ECOLOGY OF ESTUARINE FISHES

Temperate Waters
of the Western North Atlantic

ECOLOGY OF ESTUARINE FISHES

Temperate Waters
of the Western North Atlantic

KENNETH W. ABLE and MICHAEL P. FAHAY

THE JOHNS HOPKINS UNIVERSITY PRESS | *Baltimore*

The Johns Hopkins University Press
2715 North Charles Street
Baltimore, Maryland 21218-4363
www.press.jhu.edu

Library of Congress Cataloging-in-Publication Data

Able, Kenneth W., 1945–
    Ecology of estuarine fishes : temperate waters of the Western
North Atlantic / Kenneth W. Able and Michael P. Fahay.
        p. cm.
    Includes bibliographical references and index.
    ISBN-13: 978-0-8018-9471-8 (hardcover : alk. paper)
    ISBN-10: 0-8018-9471-9 (hardcover : alk. paper)
    1. Estuarine fishes—Atlantic Coast (North America) 2. Estuarine
fishes—Ecology—Atlantic Coast (North America) I. Fahay,
Michael P. II. Title.
QL621.5.A25 2010
597.177'86097—dc22   2009043635

A catalog record for this book is available from the British Library.

*Special discounts are available for bulk purchases of this book. For more
information, please contact Special Sales at 410-516-6936 or specialsales@
press.jhu.edu.*

The Johns Hopkins University Press uses environmentally friendly
book materials, including recycled text paper that is composed of at
least 30 percent post-consumer waste, whenever possible. All of our
book papers are acid-free, and our jackets and covers are printed on
paper with recycled content.

To the memory of Stacy Moore Hagan, who, for 17 years, played a central role in the collection, analysis, and management of the data that are the basis for this work. She was the mentor and matriarch for technicians, interns, and graduate students at the Rutgers University Marine Field Station. Primarily, she was a respected colleague. Her untimely death (at 36 years) has left a black hole in our scientific and personal lives. We persevered in the completion of this book, in large part, to honor her achievements.

CONTENTS

PREFACE

There are many factors that have provided the motivation and rationale for this book, most of which are associated with the dynamics of human populations and the environment. Foremost among these are the continuing and ever increasing trend for people to move to all of the coasts in the United States, with resulting implications for estuaries. This is particularly evident on the highly urbanized northeast coast and in its temperate estuaries. Such "people pressure" has elevated concern for estuarine habitats as the result of direct effects of habitat loss as well as indirect effects of eutrophication, hypoxia, contamination, and the like. At the same time the increasing realization of the effects of habitat loss and degradation has brought about, and will probably sustain, emphasis on habitat restoration. Simultaneously, the developing realization of the certainty of climate change has particular importance for shallow estuaries that are increasingly influenced by changing patterns of temperature, precipitation, sea level, and so on. All of these have increased our need to know about these ecosystems and the fish that inhabit them.

To those ends, we have brought together cumulative data from temperate estuarine fishes that result from our own intensive studies and the available literature. We present these in two formats: (1) several "synthesis" chapters that attempt to summarize the status of our understanding of estuaries as fish habitat and (2) detailed species accounts that contain much original data. We have approached these goals broadly. We do this by treating nearly 100 species, including all life history stages, as well as incorporating an improved understanding of life cycle diversity, estuarine-ocean connectivity, and the effects of climate change. We also address traditional needs, such as improved understanding of estuarine use, migrations, predator-prey interactions, and the interpretation of long-term datasets in the context of the natural history of these species.

We hope our efforts provide stimulating background as estuarine ecologists, resource managers, and fishers attempt to resolve the numerous issues facing estuaries and their ecologically and economically important fishes.

ACKNOWLEDGMENTS

An effort of this magnitude necessarily relies on the cooperation and understanding of colleagues, assistants, and other collaborators. This is especially true for those individuals from the Rutgers University Marine Field Station, both past and present. These have included volunteers, technicians, undergraduate interns, graduate students, postdoctoral fellows, and visiting scientists. Many of these contributed to the long-term collection of data that are a central portion of this book. Others contributed by allowing us to quote freely from their publications, theses, dissertations, and other documents. Two illustrators, Nancy Arthur-McGehee and Susan Kaiser, provided excellent images of juvenile fishes. Carolyn Hoss illustrated *Myrophis punctatus*. Frank Mancini was central to the effort to summarize the literature on the food habits of young-of-the-year fishes. Jen Lamonaca interpreted scale formation patterns from thousands of specimens. David Bottinelli helped to identify and summarize much of the data on carangids. Dewayne Fox shared data on gill net collections. Steve Searcy provided the specimens of *Myrophis punctatus*. Others provided insightful observations that were incorporated into the boxes scattered throughout the book, including Dave Nemerson, Dave Witting, Ed Martino, Matt Kimball, Caitlin Kennedy, and Tom Grothues. Roland Hagan helped to maintain the ichthyoplankton sampling program. Sue Able, Peter Able, and Cindy Fahay provided editorial assistance and encouragement. Fred and Judy Grassle provided frank conversation and frequent inspiration.

Several state and federal agencies provided data or logistical support. Personnel from the Jacques Cousteau National Estuarine Research Reserve provided access to environmental data. Individuals from several National Marine Fisheries Service laboratories, including those at Narragansett, Rhode Island (Jon Hare and Maureen Taylor), Beaufort, North Carolina, and Sandy Hook, New Jersey (Donna Johnson, Pete Berrien, and John Sibunka), provided data, while others assisted with library searches, access to archived data, and help with maps and other figures. Russ Brown and Don Kramer from the Woods Hole Laboratory were especially helpful with sharing decades-long trawl data from their regular seasonal surveys. Don Hoss, David Johnson, and Jeff Govoni provided space and access to the laboratory and library at the Beaufort Laboratory, where a lot of the writing by Ken Able took place over several visits. The Darling Marine Center of the University of Maine and the Belle Baruch Laboratory of the University of South Carolina also made their facilities accessible for shorter stays.

Several individuals were especially helpful in the final throes of bringing the book together. Carol Van Pelt patiently coordinated numerous revisions while keeping her sense of humor. Jim Vasslides and Stacy Hagan manipulated large amounts of data into decipherable tables and figures. Jenna Rackovan helped create and keep track of a bewildering number of figures. Jamie Caridad patiently and carefully brought the literature together. Kim Capone assisted in editing the later drafts. Dave Secor provided helpful comments on the first complete draft.

Partial funding to bring this book together was provided by the Early Life History Section of the American Fisheries Society. Other funds came from the Rutgers University Marine Field Station and the Institute of Marine and Coastal

Sciences at Rutgers University. Numerous individual grants to Ken Able and colleagues provided funding to work on estuarine fishes in many contexts. These included New Jersey Sea Grant, Hudson River Foundation, Army Corps of Engineers, Public Service Enterprise Group, and the Bluefish–Striped Bass Research Program, a Rutgers University and National Marine Fisheries Service collaborative program. The Institute of Marine and Coastal Sciences and the National Science Foundation–Research Experiences for Undergraduates (REU) (Research Internships in Ocean Science) provided support for numerous summer interns to study estuarine fishes. Another source is the Stacy Moore Hagan Endowed Scholarship through the Richard Stockton College of New Jersey in collaboration with the Rutgers University Marine Field Station.

Vincent Burke and others at the Johns Hopkins University Press provided expert assistance in the organization and publication of this complex book.

We appreciate all of the contributions of the above. Since the effort to bring this book together began in earnest in 2003, there will undoubtedly be individuals whom we have forgotten. Please forgive us for any omissions.

ECOLOGY OF
ESTUARINE FISHES

Temperate Waters
of the Western North Atlantic

1

Introduction

To appreciate how fishes use estuaries, it is important to understand estuarine landscapes. Temperate estuaries are typified by extremes in environmental conditions. In order to live in an estuarine environment, fishes must be tolerant of abrupt changes in temperature, salinity, oxygen concentrations, and turbidity levels, as well as seasonal changes in these and other physical conditions, such as opening and closing of the estuary mouth or changes in its overall size. Other environmental influences are occurring as the result of climate change. In addition to these variables, because coasts have been preferred places for human residence, industry, and recreation for centuries, human influences are profound. In recent times, the realization of the ecological importance of estuaries has prompted numerous restoration efforts. All of the above make estuaries ecological and political hot spots and increase our need to know about these landscapes and ecosystems.

The number of permanent resident fish species in estuaries is low (Day, 1981; Dando, 1984; Able and Fahay, 1998). Conversely, productivity in estuaries is extremely high, with the result that a small number of species (most of which are small in body size) are a large percentage of the ichthyofauna, in both numbers and biomass (Haedrich, 1983; Day et al., 1989; Able and Fahay, 1998). In Narragansett Bay, for example, 10 of 99 species recorded account for 90% of individuals collected (Oviatt and Nixon, 1973). Six species constitute 91% of the total in Block Island Sound (Merriman and Warfel, 1948), and 10 species comprise 90% of the total in Long Island Sound (Richards, 1963). The small number of permanent species is augmented by transient species that temporarily inhabit estuaries (Günter, 1941, 1945; McHugh, 1967; Able and Fahay, 1998). These include primarily marine species that spawn at sea, but whose young use estuaries as juvenile habitat; other species that spend much of their life history in the sea but spawn in freshwater and also use estu-

aries as juvenile habitat; and freshwater species that occasionally occur in estuaries. The estuarine fish fauna, therefore, includes both resident and transient components and a wide range of sizes and ages, thus creating a great diversity in life history stages but a more limited diversity of species.

Another way to illustrate this is that the number of species found in the estuary is many fewer than the number on the adjacent continental shelf. For example, in New England waters, the number of estuarine species is only 10% of the number of marine species (Haedrich and Hall, 1976). This relatively low species total reflects the fact that only a small number of species per family has successfully adapted to estuarine conditions, at least compared to those of the adjacent ocean. Those that have successfully invaded estuaries, however, appear to occupy a small number of broad niches (Haedrich, 1983). The sizes of estuaries may influence the number of these niches, as supported by the analysis of a number of California estuaries (Horn and Allen, 1976), which found a positive relationship between size of the estuary mouth opening and the number of species present. A recent study analyzed 28 U.S. West Coast estuaries and identified estuary mouth depth and area of the seawater zone as good predictors of species richness (Monaco et al., 1992). This relationship has not been formally assessed for estuaries of the U.S. East Coast or for any other area, however.

Our knowledge of the life history and ecology of fishes in estuaries has been improving in recent years, in part, because the information on these topics is in increasing demand by ichthyologists, estuarine ecologists, pollution biologists, and resource managers at local, state, federal, and international levels. The ichthyofauna constitutes one of the largest portions of the animal biomass and thus fishes are important to estuarine ecosystems. Sport and commercial fishermen are also becoming increasingly interested in fish life histories and ecology because they are beginning to play a larger advisory role where fish habitats and fish survival are concerned. Recently, these audiences have been further broadened by an increasingly informed general public, who are interested in (and alarmed by) the conflicting interests of aesthetic and recreational uses versus negative impacts resulting from human population pressures that bring habitat destruction and direct and indirect (non-point) sources of pollution to estuaries.

Emphases on the estuarine fish fauna of the temperate waters of the western North Atlantic (Cape Cod, Massachusetts, to Cape Hatteras, North Carolina) are lacking or site specific and scattered. The study areas of several faunal works marginally overlap the study area for this book, such as those for the Gulf of Maine (Collette and Klein-MacPhee, 2002) or Chesapeake Bay (Murdy et al., 1997). However, the spatial overlap of the fauna of the Gulf of Maine book is limited and that book does not concentrate on estuaries. The Chesapeake Bay book is only focused on that estuary and does not provide extensive detail on fishes' life history and ecology. Another set of publications (Jones et al., 1978; Hardy, 1978a, b; Johnson, 1978; Jones et al., 1978; Martin and Drewry, 1978) addresses aspects of the early life history, but these are basically compilations of all available literature up to the mid-1970s. A taxonomic guide to eggs and larvae of Middle Atlantic Bight fishes (Fahay, 1983) provides no information on the life history or ecology of these stages. An update (Fahay, 2007) concisely covers the taxonomy and description of eggs, larvae, and juvenile fish for a much expanded study area but only briefly summarizes life history and ecology. Our prior book (Able and Fahay, 1998), published more than a decade ago, was fairly narrowly focused on the early life history of fishes with emphasis on the central portion of the Middle Atlantic Bight. Thus, there is a current need for a treatment of fishes that addresses the estuarine fauna of the temperate U.S. East Coast in a comprehensive fashion. Furthermore, there are few other books that truly address temperate estuarine fishes elsewhere in the world. A treatment of southern African estuaries includes some temperate species (Whitfield, 1998). A more recent publication on estuarine fishes is focused on European temperate and subtropical estuaries (Elliott and Hemingway, 2002).

A focus on estuarine fishes is especially appropriate, largely because it is while they are in estuaries that they encounter several critical life history "bottlenecks" that can greatly affect survival rates and the resulting abundance of certain populations that we wish to harvest or conserve. Some have estimated that up to 75% of these populations are to some degree "estuarine dependent," a phrase that has become a part of resource managers' lexicons, despite a lack of critical testing or exacting definition. While these proportions are constantly evaluated and revised, there is historical and newly acquired evidence that estuaries are important, yet we lack consensus on why and to what degree. We intend to provide this missing information in a synthesis based on both historical and developing information from a variety of sources embracing both the estuary and the adjacent Atlantic Ocean, where many of these species are spawned and spend the winter.

In the western North Atlantic, the fishes of temperate estuaries, as a whole, constitute a relatively diverse fauna made up of species that occur primarily in the region between Cape Cod, Massachusetts, and Cape Hatteras, North Carolina. This fauna is enriched by boreal species from the north and tropical components from the south, thus resulting in a diverse fauna with varied life histories. Many of these species spawn over the adjacent continental shelf; larvae are then transported into estuaries, where they spend an important portion of their juvenile period growing, feeding, and avoiding predators. Eventually, these juveniles move back out of the estuary into oceanic waters, where they reach maturity and join the adult population. At the same time, larger juveniles and adults of other species are resident and transient in estuaries. Although fisheries scientists and managers rec-

ognize how critical the estuary is for a large portion of our valuable, harvestable fisheries, we currently have an incomplete understanding of these life history patterns. In order to improve that understanding, we have chosen to focus on the most abundant and important sharks and bony fish species occurring in estuaries of the region. We have provisionally chosen more than 90 of these species for treatment because they (1) occur in estuaries during some portion of their life history, (2) are important ecological components of the estuary, or (3) are of economic importance to recreational or commercial fisheries.

The initial 10 chapters of this book synthesize the results of our own studies and the available literature on temperate estuarine fishes. Chapters 2–4 provide background and methods for this approach. Characterizations of the study area in Chapter 2 include observations at the zoogeographic, oceanographic, and ecosystem levels for both the major estuaries and the adjacent continental shelf. Chapter 3 explains our materials and methods. We describe historical and current sources of information, including major surveys in estuaries and the adjacent continental shelf in our study area. Chapter 4 summarizes what is generally known about the fishes of the temperate waters off the East Coast of the United States and thus provides the context for the chapters that follow.

In the remaining synthesis chapters we describe patterns of life history and ecology. For example, fishes that use estuaries in the study area reproduce in a variety of locations and times in estuaries and the ocean with varying influences on their life histories. In Chapter 5, we summarize these patterns of reproduction and examine the significance of development based on the diverse spawning characteristic of fishes, larval duration, and location of the larval/juvenile transition. We anticipate that these differences greatly influence the spatial and temporal patterns of estuaries as fish habitat in general. In Chapter 6 we examine the occurrences of larvae and settlement in time and space and determine how these vary between resident and transient fishes as well as between cohorts of selected species. Further, we evaluate the patterns of growth, especially for juveniles, relative to species, cohorts, and season for resident and transient groups. These findings come from our original research as well as the relevant literature. These measures are especially pertinent because they may impact mortality/survival for young-of-the-year and subsequent year-class strength. Chapter 7 synthesizes our current understanding of estuarine habitat use. There have been major accomplishments since the data on estuarine habitat use for our study area were summarized (up to 1994) for eventual publication (Able and Fahay, 1998). Over the past decade more extensive qualitative and quantitative sampling for pelagic and demersal species has occurred. Equally important, there has been an increased effort to compare habitat use in estuaries and the adjacent ocean. At the same time the increased use of experimental techniques such as mark/recapture with smaller tags and ultrasonic telemetry for larger juveniles and adults have markedly enhanced our understanding of habitat use, residency, and movements. Chapter 8 summarizes the available literature on the species-specific prey and predators of young-of-the-year fishes in our study area estuaries based on available literature, as well as our own observations. We identify the dominant prey types and compare these among resident and transient species during their development. Similarly, identification of predators on young-of-the-year fishes is used to evaluate the presumed role of estuaries as refuges from predation. Other data on the food habits of larger juveniles and adults are also provided.

Transient fishes—those that use estuaries for only a portion of their early life history—make up a large component of temperate estuarine fishes. Chapter 9 focuses on the movements and migrations of these fishes, particularly as they leave estuaries with increasing size or in response to seasonal declines in temperatures, a common characteristic of temperate estuarine fishes. This examination is based on species-specific seasonal patterns of egress from estuaries and subsequent movement onto the continental shelf during the fall. In Chapter 10, we briefly review the recent literature on the evidence for climate change and discuss the influence of this change, especially that of temperature, on the distribution, behavior, and survival of estuarine fishes. The species accounts in the remaining chapters follow a common format, outlined in Chapter 3, to describe the life history (including eggs, larvae, juveniles, and adults) and ecology of selected species in temperate estuaries of the East Coast of the United States. Chapter 3 also provides the basis for many of the comments in Chapters 4 through 10. In Chapter 11, we identify and discuss the shortcomings in our present understanding of the life history and ecology of temperate estuarine fishes. Furthermore, we provide some suggestions as to how to proceed in this important endeavor.

ESTUARINE BOUNDARIES, CONNECTIVITY, AND NURSERIES

The attempt to understand fishes' use of estuaries began in the Gulf of Mexico with the observation that "The young of many animals usually thought of as marine, require areas of low salinity for nursery grounds" (Günter, 1945, 1950; Pearse and Günter, 1957). Subsequently, the "marine-estuarine life history" was considered a general law for many species of fishes (Günter, 1967). Others have referred to fishes with this life history as "marine transients" (Deegan et al., 2000) or as estuarine-catadromy (Bulger et al., 1993; Lowery et al., 1995). At the same time, several authors did not define the term but provided a list of "estuarine-dependent" species for the Gulf of Mexico and the East Coast of the United States (Clark, 1967; McHugh, 1967) or an expanded list for all coasts of the United States (Stroud, 1971). Subsequently, Darnell and Soniat (1979), while working in the Gulf of Mexico, noted the difference between "estuarine-related" species, that is, coastal marine forms that inhabit the estuary with some regularity but do not require this habitat, and "estuarine-dependent" species, which usually require the estuarine habitat for some stage of their life histories. A somewhat different interpretation, as developed from observations in South Africa, Australia, and England, is that "estuarine opportunists" or "marine estuarine opportunists" typically inhabit both estuaries and inshore marine environments (Lenanton and Potter, 1987; Potter et al., 1997). Other evaluations of estuarine use in South Africa and Australia suggested that only those taxa whose populations would be adversely affected by the loss of estuarine habitats should be considered estuarine-dependent (Blaber et al., 1989; Whitfield, 1994). Other observers have questioned the general applicability of fish "estuarine dependence" because the focus has been on sampling estuaries for fishes while there has been little effort in the adjacent ocean in both temperate (Able and Fahay, 1998; Martino and Able, 2003) and tropical (Longhurst and Pauly, 1987; Blaber, 2000) systems. Thus we do not know if estuaries and the ocean function similarly. Possible exceptions to this are evaluations from Australia (Lenanton, 1982; Blaber et al., 1989, 1995), South Africa (Lasiak, 1986; Bennett, 1989; Valesini et al., 1997; Harris et al., 2001; Strydom, 2003), and Mexico (Yáñez-Arancibia et al., 1985).

The development of ideas revolving around the concept of "estuarine dependence" is often related to our understanding of "nurseries." In fact, the first use of the term *nursery grounds* was part of the description of estuarine dependence (Pearse and Günter, 1957). Thus, the terms have had long, sometimes parallel usage, largely because early researchers considered the entire estuary to be a nursery. Lately, there has been more emphasis on individual habitats, such as salt marshes (Minello et al., 2003), sea grasses (Heck et al., 2003), and mangroves (Sheridan and Hays, 2003). A recent review of the nursery concept (Beck et al., 2001, 2003) has clarified our understanding by defining nurseries as the habitats of the juveniles of a species that contribute more, on average, to the production of individuals to the adult population than other habitats in which the juveniles occur. This definition makes a clear distinction between juvenile and nursery habitats, something many other authors have not done. This lack of differentiation accounts for the frequent use of *nursery* in the literature without any understanding of the functional significance of each habitat type. Further, the revised definition of *nursery* is based on four comparative factors: (1) density, (2) growth, (3) survival, and (4) movement to adult habitats, that is, contribution to the adult populations (Beck et al., 2001, 2003). Often only one of these factors is evaluated (most frequently density [Minello et al., 2003; Heck et al., 2003; Heupel et al., 2007] and occasionally density and growth [Ross, 2003]) but all are necessary for the identification of nurseries, including the movement to adult habitats, which is probably the least commonly studied (Gillanders, 2002; Gillanders et al., 2003).

While these reevaluations have been instructive, the nature of nurseries is the basis for continued discussion. Others have provided alternative views, such as "effective juvenile habitats" based on the relative proportion of individuals from different habitats (Dahlgren et al., 2006). Others have attempted to further define these habitats by bringing reproductive output, spatial scale, and connectivity and complexity into consideration (Sheaves et al., 2006). A further clarification of estuarine nurseries has been prompted by the questioning of estuaries as refuges from predation (Sheaves, 2001; Baker and Sheaves, 2005, 2006; Sheaves et al., 2006). A related term, *essential fish habitat*, is defined as those waters and substrate necessary for spawning, breeding, feeding, or growth to maturity (Baird, 1999). This definition and its determination is the current focus, by law, of all managed species in the United States (NOAA, 1996a; Benaka, 1999; Schmitten, 1999), and is being applied to noncommercial species as well (Able, 1999; Able and Hagan, 2003). These studies recognize the difficulties of identifying nurseries (see Layman et al., 2006; Heupel et al., 2007). Whether the focus is justified in terms of essential fish habitat or nurseries, all of these approaches have contributed to an enhanced understanding of the relationship between fishes and estuarine habitats and ecosystems.

An important consideration is how to define the borders of estuaries. Admittedly, this is a complex issue that has been frequently revisited since the earliest attempts to study estuaries in a comprehensive fashion (Cameron and Pritchard, 1963; Lauff, 1967; Hedgepeth, 1967; Day, 1981; Elliott and McLusky, 2002; Keefer et al., 2008). Although the following are not exhaustive, they do represent the spectrum (geomorphology, hydrography, sediments, fauna) of definitions found in the literature. Odum (1959) describes an estuary as a "river mouth where tidal action brings about a mixing of salt and fresh water," whereas Pritchard (1967) defines it as "a semi-enclosed body of water which has a free connection with the open sea and within which sea water is measurably diluted with fresh water derived from land drainage." Others have noted the similarities between lagoons and estuaries and define "lagoon-estuarine environments" as "shallow, semi-enclosed water bodies of variable volume, connected to the sea in a permanent or ephemeral manner, with variable temperature and salinities, permanent muddy bottoms, high turbidity, irregular topographic characteristics, and biotic elements" (Yáñez-Arancibia et al., 1994) and suggest further studies (Razinkovas et al., 2008). This definition is also sufficiently broad that it includes hypersaline estuaries/lagoons (Cowardin et al., 1997) as well as estuaries with ephemeral connections to the ocean (Lenanton and Hodgkin, 1985; Whitfield, 1998; Roy et al., 2001; James et al., 2007). These and many other definitions share a geomorphological component and are often based on a "semi-enclosed body of water." McHugh (1967) has taken a different approach and defines the nektonic estuary as "including inshore and offshore (i.e., < 33.5 salinity) components." This extends the estuary into continental shelf waters (Cameron and Pritchard,

1963) and broadens the spatial scale of the definition, but McHugh offers little explanation for the salinity boundary proposed. The offshore estuary may overlap with the shoreface entrainment volume of Ray (1991) and Ray and Hayden (1992). This same broader view is consistent with the term *estuarine zone*, that is, "an environmental system consisting of the estuary and those transitional areas consistently influenced or affected by water from the estuary" (Smith, 1966).

Of critical importance is an understanding of the connectivity between estuary and ocean and estuary and freshwater habitats. For us, connectivity refers to "the dependence of fish production and population dynamics on dispersal and migration" (Secor and Rooker, 2005) among the above habitats. Recent examples include clupeids on the northeast coast of the United States (Ray, 2005), larval fish supply to estuaries (Brown et al., 2005), the use of elemental otolith microchemistry to determine the degree of connectivity (Gillanders, 2005), commercial catches (Meynecke et al., 2008), and the relative contribution between freshwater and brackish habitats (Kraus and Secor, 2005).

In an attempt to further evaluate the degree of connectivity between estuarine and ocean habitats, we determined patterns of habitat use for young-of-the-year fishes in southern New Jersey (Able, 2005). In this estuary-ocean comparison, a consistently large proportion of the fishes used both estuarine and ocean habitats as juveniles. This also seems to be the case for many of the dominant species in the Middle Atlantic Bight. In an earlier synthesis we attempted to categorize patterns of estuarine use for 70 of the dominant species (Able and Fahay, 1998). A reevaluation of these same species suggests that although there is a large number of estuarine obligate species (e.g., cyprinodontids, fundulids), there is a similar number of transient species that use estuaries in a facultative manner (e.g., those that use both the estuary and the ocean) or estuarine use varies annually, geographically, or with some other unknown variable. Other species are insufficiently known to confidently characterize. Further, it is important to keep in mind that the patterns of habitat use are complex. For example, we now know that the spring-spawned cohort of *Scophthalmus aquosus* uses both the estuary and the inner continental shelf as juvenile habitat, while the fall-spawned cohort only uses the inner shelf (Neuman and Able, 2003). In yet another example of the difficulty of characterizing habitat use, it is clear that young-of-the-year *Chaetodon ocellatus* settle, reside, and grow in estuarine habitats in southern New Jersey, but these habitats are sinks because the young-of-the-year die when temperatures decline in the fall/winter. Thus, those that settle in temperate estuaries likely contribute nothing to the adult population (McBride and Able, 1998).

CURRENT UNDERSTANDING
Throughout the rest of this book we follow Pritchard's (1967) definition of *estuary*. There is certainly ample evidence from around the world that juvenile fishes use estuaries, regardless

of the definition one chooses. The evidence for estuaries as juvenile habitat is best documented from South Africa (Whitfield, 1998), Australia (Potter et al., 1990; Blaber, 2000), Europe (Elliott and Hemingway, 2002; Franco et al., 2008), and the United States (McHugh, 1967; Ray, 1997; Ray et al., 1997; Able and Fahay, 1998) based primarily on the abundant estuaries in each of these areas and the resulting long focus of researchers. Other documentation of the importance of estuaries is available from temperate South America (Chao et al., 1982) and from tropical estuaries in the Philippines, Mexico, and elsewhere (Yáñez-Arancibia, 1985; Longhurst and Pauly, 1987; Blaber, 2000).

In many instances the patterns of fish use of semi-enclosed estuaries are similar across the world. The categories of fish lifecycles associated with South African estuaries (Whitfield, 1998) can be found in estuaries in all of the other continents of the world. These include the following: (Ia) estuarine species that breed only in estuaries; (Ib) estuarine species that breed in estuaries and the marine environment; (IIa) euryhaline marine species that usually breed in the ocean but the juveniles are dependent on estuaries as nursery areas; (IIb) euryhaline marine species that usually breed in the ocean, with the juveniles occurring in both estuaries and the ocean; (IIc) euryhaline marine species that usually breed in the ocean, with the juveniles occurring in estuaries but being more abundant in the ocean; (III) marine stragglers not dependent on estuaries; (IV) diadromous species. These can be grouped into obligate and facultative users, a point frequently made in the literature from diverse sources and locations (Blaber, 2000). The use of estuaries may also vary over long temporal scales, especially when associated with climate change (Attrill and Power, 2002). For example, it is apparent that *Micropogonias undulatus* has expanded its use of estuaries in response to a general warming trend off the East Coast of the United States over the past decade (Hare and Able, 2007).

Physiological sources of variation, including the reduced cost of osmoregulation (Potter et al., 1990), also contribute to patterns of estuarine use by fishes. Although it is convenient to view increased food and decreased predation as advantages of estuaries (Blaber and Blaber, 1980; Yáñez-Arancibia et al., 1980; Boesch and Turner, 1984), the comparative data for oceans versus estuaries to support or refute these ideas are often unavailable. Some biotic sources of variation may also be responsible for the differences in estuarine and ocean use patterns. For example, we have yet to understand how

Estuarine Users

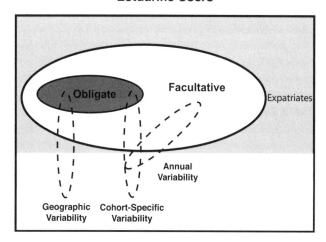

Fig. 1.1. Conceptual view of the degree of estuarine dependency (obligate, facultative, expatriate) for fishes in the Middle Atlantic Bight. Among fishes that are facultative users of estuaries are those that vary geographically, annually, or in a cohort-specific manner.

annual variation in fish abundance affects estuarine versus ocean use (Horn and Allen, 1985), or why different spawning cohorts have different patterns of estuarine versus ocean use (Neuman and Able, 2003). Further, the possibility that different contingents exist within the same population within a single estuary, for example, *Morone saxatilis* (Secor, 1999; Secor et al., 2001), can influence timing and duration of estuarine versus ocean use. In a broader sense, the diversity of lifecycles in fishes is increasingly realized (Secor and Kerr, 2009), including that of common estuarine fishes (Kraus and Secor, 2005; Able and Grothues, 2007a; Grothues et al., 2009; Kerr et al., 2009).

Perhaps the simplest approach is to acknowledge that fishes' use of estuarine and oceanic habitats is a continuum; some species have obligate life history stages (i.e., residents and diadromous species) in the estuary, others are facultative users (estuarine opportunists) of the estuary, and still others are simply strays that occasionally find their way into an estuary and may not survive and thus become expatriates (Fig. 1.1). Most important, those fishes that are facultative estuarine users may differ in their use on varying geographic, ontogenetic, annual, and cohort-specific scales. This interpretation is probably more realistic, as it reflects the diversity of estuarine use patterns and helps discourage the glib use of the term *estuarine dependence*.

2

Study Area

STUDY AREAS

We use four general study areas to relate the life history and ecology of estuarine fishes to the physical characteristics of their environment. The largest of these encompasses continental shelf waters along the East Coast of the United States between Florida and the Bahamas to the Gulf of Maine (Fig. 2.1). The focal area within this broader landscape is the temperate region, primarily between Cape Cod and Cape Hatteras, that is, the Middle Atlantic Bight (Fig. 2.2). We define the central part of this area, another area of focus, as that portion bordered by the coastlines of New York (Long Island) and New Jersey, that is, the New York Bight (Cowen et al., 1993; Pearce, 2000) (Fig. 2.2). Our most focused study area is the contiguous ecosystem that includes the Mullica River, Great Bay–Little Egg Harbor estuaries, Little Egg Inlet, and coastal ocean waters near the submerged Beach Haven Ridge (Fig. 2.3).

CHARACTERISTICS OF THE MIDDLE ATLANTIC BIGHT

The Middle Atlantic Bight is a North Atlantic Ocean neritic area delineated by Cape Cod, Massachusetts, and the adjacent Georges Bank to the north and Cape Hatteras, North Carolina, to the south (see Fig. 2.2). We chose this term because it is the same as that used by many researchers that we have cited in this book. Our coverage includes the inner estuarine borders of this bight and its continental shelf to a depth of 200 m. The width of this shelf ranges from 240 km off southern New England to 50 km at its southern extent off North Carolina. Within this area we are concerned with patterns of reproduction, largely based on gonadosomatic studies or the temporal and spatial distributions of eggs and larvae. It is also the broadest study area we have examined for describing the winter and spring distribution of young-of-the-year and large juveniles and adults after they emi-

grate from estuarine habitats in the fall and subsequently return the following spring. The South Atlantic Bight is located from Cape Hatteras to the east coast of Florida. The Gulf of Maine and Scotian Shelf are situated north and east of Georges Bank. East of the 200 m depth contour lies the Slope Sea (*sensu* Csanady and Hamilton, 1988) and, farther east, the Gulf Stream, which originates in the South Atlantic Bight and flows closely past Cape Hatteras before veering toward the northeast. The Sargasso Sea, a tropical region of the North Atlantic Ocean, is located beyond the Gulf Stream and extends south to the Bahamas and beyond. All of these neighboring oceanic, coastal, and neritic regions influence in some way the character of our study area and its fauna.

Oceanic circulation within the Middle Atlantic Bight (Fig. 2.4) is slow moving. The general transport is from northeast to southwest on the shelf and in adjacent Slope Sea waters during most of the year (Ingham, 1982; Mountain, 2003; Townsend et al., 2004). Average monthly wind conditions and their effect on surface water circulation (Ekman transport) change seasonally in the bight (Table 2.1). Superimposed on the general drift are rotary tidal currents and short-term events influenced by weather conditions (Cook, 1988). Extreme conditions, including major storms, severe droughts, oxygen depletions, major plankton blooms, or

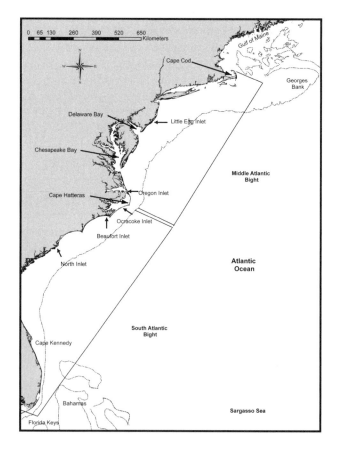

Fig. 2.1. East Coast of the United States, including temperate waters of the Middle Atlantic Bight. Limits of continental shelf indicated by the 200 m contour.

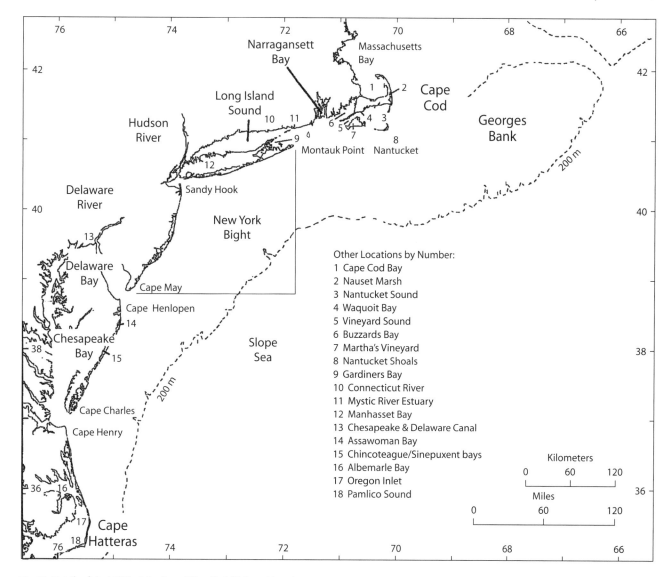

Fig. 2.2. Details of the Middle Atlantic and New York bights with major locations mentioned in the text.

ephemeral phenomena such as warm-core rings, current reversals, or coastal upwelling events, can affect both the circulation and the hydrography of the study area and its fauna.

The temperate waters of the Middle Atlantic Bight shelf are divided into three major bands: (1) a low-salinity, near-shore band affected primarily by runoff from major estuaries; (2) a mid-shelf band (between the 20 and 100 m contours); and (3) an outer shelf band near the shelf edge, which receives intrusions of warmer, more saline water from the Slope Sea (Ingham, 1982). Most of the freshwater runoff into the Middle Atlantic Bight occurs during the spring and originates from four major plumes associated with the Connecticut River, Hudson River, Delaware Bay, and Chesapeake Bay (Bigelow and Sears, 1935; Ketchum and Corwin, 1964; Charnell and Hansen, 1974; Pape, 1981; Durski, 1996; Epifanio and Garvine, 2001). Many of these plumes flow into shelf water and then south along the coastline of the bight, although this

movement varies somewhat by season (Zhang et al., 2009). Bottom currents in the bight vary in direction from region to region, but inshore-offshore movements are similar. From southern New England to the offings of Chesapeake Bay, these bottom currents move toward shore (Bumpus, 1973; Pape, 1981).

Meteorological and hydrographic conditions in the Middle Atlantic Bight are among the most variable in the world. Ocean temperatures fluctuate by more than 20 to 25°C between winter and summer (Parr, 1933; Grosslein and Azarovitz, 1982; Cook, 1988), and range seasonally from near freezing to well over tropical limits (Fig. 2.5). Sea surface temperatures vary seasonally, ranging from about 2°C along the coast during February–March (in the New York Bight) to highs near 30°C (but usually 25 to 26°C) during late summer off Cape Hatteras, North Carolina. Winter water-column temperatures off New York range from 2 to 3°C near the

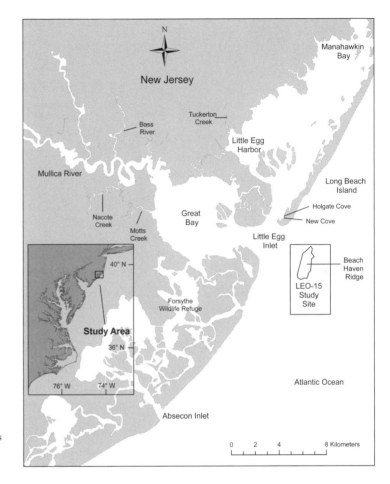

Fig. 2.3. Mullica River, Great Bay–Little Egg Harbor, Little Egg
Inlet, and Beach Haven Ridge study areas. The rectangle indicates
the Long-Term Ecosystem Observatory at 15 m (LEO-15) study
area.

coast to 7 to 11°C at the 200 m contour offshore. Comparable values for the offings of Cape May are 3 to 4°C to 9
to 10°C, and for the offings of Chesapeake Bay 5 to 5.5°C
to 9 to 10°C (Ingham, 1982). In typical winters, bottom temperatures between the 150 and 250 m contours are about 9 to
11°C throughout the bight (Cook, 1988).

Changing seasonal patterns of cooling and warming on
the bottom result in a temperature of about 10°C always
being available somewhere within the Middle Atlantic Bight,
which is important to the seasonal distributions of fishes
(Parr, 1933; Grosslein and Azarovitz, 1982). The seasonal
cycle of temperatures is pronounced and distinctive off
southern New Jersey (see Fig. 2.5). Warming begins in April,
and shelf water in the bight becomes highly stratified during
the summer and fall, when a strong thermocline separates
the warm surface mixed layer from a cold cell. This dome-
shaped cold cell forms on the shelf bottom between Georges
Bank and Cape Hatteras during the late winter months and
persists through the summer, gradually warming in late summer and fall (Bumpus, 1973; see Fig. 2.5). Its usual position is
between the 40 and 100 m isobaths, and its average thickness
(from the sea bottom to the underside of the thermocline) is
about 35 m. Its total volume represents approximately 30%
of the total water mass in the Middle Atlantic Bight during
summer (Ingham, 1982). In the summer months, surface lay-

ers (where most developing fish larvae are found) overlie this
cell and warm considerably, resulting in a strongly stratified
water column. The thermocline between these warm- and
cold-water masses occurs at depths between 10 and 25 m,
depending on the distance from shore. In nearshore areas
it intersects the bottom at depths between 10 and 15 m; the
offshore edge of this thermocline intersects the bottom at
depths between 80 and 100 m (Houghton et al., 1988). Vertical mixing accompanies the breakdown of this stratification
in the fall, and bottom temperatures reach their maximum
one or two months after surface waters reach theirs (Ketchum and Corwin, 1964). After erosion of the thermocline
during the fall, conditions are vertically isothermal during
the winter and early spring (Ingham, 1982).

Salinities in the Middle Atlantic Bight are a product of
freshwater runoff entering the bight at the surface near the
coast, combining with high-salinity slope water entering over
the bottom from the Slope Sea (Bigelow and Sears, 1935; Epifanio and Garvine, 2001; Kohut et al., 2004). Values range
from 30 ppt near shore to 35 ppt on the bottom near the
shelf edge. The average throughout the bight ranges from
32.5 ppt at the surface to 35 ppt along the bottom. Salinities are lowest during spring runoff of freshwater, increase
through summer and fall, and reach their highest levels during winter (Cook, 1988). The effect of freshwater runoff on

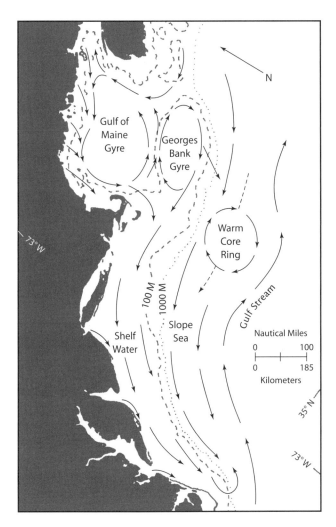

Fig. 2.4. General surface circulation in the study area (after Ingham, 1982).

Stream eddies and rings, current reversals, and coastal up-welling, may influence the transport, and subsequent distribution and abundance, of young-of-the-year fishes in the Middle Atlantic Bight. The offshore boundary of Middle Atlantic Bight shelf water is called the Shelf/Slope Front, which is usually located at the approximate location of the 200 m isobath (see Fig. 2.4). This front extends from Georges Bank to Cape Hatteras and is present year-round (Beardsley and Flagg, 1976). Offshore of this front, the slope water is warmer and more saline, and consists of a mixture of continental shelf and Gulf Stream water. Warm-core rings occasionally spin off the western edge of the Gulf Stream and slowly drift in a southwesterly direction within the slope water mass off the bight. These rings enclose highly saline, warm waters from the Gulf Stream, may measure 150 to 230 km across, and reach depths of 2000 m (Cheney, 1978). From 6 to 11 such rings are produced annually (Fitzgerald and Chamberlin, 1981). They are important because they may remove "streamers" of shelf water by "advection"; conversely, they may inject slope water and modified Gulf Stream water onto the shelf and thus modify circulation on the shelf. These rings occur most frequently in the vicinity of Georges Bank, where they are younger, larger, and stronger in rotational flow than when they enter waters off the central part of the Middle Atlantic Bight (Cook, 1988). Rings (and other features in the Slope Sea) also contain and transport fish larvae from southern waters and deliver them to shelf waters of the Middle Atlantic Bight, from whence some travel to estuarine nurseries (Cowen et al., 1993; Hare and Cowen, 1993; Epifanio and Garvine, 2001; Hare et al., 2001).

Current reversals have been identified as mechanisms allowing for the retention of blue crab larvae in the neighborhood of major Middle Atlantic Bight estuaries (Epifanio et al., 1989; Epifanio, 1995; Epifanio and Garvine, 2001), and they undoubtedly play a role in the distribution of fish larvae as well. They typically occur when prevailing winds during summer in the central Middle Atlantic Bight are from the west and southwest (see Table 2.1). When these conditions continue over a period of days, surface currents over the mid-shelf off Chesapeake and Delaware bays reverse their direction and flow to the north (Boicourt, 1982).

ocean salinities varies through the bight. The Hudson River plume can influence Middle Atlantic Bight waters depending on whether it travels along the New Jersey coast, along the Long Island coast, or along a mid-shelf offshore pathway (Zhang et al., 2009). In the process, it may lower salinities as far as 145 to 160 km offshore. The effect of the Chesapeake Bay plume rarely reaches more than 80 km (Cook, 1988).

Short-term oceanographic phenomena, such as Gulf

Table 2.1. Monthly prevailing winds and resulting Ekman transport at two locations in the study area

	Jan	Feb	Mar	Apr	May	Jun	Jul	Aug	Sep	Oct	Nov	Dec
Central Middle Atlantic Bight (39N × 72W)
Wnd	NNW	NNW	NW	W	SW	SW	SW	WSW	N	N	NW	NW
Ekm	SW	SW	SW	S	SE	SE	SE	SE	W	W	SW	SW
Southern Middle Atlantic Bight (36N × 75W)
Wnd	NW	NW	WNW	WSW	SW	SW	SW	SW	NE	NE	NW	NW
Ekm	SW	SW	SSW	SSE	SE	SSE	SSE	SE	NW	WNW	WSW	WSW

Source: Adapted from Cook, 1988.

Note: Wnd = direction from which winds blow. Ekm = direction of resultant Ekman Transport.

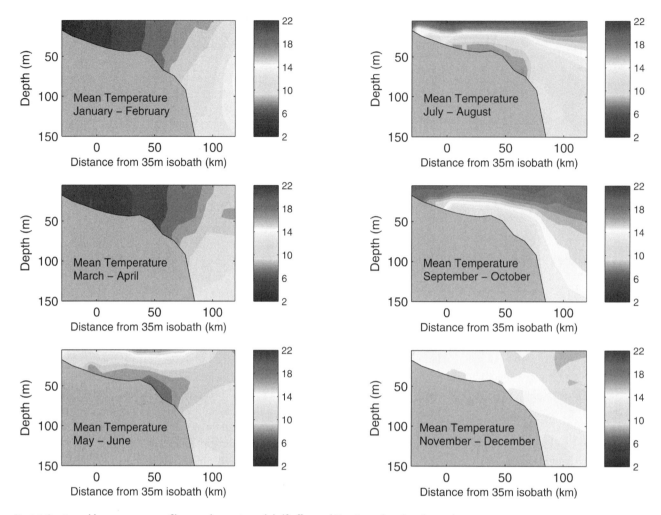

Fig. 2.5. Semimonthly temperature profiles over the continental shelf off central New Jersey based on data gathered in conjunction with ichthyoplankton collections during the MARMAP series of cruises, 1977–1987.

Another result of prevailing west and southwest summer winds is upwelling of cold bottom water. In the central part of the Middle Atlantic Bight, temperatures on the shelf bottom within the cold cell reach a minimum (3.8 to 4.7°C) during early June. Maximums occur after fall turnover, when they may reach 12 to 13°C in late November (Ingham, 1982). Nearer the coast, the front may approach (and reach) the New Jersey, New York, and Delaware shores or even enter estuaries during upwelling events precipitated by strong west or southwest winds, which are usually strongest during summer (see Table 2.1) (Scott and Csanady, 1976; Hicks and Miller, 1980; Ingham, 1982; Neuman, 1996; Münchow and Chant, 2000). In addition to lower temperatures, this inshore incursion of cold cell waters may cause stratification and result in anoxic conditions, causing mass mortalities of fishes and other marine organisms in the coastal ocean (Steimle and Sindermann, 1978; Swanson and Parker, 1988; Glenn et al., 1996). Under the opposite wind conditions, the cold cell may migrate offshore, forcing the Shelf/Slope Front to move sea-

ward and exposing the offshore shelf bottom to colder temperatures (Csanady, 1978).

Several large estuarine embayments are located along the coastline of the study area, including Narragansett Bay, Long Island Sound, Raritan–Sandy Hook Bays, Delaware Bay, Chesapeake Bay, and the Pamlico/Albermarle estuaries, as well as several smaller systems (see Fig. 2.2). The most important of these are dominated by single (or several) rivers (Table 2.2), which are framed by the Appalachian Mountains to the west and north (Patrick, 1994). The estuaries of systems to the north and south of the study area differ in geomorphology and climate (Dame et al., 2000; Roman et al., 2000). Over geological time the tectonically stable East Coast estuaries (Inman and Nordstrom, 1971), including those in the Middle Atlantic Bight, have recently formed from flooded river valleys as sea level has risen (Paul, 2001; Uncles and Smith, 2005). These estuaries vary in response to the differences in the width of the coastal plain, which is narrow in the northeast and broader in the southeast. As a result,

Table 2.2. Physical characteristics of estuarine systems considered in this study. See Able and Fahay (1998) for further details.

	Length (km)	Average Width (km)	Average Depth (m)	Mouth Dimensions		Stratification		Estuarine Drainage (km²)	Area (km²) by Salinity Zone		
				Depth (m)	Width (km)	High	Low		Tidal Fresh	Mixing	Seawater
Nauset Marsh, MA	~7	~2	~2	3.5	0.25	VH	VH	<1	0	0	10
Buzzards Bay, MA	48	11	10.6	23	10.64	VH	VH	1491	0	5	585
Narragansett Bay, RI	47	15	10.0	30	21.79	VH	VH	3444	8	52	368
Connecticut River, CN	88	0.5	3.8	7.6	1.39	HS	MS	2823	47	5	0
Long Island Sound, NY	320	20	19.5	91	34.08	VH	VH	18,725	75	427	2815
Gardiners Bay, NY	50	14	6.1	30	20.26	VH	VH	1036	0	5	505
Great South Bay, NY	116	3.5	2.2	8.5	0.76	VH	VH	2188	0	202	189
Hudson/Raritan, NY, NJ	257	2.7	4.6	15	9.67	MS	VH	21,929	96	435	241
Barnegat Bay, NJ	64	4.5	1.4	6.1	0.41	VH	VH	3496	0	78	186
Great Bay, NJ	25	4	2.5	7.6	3.57	VH	VH	1476	32	98	47
Delaware Bay, NJ, DE	222	24.1	7.4	37	17.91	VH	VH	12,302	59.6	518	1412
Chincoteague Bay, VA	51	7.2	1.3	10	1.45	VH	VH	777	0	0	355
Chesapeake Bay, VA	319	24.8	8.5	24	17.40	MS	MS	56,863	396	9226	298
Albemarle Sound, NC	172	16.9	5.8	4.6	9.57	VH	VH	15,032	1880	508	0
Pamlico Sound, NC	193	25.3	5.5	7.6	1.94	VH	VH	14,996	249	4898	104

the estuaries in the southern portion of the study area are broader, with larger volumes and gradients from full coastal salinity to freshwater. However, all of these major estuaries have roughly equivalent tidal reaches, non-tidal river lengths, and watershed sizes and are typically very turbid, with an estuarine turbidity maximum near the saltwater-freshwater interface (Paul, 2001; Uncles and Smith, 2005).

The surficial sediments of the temperate estuaries in the study area vary with estuary type. The northern estuaries are characterized by embayments that are dominated by glacial sediments. Back bay estuaries are common, especially in the areas to the south. All of these have typically muddy sediment (Holliday and Klein, 1993). Patterns of sediment distribution parallel estuarine geomorphology. The large drowned river valley estuaries, such as Chesapeake and Delaware bays, are dominated by sand bottoms on the fringes of the muddy main channels (Holliday and Klein, 1993). The river systems that contribute to these estuaries and Albermarle and Pamlico sounds are very muddy. Marine sediments are the main source for the more northern embayments and back bay estuaries. Much of the sand is concentrated in the flood tidal delta inside the inlets and behind the barrier islands as the result of washover events. In the same estuaries there are often more muddy sediments along mainland shorelines (Holliday and Klein, 1993). Much of the surficial sediment in estuaries is the result of suspended sediments, most often evident as an estuarine turbidity maximum, the resuspension of estuarine bed sediments, or sediments arriving in estuaries from rivers or the coasts (Uncles and Smith, 2005). Further, the differences in estuarine specific turbidity are largely influenced by tidal range and tidal length. All of these diverse systems have been impacted to a variable degree by a variety of human alterations (Kennish, 1992, 1998; EPA, 1998;

Pearce, 2000; Paul, 2001), reflected in eutrophication (NOAA 1996b, 1997a, b), metal contamination (Hollister et al., 2008), and increasing hypoxia and anoxia (Brush, 2009; Tyler et al., 2009). This applies to the living habitats (biotopes), including coastal intertidal wetlands (NOAA, 1990), oyster reefs, submerged aquatic vegetation (Orth and Moore, 1993; Short et al., 1993; EPA, 1998), and other estuarine habitats, such as tidal flats. Other living habitats are being created through the invasion of estuaries by *Phragmites australis*, which is prominent in wetlands throughout the study area (Saltonstall, 2002; Able et al., 2003a).

CHARACTERISTICS OF THE NEW YORK BIGHT

In this area, strong, near-bottom currents flowing toward New York Harbor and Long Island Sound have also been described (Charnell and Hansen, 1974; Hardy et al., 1976). There is an apparent divergence in bottom drift near the Hudson Canyon Shelf Valley; northeast of the valley net drift is to the north, while southwest of the valley it is westward (Charnell and Hansen, 1974; Hardy et al., 1976). Within the Hudson Canyon Shelf Valley, very strong currents flow mostly landward during winter and seaward during summer (Keller et al., 1973; Nelson et al., 1978; Ingham, 1982). All of these patterns potentially affect the distribution and drift of fish larvae (Malchoff, 1993). Two estuaries, associated with the Hudson and Delaware rivers, are major components of this region. Other important estuaries include Great South Bay and Raritan Bay in New York; Sandy Hook Bay, Barnegat Bay, Little Egg Harbor, Great Bay, and numerous sounds and bays in southern New Jersey; and Indian River Bay in Delaware (see Fig. 2.2). In this more restricted study area we emphasize the temporal and spatial distribution of late larvae and juvenile fishes, and discuss small-scale seasonal migrations associated with spawning, feeding, or other influences. This system is best described as a gradient consisting of coastal ocean waters, including a permanent sand ridge (McBride and Moslow, 1991) and surfzone, barrier beach, inlet, estuarine marsh, and tidal river. It is in this system and its varied habitats that we have the most comprehensive and long-term experience and data on ingress, habitat use, growth rates, predator-prey relationships, and egress patterns of estuarine fishes.

CHARACTERISTICS OF THE MULLICA RIVER–GREAT BAY ESTUARY AND THE ADJACENT INNER SHELF

The corridor from the Mullica River through the adjoining estuaries to the inner continental shelf is the most intensively studied portion for our purposes (see Fig. 2.3). Hydrographic conditions in our most focused study area (Mullica River–Great Bay estuary) vary considerably through the year. Mean monthly temperatures range from 2.1 to 23.9°C (see Fig. 2.6), but extremes can extend from −0.8 to 25.2°C. Salinities are usually between 28.0 and 29.8 ppt near Little Egg Inlet, but short-term events cause overall extremes of 23.6 to 34.5 ppt (Able et al., 1992). At the Long-Term Ecosystem Observatory at 15 m (LEO-15) study site, located on the Beach Haven Ridge immediately outside Little Egg Inlet, mean monthly sea surface temperatures range from 2.5 to 21.5°C and salinities from 28.8 to 30.6 ppt (Thomas and Milstein, 1973). The Mullica River and its tributaries drain an area of about 400,000 ha of relatively undeveloped pinelands in southern New Jersey. This area has been designated the Pinelands National Reserve and is relatively unaltered (Good and Good, 1984). Waters from this extensive drainage enter into Great Bay, a drowned river valley, and the adjacent Little Egg Harbor, a barrier beach estuary. Together, these constitute the Mullica River–Great Bay National Estuarine Research Reserve, one of the least disturbed estuaries in the northeastern United States (Psuty et al., 1993; Kennish, 2004). The shorelines of these relatively shallow (1.7 m average depth at mean low water) polyhaline marsh systems total 283 km and consist of extensive stands of saltmarsh cordgrass (*Spartina alterniflora*). They share qualities with many other estuaries in the Middle Atlantic Bight, including a moderate tidal range between < 0.7 m in Little Egg Harbor to 1.1 m near the mouth of Great Bay (Chizmadia et al., 1984; Durand, 1984; Able et al.,

Temp C

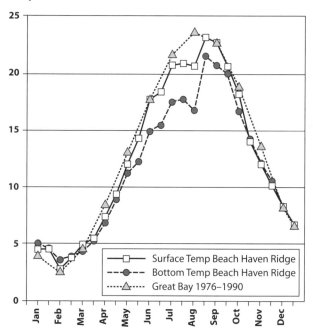

Fig. 2.6. Seasonal bottom and surface temperatures at Beach Haven Ridge and in Great Bay in southern New Jersey (from Able and Fahay, 1998).

1992; Psuty et al., 1993; Able et al., 1996a). The lower portion of the estuary incorporates an extensive salt marsh peninsula that divides an embayment (Great Bay) from an adjacent barrier beach estuary (Little Egg Harbor) (see Fig. 2.3). The peninsula is located to the west of a natural inlet (Little Egg Inlet) through which the primary source of ocean water enters the estuary. A number of thoroughfares, including Little Sheepshead Creek, run through the peninsula and connect Great Bay and Little Egg Harbor. The Beach Haven Ridge (Stahl et al., 1974) is located on the inner continental shelf just outside Little Egg Inlet. It is also the LEO-15 study site (von Alt and Grassle, 1992), and collections here and in the vicinity are the source of much of our data concerning seasonality of reproduction and settlement of fishes on the inner continental shelf. Stations throughout the estuary were regularly sampled with otter trawls between 1988 and 2006. These collections contribute to our discussions of habitat relationships across the ocean-estuary-tidal freshwater ecotone.

3

Approach

Over the past three decades, an extensive array of projects conducted with a variety of sampling gears have collected eggs, larvae, juveniles, and adult fishes in temperate estuaries and on the adjacent continental shelf of the Middle Atlantic and New York bights (see Able and Fahay, 1998, for an earlier summary). Details from the relevant literature (formally and informally published) and several new aspects of the life history, ecology, and behavior of estuarine fishes are included in our present study, based on more than 76,000 samples and millions of specimens (Table 3.1). Particular attention has been paid to the Mullica River–Great Bay–Little Egg Inlet–Inner Continental Shelf corridor, from tidal freshwater to ocean salinities. Other collections extend from the Gulf of Maine to Cape Hatteras (see Fig. 2.2). The primary consideration for this multiple sampling gear approach was to ensure that we filled in gaps relative to the size of developing fish and their temporal occurrence in estuaries, all of which might influence interpretation of their life history, ecology, and behavior, including their migrations to and from estuaries.

SAMPLING APPROACHES FOR EGGS AND LARVAE

The sources of eggs and larvae from the ocean were derived from prior studies. Eggs and larvae were collected in the ocean during the NMFS-MARMAP surveys, which provided excellent spatial and temporal coverage of continental shelf waters between Nova Scotia and Cape Hatteras, North Carolina, from 1977 through 1987 (Morse et al., 1987; Berrien and Sibunka, 1999). This sampling was focused on the entire water column and was accomplished with double-oblique tows from surface to bottom or 200 m depth, whichever was less, using 60 cm bongo plankton samplers fitted with 0.505 mm mesh nets. Data derived from these surveys provide information on the timing of reproduction, distribu-

tion and abundance of eggs and larvae, and maximum sizes attained in the pelagic larval stage.

In an attempt to discern the sources of fish larvae entering estuaries and the connectivity between the ocean and the estuary, we also sampled the surfzone and nearshore ichthyoplankton (see Table 3.1). The study area was located on a 15 km stretch of beach on the coast of New Jersey. The beaches in this region are typically high energy, exposed, and steeply sloped (Able et al., 2010). Wave heights average 0.3 to 1.2 m, with a period of 5 to 9 seconds and a tidal range of approximately 1.4 m. Within the study area are two inlets: Manasquan Inlet, at the southernmost extent, and Shark River Inlet, located roughly in the center. Sampling occurred monthly from May to July, 1995 through 1999, with each sampling event occurring over a week-long period (see Able et al., 2010, for additional details).

In a separate 18-year study, larvae were sampled weekly on night flood tides with plankton nets (1 m diameter mouth, 1.0 mm mesh) suspended in Little Sheepshead Creek behind Little Egg Inlet in southern New Jersey (see Table 3.1; Fig. 3.1) to determine the species composition and seasonal and annual variation in abundance. All fish larvae were collected from a bridge located 3 km from the Little Sheepshead Creek mouth and 2.5 km from Little Egg Inlet. Water depth at the collection site is approximately 4 m. A portion of Atlantic Ocean water flows into Little Sheepshead Creek during flood tides. These samples targeted larvae ingressing from adjacent Atlantic Ocean waters as well as larvae resulting from local estuarine spawning (Witting et al., 1999). At least three replicate half-hour sets were made on each sampling date between 1989 and 2006. Additional details of this sampling program and comparisons to a similar program inside Beaufort Inlet, North Carolina, are provided elsewhere (Warlen et al., 2002; Sullivan et al., 2006a).

In an attempt to determine the influence of upwelling on the species composition and abundance of fish larvae, we analyzed larval catches from the same long-term study site in the Great Bay–Little Egg Harbor Estuary (see Fig. 3.1) before, during, and after upwelling events (Neuman et al., 2002). We recognized the need to include additional ichthyoplankton sampling (semiweekly in July 1997 and July 1998, and daily over a two-week interval in July 1999) in order to better match the temporal scale of biological sampling with physical oceanographic monitoring. The hydrography of the study area was described by accessing data from YSI data loggers at three monitoring sites: Buoy 126, Buoy 139, and Little Sheepshead Creek Bridge (see Fig. 3.1). Upwelling events during May–September 1989–1999 were detected and monitored by gathering the following information on a daily basis prior to, during, and after upwelling events: (1) sea surface temperatures via two buoy markers approximately 30 km offshore (1989–1993) and via Advanced Very High Resolution Radiometer (AVHRR) satellite imagery along a transect that

Table 3.1. Sampling effort for larvae, juvenile, and adult fishes. See Figs. 3.1–3.4 for location of sampling sites. Datasets incorporated from Able and Fahay (1998) are also included.

Gear	Location/Habitat	Water Depth (m)	No. of Stations	Duration/ Frequency	Focus of Sampling (life history stage)	No. of Samples	Source
Bongo plankton net (0.505 mm mesh)	Middle Atlantic Bight Continental Shelf	11–1400	≈ 180	1977–1987, 6–8/year	Eggs and larvae	11,438	Sibunka and Silverman (1984, 1989); Morse et al. (1987); Berrien and Sibunka (1999)
Plankton net (1 m, 1.0 mm mesh)	Little Sheepshead Creek, Great Bay–Little Egg Harbor estuary, NJ	4	1	Feb 1989–Dec 2006/ weekly	Larvae	9222	Witting et al. (1999)
Plankton net (0.5 m, 505 μ mesh)	Ocean beach along northern New Jersey	0–2, 6–7	31	1995–1999/monthly from May to July	Larvae	465	Able et al. (2010)
Wire mesh trap (6.0 mm mesh)	RUMFS boat basin in Great Bay, NJ	2–3	1	Nov 1990–Dec 2006/daily	Juveniles	25,593	Able et al. (2005c)
Seine (30 m x 1.8 m, 6 mm mesh in wings, 2 mm in bag)	Ocean along Long Beach Island, Great Bay–Little Egg Harbor estuary, NJ	0–2	12	1998–2000	Juveniles	501	Able et al. (2003b)
Otter trawl (4.9 m, 6.0 mm mesh)	Great Bay–Little Egg Harbor–Mullica River, NJ	0.6–5.0	21	Jun 1988–Sept 2006/monthly to twice annually	Juveniles/ adults	2602	Szedlmayer and Able (1996)
Multi-mesh gill net	Delaware Bay, Mullica River–Great Bay	0–2.5	7	June–Nov 2001	Juveniles/ adults	2298	Able et al. (2009b)
Otter trawl	Gulf of Maine, Georges Bank, Middle Atlantic Bight	27–>193	—	Fall, winter, and spring, 1982–2003	Young-of-the-year/ juveniles and adults	> 16,000	Grosslein and Azarovitz (1982); Able and Brown (2005)
Tucker trawl (0.505 mm mesh)	Inner Continental Shelf (Beach Haven Ridge)	2–19	2	Jul 1991–Nov 1992/ monthly	Larvae	138	This study
Dip net (night-light)	RUMFS boat basin, Great Bay	2–3	1	Apr 1986–Aug 1992/ aperiodic	Pelagic late larvae	153	Able et al. (1997)
2 m beam trawl (3.0 mm mesh)	Beach Haven Ridge and Great Bay	2–19	30	Jul 1991–Oct 1994/ monthly	Early settled juveniles to adults	363	This study
Experimental trap	RUMFS boat basin	2–3	7	Jul 1992–Dec 1994/ daily	Juveniles	3400	This study
Throw trap (1 m²)	Great Bay–Little Egg Harbor	0.2–0.4	6	May–Sep 1988–May–Sep 1989/ biweekly	Juveniles	436	Sogard and Able (1991)
Beam trawl (3.0 mm mesh)	Great Bay–Little Egg Harbor	0.6–5.0	36	May 1992–October 1995/monthly	Juveniles/ adults	1675	This study
Gear comparison (seine, beam, otter trawls)	Great Bay–Little Egg Harbor	01.–5.0	26	May 1991–Oct 1991/ monthly	Juveniles/ adults	507	This study
Weir, seine	Tidal creek, Great Bay	0.5–2.0	3	Apr 1987–Apr 1991	Juveniles/ adults	192	Rountree et al. (1992)
Seine (4 mm mesh)	Great Bay–Little Egg Harbor	0.1–1.0	10 and 6	May 1990–Apr 1991/ monthly Jun 1994–Oct 1995/ monthly	Juveniles/ adults	408 195	This study
Throw trap	Marsh surface, Great Bay	< 1.0	30	May 1990–Jul 1991 (summer only)/ biweekly	Juveniles	840	Smith (1995)
Pop net	RUMFS boat basin, Great Bay	2–3	1	Aug 1995–May 1996/weekly	Juveniles/ adults	224	Hagan and Able (2003)

extended from the lower portion of the estuary to approximately 32 km offshore (1993–1999); (2) meteorological data (wind speed and direction) derived from a wind gauge at the Rutgers University Marine Field Station (RUMFS) (1989–1993) and from a tower located at RUMFS (1993–1999); and

(3) inshore (estuarine) water temperatures via hand-held thermometer readings at RUMFS (1989–1995), via YSI data loggers at Jacques Cousteau National Estuarine Research Reserve (JCNERR) ecological monitoring sites and an additional site at Little Sheepshead Creek Bridge (1996–1999),

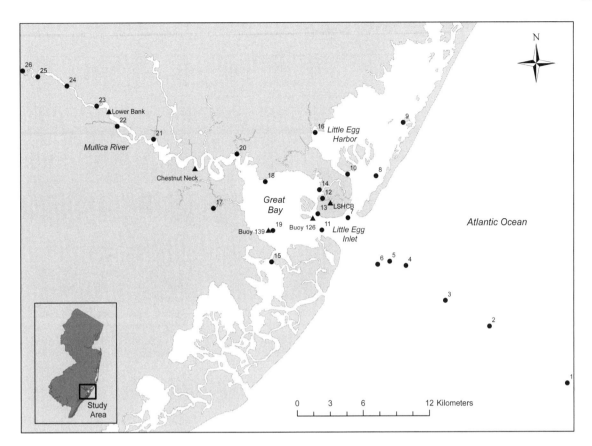

Fig. 3.1. Small otter trawl sampling locations for juvenile and adult fishes in the Ocean–Great Bay–Mullica River corridor in southern New Jersey during July and September, 1988–2006. LSHCB indicates ichthyoplankton sampling site in Little Sheepshead Creek.

and via an inshore buoy marker located approximately 5 km from RUMFS. Water temperatures derived from the four inshore and the two offshore sources were compared during a preliminary study conducted May–September 1997–1998. Upwelling events were identified by computing the difference between mean daily offshore and inshore sea surface temperatures and examining mean daily wind velocity and direction.

SAMPLING APPROACHES FOR JUVENILES AND ADULTS

We sampled recently settled juvenile fishes with a variety of gears across coastal ocean and estuarine locations (see Table 3.1; Fig. 3.1). A large number of samples was obtained from daily/weekly deployment of wire mesh traps in the RUMFS boat basin from 1990 to 2006. Other juvenile and adult fishes were sampled with a long-term otter trawl program (1988–2006) at a variety of stations/habitats located throughout the Mullica River–Great Bay–Inner Continental Shelf corridor (Table 3.2; see Fig. 3.1). These stations were chosen based on their depth, substrate type, and amount and type of structured habitat. In addition, these stations were distributed along the salinity gradient from the ocean to tidal freshwater, and thus provided an understanding of species distri-

bution across the ocean-estuary-tidal freshwater ecotone. In this sampling program, fish were collected with 3 to 4 replicate samples with a trawl (4.9 m headrope, 19 mm mesh wings, 6.3 mm mesh liner) at each station during the daytime. From each tow, all fishes were identified and counted and 20 of each were randomly measured to the nearest mm total length (TL) or fork length (FL). (See Szedlmayer and Able [1996] for additional details.) In an attempt to determine the contribution of ocean habitats to the species treated here, we sampled juvenile fishes on estuarine and ocean beaches with beach seines (30 × 1.8 m, 6 mm mesh in the wings, 2 mm mesh in the bag) during 1998–2000, 2005, and 2006 (Fig. 3.2). (See Able et al. [2003b] for additional details.)

In order to determine the estuarine distribution of larger juvenile and adult fishes in space and time, we sampled with anchored multi-mesh gill nets (15 m × 2.4 m with 5 panels of 5 mesh sizes [2.5, 3.8, 5.1, 6.4, and 7.6 cm box] and 91 m × 2.4 m with 6 panels of 3 mesh sizes [1.3, 1.9, and 2.5 cm box]) in the Mullica River–Great Bay estuary at several locations (see Tables 3.1, 3.3; Fig. 3.3). Gill nets were set at biweekly intervals during the spring, summer, and fall in upper creek, creek mouth, and nearshore bay habitats. In most instances gill nets were set during the day for approximately 60 minutes. The number of samples with each net size varied between

Table 3.2. Habitat characteristics at sampling locations along the Atlantic Ocean–Great Bay–Mullica River transects in southern New Jersey (Fig. 3.1). Estimates of habitat complexity and type are from Martino and Able (2003) and Vasslides and Able (2008).

Sampling Station	Distance from Shore (km)	Habitat Type	Depth Range (m)	Average Salinity (ppt)	Habitat Complexity Index
1	23.4	shell hash	19.9	31	2
2	14.4	shell hash	18.0	31	2
3	9.7	shell hash	16.3	31	2
4	5.5	sand/clay/*Diopatra* tubes/shell hash	13.6	30	3
5	3.9	clay/silt/sand/*Diopatra* tubes	11.5	30	3
6	3.5	sand/macroalgae/*Diopatra* tubes/clay/silt	9.4	30	3
7	0	bare sand	2.8	29	1
8	Intracoastal waterway	sand/shell	4.0–7.0	30	1
9	Marshelder Island	eelgrass/sponge	0.6–0.9	28	6
10	Marshelder Channel	shell/sand	2.7–3.7	29	3
11	Grassy Channel	sand/hydroid	1.2–4.3	29	4
12	Little Sheepshead Creek	sponge/shell/sand/silt/peat	3.7–6.1	29	5
13	Newman's Thorofare	sand/rubble	3.0–7.9	29	1
14	Cape Horn	sand/amphipod tubes	0.6–2.0	29	1
15	Little Bay	sand/sea lettuce	0.9–1.5	28	4
16	Mouth of Tuckerton Creek	silt/detritus	1.5–3.0	27	2
17	Mott's Creek	silt/peat/hydroids	1.5–2.5	22	4
18	Graveling Point	shell/silt/clam beds	2.1–2.5	24	3
19	Intracoastal waterway	silt/sand	1.7–3.7	26	1
20	Ballanger Creek	sand/silt	2.0	27	2
21	Turtle Creek	sand/silt	3.0	16	2
22	Landing Creek	silt/woody debris	3.4	13	3
23	Ed's Creek	silt/woody debris	3.2	8	3
24	Greenbank	silt/woody debris	3.0	1	3
25	Sweetwater	silt/woody debris	2.0	0	3
26	The Forks	silt/woody debris	0.6	0	3

areas. Within each area, the position in which each net was set varied such that no two locations were resampled. For each gill net sample, depth, surface temperature, salinity, and dissolved oxygen were recorded with a hand-held YSI (Table 3.3). Upon retrieval of each gill net, all fishes were identified, counted, and measured. Several of the dominant species collected were represented by multiple age classes; thus, fish were divided into two age classes: young-of-the-year (age 0) and juveniles and adults (age 1+) based on available monthly size estimates (Able and Fahay, 1998; Table 3.4).

Seasonal distributions for young-of-the-year and age 1+ juveniles and adults of estuarine species on the continental shelf were determined from the Gulf of Maine to south of Cape Hatteras by the NMFS Northeast Fisheries Science Center bottom trawl surveys (Azarovitz, 1981; Grosslein and Azarovitz, 1982). Species were selected for this study based on their frequent use of estuaries as young-of-the-year, and because all of them were transient forms that left estuaries and resided on the continental shelf for other periods of the year (Able and Fahay, 1998). The size designations for young-of-the-year and juveniles and adults are the same as those used in gill net samples (Table 3.4).

Samples were collected on the continental shelf at stratified random stations between Cape Hatteras, North Caro-

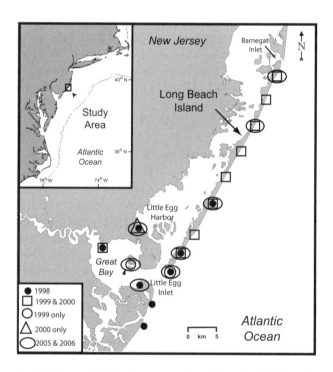

Fig. 3.2. Seine sampling locations on estuarine and ocean beaches in southern New Jersey.

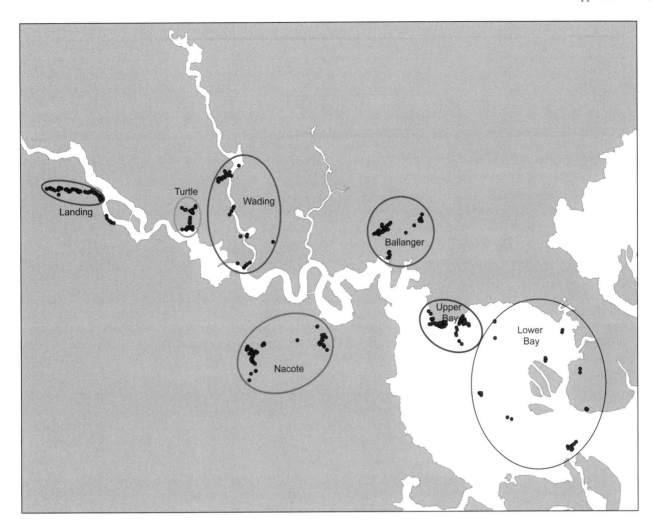

Fig. 3.3. Gill net sampling locations in the Mullica River–Great Bay estuary.

lina, and Georges Bank and the Gulf of Maine (Figs. 3.4 and 3.5) during fall (September–October), winter (January–February), and spring (March–April) (Grosslein and Azarovitz, 1982; Able and Brown, 2005). The geographical limits of the sampling program, however, varied with season and between years. Similar sampling efforts and distribution of samples occurred in the fall (7379 tows) and spring (7418 tows) over the 1982–2003 period (Fig. 3.4). The winter sampling effort was reduced in terms of number of tows (1552 tows) and geographical extent during the years in which it occurred (1992–2003). It was limited to the southern portion of Georges Bank and off Nantucket Shoals to the north and just north of Cape Hatteras to the south (see Fig. 3.4). Moreover, the number of samples in the shallow waters (depth less than 25 m) off Massachusetts and from New Jersey to North Carolina was reduced in the winter relative to the fall and spring (see Fig. 3.5). The distribution of samples varied with depth as well, with some less than 20 m (17%), a large proportion less than 100 m (81%), fewer between 100 and 250 m (16%) and fewer still in depths > 251 m (2%) (see Fig. 3.5). In addition, the distributions represented in the seasonal distri-

bution maps are composites over long periods of time and thus cannot reflect annual variation in abundance. For example, many southern species reach the northern boundaries of their distribution in the Middle Atlantic Bight and the northern limits in these maps may simply reflect one or more years when they extended their range. In the same manner, the maps for those species with more northern ranges may simply reflect a few years in which they may have been found farther south. In addition, these patterns may differ between cohorts (e.g., *Scophthalmus aquosus*) and where there are multiple stocks within the study area (e.g., *Paralichthys dentatus*). Details of the seasonal distribution for individual species are provided in the species accounts.

DEVELOPMENT

Individuals undergoing the larval/juvenile transition were selected for illustration because this portion of the early life history is often poorly described. Most of the illustrations originate from Able and Fahay (1998). Others came from recent collections of material not previously illustrated. The remainders are from the literature.

Table 3.3. Sampling effort and environmental characteristics at gill net sampling stations in the Mullica River–Great Bay estuary. See Fig. 3.3 for station locations.

Location	Distance from Little Egg Inlet (km)	Habitat Type	Average Depth (m)	Average Salinity (ppt)	Number of Samples	Number of Fish Captured
Landing Creek	30.3	Marsh creek	2.2	5.8	75	999
Turtle Creek	25.7	Marsh creek	2.2	12.7	41	662
Wading River	22.6	Marsh creek	2.5	10.7	80	737
Nacote Creek	15.5	Marsh creek	2.5	16.4	84	836
Ballanger Creek	10.6	Marsh creek	2.5	20.9	73	344
Upper Bay	5.8	Open bay	2.2	26.9	99	924
Lower Bay	3.2	Open bay	3.7	28.7	19	556

Table 3.4. Checklist of selected estuarine fishes for which seasonal distribution is mapped based on National Marine Fisheries Service (NMFS) trawl surveys. Size distinction for young-of-the-year (at 12 months), older juveniles, and adults (age 1+), and sample sizes are presented. See Figs. 3.4 and 3.5 for location and sampling depths.

Scientific Name	Common Name	Maximum Size at End of the First Year (cm)	Sample Size	
			Young-of-the-Year	Age 1+
Carcharhinidae				
Mustelus canis	smooth dogfish	< 70	6544	35,474
Congridae				
Conger oceanicus	conger eel	< 28	52	504
Engraulidae				
Anchoa hepsetus	striped anchovy	< 7	339,641	1,355,308
Anchoa mitchilli	bay anchovy	< 10	6,182,328	7920
Clupeidae				
Alosa aestivalis	blueback herring	< 15	8604	15,911
Alosa pseudoharengus	alewife	< 12	6319	62,507
Alosa sapidissima	American shad	< 15	881	3922
Brevoortia tyrannus	Atlantic menhaden	< 17	8043	1533
Clupea harengus	Atlantic herring	< 12	23,921	421,201
Synodontidae				
Synodus foetens	inshore lizardfish	< 25	2202	2389
Ophidiidae				
Ophidion marginatum	striped cusk-eel	< 10	20	919
Phycidae				
Urophycis chuss	red hake	< 17	23,480	121,518
Urophycis regia	spotted hake	< 20	207,382	62,096
Urophycis tenuis	white hake	< 35	11,550	17,934
Gadidae				
Pollachius virens	pollock	< 20	1813	13,843
Mugilidae				
Mugil cephalus	striped mullet	< 19	2	2
Mugil curema	white mullet	< 19	8	0
Atherinopsidae				
Menidia menidia	Atlantic silverside	< 12	9365	212
Gasterosteidae				
Gasterosteus aculeatus	threespine stickleback	< 6	67	115
Syngnathidae				
Hippocampus erectus	lined seahorse	< 10	67	9
Syngnathus fuscus	northern pipefish	< 19	1041	228
Triglidae				
Prionotus carolinus	northern searobin	< 8	14,530	177,193
Prionotus evolans	striped searobin	< 9	820	14,121
Cottidae				
Myoxocephalus aenaeus	grubby	< 6	370	596
Moronidae				
Morone saxatilis	striped bass	< 10	1	2132

Table 3.4 continued

Scientific Name	Common Name	Maximum Size at End of the First Year (cm)	Sample Size	
			Young-of-the-Year	Age 1+
Serranidae				
Centropristis striata	black sea bass	< 17	12,181	20,655
Pomatomidae				
Pomatomus saltatrix	bluefish	< 26	39,125	4927
Carangidae				
Caranx hippos	crevalle jack	< 20	348	15
Sparidae				
Stenotomus chrysops	scup	< 13	585,317	230,283
Sciaenidae				
Bairdiella chrysoura	silver perch	< 15	19,226	1771
Cynoscion regalis	weakfish	< 20	218,284	87,297
Leiostomus xanthurus	spot	< 15	156,927	235,564
Menticirrhus saxatilis	northern kingfish	< 23	1500	1131
Micropogonias undulatus	croaker	< 14	13,869	403,981
Pogonias cromis	black drum	< 20	33	59
Labridae				
Tautoga onitis	tautog	< 12	2	209
Tautogolabrus adspersus	cunner	< 8	49	2359
Pholidae				
Pholis gunnellus	rock gunnel	< 7	11	84
Uranoscopidae				
Astroscopus guttatus	northern stargazer	< 9	3	94
Sphyraenidae				
Sphyraena borealis	northern sennet	< 20	88	910
Stromateidae				
Peprilus tricanthus	butterfish	< 14	1,523,976	257,711
Scophthalmidae				
Scophthalmus aquosus	windowpane	< 26	36,797	39,296
Paralichthyidae				
Etropus microstomus	smallmouth flounder	< 10	9208	1080
Paralichthys dentatus	summer flounder	< 30	7867	32,457
Pleuronectidae				
Pseudopleuronectes americanus	winter flounder	< 23	5835	57,153
Tetraodontidae				
Sphoeroides maculatus	northern puffer	< 21	2516	159

In an additional attempt to improve our understanding of developmental morphology, we determined the patterns of scale formation. Specimens for this examination were collected with a variety of techniques (plankton nets, trawls, seines, dip nets), primarily in New Jersey estuaries and on the adjacent continental shelf. Effort was made to examine a length series from hatching to full coverage of scales as occurs in the adult condition. Care was taken during all stages of collection, preservation, and examination to prevent scale loss. However, some species were more likely to lose scales than others. We examined the species that tended to lose scales for the presence of scale pockets to help define their distribution. Sample sizes examined varied between species.

Individuals were stained to better visualize scales. They were preserved for at least 48 hours in either 95% ethyl alcohol (ETOH) or 10% formalin, which was switched to ETOH, then treated with a solution containing Alizarin red S to elucidate scales, as adapted from Taylor (1967) and Potthoff (1984). (See Able and Lamonaca [2006] and Able et al. [2009a] for additional details.) Subsequently, these specimens were visually examined with a stereomicroscope and patterns of scale formation were illustrated on blank templates for individual species primarily adapted from figures in Able and Fahay (1998). The illustrations were then transferred to a digital medium via Adobe Photoshop. The area covered by scales was calculated from these illustrations using the Image J software package (Rasband, 2003) and expressed as percent of body and percent of body and fins covered relative to the adult condition of scale coverage (100%). The adult condition is based on the smallest size at which scales are no longer being formed.

The illustrations of representative stages of scale formation were standardized across all species. For example, the

Sampling Locations

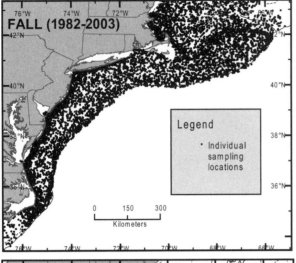

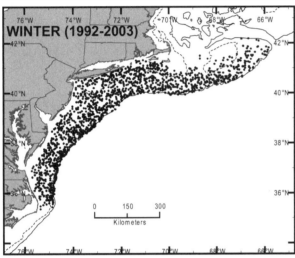

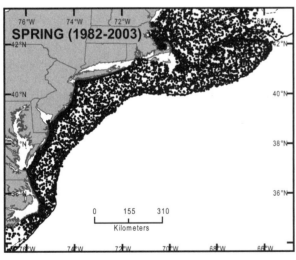

Fig. 3.4. Composite sampling locations and years sampled during fall, winter, and spring National Marine Fisheries Service surveys.

location of the lateral line is shown with a thin line, even before it is formed, to provide a local landmark. In addition, the margin of the dorsal and anal fin pterygiophores is also noted with a fine line for some flatfishes. The area behind the pectoral fins is not shaded to indicate the location of the pectoral fins in order to provide another local landmark. The darkly shaded areas indicate scale formation on the body; the gray shaded areas indicate where scales occur on the fins of some species.

BEHAVIOR

The observations of the behavior of estuarine fishes ranged from incidental ones in the field to more detailed ones in the laboratory. Observations of burying behavior were based on an assessment of burial relative to substrate grain size (sand and silt) and size of young-of-the-year cyprinodontids and fundulids. Fishes for these studies were collected from a variety of sources (seining on sandy beaches, dip netting, and wire mesh traps in marsh pools) in the vicinity of RUMFS. Once in the laboratory, these fishes were acclimated for several days in sediment-free containers in ambient flow-through water and fed flake food daily. Sediments for burying observations were gathered from the fish-collecting sites and sieved to standardized grain size for sand and silt. Cylindrical containers (20 cm in diameter) were filled with 5 cm of sediment and 2 l of water and partially submerged for 24 hours to ensure that the sediment had settled. The same containers were then reimmersed in the flow-through seawater to stabilize temperatures during the periods of observation.

Once fishes of different species and sizes were ready for testing, they were allowed to acclimate in the containers for 24 hours. Subsequently, their ability to bury was determined over 48 hours by visual observations. If burial was not observed, the fish was challenged by flicking the surface of the water with a finger. Percent burial by species and size was determined by the composite of both types of observations. Sizes of fishes tested were determined at the end of each set of observations by measuring individuals to the nearest mm total length (TL).

HABITAT USE

In an attempt to determine the dynamics of habitat use and behavior in the Mullica River–Great Bay estuary, we acoustically tagged and tracked the large juveniles and adults of several species. Wireless hydrophones were deployed as a series of gates in order to enhance detection of tagged fishes while in residence or moving along the estuarine gradient (see Fig. 3.6). At the entrance to the estuary (Little Egg Inlet), hydrophones 2, 3, and 4 (recorded as positioned at 0 km) were arranged to take advantage of local topography, such as at sand bars, channels, and the like, to detect fishes moving along several passages. The entrance to Little Egg Harbor was monitored by hydrophone 1 (considered to

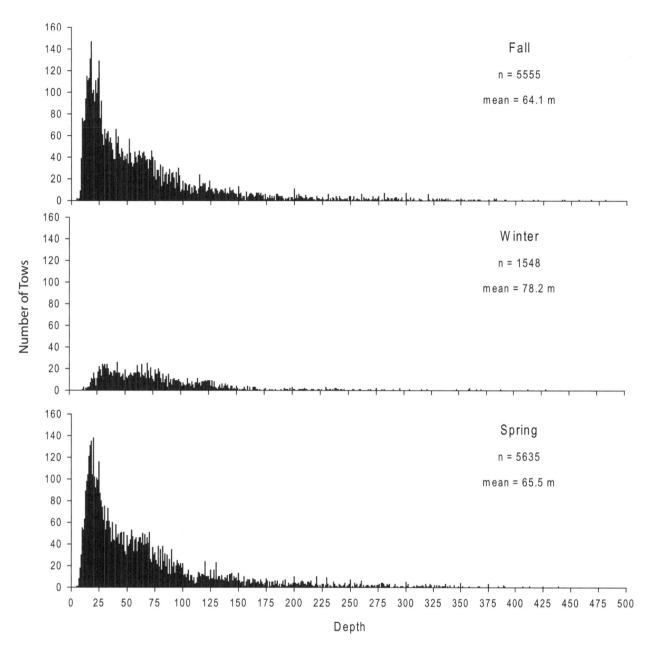

Fig. 3.5. Frequency distribution of sample depths in study area south of 42°N during fall (1982–2003), winter (1992–2003), and spring (1982–2003).

be 0 km from the inlet for the purposes of this study). This same hydrophone, along with hydrophone 5, also served to identify fishes moving through a deep (to 7 m) channel (Little Sheepshead Creek) between Little Egg Harbor and Great Bay. The channel exiting Great Bay to the south (Main Marsh Thorofare) was monitored by hydrophone 13 (4.5 km from inlet), although this hydrophone was deployed later than the others. Hydrophone 5 (4.5 km from inlet) also served to monitor fishes passing through the deepest channel in Great Bay (Newmans Thorofare). The next gate upstream was located in the Mullica River (hydrophones 6, 7, 8, and 9; approximately 18 km from the inlet). Hydrophones 6 and 8 were removed after a test period because they were

largely redundant. Farther upstream, the next gate consisted of a single hydrophone (hydrophone 10; 28.3 km from the inlet) just above the saltwater-freshwater interface. On occasion, another hydrophone (hydrophone 11; 38.1 km from the inlet) was deployed farther upstream in tidal freshwater. Additional details of this estuarine observatory are provided in Grothues et al. (2005).

Fishes bearing surgically implanted acoustic transmitters (76.8 KHz) with an individual identification code were detected when they came within range (approximately 500 m; Grothues et al., 2005) of moored wireless hydrophones (WHS-1100, Lotek Wireless, Inc., St. Johns, Newfoundland, Canada), which were suspended at a depth of

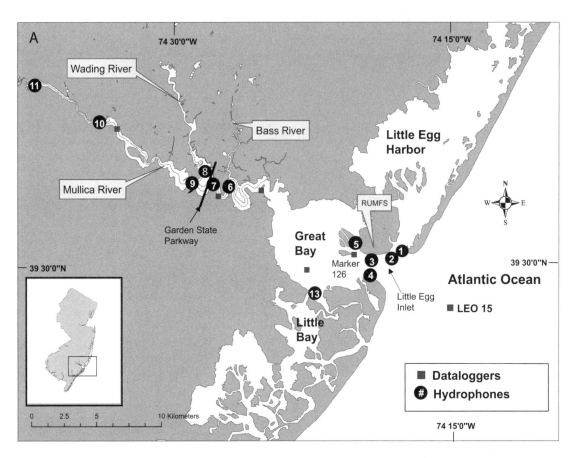

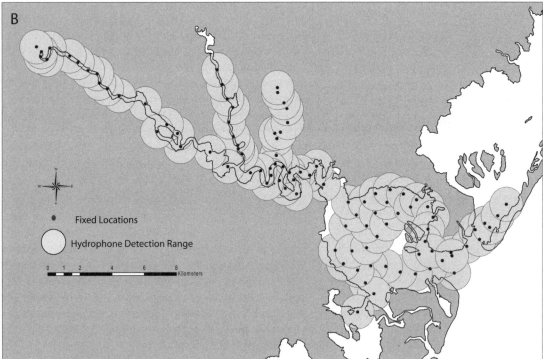

Fig. 3.6. (A) Location of hydrophones in the Mullica River–Great Bay estuary for tracking acoustically tagged fish. (B) Fixed locations (dots within circles) for weekly attempts to detect acoustically tagged fish in the Mullica River–Great Bay estuary. Circles represent the approximate detection range (500 m radius) of a directional mobile hydrophone that scanned 360 degrees while deployed from a boat.

3.2 m, where surrounding total water depth reached a depth of 10 m. Wireless hydrophones transmitted received sound in the 76.8 KHz band by VHF radio frequency unique to the unit (between 148 and 152 MHz) to shore-based receivers for interpretation and logging of the data in real time (see Grothues et al., 2005, for additional details). The JCNERR study area also provided useful infrastructure for routine environmental monitoring. Permanent instrumentation includes dataloggers used to record salinity, temperature, pH, and water depth along the estuarine gradient (see Fig. 3.1).

In addition, two mobile tracking methods were used. In the first, tracking of individual fishes occurred once a week primarily during daylight hours for 8 to 14 hours from March through December or year-round in some years. To spatially and temporally standardize tracking, 113 fixed locations encompassing an area of 150.9 km² were selected using a Geographic Information System software package (ArcView version 3.2; ESRI, Redlands, California) and visited with a directional mobile hydrophone (Lotek Wireless) (see Fig. 3.6B). Listening range resolution with the mobile hydrophone over the area varied from 0.5 to 3.0 km.

At each of the above locations, the hydrophone was lowered 1.0 m into the water and pointed at the 4 principal ordinates for 30 seconds. When a fish was detected, its position was triangulated by moving until a reading of 130 dB or above was detected at a gain of 15 or less (approximately 2 m from the hydrophone). Location was recorded with a Global Positioning System (GPS) unit in Universal Transverse Mercator (UTM) coordinates. Water temperature, salinity, and dissolved oxygen (YSI Model 85; Yellow Springs Instruments Inc., Yellow Springs, Ohio), along with date, time, tag number, tidal stage, and depth, were also recorded. Tracking was not conducted on days when the listening range was less than 0.5 km, which corresponded to wind velocities greater than 30 km/hour, or on days when there was heavy rainfall or thunderstorms.

The second tracking protocol involved following an individual for four- to five-hour periods (or until contact was lost) to further determine details of habitat use, site fidelity, and movement patterns. This procedure allowed small-scale discrimination of habitat use. Location and environmental parameters were recorded every 15 to 20 minutes, as previously described. Limited attempts at tracking at night were conducted to analyze diel variation in movement.

PREY

The prey of estuarine fishes was determined by a review of the literature and some original observations for young-of-the-year and larger juveniles and adults. More than 400 sources of published literature were reviewed on food habits of young-of-the-year of estuarine fishes in the study area (Mancini and Able, 2005). Several criteria were evaluated before a publication (Table 3.5) was included in the synthesis. First, species of interest included dominant species in estu-

aries in the study area, primarily those treated by Able and Fahay (1998). Second, we required that each study examine food habits of fishes within the size range typical of young-of-the-year within the Middle Atlantic Bight based on the size limits outlined by Able and Fahay (1998). Third, it was necessary that each study have a minimum study size (i.e., > 18 individuals/species). A published study or unpublished thesis or dissertation that met all of the above criteria was then further examined for detailed information on size and type of prey items consumed. This information was then tabulated in three ways: (1) a composite summary, consisting of a single line of information summarizing the food habits for each species; (2) a summary that provided more detailed information from each individual study on the food habits of each predator species by study; and (3) a separate summary that focused on those predator species that had fed on fish prey.

One difficulty encountered with this literature-based survey was that there was no standardization of prey categories or taxonomic levels between authors or studies; therefore, choosing prey categories was sometimes difficult. In general, if a prey item was mentioned in any of the studies we examined, it was added to our list of prey items. It was not feasible to present every prey item at the species level; therefore, many of the closely related organisms were combined into broader taxonomic levels. In addition, it was difficult to standardize the importance of different prey taxa because methods of data analysis and presentation of results were not standard across studies. As a compromise, the information we collected from each individual study was arranged into three categories: important prey, minor/rare prey, and absent. Where possible, the category assigned to each prey item was based on the original author(s)'s interpretation of the results of each study. If the study author(s) suggested a prey item was important, it was classified as important prey. All other prey items noted by the study were considered minor/rare prey. Prey items not mentioned in the study were considered absent. In studies where the author(s) did not suggest important prey items, we provided our own interpretation, considering a prey item important if it constituted greater than 20% of the diet. When we made original observations of stomach contents for young-of-the-year and larger juveniles and adults, we followed the protocols described in Nemerson and Able (2003, 2004). In other instances, for more recent literature and especially that for fishes larger than young-of-the-year, we incorporated that literature into each species account.

PREDATORS

We also reviewed the literature to identify prey of and predators on young-of-the-year of estuarine fishes (see Table 3.5). In other instances, we used directed sampling with gill nets. For these samples, all predators retained for dietary analysis were placed on ice, and their stomachs extracted within an hour after collection and then preserved in 95% ETOH. Upon return to the laboratory, contents were dissected from the stomachs and placed in vials filled with 95% ETOH. A solution of rose bengal stain was later added to the vials prior to examination to aid in the identification of the organic contents. Stomachs were initially categorized as either "empty" or "with food items," and those stomachs with food items were further classified as having parts or whole bodies of fish prey ("piscivorous") or no fish prey but other food items ("other"). Additionally, gut fullness was recorded on a qualitative scale: 0 = completely empty (the stomach lining was very thick and wrinkled), 1 = 1–25% full, 2 = 26–50% full, 3 = 51–75% full, and 4 = 76–100% full (with very thin, unwrinkled, and sometimes translucent stomach lining [adapted from Lambert, 1985]).

The proportion by weight of each prey category was determined according to the sieve fractionation method of Carr and Adams (1972) as adapted by Nemerson and Able (2003, 2004). The method was modified to focus on the fish component of the diet; thus, only three sieve sizes (2000, 850, and 75 μm) were used. The top two sieves typically retained the piscivorous components of the stomach contents. The sieve contents were then rinsed into sorting bowls and all contents were identified to the lowest practical taxonomic level. Qualitative ratios based on the sizes of constituents belonging to the same taxonomic subdivision were recorded for each sorting bowl. The contents from separate sorting bowls were then vacuum filtered onto dried, preweighed glass fiber filters, dried in a 60°C oven for 48 hours, and then weighed (± 0.0001 g). Several reference materials were used to aid in the identification of fish remains (Gregory, 1933; Brodeur, 1979; Daniels, 1996; and the RUMFS fish reference collection).

OUTLINE OF A TYPICAL SPECIES ACCOUNT
Selection of a species for a detailed account (see Table 4.4) was based on a number of criteria. Generally, the more abundant species, which are more frequently collected, are included as well as economically important species, which have often been the focus of earlier research. Other species are treated because, while they are not abundant, they may be expected to alter their life histories in response to climate change. As a result, most of the common or potentially important species in the study area have been incorporated into this synthesis. We also include documentation for rare species collected in the primary study area (see Table 4.5).

Each species account begins with the scientific name, the author of that name, and the currently used common name. Scientific nomenclature, common names, and the sequence in which families are listed generally follow Nelson et al. (2004). Departures from the latter classification include that for the Pleuronectiformes, where we follow Chapleau (1993) and Cooper and Chapleau (1998), and the Gadiformes, based on Fahay and Markle (1984). A life history calendar that summarizes the monthly occurrence of important aspects in the lives of many of these species is included and indicates principal habitats (estuary, ocean, or both) where these events occur. This interpretation is based on our original information and that from the literature. Emphasis is on the first year of life and typically represents observations in the central portion of the Middle Atlantic Bight. An illustration of an individual undergoing the larval/juvenile transition is provided.

For each species account, we endeavored to provide information in the following categories:

—distribution
—reproduction and development
—larval supply, settlement, growth, and mortality
—seasonality and habitat use
—prey and predators
—migrations

ANECDOTES
In addition to extensively documented synthesis and interpretations of large datasets, we have had the opportunity, over several decades, to make incidental observations that we think provide insights into the life history, ecology, and behavior of selected estuarine fishes. These anecdotal observations can be based on unique observations supplemented with novel interpretations. These are treated briefly, on various subjects, throughout this book in the hope that they will foster additional data collection to support or refute our suggestions or questions.

Table 3.5. Details of specific studies that identified prey for young-of-the-year estuarine fish (from Mancini and Able, 2005).

Species/Code	Location	Salinity Zone	Size Class (mm)	Method	Sample Size	Source
Mustelus canis	Little Egg Harbor–Great Bay Estuary, NJ	Poly-Euhaline	318–586 TL	%FO, Rank	85	Rountree & Able 1996 Table 3
Anguilla rostrata (A)	New Jersey Streams	N/A	145–244 TL	%FO	11	Ogden 1970 Table 3
Anguilla rostrata (B)	Delaware River Estuary	Limnetic-Euhaline	50–240 TL	%FO	43	De Sylva et al. 1962 Table 15
Anguilla rostrata (C)	James River Drainage, VA	N/A	100–249 TL	%W, %N	133	Lookabaugh & Angermeier 1992 Figures 2 & 3
Alosa aestivalis (A)	Connecticut River	N/A	5–16+	%Comp	196	Crecco & Blake 1983 Table 1
Alosa aestivalis (B)	Connecticut River	N/A	29–106 TL	%N, %V	116	Domermuth & Reed 1980 Table 1
Alosa aestivalis (C)	Hudson River Estuary, NY	Mesohaline	Mean 56.3 TL	%FO, %Comp, Signif.	149	Grabe 1996 Table 3
Alosa aestivalis (D)	James River, VA	N/A	10–90 FL	%Comp	1149	Burbidge 1974
Alosa pseudoharengus	Hudson River Estuary, NY	Mesohaline	Mean 64.8 TL	%FO, %Comp, Signif.	48	Grabe 1996 Table 3
Alosa sapidissima (A)	Connecticut River	N/A	10–16+	%Comp	372	Crecco & Blake 1983 Table 1
Alosa sapidissima (B)	Hudson River, NY	N/A	45–65 FL	%FO	55	Walburg 1957 Table 1
Alosa sapidissima (C)	Hudson River, NY	Mesohaline	Mean 64.9 TL	%FO, %Comp, Signif.	168	Grabe 1996 Table 3
Alosa sapidissima (D)	Pamunkey & Mattaponi rivers, VA	N/A	45–95 FL	%FO, %V	2049	Massmann 1963 Tables 2 & 3
Alosa sapidissima (E)	Pamunkey River, VA	N/A	47–85 FL	%FO	112	Walburg 1957 Table 1
Alosa sapidissima (F)	Neuse River, NC	N/A	35–93 FL	%FO	105	Walburg 1957 Table 1
Brevoortia tyrannus (A)	Narragansett Bay, RI	N/A	40–60	%W	30–40	Jeffries 1975 Table 5
Brevoortia tyrannus (B)	Pettaquamscutt River & Point Judith Pond, Narragansett, RI	Meso-Euhaline	10–50	%FO	300	Mulkana 1964
Brevoortia tyrannus (C)	Long Island Sound	N/A	16.2–120.1	%FO	29	Richards 1963 Appendix
Brevoortia tyrannus (D)	Indian River, DE	N/A	19–34 FL	%FO	243	June & Carlson 1971 Table 3
Brevoortia tyrannus (E)	Indian River, DE	N/A	28–94 FL	%FO	117	June & Carlson 1971 Figure 3
Brevoortia tyrannus (F)	NC Salt Marsh Ecosystems	N/A	Juveniles	%W	N/A	Lewis & Peters 1984 Table 1
Brevoortia tyrannus (G)	Newport River Estuary, NC	Polyhaline	25–32	%Comp	N/A	Peters & Kjelson 1975 Table 1
Clupea harengus	Long Island Sound	N/A	33.9–51.1	%FO	18	Richards 1963 Appendix
Anchoa mitchilli (A)	Hudson River, NY	Euryhaline	Age-0 < 40 TL	#, %FO, %W	284	Hartman et al. 2004 Table 2
Anchoa mitchilli (B)	Delaware Bay	Oligo-Mesohaline	YOY	%W	3227	Nemerson 2001 Table A-1
Anchoa mitchilli (C)	Delaware Bay	Oligo-Mesohaline	10–30	%W	—	Nemerson 2001 Figure 1–8
Anchoa mitchilli (D)	Delaware Bay	Oligo-Mesohaline	30–40	%W	—	Nemerson 2001 Figure 1–8
Anchoa mitchilli (E)	Delaware Bay	Oligo-Mesohaline	40–60	%W	—	Nemerson 2001 Figure 1–8
Anchoa mitchilli (F)	Delaware Bay	Oligo-Euhaline	15–50 SL	Rel. Abun., %#	544	Stevenson 1958 Table 19
Anchoa mitchilli (G)	Chesapeake Bay	Oligo-Polyhaline	2.5–16.4	%FO	1485	Auth 2003 Figure 32
Anchoa mitchilli (H)	York River, VA	Oligo-Polyhaline	16–60 SL	%FO	241	Smith et al. 1984 Figure 5
Anchoa mitchilli (I)	York River, VA	Oligo-Polyhaline	16–20 SL	%FO	20	Smith et al. 1984 Figure 5
Anchoa mitchilli (J)	York River, VA	Oligo-Polyhaline	21–40 SL	%FO	99	Smith et al. 1984 Figure 5
Anchoa mitchilli (K)	York River, VA	Oligo-Polyhaline	41–60 SL	%FO	80	Smith et al. 1984 Figure 5
Anchoa mitchilli (L)	Newport River Estuary, NC	N/A	20–60	%FO	398	Morgan 1990 Figure 3
Microgadus tomcod (A)	Hudson River Estuary	Meso-Polyhaline	42–75	%Comp	N/A	Metzger et al. 2001
Microgadus tomcod (B)	Haverstraw Bay, Hudson River, NY	Limnetic-Mesohaline	28.9–143.8	%FO, %Comp., Index	577	Grabe 1978 Table 2
Microgadus tomcod (C)	Haverstraw Bay, Hudson River, NY	Limnetic-Mesohaline	40–89	Importance	279	Grabe 1978 Table 3
Microgadus tomcod (D)	Haverstraw Bay, Hudson River, NY	Limnetic-Mesohaline	> 90	Importance	279	Grabe 1978 Table 3
Pollachius virens	Long Island Sound	N/A	23.1–66.2	%FO	22	Richards 1963 Appendix
Urophycis chuss	Long Island Sound	N/A	70–203.8	%FO	48	Richards 1963 Appendix
Urophycis regia (A)	Long Island Sound	N/A	48.2–195	%FO	33	Richards 1963 Appendix
Urophycis regia (B)	Magothy Bay, VA	Poly-Euhaline	34–199	Mean %V, #, FO	45	Kiel 1973 Table 15
Urophycis regia (C)	Magothy Bay, VA	Poly-Euhaline	34–80	Mean %V, #, FO	26	Kiel 1973 Table 16 & Figure 6
Urophycis regia (D)	Magothy Bay, VA	Poly-Euhaline	81–199	Mean %V, #, FO	19	Kiel 1973 Table 16 & Figure 6

Table 3.5 continued

Species/Code	Location	Salinity Zone	Size Class (mm)	Method	Sample Size	Source
Opsanus tau	Annemessex River, MD	N/A	20–100 TL	#	56	Chrobot 1951 Table 1
Cyprinodon variegatus (B)	Hereford Inlet, NJ	Poly-Euhaline	10–35	%FO	50	Clymer 1978 Table 36
Cyprinodon variegatus (C)	Hereford Inlet, NJ	Poly-Euhaline	10–15	%FO	N/A	Clymer 1978 Table 36
Cyprinodon variegatus (D)	Hereford Inlet, NJ	Poly-Euhaline	16–20	%FO	N/A	Clymer 1978 Table 36
Cyprinodon variegatus (E)	Hereford Inlet, NJ	Poly-Euhaline	21–25	%FO	N/A	Clymer 1978 Table 36
Cyprinodon variegatus (F)	Hereford Inlet, NJ	Poly-Euhaline	26–30	%FO	N/A	Clymer 1978 Table 36
Cyprinodon variegatus (G)	Hereford Inlet, NJ	Poly-Euhaline	31–35	%FO	N/A	Clymer 1978 Table 36
Cyprinodon variegatus (H)	Barnegat Bay, NJ	Poly-Euhaline	< 57	N/A	35	Pyle 1964
Cyprinodon variegatus (I)	Rehobeth & Indian River bays	Meso-Euhaline	11–50 SL	#, %FO, % V	122	Warlen 1964 Table 4
Fundulus heteroclitus (A)	Hereford Inlet, NJ	Poly-Euhaline	16–40 SL	%FO	50	Clymer 1978 Table 36
Fundulus heteroclitus (B)	Hereford Inlet, NJ	Poly-Euhaline	16–20 SL	%FO	N/A	Clymer 1978 Table 36
Fundulus heteroclitus (C)	Hereford Inlet, NJ	Poly-Euhaline	21–25 SL	%FO	N/A	Clymer 1978 Table 36
Fundulus heteroclitus (D)	Hereford Inlet, NJ	Poly-Euhaline	26–30 SL	%FO	N/A	Clymer 1978 Table 36
Fundulus heteroclitus (E)	Hereford Inlet, NJ	Poly-Euhaline	31–35 SL	%FO	N/A	Clymer 1978 Table 36
Fundulus heteroclitus (F)	Hereford Inlet, NJ	Poly-Euhaline	36–40 SL	%FO	N/A	Clymer 1978 Table 36
Fundulus heteroclitus (G)	Lower Delaware Bay	Polyhaline	6.6–30.4 SL	%W, %FO	238	Smith et al. 2000 Table 2
Fundulus heteroclitus (H)	Lower Delaware Bay	Polyhaline	6.6–20.4 SL	%W, %FO	172	Smith et al. 2000
Fundulus heteroclitus (I)	Lower Delaware Bay	Polyhaline	20.5–30.4 SL	%W, %FO	66	Smith et al. 2000
Fundulus heteroclitus (J)	Canary Creek Marsh, DE	Poly-Euhaline	12–41 SL	%FO, %W	234	Schmelz 1964 Table 1
Fundulus heteroclitus (K)	Canary Creek Marsh, DE	Poly-Euhaline	12–16 SL	%FO, %W	35	Schmelz 1964 Table 1
Fundulus heteroclitus (L)	Canary Creek Marsh, DE	Poly-Euhaline	17–21 SL	%FO, %W	21	Schmelz 1964 Table 1
Fundulus heteroclitus (M)	Canary Creek Marsh, DE	Poly-Euhaline	22–26 SL	%FO, %W	21	Schmelz 1964 Table 1
Fundulus heteroclitus (N)	Canary Creek Marsh, DE	Poly-Euhaline	27–31 SL	%FO, %W	45	Schmelz 1964 Table 1
Fundulus heteroclitus (O)	Canary Creek Marsh, DE	Poly-Euhaline	32–36 SL	%FO, %W	53	Schmelz 1964 Table 1
Fundulus heteroclitus (P)	Canary Creek Marsh, DE	Poly-Euhaline	37–41 SL	%FO, %W	59	Schmelz 1964 Table 1
Fundulus heteroclitus (Q)	NC Salt Marsh	Poly-Euhaline	< 39 SL	%FO	47	Kneib & Stiven 1978 Table 1 & Figure 3
Fundulus heteroclitus (R)	NC Salt Marsh	Poly-Euhaline	< 30 SL	%FO	N/A	Kneib & Stiven 1978 Table 1 & Figure 3
Fundulus heteroclitus (S)	NC Salt Marsh	Poly-Euhaline	30–39 SL	%FO	N/A	Kneib & Stiven 1978 Table 1 & Figure 3
Fundulus heteroclitus (T)	NC Spartina Marsh	Poly-Euhaline	< 38 SL	%FO	47	Kneib et al. 1980 Figures 2 & 3
Fundulus heteroclitus (U)	Newport River Estuary, NC	N/A	20–40	%OCC	207	Morgan 1990
Fundulus luciae (A)	Fox Creek Marsh, York River, VA	Oligo-Mesohaline	< 47 SL	%FO	303	Byrne 1978 Table 4
Fundulus luciae (B)	Salt Marsh, Tar Landing Bay, NC	Poly-Euhaline	11–34 SL	%FO	67	Kneib 1978 Figure 1
Fundulus majalis (A)	Hereford Inlet, NJ	Poly-Euhaline	26–45 SL	%FO	25	Clymer 1978 Table 36
Fundulus majalis (B)	Hereford Inlet, NJ	Poly-Euhaline	26–30 SL	%FO	N/A	Clymer 1978 Table 36
Fundulus majalis (C)	Hereford Inlet, NJ	Poly-Euhaline	31–35 SL	%FO	N/A	Clymer 1978 Table 36
Fundulus majalis (D)	Hereford Inlet, NJ	Poly-Euhaline	41–45 SL	%FO	N/A	Clymer 1978 Table 36
Fundulus majalis (E)	Delaware River Estuary	Meso-Polyhaline	30–70 SL	%FO	37	De Sylva et al. 1962 Table 19
Menidia beryllina (A)	Great South Bay, Long Island, NY	Polyhaline	40–67	%Comp, %FO	32	Grover 1982 Table 4 & Figure 4B
Menidia beryllina (B)	NJ Salt Marshes	Oligo-Polyhaline	15–40 TL	%Comp, %FO	272	Coorey 1981 Tables 10, 11 & Figure 11
Menidia beryllina (C)	Cattus Island, NJ	Oligo-Polyhaline	15–40 TL	%Comp, %FO	218	Coorey 1981 Tables 10, 11 & Figure 11
Menidia beryllina (D)	Popular Point, NJ	Mesohaline	15–40 TL	%Comp, %FO	6	Coorey 1981 Tables 10, 11 & Figure 11

Table 3.5 continued

Species/Code	Location	Salinity Zone	Size Class (mm)	Method	Sample Size	Source
Menidia beryllina (E)	Cedar Run, NJ	Polyhaline	15–40 TL	%Comp, %FO	48	Coorey 1981 Tables 10, 11 & Figure 11
Menidia beryllina (F)	Cattus Island, Upper Barnegat Bay, NJ	Oligo-Polyhaline	15–40 TL	%Comp, %FO	218	Coorey et al. 1985 Table 3
Menidia menidia (A)	Lower Pettaquamscutt River, RI	Meso-Euhaline	10–80	%FO	150	Mulkana 1964 Table V
Menidia menidia (B)	Lower Pettaquamscutt River, RI	Meso-Euhaline	10–30	%FO	69	Mulkana 1964 Table V
Menidia menidia (C)	Lower Pettaquamscutt River, RI	Meso-Euhaline	31–50	%FO	45	Mulkana 1964 Table V
Menidia menidia (D)	Lower Pettaquamscutt River, RI	Meso-Euhaline	51–80	%FO	36	Mulkana 1964 Table V
Menidia menidia (E)	Lower Point Judith Pond, RI	Euhaline	10–50	%FO	79	Mulkana 1964 Table V
Menidia menidia (F)	Lower Point Judith Pond, RI	Euhaline	10–30	%FO	45	Mulkana 1964 Table V
Menidia menidia (G)	Lower Point Judith Pond, RI	Euhaline	31–50	%FO	34	Mulkana 1964 Table V
Menidia menidia (H)	Pataguanset Estuary, CT	Limnetic-Euhaline	64–103	%W	147	Cadigan & Fell 1985 Figure 3
Menidia menidia (I)	Pataguanset Estuary, CT	Limnetic-Euhaline	64–71	%Incid	63	Cadigan & Fell 1985 Figure 4
Menidia menidia (J)	Pataguanset Estuary, CT	Limnetic-Euhaline	72–79	%Incid	106	Cadigan & Fell 1985 Figure 4
Menidia menidia (K)	Pataguanset Estuary, CT	Limnetic-Euhaline	80–87	%Incid	98	Cadigan & Fell 1985 Figure 4
Menidia menidia (L)	Pataguanset Estuary, CT	Limnetic-Euhaline	88–95	%Incid	64	Cadigan & Fell 1985 Figure 4
Menidia menidia (M)	Pataguanset Estuary, CT	Limnetic-Euhaline	96–103	%Incid	22	Cadigan & Fell 1985 Figure 4
Menidia menidia (N)	Pataguanset Estuary, CT	Limnetic-Polyhaline	64–103	N/A	N/A	Cadigan & Fell 1985
Menidia menidia (O)	Pataguanset Estuary, CT	Euhaline	64–103	N/A	N/A	Cadigan & Fell 1985
Menidia menidia (P)	Great South Bay, Long Island, NY	Polyhaline	30–120	%FO, %Comp	200	Grover 1982 Table 4 & Figure 4A
Menidia menidia (Q)	Long Island Sound	N/A	47–116.9	%FO	54	Richards 1963 Appendix
Menidia menidia (R)	Delaware River Estuary	Oligo-Euhaline	50–120	%FO	193	De Sylva et al. 1962 Table 42
Menidia menidia (S)	Magothy Bay, VA	Poly-Euhaline	16–119	Mean %V, #, FO	47	Kiel 1973 Table 24
Menidia menidia (T)	Magothy Bay, VA	Poly-Euhaline	16–69	Mean %V, #, FO	19	Kiel 1973 Table 25 & Figure 9
Menidia menidia (U)	Magothy Bay, VA	Poly-Euhaline	71–118	Mean %V, #, FO	28	Kiel 1973 Table 25 & Figure 9
Menidia menidia (V)	Newport River Estuary, NC	N/A	20–100	%FO	648	Morgan 1990 Figure 3
Hippocampus erectus (A)	Chesapeake Bay	Meso-Euhaline	23–126 TL	%FO, %#	136	Teixeira 1995 Table 5.1
Hippocampus erectus (B)	Chesapeake Bay	Meso-Euhaline	< 60 TL	%FO	26	Teixeira 1995 Table 5.5
Hippocampus erectus (C)	Chesapeake Bay	Meso-Euhaline	60–99 TL	%FO	67	Teixeira 1995 Table 5.5
Hippocampus erectus (D)	Chesapeake Bay	Meso-Euhaline	> 99 TL	%FO	21	Teixeira 1995 Table 5.5
Syngnathus fuscus (A)	Long Island Sound	N/A	88–177.6	%FO	13	Richards 1963 Appendix
Syngnathus fuscus (B)	Lower Chesapeake Bay	N/A	50–200 TL	%W	136	Ryer & Orth 1987 Table 1
Syngnathus fuscus (C)	Lower Chesapeake Bay	N/A	50–99 TL	%W	73	Ryer & Orth 1987 Table 1
Syngnathus fuscus (D)	Lower Chesapeake Bay	N/A	100–149 TL	%W	52	Ryer & Orth 1987 Table 1
Syngnathus fuscus (E)	Lower Chesapeake Bay	N/A	150–200 TL	%W	11	Ryer & Orth 1987 Table 1
Syngnathus fuscus (F)	Lower York River, VA	Oligo-Polyhaline	28–193	#	1905	Mercer 1973 Table 4
Syngnathus fuscus (G)	Lower York River, VA	Meso-Polyhaline	32–209 TL	%W, %#	3488	Teixeira 1995 Table 3.2
Syngnathus fuscus (H)	Lower York River, VA	Meso-Polyhaline	30–89	%W	607	Teixeira 1995 Figure 3.3
Syngnathus fuscus (I)	Lower York River, VA	Meso-Polyhaline	90–119	%W	1252	Teixeira 1995 Figure 3.3
Syngnathus fuscus (J)	Lower York River, VA	Meso-Polyhaline	120–139	%W	884	Teixeira 1995 Figure 3.3
Syngnathus fuscus (K)	Lower York River, VA	Meso-Polyhaline	140–209	%W	745	Teixeira 1995 Figure 3.3
Prionotus carolinus (A)	Woods Hole, MA	N/A	20–69 SL	#, %#	42	Marshall 1946 Table III
Prionotus carolinus (B)	Long Island Sound, CT	N/A	37–119 SL	#, %#, %FO	120	Richards et al. 1979 Table 5
Prionotus carolinus (C)	Long Island Sound, CT	N/A	21–159.7	%FO	126	Richards 1963 Appendix
Prionotus carolinus (D)	Magothy Bay, VA	Poly-Euhaline	22–115	Mean %V, #, FO	48	Kiel 1973 Table 48
Prionotus carolinus (E)	Magothy Bay, VA	Poly-Euhaline	22–54	Mean %V, #, FO	24	Kiel 1973 Table 49 & Figure 18
Prionotus carolinus (F)	Magothy Bay, VA	Poly-Euhaline	57–115	Mean %V, #, FO	24	Kiel 1973 Table 49 & Figure 18
Prionotus evolans (A)	Great Harbor, Woods Hole, MA	N/A	33–60 SL	#, %#, FO	10	Marshall 1946 Table V
Prionotus evolans (B)	Morris Cove, CT	N/A	YOY	FO	28	Marshall 1946

continued

Table 3.5 continued

Species/Code	Location	Salinity Zone	Size Class (mm)	Method	Sample Size	Source
Prionotus evolans (C)	Long Island Sound	N/A	40–150 SL	#, %#, %FO	271	Richards et al. 1979 Table 5
Myoxocephalus aenaeus (A)	Nauset Marsh Estuary, Cape Cod, MA	Poly-Euhaline	37–109 SL	%V, %FO	168	Lazzari et al. 1989 Table 1
Myoxocephalus aenaeus (B)	Nauset Marsh Estuary, Cape Cod, MA	Poly-Euhaline	< 60 SL	%V, %FO	63	Lazzari et al. 1989 Table 1
Myoxocephalus aenaeus (C)	Nauset Marsh Estuary, Cape Cod, MA	Poly-Euhaline	> 60 SL	%V, %FO	105	Lazzari et al. 1989 Table 1
Myoxocephalus aenaeus (D)	Long Island Sound	N/A	76.7–134.9	%FO	28	Richards 1963 Appendix
Morone americana (A)	Connecticut River	N/A	0–100 TL	%FO	759	Marcy 1976a Table 24
Morone americana (B)	Connecticut River	N/A	0–40 TL	%FO	245	Marcy 1976a Table 25
Morone americana (C)	Connecticut River	N/A	41–60 TL	%FO	231	Marcy 1976a Table 25
Morone americana (D)	Connecticut River	N/A	61–100 TL	%FO	99	Marcy 1976a Table 25
Morone americana (E)	Hudson River, NY	Oligohaline	< 110 SL	%FO	258	Bath & O'Connor 1985 Figure 2
Morone americana (F)	Delaware River Estuary	Oligo-Polyhaline	40–109 SL	%FO	85	Miller 1963 Table 3
Morone americana (G)	Delaware Bay Marsh Creeks	Limnetic-Polyhaline	60–139	%W	1720	Nemerson 2001 Table 2–3
Morone americana (H)	Delaware Bay Marsh Creeks, Upper Bay	Limnetic-Mesohaline	60–139	%W	1623	Nemerson 2001 Table 2–3
Morone americana (I)	Delaware Bay Marsh Creeks, Lower Bay	Meso-Polyhaline	60–139	%W	97	Nemerson 2001 Table 2–3
Morone americana (J)	Delaware River Estuary	Limnetic-Polyhaline	30–109	FO	65	De Sylva et al. 1962 Table 22
Morone americana (K)	York River, VA	Oligo-Polyhaline	21–100 SL	%W	84	Smith et al. 1984 Figure 5
Morone saxatilis (A)	Hudson River Estuary	N/A	< 150	%FO	273	Gardinier & Hoff 1982 Table 2
Morone saxatilis (B)	Hudson River Estuary	N/A	< 76	%FO	67	Gardinier & Hoff 1982 Table 2
Morone saxatilis (C)	Hudson River Estuary	N/A	76–150	%FO	206	Gardinier & Hoff 1982 Table 2
Morone saxatilis (D)	Hudson River	N/A	88–150 TL	%FO, %W	> 282	Hurst & Conover 2001 Table 1
Morone saxatilis (E)	Delaware River Estuary	Limnetic-Polyhaline	40–149	FO	262	De Sylva et al. 1962 Table 24
Morone saxatilis (F)	Delaware Bay Marsh Creeks	Oligo-Polyhaline	< 99	%W	80	Nemerson & Able 2003 Figure 9
Morone saxatilis (G)	Delaware Bay Marsh Creeks, Upper Bay	Oligohaline	< 99	%W	68	Nemerson & Able 2003 Figure 10
Morone saxatilis (H)	Delaware Bay Marsh Creeks, Lower Bay	Mesohaline	< 99	%W	12	Nemerson & Able 2003 Figure 10
Morone saxatilis (I)	Chesapeake Bay	N/A	Age 0	%W	293	Hartman & Brandt 1995 Table 1
Morone saxatilis (J)	Potomac Estuary	Limnetic-Mesohaline	25–100	%W	703	Boynton et al. 1981 Figure 2
Morone saxatilis (K)	Potomac Estuary	Limnetic-Oligohaline	25–100	%W	N/A	Boynton et al. 1981 Figure 2
Morone saxatilis (L)	Potomac Estuary	Mesohaline	25–100	%W	N/A	Boynton et al. 1981 Figure 2
Morone saxatilis (M)	Virginia Rivers	Oligo-Mesohaline	YOY	%FO, %V	297	Markle & Grant 1970 Table 1
Morone saxatilis (N)	James River, VA	Oligohaline	YOY	%FO, %V	103	Markle & Grant 1970 Table 1
Morone saxatilis (O)	Rappahannock River, VA	Oligo-Mesohaline	YOY	%FO, %V	58	Markle & Grant 1970 Table 1
Morone saxatilis (P)	York River, VA	Mesohaline	YOY	%FO, %V	136	Markle & Grant 1970 Table 1
Morone saxatilis (Q)	Albemarle Sound, NC	Limnetic-Oligohaline	< 160	%FO	467	Cooper et al. 1998 Table 2
Centropristis striata (A)	Long Island Sound	N/A	19.1–49.5	%FO	28	Richards 1963 Appendix
Centropristis striata (B)	Magothy Bay, VA	Poly-Euhaline	30–146	Mean %V, #, FO	48	Kiel 1973 Table 29
Centropristis striata (C)	Magothy Bay, VA	Poly-Euhaline	30–91	Mean %V, #, FO	28	Kiel 1972 Table 30 & Figure 11
Centropristis striata (D)	Magothy Bay, VA	Poly-Euhaline	92–146	Mean %V, #, FO	20	Kiel 1972 Table 30 & Figure 11
Pomatomus saltatrix (A)	Great South Bay, Long Island, NY	N/A	61–206	%FO, %W	79	Juanes et al. 2001 Table 3 & Figure 3A
Pomatomus saltatrix (B)	Great South Bay, Long Island, NY	Meso-Polyhaline	< 90–> 180 TL YOY	%FO, %W	841	Juanes & Conover 1995 Tables 2–4 & Figure 4
Pomatomus saltatrix (C)	Great South Bay, Long Island, NY	Meso-Polyhaline	< 90–> 180 TL YOY Spring Spawned	%FO, %W	710	Juanes & Conover 1995 Tables 2, 3 & Figure 4

Table 3.5 continued

Species/Code	Location	Salinity Zone	Size Class (mm)	Method	Sample Size	Source
Pomatomus saltatrix (D)	Great South Bay, Long Island, NY	Meso-Polyhaline	< 90–119 TL YOY Summer Spawned	%FO, %W	131	Juanes & Conover 1995 Table 4 & Figure 4
Pomatomus saltatrix (E)	Great South Bay, Long Island, NY	N/A	< 90 TL	%W	157	Juanes et al. 1994 Table I
Pomatomus saltatrix (F)	Great South Bay, Long Island, NY	N/A	YOY Spring Spawned	%W	256	Juanes & Conover 1994 Figure 1
Pomatomus saltatrix (G)	Lower Hudson River	N/A	< 110 TL	%W	163	Juanes et al. 1994 Table II
Pomatomus saltatrix (H)	Lower Hudson River	Oligo-Mesohaline	78–278 TL Age-0	%FO, %W	374	Juanes et al. 1993 Table 1 & Figure 2
Pomatomus saltatrix (I)	Lower Hudson River	Oligo-Mesohaline	78–278 TL Age-0 Spring Spawned	%FO, %W	341	Juanes et al. 1993 Table 1 & Figure 2
Pomatomus saltatrix (J)	Lower Hudson River	Oligo-Mesohaline	47–138 TL Age-0 Summer Spawned	%FO, %W	33	Juanes et al. 1993 Table 1 & Figure 2
Pomatomus saltatrix (K)	Hudson River	N/A	YOY	%W	1585	Buckel et al. 1999A Figure 1
Pomatomus saltatrix (L)	Hudson River Estuary	N/A	YOY	%F, %W	1338	Buckel & Conover 1997 Tables 1 & 2
Pomatomus saltatrix (M)	Hudson River Estuary	N/A	YOY Spring Spawned	%F, %W	1274	Buckel & Conover 1997 Tables 1 & 2
Pomatomus saltatrix (N)	Hudson River Estuary	N/A	YOY Summer Spawned	%F, %W	64	Buckel & Conover 1997 Table 1
Pomatomus saltatrix (O)	Sandy Hook Bay, NJ	Polyhaline	60–180 TL	%FO, %#, %W	1078	Friedland et al. 1988 Table 1
Pomatomus saltatrix (P)	Great Bay–Little Egg Harbor, NJ	Poly-Euhaline	20–200 FL	#, %FO	72	Able et al. 2003 Table 2
Pomatomus saltatrix (Q)	Delaware River Estuary	Oligo-Polyhaline	40–230 TL	FO	152	De Sylva et al. 1962 Table 26
Pomatomus saltatrix (R)	Indian River, DE	N/A	34–240 FL	#, FO, %FO, V, %V	262	Grant 1962 Table 1
Pomatomus saltatrix (S)	Chesapeake Bay	N/A	Age-0	%W	100	Hartman & Brandt 1995 Table A.3
Pomatomus saltatrix (T)	Lower Chesapeake Bay	N/A	33–290 FL	%IRI, %F, %N, %W	406	Gartland 2002 Table 2, Figures 6 & 7
Caranx hippos	Delaware River Estuary	N/A	30–160 TL	FO	40	De Sylva et al. 1962 Table 28
Stenotomus chrysops	Long Island Sound	N/A	20–146	%FO	181	Richards 1963 Appendix
Bairdiella chrysoura (A)	Delaware River Estuary	Oligo-Euhaline	20–159 TL	FO	218	De Sylva et al. 1962 Table 30
Bairdiella chrysoura (B)	Lower Delaware River	Limnetic-Mesohaline	65–135	%FO	214	Thomas 1971 Table 75
Bairdiella chrysoura (C)	York River Estuary, VA	Limnetic-Polyhaline	57–153 TL	%FO	68	Chao & Musick 1977 Table 11
Bairdiella chrysoura (D)	Cape Charles, VA	N/A	60–82 SL	%V	21	Welsh & Breder 1923
Bairdiella chrysoura (E)	Beaufort, NC	N/A	7–80 SL	N/A	109	Hildebrand & Cable 1930
Bairdiella chrysoura (F)	Beaufort, NC	N/A	7–20	N/A	30	Hildebrand & Cable 1930
Bairdiella chrysoura (G)	Beaufort, NC	N/A	25–50	N/A	64	Hildebrand & Cable 1930
Bairdiella chrysoura (H)	Beaufort, NC	N/A	50–80	N/A	15	Hildebrand & Cable 1930
Cynoscion regalis (A)	Acushnet River, MA	N/A	70–100 SL	%FO	28	Welsh & Breder 1923
Cynoscion regalis (B)	Long Island Sound	N/A	37–99	%FO	71	Richards 1963 Appendix
Cynoscion regalis (C)	Cape May, NJ	N/A	26–78 SL	%FO	32	Welsh & Breder 1923
Cynoscion regalis (D)	Delaware Bay Marsh Creeks	Limnetic-Mesohaline	< 200 TL	%W	822	Nemerson 2001 Table A-2
Cynoscion regalis (E)	Delaware River Estuary	Oligo-Polyhaline	10–140 TL	FO	220	De Sylva et al. 1962 Table 32
Cynoscion regalis (F)	Delaware Bay	N/A	< 3.5– > 7.55 NL	%#, %W	354	Goshorn 1990 Table 4.1
Cynoscion regalis (G)	Delaware Bay	Meso-Euhaline	10–89	%W	146	Grecay 1990 Figure 2.10
Cynoscion regalis (H)	Lower Delaware River	Limnetic-Mesohaline	< 180 FL	%FO	558	Thomas 1971 Table 75

continued

Table 3.5 continued

Species/Code	Location	Salinity Zone	Size Class (mm)	Method	Sample Size	Source
Cynoscion regalis (I)	Chesapeake Bay	N/A	Age-0	%W	564	Hartman & Brandt 1995 Table A.2
Cynoscion regalis (J)	York River Estuary, VA	Limnetic-Polyhaline	70–183	%FO	36	Chao & Musick 1977 Table 10
Cynoscion regalis (K)	Magothy Bay, VA	Poly-Euhaline	15–151	Mean %V, #, FO	27	Kiel 1973 Table 37
Cynoscion regalis (L)	Cape Charles, VA	N/A	43–115 SL	%FO	45	Welsh & Breder 1923
Cynoscion regalis (M)	Pamlico Sound, NC	N/A	Age-0 Mean 151 SL	%FO	26	Merriner 1975 Table 2
Cynoscion regalis (N)	Morehead City, NC	N/A	Age-0 Mean 152 SL	%FO	26	Merriner 1975 Table 3
Leiostomus xanthurus (A)	Delaware Bay Marsh Creeks	Limnetic-Polyhaline	20–140	%W	1159	Nemerson 2001 Table A-3
Leiostomus xanthurus (B)	Rehoboth Bay, DE	Meso-Polyhaline	Avg. 92	FO, #, %V	20	Tions 1995 Table 1.6
Leiostomus xanthurus (C)	Rhode River, Chesapeake Bay	Mesohaline	68–124	%Comp	187	Hines et al. 1990 Appendix 1
Leiostomus xanthurus (D)	York River, VA	Oligo-Polyhaline	16–125 SL	%W	1753	Smith et al. 1984 Figure 4
Leiostomus xanthurus (E)	York River, VA	Oligo-Polyhaline	16–20 SL	%W	77	Smith et al. 1984 Figure 5
Leiostomus xanthurus (F)	York River, VA	Oligo-Polyhaline	21–125 SL	%W	1645	Smith et al. 1984 Figure 5
Leiostomus xanthurus (G)	York River Estuary, VA	Oligo-Polyhaline	16–> 100	%W	1748	O'Neil & Weinstein 1987 Figure 3
Leiostomus xanthurus (H)	York River Estuary, VA	Oligo-Polyhaline	16–25	%W	207	O'Neil & Weinstein 1987 Figure 3
Leiostomus xanthurus (I)	York River Estuary, VA	Oligo-Polyhaline	26–40	%W	235	O'Neil & Weinstein 1987 Figure 3
Leiostomus xanthurus (J)	York River Estuary, VA	Oligo-Polyhaline	41–100	%W	1245	O'Neil & Weinstein 1987 Figure 3
Leiostomus xanthurus (K)	York River Estuary, VA	Oligo-Polyhaline	> 100	%W	61	O'Neil & Weinstein 1987 Figure 3
Leiostomus xanthurus (L)	Goalders Creek, York River Estuary, VA	Oligo-Mesohaline	16–> 100	%W	881	O'Neil & Weinstein 1987 Figure 5
Leiostomus xanthurus (M)	Blevins Creek, York River Estuary, VA	Polyhaline	16–> 100	%W	872	O'Neil & Weinstein 1987 Figure 5
Leiostomus xanthurus (N)	Magothy Bay, VA	Poly-Euhaline	11–159	Mean %V, #, FO	50	Kiel 1973 Table 38
Leiostomus xanthurus (O)	Magothy Bay, VA	Poly-Euhaline	16–100	Mean %V, #, FO	13	Kiel 1973 Table 39 & Figure 15
Leiostomus xanthurus (P)	Magothy Bay, VA	Poly-Euhaline	101–159	Mean %V, #, FO	37	Kiel 1973 Table 39 & Figure 15
Leiostomus xanthurus (Q)	Newport River Estuary, NC	Polyhaline	17–24	%Comp	N/A	Kjelson et al. 1975 Table 1
Leiostomus xanthurus (R)	Beaufort, NC	N/A	15–100	%FO	135	Hildebrand & Cable 1930
Leiostomus xanthurus (S)	NC Estuaries	N/A	63.5–152.4	%FO	73	Roelofs 1954 Table 1
Leiostomus xanthurus (T)	Cape Fear River Estuary, NC	Meso-Euhaline	9–124 SL	%FO, %V	1026	Hodson et al. 1981 Table 1
Menticirrhus saxatilis (A)	Delaware River Estuary	Meso-Euhaline	10–140	%FO	119	De Sylva et al.1962 Table 34
Menticirrhus saxatilis (B)	York River Estuary, VA	N/A	37–118 TL	%FO	20	Chao & Musick 1977 Table 13
Micropogonias undulatus (A)	Delaware Bay Marsh Creeks	Limnetic-Polyhaline	YOY	%W	2002	Nemerson 2001 Table A-4
Micropogonias undulatus (B)	Lower Delaware River	Limnetic-Mesohaline	23–142	%FO	27	Thomas 1971 Table 75
Micropogonias undulatus (C)	Rhode River, Chesapeake Bay	Mesohaline	112–119 `	%W	40	Hines et al. 1990 Appendix 1
Micropogonias undulatus (D)	Chesapeake Bay	N/A	17–42	%V	45	Welsh & Breder 1923
Micropogonias undulatus (E)	NC Estuaries	N/A	63.5–152.4	%FO	159	Roelofs 1954 Table 1
Pogonias cromis (A)	Delaware River Estuary	Meso-Euhaline	30–190	FO	79	De Sylva et al. 1962 Table 36
Pogonias cromis (B)	Lower Delaware River	Limnetic-Mesohaline	101–211	%FO	189	Thomas 1971 Table 75
Sphyraena borealis	Delaware River Estuary	N/A	50–110	FO	85	De Sylva et al. 1962 Table 38
Tautoga onitis	Great South Bay, Long Island, NY	Polyhaline	31–71	%Comp, %FO	36	Grover 1982 Figure 4C & Table 4
Tautogolabrus adspersus	Great South Bay, Long Island, NY	Polyhaline	11–64	%Comp, %FO	54	Grover 1982 Figure 4D & Table 4

Table 3.5 continued

Species/Code	Location	Salinity Zone	Size Class (mm)	Method	Sample Size	Source
Ammodytes americanus (A)	Long Island Sound	N/A	< 80 SL	%FO	143	McKown 1984 Table 7
Ammodytes americanus (B)	Long Island Sound	N/A	< 8--> 24	W	165	Monteleone & Peterson 1986 Table 2
Ammodytes americanus (C)	Long Island Sound	N/A	< 8.0	W	15	Monteleone & Peterson 1986 Table 2
Ammodytes americanus (D)	Long Island Sound	N/A	8–11.9	W	37	Monteleone & Peterson 1986 Table 2
Ammodytes americanus (E)	Long Island Sound	N/A	12–18.9	W	84	Monteleone & Peterson 1986 Table 2
Ammodytes americanus (F)	Long Island Sound	N/A	19–23.9	W	16	Monteleone & Peterson 1986 Table 2
Ammodytes americanus (G)	Long Island Sound	N/A	> 24	W	13	Monteleone & Peterson 1986 Table 2
Ammodytes americanus (H)	Great South Bay, Long Island, NY	Polyhaline	45–159	%FO	142	Grover 1982 Table 4
Ammodytes americanus (I)	Long Island Sound	N/A	74.3–132.1	%FO	290	Richards 1963 Appendix
Gobiosoma ginsburgi	Sakonnet River, RI	Poly-Euhaline	18–42 SL	FO, %FO, #, %#	49	Munroe & Lotspeich 1979 Table 2
Peprilus triacanthus	Long Island Sound	N/A	21.6–100	%FO	19	Richards 1963 Appendix
Scophthalmus aquosus (A)	Long Island Sound	N/A	31–131.8	%FO	74	Richards 1963 Appendix
Scophthalmus aquosus (B)	Great Bay–Little Egg Harbor Estuary, NJ	Polyhaline	2.8–89.7 TL	%W	415	Haberland 2002 Figure 4
Scophthalmus aquosus (C)	Great Bay–Little Egg Harbor Estuary, NJ	Polyhaline	0–5 TL	%W	103	Haberland 2002 Figure 4
Scophthalmus aquosus (D)	Great Bay–Little Egg Harbor Estuary, NJ	Polyhaline	6–40 TL	%W	173	Haberland 2002 Figure 4
Scophthalmus aquosus (E)	Great Bay–Little Egg Harbor Estuary, NJ	Polyhaline	40–70 TL	%W	69	Haberland 2002 Figure 4
Scophthalmus aquosus (F)	Great Bay–Little Egg Harbor Estuary, NJ	Polyhaline	70–90 TL	%W	38	Haberland 2002 Figure 4
Scophthalmus aquosus (G)	Delaware River Estuary	Meso-Polyhaline	50–170	%FO	31	De Sylva et al. 1962 Table 46
Paralichthys dentatus (A)	Great Bay–Little Egg Harbor Estuary, NJ	N/A	8.1–14.6 SL	%#, %V, %FO, %IRI	119	Grover 1998 Table 5
Paralichthys dentatus (B)	Great Bay–Little Egg Harbor, NJ	Polyhaline	167–305	%FO, Mean #, Mean W	137	Rountree & Able 1992 Table 4
Paralichthys dentatus (C)	Delaware River Estuary	Meso-Polyhaline	50–210	FO	26	De Sylva et al. 1962 Table 44
Paralichthys dentatus (D)	Northeastern Cove, Rehoboth Bay, DE	Meso-Polyhaline	AVG 141 TL	FO, #, %V	23	Tions 1995 Tables 1.5A&B
Paralichthys dentatus (E)	Lower Chesapeake Bay	N/A	Avg. 274 SL	%W, %FO	45	Lascara 1981
Paralichthys dentatus (F)	York River, VA	Oligo-Polyhaline	81–160 SL	%W	28	Smith et al. 1984 Figure 4
Paralichthys dentatus (G)	Magothy Bay, VA	Poly-Euhaline	42–198	Mean %V, #, FO	56	Kiel 1973 Table 54 & Figure 20
Paralichthys dentatus (H)	Newport & North Rivers, NC	Poly-Euhaline	20–60 SL	%IRI	63	Burke 1995 Figure 7
Paralichthys dentatus (I)	Newport & North Rivers, NC	Poly-Euhaline	9–20 SL	%IRI	173	Burke 1995 Figure 5
Paralichthys dentatus (J)	Newport & North Rivers, NC	Poly-Euhaline	9–10 SL	%IRI	41	Burke 1995 Figure 5
Paralichthys dentatus (K)	Newport & North Rivers, NC	Poly-Euhaline	11–12 SL	%IRI	69	Burke 1995 Figure 5
Paralichthys dentatus (L)	Newport & North Rivers, NC	Poly-Euhaline	13–14 SL	%IRI	43	Burke 1995 Figure 5
Paralichthys dentatus (M)	Newport & North Rivers, NC	Poly-Euhaline	15–16 SL	%IRI	10	Burke 1995 Figure 5
Paralichthys dentatus (N)	Newport & North Rivers, NC	Poly-Euhaline	17–18 SL	%IRI	6	Burke 1995 Figure 5
Paralichthys dentatus (O)	Newport & North Rivers, NC	Poly-Euhaline	19–20 SL	%IRI	4	Burke 1995 Figure 5
Paralichthys dentatus (P)	Pamlico Sound, NC	N/A	100–300 TL	%V, %FO	528	Powell & Schwartz 1979 Table 1

continued

Table 3.5 continued

Species/Code	Location	Salinity Zone	Size Class (mm)	Method	Sample Size	Source
Paralichthys dentatus (Q)	Pamlico Sound, NC	N/A	100–200 TL	%V, %FO	470	Powell & Schwartz 1979 Table 1
Paralichthys dentatus (R)	Pamlico Sound, NC	N/A	201–300 TL	%V, %FO	49	Powell & Schwartz 1979 Table 1
Pseudopleuronectes americanus (A)	Lower Pettaquamscutt River, RI	Meso-Euhaline	10–80	Avg. #, %FO	123	Mulkana 1964 Table III
Pseudopleuronectes americanus (B)	Lower Pettaquamscutt River, RI	Meso-Euhaline	10–30	Avg. #, %FO	58	Mulkana 1964 Table III
Pseudopleuronectes americanus (C)	Lower Pettaquamscutt River, RI	Meso-Euhaline	31–60	Avg. #, %FO	50	Mulkana 1964 Table III
Pseudopleuronectes americanus (D)	Lower Pettaquamscutt River, RI	Meso-Euhaline	61–80	Avg. #, %FO	15	Mulkana 1964 Table III
Pseudopleuronectes americanus (E)	Lower Point Judith Pond, RI	Euhaline	10–80	Avg. #, %FO	76	Mulkana 1964 Table III
Pseudopleuronectes americanus (F)	Lower Point Judith Pond, RI	Euhaline	10–30	Avg. #, %FO	5	Mulkana 1964 Table III
Pseudopleuronectes americanus (G)	Lower Point Judith Pond, RI	Euhaline	31–60	Avg. #, %FO	59	Mulkana 1964 Table III
Pseudopleuronectes americanus (H)	Lower Point Judith Pond, RI	Euhaline	61–80	Avg. #, %FO	12	Mulkana 1964 Table III
Pseudopleuronectes americanus (I)	Mystic River Estuary, CT	Limnetic-Euhaline	3–8	%FO	140	Pearcy 1962 Table I
Pseudopleuronectes americanus (J)	Mystic River Estuary, CT	Limnetic-Euhaline	6–165	%FO	337	Pearcy 1962 Table II
Pseudopleuronectes americanus (K)	New Haven Harbor, CT	Polyhaline	100–300 TL	%W, %#, %FO, IRI	151	Carlson et al. 1997 Table 1
Pseudopleuronectes americanus (L)	New Haven Harbor, CT	Meso-Polyhaline	50–200 TL	%W, %#, %FO, IRI	29	Carlson et al. 1997 Table 2
Pseudopleuronectes americanus (M)	Long Island Sound	N/A	37–165	%FO	373	Richards 1963 Appendix
Pseudopleuronectes americanus (N)	Hudson River Estuary, NJ	N/A	30–61 TL	%Comp	30	Vivian et al. 2000 Table 1
Pseudopleuronectes americanus (O)	Hudson-Raritan River Estuary	Limnetic-Euhaline	15–299 TL	%V	< 1291	Stehlik & Meise 2000 Figure 2
Pseudopleuronectes americanus (P)	Hudson-Raritan River Estuary	Limnetic-Euhaline	15–49 TL	%V	N/A	Stehlik & Meise 2000 Figure 2
Pseudopleuronectes americanus (Q)	Hudson-Raritan River Estuary	Limnetic-Euhaline	50–299 TL	%V	N/A	Stehlik & Meise 2000 Figure 2
Pseudopleuronectes americanus (R)	Delaware River Estuary	N/A	30–130 TL	FO	95	De Sylva et al. 1962 Table 48
Pseudopleuronectes americanus (S)	Herring Creek, Rehoboth Bay, DE	Meso-Polyhaline	Mean 100 TL	FO, Avg. #, %V	36	Tions 1995 Table 1.4A
Trinectes maculatus	York River, VA	Oligo-Polyhaline	21–60 SL	%W	83	Smith et al. 1984 Figure 4
Sphoeroides maculatus	Long Island Sound	N/A	29.4–89.5	%FO	22	Richards 1963 Appendix

4

General Characteristics of the Temperate Ichthyofauna

The region of the U.S. East Coast between Cape Cod and Cape Hatteras is variously referred to as a warm temperate region (Ekman, 1953) or the Virginian province (Engle and Summers, 1999). This designation is based on a near consensus by numerous investigators focusing on different components of the fauna and the temperature regime. Furthermore, portions of this interpretation are based on early syntheses (Ekman, 1953; Stephenson and Stephenson, 1954; Hedgepeth, 1957; Hazel, 1970), more recent cladistic biogeography for sponges and fishes (Humphries and Parenti, 1999), and, to some extent, benthic invertebrates (Engle and Summers, 1999). More current evaluation is based exclusively on fishes (Grosslein and Azarovitz, 1982; Colvocoresses and Musick, 1984; Ray et al., 1997; Briggs and Waldman, 2002), including those found in estuaries (Ayvazian et al., 1992; Able et al., 2002; Nordlie, 2003). The only major recent exception to the designation of Cape Cod–Cape Hatteras as a separate region is the interpretation by Briggs (1974). While recognizing the biogeographic distinctiveness of areas north (boreal or cold-temperate Acadian province) and south of Cape Cod, he did not consider Cape Hatteras as a distinct barrier because of the lack of endemism north of it and the large number of subtropical or Carolinian Province forms that occur there. Subsequently Briggs (1995) considered that the Western Atlantic Boreal Region extends from the northern entrance of the Gulf of St. Lawrence to Cape Hatteras. While admitting there were many other opinions in conflict with his regarding Cape Cod as a faunal barrier, he doubted its significance because of the large number of boreal species and little endemism south of it. More recently, others have questioned the reliance on endemism as a means of delineating biogeographic regions (Adey and Steneck, 2001). Others have recognized that the region between Cape Cod and Cape Hatteras is dominated by seasonal transients from the north

in winter and the south in summer (Grosslein and Azarovitz, 1982; see Chapter 9), as a result of the extensive seasonal temperature range that occurs in the area (Parr, 1933; Grosslein and Azarovitz, 1982). This broad range of temperatures occurs for many temperate regions worldwide (Tyler, 1971; Ray et al., 1997). The region between Cape Cod and Cape Hatteras is referred to as the Middle Atlantic Bight by numerous authors (Able and Fahay, 1998), but the term is objected to by some (Richards, 1999). We use the term throughout this text for convenience and because it occurs so frequently in the studies we cite.

Typical of the bight is a group of about 60 migratory species, many of which are important to both commercial and recreational interests. This group is dominated by species that tend to occur near the coast and in northern parts of their range during summer and offshore and in the southern part of their range during winter (Grosslein and Azarovitz, 1982). Some spend the colder months south of the Middle Atlantic Bight, or south of Cape Hatteras and as far south as Florida. The estuarine fauna has a large migratory component as well. Some of these migrations are between estuaries and the continental shelf; some occur seasonally across the shelf from estuaries to the continental shelf edge; and some involve migrations out of the Middle Atlantic Bight, either to the south or to the north and east (see Chapter 9). The life histories of fishes from the Middle Atlantic Bight are complex, in large part because of these migrations (Grosslein and Azarovitz, 1982; Able and Fahay, 1998). Often, dissimilar migration patterns are exhibited by different-aged fishes, such that wintering areas for young-of-the-year will differ from those of older, adult fishes. The mobility of certain species is further complicated by increased movement with age, with the result that older and younger adults will dominate in different parts of the range. For some species, such as *Acipenser* spp., many aspects of their distribution and migrations are still poorly known.

A checklist that includes all fish species that occur in the northern part of the western North Atlantic Ocean (north of the latitude of Cape Hatteras) is available (Fahay, 2007). This list contains many range extensions, often based on early stages, and should be consulted for occurrences that are not published elsewhere. The ichthyofauna of the central part of the Middle Atlantic Bight is comprised of numerous marine and estuarine species (Able, 1992; Fahay, 1993; Briggs and Waldman, 2002). Included in the estuaries of this region are species-rich families such as Clupeidae (herrings), Fundulidae (killifishes), and Sciaenidae (drums). The ontogenetic composition of this ichthyofauna is also well described (Fahay, 1983, 1993, 2007), so that we can begin to measure the importance of various habitats to life history stages. The seasonality of reproduction and egg and larval occurrences on the continental shelf and in estuaries of the region is treated in Chapter 5. Comparison of faunal works to the north (Col-

The Importance of Taxonomic Precision

The present study includes examples of several taxa that are difficult to identify in their early life history stages. This difficulty is reflected when investigators refer to them at the generic level, followed by the epithet "sp." Use of this label implicitly suggests that more than one species was involved in the collection, or, if only a single species, one that cannot be identified. In our study area, such "lumping" is often reported for the genera *Alosa, Anchoa, Urophycis, Fundulus, Ammodytes,* and *Mugil*. As adults, some of these lumped taxa are sometimes managed together, which may suggest that their life histories are similar. However, early life history patterns that emerge from such "assemblages" are likely to be averages of different patterns, and their values are therefore somewhat questionable.

For the genus *Urophycis*, for example, the present study outlines three very distinct life histories for three species (*Urophycis chuss, U. regia,* and *U. tenuis*). All three species utilize estuarine habitats, but at different times of year and at various stages in their development. The three species also demonstrate diverse responses to estuaries geographically, and in the case of *U. regia*, different cohorts exhibit differing patterns. Of the three species, only *U. regia* might be considered to be an obligate user of estuaries, and that is truer in the southern part of its range than in the northern. Combining the evidence for all three species, therefore, is likely to result in a generalized picture that bears little resemblance to the pattern displayed by any one species.

Table 4.1. Ranking of monthly species composition of fish eggs collected in ocean waters from Cape Hatteras, North Carolina, to the Gulf of Mexico during NMFS-MARMAP surveys, 1977–1987 (modified from Berrien and Sibunka, 1999). * indicates species that use estuaries in the study area. Dashes indicate few or no collections of that species in that month.

Taxon	Jan	Feb	Mar	Apr	May	Jun	Jul	Aug	Sep	Oct	Nov	Dec
*Pollachius virens**	1	2	9	—	—	—	—	—	—	—	2	1
Gadus morhua	2	1	1	3	—	—	—	—	—	—	3	2
*Urophycis spp.**	3	4	4	9	—	5	1	1	1	1	1	3
Melanogrammus aeglefinus	4	3	2	1	10	—	—	—	—	—	—	—
Merluccius bilinearis	5	6	5	8	4	4	3	2	3	2	6	4
*Prionotus spp.**	6	—	8	—	—	—	10	4	2	4	—	—
Maurolicus weitzmani	7	—	—	—	—	—	—	—	8	9	7	6
*Citharichthys/Etropus**	8	8	—	—	9	9	4	5	4	5	—	—
Merluccius albidus	9	7	6	6	—	—	—	—	10	8	10	5
Enchelyopus cimbrius	10	—	—	7	5	8	—	9	6	10	—	7
Hippoglossoides platessoides	—	5	7	5	—	—	—	—	—	—	—	9
*Scophthalmus aquosus**	—	9	9	10	7	—	9	8	5	3	4	—
Glyptocephalus cynoglossus	—	10	10	—	—	—	—	—	—	—	—	—
Pleuronectes ferrugineus	—	—	3	2	3	6	—	—	—	—	—	—
Scomber scombrus	—	—	—	4	1	2	—	—	—	—	—	—
*Anchoa mitchilli**	—	—	—	—	2	1	2	3	—	—	—	—
*Peprilus triacanthus**	—	—	—	—	6	10	8	6	—	—	—	—
Paralichthys oblongus	—	—	—	—	8	7	6	7	—	—	—	—
*Tautogolabrus adspersus**	—	—	—	—	—	3	5	—	—	—	—	—
*Pomatomus saltatrix**	—	—	—	—	—	—	7	—	—	—	—	—
*Micropogonias undulatus**	—	—	—	—	—	—	—	—	7	7	9	—
*Centropristis striata**	—	—	—	—	—	—	—	—	9	—	—	—
*Paralichthys dentatus**	—	—	—	—	—	—	—	—	—	6	5	8
*Brevoortia tyrannus**	—	—	—	—	—	—	—	—	—	—	8	—

lette and Klein-McPhee, 2002) and to the south (Carpenter, 2001) provides further clarification regarding the Middle Atlantic Bight fish fauna.

Eggs and larvae are an important but seasonally variable component of the water column in the Middle Atlantic Bight (Tables 4.1 and 4.2). Many of these eventually use the adjacent estuaries. Typically, eggs in winter are primarily those of gadids (*Pollachius virens, Gadus morhua,* and *Melanogrammus*

aeglefinus), phycids (*Urophycis* spp.), and merluciids (*Merluccius bilinearis* and *M. albidus*), while the winter larvae are dominated by *Ammodytes* spp., *Gadus morhua, Pholis gunnellus,* and *Paralichthys dentatus*, at least based on the extensive sampling during 1977–1987. Both egg and larval abundance and diversity increase during the early spring and include such species as *Glyptocephalus cynoglossus* and *Scophthalmus aquosus*. *Ammodytes* spp. larvae remain abundant but are joined by such spe-

Table 4.2. Ranking of monthly species composition of fish larvae collected in ocean waters in the central part of the Middle Atlantic Bight, during NMFS-MARMAP Surveys, 1977–1987 (from Able and Fahay, 1998). * indicates those species that typically use estuaries in the study area. Dashes indicate few or no collections of that species in that month.

Taxon	Jan	Feb	Mar	Apr	May	Jun	Jul	Aug	Sep	Oct	Nov	Dec
Ammodytes spp.	1	1	1	1	2	—	—	—	—	—	—	4
Gadus morhua	2	3	2	3	7	—	—	—	—	—	—	7
*Paralichthys dentatus**	3	2	9	—	—	—	—	—	—	5	2	1
*Brevoortia tyrannus**	4	—	—	—	—	—	—	—	—	—	5	5
Merluccius bilinearis	5	9	—	—	—	7	9	9	—	6	4	2
Maurolicus muelleri	6	—	—	—	—	—	—	—	—	—	—	8
*Leiostomus xanthurus**	7	—	—	—	—	—	—	—	—	—	—	—
*Pollachius virens**	8	4	7	—	—	—	—	—	—	—	—	—
Gobiidae	9	—	—	—	—	—	—	—	—	—	8	—
*Clupea harengus**	10	—	—	—	—	—	—	—	—	—	—	—
*Micropogonias undulatus**	11	—	—	—	—	—	—	—	—	—	—	—
*Pholis gunnellus**	—	5	4	11	—	—	—	—	—	—	—	—
Myoxocephalus octodecemspinosus	—	6	3	10	—	—	—	—	—	—	—	—
Paralepididae	—	7	—	—	—	—	—	—	—	—	—	—
*Anguilla rostrata**	—	8	—	—	—	—	—	—	—	—	—	—
Notolepis rissoi	—	10	10	—	—	—	—	—	—	—	—	—
*Pseudopleuronectes americanus**	—	—	5	5	—	—	—	—	—	—	—	—
*Myoxocephalus aenaeus**	—	—	6	9	—	—	—	—	—	—	—	—
Cottidae	—	—	8	7	—	—	—	—	—	—	—	—
Benthosema glaciale	—	—	—	2	9	—	—	—	—	—	—	—
Limanda ferruginea	—	—	—	4	1	1	10	—	—	—	—	—
Liparis spp.	—	—	—	6	5	—	—	—	—	—	—	—
Melanogrammus aeglefinus	—	—	—	8	10	—	—	—	—	—	—	—
Scomber scombrus	—	—	—	—	3	3	—	—	—	—	—	—
Enchelyopus cimbrius	—	—	—	—	4	2	—	—	—	—	10	—
*Scophthalmus aquosus**	—	—	—	—	6	6	11	—	—	3	3	3
Glyptocephalus cynoglossus	—	—	—	—	8	5	—	—	—	—	—	—
Lophius americanus	—	—	—	—	—	4	7	—	—	—	—	—
*Tautogolabrus adspersus**	—	—	—	—	—	8	3	7	—	—	—	—
Hippoglossina oblonga	—	—	—	—	—	9	1	2	4	8	—	—
*Urophycis chuss**	—	—	—	—	—	10	4	—	—	—	—	—
*Peprilus triacanthus**	—	—	—	—	—	—	2	1	8	—	—	—
*Pomatomus saltatrix**	—	—	—	—	—	—	5	6	—	—	—	—
Engraulidae	—	—	—	—	—	—	6	8	10	—	—	—
Citharichthys arctifrons	—	—	—	—	—	—	8	4	2	2	6	—
Urophycis spp.	—	—	—	—	—	—	—	3	1	—	—	—
*Etropus microstomus**	—	—	—	—	—	—	—	5	3	7	—	—
*Prionotus carolinus**	—	—	—	—	—	—	—	10	5	9	—	—
*Ophidion marginatum**	—	—	—	—	—	—	—	—	6	—	—	—
Lepophidium profundorum	—	—	—	—	—	—	—	—	7	—	9	—
*Centropristis striata**	—	—	—	—	—	—	—	—	9	—	—	—
Ophidiidae	—	—	—	—	—	—	—	—	—	4	—	—
Bothus spp.	—	—	—	—	—	—	—	—	—	10	11	—
*Urophycis regia**	—	—	—	—	—	—	—	—	—	1	1	6
Ceratoscopelus maderensis	—	—	—	—	—	—	—	—	—	—	7	—
Diaphus spp.	—	—	—	—	—	—	—	—	—	—	—	9

cies as *Scomber scombrus, Limanda ferruginea* in the northern part, and *Cynoscion regalis* in the southern part of the bight. Abundance and diversity of eggs and larvae reach a peak during mid- to late summer, when many species reproduce. Several other taxa become abundant, including the eggs and larvae of *Enchelyopus cimbrius, Hippoglossina oblonga, Peprilus triacanthus, Tautogolabrus adspersus,* and a mixture of *Urophycis* spp. Included in this group are many species that use estuaries (e.g., *Anchoa mitchilli, Scophthalmus aquosus, Tautoga onitis, Prionotus carolinus,* and *Pomatomus saltatrix*). Spawning is somewhat reduced during the fall, but the eggs of *Urophycis* spp., *Merluccius bilinearis, Maurolicus weitzmani,* and estuarine users such as *Micropogonias undulatus* are common. Larvae of *Urophycis regia* are abundant then, along with those of *Citharichthys arctifrons, Etropus microstomus,* and *Scophthalmus aquosus.* Also, *Paralichthys dentatus* eggs reach peak abundance in northern bight waters and *Brevoortia tyrannus* and *Prionotus* spp. egg concentrations occur in the southern bight at this time (Smith and Morse, 1988). Further details for the above, and other, estuarine species are provided in the species accounts.

Patterns of estuarine use by all life history stages of fishes vary spatially in the Middle Atlantic Bight. We have plotted these reported occurrences within 16 estuaries to demonstrate some of this variation (Table 4.3). We caution, however, that these data are not quantitative values, and in some cases, a positive occurrence may be based on the collection of a single specimen of a certain life history stage. We have endeavored to include evaluations of these occurrences in the appropriate species accounts to correct misleading generalizations. Some obvious trends in these distributions are due solely to latitudinal differences in the fauna. Some species (e.g., *Clupea harengus*, *Pollachius virens*, *Stenotomus chrysops*, and *Tautogolabrus adspersus*) reach the southern limit of their range in the study area, extending as far as the vicinity of Chesapeake Bay. Other species with northern affinities

(e.g., *Osmerus mordax*, *Microgadus tomcod*, *Pholis gunnellus*, and *Myoxocephalus aenaeus*) are found only in the northernmost estuaries of the bight. The former two occur as far south as the Hudson–Raritan estuary, whereas *P. gunnellus* and *M. aenaeus* currently extend their ranges to Great Bay and Delaware inland bays, respectively. Conversely, those with more southern affinities (*Urophycis regia*, *Opsanus tau*, *Lucania parva*, *Cynoscion regalis*, *Leiostomus xanthurus*, *Menticirrhus saxatilis*, *Chaetodon ocellatus*, *Mugil cephalus*, *Sphyraena borealis*, *Trinectes maculatus*, and *Sphoeroides maculatus*) reach the northern limit of their range in estuaries near Cape Cod. Other species (*Membras martinica*, *Astroscopus guttatus*, *Hippocampus erectus*, and *Hypsoblennius hentz*) are found only in estuaries in the southern part of the bight. In certain species (*Pollachius virens*, *Urophycis chuss*, and *Stenotomus chrysops*), eggs, larvae, and juveniles are found in northern estuaries, whereas only juveniles are found in southern estuaries. The opposite trend is found in *Bairdiella chrysoura* and *Pogonias cromis*.

Table 4.3. Distribution of eggs (E), larvae (L), juveniles (J), and adults (A) of selected fishes for representative estuaries in the Middle Atlantic Bight.

Species	Pam Alb NC	Ches Bay VA/ MD	East Shore VA/ MD	Inlnd Bays DE	Del Bay DE/ NJ	South Inlnd NJ	Great Bay NJ	Barn Bay NJ	Hudsn Raritn NJ	Great South NY	Gard Bay NY	L.I. Sound NY	Conn River CT	Narr Bay RI	Buzz Bay MA	Nauset Marsh MA
Mustelus canis			J				J		J		J				J	
Anguilla rostrata	J	J	J	J	J	J	J	J	J	J	J	J	J	J	J	J
Conger oceanicus	J		J	J	J	J	J		J	J		J				J
Alosa aestivalis	JA	ELJA	JA	ELJA	ELJA	JA		ELJA	ELJA	JA	JA	LJA	ELJA	ELJA	ELJ	
Alosa mediocris		ELJ	J		J		A		J						ELJ	
A. pseudoharengus	JA	ELJA	JA	ELJA	ELJA	ELJA	ELJA	ELJA	ELJA	JA	JA	ELJA	ELJA	ELJA	ELJA	J
A. sapidissima	JA	ELJA	J	ELJA	ELJA	JA		JA	ELJA	JA	JA	ELJA	ELJA	JA	JA	
Brevoortia tyrannus	LJA	ELJA	LJA	LJA	ELJA	ELJA	ELJA	ELJA	ELJA	ELJA	ELJA	ELJA	LJA	ELJA	ELJA	LJA
Clupea harengus		JA	JA	LJA	LJA	LJA	LJA	LJA	LJA	JA	JA	LJA	JA	LJA	LJA	LJ
Anchoa hepsetus	ELJ	LJ	J	EJ	J	ELJ	ELJA		E	EJ				J		
Anchoa mitchilli	LJA	ELJA	ELJA	ELJA	ELJA	ELJA	ELJA	ELJA	ELJA	ELJA	ELJA	ELJA	ELJA	ELJA	ELJA	J
Osmerus mordax									ELJA		JA	ELJA	ELJA	ELJA	ELJA	
Synodus foetens	J	J	J	J	J	J	LJ	J	LJ			LJ				
Microgadus tomcod			J						ELJA	ELJA	ELJA	ELJA	ELJA	ELJA	ELJA	J
Pollachius virens		J	J	J	J	J	J	J	JA	J		LJA	LJA	LJ	LJA	ELJ	J
Urophycis chuss		JA	J	J	JA	JA	EJ	J	LJA	EJA	LJA	JA	JA	ELJA	ELJA	J
U. regia	J	J	J	J	LJ	J	J	J	J	J		J			J	
U. tenuis						J	J			J				J		J
Ophidion marginatum		J	LJ	LJ	LJ	LJ	J		L							
Opsanus tau	LJ	ELJA	ELJA	ELJA	ELJA	ELJA	ELJA	ELJA	ELJA	ELJA	ELJA	ELJA	ELJA	ELJA	ELJA	
Strongylura marina	J	LJ	J	J	J	J	LJ	J	J	LJ						
Cyprinodon variegatus	JA	ELJA	ELJA	ELJA	ELJA	ELJA	ELJA	ELJA	ELJA	ELJA	ELJA	ELJA	ELJA	ELJA	ELJ	
Fundulus heteroclitus	JA	ELJ	ELJ	ELJ	ELJ	ELJ	ELJ	ELJ	ELJ	ELJ	ELJ	ELJ	J	ELJ	ELJ	ELJA
F. luciae	ELJ	ELJ	ELJ	ELJ	ELJ	ELJ	ELJ	ELJ								
F. majalis	LJ	ELJ	ELJ	ELJ	ELJ	ELJ	ELJ	ELJ		J	ELJ		J		ELJ	ELJA
Lucania parva	J	ELJ	ELJ	ELJ	ELJ	ELJ	ELJ	ELJ	ELJ	J					ELJ	
Membras martinica	J	ELJ	LJ	ELJ	ELJ	ELJ	ELJ	J	ELJ							
Menidia beryllina	J	LJ	ELJ	ELJ	ELJ	ELJ	ELJ	ELJ	ELJ	J		J			ELJ	JA
M. menidia	LJ	LJ	ELJ	ELJ	ELJ	ELJ	ELJ	ELJ	ELJ	ELJ	ELJ	ELJ	LJ	L	ELJ	LJA
Apeltes quadracus		ELJ	J	ELJ	ELJ	ELJ	ELJ	ELJ	ELJ	LJ	ELJ	ELJ	J		ELJ	ELJA
Gasterosteus aculeatus		ELJ		ELJ	ELJ	ELJ	ELJ	ELJ		LJ		LJ		L	ELJ	LJA
Hippocampus erectus	J	J	J	LJ		J	LJ	LJ		J		J				
Syngnathus fuscus	J	LJA	LJA	LJA	LJA	LJA	LJA	LJA	LJA	LJA	LJA	LJA	LJA	LJA	LJA	LJA

Table 4.3 continued

Species	Pam Alb NC	Ches Bay VA/MD	East Shore VA/MD	Inlnd Bays DE	Del Bay DE/NJ	South Inlnd NJ	Great Bay NJ	Barn Bay NJ	Hudsn Raritn NJ	Great South NY	Gard Bay NY	L.I. Sound NY	Conn River CT	Narr Bay RI	Buzz Bay MA	Nauset Marsh MA
Prionotus evolans	J	J	J	J	J	ELJ	LJ	J	J			EL		E	J	
P. carolinus	J	JA	ELJA	ELJA	ELJA	ELJA	LJA	JA	ELJA	ELJA	ELJA	ELJA	ELJA	ELJA	ELJA	
Myoxocephalus aenaeus				J	ELJ	J	LJ	ELJ	ELJ	L	ELJ	ELJ			ELJ	ELJA
Morone americana	JA	ELJA	ELJA	JA	ELJA	ELJA	LJA	ELJA	ELJA	JA	ELJA	ELJA	ELJA	ELJA	ELJA	
M. saxatilis	JA	ELJA	J	JA	ELJA	JA	ELJA	JA	ELJA	JA	JA	JA	JA	JA	JA	
Centropristis striata	LJ	JA	JA	JA	JA	LJA	LJA	LJA	JA	A	LJA	LJA		LJA	ELJA	
Pomatomus saltatrix	JA	JA	LJA	JA	LJA	JA	LJA	LJA	ELJA	JA	ELJA	JA	JA	LJA	LJA	
Caranx hippos	J	J		J	J	J	J	J	J		J	J			J	
Lutjanus griseus	J			J	J	J	J		J	J					J	
Lagodon rhomboides	LJA	A		A		A	J		A					A		
Stenotomus chrysops		JA	J	JA	JA	JA	LJ	JA	ELJA	JA	ELJA	ELJA	ELJA	ELJA	ELJA	
Bairdiella chrysoura		LJ	ELJ	EJ	LJ	J	ELJ	LJ	J	LJ	LJ		LJ			
Cynoscion regalis	LJA	ELJA	ELJA	ELJA	ELJA	ELJA	LJA	ELJA	ELJA	ELJA	ELJA	ELJA	JA	ELJA	ELJA	
Leiostomus xanthurus	LJA	LJA	LJA	ELJA	LJA	JA	LJ	LJA	LJA	JA	JA	JA	JA	ELJA	J	
Menticirrhus saxatilis	LJ	JA	J	LJA	ELJA	LJA	JA	ELJA	ELJA	ELJA	ELJA	A	ELJA	ELJA		
Micropogonias undulatus	LJA	LJA	LJ	ELJA	LJA	LJA	LJA	LJA	LJA		JA					
Sciaenops ocellatus	LJA	A			A	A		A								
Pogonias cromis	JA	ELJA	EJ	LJA	ELJA	JA	JA	JA	JA					JA		
Chaetodon ocellatus		J	J	J		J	J		J					J		
Mugil cephalus	JA	J	J	J	J	J	J	J	J	J				J	J	
M. curema	J	J	J	J	J	J	J	J	J	J	J	J			J	J
Sphyraena borealis		J	J	J	J	J	J		J	J	J	J			J	
Tautoga onitis		ELJA	JA	ELJA	ELJA	ELJA	ELJA	ELJA	ELJA	ELJA	ELJA	ELJA	ELJA	ELJA	ELJA	J
Tautogolabrus adspersus		J		ELJA	ELJA	ELJA	ELJA	ELJA	ELJA	ELJA	ELJA	ELJA	ELJA	ELJA	LJA	
Pholis gunnellus							LJ				E	EJ		L	J	LJ
Astroscopus guttatus		LJ		J	J	J	LJ	J	J							
Hypsoblennius hentz	LJ	LJ	LJ	LJ	L		LJ	ELJ								
Ammodytes americanus		L	LJA	ELJ	ELJA	ELJA	ELJA	ELJA	ELJA	ELJA	ELJA	ELJA	ELJA	ELJA	ELJA	LJA
Gobionellus boleosoma	LJ			L		LJ										
Gobiosoma bosc	LJ	ELJ	ELJ	ELJ	ELJ	ELJ	ELJ	ELJ	J			J			J	
G. ginsburgi	L	LJ	LJ	LJ	LJ	ELJ	LJ		J		L	J		L		
Peprilus triacanthus	LJ	ELJA	J	JA	ELJA	LJA	LJ	LJA	ELJA	ELJA	ELJA	ELJA	JA	ELJA	ELJA	J
Scophthalmus aquosus	LJ	ELJA	LJA	ELJA	JA	ELJA	ELJA	ELJA	ELJA	ELJA	ELJA	ELJA	ELJA	ELJA	ELJA	JA
Etropus microstomus	L		J	J	LJ	ELJ	LJ	J	J	E		J				J
Paralichthys dentatus	LJA	LJA	JA	LJA	LJA	LJA	LJA	LJA	ELJA	JA	JA	LJA	JA	ELJA	ELJA	J
Pseudopleuronectes americanus		ELJA	LJA	ELJA	ELJA	ELJA	ELJA	ELJA	ELJA	ELJA	ELJA	ELJA	ELJA	ELJA	ELJA	LJA
Trinectes maculatus	LJ	ELJA	ELJA	ELJA	ELJA	ELJA	ELJA	ELJA	ELJA	ELJA	ELJA	ELJA	ELJA	ELJA	ELJA	
Sphoeroides maculatus	LJ	LJ	J	LJ	J	LJ	LJ	ELJ	J	LJ	J	LJ		L	LJ	

Sources: Pamlico–Albemarle Sounds, North Carolina—45–47; Chesapeake Bay and tributaries, Maryland—1, 20, 23, 50–59; eastern shore, Virginia and Maryland—1, 12, 18, 20, 34, 36, 53–59; Delaware inland bays, Delaware—1, 8, 20, 21, 22, 34, 38; Delaware Bay, Delaware and New Jersey—1, 4, 7, 10, 20, 33; Great Bay, New Jersey—20, 26, 27, 30, 35, 48; Barnegat Bay, New Jersey—1, 3; Hudson–Raritan–Sandy Hook Bays, New Jersey—1, 6, 14, 16, 17, 19, 20, 24, 60; Great South Bay, New York—1, 20, 21, 28, 39; Gardiners Bay, New York—1, 9, 20, 21, 24; Long Island Sound, New York—1, 11, 13, 21, 25, 32; Connecticut River, Connecticut—1, 42, 43; Narragansett Bay, Rhode Island—1, 11, 20, 29, 31, 40; Buzzards Bay, Massachusetts—1, 5, 20, 24, 37, 41; Nauset Marsh, Massachusetts—44.

Table sources listed by number:

1. Stone et al., 1994
2. Musick, 1972
3. Tatham et al., 1984
4. Allen et al., 1978
5. Lux and Nichy, 1971
6. Smith, 1985
7. Bean, 1887
8. Wang and Kernehan, 1979 (includes de Sylva et al., 1962)
9. Hickey et al., 1975
10. Himchak, 1982b
11. Roberts, 1978
12. Schwartz, 1961
13. Richards, 1959
14. Kahnle and Hattala, 1988
15. Weisberg and Burton, 1993
16. Wilk and Silverman, 1976
17. Berg and Levinton, 1985
18. Cowan and Birdsong, 1985
19. Dovel, 1981
20. Pacheco, 1973
21. Perlmutter, 1939
22. Pacheco and Grant, 1973
23. Olney and Boehlert, 1988
24. Nichols and Breder, 1927
25. Wheatland, 1956
26. Rountree and Able, 1992a
27. Szedlmayer and Able, 1996
28. Monteleone, 1992
29. Herman, 1963
30. Witting, 1995; Witting et al., 1999
31. Bourne and Govoni, 1988
32. Pearcy and Richards, 1962
33. McDermott, 1971
34. Schwartz, 1964
35. Swiecicki and Tatham, 1977
36. Richards and Castagna, 1970
37. Summer et al., 1913
38. Derickson and Price, 1973
39. Briggs and O'Connor, 1971
40. Oviatt and Nixon, 1973
41. Lux and Wheeler, 1992
42. Marcy, 1976
43. Marcy, 1976
44. Able and Fahay, unpubl. observ.
45. Tagatz and Dudley, 1961
46. Ross and Epperly, 1985
47. Hettler and Barker, 1993
48. Fahay and Able, 1989
49. Barans, 1972
50. Virginia Institute of Marine Sciences, 1962
51. Orth and Heck, 1980
52. Hildebrand and Schroeder, 1928
53. Jones et al., 1978
54. Hardy, 1978b
56. Johnson, 1978
57. Fritzsche, 1978
58. Martin and Drewry, 1978
59. Joseph et al., 1964
60. Croker, 1965.

Note: See Figures 2.1, 2.2, and 2.3 for location of estuaries and Table 2.3 for physical characteristics of these estuaries. Occurrence indications are based on published or unpublished sources and are not quantitative; thus the collection of a single specimen results in a positive indication. See individual species accounts for evaluations of certain of these generalizations. *Gambusia holbrooki* is not included because of frequent introduction of congeners.

Table 4.4. Checklist of Middle Atlantic Bight estuarine fishes treated in this study. *Species treated in individual chapters (n = 93). Details of the remaining, rarely collected, species (n = 39) are in Table 4.5.

Scientific Name (Author)	Common Name	Scientific Name (Author)	Common Name
Carcharhinidae		Cyprinodontidae	
Carcharhinus plumbeus (Nardo)★	sandbar shark	*Cyprinodon variegatus* Lacepède★	sheepshead minnow
Mustelus canis (Mitchill)★	smooth dogfish	Fundulidae	
Rajidae		*Fundulus confluentus* Goode and Bean★	marsh killifish
Raja eglanteria Bosc	clearnose skate	*F. diaphanus* (Lesueur)★	banded killifish
Myliobatidae		*F. heteroclitus* (Linnaeus)★	mummichog
Rhinoptera bonasus (Mitchill)★	cownose ray	*F. luciae* (Baird)★	spotfin killifish
Acipenseridae		*F. majalis* (Walbaum)★	striped killifish
Acipenser brevirostrum (Lesueur)★	shortnose sturgeon	*Lucania parva* (Baird and Girard)★	rainwater killifish
Acipenser oxyrinchus (Mitchill)★	Atlantic sturgeon	Poeciliidae	
Elopidae		*Gambusia holbrooki* Girard★	eastern mosquitofish
Elops saurus Linnaeus	ladyfish	Atherinopsidae	
Albulidae		*Membras martinica* (Valenciennes)★	rough silverside
Albula vulpes (Linnaeus)	bonefish	*Menidia beryllina* (Cope)★	inland silverside
Anguillidae		*M. menidia* (Linnaeus)★	Atlantic silverside
Anguilla rostrata (Lesueur)★	American eel	Gasterosteidae	
Muraenidae		*Apeltes quadracus* (Mitchill)★	fourspine stickleback
Gymnothorax sp. Jordan and Davis	honeycomb moray	*Gasterosteus aculeatus* Linnaeus★	threespine stickleback
Ophichthidae		*G. wheatlandi* Putnam	blackspotted stickleback
Myrophis punctatus Lütken★	speckled worm eel	*Pungitius pungitius* (Linnaeus)	ninespine stickleback
Ophichthus gomesi (Castelnau)	shrimp eel	Fistulariidae	
Congridae		*Fistularia tabacaria* Linnaeus	bluespotted cornetfish
Conger oceanicus (Mitchill)★	conger eel	Syngnathidae	
Clupeidae		*Hippocampus erectus* Perry★	lined seahorse
Alosa aestivalis (Mitchill)★	blueback herring	*Syngnathus fuscus* Storer★	northern pipefish
A. mediocris (Mitchill)★	hickory shad	Triglidae	
A. pseudoharengus (Wilson)★	alewife	*Prionotus carolinus* (Linnaeus)★	northern searobin
A. sapidissima (Wilson)★	American shad	*P. evolans* (Linnaeus)★	striped searobin
Brevoortia tyrannus (Latrobe)★	Atlantic menhaden	Cottidae	
Clupea harengus Linnaeus★	Atlantic herring	*Myoxocephalus aenaeus* (Mitchill)★	grubby
Dorosoma cepedianum (Lesueur)★	gizzard shad	Percichthyidae	
Opisthonema oglinum (Lesueur)★	Atlantic thread herring	*Morone americana* (Gmelin)★	white perch
Sardinella aurita Valenciennes	Spanish sardine	*M. saxatilis* (Walbaum)★	striped bass
Engraulidae		Serranidae	
Anchoa hepsetus (Linnaeus)★	striped anchovy	*Centropristis striata* (Linnaeus)★	black sea bass
A. mitchilli (Valenciennes)★	bay anchovy	*Epinephelus niveatus* (Valenciennes)	snowy grouper
Engraulis eurystole (Swain and Meek)★	silver anchovy	*E. striatus* (Bloch)	Nassau grouper
Osmeridae		*Mycteroperca microlepis* (Goode and Bean)★	gag
Osmerus mordax (Mitchill)★	rainbow smelt	Apogonidae	
Synodontidae		*Phaeoptyx pigmentaria* (Poey)	dusky cardinalfish
Synodus foetens (Linnaeus)★	inshore lizardfish	Pomatomidae	
Phycidae		*Pomatomus saltatrix* (Linnaeus)★	bluefish
Enchelyopus cimbrius (Linnaeus)★	fourbeard rockling	Carangidae	
Urophycis chuss (Walbaum)★	red hake	*Caranx crysos* (Mitchill)	blue runner
U. regia (Walbaum)★	spotted hake	*C. hippos* (Linnaeus)★	crevalle jack
U. tenuis (Mitchill)★	white hake	*Decapterus macarellus* (Cuvier)	mackerel scad
Merlucciidae		*Selene vomer* (Linnaeus)	lookdown
Merluccius bilinearis (Mitchill)	silver hake	*Seriola zonata* (Mitchill)	banded rudderfish
Gadidae		*Trachinotus carolinus* (Linnaeus)★	Florida pompano
Gadus morhua Linnaeus	Atlantic cod	*T. falcatus* (Linnaeus) ★	permit
Microgadus tomcod (Walbaum)★	Atlantic tomcod	*T. goodei* Jordan and Evermann	Palometa
Pollachius virens (Linnaeus)★	pollock	Lutjanidae	
Ophidiidae		*Lutjanus griseus* (Linnaeus)★	gray snapper
Ophidion marginatum (DeKay)★	striped cusk-eel	Gerreidae	
O. welshi (Nichols and Breder)	crested cusk-eel	Unidentified species	mojarras
Batrachoididae		Haemulidae	
Opsanus tau (Linnaeus)★	oyster toadfish	*Orthopristis chrysoptera* (Linnaeus)	pigfish
Hemiramphidae		Sparidae	
Hyporhamphus meeki Banford and Collette	silverstripe halfbeak	*Lagodon rhomboides* (Linnaeus)★	pinfish
Belonidae		*Stenotomus chrysops* (Linnaeus)★	scup
Strongylura marina (Walbaum)★	Atlantic needlefish		

Table 4.4 continued

Scientific Name (Author)	Common Name	Scientific Name (Author)	Common Name
Sciaenidae		Sphyraenidae	
Bairdiella chrysoura (Lacepède)*	silver perch	Sphyraena barracuda (Edwards)	great barracuda
Cynoscion nebulosus (Cuvier)*	spotted seatrout	S. borealis DeKay*	northern sennet
Cynoscion regalis (Bloch and Schneider)*	weakfish	Scombridae	
Leiostomus xanthurus Lacepède*	spot	Scomber japonicus Houttuyn	chub mackerel
Menticirrhus saxatilis (Bloch and Schneider)*	northern kingfish	S. scombrus Linnaeus	Atlantic mackerel
Micropogonias undulatus (Linnaeus)*	Atlantic croaker	Scomberomorus maculatus (Mitchill)	Spanish mackerel
Pogonias cromis (Linnaeus)*	black drum	Stromateidae	
Sciaenops ocellatus (Linnaeus)*	red drum	Peprilus paru (Linnaeus)	harvestfish
Chaetodontidae		P. triacanthus (Peck)*	butterfish
Chaetodon capistratus Linnaeus*	foureye butterflyfish	Bothidae	
C. ocellatus Bloch*	spotfin butterflyfish	Bothus sp.	
Mugilidae		Scophthalmidae	
Mugil cephalus Linnaeus*	striped mullet	Scophthalmus aquosus (Mitchill)*	windowpane
M. curema Valenciennes*	white mullet	Paralichthyidae	
Labridae		Citharichthys spilopterus Günther	bay whiff
Tautoga onitis (Linnaeus)*	tautog	Etropus crossotus Jordan and Gilbert	fringed flounder
Tautogolabrus adspersus (Walbaum)*	cunner	E. microstomus (Gill)*	smallmouth flounder
Stichaeidae		Paralichthys dentatus (Linnaeus)*	summer flounder
Lumpenus lumpretaeformis (Walbaum)	snakeblenny	Pleuronectidae	
L. maculatus (Fries)	daubed shanny	Pseudopleuronectes americanus Walbaum*	winter flounder
Ulvaria subbifurcata (Storer)	radiated shanny	Achiridae	
Pholidae		Trinectes maculatus (Bloch and Schneider)*	hogchoker
Pholis gunnellus (Linnaeus)*	rock gunnel	Cynoglossidae	
Ammodytidae		Symphurus plagiusa (Linnaeus)*	blackcheek tonguefish
Ammodytes americanus DeKay*	American sand lance	Monacanthidae	
Uranoscopidae		Aluterus schoepfii (Walbaum)	orange filefish
Astroscopus guttatus Abbott*	northern stargazer	A. scriptus (Osbeck)	scrawled filefish
Blenniidae		Stephanolepis hispidus (Linnaeus)	planehead filefish
Chasmodes bosquianus (Lacepède)*	striped blenny	Ostraciidae	
Hypleurochilus geminatus (Wood)	crested blenny	Lactophrys sp.	unidentified boxfish
Hypsoblennius hentz (Lesueur)*	feather blenny	Tetraodontidae	
Gobiesocidae		Chilomycterus schoepfii (Walbaum)*	striped burrfish
Gobiesox strumosus Cope*	skilletfish	Sphoeroides maculatus (Bloch and Schneider)*	northern puffer
Gobiidae			
Ctenogobius boleosoma (Jordan and Gilbert)*	darter goby		
Gobiosoma bosc (Lacepède)*	naked goby		
G. ginsburgi Hildebrand and Schroeder*	seaboard goby		
Microgobius thalassinus (Jordan and Gilbert)*	green goby		

Several species (e.g., *Caranx hippos*, *Lutjanus griseus*, *Chaetodon ocellatus*, *Mugil cephalus*, and *M. curema*) occur in Middle Atlantic Bight estuaries as juveniles only, and these are all products of spawning in the South Atlantic Bight or oceanic waters east of the Bahamas (Sargasso Sea) as is true of *Anguilla rostrata* and *Conger oceanicus*. Other species (*Urophycis tenuis* and a spring cohort of *Urophycis regia*) spawn in Slope Sea waters beyond the continental shelf, then appear in estuaries as juveniles. In some species (e.g., *Ophidion marginatum*), larvae and juveniles are simply found in estuaries adjacent to continental shelf areas where they spawn. For other species, there are curious omissions, such as the lack of reports of *Sphoeroides maculatus* eggs, although their larvae and juveniles occur throughout the bight. We conclude that our understanding of some of this information is influenced somewhat by sampling deficiencies. Other apparent differences in distribution may be due to variations in the charac-

teristics of individual estuaries. For example, eggs and larvae of *Alosa* spp. are absent in systems lacking a freshwater tidal component (e.g., eastern shore of Virginia, Gardiners Bay, Great South Bay). However, despite different estuarine characteristics, all early stages of several species (e.g., *Brevoortia tyrannus*, *Anchoa mitchilli*, *Opsanus tau*, *Fundulus heteroclitus*, *Menidia menidia*, *Pseudopleuronectes americanus*, and *Trinectes maculatus*) are ubiquitous in their occurrences, perhaps attesting to their adaptability to a wide range of conditions.

The Mullica River–Great Bay estuary has been the site of extensive sampling for early stages of fishes during the past two decades, and this has prompted several analyses, including the present one. A summary of specific aspects of the fish fauna within this system has been published, mostly focusing on young-of-the-year (Able and Fahay, 1998). The numerically dominant species are members of the families Engraulidae, Clupeidae, Fundulidae, Gobiidae, Syngnathidae,

Atherinopsidae, and Sciaenidae. Seasonal and annual variation in abundance of many species is pronounced. An analysis suggests a stable ichthyoplankton assemblage with marked seasonal components that occur in a regular progression (Witting et al., 1999). Important annual variations in abundance, measured in regular and repeated sampling programs, can be influenced by long-term trends, as for recovering and then declining stocks of *Clupea harengus*. Variations in delivery systems to this estuary may be influenced by hy-

drographic perturbations, such as locally produced upwelling along the coast or features associated with the Slope Sea or Gulf Stream near the edge of the continental shelf (Epifanio and Garvine, 2001; Hare et al., 2001; see Chapter 6). The effects of climate change are having a demonstrable influence on the fauna here (Hare and Able, 2007; see Chapter 10). Other components of the estuarine fish fauna are derived from adjacent, upstream freshwaters (Arndt, 2004). Together, these represent a diverse fauna under multiple influences. Of

Table 4.5. Summary of collections of rare fish species in the Mullica River–Great Bay estuary.

Species	No. Collected	Years	Primary Collecting Gear	Length Range	Seasonal Occurrence*
Rajidae					
Raja eglanteria	79	1988–1990, 1994, 1997–2000, 2003, 2005–2006	Otter trawl	4–68 cm BW	Sp–Su
Elopidae					
Elops saurus	3	1991, 1995, 2000	Plankton net	3 cm SL	Fa
Albulidae					
Albula vulpes	3	1992, 1994, 1999	Plankton net / Dip net	27 FL–60.0 mmTL	Su–Fa
Muraenidae					
Gymnothorax sp.	2	1989–1990	Misc.	7–9 cm TL	Su
Ophichthidae					
Ophichthus gomesi	9	1989, 1991–1992	Plankton net / Dip net	7–9 cm TL	Su–Fa
Clupeidae					
Sardinella aurita	95	1988–1990, 1998	Weir / Seine / Methot trawl	12.3 SL–80 mm TL	Su
Gadidae					
Gadus morhua	2	1992, 2003	Plankton net	1–3 cm TL	Sp
Merlucciidae					
Merluccius bilinearis	17	1988–1990, 1999, 2003, 2006	Otter trawl	8–20 cm TL	Su–Wi
Ophidiidae					
Ophidion welshi	1	1989	Plankton net	3 cm SL	Fa
Hemiramphidae					
Hyporhamphus meeki	4	1987, 1990, 2005	Killitrap / Dip net	1–6 cm TL	Sp–Fa
Gasterosteidae					
Gasterosteus wheatlandi	1	1992	Dip net	3 cm TL	Sp
Pungitius pungitius	2	1979, 1982	Dip net	5 cm TL	Sp
Fistulariidae					
Fistularia tabacaria	20	1990, 1998–2000, 2002, 2006	Seine	94–520 mm TL	Su–Fa
Serranidae					
Epinephelus niveatus	3	1991, 1995, 1999	Misc	3–15 cm TL	Su
Epinephelus striatus	1	1992	Trap	8 cm TL	Fa
Apogonidae					
Phaeoptyx pigmentaria	1	1992	Trap	4 cm FL	Fa
Carangidae					
Caranx crysos	20	1992, 1998–2000, 2006	Methot trawl / Otter trawl / Seine	4mm SL–15cm FL	Su
Decapterus macarellus	1	1989	Dip net	10 cm TL	Su
Selene vomer	80	1988–1990, 1998–2000, 2002–2005	Otter trawl / Seine	28 mm SL–117 mm FL	Su–Fa
Seriola zonata	14	1986, 1992–1993, 1997, 2000, 2005–2006	Dip net / Trap / Seine	5–18 cm FL	Su
Trachinotus goodei	115	1996, 1996–2000	Seine	20–145 mm FL	Su–Fa
Gerreidae	89	1989–1995, 1997–2003, 2006	Beam trawl / Otter trawl / Plankton net / Seine	1–6 cm SL	Sp–Fa
Haemulidae					
Orthopristis chrysoptera	42	1996–97, 2006	Trap	5–11 cm FL	Su–Fa
Sphyraenidae					
Sphyraena barracuda	1	1989	Plankton net	3 cm SL	Su
Stichaeidae					
Lumpenus lumpretaeformis	1	1990	Plankton net	2 cm SL	Su
L. maculatus	2	1989	Plankton net	2–3 cm SL	Su
Ulvaria subbifurcata	28	1989–1990, 1992, 1999, 2001	Plankton net	1–2 cm SL	Sp

Table 4.5 continued

Species	No. Collected	Years	Primary Collecting Gear	Length Range	Seasonal Occurrence*
Blenniidae					
Hypleurochilus geminatus	2	1992	Killitrap/Dip net	6 cm TL	Su–Fa
Scombridae					
Scomber japonicus	1	1990	Dip net	9 cm TL	Su
S. scombrus	40	1996, 1998–1999, 2001–2005	Plankton net	4 mm SL–85 cm FL	Sp–Fa
Scomberomorus maculatus	5	1989, 1998, 2000	Methot trawl/Otter trawl	31 mm SL–140 mm TL	Su
Stromateidae					
Peprilus paru	1	1988	Otter trawl	2 cm FL	Su
Bothidae					
Bothus sp.	5	1990, 1991, 2004, 2006	Plankton net	8 mm SL–25 mm TL	Su–Fa
Paralichthyidae					
Citharichthys spilopterus	2	1993–1994	Plankton net	1 cm SL	Fa
Etropus crossotus	2	1990	Plankton net	4–7 cm SL	Fa
Monacanthidae					
Aluterus schoepfii	2	1989	Plankton net/Otter trawl	8–27 cm TL	Su
Aluterus scriptus	7	1989, 1994	Otter trawl	6–34 cm TL	Su–Fa
Stephanolepis hispidus	20	1989, 1992–1994, 1999, 2000	Otter trawl	2–15 cm TL	Su–Fa
Ostraciidae					
Lactophrys sp.	2	1990, 1993	Misc.	8 mm SL	Su

*Spring (Sp) = March–May, Summer (Su) = June–August, Fall (Fa) = September–November, Winter (Wi) = December–February

this fauna we treat 93 species in detail (Table 4.4). Several rare species occur irregularly in the Great Bay–Little Egg Harbor system in young-of-the-year stages, and many of these may be more common in estuaries of the southern or northern parts of the Middle Atlantic Bight (Table 4.5).

To summarize this chapter on the ichthyofauna of our study area, estuarine fishes are not always impeded by biogeographic barriers at Cape Cod and Cape Hatteras, and the extreme temperatures within this temperate region make it possible for both northern and southern species to occur here, albeit only seasonally. This dilution can occur as a result of larval dispersal and/or seasonal migrations by adults. The degree to which both dispersal and migration influence the fauna of the region may be increasingly influenced by climate change (see Chapter 10).

5

Reproduction and Development

For many fish species that use temperate estuaries, reproduction occurs and development begins in the ocean. For this reason, we have assembled much of the available data from the ocean in order to discern patterns in their seasonality, which is so marked in this region (see Chapter 4). In this chapter we also bring together aspects of development that occur while these fishes are in estuaries, and for the first time, include some of our observations of scale formation in these fishes.

REPRODUCTION

Fishes that use temperate estuaries spawn in a variety of locations, some quite distant from the estuaries where the early stages spend much of their first year. Most spawning occurs in the Middle Atlantic Bight on the continental shelf, in estuaries or freshwaters (Table 5.1; Fig. 5.1). A relatively large number of species (e.g., *Mugil curema, Caranx hippos, Lutjanis griseus,* and *Myrophis punctatus*) originate from spawning that occurs in the South Atlantic Bight. A smaller number (*Pomatomus saltatrix* and *Peprilus triacanthus*) may come from a combination of South Atlantic Bight and Middle Atlantic Bight spawning. This combination pattern is best demonstrated by *Pomatomus saltatrix,* in which individuals spawning in both bights use the temperate estuaries of the Middle Atlantic Bight (Hare and Cowen, 1993). The same may be true for *Brevoortia tyrannus,* based on the timing of spawning, larval ingress, and back-calculation of hatching dates (Warlen et al., 2002; see species account). The most distant spawning site of which we are aware for species using temperate estuaries is the Sargasso Sea (*Anguilla rostrata* and *Conger oceanicus*) to the south and Georges Bank / Nantucket Shoals (*Clupea harengus*) to the north.

Most resident fishes in temperate estuaries spawn during the spring and summer, whereas transients in these same estuaries are spawned during every season (see Table 5.1), per-

haps so they can take advantage of warm temperatures in other regions. The timing of spawning may also vary within a species. Some species may spawn in a distinctly bimodal temporal pattern. Examples from the study area include *Urophycis regia* and *Scophthalmus aquosus.* Others may spawn in different areas, and the progeny produce a bimodal size pattern. The clearest example is *Pomatomus saltatrix,* which spawns continuously from the South Atlantic to Middle Atlantic bights; the resulting cohorts demonstrate multi- or bimodal size distributions (Hare and Cowen 1993; Smith et al., 1994; Able et al., 2003b; Wuenschel et al., 2009).

These patterns of reproduction within temperate waters of the U.S. East Coast are exemplified by the seasonal occurrence of eggs and larvae (Morse et al., 1987; Smith and Morse, 1988; Able and Hagan, 1995; Berrien and Sibunka, 1999). Larvae of species that eventually use estuaries in the study area are produced throughout the year, and many of these are among the most abundant taxa during certain months (see Tables 4.1 and 4.2). Spawning over the continental shelf is at its lowest during winter, when few species spawn, although an explosion of the *Ammodytes dubius* population in the mid-1970s resulted in an abundance of larvae during winter throughout the Middle Atlantic Bight (Sherman et al., 1984).

The spatial and temporal patterns we have observed in estuarine use by temperate fishes during their first year of life (see Able and Fahay, 1998; species accounts) vary among and within resident and transient species and cohorts and often change as we gain an improved understanding of fish lifecycle diversity (Secor and Kerr, 2009). Another difficulty in assigning fishes to these categories is that these patterns may vary from northern to southern estuaries. The clearest difference in the estuarine fish fauna is that between residents (species that spend their entire lifespans in estuaries) and transients (species that spend only a portion of their lives there). Evidence for assignment to specific groups draws on our prior attempts to synthesize this information (Able and Fahay, 1998), is in the individual species accounts, or is largely based on spawning time and location and, as a result, the pattern of estuarine use. Consequently, it appears that most Middle Atlantic Bight estuarine species for which we have good information are transients (75%), and fewer are residents (25%). A surprisingly large number are facultative estuarine users because they use both the estuary and ocean. Other species are represented by multiple cohorts and these belong to different groups (Table 5.2).

Group I includes several species that typically spawn during summer. They are facultative users of estuaries because the juvenile habitats are in estuaries and/or the inner continental shelf. The larvae and early juveniles can be found in both estuaries and the coastal ocean for the remainder of the summer before they emigrate from these habitats into deeper or more southern ocean waters in the fall. The most completely documented example is *Centropristis striata,* but

Table 5.1. Summary of selected reproduction characteristics of estuarine fishes for temperate estuaries in the Middle Atlantic Bight. Sources include Able and Fahay 1998 and this study.

Species	Size at First Reproduction	Spawning Time[†]	Spawning Location[‡]	Egg Type	Egg Size (mm)
Carcharhinidae					
Carcharhinus plumbeus	120–180 cm	Sp, Su	Estuary / MAB / SAB	Live	—
Mustelus canis	86 cm M, 102 cm F	Sp	Estuary / MAB	Live	—
Myliobatidae					
Rhinoptera bonasus	90–96 cm	Su	Estuary / MAB / SAB	Live	—
Acipenseridae					
Acipenser brevirostrum	—	Sp	Freshwater	Benthic	3.0
Acipenser oxyrinchus	122 cm M, 183 cm F	Sp	Freshwater	Benthic	2.9
Anguillidae					
Anguilla rostrata	28 cm M, 46 cm F	Sp	SS	?	0.6–1.2*
Ophichthidae					
Myrophis punctatus	?	Fa	SAB	?	?
Congridae					
Conger oceanicus	?	Su–Wi	SS	?	?
Clupeidae					
Alosa aestivalis	< 250 mm	Sp	FW	Pelagic	0.8–1.1
Alosa mediocris	287 mm TL M, 320 mm TL F	Sp	FW	Demersal / Pelagic	0.9–1.6
Alosa pseudoharengus	265–278 mm M, 284–308 mm F	Sp	FW	Pelagic	0.8–1.3
Alosa sapidissima	305–447 mm FL M, 383–485 mm FL F	Sp	FW	Demersal / Pelagic	2.5–3.8
Brevoortia tyrannus	180–200 mm	Fa, Sp	MAB / SAB	Pelagic	1.3–1.9
Clupea harengus	181–185 mm SL	Sp	MAB	Demersal	1.0–1.4
Dorosoma cepedianum	?	Sp	FW	Demersal	0.75
Opisthonema oglinum	?	Sp–Su	MAB? / SAB	Pelagic	1.08–1.31
Engraulidae					
Anchoa hepsetus	< 75 mm	Su	MAB	Pelagic	1.2–1.7 long axis
Anchoa mitchilli	31–45 mm FL	Su	Estuary / MAB	Pelagic	0.8–1.1 long axis
Engraulis eurystole	?	Sp–Fa	SAB?	Pelagic	1.02–1.25 long axis
Osmeridae					
Osmerus mordax	?	Sp	FW	Demersal	≈ 1.0
Synodontidae					
Synodus foetens	?	?	SAB	?	?
Phycidae					
Enchelyopus cimbrius	?	Sp–Fa	MAB	Pelagic	0.66–0.98
Urophycis chuss	240–300 mm M, 270–300 mm F	Su	MAB	Pelagic	0.6–1.0
Urophycis regia	210 mm M, 310 mm F	Su–Fa, Sp	MAB	Pelagic	0.6–1.0
Urophycis tenuis	400 mm M, 480 mm F	Sp	MAB (Slope Sea)	Pelagic	0.7–0.8
Gadidae					
Microgadus tomcod	170 mm F	Wi	FW	Demersal	1.4–1.7
Pollachius virens	~ 410 mm M, ~ 390 mm F	Fa–Wi	MAB	Pelagic	1.0–1.2
Ophidiidae					
Ophidion marginatum	160 mm TL F	Su–Fa	Middle Atlantic Bight	Pelagic	Oblong 0.8–1.0 × 0.9–1.1
Batrachoididae					
Opsanus tau	?	Sp–Su	Estuary	Demersal	5.0–5.5
Belonidae					
Strongylura marina	205 mm	Sp	Estuary	Demersal	3.0
Cyprinodontidae					
Cyprinodon variegatus	?	Sp–Su	Estuary	Demersal	1.2–1.4
Fundulidae					
Fundulus confluentes	35 mm	Sp–Fa	Estuary	Demersal	1.6–1.8
Fundulus diaphanus	55 mm	Sp–Su	Estuary	Demersal	1.5–2.0
Fundulus heteroclitus	23 mm SL M, 28 mm SL F	Sp–Su	Estuary	Demersal	1.7–2.0
Fundulus luciae	24–27 mm M, 28–30 mm F	Sp–Su	Estuary	Demersal	1.7–2.2
Fundulus majalis	63.0 mm M, 76.0 mm F	Sp–Su	Estuary	Demersal	2.0–3.0
Lucania parva	25.0 mm TL	Sp–Su	Estuary	Demersal	1.0–1.3
Poeciliidae					
Gambusia holbrooki	?	Su	FW	Live	—
Atherinopsidae					
Membras martinica	?	?	?	Demersal	0.7–0.8

continued

Table 5.1 continued

Species	Size at First Reproduction	Spawning Time[†]	Spawning Location[‡]	Egg Type	Egg Size (mm)
Menidia beryllina	33 mm SL	Sp–Su	Estuary	Demersal	0.8–0.9
Menidia menidia	50–80 mm	Sp–Su	Estuary	Demersal	1.0–1.5
Gasterosteidae					
Apeltes quadracus	27 mm M, 33 mm F	Sp	Estuary	Demersal	1.3–1.6
Gasterosteus aculeatus	50 mm	Sp	Estuary	Demersal	1.5–1.9
Syngnathidae					
Hippocampus erectus	?	Sp–Su	Estuary/MAB	Live	—
Syngnathus fuscus	83 mm	Su	Estuary	Live	—
Triglidae					
Prionotus carolinus	140–200 mm	Su–Fa	MAB (Estuary?)	Pelagic	0.9–1.5
Prionotus evolans	> 200 mm	Su–Fa	MAB (Estuary?)	Pelagic	1.0–1.2
Cottidae					
Myoxocephalus aenaeus	?	Wi	Estuary/MAB	Demersal	1.5–1.7
Percichthyidae					
Morone americana	72 mm M, 98 mm F	Sp	FW	Demersal/Pelagic	0.7–1.0
Morone saxatilis	174 mm M, 432 mm F	Sp	FW	Pelagic	1.3–4.6
Serranidae					
Centropristis striata	190 mm TL	Sp–Fa	MAB	Pelagic	0.9–1.0
Mycteroperca microlepis	?	Wi–Sp	SAB	Pelagic	?
Pomatomidae					
Pomatomus saltatrix	?	Sp, Su	South Atlantic Bight/ MAB	Pelagic	1.0
Carangidae					
Caranx hippos	?	?	SAB	Pelagic	0.7–0.9
Trachinotus carolinus	275 mm TL	Sp–Fa	SAB	Pelagic	0.7
Trachinotus falcatus	486 mm M, 547 mm F	Sp–Su	SAB	?	?
Lutjanidae					
Lutjanus griseus	185 mm SL M, 195 mm SL F	Su	SAB	Pelagic	0.7–0.8
Sparidae					
Lagodon rhomboides	?	Fa–Sp	SAB	Pelagic	0.99–1.05
Stenotomus chrysops	150 mm	Sp–Su	Estuary[||]	Pelagic	0.8–1.0
Sciaenidae					
Bairdiella chrysoura	130 mm	Su	?	Pelagic	0.7–0.8
Cynoscion nebulosis	292 mm	Sp–Su	Estuary/SAB	Pelagic	0.7–0.8
Cynoscion regalis	?	Sp–Su	Estuary/MAB	Pelagic	0.8–0.9
Leiostomus xanthurus	170–210 mm	Wi	MAB**	Pelagic	0.7–0.9
Menticirrhus saxatilis	?	Su	MAB	Pelagic	0.8–0.9
Micropogonias undulatus	140–180 mm	Su–Fa	MAB**	Pelagic	?
Pogonias cromis	320 mm	Su	MAB	Pelagic	0.8–1.0
Sciaenops ocellatus	621–740 mm M, 801–860 mm F	Su–Fa	Estuary/Ocean/ MAB/SAB	Pelagic	0.86–0.98
Chaetodontidae					
Chaetodon capistratus	?	Wi–Sp	SAB?	?	0.7
Chaetodon ocellatus	?	?	SAB?	Pelagic	0.60–0.75
Mugilidae					
Mugil cephalus	230–534 mm FL M, 240–614 mm FL F, ~ 200 mm for MAB	Wi	SAB	Pelagic	0.6–0.9?
Mugil curema	189 mm SL M, 209 mm SL F	Sp	SAB	Pelagic	0.9
Labridae					
Tautoga onitis	?	Sp–Fa	Estuary/MAB	Pelagic	0.9–1.0
Tautogolabrus adspersus	50–60 mm SL	Sp–Fa	MAB	Pelagic	0.8–0.9
Pholidae					
Pholis gunnellus	?	Wi	Estuary/MAB	Demersal	1.7–2.2
Ammodytidae					
Ammodytes americanus	?	Wi	Ocean/Estuary	Demersal	0.9–1.0
Uranoscopidae					
Astroscopus guttatus	> 381 mm	Su	Estuary/MAB	?	?
Blenniidae					
Chasmodes bosquianus	?	Sp–Su	Estuary	Demersal	0.92–1.10
Hypsoblennius hentz	?	Su	Estuary	Demersal	0.7–0.8

Table 5.1 continued

Species	Size at First Reproduction	Spawning Time[†]	Spawning Location[‡]	Egg Type	Egg Size (mm)
Gobiesocidae					
Gobiesox strumosus	40–45 mm	Sp–Su	Estuary	Demersal	0.75–0.94
Gobiidae					
Ctenogobius boleosoma	25–30 mm M, 18 mm F	Su	Estuary	Demersal	0.3
Gobiosoma bosc	?	Sp–Su	Estuary	Demersal	1.4–2.0
Gobiosoma ginsburgi	?	Su	Estuary	Demersal	?
Microgobius thalassinus	?	Su–Fa	Estuary	?	?
Sphyraenidae					
Sphyraena borealis	?	Sp	SAB	Pelagic	1.2
Stromateidae					
Peprilus triacanthus	114 mm M, 120 mm F	Sp, Su	Estuary/MAB	Pelagic	0.7–0.8
Scophthalmidae					
Scophthalmus aquosus	210–230 mm TL	Sp, Fa	Estuary/ MAB	Pelagic	0.9–1.4
Paralichthyidae					
Etropus microstomus	59.9 mm TL	Sp–Fa	MAB	Pelagic	?
Paralichthys dentatus	250–280 mm	Fa–Wi	MAB	Pelagic	0.9–1.1
Pleuronectidae					
Pseudopleuronectes americanus	200–250 mm TL	Wi	Estuary/MAB	Demersal	0.7–9.0
Achiridae					
Trinectes maculatus	50 mm	Sp–Fa	Estuary	Pelagic	0.6–0.9
Cynoglossidae					
Symphurus plagiusa	91 mm M, 101 mm F	Su–Fa	Estuary/Ocean	Pelagic	?
Tetraodontidae					
Chilomycterus schoepfii	?	Sp–Fa	Ocean	?	1.8
Sphoeroides maculatus	88, 70–100 mm SL	Sp–Su	Estuary	Demersal	0.8–0.9

[†] Spring, summer, fall, winter—Sp. = March–May, Su = June–August, Fa = September–November, Wi = December–February.

[‡] Estuarine, freshwater, Middle Atlantic Bight continental shelf, South Atlantic Bight continental shelf, Sargasso Sea.

[§] Unresolved for young-of-the-year.

[||] Large bays and inner continental shelf.

[**]Southern only.

[*] Ovarian eggs.

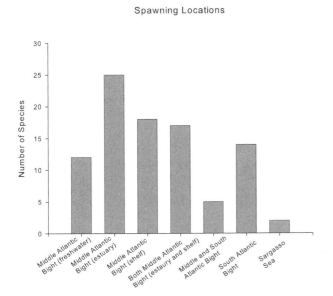

Fig. 5.1. Spawning locations for selected estuarine species in the Middle Atlantic Bight.

this group also includes such species as the spring-spawned cohort of *Scophthalmus aquosus*.

Group II includes species whose adults migrate into estuaries to spawn in spring or summer. Their progeny remain in the estuaries before emigrating to the ocean in the fall. The best documentation of this pattern of obligatory spawning in estuaries is for *Menidia menidia* and *Syngnathus fuscus*, but the group also includes *Hippocampus erectus, Strongylura marina,* and *Sphoeroides maculatus. Gasterosteus aculeatus* follows a similar pattern but enters the estuary in the winter and spawns in early spring. These juveniles then emigrate out of the estuary in early summer.

Group III includes anadromous species, such as *Morone saxatilis* and *Alosa* spp. In this group, adults migrate through estuaries in order to spawn in freshwaters. Young-of-the-year use freshwaters as well as saline portions of estuaries as nurseries, and most emigrate before winter, although those of *Morone saxatilis* remain in the estuary during their first few winters.

Groups IV to VI spawn exclusively in the ocean, but the location, timing, and manner in which young-of-the-year use

Table 5.2. Groupings of transient and resident estuarine bony fishes in the Middle Atlantic Bight according to temporal patterns of estuarine use by young-of-the-year and spawning site.

TRANSIENTS

Group I: Facultative Estuarine Users
  Ammodytes americanus?, Astroscopus guttatus?, Brevoortia tyrannus II, Centropristis striata, Chilomycterus schoepfii?, Ctenogobius boleosoma?, Cynoscion regalis?, Enchelyopus cimbrius, Engraulis eurystole, Etropus microstomus, Gobiosoma ginsburgi, Menticirrhus saxatilis?, Myoxocephalus aenaeus, Opisthonema oglinum, Peprilus triacanthus II, Pholis gunnellus, Pomatomus saltatrix II, Prionotus evolans, Scophthalmus aquosus I, Symphurus plagiusa, Tautogolabrus adspersus

Group II: Seasonal Residents
  Anchoa mitchilli, Bairdiella chrysoura, Chasmodes bosquianus, Cynoscion regalis?, Gasterosteus aculeatus, Hippocampus erectus, Menidia menidia, Micropogonias undulatus, Mustelus canis, Pogonias cromis, Pseudopleuronectes americanus, Sciaenops ocellatus, Sphoeroides maculatus, Strongylura marina, Syngnathus fuscus, Tautoga onitis

Group III: Anadromous
  Acipenser oxyrinchus, Alosa aestivalis, Alosa mediocris, Alosa pseudoharengus, Alosa sapidissima, Morone saxatilis

Group IV: Early Users
  Clupea harengus, Leiostomus xanthurus, Paralichthys dentatus, Pollachius virens, Urophycis regia I and II, Urophycis tenuis

Group V: Delayed Users
  Ophidion marginatum, Prionotus carolinus, Scophthalmus aquosus II, Urophycis chuss

Group VI: Distant Spawners
  Anguilla rostrata, Brevoortia tyrannus I, Caranx hippos, Conger oceanicus, Mugil cephalus, Mugil curema, Myrophis punctatus, Peprilus triacanthus I, Pomatomus saltatrix I, Sphyraena borealis, Trachinotus falcatus

Group VII: Expatriates
  Chaetodon capistratus, Chaetodon ocellatus, Lagodon rhomboides, Lutjanus griseus, Microgobius thalassinus, Monacanthus hispidus, Mycteroperca microlepis, Sardinella aurita?, Synodus foetens?

RESIDENTS

Group VIII: Summer Spawners
  Apeltes quadracus, Cyprinodon variegatus, Dorosoma cepedianum, Fundulus diaphanus, Fundulus heteroclitus, Fundulus luciae, Fundulus majalis, Gambusia affinis, Gobiesox strumosus, Gobionellus boleosoma?, Gobiosoma bosc, Lucania parva, Menidia beryllina, Opsanus tau

Group IX: Winter Spawners
  Ammodytes americanus

Group X: Migrating Spawners
  Microgadus tomcod, Morone americana, Osmerus mordax, Trinectes maculatus, Acipenser brevivostrum, Acipenser oxyrinchus

Note: Roman numerals for some species refer to different cohorts. *I* indicates the products of an early spawning; *II* indicates those of a later spawning.

estuaries vary. Most species in Group IV (e.g., *Urophycis regia, U. tenuis,* and *Pollachius virens*) spawn in the ocean during the winter or spring (see Table 4.1), but at least some of their off-spring enter estuaries, stay only a short time, and then leave to spend the remainder of their first year on the continental shelf. The young-of-the-year of *Leiostomus xanthurus* enter estuaries in early spring (or earlier) but remain through the fall before emigrating. In Group V, spawning occurs in the ocean during summer and fall, and transforming larvae settle and spend their first winter on the continental shelf. They then enter estuaries for the first time the following spring,

remain through the summer, and then emigrate in the fall. Examples include *Ophidion marginatum, Prionotus carolinus,* and the fall-spawned cohort of *Scophthalmus aquosus.* This pattern is slightly different for *Urophycis chuss:* some enter estuaries in the winter or spring after an initial stage involving a commensal association with sea scallops (*Placopecten megallancius*) on the continental shelf, but they do not stay through the summer.

Species in Group VI spawn outside the Middle Atlantic Bight, yet their progeny regularly occur in the Gulf Stream, then ultimately use estuaries in the Middle Atlantic Bight. Certain of these (e.g., *Mugil curema, Caranx hippos, Sphyraena borealis,* and the spring-spawned cohort of *Pomatomus saltatrix*) spawn in the South Atlantic Bight. Others (*Anguilla rostrata* and *Conger oceanicus*) reproduce in the Sargasso Sea. Larvae of all these species eventually arrive (by either passive transport, active swimming, or a combination of both) in waters of the Middle Atlantic Bight, from which they migrate into estuarine nurseries. Some of these, such as *Myrophis punctatus,* may survive in southern estuaries but not in northern ones (Able et al., in prep.).

In Group VII, larvae result from distant spawning, usually on the continental shelf south of Cape Hatteras. They arrive in Middle Atlantic Bight estuaries in the summer, but do not successfully emigrate back to the South Atlantic Bight and succumb to falling temperatures during fall and winter. Examples include *Chaetodon ocellatus, Lutjanus griseus* (see species accounts), and *Monacanthus hispidus* (Moss, 1973). The evidence is less clear regarding the fate of several other regularly occurring transients, especially in response to warming winter temperatures as the result of climate change (see Chapter 10).

Resident estuarine fishes in the central part of the Middle Atlantic Bight exhibit less diverse patterns of estuarine use than transients, because their entire life histories are limited to the estuary. Despite this spatial constraint, there are differences.

Group VIII has the most residents and these are represented by primarily shallow-water species that spawn in the summer and develop in the immediate vicinity of their spawning sites. These include *Cyprinodon variegatus, Menidia beryllina,* and *Apeltes quadracus.* Group IX constitutes a few resident species that are spawned in the winter and spring (e.g., *Ammodytes americanus*). Group X differs from the above groups of residents because its members undergo a spawning migration within the estuary. One of these, *Trinectes maculatus,* appears to be semicatadromous because it leaves freshwater or low-salinity portions of the estuary and moves downstream into higher-salinity areas to spawn pelagic eggs. Others in this group are semianadromous because they leave the saltier portions of the estuary and move into freshwater to spawn. These include *Morone americana, Microgadus tomcod,* and *Osmerus mordax.* Others, such as *Acipenser brevirostrum,* remain primarily in freshwaters.

DEVELOPMENT

Most resident temperate fishes (85%) have demersal eggs, while the majority of transients (70%) have pelagic eggs (see Table 5.1; Fig. 5.2). Demersal eggs may be typical of most resident estuarine fish assemblages (Pearcy and Richards, 1962; Able, 1978; Haedrich, 1983; Dando, 1984) because this characteristic increases retention in the estuary. Hatching in temperate estuarine fishes occurs over a relatively broad range of sizes, from 1.2 to 14.0 mm with most individuals < 6 mm (Table 5.3; Fig. 5.3). We have information on larval stage duration for only 16% of the species we treat, and these range from less than 11 to 82 days, with an exceptional few that are larvae for from 100 days to one year. Settlement size ranges from less than 17 to 100 mm, with size at settlement evenly distributed over this size range.

It is during the larval stage that dispersal typically occurs. While this was originally considered to be largely passive

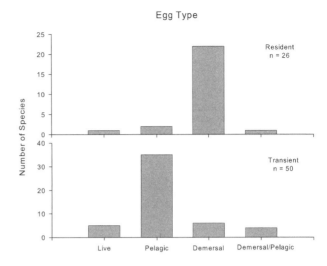

Fig. 5.2. Egg types in resident and transient components of the estuarine fish fauna in the central part of the Middle Atlantic Bight.

Table 5.3. Summary of selected developmental characteristics for temperate estuarine fishes in the Middle Atlantic Bight.

Species	Hatching/ Birth Size (mm)	Completion of Fin Rays (mm)	Larval Duration (mo, d)	Settlement/ Transformation Size (mm)	Onset of Scale Formation	Completion of Scale Formation	Juvenile Pigmentation
Carcharhinidae							
Carcharhinus plumbeus	50–60 cm	—	—	—	Before birth	Before birth	Before birth
Mustelus canis	28–39 cm	—	—	—	Before birth	Before birth	Before birth
Myliobatidae							
Rhinoptera bonasus	?	—	—	—	Before birth	Before birth	Before birth
Acipenseridae							
Acipenser brevirostrum	7.3–11.3	?	?	57–67	20	?	?
Acipenser oxyrinchus	7–9	60–70	?	116–136	?	?	?
Anguillidae							
Anguilla rostrata	6	?	175–209 d	56–61	45	?	≈ 60
Ophichthidae							
Myrophis punctatus	?	?	?	60–75	?	?	?
Congridae							
Conger oceanicus	?	?	3–6 mo	100	—	—	≈ 60
Clupeidae							
Alosa aestivalis	3.1–5.0	18.0	25–35 d	20	25–29	45	25–40
Alosa mediocris	5.2–6.5	17.5	?	≈ 18?	35	?	?
Alosa pseudoharengus	2.5–5.0	20.0	?	20	25–29	?	30–40
Alosa sapidissima	5.7–10.0	30.0	21–28 d	≈ 27	34	51.8	?
Brevoortia tyrannus	2.4–4.5	23.0	?	30–38	> 33	40–45	?
Clupea harengus	4.0–10.0	35.0	5–7 mo	30–42	?	?	45
Dorosoma cepedianum	3.2	?	?	—	?	50	?
Opisthonema oglinum	3.0	15–25	?	—	?	?	?
Engraulidae							
Anchoa hepsetus	3.6–4.0	13.0	?	?	19.1 FL	46.9 FL	?
Anchoa mitchilli	2.0	7–10 SL	?	?	10.0 FL	75.8 FL or less	?
Engraulis eurystole	2.0–3.0	?	?	—	?	?	?
Osmeridae							
Osmerus mordax	5.5–6.0	?	?	?	36+	?	?
Synodontidae							
Synodus foetens	2.5	30.0	?	30	?	?	30–50
Phycidae							
Enchelyopus cimbrius	1.6–2.4	?	?	?	?	?	?
Urophycis chuss	< 2.0	22.0 TL	≈ 60 d?	23–49	25	35–50	?

continued

Table 5.3 continued

Species	Hatching/ Birth Size (mm)	Completion of Fin Rays (mm)	Larval Duration (yr, mo, d)	Settlement/ Transformation Size (mm)	Size range (TL mm) at		
					Onset of Scale Formation	Completion of Scale Formation	Juvenile Pigmentation
Urophycis regia	< 2.0	15.0	?	25–30	?	?	35–50
Urophycis tenuis	< 2.0	?	50–60 d	50–80	?	?	67–76
Gadidae							
Microgadus tomcod	5.0–7.0	?	?	12–23	?	?	?
Pollachius virens	3.0–4.0	≈ 25	?	> 25 TL	≈ 40	> 95	≈ 25
Ophidiidae							
Ophidion marginatum	≈ 2.0	?	?	≈ 22	≈ 30.0	≈ 60	?
Batrachoididae							
Opsanus tau	≈ 7	?	—	—	—	—	?
Belonidae							
Strongylura marina	14.0	?	?	?	87	≈ 200	?
Cyprinodontidae							
Cyprinodon variegatus	4.2–5.2	12.0	?	—	8.5	18.5	?
Fundulidae							
Fundulus confluentus	4.0–5.6	?	?	?	9.7	20.2	?
Fundulus diaphanus	5.0–5.5	12	?	?	?	?	?
Fundulus heteroclitus	4.8–5.5	20	?	—	9.3	20.0	≈ 20
Fundulus luciae	5.3–6.0	13	?	—	10.1–11.5	18.3	11
Fundulus majalis	7.0	20.0 TL	?	—	13.4–13.8	21.6–23.5	14
Lucania parva	4.0–5.5	15–20 TL	?	—	10.1–10.9	18.0	?
Poeciliidae							
Gambusia holbrooki	7–10	?	—	—	≈ 6	15.7	9–10
Atherinopsidae							
Membras martinica	3.0–5.0	15.0	?	?	< 20	30–40	< 30
Menidia beryllina	3.5–4.0	20.0	?	?	< 20	27.7	?
Menidia menidia	3.8–5.0	≈ 13.0–15.0	?	?	≤ 14	25.0	?
Gasterosteidae							
Apeltes quadracus	4.2–4.5	14.0	?	?	—	—	?
Gasterosteus aculeatus	4.2	15	?	?	12	25–55	?
Syngnathidae							
Hippocampus erectus	?	?	?	?	?	?	?
Syngnathus fuscus	10.0–12.0 at birth	13.0 TL	?	40?	?	?	?
Triglidae							
Prionotus evolans	2.8	7.8	18–19 d	7–12	> 10.6	18.2	?
Prionotus carolinus	2.6–2.8	7.0 SL	18–19 d	7–12	≈ 12–20	30–35	?
Cottidae							
Myoxocephalus aenaeus	5.4	?	55 d	9–10?	?	?	?
Percichthyidae							
Morone americana	≈ 2.6	30 TL	≈ 42 d	20–30?	16.2 FL	34 FL	?
Morone saxatilis	2.0–3.7	20 TL	?	25–36	16 FL	56.9 FL	≈ 40?
Serranidae							
Centropristis striata	1.5–2.0	9.0	?	10–16	≈ 15	71.2	?
Mycteroperca microlepis	?	?	44 d	?	?	?	?
Pomatomidae							
Pomatomus saltatrix	2.0–2.4	14	60 d	24	11.8	24.0	?
Carangidae							
Caranx hippos	1.7	20–21	?	?	< 26 FL	> 40 FL	?
Trachinotus carolinus	?	?	?	?	≈ 25 FL	≈ 60 FL	?
Trachinotus falcatus	?	?	?	?	≈ 30 FL	51 FL	?
Lutjanidae							
Lutjanus griseus	?	17.0	?	10–15	< 18	28	?
Sparidae							
Lagodon rhomboides	?	?	?	10–12	?	?	?
Stenotomus chrysops	2.6 SL	14.0	?	15–30	14.5	?	?
Sciaenidae							
Bairdiella chrysoura	1.5–1.9	10–12	?	?	19.2	56.7	?
Cynoscion nebulosis	1.3–1.6	7	?	4–8	?	?	25–30
Cynoscion regalis	1.5–1.7	8.0–10.0	?	?	14.9	53.3	?

Table 5.3 continued

Species	Hatching/ Birth Size (mm)	Completion of Fin Rays (mm)	Larval Duration (yr, mo, d)	Settlement/ Transformation Size (mm)	Onset of Scale Formation	Completion of Scale Formation	Juvenile Pigmentation
Leiostomus xanthurus	1.6–1.7	10–12	82 d	9–20	19.6	69.4	?
Menticirrhus saxatilis	2.0–2.5	10	?	?	14.8	28.8	?
Micropogonias undulatus	?	11–12	?	8–20	14.5	46.0	?
Pogonias cromis	1.9–2.4	8–10	?	?	18.7	41.8	?
Sciaenops ocellatus	1.7–1.8	10	?	?	?	?	?
Chaetodontidae							
Chaetodon capistratus	?	?	?	?	?	?	?
Chaetodon ocellatus	?	31.0 TL	?	≈20 TL	< 19	≈ 40	?
Mugilidae							
Mugil cephalus	2.4	12–16 TL	20–24 d	?	< 19 FL	≈ 54 FL	8–15
Mugil curema	2.6	14.5 TL	?	?	< 20 FL	≈ 30 FL	?
Labridae							
Tautoga onitis	2.2	> 10	11–23 d	8–13	13–17	28.5	?
Tautogolabrus adspersus	≈ 2.2	8–10	?	≈ 10	14.8	32.0	?
Pholidae							
Pholis gunnellus	9.0	?	?	30–40	> 29	≈ 50	40–50
Ammodytidae							
Ammodytes americanus	5.7–6.3	?	?	?	?	?	?
Uranoscopidae							
Astroscopus guttatus	?	29.0 TL	?	23?	> 60	?	?
Blenniidae							
Chasmodes bosquianus	?	?	?	?	?	?	?
Hypsoblennius hentz	2.6–2.8	8.0	?	?	13.9 SL	?	?
Gobiesocidae							
Gobiesox strumosus	2.4–3.4	?	?	9.0	—	—	?
Gobiidae							
Ctenogobius boleosoma	1.2	13.0	?	?	> 9 SL	?	?
Gobiosoma bosc	2.0	5.0–6.8	?	8–13	—	—	?
Gobiosoma ginsburgi	?	10–11	?	9–12	?	?	?
Microgobius thalassinus	2.0	?	?	?	18	23	?
Sphyraenidae							
Sphyraena borealis	2.6 SL	14.0	?	?	14.5	?	> 14.5
Stromateidae							
Peprilus triacanthus	1.7	21.0	?	—	< 30	?	?
Scophthalmidae							
Scophthalmus aquosus	2.0	8.5 TL	?	10–20	27.0	80.0	?
Paralichthyidae							
Etropus microstomus	2.0	11–12 TL	?	< 20	9.0–11.0	43.6	?
Paralichthys dentatus	2.4–2.8	8–12	?	11–16	16.7	88.5	< 27
Pleuronectidae							
Pseudopleuronectes americanus	3.0–3.5	13.0	?	9–13	13.5	68.2	?
Achiridae							
Trinectes maculatus	1.7–2.0	3.8	?	?	< 25	56.5	?
Cynoglossidae							
Symphurus plagiusa	< 1.3	?	?	?	?	?	?
Tetraodontidae							
Chilomycterus schoepfii	?	?	?	?	?	?	?
Sphoeroides maculatus	≈ 2.4	?	?	?	3.7	27	?

drift (Roberts, 1997), recent authors have suggested behavior, including vertical and horizontal movements, during late larval stages greatly influences transport (Cowen et al., 1993; Leis et al., 1996; Stobutzki and Bellewood, 1997). Moreover, the swimming abilities of early-stage larvae may contribute to dispersal as well (Fisher et al., 2000). In fact, a recent review (Leis, 2006) indicates that the larvae of many fishes are capable of swimming for long periods for much of the larval stage at speeds greater than ambient currents. Further, it has been demonstrated that larvae and pelagic juveniles can use sound for orientation (Montgomery et al., 2006). Thus, we are only beginning to understand the locomotory and sensory capabilities of fishes in their early life history stages (Fuiman and Cowan, 2003).

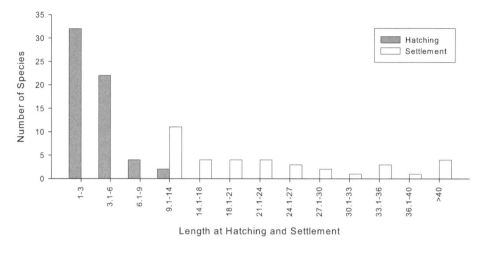

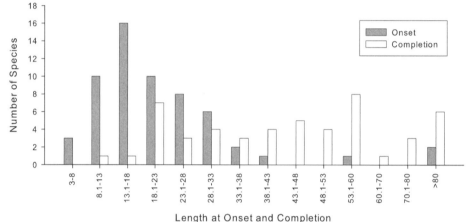

Fig. 5.3. Length at hatching, settlement, and onset and completion of scale formation in temperate estuarine species.

Development during settlement warrants further study because important morphological, physiological, and behavioral transitions occur while fishes are undergoing habitat transitions (Moser, 1981; Balon, 1984; Chambers et al., 1988; Youson, 1988; Levin, 1991; Kaufman et al., 1992). For many, this occurs as they leave the ocean and enter estuaries. The small size of recently settled fishes may make them especially susceptible to predators, and several reviews and species-specific studies have clearly identified high mortality rates during these early juvenile periods (Gulland, 1965; Cushing, 1974; Sissenwine, 1984; Sissenwine et al., 1984; Smith, 1985; Houde, 1987; Elliott, 1989; Doherty, 1991; Beverton and Iles, 1992; Tupper and Boutilier, 1995a, b; Cushing, 1996). The period of transformation and settlement may be especially important to fishes entering estuaries because they also encounter osmoregulatory challenges during these transitions. The coincidence between transformation and settlement is evident in many temperate estuarine fishes (see species accounts) as well as estuarine species in the South Atlantic Bight of the United States (Hoss and Thayer, 1993), the Gulf of Mexico (Yáñez-Arancibia, 1985), Spain (Arias and Drake, 1990), South Africa (Day, 1981; Beckley, 1984, 1985, 1986), and Australia (West and King, 1996). It does not apply to all species, however. All of the *Urophycis* spp. in the Middle Atlantic

Bight have a pelagic-juvenile stage, and they do not settle to the bottom until they are older and larger than most other species (Fahay and Able, 1989).

While the transition between larvae and juveniles is still the most poorly known stage in the life history of fishes (Hempel, 1965; Houde, 1987; de Lafontaine et al., 1992), it might be vital to understanding the variation in annual recruitment (but see Houde, 2008). It is longer in duration than typically perceived, at least in the morphological sense, because it does not end until the major sensory and organ systems are complete. For many species, one convenient morphological end point is completion of scale formation.

PATTERNS OF SCALE FORMATION

Scales of teleost fishes are formed by the multiplication of fibroblasts (Elson, 1939; van Oosten, 1957) that develop into papillae (Elson, 1939; Waterman, 1970; Lindsey, 1988). These originate in one or several loci on the surface of the body (Andrews, 1970; Armstrong, 1973; Sire, 1981; Sire and Arnulf, 1990). Eventually, they become flexible, calcified plates lying within shallow envelopes, or scale pockets, in the upper layers of the dermis (Bullock and Roberts, 1974). For many species, scale formation begins at a single locus on the lateral midline of the caudal peduncle (Andrews, 1970; White, 1977)

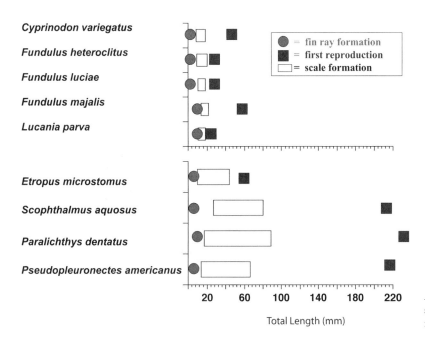

Fig. 5.4. Relationship between length at fin ray formation, scale formation, and first reproduction for selected estuarine species in the Middle Atlantic Bight.

and often occurs between the larval and juvenile periods (Ahlstrom et al., 1976; Kendall et al., 1984; Gozlan et al., 1999; Webb, 1999; Urho, 2002). As a result, the completion of scale formation may be another useful indicator of the conclusion of the larval period and the beginning of the juvenile period (Fuiman, 1997; Fuiman et al., 1998). This end point has been used by others (Copp and Kovac, 1996) and can be applied to species with "direct" or "indirect" development (Balon, 1990).

Scale formation patterns of onset, spreading over the surface of the body, and completion in the species examined in this study vary between families and species, but remain fairly consistent within species (see Table 5.3; Fig. 5.2; Able and Lamonaca, 2006; Able et al., 2009a, b). The origin of scales on the caudal peduncle is typical for many freshwater families of fishes (Andrews, 1970; White, 1977; Sire and Arnulf, 1990; Penaz, 2001), several families of flatfishes (Able and Lamonaca, 2006), and cyprinodontids and fundulids (Able et al., 2009a), although numerous other sites (up to seven) have been identified (see Sire and Arnulf, 1990; Able et al., 2009b). In the temperate estuarine species examined for this study, for which we have data, scales mostly originate on the caudal peduncle (see species accounts). An unusual pattern occurs in *Strongylura marina*. Scales begin on the caudal peduncle and proceed forward as for other species. However, they soon begin to develop faster on the dorsal and ventral surfaces than in the mid-flank region (see species account). Another unusual pattern is found in *Sphoeroides maculatus*. Onset of the formation of scales occurs first on the ventral surface of the trunk, then on the dorsal surface. The caudal peduncle is the last area where scales develop (see species account).

The size of fishes at the onset of scale formation may be related to the beginning of burial behavior, as in *Scophthalmus aquosus* (Neuman and Able, 1998). However, other species, such as the cyprinodontids and fundulids, are capable of

burial before scale formation (see species accounts). Among the species studied here, the onset of scale formation occurs at sizes as small as 6 mm for *Gambusia holbrooki* to as large as 45 mm for *Anguilla rostrata* and > 60 mm for *Astroscopus guttatus*. For most species for which we have data, it occurs at sizes < 33 mm. The completion of scale formation occurs over a broad range of sizes and for some, such as *Strongylura marina*, it occurs at approximately 200 mm. In some instances, size at onset of scale formation appears to be related to the size of the adults, as in fundulids and flatfishes (Fig. 5.4; Able and Lamonaca, 2006; Able et al., 2009a, b).

The time duration over which scales form is seldom known because daily ages are not available for most species. However, we can estimate the duration of scale formation by comparing the sizes at onset and completion of scales based on our observations (Fig. 5.5). For many temperate estuarine fishes, for which we have data, scale formation occurs in fishes measuring < 10–40 mm. These include many small species, such as fundulids, atherinids, and engraulids (see Table 5.3). Some species are much longer (e.g., *Strongylura marina*, 200 mm). For the smaller species, the size at the completion of scale formation is near to the size at first reproduction, while for others scale formation is completed long before reproduction. Fully formed scales cover a variable amount of the body in the adult (Fig. 5.6). For most species, this value is > 80% but can be much less in others.

The terminology for the development of fishes has been discussed frequently (Balon, 1981, 1999; Kendall et al., 1984; Urho, 2002) with a particular focus on the larval/juvenile transition (Balon, 1985; Copp and Kovac, 1996; Penaz, 2001). In many of these treatments scale formation has been included or is central to the discussion (Fuiman, 1997; Fuiman et al., 1998; Urho, 2002; Able and Lamonaca, 2006), although Balon (1999) considered scales "less vital characters" and the

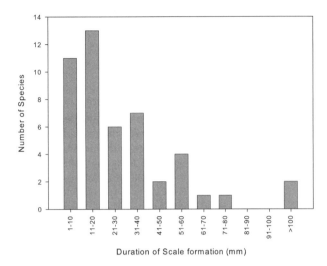

Fig. 5.5. Duration over which scale formation occurs for selected estuarine species in the Middle Atlantic Bight.

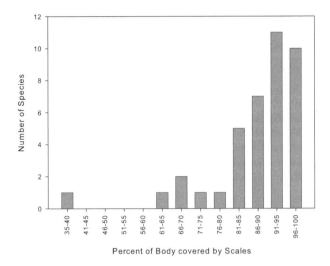

Fig. 5.6. Percent of body covered by scales for selected estuarine species in the Middle Atlantic Bight.

onset of scale formation has been regarded as not very valuable by others (Copp and Kovac, 1996). However, in some of these discussions the beginning of scale formation is one of the characters associated with the onset of the juvenile period (Penaz, 1974; Ahlstrom et al., 1976). One advantage of the use of scale formation as a means of distinguishing ontogenetic periods is that it is an external character that is relatively easy to observe. Also, it is length-based, and thus has several benefits relative to an age-based character (Armstrong, 1973; Fuiman, 1997). Alternatively, scale formation as a measure of early ontogeny occurs later in development (e.g., during or after the completion of median fin rays, a common character for discerning the end of the larval period

[Kendall et al., 1984; Fuiman, 1994, 1997; Fuiman et al., 1998; Urho, 2002]), than conventionally considered (Penaz, 2001; Able and Lamonaca, 2006; Able et al., 2009a; see also Fig. 5.4). However, as has been pointed out, the size at completion of fin rays is still somewhat arbitrarily reported because development of this character varies depending on whether fin rays are counted as soon as they are visible or whether they are considered complete when they are ossified or segmented (Mansueti and Hardy, 1967; Urho, 2002). In consideration of the above aspects of development, we suggest that it is important to recognize the longer and complex nature of the larval/juvenile transition, as supported by others (Penaz, 2001; Urho, 2002).

6

Larval Supply, Settlement, Growth, and Mortality

Some of the most important events in the life of estuarine resident and transient fishes occur during the larval/juvenile transition. Important questions during this life history transition include: How many and what kinds of larvae are supplied to the estuary? When and where do they settle or undergo transformation? How fast do they grow? What affects their survival rate? This chapter attempts to answer these questions for temperate estuarine fishes on the East Coast of the United States, based on the literature and our own findings. Wherever possible, we compare events in the estuary with those in the ocean for the same species to provide a broader perspective.

LARVAL SUPPLY TO ESTUARIES

The larval fish fauna in temperate estuaries is strongly influenced by the timing and location of spawning by transient fishes. However, the distribution of larvae on the inner shelf and on ocean beaches can provide insights into those components of the larval fauna that may spawn in both the ocean and the estuary and are transported into and, perhaps, out of estuaries through inlets (Boehlert and Mundy, 1988; Able et al., 2010). In this chapter, we evaluate the occurrence and abundance of larval fishes across the inner shelf–ocean beach–estuary ecocline to address these issues. Sampling for this evaluation comes from several different sampling programs (see Chapter 3) but together they make such an evaluation possible.

Numerous species that spawn in the ocean use estuaries after traveling long distances from tropical to temperate waters (Booth et al., 2007). These transients are especially evident in temperate estuaries on the East Coast of the United States (Able and Fahay, 1998; Witting et al., 1999; Hare et al., 2002) (Table 6.1) and other parts of the world (Keane and Neira, 2008). Representative taxa that supply larvae after spawning in the Sargasso Sea include anguillid and congrid

eels; numerous other species are derived from spawning in the South Atlantic Bight and the Bahamas (and perhaps the Gulf of Mexico), including representatives from the clupeids, synodontids, pomatomids, carangids, lutjanids, chaetodontids, and mugilids (see Table 5.1). In addition, their numbers and diversity appear to be increasing as a result of climate change (see Chapter 10). Larvae of many other species, including engraulids, gadids, ophidiids, triglids, serranids, pomatomids, sciaenids, labrids, stromateids, and others (see Table 5.1), result from spawning on the continental shelf of the Middle Atlantic Bight. All the available evidence suggests that the only species that spawn above estuaries in freshwater and subsequently use estuaries to any extent are the anadromous species, represented by clupeids, moronids, osmerids, and a gadid (see Table 5.1). Many other larval taxa are the result of local estuarine spawning by transients that overwinter on the continental shelf or by those that are estuarine residents (see Table 5.1).

LARVAL SUPPLY FROM THE OCEAN

At the geographical extremes of the study area, larval ingress into temperate estuarine inlets is believed to be the result of Gulf Stream transport from the south and subsequent movement inshore (Hare et al., 1999; Grothues et al., 2002) or the Labrador Current from more northern areas (Epifanio and Garvine, 2001). We do not know the source of these larvae in many temperate estuaries. One of the most direct ways to identify sources of larvae, and how they vary along the coast, is to measure larval characteristics at ingress into inlets (Hare et al., 2005a, b; Love et al., 2009). In an attempt to do this we have summarized, from the available literature and our own sampling, species composition and abundance from southern New Jersey, Delaware, and North Carolina to just south of Cape Hatteras (Table 6.2). As expected, there are several species (e.g., *Syngnathus floridae*, *S. louisianae*, *Cynoscion nebulosus*, *Sciaenops ocellatus*, and *Gobiesox strumosus*) that are only represented from North Carolina inlets. Others (*Menidia beryllina*, *Tautoga onitis*, *Ammodytes* spp., and *Pseudopleuronectes americanus*) are found only in the two northernmost inlets in New Jersey and Delaware. There is a large number of species represented in all five inlets. These include several that are spawned at distant sites to the south (Sargasso Sea, e.g., *Anguilla rostrata* and *Conger oceanicus*) and others that are widely distributed and presumably spawn near or in all of these estuaries (e.g., *Anchoa mitchilli*, *Brevoortia tyrannus*, *Bairdiella chrysoura*, *Menidia menidia*, and *Syngnathus fuscus*).

LARVAL SUPPLY FROM THE SURFZONE AND NEARSHORE COASTAL OCEAN

A unique set of collections identified the larvae available for ingress into the estuary from the surfzone and nearshore (Able et al., 2010). The summer (May–July) sampling for this study was near or during the peak in larval abundance as a result of a typical mid- to late summer peak in reproduction

Subtle Effects of Vacillating Freshwater Input on Recruitment Success

The human-induced reduction of freshwater flow into estuaries is a common concern for resource managers and estuarine ecologists. Often these impacts—dams or withdrawal of freshwater for irrigation, for example—are overt. The response of many estuarine-dependent fishes can be very negative, as that of the sciaenid *Totoaba macdonaldi*, whose populations have been reduced in the Gulf of California as the result of freshwater withdrawal from the upstream source, the Colorado River (Lercari and Chávez, 2007; Rowell et al., 2008). The same type of response may be occurring in many other estuaries, only at a more subtle level that is more difficult to detect.

We suspect that reduced freshwater input into estuaries may cause variation in the ability of some species to find estuarine nurseries. This occurs because many species that spawn in the ocean but use estuaries as juvenile habitat may be affected when freshwater cues to estuarine location are diminished. Circumstantial evidence for this kind of subtle response is evident from the examination of the time series of *Anguilla rostrata* ingress into two temperate estuaries. These data from Little Egg Inlet, New Jersey (16 years), and Beaufort Inlet, North Carolina (18 years), demonstrated no synchrony in timing and size at ingress (Sullivan et al., 2006a). Yet there was a clear response at both inlets to precipitation during the winter. This suggests that freshwater flow, and its detection by ingressing glass eels, has a positive influence on their abundance. Thus, it follows that decreased precipitation or freshwater input may reduce the ability of glass eels to find estuarine plumes and thus estuarine nurseries. While eels are well known for their olfactory capabilities, this same possibility may be true for the larvae and pelagic juveniles of other estuarine-dependent species and thus reduced freshwater input could influence subsequent recruitment of these species to the adult population as well. We could determine if this is likely with a better understanding of the sensory capabilities of the early life history stages of estuarine-dependent fishes and an improved identification of the influences of variable freshwater input on estuaries and subsequent recruitment success, especially as we attempt to ascertain the effect of climate change on estuaries.

Table 6.1. General life history and ecological characteristics for young-of-the-year and other life history stages.

Species	General Pattern of Utilization*	Young-of-the-Year Season in Estuary†	Young-of-the-Year Habitat		Young-of-the-Year Growth Rate (mm/day)		Fall Migration or End of Summer (Oct) (cm)	End of Winter (Apr) (cm)	End of First Year (12 mo) (cm)	Maximum Size Attained
			Summer	Winter	Summer	Winter				
Carcharhinidae										
Carcharhinus plumbeus	Transient	Sp–Su–Fa	Estuary	Ocean	—	—	—	—	—	225–250 cm
Mustelus canis	Transient	Sp–Su	Estuary	Ocean	1.9	?	55–70	?	?	1.5 m TL
Myliobatidae										
Rhinoptera bonasus	Transient	Su	Estuary	Ocean	—	—	47–59	—	—	107 cm
Acipenseridae										
Acipenser brevirostrum	Resident	All	Estuary	Estuary/ River	0.3	—	14–30	?	?	1.4 m TL
Acipenser oxyrinchus	Transient	All	Estuary	Estuary	?	?	?	?	?	4.3 m TL
Anguillidae										
Anguilla rostrata	Transient	All	Estuary	Estuary	?	?	10–15	?	≈ 30	61 cm TL M, 2.2 m TL F
Ophichthidae										
Myrophis punctatus	Transient	Wi–Su	Estuary	?	?	?	?	?	?	?
Congridae										
Conger oceanicus	Transient	?	Estuary	?	?	?	10–15?	15–20	20?	76 cm TL M, 2.2 m TL F
Clupeidae										
Alosa aestivalis	Transient	Sp–Fa	Estuary	Ocean	0.4	?	5–10	5–10	10–15	38 cm SL
Alosa mediocris	Transient	?	?	?	?	?	?	?	15–20	60 cm SL
Alosa pseudoharengus	Transient	All	Estuary	Ocean	0.6	0	9–11	10–15	10–15	38 cm SL
Alosa sapidissima	Transient	Sp–Fa	FW/ Estuary	Ocean	0.5	0	7–12	10–15	10–15	60 cm SL

Table 6.1 continued

Species	General Pattern of Utilization*	Young-of-the-Year Season in Estuary†	Young-of-the-Year Habitat		Young-of-the-Year Growth Rate (mm/day)		Young-of-the-Year Size Range (TL) at:			
			Summer	Winter	Summer	Winter	Fall Migration or End of Summer (Oct) (cm)	End of Winter (Apr) (cm)	End of First Year (12 mo) (cm)	Maximum Size Attained
Brevoortia tyrannus	Transient	Sp–Su	Estuary	Ocean	1.0	?	8–17	?	8–18	51 cm TL
Clupea harengus	Transient	Sp	Estuary	Ocean	?	?	9–17	?	?	54 cm TL
Dorosoma cepedianum	Transient	Su–Fa	Estuary	FW	?	?	109–160	109–160	?	50 cm SL
Opisthonema oglinum	Transient	Su	Estuary/Ocean	Ocean	?	?	60	?	?	30 cm SL
Engraulidae										
Anchoa hepsetus	Transient	Su–Fa	Estuary/Ocean	Estuary/Ocean	?	?	9–14	?	?	150 mm SL
Anchoa mitchilli	Transient	Sp–Fa	Estuary	Ocean	0.3	?	2–10	5–10	5–10	110 mm SL
Engraulis eurystole	Stray	Su–Fa	Estuary/Ocean	Ocean	?	?	?	?	?	?
Osmeridae										
Osmerus mordax	Resident	Sp–Su	Brackish	Estuary	< 0.5	?	?	?	?	35 cm
Synodontidae										
Synodus foetens	Transient	Su–Fa	Estuary	Ocean	?	?	25	?	?	40 cm SL
Phycidae										
Enchelyopus cimbrius	Stray	Su–Fa	Estuary/Ocean	Ocean	?	?	?	?	?	32.8 cm
Urophycis chuss	Stray	Sp	Ocean	Ocean	1.0+	?	1–15	5–15	5–20	52 cm TL
Urophycis regia	Transient	Sp	Ocean	Ocean	1.0+	0?	1–7	3–17	15–20	41 cm TL
Urophycis tenuis	Transient	Sp	Ocean	Ocean	1.0	0	20–30	?	25–30	135 cm
Gadidae										
Microgadus tomcod	Resident	Sp–Su	Estuary/FW	FW	< 0.5?	?	10–15	15–22	15–22	44.7 cm
Pollachius virens	Transient	Sp–Su	Estuary	Ocean	0.8	?	?	?	?	130 cm TL
Ophidiidae										
Ophidion marginatum	Transient	Sp–Su	Estuary/Ocean	Ocean	?	?	1–2	?	8–12	250 mm TL
Batrachoididae										
Opsanus tau	Resident	All	Estuary	Estuary	0.3	0	4–9	4–9	6–12	38 cm TL
Belonidae										
Strongylura marina	Transient	Su	Estuary	?	2.3	?	35–40	?	?	64 cm SL
Cyprinodontidae										
Cyprinodon variegatus	Resident	All	Marsh	Estuary	0.3	0	2–6	3–6	3–6	93 mm
Fundulidae										
Fundulus confluentus	Resident	All	Marsh	Marsh	?	?	?	?	?	60 mm TL
Fundulus diaphanus	Resident	All	Estuary	Estuary	?	?	24–58	?	35–65	110 mm TL
Fundulus heteroclitus	Resident	All	Marsh	Estuary	0.3	0	3–7	3–7	3–8	130 mm
Fundulus luciae	Resident	All	Marsh	Estuary	0.2	0	2–5	2–5	3–5	56 mm
Fundulus majalis	Resident	All	Creeks/Shores	Estuary	0.5	0	3–9	4–9	5–10	200 mm TL
Lucania parva	Resident	All	Marsh	Estuary	0.3	0	1–4	3–4	3–4	58 mm SL
Poeciliidae										
Gambusia holbrooki	Resident	All	FW/Estuary	FW/Estuary	0.1	0	2–3	3	3–4	63 mm TL
Atherinopsidae										
Membras martinica	?	?	?	?	?	?	?	?	?	110 m
Menidia beryllina	Resident	All	Marsh	Estuary	0.3	0	2–6	3–6	5–6	100 mm SL
Menidia menidia	Transient	Sp–Fa	Estuary	Ocean	0.6	0	1–9	4–13	5–15	125 mm SL
Gasterosteidae										
Apeltes quadracus	Resident	All	Eelgrass	Estuary	0.1	0	3–6	3–6	3–6	60 mm SL
Gasterosteus aculeatus	Transient	Wi–Sp	Marsh	Ocean	?	?	?	4–6	4–7	70 mm SL
Syngnathidae										
Hippocampus erectus	Transient	Sp–Fa	Estuary	Ocean	0.5	0	1–10	4–10	7–10	170 mm TL
Syngnathus fuscus	Transient	Sp–Fa	Estuary	Ocean	?	0	6–20	8–20	7–21	305 mm TL

continued

Table 6.1 continued

Species	General Pattern of Utilization*	Young-of-the-Year Season in Estuary†	Young-of-the-Year Habitat		Young-of-the-Year Growth Rate (mm/day)		Young-of-the-Year Size Range (TL) at:			
			Summer	Winter	Summer	Winter	Fall Migration or End of Summer (Oct) (cm)	End of Winter (Apr) (cm)	End of First Year (12 mo) (cm)	Maximum Size Attained
Triglidae										
Prionotus carolinus	Transient	Sp–Su	Estuary/Ocean	Ocean	0.5	0	2–10	?	5–15	450 mm TL
Prionotus evolans	Transient	Su–Wi	Estuary/Ocean	Ocean	0.5	0	1–10	?	?	~ 480 mm TL
Cottidae										
Myoxocephalus aenaeus	Resident	All	Estuary	Estuary/Ocean	0.3	0	3–10	3–10	3–10	460 cm TL
Percichthyidae										
Morone americana	Resident	All	Estuary/FW	Estuary	0.3	0	4–9	5–11	5–11	480 mm TL
Morone saxatilis	Transient	All	Estuary/FW	Estuary/Ocean	0.4	0	6–11	?	?	1.8 m TL
Serranidae										
Centropristis striata	Transient	Sp–Fa	Estuary/Ocean	Ocean	0.6	0	2–11	6–12	6–16	61 cm TL
Mycteroperca microlepis	Transient	Su	Estuary	Ocean	?	?	10–15	?	?	1.2 m TL
Pomatomidae										
Pomatomus saltatrix	Transient	Sp–Su	Estuary/Ocean	Ocean	1.0	?	18–30	?	30–40	150 cm
Carangidae										
Caranx hippos	Transient	Su	Estuary	?	1.3	?	2–20	?	?	1000 mm TL
Trachinotus carolinus	Transient	Su–Fa	Estuary	?	0.9	?	12–15	?	?	60 cm FL
Trachinotus falcatus	Transient	Su–Fa	Estuary/Ocean	?	?	?	8	?	?	1.1 m FL
Lutjanidae										
Lutjanus griseus	Stray	Su–Fa	Estuary	?	?	?	1–11	?	?	60 cm
Sparidae										
Lagodon rhomboides	Transient	Sp–Fa	Estuary	?	?	?	?	?	?	40 cm TL
Stenotomus chrysops	?	Sp–Su	Estuary	Ocean	0.5	0	3–15	?	11–17	45 cm TL
Sciaenidae										
Bairdiella chrysoura	Transient	Su–Fa	Estuary	?	0.6	?	3–15	8–17	?	30 cm TL
Cynoscion nebulosus	Transient	Sp–Fa	Estuary	Ocean	1.4	?	?	?	?	90 cm TL
Cynoscion regalis	Transient	Sp–Fa	Estuary	Ocean	1.4	0	2–25	?	10–30	1 m TL
Leiostomus xanthurus	Transient	Sp–Fa	Estuary	Ocean	0.6	?	5–20	9–25	9–25	34 cm TL
Menticirrhus saxatilis	Transient	Su–Fa	Ocean/Estuary	Ocean	?	?	2–24	11–30	20–30	55 cm TL
Micropogonias undulatus	Transient	All	Estuary	Estuary	0.8–1.4	0	1–10	3–10	8–22	50 cm TL
Pogonias cromis	Transient	Su–Fa	Estuary	Ocean	1.1	?	9–23	?	?	1.7 m TL
Sciaenops ocellatus	Transient	Su–Fa	Estuary	Estuary/Ocean	0.45–0.75	?	?	?	?	1.5 m TL
Chaetodontidae										
Chaetodon capistratus	Transient	Su–Fa	Estuary	?	?	?	?	?	?	?
Chaetodon ocellatus	Stray	Su–Fa	Estuary	?	0.1–0.2	?	1–4	2–6	?	15 cm TL
Mugilidae										
Mugil cephalus	Transient	Sp–Su	Estuary/FW	Ocean	?	?	19–25	?	?	1.2 m TL
Mugil curema	Transient	Sp–Su	Estuary	Ocean	1.1	?	3–16	?	?	45 cm TL
Labridae										
Tautoga onitis	Transient	All	Estuary	Estuary	0.5	0	3–20	4–20	4–18	95 cm SL
Tautogolabrus adspersus	Resident	All	Estuary	Estuary/Ocean	0.3	0	3–7	3–10	5–12	43 cm TL

Table 6.1 continued

Species	General Pattern of Utilization*	Young-of-the-Year Season in Estuary†	Young-of-the-Year Habitat Summer	Young-of-the-Year Habitat Winter	Young-of-the-Year Growth Rate (mm/day) Summer	Young-of-the-Year Growth Rate (mm/day) Winter	Fall Migration or End of Summer (Oct) (cm)	End of Winter (Apr) (cm)	End of First Year (12 mo) (cm)	Maximum Size Attained
Pholidae										
Pholis gunnellus	Stray	Sp–Su–Fa	Estuary	Ocean	?	?	?	?	6–7	30 cm
Ammodytidae										
Ammodytes americanus	Resident	Fa–Sp	Estuary	Estuary	0.6	0	8–15	?	10–17	22 cm SL
Uranoscopidae										
Astroscopus guttatus	Stray	Sp–Fa	Estuary/Ocean	Ocean	?	?	?	?	?	31 cm SL
Blenniidae										
Chasmodes bosquianus	Resident	All	Estuary	Estuary	?	?	7	?	?	8 cm SL
Hypsoblennius hentz	Stray	Su–Wi	Estuary	Estuary	0.5	0	4–9	?	?	10 cm SL
Gobiesocidae										
Gobiesox strumosus	Resident	All	Estuary	Estuary	?	?	?	?	?	75 mm SL
Gobiidae										
Ctenogobius boleosoma	Resident	All	Estuary	Estuary	?	?	1–4	3–4	?	62 mm TL
Gobiosoma bosc	Resident	All	Estuary	Estuary	0.2	0	2–5	3–5	3–6	60 mm TL
Gobiosoma ginsburgi	Resident	All	Estuary/Ocean	?	0.2	?	1–4	2–4	?	52 mm TL
Microgobius thalassinus	Stray	Su	Estuary	?	?	?	?	?	4–5	40 mm SL
Sphyraenidae										
Sphyraena borealis	Transient	Su–Fa	Estuary/Ocean	?	1.5	?	15–30	?	?	46 cm
Stromateidae										
Peprilus triacanthus	Stray	Su	Estuary/Ocean	Ocean	0.2	?	?	?	?	30 cm TL
Scophthalmidae										
Scophthalmus aquosus	Transient	Sp–Fa	Estuary/Ocean	Ocean	1.1 (Sp) ? (Fa)	? (Sp) ? (Fa)	11–19 (spring) 4–7 (fall)	> 16 (spring) 4–8 (fall)	? (spring) 18–26 (fall)	51 cm
Paralichthyidae										
Etropus microstomus	Transient	Sp–Fa	Estuary/Ocean	Ocean	0.2	0	2–11	3–12	4–12	15 cm TL
Paralichthys dentatus	Transient	All	Estuary	Estuary/Ocean	1.5–1.9	0	20–30	?	20–30	95 cm TL
Pleuronectidae										
Pseudopleuronectes americanus	Resident	All	Estuary/Ocean	Estuary/Ocean	0.5	0.3	3–17	5–20	5–22	64 cm TL
Achiridae										
Trinectes maculatus	Resident	All	Estuary	Estuary	0.3	0	1–6	2–6	3–8	20 cm TL
Cynoglossidae										
Symphurus plagiusa	Stray	Su	Estuary	?	?	?	?	?	76–79	20 cm TL
Tetraodontidae										
Chilomycterus schoepfii	Transient	Sp–Fa	Estuary	?	?	?	?	?	?	330 mm
Sphoeroides maculatus	Transient	Su–Fa	Estuary	Ocean	1.0	0	1–16	?	11–26	360 mm TL

* Resident, Stray, Transient (nursery, seasonal resident).

† Spring, summer, fall, winter—Sp = March–May, Su = June–August, Fa = September–November, Wi = December–February.

for many species in the Middle Atlantic Bight (Able and Fahay, 1998; see Table 5.1). The density of larvae in the northern New Jersey surfzone and nearshore study area varied across years over the area from Long Branch to Manasquan Inlet during 1996–1999 (Able et al., 2010). There was no obvious correspondence between larval densities in nearshore versus surfzone samples in any given year. In addition, there was no general pattern of overall greater abundance at any one location, for either nearshore or surfzone collections (Able et al., 2009c). The lack of consistency in the spatial occurrence and abundance of larvae in surfzone versus nearshore samples at the various sampling locations may be due

Table 6.2. Larval species composition and abundance (number of individuals/100 m³) at several temperate estuarine inlets on the East Coast of the United States. See Fig. 2.1 for location of inlets. Sources are Hettler and Chester, 1990; Hettler and Barker, 1993; Witting et al. 1999; Rhodes, 2008. — Indicates not reported.

Family / Species	Common Name	Little Egg Inlet, NJ	Indian River Inlet, DE	Oregon Inlet, NC	Ocracoke Inlet, NC	Beaufort Inlet, NC
Elopidae						
Elops saurus	Lady fish	0.00025	—	0.90	0.70	0.024
Megalopidae						
Megalops atlanticus	Tarpon	—	—	—	—	0.007
Anguillidae						
Anguilla rostrata	American eel	0.345	0.362	11.14	1.55	0.044
Ophichthidae						
Myrophis punctatus	Speckled worm eel	0.013	0.003	2.23	3.79	2.271
Ophichthus cruentifer	Margined snake eel	—	—	—	—	0.003
Ophichthus gomesi	Shrimp eel	0.00042	—	—	—	0.027
Congridae						
Conger oceanicus	Conger eel	0.50	0.032	0.27	0.85	0.078
Engraulidae						
Anchoa hepsetus	Striped anchovy	0.061	—	70.11	269.80	8.255
Anchoa mitchilli	Bay anchovy	8.14	10.01	213.90	27.50	0.450
Anchoa sp.	(unidentified anchovy)	2.66	15.08	1004.90	1237.40	121.262
Clupeidae						
Brevoortia tyrannus	Atlantic menhaden	1.82	1.17	405.14	146.11	2.781
Harengula jaguana	Scaled sardine	—	—	—	0.42	—
Opisthonema oglinum	Atlantic thread herring	0.024	—	55.72	4.14	0.780
Sardinella aurita	Spanish sardine	—	—	—	—	0.396
Synodontidae						
Synodus foetens	Inshore lizard fish	0.003	0.003	6.39	3.13	0.044
Synodus poeyi	Offshore lizard fish	—	—	—	0.12	—
Myctophidae						
(genus unknown)	(unidentified lantern fish)	—	—	0.12	—	—
Ophdiidae						
Ophidion sp.	(unidentified cusk-eel)	0.00008	—	1.80	0.41	0.002
Ophidion marginatum	Striped cusk-eel	0.0002	0.005	—	—	—
Phycidae						
Urophycis chuss	Red hake	0.0009	—	—	—	—
Urophycis regia	Spotted hake	0.0017	—	—	0.22	0.012
Echelyopus cimbrius	Fourbeard rockling	0.010	—	—	0.10	—
Batrachoididae						
Opsanus tau	Oyster toadfish	0.005	0.01	—	0.21	—
Atherinopsidae						
Membras martinica	Rough silverside	0.0003	—	—	—	—
Menidia beryllina	Inland silverside	0.0005	0.087	—	—	—
Menidia menidia	Atlantic silverside	0.257	0.25	5.05	6.87	0.039
Hemiramphidae						
Hyporhamphus meeki	Atlantic silverstripe halfbeak	0.00008	—	—	—	0.002
Fundulidae						
Fundulus heteroclitus	Mummichog	0.055	0.003	—	—	—
Fistularidae						
Fistularia tabacaria	Bluespotted cornetfish	—	—	—	0.11	—
Syngathidae						
Hippocampus erectus	Lined seahorse	0.013	0.003	—	3.56	0.100
Hippocampus reidi	Longsnout seahorse	—	—	—	0.10	—
Syngnathus floridae	Dusky pipefish	—	—	9.84	20.93	0.008
Syngnathus fuscus	Northern pipefish	1.56	0.063	7.47	1.66	0.082
Syngnathus louisianae	Chain pipefish	—	—	7.24	22.03	0.727
Scorpaenidae						
(genus unknown)	(unidentified scorpenid)	—	—	0.11	0.12	—
Triglidae						
Prionotus carolinus	Northern searobin	0.003	—	0.06	—	—
Prionotus evolans	Striped searobin	0.011	—	—	—	—
Prionotus sp.	(unidentified searobin)	0.0003	—	12.48	10.95	0.528

Table 6.2 continued

Family / Species	Common Name	Little Egg Inlet, NJ	Indian River Inlet, DE	Oregon Inlet, NC	Ocracoke Inlet, NC	Beaufort Inlet, NC
Serranidae						
Centropristis striata	Black sea bass	0.013	0.011	2.39	1.01	0.012
Epinephelus spp.	Groupers	0.00008	—	—	—	—
Mycteroperca microlepis	Gag	—	—	—	—	0.008
Pomatomidae						
Pomatomus saltatrix	Bluefish	0.002	—	0.73	0.65	0.017
Carangidae						
Decapterus punctatus	Round scad	—	—	—	0.12	0.129
Trachinotus falcatus	Permit	—	—	—	0.25	0.003
Lutjanidae						
Lutjanus griseus	Gray snapper	0.0002	—	0.09	2.02	—
Lobotidae						
Lobotes surinamensis	Tripletail	—	—	0.11	—	—
Gerreidae						
Eucinostomus argenteus	Spotfin mojarra	0.00008	—	—	—	—
Eucinostomus sp.	(unidentified mojarra)	0.003	—	3.87	4.18	0.034
Haemulidae						
Orthopristis chrysoptera	Pigfish	0.0003	0.029	15.79	27.64	3.269
Sparidae						
Lagodon rhomboides	Pinfish	0.0004	—	0.23	17.09	0.620
Archosargus probatocephalus	Sheepshead	—	—	—	0.82	0.015
Sciaenidae						
Sciaenidae spp.	Drums and croakers	0.012	0.193	—	—	—
Bairdiella chrysoura	Silver perch	0.0007	0.3373	13.67	3.94	10.581
Cynoscion nebulosus	Spotted seatrout	—	—	0.49	0.73	0.202
Cynoscion regalis	Weakfish	0.185	0.177	52.37	16.59	5.335
Leiostomus xanthurus	Spot	0.009	0.037	86.33	56.39	3.420
Menticirrhus sp.	(unidentified kingfish)	0.005	—	4.19	1.94	0.336
Menticirrhus americanus	Southern kingfish	—	0.040	—	—	—
Menticirrhus saxatilis	Northern kingfish	0.007	—	—	—	—
Micropogonias undulatus	Atlantic croaker	1.46	9.10	782.31	77.10	3.916
Pogonias cromis	Black drum	—	—	—	—	0.007
Sciaenops ocellatus	Red drum	—	—	29.18	3.57	0.005
Mugilidae						
Mugil cephalus	Striped mullet	—	—	—	0.76	0.238
Mugil curema	White mullet	—	—	0.10	1.92	0.007
Pomacentridae						
Abudefduf saxatilis	Sergeant major	—	—	—	0.13	0.002
Labridae						
Tautoga onitis	Tautog	0.038	0.010	—	—	—
Ammodytidae						
Ammodytes spp.	Sand lances	0.310	0.003	—	—	—
Uranoscopidae						
Astroscopus guttatus	Northern stargazer	0.005	0.003	—	—	0.002
Astroscopus y-graecum	Southern stargazer	—	—	0.40	—	—
Dactyloscopidae						
Dactyloscopus spp.	Stargazer	—	—	—	—	0.002
Blenniidae						
Chasmodes bosquianus	Striped blenny	—	—	—	—	0.088
Hypleurochilus germinatus	Crested blenny	—	0.003	—	—	—
Hypsoblennius hentz	Feather blenny	0.003	0.013	3.85	9.59	0.489
Gobiesocidae						
Gobiesox strumosus	Skilletfish	—	—	0.93	0.45	0.008
Eleotridae						
Dormitator maculatus	Fat sleeper	0.0005	—	—	0.43	0.003
Gobiidae						
Ctenogobius bolesoma	Darter goby	0.051	0.830	31.32	46.49	0.722
Gobiidae spp.	Gobies	0.014	0.003	—	—	—
Gobionellus hastatus	Sharptail goby	—	—	0.41	2.74	0.046
Gobiosoma bosc	Naked goby	0.286	1.45	35.01	72.70	0.138

continued

Table 6.2 continued

Family / Species	Common Name	Little Egg Inlet, NJ	Indian River Inlet, DE	Oregon Inlet, NC	Ocracoke Inlet, NC	Beaufort Inlet, NC
Gobiosoma ginsburgi	Seaboard goby	0.069	0.886	3.69	0.79	0.246
Gobiosoma sp.	(unidentified goby)	0.057	0.894	2.59	0.85	2.042
Microgobius gulosus	Clown goby	—	—	0.12	1.13	0.530
Microgobius thalassinus	Green goby	0.006	0.872	—	—	—
Microdesmidae						
Microdesmus longipinnis	Pink wormfish	0.00008	—	0.22	0.68	—
Sphyraendiae						
Sphyraena barracuda	Great barracuda	0.00008	—	0.13	0.12	—
Sphyraena borealis	Northern sennet	0.00008	—	—	—	0.003
Scombridae						
Scomberomorus spp.	Mackerels	—	—	—	—	—
Scomberomorus maculatus	Spanish mackerel	0.00008	—	13.35	0.42	0.163
Stromateidae						
Peprilus alepidotus	Harvestfish	—	—	0.45	—	0.005
Peprilus triacantus	Butterfish	0.004	—	1.81	0.35	0.032
Bothidae						
Bothus ocellatus	Eyed flounder	0.00008	—	0.85	—	—
Scophthalmidae						
Scophthalmus aquosus	Windowpane	0.943	0.026	0.09	0.09	0.019
Paralichthydae						
Citharichthys arctifrons	Gulf Stream flounder	—	—	0.12	—	—
Citarichthys spilopterus	Bay whiff	0.0002	0.005	—	—	—
Citharichthys sp.	(unidentified flounder)	0.0008	—	0.10	4.77	0.168
Etropus crossotus	Fringed flounder	—	—	0.22	0.08	0.048
Etropus microstomus	Smallmouth flounder	0.09	0.016	7.50	5.23	0.041
Paralichthys albigutta	Gulf flounder	—	—	0.21	2.77	0.826
Paralichthys dentatus	Summer flounder	0.34	0.568	1.28	2.41	1.351
Paralichthys lethostigma	Southern flounder	—	—	—	1.50	0.027
Pleuronectidae						
Etropus spp.	Flounder	—	—	—	—	0.010
Pseudopleuronectes americanus	Winter flounder	1.62	1.41	—	—	—
Achiridae						
Trinectes maculatus	Hogchoker	0.0003	—	0.97	0.29	0.234
Cynoglossidae						
Symphurus civitatus	Offshore tonguefish	—	—	0.42	1.24	0.024
Symphurus plagiusa	Blackcheek tonguefish	0.00008	0.045	51.13	37.86	0.583
Monocanthidae						
Aluterus schoepfii	Orange filefish	—	—	—	0.47	0.029
Stephanolepis hispidus	Planehead filefish	—	—	0.42	13.74	0.048
Tetradontidae						
Chilomycterus schoepfii	Striped burrfish	0.00008	—	—	—	—
Sphoeroides maculatus	Northern puffer	0.010	—	0.57	3.60	0.063

to complex circulation or species-specific differences in the timing of larval occurrence and transport or retention. However, there were pronounced temporal differences in larval occurrence and abundance that varied between months and years (Able et al., 2010).

While the sampling nets in the nearshore and surfzone collections in this study differed in size and mode of deployment, the size and stage of larval development appeared to be similar across years. The composite and annual size frequency distributions (Able et al., 2010) were similar (mean 5.7 mm in the surfzone collections and 4.8 in the nearshore collections), with most individuals < 5 mm (however, individuals up to 15 mm were fairly equally represented in both

nets, and some individuals > 20 mm were also collected). The relatively small size across habitats and gears is consistent with the relatively large numbers of preflexion larvae in almost every year. Also, for many of the species, the average size at collection (both gears combined) approximated the size at hatching; thus spawning probably occurred nearby (Able et al., 2009c). This was evident for *Brevoortia tyrannus*, *Cynoscion regalis*, *Lophius americanus*, *Scophthalmus aquosus*, *Sphoeroides maculatus*, and *Tautogolabrus adspersus*. The possibility that many of these species spawn in relatively shallow waters on the inner continental shelf is supported by prior studies (Able and Fahay, 1998; Berrien and Sibunka, 1999). Spawning by *Lophius americanus* in these nearshore waters has not previously been reported, but was evident in a number of years. For other species that were represented by two size modes, including *Hippoglossina oblonga, Pomatomus saltatrix, Pseudopleuronectes americanus,* and *Tautoga onitis,* the smallest mode also represented the size at hatching. Most of these species are reported to reproduce in coastal waters (Grosslein and Azarovitz, 1982). The exception is *Pseudopleuronectes americanus,* which is presumed to spawn in adjacent estuaries (Able and Fahay, 1998), but some of the larvae may be transported out of the estuary soon after hatching (Able and Fahay, 1998; Chant et al., 2000) or spawning may occur in the ocean (Wuenschel et al., 2009). In other instances, the smallest individuals collected were larger than the size at hatching. This may be due to spawning that occurs farther offshore over the continental shelf, as is true of the *Prionotus* species (McBride and Able, 1994), or spawning that takes place during the previous fall (*Etropus microstomus*).

The composite nearshore and surfzone larval collection was dominated by a diverse assemblage of fishes from 33 families (Able et al., 2010). Of these, the sciaenids, gadids, engraulids, scombrids, and bothids were the most speciose. The sciaenids included four species (*Bairdiella chrysoura, Cynoscion regalis, Menticirrhus saxatilis,* and *Micropogonias undulatus*); the phycids and gadids, four (*Enchelyopus cimbrius, Pollachius virens, Urophycis chuss,* and *U. regia*); the engraulids, three (*Anchoa hepsetus, A. mitchilli,* and *Engraulis eurystole*); the scombrids, three (*Auxis* sp., *Scomber scombrus,* and *Scomberomorus maculatus*); and the paralichthyids and bothids, three (*Bothus ocellatus, Etropus microstomus,* and *Hippoglossina oblonga*). Several species were among the most abundant in both nearshore and surfzone collections, including *Anchoa* sp. (most likely *Anchoa mitchilli,* given the dominance of this form over other engraulids among those identified to species), *Brevoortia tyrannus, Scomber scombrus, Scophthalmus aquosus,* and *Lophius americanus.* Species richness, based on combined nearshore and surfzone collections, was relatively high, with 51 species represented over the 5 years from 1995 to 1999. The number of species also varied between habitats, with a larger number of species in nearshore (46) than in surfzone (38) collections.

Some relatively abundant (> 10 in total catch) species were restricted to the habitats in which they occurred (Able et al., 2009c, 2010). Those species that were collected only in the nearshore samples included *Anchoa hepsetus, Opisthonema oglinum, Engraulis eurystole, Urophycis regia,* and *Gasterosteus aculeatus.* Species that appeared disproportionately abundant in the nearshore were *Anchoa* sp., *Tautoga onitis,* and *Pomatomus saltatrix.* There were no abundant species that were found only in the surfzone; however, several were relatively more abundant there, including *Menidia menidia, Scophthalmus aquosus,* Atherinopsidae, and *Peprilus triacanthus.* Many of these species are typical larvae from the central part of the Middle Atlantic Bight based on an 11-year dataset from NMFS-MARMAP surveys (Able and Fahay, 1998; see Table 4.1), and

with the inshore larval assemblage reported from slightly deeper water in the New York Bight (Cowen et al., 1993).

The sources of the larvae in the coastal ocean, that is, nearshore and surfzone combined, often included estuarine species or those that occur in both the estuary and the ocean (Able et al., 2009c, 2010). These included many abundant species, for example, *Anchoa* sp. (probably *A. mitchilli*), *Tautoga onitis, Scophthalmus aquosus, Cynoscion regalis,* and *Hypsoblennius hentz.* Most of the estuarine species (e.g., *Menidia menidia*) were collected exclusively in the surfzone.

While there was considerable overlap in presumed estuarine and ocean species in the surfzone/nearshore, it is also worth noting that there are some species from these habitats that seldom occur in the estuary based on our extensive collections at Little Egg Inlet (see below). These include such species as *Scomber scombrus* and *Lophius americanus.* The implication is that those transient species that do enter the estuary may do so as the result of behavioral changes, especially around the larval/juvenile transition. The most frequently cited behavior to explain this is selective tidal stream transport (Forward and Tankersley, 2001), but multiple mechanisms are likely important and may differ among species and ontogenetic stages (Joyeux, 1998; Churchill et al., 1999a, b; Hare et al., 2005a, b). While there are some differences in larval species composition between the ocean and just inside the estuary, the overlap between larval abundance and settlement in both of these habitats suggests that they usually function in the same way for many species (Able et al., 2006a). The overlap between continental shelf, nearshore and surfzone, and adjacent estuarine larval faunas is due, in large part, to several common life history patterns. First, many species spawn in the ocean and the larvae are transported to estuaries (Able and Fahay, 1998); thus they could be collected anywhere along that continuum. Second, other species spawn in both the inner continental shelf and the estuary (Able et al., 2006a); thus the larvae are available in both habitats. Third, the broad seasonal spawning migrations of many species in the region (Grosslein and Azarovitz, 1982; Able and Fahay, 1998) ensure that the eggs and larvae are broadly distributed and mixed across habitats.

LARVAL SUPPLY THROUGH LITTLE EGG INLET, NEW JERSEY

The intensive sampling in Little Egg Inlet in southern New Jersey clearly indicates the seasonal pattern of larval abundance based on weekly sampling over 17 years (Fig. 6.1). The total abundance of larvae is greatest from June through October, when water temperatures are typically warmest (see Fig. 2.6). The periodic peaks in abundance during June through August are primarily the result of large collections of *Anchoa mitchilli* (and *Anchoa* spp.). When *A. mitchilli* is removed from the collections, there are higher abundances of other larvae in the spring (weeks 16–20 and 24). The lowest values for larval abundance typically occur in January

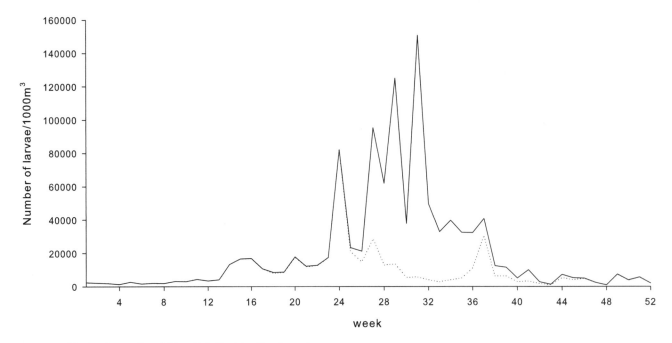

Fig. 6.1. Weekly abundance of larval fish at Little Egg Inlet, New Jersey, during 1989–2006. Solid line = total larvae. Dotted line = total larvae without *Anchoa* spp.

through March. The larvae are consistently dominated by postflexion stages, probably because of the larger mesh (1.0 mm) and the sizeable number of transient species that typically enter at bigger sizes and later in development (Fig. 6.2). The highest proportion of preflexion stage larvae occurs during April through August, when many resident and seasonally resident species are spawning in the estuary (see Chapter 5).

The abundance of total larvae at Little Egg Inlet varied annually over this 17-year time series as well (Fig. 6.3). Overall, the peaks in larval abundance were greater, relative to the long-term average, starting in 2002. Over this same time period, the species ranking, by abundance, varied as well (Table 6.3). *Anchoa mitchilli* was consistently the most abundant species in most years with a top ranking (1) in 8 of the 18 years (mean 2.1, range 1–7). The species with the next highest overall ranking were *Syngnathus fuscus* (mean 5.5, range 1–8) and *Pseudopleuronectes americanus* (mean 3.0, range 1–6). All of these were products of reproduction in the estuary or in adjacent inner continental shelf waters, or both (see species accounts). The next most abundant species were much more variable in ranking. These included *Brevoortia tyrannus* (mean 7.5, range 1–14), *Anguilla rostrata* (mean 8.7, range 4–15), and *Micropogonias undulatus* (mean 10.1, range 1–30). Another estuarine spawner, *Menidia menidia*, was highly ranked (mean 12.6, range 4–28). This species may be more abundant than indicated because the smallest individuals of this species are difficult to separate from *Menidia beryllina* or other atherinids (possibly *Membras martinica*, but this species was seldom collected). The next highest-ranked, *Ammodytes* spp. (mean 10.4, range 2–18), may spawn in the estuary or the ocean. Other relatively abundant species were *Clupea harengus* (mean 12.2,

range 1–30), *Gobiosoma bosc* (mean 15.2, range 5–60), and *Paralichthys dentatus* (mean 12.2, range 2–24). These patterns of larval supply, in some cases, differ from the earlier portion of the time series (Witting et al., 1999). The explanations for these can be found in the species accounts and in Chapter 10.

The seasonal patterns of abundance are also reflected in the ranking by month (Table 6.4). In winter (December–February), some of the highest-ranking species include several that spawn elsewhere, either to the north (*Clupea harengus*), to the south (*Anguilla rostrata*, *Brevoortia tyrannus*, and *Myrophis punctatus*), or in the vicinity of the inlet (*Paralichthys dentatus* and *Micropogonias undulatus*). In spring (March–May), many of these species were still highly ranked, but others that were ranked high for this season (*Ammodytes americanus*, *Pseudopleuronectes americanus*, *Myoxocephalus aenaeus*, and *Pholis gunnellus*) were likely the result of local spawning. In summer (June–August), the dominant species represent those less typically seen in the other months/seasons. Most of these result from spawning in or near the estuary, including a different cohort of *Brevoortia tyrannus* from that collected in the winter, *Menidia menidia*, *Syngnathus fuscus*, *Anchoa mitchilli*, *Gobiosoma bosc*, *Cynoscion regalis*, and *Opisthonema oglinum*. Many of these species also occur in fall (September–November), but some appear for the first time this season (*Ctenogobius boleosoma*, *Anchoa hepsetus*, and *Symphurus plagiusa*).

One of the most distinct seasonal differences for larvae in this sampling program is the size at capture (Fig. 6.4). Typically, larvae collected in the fall and winter months are considerably larger (> 40 mm), and there are fewer small larvae (< 10 mm). In the summer, the collections are dominated by

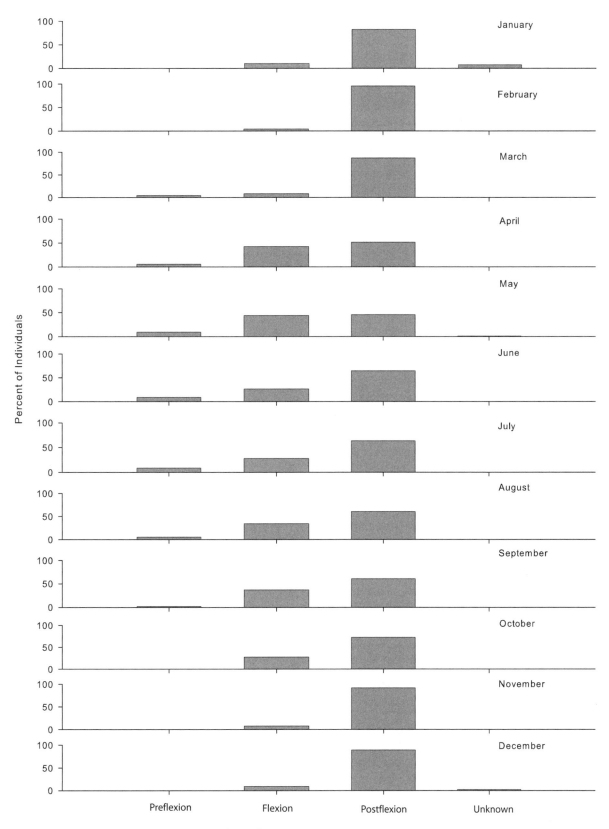

Fig. 6.2. Monthly variation in larval development stage (preflexion, flexion, postflexion) at Little Egg Inlet, New Jersey, during 1989–2006.

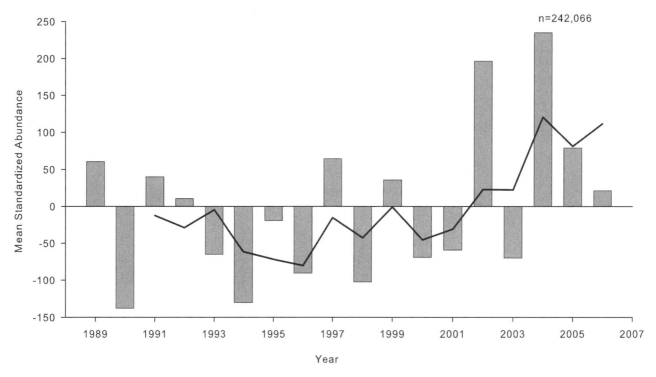

Fig. 6.3. Annual variation in abundance of all larvae (n = 242,066) in Little Sheepshead Creek Bridge 1 m plankton net collections, 1989–2006. Data were standardized by subtracting the overall mean from annual abundance. Vertical bars represent standardized annual abundance; the solid line is a three-year running mean that was calculated by taking the mean of the standardized values from the previous two years and the year in which the data are plotted.

smaller larvae (< 20 mm) with fewer of the larger individuals (> 40 mm). Presumably, the difference results from local spawning by residents and distant spawning and subsequent growth required of the transient species that move into the estuary.

ENVIRONMENTAL EFFECTS ON LARVAL SUPPLY

The interaction between physical and biological processes has yielded important insights into the movement of larvae into estuaries and their subsequent retention within them (Weinstein et al., 1980; Fortier and Leggett, 1983; Boehlert and Mundy, 1988; Hare et al., 1999; Epifanio and Garvine, 2001). In temperate estuaries, temperature has a profound effect on larval supply. The larvae of both resident and transient species are most abundant during spring through fall and especially in the summer (see Fig. 6.1), presumably in response to the temperature effects on the timing of reproduction but also because transport may not be possible at low winter temperatures. Certainly, in our experience at Little Egg Inlet, larvae of transient forms (e.g., *Anguilla rostrata* and *Paralichthys dentatus*) are less available at low temperatures (< 2°C), perhaps because they are inactive and less capable of vertical migrations associated with tidal stream transport into the estuary.

In some instances precipitation influences larval catches. For example, there is a clear, positive association between precipitation and abundance of *Anguilla rostrata*; greater freshwater flow might allow the ingressing glass eels to sense estuarine plumes and orient to the estuary more easily (Sullivan et al., 2006a). This relationship could be negatively affected by reduced freshwater flows to estuaries due to human impacts (see Chapter 7).

Few researchers have examined the effect of coastal upwelling, an important physical oceanographic event that occurs during the summer months along the east coast of Middle Atlantic Bight states (Walford and Wicklund, 1968; Crowley and Glenn, 1994; Neuman, 1996; Reiss and McConaugha, 1999), on estuarine larval fish communities. During the summer months along the New Jersey coast, an alongshore southerly wind stress could cause the transport of cold, nutrient-rich bottom water onshore and warm, nutrient-poor surface water offshore, resulting in coastal upwelling conditions (Hicks and Miller, 1980). As a result, coastal upwelling may influence the timing of ingress of estuarine-dependent larvae occurring offshore in cold bottom water into estuaries or may result in the loss of larval fish that inhabit warm, estuarine surface waters (Fig. 6.5). Previously, Witting et al. (1999) concluded that the variation in the density of the dominant species comprising the ichthyoplankton assemblage at Little Egg Inlet was primarily explained by intra-annual rather than inter-annual variation. The timing and duration of coastal upwelling events also varies intra-annually (Neuman, 1996), and thus a physical oceanographic process, such as upwelling, may explain a portion of the intra-annual variation in

Table 6.3. Annual variation in species rank abundance for larvae at Little Egg Inlet, New Jersey, 1989–2006. 1 = most abundant.

Genus	Species	1989	1990	1991	1992	1993	1994	1995	1996	1997	1998	1999	2000	2001	2002	2003	2004	2005	2006
Anchoa	*mitchilli*	1	1	2	2	2	3	1	4	1	1	1	3	2	4	7	1	1	2
Pseudopleuronectes	*americanus*	3	4	3	3	4	6	2	2	4	2	4	1	3	3	1	4	3	3
Syngnathus	*fuscus*	6	6	5	4	6	1	6	3	6	5	8	7	4	7	4	7	7	8
Brevoortia	*tyrannus*	9	8	14	9	14	14	10	5	3	13	7	6	7	1	3	5	4	4
Anguilla	*rostrata*	4	5	6	8	7	4	9	10	8	8	15	13	11	12	9	13	9	7
Micropogonias	*undulatus*	14	11	30	19	17	27	3	11	10	3	9	5	1	6	11	2	2	1
Ammodytes	sp.	5	2	7	12	13	2	8	14	14	9	13	17	6	13	18	14	8	12
Paralichthys	*dentatus*	12	10	9	14	16	12	15	24	11	14	17	8	18	2	5	12	14	6
Clupea	*harengus*	30	18	12	5	3	8	4	1	5	4	5	22	8	26	14	16	18	21
Menidia	*menidia*	28	16	4	13	15	22	11	8	9	15	11	15	13	8	6	9	13	11
Anchoa	sp.	60	59	1	1	1	25	18	12	2	2	26	2	2	5	11	2	3	5
Menidia	sp.	2	9	34	11	11	5	46	7	43	10	6	4	9	5	8	6	12	14
Scophthalmus	*aquosus*	8	33	25	20	9	9	16	6	21	7	16	18	19	9	12	8	15	13
Gobiosoma	*bosc*	60	59	8	6	5	7	5	9	13	6	19	10	10	10	10	11	6	19
Etropus	*microstomus*	10	17	13	10	8	13	23	25	15	11	18	26	24	22	13	17	16	18
Cynoscion	*regalis*	11	59	20	15	18	15	12	44	7	12	3	9	23	17	16	10	10	51
Fundulus	*heteroclitus*	17	20	31	18	19	16	7	15	16	21	30	21	14	35	19	18	34	34
Anchoa	*hepsetus*	40	15	17	34	12	29	17	18	12	23	10	12	20	20	28	24	46	15
Conger	*oceanicus*	16	13	22	23	10	10	33	19	23	32	32	24	27	15	23	29	31	20
Gobiosoma	*ginsburgi*	60	59	10	17	30	40	13	16	17	16	14	38	15	16	21	15	21	9
Tautoga	*onitis*	21	27	48	7	29	18	41	35	18	22	34	14	32	30	20	23	20	26
Gasterosteus	*aculeatus*	22	12	26	30	25	23	21	22	25	39	28	30	33	27	31	26	39	27
Ctenogobius	*boleosoma*	60	59	11	54	28	21	14	44	30	20	20	16	22	25	25	22	17	10
Hippocampus	*erectus*	24	21	18	47	20	24	19	20	28	25	27	44	25	41	22	52	28	25
Prionotus	*evolans*	25	30	27	25	21	30	44	38	19	17	25	28	40	34	35	28	27	35
Sphoeroides	*maculatus*	32	59	29	26	41	42	22	27	29	29	31	27	17	33	26	36	29	39
Centropristis	*striata*	42	35	60	31	34	46	46	44	20	30	22	25	16	21	29	30	22	23
Myrophis	*punctatus*	60	36	21	40	22	31	25	26	39	40	23	19	21	14	41	41	54	36
Myoxocephalus	*aenaeus*	18	44	41	28	37	31	20	31	26	32	43	47	26	52	32	25	31	36
Pholis	*gunnellus*	15	29	36	21	47	11	28	23	32	27	42	40	36	52	36	27	54	51
Opisthonema	*oglinum*	60	59	60	54	47	46	33	44	22	24	36	23	12	19	15	21	19	17
Opsanus	*tau*	23	38	33	39	24	46	30	29	54	38	39	41	38	36	34	37	30	45
Tautogolabrus	*adspersus*	31	59	49	22	47	17	46	44	44	37	45	32	51	28	17	19	24	44
Enchelyopus	*cimbrius*	29	14	47	35	47	28	46	44	54	19	36	39	30	43	27	38	54	33
Astroscopus	*guttatus*	60	32	39	54	26	26	46	31	33	28	26	36	30	37	41	35	35	51
Clupeidae	sp.	60	24	24	24	47	41	29	28	31	51	58	11	37	47	49	33	54	51
Leiostomus	*xanthurus*	60	22	19	54	47	19	46	21	54	36	33	31	49	23	55	52	54	31
Menticirrhus	*saxatilis*	60	59	28	40	47	46	27	31	23	40	58	29	43	52	30	20	25	51
Menticirrhus	sp.	34	28	22	28	26	46	46	44	39	40	24	47	43	52	41	52	54	51
Engraulis	*eurystole*	13	25	51	27	31	46	42	30	35	51	58	34	39	52	55	52	47	32
Peprilus	*triacanthus*	26	44	60	32	37	31	33	44	54	18	35	36	43	43	55	31	54	51
Gobiidae	sp.	60	59	15	40	47	35	31	17	26	40	40	47	57	43	55	52	54	22
Gobiosoma	sp.	7	7	52	54	47	46	24	44	45	51	21	47	57	52	51	52	33	51
Hypsoblennius	*hentz*	60	59	41	35	33	46	33	44	39	32	29	47	57	29	55	38	36	41
Prionotus	*carolinus*	33	31	46	38	43	46	32	44	48	31	51	42	27	40	55	43	54	51
Gerreidae	sp.	37	40	35	54	47	46	25	44	36	51	46	47	34	24	41	52	54	51
Fundulus	sp.	60	3	60	33	23	46	46	36	54	51	58	47	50	52	37	46	54	51
Synodus	*foetens*	20	59	37	46	35	34	46	44	46	51	58	35	52	52	55	34	54	51
Scomber	*scombrus*	60	59	60	54	47	46	46	13	54	32	40	47	34	38	23	52	54	51
Symphurus	sp.	60	23	16	16	32	46	46	44	54	46	58	47	57	52	55	52	54	51
Urophycis	*regia*	58	26	50	49	36	46	46	40	49	51	52	47	41	39	40	48	49	46
Strongylura	*marina*	60	39	32	54	47	46	43	37	47	47	58	47	53	52	38	32	48	51
Microgobius	*thalassinus*	60	59	60	54	47	46	46	44	54	51	58	47	57	18	55	44	23	16
Sciaenidae	sp.	60	59	41	54	47	46	46	44	54	51	46	47	57	52	55	52	11	28
Ulvaria	*subbifurcata*	44	19	60	40	47	46	46	44	54	51	46	47	42	52	55	52	54	51
Gobionellus	sp.	19	44	40	54	47	46	46	44	54	51	43	47	57	52	55	52	54	51
Pomatomus	*saltatrix*	36	34	54	54	44	20	46	44	54	51	58	47	57	52	52	52	54	51
Clupeiformes	sp.	60	59	60	54	47	46	46	44	54	51	58	20	29	30	55	52	54	51
Ophichthidae	sp.	60	59	47	35	47	35	46	44	54	51	58	47	43	52	41	52	54	51
Urophycis	*chuss*	27	42	56	54	47	46	46	39	54	51	58	47	57	52	55	52	54	40
Apeltes	*quadracus*	57	59	57	54	46	43	46	42	38	50	53	47	57	50	50	40	40	48
Menidia	*beryllina*	60	57	45	51	47	46	45	44	50	49	54	45	55	42	39	50	51	51

continued

Table 6.3 continued

Genus	Species	1989	1990	1991	1992	1993	1994	1995	1996	1997	1998	1999	2000	2001	2002	2003	2004	2005	2006
Pollachius	virens	44	59	60	40	47	46	46	31	54	51	58	47	57	52	55	52	41	51
Gobionellus	oceanicus	60	59	60	54	47	46	46	44	54	51	58	47	57	52	55	52	26	24
Mugil	sp.	60	44	47	54	47	46	46	44	54	51	58	47	57	52	55	52	54	28
Fundulus	majalis	41	55	58	50	47	45	46	44	52	51	55	47	54	48	53	49	54	49
Atherinopsidae	sp.	60	59	60	54	47	46	46	44	54	51	12	47	57	52	55	52	54	51
Stenotomus	chrysops	60	44	60	54	47	46	46	44	54	51	58	47	57	52	32	44	54	51
Bairdiella	chrysoura	60	59	53	54	42	44	46	44	54	51	38	47	57	52	55	47	50	51
Trinectes	maculatus	44	44	60	54	37	46	46	44	54	51	58	47	57	52	55	52	54	51
Chilomycterus	sp.	34	59	47	54	47	46	46	44	54	51	58	47	57	52	55	52	54	51
Ophichthus	gomesi	44	59	38	54	47	46	46	44	54	51	58	47	57	52	55	52	54	51
Prionotus	sp.	60	59	60	54	47	46	46	44	54	51	58	47	43	43	41	52	54	51
Urophycis	sp.	60	44	44	48	47	46	46	44	54	51	58	47	57	52	55	52	54	51
Dormitator	maculatus	60	59	60	54	47	46	46	44	54	51	58	47	57	52	55	52	41	28
Membras	martinica	60	59	60	54	47	46	46	44	36	51	58	47	57	52	55	52	36	51
Cyprinodon	variegatus	43	56	60	54	45	46	46	41	54	48	58	47	57	49	55	52	54	47
Muraenidae	sp.	44	40	60	54	47	46	46	44	54	51	58	47	57	52	55	52	54	51
Clupea	sp.	60	59	60	54	47	46	46	44	54	51	58	47	57	52	55	52	36	36
Elops	saurus	60	59	47	54	47	46	33	44	54	51	58	43	57	52	55	52	54	51
Ophidion	marginatum	59	43	59	52	47	46	46	44	53	51	57	46	56	51	54	51	53	50
Chaetodon	sp.	44	59	60	54	47	46	46	44	54	51	58	47	57	52	55	52	41	51
Hippoglossina	oblonga	60	59	60	54	47	46	46	44	54	51	58	33	57	52	41	52	54	51
Gadus	morhua	60	59	60	40	47	46	46	44	54	51	58	47	57	52	41	52	54	51
Bothus	sp.	60	44	60	54	47	46	46	44	54	51	58	47	57	52	55	41	54	51
Lutjanus	griseus	60	59	47	54	47	35	46	44	54	51	58	47	57	52	55	52	54	51
Blenniidae	sp.	37	59	60	54	47	46	46	44	54	51	58	47	57	52	55	52	54	51
Lumpenus	maculatus	37	59	60	54	47	46	46	44	54	51	58	47	57	52	55	52	54	51
Lagodon	rhomboides	60	59	60	54	47	46	46	44	54	51	58	47	57	30	55	52	54	51
Tetraodontidae	sp.	60	37	60	54	47	46	46	44	54	51	58	47	57	52	55	52	54	51
Citharichthys	spilopterus	60	59	60	54	37	35	46	44	54	51	58	47	57	52	55	52	54	51
Orthopristis	chrysoptera	60	59	60	54	47	46	46	44	34	51	58	47	57	52	55	52	54	51
Aluterus	schoepfii	44	59	60	54	47	46	46	44	54	51	58	47	57	52	55	52	54	51
Citharichthys	sp.	44	59	60	54	47	46	46	44	54	51	58	47	57	52	55	52	54	51
Gymnothorax	sp.	44	59	60	54	47	46	46	44	54	51	58	47	57	52	55	52	54	51
Monacanthus	hispidus	44	59	60	54	47	46	46	44	54	51	58	47	57	52	55	52	54	51
Scomberomorus	maculatus	44	59	60	54	47	46	46	44	54	51	58	47	57	52	55	52	54	51
Serranidae	sp.	44	59	60	54	47	46	46	44	54	51	58	47	57	52	55	52	54	51
Sphyraena	barracuda	44	59	60	54	47	46	46	44	54	51	58	47	57	52	55	52	54	51
Ophidion	sp.	60	58	55	53	47	46	46	43	51	51	56	47	57	52	55	52	52	51
Haemulidae	sp.	60	59	60	54	47	46	46	44	39	51	58	47	57	52	55	52	54	51
Engraulidae	sp.	60	44	60	54	47	46	46	44	54	51	58	47	57	52	55	52	54	51
Lactophrys	sp.	60	44	60	54	47	46	46	44	54	51	58	47	57	52	55	52	54	51
Lumpenus	lumpretaeformis	60	44	60	54	47	46	46	44	54	51	58	47	57	52	55	52	54	51
Chilomycterus	schoepfii	60	59	60	54	47	46	46	44	54	51	58	47	43	52	55	52	54	51
Epinephelus	sp.	60	59	60	54	47	46	46	44	54	51	58	47	57	52	55	52	41	51
Hyporhamphus	meeki	60	59	60	54	47	46	46	44	54	51	58	47	57	52	55	52	41	51
Lucania	parva	60	59	60	54	47	46	33	44	54	51	58	47	57	52	55	52	54	51
Rachycentron	canadum	60	59	60	54	47	46	33	44	54	51	58	47	57	52	55	52	54	51
Symphurus	plagiusa	60	59	60	54	47	46	33	44	54	51	58	47	57	52	55	52	54	51
Anguilliformes	sp.	60	59	47	54	47	46	46	44	54	51	58	47	57	52	55	52	54	51
Synodontidae	sp.	60	59	47	54	47	46	46	44	54	51	58	47	57	52	55	52	54	51
Lophius	americanus	60	59	60	54	47	46	46	44	54	51	46	47	57	52	55	52	54	51
Sphyraena	borealis	60	59	60	54	47	46	46	44	54	51	46	47	57	52	55	52	54	51
Bothus	ocellatus	60	59	60	54	47	46	46	44	54	40	58	47	57	52	55	52	54	51
Limanda	ferruginea	60	59	60	54	47	46	46	44	54	40	58	47	57	52	55	52	54	51
Albula	vulpes	60	59	60	54	47	35	46	44	54	51	58	47	57	52	55	52	54	51
Eucinostomus	argentus	60	59	60	54	47	46	46	44	54	51	58	47	57	52	55	52	54	41
Microdesmus	longipinnis	60	59	60	54	47	46	46	44	54	51	58	47	57	52	55	52	54	41

Table 6.4. Variation in larval rank abundance by month for the top 10 taxa at Little Egg Inlet, New Jersey, 1989–2000. 1 = most abundant.

Genus	Species	Mean	Jan	Feb	Mar	Apr	May	Jun	Jul	Aug	Sep	Oct	Nov	Dec
Clupea	*harengus*	21.6	1	2	1	3	6	—	—	—	—	—	—	9
Anguilla	*rostrata*	19.3	2	1	2	4	—	—	—	—	—	—	8	4
Brevoortia	*tyrannus*	5.3	3	3	3	5	9	1	5	—	5	3	3	1
Paralichthys	*dentatus*	23.3	4	4	4	8	—	—	—	—	—	5	1	2
Micropogonias	*undulatus*	17.6	5	5	—	—	—	—	—	3	2	1	2	3
Myrophis	*punctatus*	25.8	6	—	—	—	—	—	—	—	—	—	7	7
Anchoa	*mitchilli*	8.6	7	—	—	—	—	5	1	1	1	2	4	5
Clupeiformes	sp.	26.6	8	—	—	—	—	—	—	—	—	—	—	—
Clupeidae	sp.	26.6	9	9	10	—	—	—	—	—	—	—	10	6
Pholis	*gunnellus*	35.8	10	7	7	6	—	—	—	—	—	—	—	—
Ammodytes	sp.	26.9	10	8	6	2	5	—	—	—	—	—	—	—
Leiostomus	*xanthurus*	31.0	—	6	9	10	—	—	—	—	—	—	—	—
Myoxocephalus	*aenaeus*	33.7	—	9	8	7	—	—	—	—	—	—	—	—
Ctenogobius	*boleosoma*	21.0	—	—	—	—	—	—	—	—	6	4	5	8
Anchoa	sp.	14.7	—	—	—	—	—	6	2	2	3	—	—	10
Syngnathus	*fuscus*	13.3	—	—	—	—	2	2	6	—	10	10	—	—
Engraulis	*eurystole*	23.3	—	—	—	—	—	—	—	—	9	—	6	—
Scophthalmus	*aquosos*	21.3	—	—	—	—	4	7	—	—	—	—	9	—
Anchoa	*hepsetus*	16.8	—	—	—	—	—	—	—	9	7	6	—	—
Gobiosoma	sp.	17.2	—	—	—	—	—	—	9	7	—	—	—	—
Symphurus	sp.	26.3	—	—	—	—	—	—	—	—	—	7	—	—
Etropus	*microstomus*	18.0	—	—	—	—	—	—	—	4	4	8	—	—
Gobiosoma	*bosc*	15.3	—	—	—	—	—	—	3	6	—	—	—	—
Gobiosoma	*ginsburgi*	17.1	—	—	—	—	—	—	10	10	8	—	—	—
Sciaenidae	sp.	26.0	—	—	—	—	—	—	—	—	9	—	—	—
Menidia	*menidia*	18.3	—	—	—	—	10	4	8	—	—	—	—	—
Cynoscion	*regalis*	19.7	—	—	—	—	—	—	4	5	—	—	—	—
Opisthonema	*oglinum*	25.3	—	—	—	—	—	—	—	8	—	—	—	—
Fundulus	*heteroclitus*	21.8	—	—	—	—	—	8	—	—	—	—	—	—
Fundulus	sp.	25.9	—	—	—	—	—	—	7	—	—	—	—	—
Menidia	sp.	21.2	—	—	—	9	3	3	—	—	—	—	—	—
Conger	*oceanicus*	26.8	—	—	—	—	8	10	—	—	—	—	—	—
Atherinopsidae	sp.	31.3	—	—	—	—	—	9	—	—	—	—	—	—
Pseudopleuronectes	*americanus*	27.3	—	—	5	1	1	—	—	—	—	—	—	—
Gasterosteus	*aculeatus*	33.4	—	—	—	—	7	—	—	—	—	—	—	—

the timing of occurrence and densities of larval fishes. In addition, Glenn et al. (1996) identified a bathymetrically generated recurrent coastal upwelling center adjacent to Little Egg Inlet. This, together with the fact that ichthyoplankton samples have been consistently collected at this inlet on a weekly basis from 1989 to 2006, make this an ideal location to determine if upwelling influences larval abundance or species composition (Neuman et al., 2002).

We focused our upwelling study on a number of target species (*Conger oceanicus, Etropus microstomus, Cynoscion regalis, Scophthalmus aquosus, Tautoga onitis, Anchoa* spp., *Gobiosoma* spp., *Menidia* spp., and *Syngnathus fuscus*), for the following reasons: (1) these species are present and abundant in samples collected during the summer months (May–September) when upwelling events are known to occur frequently (Glenn et al., 1996); and (2) the spawning areas for these species range from the estuary to the continental shelf; therefore, the origin of larval supply differs among them.

This variation allows us to compare how different spawning strategies might be affected by upwelling events.

Based on the information gathered during preliminary analyses (Neuman et al., 2002), we arrived at the following conclusions that helped us establish guidelines for the examination of the long-term dataset: (1) a $\geq 4°C$ difference between inshore and offshore water temperatures was required for an upwelling event to significantly affect inshore, estuarine waters; (2) wind velocities of ≥ 4 m/s from the south to southwest had to continue for at least two days in order for cold, upwelled water to break the surface inshore and be detected with satellite imagery; however, wind stress of this magnitude and direction did not always result in a $\geq 4°C$ difference between inshore and offshore water temperatures; (3) the strongest upwelling signal (i.e., the coolest water temperatures) at Little Egg Inlet was evident on incoming and high tides, the time period during which ichthyoplankton sampling was conducted; (4) inshore water

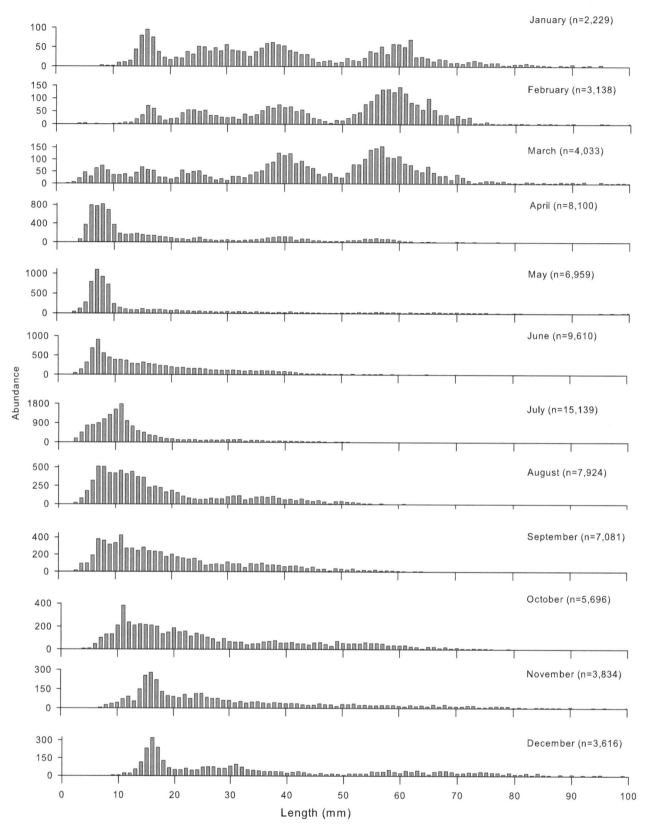

Fig. 6.4. Composite monthly size frequency for larvae collected at Little Egg Inlet, New Jersey, during 1989–2006.

a. No wind

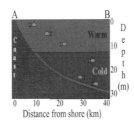

b. Southwesterly wind for 8 days

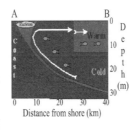

c. Wind stops for 3 days

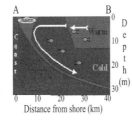

d. Wind stops for 8 days

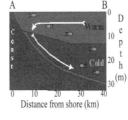

Fig. 6.5. Lifecycle of a summer, coastal upwelling event in southern New Jersey at inshore (A) and offshore (B) stations and its potential effect on larval fishes. *a.* Warming of coastal surface waters by the sun causes stratification (warm surface / cold bottom). *b.* Persistent southwesterly winds blow the surface water to the southeast. This 90 degree difference in wind and water current direction is due to earth's Coriolis force. Cold bottom water, rich in nutrients, rises to the surface along the coast and phytoplankton blooms may result. *c.* Upwelling relaxes after southwesterly winds cease. *d.* Downwelling occurs and typical summer conditions return.

temperature data from a variety of sources (sampling site, Buoy 126, hand-held thermometer readings at Rutgers University Marine Field Station [RUMFS], and an inshore buoy at Long-Term Ecosystem Observatory at 15 m [LEO-15]) were similar during upwelling events. Given this similarity, we pieced together data from RUMFS (1989–1992) together with the inshore buoy (1993–1999) to cover the entire 11-year period of interest (1989–1999). Therefore, the long-term data were examined on a species-specific basis in order to determine how fishes with different life histories might be affected by upwelling events. While the strongest and most persistent upwelling events occur during July, they also take place during other summer months. Therefore, the long-term data were examined from May to September.

As a result of this analysis, the patterns indicate that coastal upwelling events occurred during every year of the study and ranged from one event in 1990, 1991, 1992, and 1995 to five events in 1994 (Neuman et al., 2002). The duration of upwelling events ranged from 2 to 12 days. The total number of upwelling days ranged from 5 in 1992 to 23 in 1994. In 7 out of the 11 years, total abundance of ichthyoplankton (mean number/1000 m³ ± SE) was greater than mean abundance of ichthyoplankton during upwelling events. Total number of species captured ranged from 27 in 1996 to 49 in 1989, and the mean number of species captured during upwelling events ranged from 4 in 1992 to 18 in 1993 (Table 6.5). There was a significant, positive correlation between: (1) the number of upwelling events in a given year and species diversity during upwelling events; and (2) the total number of upwelling days in a given year and species diversity during upwelling events.

Species diversity increased as the number of upwelling events and the total number of upwelling days increased (p < 0.05; Fig. 6.6).

Corresponding larval abundances before, during, and after each upwelling event were examined for species that ranked among the top 15 in abundance for a given year during the time period examined (Neuman et al., 2002). Nine species (*Conger oceanicus, Ammodytes* spp., *Etropus microstomus, Cynoscion regalis, Scophthalmus aquosus, Gobiosoma* sp., *Anchoa* spp., *Menidia* spp., and *Syngnathus fuscus*) ranked among the top 15 in abundance in ≥ 8 of the 11 years examined (> 70% of the years examined). Three species (*Tautoga onitis, Brevoortia tyrannus,* and *Pseudopleuronectes americanus*) ranked among the top 15 in abundance in > 45% of the years examined, and 7 species (*Pomatomus saltatrix, Clupea harengus, Leiostomus xanthurus, Micropogonias undulatus, Scomber scombrus, Centropristis striata,* and *Anguilla rostrata*) occurred during < 30% of the years examined. Species were categorized by spawning location into subgroups (estuary, estuary/shelf, continental shelf, Sargasso Sea, incidental stray; Able and Fahay, 1998; Witting et al., 1999) to determine if these affected the response to the upwelling (Fig. 6.7). Across most years, two interesting patterns were noted: (1) each group exhibited similar trends in abundance over time regardless of spawning strategy (i.e., more often than not, all of the groups peaked in abundance at the same time); and (2) the most abundant group was the estuary/shelf spawners, followed by estuarine spawners, and finally the shelf spawners. However, if *Anchoa* spp., considered to be estuary/shelf spawners, were removed from the dataset, estuarine spawners were most abundant,

Table 6.5. Summary of frequency and duration (d) of upwelling events at Little Egg Inlet, May–September 1989–1999. Abundance (mean number/1000 m³) and diversity (mean number of species/tow) of ichthyoplankton at Little Sheepshead Creek Bridge during the entire study period (total) and only during upwelling events are given for each year.

Sampling Duration	Number of Events	Mean Duration of Events (d)	Total Upwelling Days (d)	Total Abundance	Mean Abundance during Upwelling	Total Diversity	Mean Diversity during Upwelling
May–Sep 1989	2	7.5 [5–10]	15	821 ± 2425.9 N = 185	69 ± 70.2 N = 18	49	11
May–Sep 1990	1	6	6	235 ± 579.8 N = 121	NA N = 0	38	N/A
May–Sep 1991	1	7	7	776 ± 1129.8 N = 94	1139 ± 1229.6 N = 6	43	7
May–Sep 1992	1	5	5	967 ± 1996.7 N = 49	374 ± 382.3 N = 3	35	4
May–Sep 1993	2	6.3 [6–7]	19	369 ± 618.3 N = 51	1183 ± 1099.4 N = 6	34	18
May–Sep 1994	5	4.6 [2–10]	23	258 ± 323.8 N = 53	412 ± 463.0 N = 9	31	15
May–Sep 1995	1	12	12	1031 ± 3623.1 N = 48	166 ± 102.9 N = 3	32	6
May–Sep 1996	2	6	12	328 ± 472.7 N = 60	125 ± 104.2 N = 6	27	7
May–Sep 1997	2	7.5 [7–8]	15	809 ± 1715.2 N = 54	574 ± 450.6 N = 3	29	11
May–Sep 1998	3	6 [5–7]	18	193 ± 190.8 N = 77	149 ± 120.8 N = 15	37	17
May–Sep 1999	2	6.5 [6.7]	13	1386 ± 3931.3 N = 92	136 ± 78.4 N = 6	44	11

followed by estuary/shelf spawners, and finally shelf spawners. This more detailed analysis did not reveal an association between the occurrence of upwelling events and a response in the larval assemblage by spawning strategy.

Two Sargasso Sea spawners (*Conger oceanicus* and *Anguilla rostrata*), four continental shelf spawners (*Ammodytes* spp., *Pomatomus saltatrix*, *Clupea harengus*, and *Leiostomus xanthurus*), one estuarine and continental shelf spawner (*Scophthalmus aquosus*), and three estuarine spawners (*Pseudopleuronectes americanus*, *Menidia* spp., and *Syngnathus fuscus*) generally peaked in abundance prior to the onset of upwelling events. Two continental shelf spawners (*Etropus microstomus* and *Micropogonias undulatus*) peaked in abundance after the completion of upwelling events. A few species (*Cynoscion regalis* in 1991, *Anchoa* spp. in 1993, and *Gobiosoma* sp. in 1996 and 1998) occasionally peaked during upwelling events (Neuman et al., 2002). In addition, a few species exhibited higher abundances with increasing number or duration of upwelling events in a given year. A number of recent studies have shown that fluctuations in larval abundance are related not only to wind stress, but also to lunar cycles (in tropical systems), tidally dependent hydrological fronts, or low barometric pressure (Cowen et al., 1993; Jenkins et al., 1997; Dixon et al., 1999; Reiss and McConaugha, 1999; Smith and Suthers, 1999). Species-specific biological responses to these physical processes with respect to spawning time, location, behavior, and ontogenetic development of larvae are what may ulti-

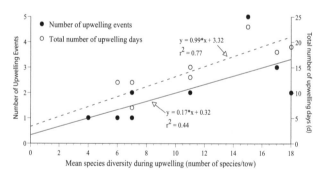

Fig. 6.6. The number of upwelling events (solid circle) and the number of upwelling days (open circle) regressed against mean species diversity during upwelling (number of species/tow) for each year of the study, 1989–1999. The regression for the number of upwelling events is shown by the solid line and the regression for the number of upwelling days is shown by the broken line. The equations and r² values for the linear regressions are given in the figure.

mately mediate fluctuations in larval supply (Cowen et al., 1993; Dixon et al., 1999).

Overall, our results suggest that upwelling events off the coast of New Jersey are most prominent during the summer months (May–September), especially July. Furthermore, these episodic events can potentially affect water temperatures within adjacent estuaries when relatively strong south to southwest winds persist for at least two days. These findings agree with previous research conducted in the region (Crowley and Glenn, 1994; Neuman, 1996).

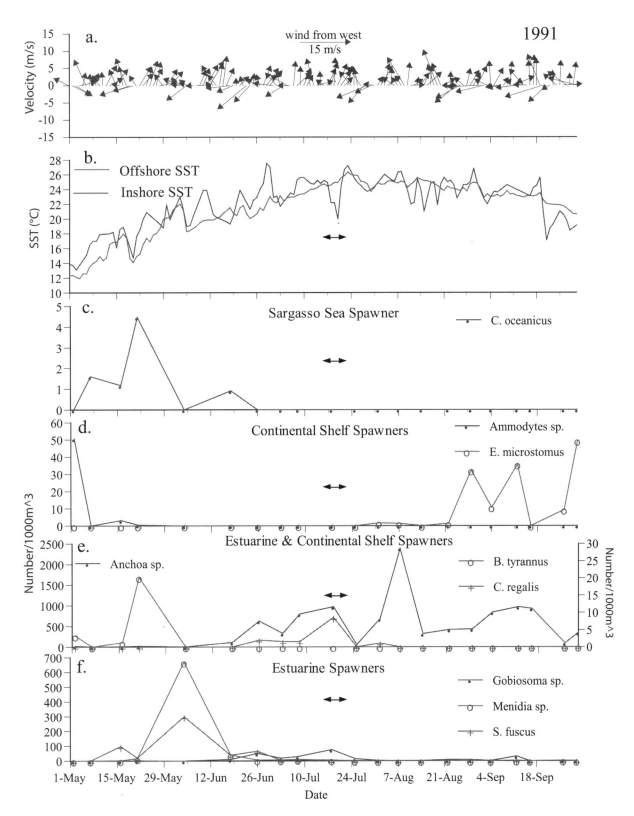

Fig. 6.7. Meteorological and larval fish characterization for the study area in 1991. *a.* Daily mean wind velocity (m/s) and direction; *b.* inshore and offshore sea surface temperatures (SST, °C); and mean abundance (number/1000 m³) of *c.* Sargasso Sea spawners, *d.* continental shelf spawners, *e.* estuarine and continental shelf spawners, and *f.* estuarine spawners. The species groups shown were ranked among the top 15 in abundance, by month, in 1991. Winds that resulted in upwelling conditions are circled, while the arrows denote the time period over which offshore and inshore water temperatures differed by ≥ 4 °C (indicative of upwelling).

A study conducted in Chesapeake Bay suggested that larval stages of *Anchoa* spp. and the earliest stages (preflexion) of *Micropogonias undulatus,* both considered to be part of the estuarine plume assemblage (as opposed to considering *M. undulatus* as part of the shelf assemblage and *Anchoa* spp. as part of the estuary/shelf assemblage; this study; Able and Fahay, 1998; Witting et al., 1999), were transported rapidly offshore due to a combination of upwelling favorable conditions and plume dynamics (Reiss and McConaugha, 1999). Our findings differed from those of Reiss and McCounaugha (1999) in that we did not identify any negative correlations between the abundance of estuarine spawners and upwelling events. In their study, *M. undulatus* exhibited higher abundances during upwelling events, suggesting that larvae may have been transported into the estuary from the shelf. A possible explanation for this discrepancy is that during the time that Reiss and McCounaugha (1999) conducted their study in Chesapeake Bay, *M. undulatus* were preflexion-stage, planktonic, estuarine plume-inhabiting larvae. At the time that this study was conducted off New Jersey, *M. undulatus* were later-stage (85% were postflexion; Witting et al., 1999), settling larvae and may have been transported inshore with cold bottom water from offshore in association with upwelling events.

SETTLEMENT

Much of the emphasis on fish recruitment since the pioneering work of Hjort (1914, 1926; recently summarized in May, 1974; Blaxter, 1988; Pepin, 1991; Miller, 1994; Chambers and Trippel, 1997) has focused on the search for a "critical period" of mortality during egg and larval development that might explain the extreme population fluctuations so frequently observed. To date, this search for a single critical period has been unsuccessful (Leggett, 1986; Houde, 1987; Blaxter, 1988; Browman, 1989; Houde, 2008). Recent efforts have considered the possibility that more than one period of increased mortality may occur during the early life history of fishes (Graham et al., 1984; Walline, 1985; Victor, 1986; Veer and Bergman, 1987; Campana, 1996; thoroughly summarized by Houde [2008]). An emerging view is that morphological development of transitory characters in late larval stages is related to critical survival and represents another in a long list of life history "bottlenecks" (M. P. Fahay, unpubl. observ.).

Much of the use of estuarine nurseries by fishes is begun during a transitional morphological period. During this time, a fish undergoes anatomical alterations associated with allometric growth and profound changes involving physiology and development. Shifts in habitats are often concurrent with these changes (Able and Fahay, 1998; Montgomery et al., 2001; Gagliano et al., 2007a; Juanes, 2007). The extent to which concurrence of developmental change and environmental shift affects settlement is largely unknown, especially for estuarine fishes. This is also true for demersal species with pelagic larvae that may face the added stress associated with settlement to the bottom (McGurk, 1984; Houde, 1987;

Doherty et al., 2004). The statement by Hempel (1965) that we know less about the larval/juvenile transition (Bailey and Houde, 1989) than any other part of the life history remains true decades later.

There is an increasing realization that processes that occur immediately before and after juveniles settle to the bottom (following a pelagic larval stage) may influence recruitment (McGurk, 1984; Houde, 1987, Cushing, 1996). This idea is best developed for tropical and temperate reef fishes (Jones, 1987a, b, 1990; Shulman and Ogden, 1987; Robertson, 1988a, b; Forrester, 1990; Carr, 1991; Hunt von Herbing and Hunte, 1991; Jones, 1991; Levin, 1991, 1993; Tupper and Hunte, 1994; Williams et al., 1994; Booth, 1995; Carr and Hixon, 1995; Cowen and Sponaugle, 1997; Leis and Carson-Ewant, 2002; Doherty et al., 2004) and for fishes in temperate habitats as well (Breitburg, 1991; Malloy and Targett, 1991; Szedlmayer et al., 1992; Tupper and Boutilier, 1995a, b; Campana, 1996; Juanes, 2007). In many cases, the search for the mortality that fuels year-class variation is expected in post-settlement young-of-the-year juveniles (Sissenwine, 1984; Smith, 1985; Houde, 1987; Elliott, 1989; Doherty, 1991; Beverton and Iles, 1992) and extends to the first winter (Post and Evans, 1989; Conover and Present, 1990; Hales and Able, 2001; Houde, 2008). The lack of adequate research on mortality during settlement of benthic fishes may be due to the difficulty in sampling these small individuals quantitatively (Able and Fahay, 1998).

JUVENILE GROWTH

The growth of juveniles of temperate estuarine fishes has received frequent attention and, as a result, some generalities are becoming apparent. Many of our growth estimates in this synthesis are based on modal length-frequency progressions, which provide reasonable estimates when compared with other aging techniques. These estimates range from 0 to 2.3 mm per day (see Table 6.1; Fig. 6.8). Rates observed in some transients (typically greater than for residents) are among the fastest reported for the young-of-the-year of any fish species, and this is consistent with the prediction that migratory species will have quicker growth rates (Roff, 1988). Slower growth rates in resident species in a Massachusetts estuary were attributed to their using more predictable, but lower-quality food sources (Werme, 1981; Teal, 1986). Growth for these fishes may appear to be slow because many residents (e.g., fundulids, gasterosteids, soleids, etc.) are small as adults (see Table 6.1). If we express growth as a function of adult size, these estimates would be more similar to those of transient species. During the winter, growth rates are consistently low or nil for both residents and transients, even for different cohorts (Fig. 6.8). The slower estimated growth rates for many residents is reflected in the smaller sizes they reach during the fall, during the following spring, and at the end of the first year (see Table 6.1; Fig. 6.9). In species that use both estuaries and coastal ocean habitats during the first year, growth rates can be similar in both habitats, as in *Cen-*

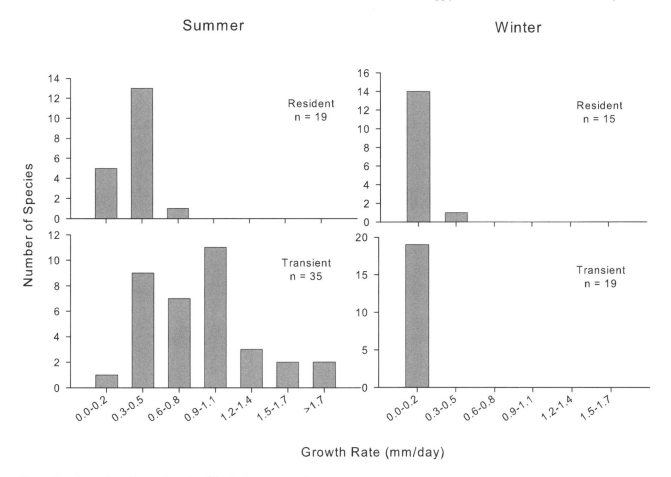

Fig. 6.8. Growth rates in resident and transient fishes in the summer and winter in the central part of the Middle Atlantic Bight.

tropristis striata (Able et al., 1995a), or higher in the estuary, as in *Urophycis regia* (Barans, 1972).

Some evidence shows that growth is fastest immediately following settlement. Growth rates from caged *Pseudopleuronectes americanus* in different estuarine habitats were quickest in the smallest sizes (< 50 mm TL), and this pattern was similar for uncaged individuals (Sogard, 1991, 1992; Able et al., 1996b). Other studies of recently settled fishes have also determined that fast growth rates are essential for survival (Campana et al., 1989; Tupper and Boutilier, 1995a, b; Campana, 1996). This result is consistent with the hypothesis that growth rate and stage duration are important determinants of subsequent survival and year-class strength (Houde, 1987, 2008).

One reason that the growth of estuarine fishes is receiving increasing attention is because it is an important component of essential fish habitat (EFH). This U.S. legislative mandate requires the EFH, or the water and substrate necessary for spawning, breeding, feeding, or growth to maturity, be considered in fisheries management (NOAA, 1996a; Mangel et al., 2006). As such, growth is considered as one indication of habitat quality (EFH) because it implies that (1) there is sufficient food available; (2) individuals may achieve a size refuge from predation; and (3) individuals may be larger at

the end of the growing season, thus enhancing the chances of overwinter survival (Sogard, 1997; Able, 1999). This focus on growth and an essential fish habitat approach is being used in other countries as well (Le Pape et al., 2003; Gilliers et al., 2006). Based on this rationale, several studies have examined growth rates of caged fish to evaluate habitat quality in the study area (see Duffy-Anderson and Able, 1999; Phelan et al., 2000; Meng et al., 2008, for recent reviews; see Chapter 7 for application of this approach due to anthropogenic effects). More recently, growth has been reevaluated for *Micropogonias undulatus* in order to understand how these rates change in response to habitat-specific environmental conditions, size-selective mortality, previous growth history, and density dependence (Searcy et al., 2007a). However, it is also clear that including growth with other measures of habitat quality (i.e., habitat-specific density and production) is a more robust measure than a single index (Searcy et al., 2007b).

MORTALITY

One of the most important shortcomings in our knowledge of estuarine fishes is the lack of estimates of the sources and rates of mortality for any life history stage (Houde, 2008). This is an especially important issue given the high rates typically estimated (Roff, 1992; Bradford and Cabana, 1997), al-

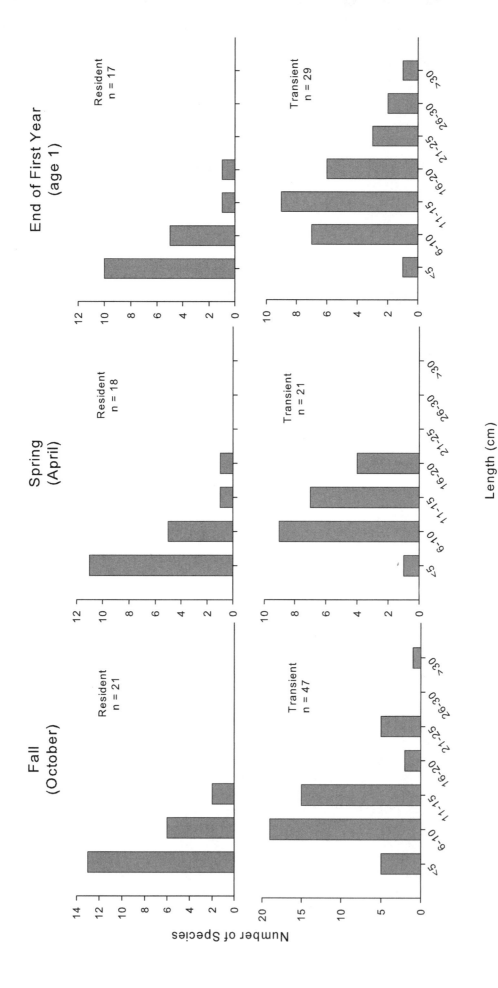

Fig. 6.9. Size attained by the fall, spring, and end of first year for resident and transient fishes in the central part of the Middle Atlantic Bight.

though it is clear that it is the variability in mortality that is the most important factor (Chambers and Trippel, 1997). It has been suggested (Günter, 1961) and often repeated that there may be relaxed predation pressure in estuaries relative to the ocean, and this may account, in part, for the large number of juvenile fishes surviving the first year in estuaries (Joseph, 1972; Whitfield and Blaber, 1978; Blaber, 1980; Blaber and Blaber, 1980). Recent studies have frequently questioned that interpretation and have instead documented the frequent occurrence of piscivory on the part of small resident fishes, juveniles of other species (including cannibalism), and abundant large piscivores in estuaries. The same applies to avian predators on temperate estuarine fishes. This issue is treated in more detail in Chapter 9. Even where mortality estimates have been made for estuarine species, the influence of confounding factors (i.e., gear avoidance, inaccessible habitats, etc.) makes it difficult to determine mortality. This makes it even harder to evaluate the relative contribution of mortality among different early life history stages or cohorts and the role of growth, habitat, and the like, and their influence on recruitment to the adult stage. Another important contributor to this problem is that the mobility of the juveniles of many estuarine forms makes it difficult to separate emigration from actual mortality when calculating loss rates (Herke, 1977; Weinstein, 1983; Sogard, 1989). The exceptions may be for some pelagic species such as *Anchoa mitchilli*, where gear avoidance is less of an issue (Houde, 1987) or for individual demersal fish, which can be tagged and followed with acoustic techniques (Szedlmayer and Able, 1993).

The role of diseases and parasites on mortality is virtually unknown for most estuarine species. There is increasing evidence that parasites and diseases, including mobile temporary parasites or micropredators (Penfold et al., 2008), play an important role in the life history, ecology, and behavior of fishes (Sindermann, 1966; Poulin, 1995; Harvell et al., 1999; Barber et al., 2000; Lafferty et al., 2008). One of the clearest examples is *Pseudopleuronectes americanus,* which occurs in heavily urbanized polluted estuaries where "fin rot" disease can be common (Murchelano and Ziskowski, 1982). Several species of bacteria are associated with this condition, which also produces changes in blood characteristics (Mahoney and McNulty, 1992). Broader surveys of disease in commercially important fishes, including some estuarine fishes, found overall disease prevalence to be relatively low but with increasing incidence for flatfishes in nearshore portions of the continental shelf and in the vicinity of major cities (Ziskowski et al., 1987). More recently, the frequent, seasonal occurrence of *Myxosporidium* in *Morone saxatilis*, particularly in Chesapeake Bay, is characterized by the factors associated with the disease, including reduced body condition associated with an insufficient prey supply and high population densities of the hosts, but the exact mechanism(s) responsible for its increasing frequency are unknown.

In estuaries, parasites are pervasive. Cestode parasites are

likely to influence larval survival for *Osmerus mordax* (Sirois and Dodson, 2000), and hemoflagellates cause mortalities in juvenile *Paralichthys dentatus* (Burreson and Zwerner, 1984). More subtle but perhaps just as important effects are manifested in fish morphology, swimming ability, and reproduction. The gills of *Fundulus heteroclitus* can become abnormal, with multiple branches of gill filaments appearing, when infected with digenean metacercariae endoparasites (Bass et al., 2007). The heart of *Cyprinodon variegatus* can become infected with the cysts of another digenean parasite that influences the population dynamics of this species by increasing winter mortality (Coleman and Travis, 1998). The fast-start response by *Gasterosteus aculeatus*, a behavior used to avoid predators and capture prey, varies with the degree of infection and the species of cestode (Blake et al., 2006). In addition, the more harmful cestodes were also found to reduce the ability of the same host to produce gametes and spawn (Heins and Baker, 2008). In other instances, as for an isopod ectoparasite of young-of-the-year *Pomatomus saltatrix*, it was determined that the parasite does not cause a serious threat to individual survival (Marks et al., 1996). The same may not be true for an introduced nematode (*Anguillicola crassus*) that is found, with increasing frequency, in the swimbladder of *Anguilla rostrata* (Barse and Secor, 1999). The parasite, native to Japan, where it naturally infects *Anguilla japonica*, has been introduced into Europe, where it may severely impair the swimming ability of the silver eel stage of *Anguilla anguilla* such that infected individuals may be unable to reach the Sargasso Sea spawning grounds (Palstra et al., 2007). A similar effect is predicted for *Anguilla rostrata* (Barse and Secor, 1999; Barse et al., 2001; Sures and Knopf, 2004) and may be a major factor causing the decline of this species in recent years. The ultimate effect of this parasite on the status of *Anguilla rostra* is being determined as we write these words.

Another source of mortality in temperate estuarine fishes, particularly small, relatively immobile species, is winter, with its low temperatures and an accompanying loss of energy reserves during the period of little or no feeding (Conover and Ross, 1982; Henderson et al., 1988; Conover and Present, 1990; Shuter and Post, 1990; Hurst and Conover, 1998; Hurst et al., 2000; Hales and Able, 2001; Hurst, 2007). Mortality may occur, especially in the first winter of life, because individual estuarine temperate fishes may experience net energy deficits due to low temperature and food scarcity and both of these factors result in decreased feeding (see Sogard, 1997, for review). Small individuals may also accrue an energy deficit due to their relatively low weight-specific standard metabolic rates, relatively low weight-specific energy reserves, and osmoregulatory stress (Calder, 1984; Thompson et al., 1991; Johnson and Evans, 1996; Hurst et al., 2000; Hales and Able, 2001). This size selectivity may be especially important because individuals may grow slowly due to a variety of possible factors (i.e., poor food resources, threat of predation, reduced access to food, disease, parasitism, etc.).

Size-dependent winter mortality may also affect many life history traits such as migration, growth, and reproduction (Fox and Keast, 1990, 1991; Snyder and Dingle, 1990; Conover, 1992; Lankford, 1997) and could impact habitat-specific species composition, trophic structure, and other aspects of poikilotherm ecology (Persson, 1986a; Hall and Ehlinger, 1989; Shuter and Post, 1990; Atkinson, 1994; Johnson and Evans, 1996). Effects of mortality may be especially pronounced along the Atlantic coast of North America, where average winter water temperatures decline about 1°C with each degree of latitude and the annual range in sea surface temperatures exceeds 20°C (Schroeder, 1966). This especially applies to shallow estuaries (see below). However, much of the information available on winter mortality, growth, and behavior of estuarine and marine fishes is restricted to subtropical faunas and is concentrated on severe storm impacts (Günter, 1947; Hoff, 1971; Gilmore et al., 1978; Mora and Ospina, 2002; Hsieh et al., 2008), although an improved understanding is developing for temperate estuarine fishes in the study area (Schwartz, 1964; Moss, 1973; Able and Fahay, 1998; Hurst and Conover, 1998; Hurst et al., 2000; Lankford and Targett, 2001; Hurst, 2007; Able and Curran, 2008) and elsewhere (Kelley, 2002).

Additional evidence for overwinter mortality is accumulating based on laboratory observations of several obligate and facultative estuarine users and comparisons of estuarine year-round residents, seasonal summer residents, and winter residents (see Table 10.2). As expected, those species that are year-round residents had the highest tolerance (0 to 36% mortality), which was similar to that of a winter resident (*Pseudopleuronectes americanus*, 25%). Those summer residents that migrated out of the estuary for a winter thermal refuge had higher mortality (33 to 100%), with most of these species experiencing total mortality. The presumed expatriate (*Lutjanus griseus*) also experienced 100% mortality. (See species accounts for additional details.) The response to winter temperatures can also vary between cohorts, as for *Scophthalmus aquosus*, which had high (75%) mortality for the smaller fall cohort and lower mortality (31%) for the larger spring cohort (Neuman and Able, 2009).

Besides those species already indicated in the literature as being susceptible to overwinter mortality (*Centropristis striata, Menidia menidia,* and *Morone saxatilis;* Conover and Present, 1990; Hare and Cowen, 1996; Hurst, 2007), we would add other species (*Hippocampus erectus, Syngnathus fuscus, Menidia beryllina, Hypsoblennius hentz, Gobiosoma bosc,* and *Gobiosoma ginsburgi*) because of the absence of small individuals in early spring relative to the previous fall (Able and Fahay, 1998; see species accounts). We suggest this differential is due to size-selective mortality in the absence of growth during the winter.

Winter mortality is likely most pronounced for southern species that typically enter estuaries in the fall, when temperate estuarine waters are warmest, followed by cold winter temperatures. The extremes of temperature could selectively cause mortality and, as a result, differentially influence those species affected in cold versus relatively warm winters. For species entering temperate estuaries, there are three potential responses: (1) those that die during all winters, (2) those that die only in cold winters, and (3) those that are seldom influenced by winter temperatures (see Chapter 10).

Because fish have complex lifecycles, events during one life history stage can influence subsequent stages (see Houde, 1987; Gagliano et al., 2007b). Thus, it is important to understand the relationship between growth and mortality, especially during subsequent life history stages. Some authors suggest that faster-growing fish pass through size-dependent predation periods faster than slower-growing conspecifics (Werner and Gilliam, 1984; Sogard, 1997; but see alternative interpretations in Walters and Juanes, 1993). These types of interactions between growth and mortality are evident for temperate estuarine fishes such as that between a predator, *Paralichthys lethostigma,* and a prey, *Leiostomus xanthurus* (Craig et al., 2006). However, despite recent progress in identifying some of the mechanisms contributing to growth and mortality and thus recruitment variability, accurate forecasting remains a formidable challenge. This challenge is not likely to be easily met because recruitment variability is likely the result of complex interactions between abiotic and biotic processes over multiple temporal and spatial scales (Houde, 2008).

7

Habitat Use

Our knowledge of estuarine habitat use by fishes is incomplete partly because of the confusion regarding the definition of habitat (Hall et al., 1997; Mitchell, 2005). For our purposes, we define *habitat* as the place where a population or a life stage of a species lives at any particular time (see Minello et al., 2003, for further review). Some of the uncertainty regarding habitat use revolves around the degree of dependence on estuarine habitat use for many fishes (see Chapter 1). Also, much of the current level of understanding is based on non-quantitative collection techniques that can provide only a snapshot of actual habitat use (Able and Fahay, 1998; Able, 1999; Johnson et al., 2008). Additionally, most of the habitat use data are based solely on daytime collecting. Where comparisons between day and night collections have been made, they provide a very different view of habitat use (Rountree and Able, 1993; Sogard and Able, 1994; Hagan and Able, 2008). Fishes are also highly mobile, and their capture may indicate only a fraction of the habitats they use. This may be especially true for pelagic fishes (Hagan and Able, 2003). Such dynamic use of fluctuating estuarine habitats (Deegan and Day, 1986) might require a broader, landscape approach to the study of habitats for both pelagic and benthic fishes (Ray, 1991, 1997; Hoss and Thayer, 1993). As a result, connectivity between habitats has received increasing attention (Able, 2005; Gillanders, 2005; Herzka, 2005; Ray, 2005; Secor and Rooker, 2005).

Several general characteristics of estuaries may make them better habitats for young-of-the-year and larger juvenile and adult fishes than open ocean coasts. Much of the literature on estuaries refers to them as refuges from predation. For example, it has been suggested that they provide structurally complex shelters (Webster and Paul, 2004; Caddy, 2008), such as *Zostera* beds and *Spartina* marshes (Day et al., 1989; Raposa and Oviatt, 2000). The general and usually extensive shallowness of estuaries may result in an increase in the area of potential refuge (Nixon, 1980; Kneib, 1984; Ruiz et al., 1993; see Chapter 8 for a more thorough treatment of predation on fishes in estuaries). In terms of physical refuge, estuaries may offer more protection by reducing wave surge. Certainly, estuaries are used by adults of many anadromous species as spawning areas as well as habitat for their young (Dadswell et al., 1987). The usually turbid waters of estuaries might also provide concealment from predators (Livingston, 1975; Johnston et al., 2007), or may favor either the predator or the prey (Minello et al., 1987; Grecay and Targett, 1996), or may affect the ability to feed (which may vary between species; Ljunggren and Sandstrom, 2007). Some estuarine predators may have adapted to turbid conditions or rely on other sensory modalities (Cyrus and Blaber, 1992). In recent years, there has been an increased focus on the estuary turbidity maximum and its effects on feeding and survival of estuarine fishes (Sirois and Dodson, 2000; North and Houde, 2001, 2003, 2006; Islam et al., 2006; Martino, 2008). The young-of-the-year of many other species take advantage of the habitat in estuaries after spawning in the ocean (Able and Fahay, 1998). Other fishes, both young-of-the-year and larger juveniles and adults, presumably use estuarine habitats because of access to food (see Chapter 8).

In this chapter we concentrate on several issues that influence habitat use in temperate estuaries, with particular focus on seasonality in the estuary for both pelagic and benthic fishes, new data on the spatial distribution of larger juveniles and adults, how they interact with the substrate, the dynamics of habitat use, habitat quality and anthropogenic influences on the latter, as well as the increasingly important issue of habitat restoration for estuarine fishes.

PELAGIC AND BENTHIC FISHES

One of the major ecological distinctions for estuarine fishes is that between those that use pelagic habitats and those that use benthic habitats. Pelagic fishes in estuaries are typically those that swim in the water column and away from the bottom. These fishes are often extremely abundant in estuaries and are an integral link in estuarine food webs because they are important predators of zooplankton (Blaxter and Hunter, 1982; Baird and Ulanowicz, 1989; Johnson et al., 1990; Durbin and Durbin, 1998; see Chapter 8). These same pelagic fishes are also important prey for many species of estuarine fishes, birds, and marine mammals (see Chapter 8).

To date, we know that pelagic fishes are seasonally abundant in the study area if we assume that the pattern observed in the Mullica River–Great Bay estuary in New Jersey is typical. The pelagic species in this system are dominated by clupeids (*Clupea harengus, Alosa pseudoharengus,* and *Brevoortia tyrannus*), engraulids (*Anchoa mitchilli, Anchoa hepsetus,* and *Engraulis eurystole*), and an atherinid (*Menidia menidia*) (Hagan and Able, 2003). There is a well-defined seasonal structure to the pelagic fish assemblage, with five seasonal groups (winter, spring, early summer, late summer, and fall),

all of which are dominated by young-of-the-year. Both temperature and salinity contribute to this variation (Hagan and Able, 2003), but their influence fluctuates somewhat between night and day (Hagan and Able, 2008).

For benthic fishes, that is, those resting on or in the substrate or swimming just above it, we have chosen to focus on several discrete aspects of their habitat use. We review the evidence for specific habitat associations for temperate estuaries, look specifically at the underappreciated interaction with the substrate, and reexamine the evidence for the role of estuaries as non-nursery habitats for fish, especially along environmental gradients. Additionally, we detail the available literature and our other studies on habitat dynamics during ontogeny and at tidal, diel, and seasonal scales and during episodic events.

SEASONALITY IN THE ESTUARY

The occurrence of many species of pelagic and benthic fishes in temperate estuaries is influenced by pronounced seasonal temperature changes that occur in these estuaries and the adjacent ocean (Parr, 1933; Grosslein and Azarovitz, 1982; Able and Fahay, 1998). For example, this pattern in temperature was evident in the Delaware Bay, in both deep and shallow waters, across all the habitats studied during April through November (Fig. 7.1). This was most evident in the shallower waters of the bay, where temperatures reached a peak in July–August near 25°C. Deep habitats of the bay followed the same seasonal pattern in temperatures—highest during the same months although slightly lower, overall, than in shallow waters. The same general pattern in temperature is evident in Great Bay, New Jersey (see Fig. 2.6).

The seasonal pattern of occurrence and abundance by habitat for most of the transient (both facultative and obligate; see Able, 2005) fish species in Delaware Bay mirrored that for temperature, with low values in the spring and fall and higher values in mid-summer (see Fig. 7.1). The possible exceptions are for the bay-water column (pelagic trawl) and the bay-deep (small trawl) habitats, where the abundance increased in the fall as the result of high abundance of young-of-the-year *Anchoa mitchilli* and *Micropogonias undulatus* in both habitats. The same general seasonal pattern is evident for resident species (see Fig. 7.1), and also holds true for adjacent estuaries in New Jersey and other temperate estuaries (Able et al., 1996a; Able and Fahay, 1998; Able, 1999). A recent review for flatfishes, including estuarine species, indicates seasonal changes in habitat use are common (Able et al., 2005a). Other studies suggest that conditions in the estuary during the winter, especially temperature and flow, can influence subsequent events, such as fish assemblages in the summer and fall (Wingate and Secor, 2008). While seasonal differences in abundance are clear, responses to temperature can extend to the longer temporal scales at annual and decadal levels. For example, annual variation in abundance of transient and even resident species in Delaware Bay is com-

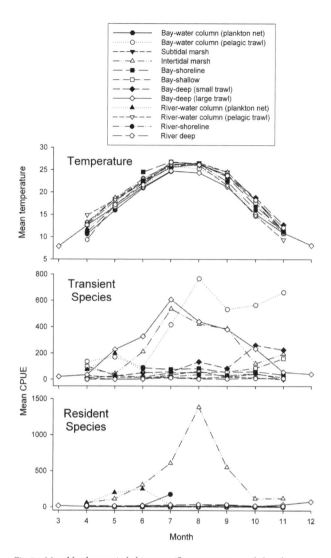

Fig. 7.1. Monthly changes in habitat-specific temperature and abundance (catch per unit effort [CPUE]) for transient and resident fishes during 1998–2004 in Delaware Bay (based on Able et al., 2007a).

mon (Able et al., 2007a). Certainly, changes in abundance at longer scales are also evident in the bay, as for *Micropogonias undulatus* (Hare and Able, 2007), but we often lack data on this for many species (see species accounts).

ASSEMBLAGES ALONG ESTUARINE ENVIRONMENTAL GRADIENTS

Many mechanisms influence the distribution and size of fishes within estuarine ecosystems. The general composition of transient fishes in estuaries originates from two major sources: those that enter the estuary from the ocean and those that penetrate the estuary from freshwaters (Attrill and Rundle, 2002; Greenwood, 2007). Thus, some estuaries represent double ecoclines (Attrill and Rundle, 2002). Many of the individuals collected in the Mullica River–Great Bay estuary, for example, are marine species that use the estuary as young-of-the-year (Able and Fahay, 1998) and/or as a feeding area for larger juveniles or adults (e.g., *Morone saxatilis*, *Mugil* curema, *Leiostomus xanthurus, Brevoortia tyrannus,* and *Pomatomus saltatrix).* The dominance by marine species, as represented by *Menticirrhus saxatilis, Pogonias cromis, Cynoscion regalis,* and *Stenotomus chrysops,* is especially evident in the lower, high-salinity portions of the estuary. The same pattern is evident in Delaware Bay (Able et al., 2007a) and elsewhere in southern New Jersey (Rountree and Able, 1992a). What is less frequently recognized is the contribution of freshwater forms to the estuarine fauna (Barthem and Goulding, 1997; Attrill and Rundle, 2002; Bate et al., 2002; Greenwood, 2007). Many of these appear to be larger individuals that may result from being spawned in freshwater and then moving into the estuary as they develop the ability to tolerate higher salinity.

The distribution of some of the dominant large fishes as collected by gill net in the Mullica River–Great Bay estuary was clearly reflected by the salinity gradient (Table 7.1; Figs. 7.2 and 7.3). Many species (e.g., *Morone americana, M. saxatilis, Mugil curema, Leiostomus xanthurus, Brevoortia tyrannus,* and *Pomotomus saltatrix*) were collected at most stations represented by the full range of salinities. Others (*Menticirrhus saxatilis, Pogonias cromis, Cynoscion regalis,* and *Stenotomus chrysops*) were captured only at the highest salinities in the lower estuary. Other species, such as *Ameiurus catus, Dorosoma cepedianum,* and *Notemigonus crysoleucas,* were captured only at the lowest salinities in the upper estuary (Fig. 7.3). This response to salinity is evident across the ocean to tidal freshwater spectrum as demonstrated by earlier sampling in the same estuary (Martino and Able, 2003) and complementary sampling that compares the inner continental shelf fauna to that in the estuary (Vasslides and Able, 2008).

There is a preponderance of small fishes in the Delaware Bay estuary (Fig. 7.4). In this system, most individuals collected were < 100 mm. Admittedly, some of this is the result of gear bias, but nonetheless multiple gears across 12 habitat types are consistently dominated by small fishes (although larger individuals [> 100 mm] did occur). For larger fishes, as collected by gill nets in the Mullica River–Great Bay estuary, the size distribution of some of the dominant species varies across the salinity gradient as well (Fig. 7.5). For *Leiostomus xanthurus,* the size was similar across all locations along the salinity gradient, with most individuals dominated by apparent young-of-the-year based on prior analysis (Able and Fahay, 1998). There was no apparent pattern for *Pomotomus saltatrix,* which was represented by young-of-the-year and apparent age 1 individuals. Alternatively, for the resident species *Morone americana,* the smallest average size individuals were found farthest up the estuary, while the average size increased with distance down the estuary, with the larger representing age 1+ individuals based on examinations elsewhere (Able and Fahay, 1998; Jones and Able, in review). This focus on salinity is appropriate given its frequently documented effect (Günter, 1956; Weinstein et al., 1980; Peterson and Ross, 1991; Rakocinski et al., 1992; Wagner and Austin, 1999). However, numerous other factors, both physical and

Table 7.1. Species composition, frequency of occurrence (Frq), CPUE (number of fish per tow), and standard error (SE) of CPUE of fish collected by gill net in marsh creeks tributary to the Mullica River (River) and in the Great Bay estuary (Bay) in 2002.

Species	Bay			River		
	Frq	CPUE	SE	Frq	CPUE	SE
Alosa aestivalis	2.54	0.16	0.12	—	—	—
Alosa mediocris	2.54	0.03	0.02	0.28	—	—
Alosa pseudoharengus	3.39	0.04	0.02	0.28	—	—
Alosa sp.	2.54	0.06	0.04	0.28	—	—
Ameiurus catus	—	—	—	11.33	0.87	0.21
Ameiurus nebulosus	—	—	—	1.7	0.02	0.01
Brevoortia tyrannus	23.73	9.91	3.88	8.7	3.31	0.87
Caranx hippos	0.85	0.01	0.01	0.57	0.01	0.01
Caranx sp.	—	—	—	0.28	—	—
Carcharhinus plumbeus	0.85	0.01	0.01	—	—	—
Catostomus commersoni	—	—	—	1.7	0.03	0.01
Centropristis striata	1.69	0.02	0.01	0.28	—	—
Cynoscion regalis	4.24	0.11	0.07	0.28	—	—
Dasyatis americana	0.85	0.01	0.01	—	—	—
Dasyatis sp.	0.85	0.01	0.01	—	—	—
Dorosoma cepedianum	—	—	—	5.38	0.1	0.03
Esox niger	—	—	—	3.68	0.05	0.02
Fundulus heteroclitus	—	—	—	0.28	—	—
Ictalurus punctatus	—	—	—	1.42	0.03	0.01
Lagodon rhomboides	5.08	0.13	0.06	0.28	—	—
Leiostomus xanthurus	7.63	0.18	0.07	21.81	1.13	0.3
Lepomis auritus	—	—	—	0.28	—	—
Lepomis gibbosus	—	—	—	0.28	—	—
Menticirrhus saxatilis	6.78	0.1	0.04	—	—	—
Micropogonias undulatus	—	—	—	0.85	0.01	0.01
Morone americana	0.85	0.01	0.01	50.14	3.75	0.47
Morone saxatilis	—	—	—	7.08	0.1	0.02
Mugil cephalus	—	—	—	1.13	0.01	0.01
Mugil curema	1.69	0.09	0.08	4.25	0.06	0.02
Mustelus canis	11.02	0.27	0.11	—	—	—
Notemigonus crysoleucas	—	—	—	3.97	0.14	0.07
Opsanus tau	—	—	—	0.28	—	—
Paralichthys dentatus	3.39	0.03	0.02	0.85	0.01	—
Pogonias cromis	—	—	—	1.13	0.02	0.01
Pomatomus saltatrix	23.73	1.19	0.34	12.75	0.46	0.11
Prionotus carolinus	0.85	0.01	0.01	—	—	—
Sphoeroides maculatus	2.54	0.03	0.01	—	—	—
Stenotomus chrysops	6.78	0.09	0.03	—	—	—
Strongylura marina	1.69	0.03	0.03	0.28	—	—
Synodus foetens	0.85	0.01	0.01	—	—	—
Tautoga onitis	2.54	0.03	0.01	—	—	—

biological, and their interactions can influence the distribution of fishes in estuaries (Rowe and Dunson, 1995; Martino and Able, 2003, Sackett et al., 2008).

Prior efforts to resolve fish distributions in the Mullica River–Great Bay estuary and place those observations in perspective relative to other estuaries found that large-scale patterns in the structure of estuarine fish assemblages are the result of individual species responses to the dominant environmental gradients (e.g., salinity, turbidity, temperature, and, perhaps somewhat uniquely, pH in this system) (Able

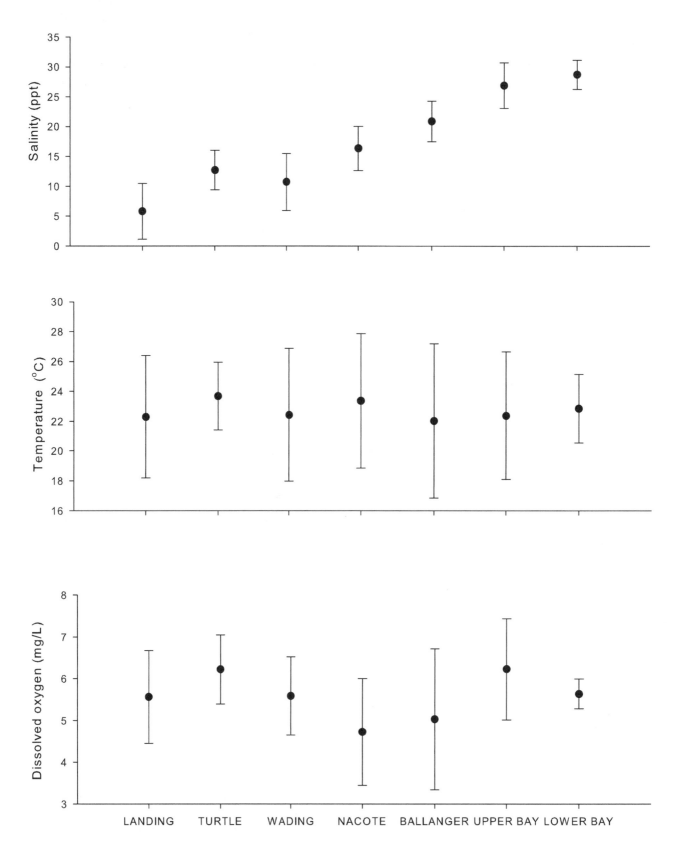

Fig. 7.2. Environmental variables across the Mullica River–Great Bay gradient. See Fig. 3.3 for location of sampling areas.

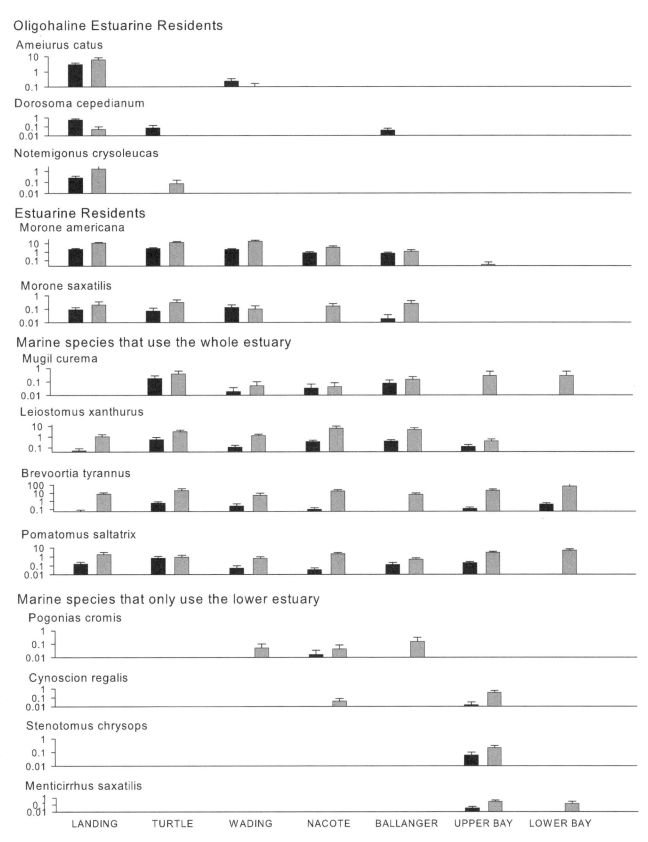

Fig. 7.3. Spatial distribution of selected large nektonic fishes in Mullica
River–Great Bay. Black bars = 15 m and grey bars = 90 m gill net samples.
See Fig. 3.3 for location of sampling areas.

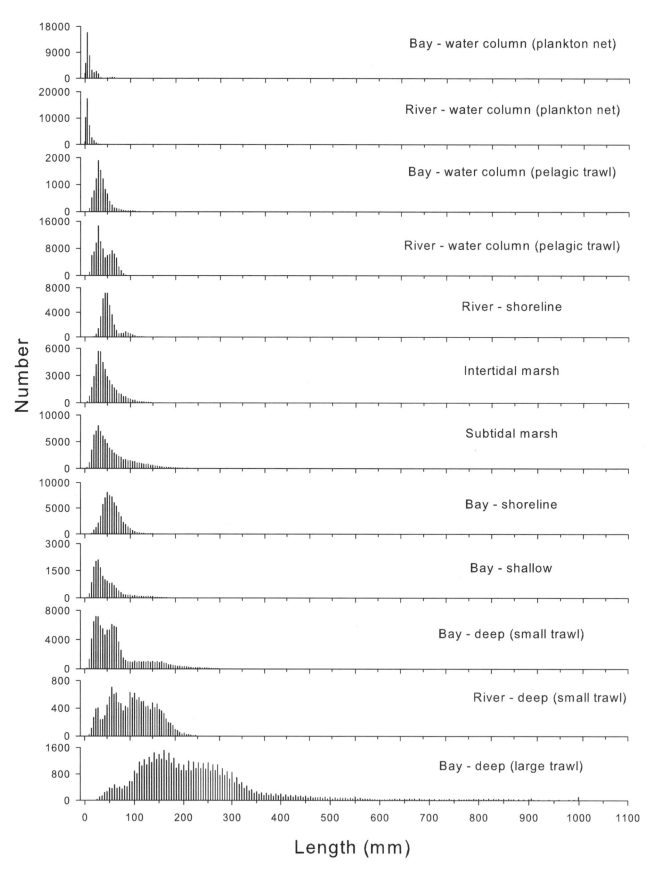

Fig. 7.4. Size of fishes in Delaware Bay estuary across 12 habitat / gear types
(based on samples collected from Able et al., 2007a).

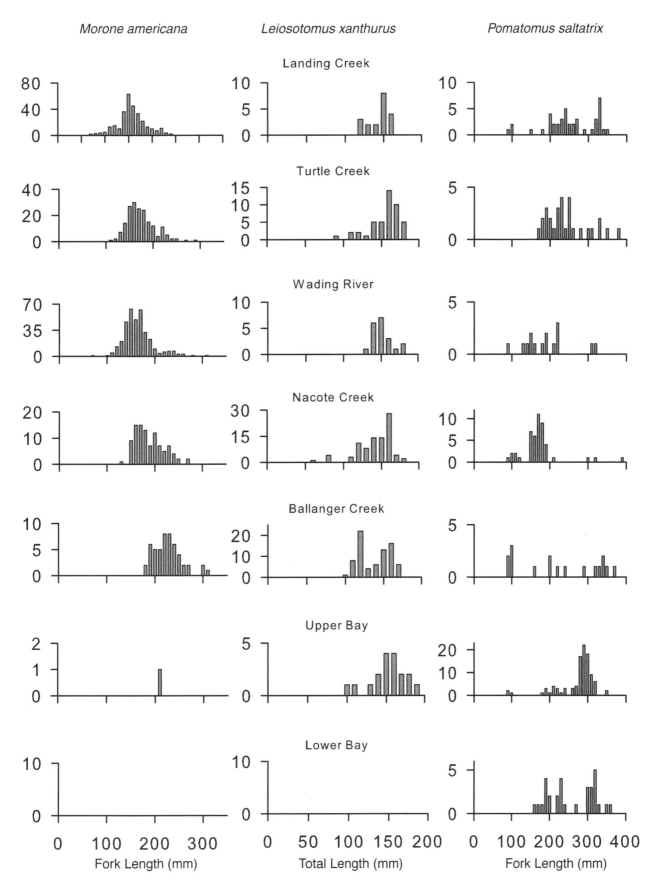

Fig. 7.5. Length-frequency distribution from gill net samples relative to location along the salinity gradient in Mullica River–Great Bay. See Fig. 3.3 for location of sampling areas.

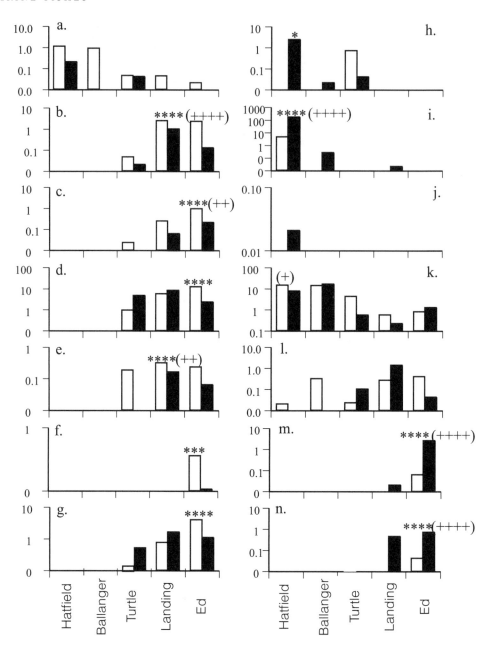

Fig. 7.6. Abundance of fish (CPUE, mean number per tow) by location and area in the Mullica River–Great Bay estuary for otter trawl sampling during April through November in 1998 and 1999. Tidal creek areas are represented by solid bars and adjacent channel areas by unshaded bars. *a. Anchoa mitchilli* 1+ *b. Ameiurus catus c. Ameiurus nebulosus d. Morone americana* 1+ *e. Trinectes maculatus* 1+ *f. Trinectes maculatus* young-of-the-year *g. Morone americana* young-of-the-year *h. Clupea harengus i. Menidia menidia* young-of-the-year *j. Menidia menidia* 1+ *k. Anchoa mitchilli* young-of-the-year *l. Clupeidae spp. m. Notemigonus crysoleucas n. Etheostoma olmstedi.* Significantly different among locations at *p < 0.05 **p < 0.01 ***p < 0.001 ****p < 0.0001. Significantly different among areas at +p < 0.05 ++p < 0.01 +++p < 0.001 ++++p < 0.0001. (Modified from Martino, 2001.)

et al., 1996b) as well as ontogenetic habitat shifts (Martino and Able, 2003; Sackett et al., 2008). In addition, others have attributed fish response to a variety of specific abiotic variables, including salinity (Günter, 1956; Weinstein et al., 1980; Peterson and Ross, 1991; Rakocinski et al., 1992; Szedlmayer and Able, 1996; Wagner and Austin, 1999; Able et al., 2001; Greenwood, 2007), temperature (Peterson and Ross, 1991; Rakocinski et al., 1992; Szedlmayer and Able, 1996), turbidity (Peterson and Ross, 1991), dissolved oxygen (Fraser, 1997; Sackett et al., 2008), freshwater inflow (Rogers et al., 1984; Fraser, 1997), structural attributes of the habitat (Weinstein et al., 1980; Thorman, 1986; Sogard and Able, 1991; Everett and Ruiz, 1993; Szedlmayer and Able, 1996; Wagner and Austin, 1999), depth (Rakocinski et al., 1992; Ruiz et al., 1993), stream order/geomorphology (Granados-Dieseldorff and

Baltz, 2008), and the interactions between abiotic and biotic factors (Rowe and Dunson, 1995). However, small-scale patterns appear to be the result of habitat associations that are most likely driven by food resources, competition, and predator avoidance. The biotic influences are evident for fishes in other estuaries as well (Ross and Epperly, 1985; Holbrook and Schmitt, 1989; Ogburn-Matthews and Allen, 1993; Lankford and Targett, 1994; Able et al., 2001), but this area needs further attention.

Additional attempts to resolve the factors responsible for estuarine habitat use compared the distributions of fishes in deeper channels and adjacent shallow tidal marsh tributary creeks along the gradient of salinity, temperature, pH, and other variables in the Mullica River–Great Bay estuary (Martino, 2001). Some species were collected exclusively in

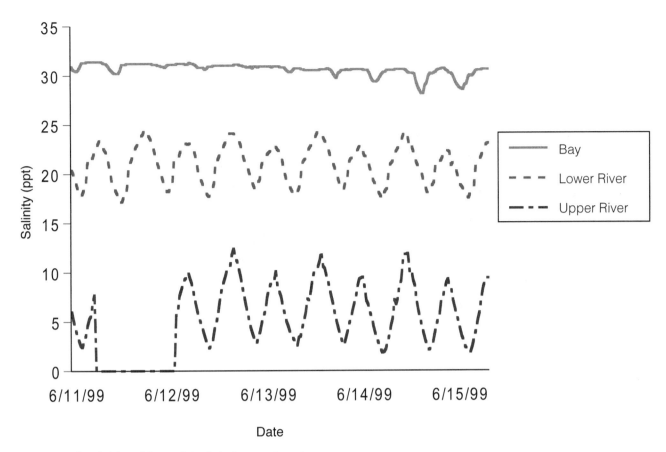

Fig. 7.7. Examples of tidal variability in salinity during June 11 to June 15, 1999, in the bay, lower river, and upper river of the Mullica River–Great Bay estuary. (Modified from Martino, 2001.)

marsh creeks (*Gobiosoma bosc, Apeltes quadracus,* and *Fundulus heteroclitus*) or in the channels (*Raja eglanteria, Hippocampus erectus,* and *Cynoscion regalis*). Also, the abundance of several dominant taxa, which occurred in both habitats, varied significantly between marsh creeks and channels, with *Menidia menidia, Etheostoma olmstedi,* and *Notemigonus crysoleucas* more abundant in the creeks and young-of-the-year *Anchoa mitchilli, Ameiurus catus, A. nebulosus,* and *Trinectes maculatus* more plentiful in the adjacent channel areas (Fig. 7.6). While salinity is an important mechanism influencing estuarine fish assemblages in general, it should be remembered that the variability in salinity differs across the estuary, from channels to tributary creeks, and may influence local distribution, especially in the upper estuary, where it can be more variable on scales of hours to days in this system (Fig. 7.7). As a further source of variation, many of the species in both channel and marsh creeks originated from diverse sources, both distant (Sargasso Sea, South Atlantic Bight, and Mid-Atlantic Bight) and local (estuarine and freshwaters) (Fig. 7.8).

ASSOCIATIONS WITH STRUCTURED HABITATS

In general, there do not appear to be many habitat-specific use patterns for estuarine fishes in the study area, but it is just as likely that they do exist and we simply lack the *in situ*

observations necessary to detect them in typically turbid estuaries. Possible habitat associations in temperate estuaries have often been studied in the Mullica River–Great Bay estuary. One of the most frequently investigated is the salt marsh habitat, including marsh surface (also marsh pools), intertidal creeks, and subtidal creeks. A particularly obvious example of marsh use is *F. luciae,* which spends its entire life in shallow depressions, puddles, and occasionally pools on the marsh surface, often in < 100 mm of water. A number of other residents (several fundulids, *Menidia beryllina,* and *Cyprinodon variegatus*) and transients (*Menidia menidia* and *Gasterosteus aculeatus*) use the intertidal marsh surface, including pools located there (Able et al., 1990; Able et al., 2005b; Adamowicz and Roman, 2005), for spawning (see species accounts). The marsh pools provide spawning habitat (e.g., *F. heteroclitus;* Hunter et al., 2007, 2009), but are also used as larval and juvenile habitat by *Gasterosteus aculeatus* (Able and Fahay, 1998). Pools are also overwinter habitat for *Fundulus heteroclitus* because temperatures are warmer there than in adjacent subtidal creek waters (Smith and Able, 1994). Use by some species varies between pools, with *Cyprinodon variegatus* and *Lucania parva* preferring vegetated (*Ruppia maritima*) to non-vegetated pools (Smith, 1995).

Of salt marsh species, *Fundulus heteroclitus heteroclitus*

Channel

Tidal Creek

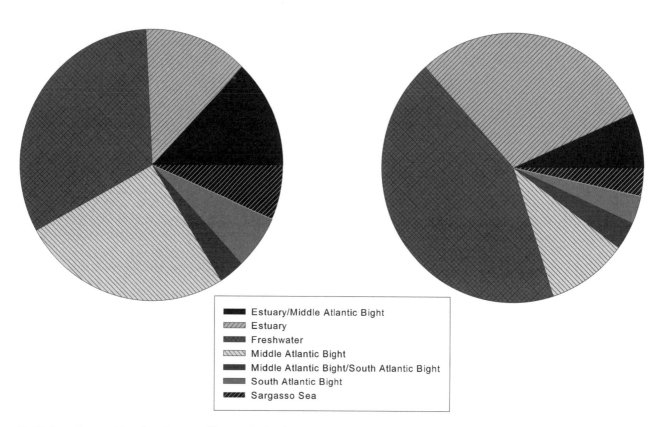

Legend:
- ■ Estuary/Middle Atlantic Bight
- ▨ Estuary
- ▨ Freshwater
- ▨ Middle Atlantic Bight
- ▨ Middle Atlantic Bight/South Atlantic Bight
- ▨ South Atlantic Bight
- ▨ Sargasso Sea

Fig. 7.8. Percent composition of species grouped by spawning location by habitat in the Mullica River–Great Bay estuary during April through November, 1998 and 1999. (Modified from Martino, 2001.)

shows the greatest degree of adaptation to this habitat, as it consistently deposits its eggs along intertidal and subtidal creeks, either at the base of blades of *Spartina alterniflora,* in broken, hollow stems of *Phragmites,* or in the valves of *Geukensia demissa*; this behavior may not occur in *F. h. macrolepidotus* (Able and Felley, 1986; Able and Hagan, 2003). Based on extensive tag/recapture studies (Able et al., in prep.), it has been determined that this species also uses other components of salt marshes, including subtidal and intertidal creeks and marsh pools. Spawning on the vegetated marsh surface is typical for *Menidia menidia*. The fauna of intertidal creeks are relatively depauperate in this and other temperate estuaries, but fishes can be quite abundant (Rountree and Able, 1993; Able et al., 1996b; Kimball and Able, 2007a, b), with species composition dependent on sampling approach (Kimball and Able, in review). For obvious reasons, members of the fauna are greatly influenced by tidal and diel variation (Kimball and Able, 2007a, b) as well. For subtidal creeks, at least in southern New Jersey, the fish fauna can be relatively rich (64 species). There is a pronounced seasonal variation in abundance, with peaks during spring and summer (Rountree and Able, 1992a).

When abundance and species composition were com-

pared between and within subtidal and intertidal creeks in the same marsh in Great Bay, New Jersey, there were pronounced differences, which included changes in assemblages as the result of diel fish movements between subtidal and intertidal creeks and the marsh surface (Rountree and Able, 1993). Subsequent, more detailed, analysis found strong differences between species relative abundances among and within subtidal and intertidal creeks. Among-creek variations appear to be more closely related to creek characteristics (length, depth, area, volume, percent tidal exchange in water volume) than to creek location (Rountree, 1992). Observed among-creek patterns in weir and seine catches suggest that creek size primarily influences catches through the volume of tidal exchange. As creek size decreases and percent tidal exchange increases, more species are concentrated near the creek mouth or are forced out of the creeks at low tide, resulting in both increasing catch diversity in the subtidal creek and increasing divergence of faunal assemblages between subtidal creeks and intertidal creeks. Within-creek patterns suggest that community structure differs between upper and lower regions, and between intertidal and subtidal habitats, in response to hypothetical gradients in physical conditions from the creek mouth to headwaters along the creek gradi-

In Situ Video Identifies Unique Tidal Behaviors

During the fall of 1998, we set up a video camera at the edge of high-salinity *Spartina alterniflora* marshes in order to complement our fish sampling techniques (e.g., weirs, seines, pit traps) (Rountree and Able 1992a, 1993). These observations took place in Schooner Creek, immediately adjacent to the Rutgers University Marine Field Station in Great Bay, and in a low-salinity marsh at Hog Islands, farther up the Mullica River. In order to overcome the short viewing distance in these typically turbid waters, we built an X-shaped flume of transparent plastic that concentrated fish moving from the creek, a low-tide refuge, onto the marsh surface on flood tides. These video observations in the flume allowed us to identify several species that we had not collected on the marsh surface with other gears. At the high-salinity marsh, these included *Menidia* (probably *M. menidia*), which occurred at the highest tides and was feeding in the water column but may also have been avoiding predators in the creek such as *Pomotomus saltatrix* (Rountree and Able, 1992) and *Paralichthys dentatus* (Szedlmayer et al., 1992; Rountree and Able, 1992b). Another species that is commonly collected and seen on the marsh surface, *Fundulus heteroclitus*, was observed feeding in algal mats, presumably on associated invertebrates, at the base of the *Spartina* plants through both early and late stages of the tide. Later in the flood tidal cycle, *Gobiosoma bosc*, a typical creek resident, was observed moving onto the marsh surface. Another species that we did not anticipate and had not previously collected was *Fundulus majalis*. In the low-salinity marsh we saw *Anguilla rostrata* and *Morone americana*. Thus, these types of observations broaden our understanding of the role of the marsh surface in the ecology of marsh fishes.

One wonders if the more frequent deployment of video cameras on the marsh surface and in marsh pools would not be relatively easy and productive. In these typically shallow waters, light penetration helps to overcome some of the difficulties associated with high turbidity and provides the visual enhancement that helps us to interpret observations from traditional sampling approaches.

ent (Rountree and Able, 2007). Similar findings are reported for Georgia (Hackney et al., 1976), North Carolina (Hettler, 1989a), and Virginia (Weinstein and Brooks, 1983; Smith et al., 1984; Rozas and Odum, 1987) marsh creeks. At least some young-of-the-year *Paralichthys dentatus* show continuous use of salt marsh creeks during the summer (Rountree and Able, 1992b; Szedlmayer and Able, 1993). Other fish use of salt marsh creeks or open water adjacent to salt marshes may be because of an association with "peat reefs" (Able et al., 1988), as apparently occurs for *Tautoga onitis* (K. W. Able, pers. observ.).

Coves may provide unique estuarine habitats as well. For example, coves near an inlet in Great Bay–Little Egg Harbor annually have very high abundances of recently settled *Pseudopleuronectes americanus* (Witting, 1995; Curran and Able, 2002; see species account), although the importance of this habitat type has not been verified for other estuaries where deep water settlement areas occur (Manderson et al., 1997; Stoner et al., 2001). Perhaps both areas share retentive water masses that serve to focus settlement.

Other species are consistently found in eelgrass (*Zostera marina*) (Hunter-Thomson, 2002; Lazzari et al., 2003), including *Apeltes quadracus*, which builds its nests there (Reisman, 1963), and *Syngnathus fuscus, Hippocampus erectus, Opsanus tau,* and *Tautoga onitis*, which use these habitats during the summer (Szedlmayer and Able, 1996; Able and Fahay, 1998). This preference for eelgrass is also demonstrated by *Bairdiella chrysoura*, but in years when it is abundant it can also be found in many other habitats. In addition, several caging experiments have been designed to determine habitat quality as measured by its effect on growth of young-of-the-year of

benthic fishes. Studies comparing vegetated to unvegetated substrates indicated that the growth response is species specific, with *Tautoga onitis* and *Pseudopleuronectes americanus* clearly benefiting from vegetated sites, although this varies somewhat with location (Sogard, 1992). Another species in this study (*Gobiosoma bosc*) was most abundant in *Zostera marina* beds but did not grow well there, indicating a possible tradeoff between slower growth and enhanced protection from predation. When growth rates of caged young-of-the-year *Pseudopleuronectes americanus* and *Tautoga onitis* were compared across vegetated (eelgrass and sea lettuce [*Ulva lactuca*]) habitats, growth was often higher but not consistently so across different estuaries (Phelan et al., 2000). At the other extreme, growth for *Pseudopleuronectes americanus* in sea lettuce was low when dissolved oxygen was also low. Despite these differences, sea lettuce often supports high densities of fishes, especially in areas lacking eelgrass (Sogard and Able, 1991).

A few species are consistently collected over sand substrates in the lower portion of the estuary, including *Ammodytes americanus* (Haroski, 1998; Able et al., 2002) and *Scophthalmus aquosus* (Neuman and Able, 1998). In another study, *Pseudopleuronectes americanus* was often most abundant in unvegetated areas adjacent to eelgrass (Goldberg et al., 2002). Other species have specific habitat preferences, such as *Gobiosoma bosc* in association with oyster reefs (Breitburg, 1991) and *Centropristis striata* in these and other hard substrates (Able et al., 1995a). Others have found that the percent silt across a variety of estuarine habitats is a good predictor of fish assemblage structure (Szedlmayer and Able, 1996). Because estuaries consist of multiple habitats, it is not unexpected that multiple habitats influence use by mobile fishes. As an example from a temperate estuary (Great South Bay, New York), abundance of some species was significantly higher in eelgrass beds adjacent to salt marshes; others were more plentiful in eelgrass near beaches. In some instances, distance from shore was important (Raposa and Oviatt, 2000).

Oyster reefs are common to many temperate estuaries and can be important for many fishes (Wells, 1961; Breitburg et al., 1995; Breitburg, 1999; Coen et al., 1999; Harding and Mann, 1999). It has been suggested that three types of fishes—reef residents, facultative residents, and reef transients (Breitburg, 1999)—are associated with subtidal reefs in Chesapeake Bay. However, our understanding of the value of these habitats is often incomplete, in part because of the difficulty in sampling these complex habitats (Wenner et al., 1996) and the overall decline of oyster reefs in general (Breitburg et al., 2000).

Telemetry studies have provided unique insights into habitat use by larger estuarine fishes. We documented that the large juveniles and adults of *Paralichthys dentatus* have annual fidelity to the Mullica River–Great Bay estuary, and in some instances revisits occurred within 500 m of the habitat occupied in the previous year (Sackett et al., 2007). In other telemetry studies of *Morone saxatilis* in the same estuary, we determined that large juvenile and adults consistently used the deeper water of the estuary, whether near the inlet to the ocean or farther up in the river. In these deep areas they were typically found along marsh edges with abrupt dropoffs or in the middle of channels with variable bottom topography (Able and Grothues, 2007b; Ng et al., 2007). Some of these individual locations were occupied by individual fish from weeks to months and in subsequent years (Ng et al., 2007).

INTERACTING WITH THE SUBSTRATE

The available literature on habitat use for fishes, including those in estuaries, often focuses on structured habitats that occur above the substrate, for example. eelgrass beds, sea lettuce, and oyster beds (see above). Often the role of substrate use and the particular mode of use are underappreciated. We suggest that this interaction is often more frequent, of longer duration, and more direct than reported. The manner of substrate use also varies and warrants clear distinction (see Atkinson and Taylor, 1991; Gibson and Robb, 2000). Thus, there are species that burrow, that is, create temporally stable structures (burrows), that they may use over and over again. Many other fishes bury by periodically covering themselves with the surrounding substrate.

The documented incidences of temperate estuarine fishes interacting with the substrate in the study area are surprisingly common, but details of the structural nature of these interactions and their functional significance are poorly understood. Of the taxa treated in this book, 28 species in 19 families are known to either burrow or bury based on the literature or our own observations (Table 7.2). Of these, both

types of substrate use occur in muddy and sandy substrates. Burying, both shallow and deep, is more frequent and typically occurs in juveniles and adults, but there is relatively little information available on the ontogeny of burial or burrowing (but see Tanda, 1990; Stoner and Ottmar, 2003; Morioka, 2005). An exception, for temperate estuaries, is that for burial by *Paralichthys dentatus* (Keefe and Able, 1993, 1994). As this species is entering estuaries, it is undergoing transformation (eye migration stages G through H+; 10.0–15.6 mm SL) and is capable of partial burial (sediment covering portions of the body but never the head). At later stages (stage H+), shallow burial is more complete (the entire body covered, except for eyes and gill openings). Burial is not related to scale formation, a potential means of protecting the body from damage during burial, because burial occurs at sizes before scale formation begins (see species account). Burial by these small individuals may be related to predator (estuarine decapod crustaceans) avoidance (Witting and Able, 1993; Keefe and Able, 1994; Barbeau, 2000). Burying for *P. dentatus* undergoing transformation has a distinct diel periodicity, with most of it occurring during the day (Keefe and Able, 1994); the same applies to juveniles (Klein-MacPhee, 1979). Burial by *Paralichthys dentatus* continues in juveniles and adults (Olla et al., 1972). At these larger sizes, burial may be used as a means of ambushing prey (Olla et al., 1972). Other species that bury in the substrate include the cyprinodontid (*Cyprinodon variegatus*) and all of the fundulids examined (*Fundulus heteroclitus, F. luciae, F. majalis,* and *Lucania parva*) (see species accounts). For all these species, burial occurred as the fish attempted to "swim" into the substrate with exaggerated movements of the posterior portion of the body. For some, such as *Fundulus diaphanus*, burial behavior may be quite flexible, based on prior exposure to different substrate types (Colgan and Costeloe, 1980).

Both burying and burrowing offer presumed advantages and involve potential costs. All types of burial that make fishes more cryptic or less visible presumably provide reduced accessibility to visual predators for species from a variety of families, including paralichthyids (Olla et al., 1972; Tanda, 1990; Keefe and Able, 1994), labrids (Nemtzov, 1994), soleids (Ellis et al., 1997), a number of other flatfishes (Stoner and Ottmar, 2003), a cyprinodontid (Colgan and Costeloe, 1980), and ammodytids (Meyer et al., 1979). Thus, it is not surprising that many of these same species demonstrate a preference for different types of sediment (Meyer et al., 1979; Tanda, 1990; Burke et al., 1991; Gibson and Robb, 1992; Moles and Norcross, 1995; Wright et al., 2000; Neuman and Able, 2003; Stoner and Ottmar, 2003). Burial may also be a means to deal with stressful temperatures (Olla et al., 1972). When burial is long term, that is, seasonal, as for *Ammodytes* spp., including *A. americanus* (Able et al., 2002), these antipredator advantages may extend over long periods when the animals are inactive due to low winter temperatures (Winslade, 1974a, b, c; Freeman et al., 2004). This advantage may

Table 7.2. Occurrence and type of substrate use by temperate estuarine fishes.

Family/Species	Type of Substrate Use	Type of Substrate	Life History Stage	Source
Raja eglanteria	Bury	Sand	Adults	Pers. observ.
Anguilla rostrata	Burrow	Silt	Juveniles	Able and Fahay (1998), pers. observ.
Myrophis punctatus	Burrow?	Silt	Juveniles	Able et al. in prep.
Conger oceanicus	Hide	Under peat reefs	Juveniles/Adults	Pers. observ.
Synodus foetens	Bury	Sand	Juveniles	Breder (1962), Hoese (1965)
Urophycis regia	Bury	Sand	Juveniles/Adults	Barans (1969), pers. observ.
Urophycis tenuis	Bury	Sand	Juveniles	McAllister (1960)
Ophidion marginatum	Burrow	Sand	Juveniles/Adults	Bowers-Altman (1993), this study
Cyprinodon variegatus	Bury	Sand/silt	Juveniles/Adults	This study
Fundulus diaphanus	Bury	Sand	Juveniles	Colgan and Costeloe (1980)
Fundulus heteroclitus	Bury	Silt	Juveniles/Adults	Chittendon (1980), this study
Fundulus luciae	Bury	Silt	Juveniles	This study
Fundulus majalis	Bury	Sand/silt	Juveniles	This study
Lucania parva	Bury	Silt	Juveniles	This study
Syngnathus fuscus	Bury	Sand	Adults	Wicklund et al. (1968)
Prionotus carolinus	Bury	Sand	Adults	Pers. observ.
Prionotus evolans	Bury	Sand	Adults	Pers. observ.
Morone saxatilis	Bury	Sand	Juveniles	J. Manderson, pers. commun.
Tautoga onitis	Hide	Under rocks	Juveniles	Olla et al. (1979)
Tautogolabrus adspersus	Hide	Under rocks	Juveniles	Curran (1992)
Astroscopus guttatus	Bury	Sand	Juveniles	R. McBride, pers. observ.
Gobiosoma bosc	Hide	Oyster shells	Juveniles/Adults	Pers. observ.
Ammodytes americanus	Bury	Sand	Juveniles/Adults	Able et al. (2002)
Scophthalmus aquosus	Bury	Sand	Juveniles	Neuman and Able (1998)
Etropus microstomus	Bury	Sand	Juveniles	Pers. observ.
Paralichthys dentatus	Bury	Sand	Juveniles/adults	Olla et al. (1979), R. Rountree, pers. observ.
Pseudopleuronectes americanus	Bury	Silt	Juveniles/Adults	C. Curran, pers. observ., Fletcher (1977)
Sphoeroides maculatus	Bury	Sand	Juveniles	Pers. observ.

co-occur with physiological costs associated with burial, including reduced access to respiratory currents and appropriate levels of dissolved oxygen (Pelster et al., 1988; Behrens et al., 2007). Burrowing may offer similar advantages and disadvantages. Avoidance of visual predators may be the primary reason for burrowing, because many of those species that burrow do so during the day and are active at night (see species account of *Ophidion marginatum*). The respiratory constraints and adaptations have been addressed for a few burrowing species (Pelster et al., 1988; Behrens et al., 2007) but not in sufficient detail to draw conclusions.

THE NON-NURSERY ROLES OF ESTUARIES

Much of the emphasis on fishes and their habitats has focused on young-of-the-year (Able and Fahay, 1998), as we have in the preceding section, because of the assumed importance of estuaries as "nurseries." As a result, we know less about larger fishes that use estuaries. This is compounded by the lack of effective means of sampling these frequently active fishes. In an attempt to resolve some of these shortcomings, we have recently emphasized sampling with gill nets in two temperate estuaries. In Delaware Bay nearshore waters and marsh creeks the larger pelagic and benthic fishes (mean 261.4 mm TL, with some > 600 mm TL) were represented by 22 species in 15 families, with sciaenids (*Cynoscion*

regalis, Leiostomus xanthurus, Micropogonias undulatus, and *Pogonias cromis*), clupeids (*Alosa aestivalis, A. mediocris, A. pseudoharengus,* and *Brevoortia tyrannus*), and ictalurids (*Ameiurus catus, A. nebulosus,* and *Ictalurus punctatus*) the most speciose (Able et al., 2009b). The most abundant species was *Morone americana* (36% of total) but an equally abundant species was *Pomotomus saltatrix*. Together these constituted 85.6% of the total catch. While young-of-the-year fishes were represented in these samples, larger, older fishes made up a large percentage of the catch.

Habitat use based on the same gill net study varied between species (see species accounts); in general, the patterns of assemblage structure of these fishes were influenced by spatial gradients in salinity and dissolved oxygen and temporal changes in temperature during the June–November sampling period. A similar effort, using two gill nets (one the same type as in the above study), was based on an extensive effort in the Mullica River–Great Bay estuary (see Table 7.1; Fig. 7.3). The average size of the fishes captured was similar (mean 181 mm, range 64–800 mm) to that in the Delaware Bay study and larger than most fishes sampled there with other techniques (see Fig. 7.4) and in other estuaries. In addition, the same families of fishes, including adults of sciaenids, clupeids, and ictalurids as well as other abundant species, dominated the catches (see Table 7.1).

Another instance where we know about the occurrence and abundance of adult fishes is the anadromous species that swim through estuaries on the way to freshwater during the spawning season. These include *Morone saxatilis* and some species of *Alosa* (*A. sapidissima, A. aestivalis,* and *A. pseudoharengus*).

Other evidence is accumulating that large predatory fishes are more common in temperate estuaries than previously accepted (see Chapter 8). Patterns of habitat use by these predators are being elucidated with a variety of techniques, including telemetry. As a result, we know that some predators are only seasonal visitors to the estuary. This is evident for large juvenile and adult *Morone saxatilis* that make seasonal coastal migrations from wintering areas off Virginia and North Carolina to the north in spring and back south again in the fall (Boreman and Lewis, 1987). In the process, ultrasonically tagged fishes enter estuaries, where they stay from days to months. Some of these fishes enter estuaries and move upstream to low-salinity water or freshwater for spawning, especially in larger estuaries and portions of Chesapeake Bay (Secor et al., 2001; Able and Grothues, 2007b; Grothues et al., 2009). Others, especially in smaller estuaries, enter these systems in the spring and fall during the coastal migrations, while still others may become resident during much of the summer (Ng et al., 2007). This was especially evident in the Mullica River–Great Bay estuary, where fish originally tagged in this system often revisited it for variable durations (from days to weeks). In some instances these included rapid movements upstream to potential spawning areas, but there is little evidence of successful spawning and survival of eggs and larvae due to the paucity of young-of-the-year in the system over many years of sampling (Able and Grothues, 2007b). Based on both active and passive tracking techniques, it was determined that other individuals remained resident in the estuary (Able and Grothues, 2007b; Ng et al., 2007; Grothues et al., 2009). Gill net catches and ultrasonic telemetry demonstrate that another presumed coastal migrant, *Pomatomus saltatrix*, has a seasonal periodicity in estuaries (Grothues and Able, 2007; Able et al., 2009b). Large juveniles and adults of this species enter estuaries with rising temperatures in the spring, feed through the summer, and depart as temperatures cool in the fall.

Another species with a different life history, *Paralichthys dentatus*, migrates inshore in the spring, is typically abundant in estuaries from spring through fall, and then migrates offshore to spawn over the continental shelf (Able et al., 1990; Able and Kaiser, 1994; Packer and Hoff, 1999; Sackett et al., 2007). The timing of estuarine ingress and egress by large juveniles and adults is variable and dependent on ambient conditions (Szedlmayer and Able, 1993; Sackett et al., 2007); this is true for the piscivorous young-of-the-year as well (Rountree and Able, 1992a, b). This is especially evident for egress in the Mullica River–Great Bay estuary, where it can be influenced by storms and delayed until December in some warmer years (Sackett et al., 2007).

SITE FIDELITY/RESIDENCY

The importance of estuaries to some of these species is supported by the high degree of inter-annual site fidelity. For large juvenile and adult *Morone saxatilis* there are ample data to indicate that acoustically tagged individuals returned to the Mullica River–Great Bay estuary on numerous occasions (Able and Grothues, 2007b; Grothues et al., 2009) with evidence for movement between New Jersey and Maine estuaries (Grothues et al., 2009). Further, some individuals, that had left and returned to the estuary in subsequent years, came back to the same location/habitat they had occupied previously (Ng et al., 2007). The same estuarine-specific site fidelity has also been reported in the same estuary at very high return rates (39%) for *Paralichthys dentatus* (Sackett et al., 2007), a pattern evident for this species for other temperate estuaries (Hamer and Lux, 1962; Poole, 1962; Jesien et al., 1992).

The young-of-the-year of several temperate estuarine species tagged with coded wire tags have demonstrated a high degree of local site fidelity, on a scale of meters, based on tag/recapture studies. This has been shown for *Centropristis striata* (Able and Hales, 1997), *Tautoga onitis* and *Tautogolabrus adspersus* (Able et al., 2005c), *Chaetodon ocellatus* (McBride and Able, 1998), and *Pseudopleuronectes americanus* (Saucerman and Deegan, 1991). Because of the mobility of many young-of-the-year fishes, it is often difficult to determine residence time in discrete habitats (Herke, 1977; Sogard, 1989; Sogard and Able, 1994). One example where this was possible is based on studies of *Leiostomus xanthurus* in Chesapeake Bay salt marsh creeks. In this instance, the marked population was resident for up to 181 days with an average duration of 91 days (Weinstein et al., 1984). Additional studies in the Mullica River–Great Bay estuary and other temperate estuaries indicate a variable period of residency. A number of tag/recapture or ultrasonic telemetry studies in the Mullica River–Great Bay estuary have demonstrated that the residency time in any estuarine habitat varies between species (Fig. 7.9). For most species, maximum residency was < 130 days and quite frequently much shorter. Of course, for species that are primarily in estuaries only for the summer, as for young-of-the-year, the duration could be a function of when in the summer they were tagged. The short residency of *Pomatomus saltatrix* is not surprising for this fast-swimming pelagic fish. The short duration of residency by estuarine *Menticirrus saxatilis* and *Paralichthys dentatus* probably reflects the late summer tagging and departure from the estuary with cooling temperatures in the fall. At the other extreme, the long period of maximum residency by *Fundulus heteroclitus* demonstrates that this species consistently shows high site fidelity to individual salt marshes in this (Able et al., 2006b;

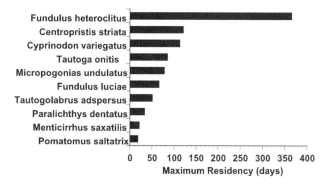

Fig. 7.9. Duration of residency for a variety of estuarine fishes based on tag/recapture studies.

Able et al., in prep.) and other (Meredith and Lotrich, 1979; Teo and Able, 2003) estuaries.

DYNAMIC HABITAT USE

Assessing dynamic habitat use for estuarine fishes is often difficult because of the variation that occurs at several temporal scales (seasonal, ontogenetic, tidal, diel) and because of ineffective sampling at these multiple scales (Able, 1999; Hagan and Able, 2008; Rotherham et al., 2008). We examined these patterns for temperate estuarine fishes based on the available literature and our own research. Some of the clearest examples of dynamic habitat use during ontogeny are associated with settlement. Settlement patterns for benthic fishes may be largely influenced by larval supply, but can be modified by subsequent habitat selection (see Chapter 6). Some species settle and stay (Bell et al., 1985, 1987), while others settle but do not stay because the use of settlement habitats is only temporary (Sogard, 1989), as can occur for *Pseudopleuronectes americanus* (Curran and Able, 2002). Other examples of ontogenetic changes in habitat have been reviewed for flatfishes from a variety of families (Able et al., 2005a). Other species settle in very different habitats from the estuary to the ocean (Able et al., 2006a).

Tidal movements are among the most common type of estuarine movements (Rountree and Able, 2007). They have been observed in fishes in salt marshes in Massachusetts (Werme, 1981), New Jersey (Rountree and Able, 1992a), Virginia (McIvor and Odum, 1988), and North Carolina (Hettler, 1989a). Extensive movements, over hundreds of meters in marsh creeks, have been recorded in a single tidal stage for acoustically tagged young-of-the-year *Paralichthys dentatus* (Szedlmayer and Able, 1993). These movements could be associated with the movements of prey fish species that have pronounced cycles with tides and diel changes in temperate estuaries. This movement of potential prey occurs, for example, in subtidal (Rountree and Able, 1992a) and intertidal (Kimball and Able, 2007a, b, and references cited therein) creeks and other habitats (Sogard and Able, 1994).

A summary that includes diel and tidal patterns for sev-

eral families of flatfishes, some of which can occur in estuaries, points out the importance of these variables (Able et al., 2005a). The demonstration of diel differences in behavior often relies on different catch rates in a variety of habitats. as for subtidal creeks, seagrass beds, open sand, and macroalgae-dominated areas (Stoner, 1991; Rountree and Able, 1993; Sogard and Able, 1994; Gibson et al., 1996; Gray et al., 1998). Other studies have focused on diel variation of a single species of estuarine pelagic fish (*Anchoa mitchilli*: Reis and Dean, 1981; Castillo-Rivera et al., 1994; *Menidia menidia*: Middaugh and Takita, 1983; Conover and Kynard, 1984; *Pomatomus saltatrix*: Buckel and Conover, 1997) and note that it usually results in movements between habitats. A detailed examination of diel changes in habitat use for pelagic fishes in the Great Bay estuary demonstrated strong differences in species composition, such that some species only occurred during the day or at night (Hagan and Able, 2008). As a result, the abundance of some species would have been drastically underestimated by sampling only during the day. In addition, diel differences change with season, reflecting the mobility of these fishes and how estuarine fish habitat preferences change from day to night. Seasonal cycles in habitat use are most pronounced during estuarine ingress and egress. These patterns are dealt with in the chapters on larval supply (Chapter 6) and migration (Chapter 9).

In addition, habitat use by estuarine fishes can be very dynamic in response to episodic events (Childs et al., 2008). For example, during the summer of 1976 low dissolved oxygen conditions on the adjacent continental shelf forced *Paralichthys dentatus* into adjacent estuaries of the New York Bight (Freeman and Turner, 1977; Swanson and Sindermann, 1979). An alternative movement of summer flounder occurred out of an estuary during a period of low barometric pressure during a storm event (Sackett et al., 2007). However, upwelling of ocean bottom waters into an estuary did not produce a major change in larval ingress (see Chapter 6).

ANTHROPOGENIC INFLUENCES

Many of the temperate estuaries that are situated in the densely human-populated portions of the northeastern United States have relatively diverse fish faunas despite being profoundly impacted. This occurs in the Bronx River and Hudson River in New York (Able and Duffy-Anderson, 2006; Rachlin et al., 2007), Hackensack Meadowlands in New Jersey (Neuman et al., 2004), and the Mystic River in Massachusetts (Haedrich and Haedrich, 1974). However, anthropogenic impacts on estuarine fishes are common. We comment on the detection and expression of these impacts as the result of our own studies and those of others. Our emphasis is on sub-lethal effects because the topic of fish kills in temperate estuaries has been frequently covered (Biernacki, 1979; Lowe et al., 1991; Burkholder et al., 1995), while that is not the case for more sub-lethal impacts (Rose, 2000; O'Connell et al.,

2004). This brief review covers several sub-lethal metrics of anthropogenic effects and focuses particularly on some of the most common impacts in estuaries, including those at the landscape level (reduction in freshwater flow, hypoxia, urbanization) and at the level of individual habitats. However, we must recognize that, in general, degradation of estuaries and the adjacent coastal ocean (Edgar et al., 2000; Levin and Stunz, 2005; Kleppel et al., 2006; Lotze et al., 2006; Saenger et al., 2008; Courrat et al., 2009), as well as fishing pressure (Jackson et al., 2001), may simultaneously influence these same fish populations and thus their capacity to use some estuarine habitats, and our ability to detect anthropogenic effects.

It has long been understood that freshwater inflow into estuaries is central to estuarine function (Livingston et al., 1997; Grange et al., 2000; Niklitschek and Secor, 2005; Ritter et al., 2008), and that dams and other types of freshwater withdrawals can have profound effects on the physical and chemical aspects of watersheds (Pringle, 2000), but their ecological effects on watersheds are less well understood (Tolan, 2008). An examination of the literature supports the likelihood that freshwater removal can change trophic structure, including that for fish assemblages and their distribution (Ter Morshuizen et al., 1996; Livingston et al., 1997; Tsou and Matheson, 2002; Rubec et al., 2006; Sheaves and Johnston, 2008), migrations (Yako et al., 2002), and year-class strength (Strydom et al., 2002; Staunton-Smith et al., 2004; Halliday et al., 2008). While these examples range from all over the world, there are few clear ones from temperate estuaries perhaps because fish assemblages have been altered for such a long time that their current states are often considered the norm. One of the common causes of these alterations are dams in the freshwater portions of estuaries. Dams clearly affect estuarine habitat use by limiting access to spawning areas for anadromous species (Beasley and Hightower, 2000).

At a broader level, the urbanization of entire watersheds is commonplace in the northeastern United States, but again, the ecological impacts are poorly known. Exceptions include those for detecting differences in anadromous fish spawning in the Hudson River estuary (Limburg and Schmidt, 1990) and various types of altered shorelines in the James River estuary (Bilkovic and Roggero, 2008) and in New England (Bertness et al., 2009). In the first two instances there was a negative response by the fish fauna to urbanization.

Another common manifestation of urbanization and eutrophication of many estuaries, including temperate estuaries, is the development of hypoxia (Wannamaker and Rice, 2000; Buzzelli et al., 2002; Gray et al., 2002; Baird et al., 2004; Long and Seitz, 2009). While the phenomenon has likely occurred over thousands of years (Cooper and Brush, 1991), the frequency, duration, and spatial extent are increasing in recent decades (see above references), with numerous and sometimes complex impacts on estuarine fishes. In some estuaries, hypoxia has resulted in changes in the distribution and habitat shifts for larvae and juveniles (Taylor and Rand, 2003; Tyler and Targett, 2007), as noted specifically for fishes associated with oyster reefs (Lenihan and Peterson, 1998; Lenihan et al., 2001); hypoxia also adversely affects food habits (Pihl et al., 1992) and growth (McNatt and Rice, 2004; Stierhoff et al., 2006). However, in some instances hypoxia may facilitate feeding (Diaz et al., 1992; Long and Seitz, 2008). In other estuaries hypoxia may enhance predation on fish larvae (Breitburg, 1992) and increase direct mortality as well (Secor and Gunderson, 1998; Shimps et al., 2005), although the response may vary with the type of predator (Brietburg et al., 1994). At the opposite extreme, hypoxia/anoxia on the continental shelf has forced fishes into adjacent, better-oxygenated estuaries, thus increasing their abundance there and influencing catch rates for species such as *Paralichthys dentatus* (Freeman and Turner, 1977; Swanson and Sindermann, 1979). Together, these examples demonstrate the often negative and typically complex response of temperate estuarine fishes to hypoxia.

In addition to these types of impacts, there are frequently habitat-specific responses by fishes to degradation in temperate estuaries, including the effect of ditching of salt marshes for mosquito control (Adamowicz and Roman, 2005; Crain et al., 2009) and harvesting of oyster reefs (Lenihan and Peterson, 1998; Lenihan et al., 2001) and their associated fishes. In a clear example of anthropogenic effects, a series of habitat-specific growth experiments in the heavily impacted Hudson River estuary demonstrated the negative effects of shading by large piers on young-of-the-year *Pseudopleuronectes americanus* and *Tautoga onitis* abundance and growth, across open water, pier edge, and under-pier areas (Able et al., 1998, 1999; Duffy-Anderson and Able, 1999, 2001; Metzger et al., 2001; Duffy-Anderson et al., 2003; Able and Duffy-Anderson, 2006). In a further attempt to compare the values for juvenile fish growth for the heavily impacted Hudson River estuary to the relatively unaltered Great Bay estuary in southern New Jersey, it was demonstrated that the former had slower average growth rates (Fig. 7.10). These experiments with young-of-the-year of the same species in open water habitats in each estuary assumed that growth was a good indicator of habitat quality. In each instance, during 1994 both species grew significantly faster in Great Bay. This strongly suggests that the cumulative effects of anthropogenic influences on the Hudson River diminish growth rates. As a result, these smaller fish might be exposed to predation for a longer period of time and thus have reduced survival rates. Another possibility is that the slower growth in the Hudson River estuary might cause individuals to enter the winter at smaller sizes that would be susceptible to size-selective winter mortality (see Chapter 6).

Estuarine shorelines, in large part due to the extensive shallow waters, provide important habitat for many fishes and their prey (Boesch and Turner, 1984; Ruiz et al., 1993; Rypel et al., 2007). Human modification of these shore-

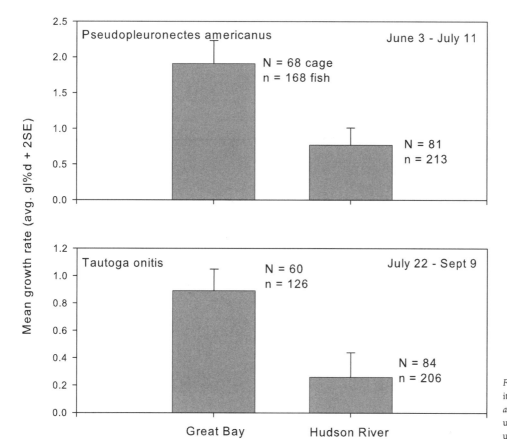

Fig. 7.10. Comparison of habitat quality (growth of *Pseudopleuronectes americanus* and *Tautoga onitis*) between urban (Hudson River) and relatively unaltered (Great Bay) estuaries.

lines alters physical properties at this land-water interface, introduces contaminants, and may impact the structure and function of this habitat (Anthony, 1994; Dugan et al., 2008; Beck et al., 2009; Bertness et al., 2009). These modifications often take two forms (Cox et al., 1994): (1) barriers between water and land (e.g., seawalls, groins, rip-rap, etc.); and (2) overwater structures (e.g., piers). For the former, the assumption is that one advantage of shallow water in estuaries is that it provides a refuge from predation (Blaber and Blaber, 1980; Boesch and Turner, 1984; Paterson and Whitfield, 2000); however, tethering studies in a tropical estuary did not support this assumption (Baker and Sheaves, 2007). For pier habitats, a series of studies in the Hudson River estuary (see Able and Duffy-Anderson, 2006, for review) suggests that fish species diversity and species abundance were depressed under piers relative to nearby habitats. The only species that were routinely collected from under piers were those that do not appear to rely solely on the use of vision to forage (*Anguilla rostrata*, *Gobiosoma bosc*, and *Microgadus tomcod*). Results from studies of the distribution of benthic invertebrate prey for fishes around piers suggest that prey abundances under piers are more than sufficient to support fish growth (Able and Duffy-Anderson, 2006). However, results of directed growth studies indicate that feeding and growth rates of visually feeding fish species (*Pseudopleuronectes americanus* and *Tautoga onitis*) are negative under piers (i.e., fish lose weight). It is not likely that factors associated with pier

pilings, such as reduced flow or sedimentation, affect feeding, since studies of fish growth in pile fields (piers without the decking) indicate that fish grow well in that habitat. Rather, it appears that the decking associated with piers creates conditions of intense shading that impede foraging activities, at least for the species examined to date. We propose that large under-pier areas, and potentially any areas that significantly reduce light penetration to depth in nearshore areas, are poor habitats for fishes.

While plant invasions of estuaries are common, few are known to affect fish populations in temperate estuaries. An exception is the invasion of *Phragmites australis* into salt marshes across the northeastern United States in the past 100 years (Chambers et al., 1999). This introduced European lineage has aggressively displaced the native marsh grasses (Saltonstall, 2002; Myerson et al., 2009). As a result, the marsh surface, particularly in brackish marshes, has been altered such that there is a loss of surface standing water due to increased marsh elevation and reduced drainage area (Windham and Lathrop, 1999; Lathrop et al., 2003; Osgood et al., 2003). These alterations become more pronounced as the invasion progresses (Able et al., 2003; Lathrop et al., 2003) (Fig. 7.11).

The fish response, particularly for marsh surface resident species such as *Fundulus heteroclitus* (and *F. luciae*), is to reduce survival of larvae and juveniles and relative abundance of adults in *Phragmites*-invaded marshes relative to un-

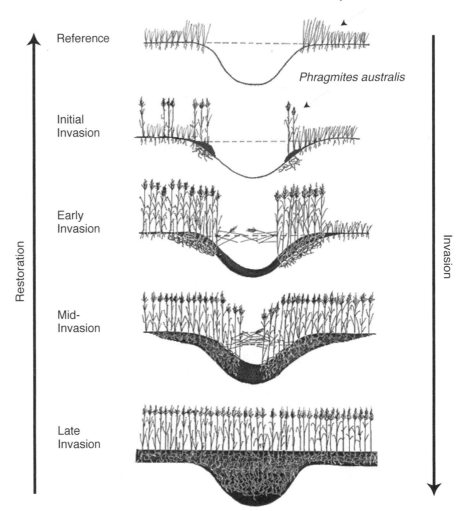

Fig. 7.11. Diagram of changes in marsh surface topography and elevation over the period of the invasion by *Phragmites australis* and subsequent restoration to *Spartina* (from Able et al., 2003a).

altered *Spartina* marshes (Table 7.3; Fig. 7.12). This response is evident in marshes in New Jersey, Delaware, and Maryland (Hunter et al., 2006). This response among marsh types is also evident in intertidal creek fish assemblages (Kimball and Able, 2007a, b). More extensive studies in Delaware Bay marshes have shown that the *Phragmites* invasion has dramatically reduced fish production by three orders of magnitude lower than that for unaltered *Spartina* marshes (Hagan et al., 2007).

Dredging is one of the most common forms of human influence in estuaries (Kennish, 1992, 1998; Elliott and Hemingway, 2002; Wilber and Clarke, 2002), yet we still know relatively little about its impact on fishes (Wilber and Clarke, 2002). A recent study suggested that there was relatively little impact of the dredging of a boat basin on *Fundulus heteroclitus* and associated species (Able et al., in review).

HABITAT RESTORATION

Restoration is becoming one of the most important components of environmental sciences and ecosystem management and will continue to influence these sciences for the foresee-

able future (National Research Council, 1992; Young, 2000; Hobbs and Harris, 2001; Thayer et al., 2005). Accordingly, there is growing and dynamic literature on restoration that has recently focused on issues such as the conceptual basis for restoration ecology, with considerable emphasis on salt marsh systems (Zedler, 2000; Boumans et al., 2002; Neckles et al., 2002; Warren et al., 2002; Callaway, 2005; Able et al., 2008). Further, there is an increasing realization of the need to understand both the structure and function of restored habits (Zedler, 1995; Mitsch and Wilson, 1996; Zedler and Callaway, 1999; van Diggelen et al., 2001; Rozas et al., 2005). Just as human-induced anthropogenic changes in temperate estuarine habitats have caused declines or alterations in some habitat types and quality, the restoration of selected estuaries and habitats has had various effects on the fish fauna.

In some instances, this has occurred as the result of broad U.S. programs, such as the Clean Water Act (National Research Council, 2001). For example, the Hudson River estuary is recovering from earlier human impacts (Brosnan and O'Shea, 1996; Levinton and Waldman, 2006), resulting in restoration of fish diversity and abundance (Daniels et al., 2005).

Table 7.3. Structural and functional responses of fishes to restoration of *Phragmites*-invaded marshes.

	Before Restoration	After Restoration	Sources
Structural Attributes			
Fish abundance			
Marsh surface	−	+ (−)	Able and Hagan (2000); Able and Hagan (2003); Able et al. (2003a); Raichel et al. (2003); Hagan et al. (2007)
Intertidal creeks	−+	++	Hagan et al. (2007); Kimball and Able (2007a, b)
Subtidal creeks	+	+	Grothues and Able (2003a, b); Hagan et al. (2007); Jones and Able (in review)
Fish diversity			
Marsh surface	−	+	Able and Hagan (2000); Able and Hagan (2003); Able et al. (2003a); Raichel et al. (2003)
Intertidal creeks	−	−+	Kimball and Able (2007a, b)
Subtidal creeks	+	+	Grothues and Able (2003a, b)
Functional Attributes			
Marsh surface			
reproduction	+	−	Able and Hagan (2003)
feeding	−	+	Raichel et al. (2003)
growth	−	+	Grothues and Able (2003a)
survival	−	+	Hagan et al. (2007)
production	−	+	Hagan et al. (2007)
Intertidal creeks			
growth	−	+	Hagan et al. (2007)
Subtidal creeks			
feeding	+	+	Neuman et al. (2004); Jones and Able (in review)
growth	−	+	Hagan et al. (2007)

Some specific habitats, including salt marshes, eelgrass beds, and oyster beds, have received more focused attention. Salt marshes have been recipients of more restoration efforts than any other temperate estuarine habitat because they are very productive, including for fishes (National Research Council, 1992; Zedler, 2001; Laegdsgaard, 2006; Hagan et al., 2007), and because they have been altered in numerous ways over the past 200+ years (Philipp, 2005; Lotze et al., 2006; Gedan and Silliman, 2009; Stevenson and Kearney, 2009). Because marshes have often been diked for a variety of reasons, removing dikes and restoring tidal flow is a high priority in many marsh restoration efforts. In an extensive effort over nine years, we evaluated numerous characteristics of former salt hay farms that had been restored to tidal flow relative to reference marshes in Delaware Bay with emphasis on the fish response (Able et al., 2008, and references cited therein). More specifically, we compared distribution, abundance, feeding, growth, survival, reproduction, and production to evaluate restoration success. The fish fauna responded positively in one or two years, once tide had been restored and before vegetation had recovered. This suggested that access to the marsh surface and intertidal and subtidal creeks was the key to the success of the restoration. Subsequent evaluations have focused on how tide influences the fish assemblages in these restored and reference marshes (Kimball and Able, 2007a, b). In other temperate marshes the restoration of tidal flow has proven effective in restoring fish populations in Massachusetts marshes (Raposa and Roman, 2001; Roman et al., 2002) and reestablishing *Fundulus heteroclitus*'s normal diet after marsh restoration (James-Pirri et al., 2001).

Another type of salt marsh (*Spartina alterniflora* and *S. patens*) restoration involves activities associated with biological control of mosquitoes by habitat alteration. These efforts are designed to increase habitat for fish predators of mosquitoes by converting salt marsh grid-ditched drainage patterns by alterations collectively known as open marsh water management (OMWM). In this approach, a system of ponds and ditches is dug to connect small isolated mosquito breeding depressions to tidal or larger non-tidal water bodies. These provide more, diverse habitats and result in greater abundance of fishes and greater fish diversity than grid-ditched marshes (Talbot et al., 1986). In fact, the fish assemblages in OMWM-modified marshes often resemble those of unaltered salt marshes (K. W. Able, unpubl. data). Created marshes in temperate estuaries can also provide fish habitat even in a heavily urbanized estuary such as the Hackensack Meadowlands in northern New Jersey (Neuman et al., 2004). In this instance, the dredged creeks provided habitat for typical large juvenile and adult marsh fishes, including *Morone saxatilis*, *M. americana*, and *Pomatomus saltatrix*, although there were no comparisons possible with natural marshes in this heavily urbanized estuary.

Just as the *Phragmites* invasion of temperate estuarine salt marshes has a negative effect on the resident marsh surface fishes, it appears that elimination of *Phragmites* and restoration of *Spartina* marshes, including standing water in surface pools and puddles, also restores the fishes. In this process, once the *Phragmites* has been treated and eliminated, the marsh surface becomes wetter, with more pools and water-filled depressions (see Fig. 7.11). In response, fishes (primarily

**Larval
Habitat**

**Juvenile/Adult
Habitat**

Spartina

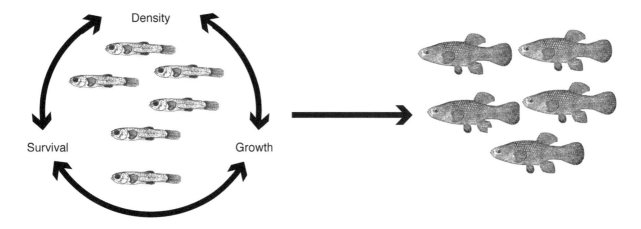

Density

Survival

Growth

Phragmites

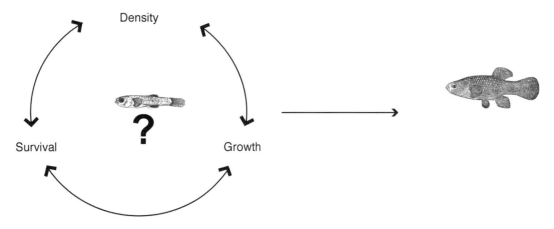

Density

?

Survival

Growth

Fig. 7.12. Relationship between young-of-the-year density, growth, and presumed survival of *Fundulus heteroclitus* in *Spartina alterniflora* and *Phragmites alterniflora*. See Table 7.3 for related information.

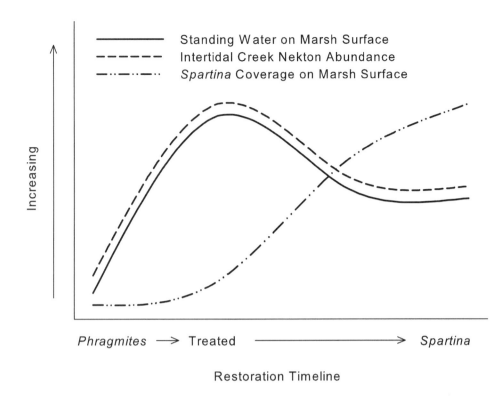

Fig. 7.13. Conceptual relationship between the amount of standing water on the marsh surface, *Spartina* coverage, and intertidal creek nekton abundance for salt marsh restorations. The restoration timeline is depicted on the x-axis beginning with *Phragmites*-dominated marshes followed by the transition from treatment for *Phragmites* removal to *Spartina*-dominated marshes (Kimball and Able, in review and pers. observ.).

Fundulus heteroclitus, but also including other species) are more abundant on the marsh surface and intertidal creeks; feeding, growth, and survival are enhanced; and, as a result, production is increased (Table 7.3). In a cascading effect, feeding and growth of some fishes are enhanced because of enhanced prey availability in intertidal and subtidal creeks. These findings are from extensive studies in Delaware Bay, Chesapeake Bay, and the Hackensack Meadowlands of Newark Bay (see Table 7.3). Further, the impact of the restoration is not linear but rather changes over time, with an ini-

tial increase in intertidal marsh nekton abundance followed by decreased abundance as the marsh surface becomes more vegetated (Fig. 7.13).

Restoration of oyster reefs is an important activity in many temperate estuaries (Breitburg et al., 2000; Rodney and Paynter, 2006), but little attention is devoted to the fish fauna of these reefs (Coen and Luckenbach, 2000). This lack of information is particularly an issue when multiple stressors (e.g., hypoxia and fishery-caused degradation) can interact to adversely affect fishes using oyster reefs (Lenihan et al., 2001).

8

Prey and Predators

One of the frequently cited assumptions about the functional significance of estuaries to young-of-the-year fishes is that estuarine habitats provide abundant food resources (Blaber and Blaber, 1980; Yáñez-Arancibia et al., 1980; Blaber, 1981; Boesch and Turner, 1984; Livingston, 2003) that contribute to the presumed habitat value (Beck et al., 2001; Layman et al., 2006; see Chapter 1). By reviewing and synthesizing the available literature, we hoped to pose, and answer, several questions: What are the general patterns of food habits? What role does piscivory play in diet during the first year? Are estuaries refuges from predation, as is often assumed?

FISH PREY

More than 400 papers published during 1907–2004 were utilized for this synthesis of the literature on food habits of young-of-the-year estuarine fish of 70 target species from 42 families in temperate estuaries. Of these, 91 papers contained food habit information for 47 (67%) species. As a result, we have examined published data for more than 54,000 stomachs. The number of studies varied between species, with several species (*Pseudopleuronectes americanus*, *Leiostomus xanthurus*, *Cynoscion regalis*, and *Pomatomus saltatrix*) having more than 10 published papers on food habits (Fig. 8.1). The individual sample sizes varied between each study; some were represented by > 3000 stomachs while most were represented by < 400 stomachs (Fig. 8.2). Most fishes studied were < 300 mm in length, with the exception of *Mustelus canis* (318 to 586 mm).

Prey were divided into 47 categories (Table 8.1) based on the available literature. Seven categories of invertebrate prey were dominant (important prey of > 20% of species examined) across many species, including copepods, amphipods, mysids, decapod shrimp, polychaetes, crabs, and insects/ arachnids. Copepods, for example, were found to be important prey for 33 species (70% of those for which there are data

available), amphipods for 27 species (57%), mysids for 26 species (55%), decapod shrimp for 21 species (45%), polychaetes for 17 species (36%), crabs for 13 species (28%), and insects/ arachnids for 11 species (23%). In addition, these same prey categories were commonly consumed by all species studied, as indicated by their frequency of occurrence, which was 81%, 91%, 74%, 72%, 79%, 68%, and 51%, respectively. Other prey categories that were also frequently encountered (consumed by > 30% of species examined) included isopods (74%), bivalves (62%), gastropods (55%), ostracods (51%), invertebrate eggs (47%), nematodes (38%), cnidarians (34%), algae (32%), and cladocerans (32%). Detritus and sediment were also frequently encountered (each 62%). Additional breakdown of this literature, by species, is summarized in the species accounts and available in Mancini and Able (2005). We also examined the food habits for young-of-the-year of an additional eight species (*Astroscopus guttatus*, *Anchoa hepsetus*, *Apeltes quadracus*, *Clupea harengus*, *Gasterosteus aculeatus*, *Mugil curema*, *Ophidion marginatum*, and *Strongylura marina*, total n = 442), which are not well represented in the literature for the study area. In these instances, food habits were reported in broad categories (annelids, crustaceans, insects, meiofauna, mollusks, mysids, phytoplankton and zooplankton, fish, sand, detritus, and unidentified material).

The importance of copepods, amphipods, mysids, decapod shrimp and crabs, polychaetes, and insects/arachnids, as found in this review of the literature (Table 8.1) and our examination of stomach contents (see species accounts), is typical for young-of-the-year in other estuarine studies in other regions (Festa, 1979; DeMorais and Bodiou, 1984; Fitzhugh and Fleeger, 1985; Gee, 1989; Coull, 1990; Feller et al., 1990; Coull et al., 1995). The importance of small prey, as indicated by the dominance of the above groups, likely reflects the morphological influence of gap limitations (see Karpouzi and Stergiou, 2003) of the small, young-of-the-year fishes (most < 300 mm) that occurs across all taxa, regardless of habitat.

There were numerous sources of variation in the prey consumed by young-of-the-year fishes in the study area based on the available literature (Table 8.2). One of the most common, despite the fact that many of the species had been poorly studied, was the frequency of ontogenetic variation in the diet, as occurs in other systems (Livingston, 1988). At least 23 species in 18 families of fishes exhibited a change in diet with increasing development/size. Much of this variation can be attributed to a switch in diet that takes place during the transition from the larval to the juvenile stage and accounts, to some degree, for the different feeding modes exhibited by young-of-the-year estuarine fishes, including herbivory, carnivory (including cannibalism), omnivory, and scavenging. This transformation is often associated with a switch to larger prey, as in those species that become piscivorous (e.g., *Anchoa hepsetus* and *Urophycis regia*). However, for some species, such as *Brevoortia tyrannus* and *Cyprinodon*

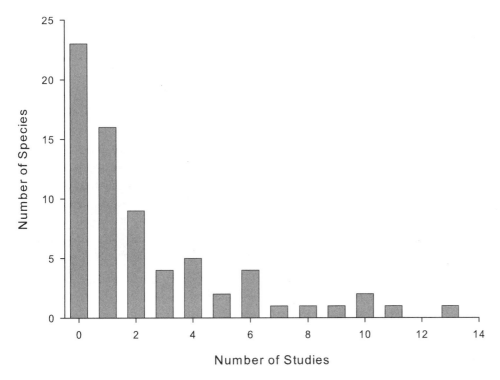

Fig. 8.1. Frequency distribution of number of food habit studies by fish species for those young-of-the-year in the study area based on the available literature.

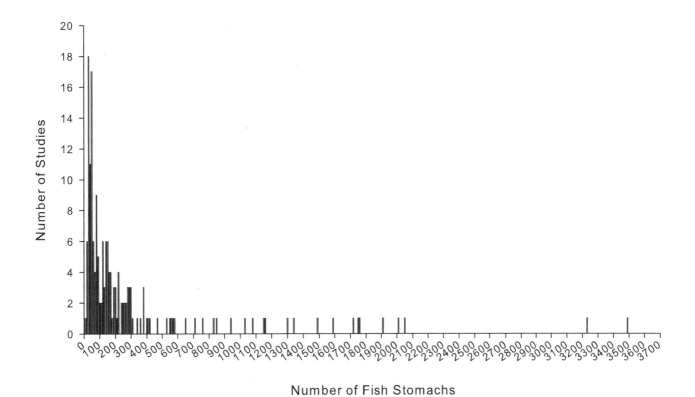

Fig. 8.2. Frequency distribution of number of young-of-the-year fish stomach samples per study based on the available literature.

Table 8.1. Identification of prey ranking (important = black-filled box, minor/rare = gray-filled box, absent = open box) by young-of-the-year estuarine fish based on a composite of studies examined (modified from Mancini and Able, 2005).

Prey Items (column headers): Phytoplankton, Zooplankton, Meiofauna, Algae, Diatoms, Ciliates, Foraminiferans, Flagellates, Porifera, Cnidarians, Platyhelminthes, Turbellarians, Cestoda, Trematodes, Nemertines, Nematodes, Rotifers, Polychaetes, Oligochaetes, Gastropods, Bivalves, Cephalopods, Tunicates, Insects/arachnids, Pycnogonids, Mysid Shrimps, Cumaceans, Tanaids, Isopods, Amphipods, Decapod Shrimps, Crabs, Astacids, Ostracods, Copepods, Cladocerans, Cirripedes, Limulus, Tardigrades, Bryozoans, Chaetognaths, Larvaceans, Invert Eggs, Fish Eggs, Fish, Detritus, Sediment

Species	Size Class (mm)	N
Mustelus canis	318–586 TL	85
Anguilla rostrata	50–249 TL	187
Conger oceanicus		
Alosa aestivalis	5–106 TL	1610
Alosa mediocris		
Alosa pseudoharengus	mean 64.8 TL	48
Alosa sapidissima	10–95 FL	2861
Brevoortia tyrannus	10–120.1	719
Clupea harengus	33.9–51.1	18
Anchoa hepsetus		
Anchoa mitchilli	2.5–60	6179
Osmerus mordax		
Synodus foetens		
Microgadus tomcod	28.9–143.8	577
Pollachius virens	23.1–66.2	22
Urophycis chuss	70–203.8	48
Urophycis regia	34–199	78
Urophycis tenuis		
Ophidion marginatum		
Opsanus tau	20–100 TL	56
Strongylura marina		
Cyprinodon variegatus	10–57 SL	1147
Fundulus heteroclitus	6.6–41 SL	828
Fundulus luciae	11–47 SL	370
Fundulus majalis	26–70 SL	62
Lucania parva		
Gambusia holbrooki		
Membras martinica		
Menidia beryllina	15–67 TL	522
Menidia menidia	10–120	1439
Apeltes quadracus		
Gasterosteus aculeatus		
Hippocampus erectus	23–126 TL	136
Syngnathus fuscus	28–209 TL	5542
Prionotus carolinus	20–159.7	336
Prionotus evolans	33–150 SL	309
Myoxocephalus aenaeus	37–134.9	196

Species	Size range	Count
Morone americana	<139 SL	2971
Morone saxatilis	<160	2657
Centropristis striata	19.1–146	76
Pomatomus saltatrix	20–290 FL	6863
Caranx hippos	30–160 TL	40
Lutjanus griseus		
Stenotomus chrysops	20–146	181
Bairdiella chrysoura	7–159	630
Cynoscion regalis	3.5–200	2989
Leiostomus xanthurus	9–159	6151
Menticirrhus saxatilis	10–140	139
Micropogonias undulatus	17–152.4	2273
Pogonias cromis	30–211	268
Chaetodon ocellatus		
Mugil cephalus		
Mugil curema		
Sphyraena borealis	50–110	85
Tautoga onitis	31–71	36
Tautogolabrus adspersus	11–64	54
Pholis gunnellus		
Astroscopus guttatus		
Hypsoblennius hentz		
Ammodytes americanus	8–159	740
Gobionellus boleosoma		
Gobiosoma bosc		
Gobiosoma ginsburgi	18–42 SL	49
Peprilus triacanthus	21.6–100	19
Scophthalmus aquosus	2.8–170	520
Etropus microstomus		
Paralichthys dentatus	8.1–300	1211
Pseudopleuronectes americanus	3–300	2681
Trinectes maculatus	21–60 SL	83
Sphoeroides maculatus	29.4–89.5	22

Table 8.2. Documented sources of variation in diet of young-of-the-year fishes in temperate estuaries based on species accounts and literature (see Chapter 3).

Species	Ontogeny	Annual	Geographical	Cannibalism
Ammodytes americanus	X	X		X
Anchoa mitchilli	X			X
Anguilla rostrata	X			
Bardiella chrysoura	X			
Brevoortia tyrannus	X			
Centropristis striata	X	X		
Cynoscion regalis	X		X	
Fundulus heteroclitus		X		
Fundulus luciae		X		
Hippocampus erectus	X			
Leiostomus xanthurus	X	X	X	
Menidia beryllina		X	X	
Menidia menidia	X		X	X
Micogadus tomcod	X			
Micropogonias undulatus	X	X	X	
Morone americana	X	X	X	
Morone saxatilis	X	X		
Myoxocephalus aenaeus	X			
Opsanus tau	X			X
Paralichthys dentatus	X			
Pogonias cromis			X	
Pomatomus saltatrix	X	X	X	X
Prionotus carolinus	X			
Pseudopleuronectes americanus	X			
Scophthalmus aquosus	X			
Stenotomus chrysops	X			
Syngnathus fuscus	X	X		X

variegatus, the switch is from carnivory to herbivory with ontogeny, while for others, like Mugil spp., it is from carnivory to detritivory. Still others, such as Fundulus heteroclitus, change from carnivory (including piscivory) to omnivory and, in some instances, scavenging (Able et al., 2007b).

There are other sources of variation as well. Seasonal change in diet is commonplace but difficult to separate from the co-occurring changes in size/ontogeny. Change also occurs at intra-annual (12 species in 8 families) as well as geographical (8 species in 4 families) scales. Not unexpectedly, differences in salinity are believed to be a source of variation in available prey and thus diet (3 species). Other differences may be the result of episodic sources of prey. A good example in many temperate estuaries is the early summer occurrence of Limulus polyphemus (horseshoe crab) eggs during the early summer spawning season. At least four common species are reported to take advantage of this resource (Nemerson, 2001). As in many studies of fishes, the plasticity in diet of estuarine fishes is commonplace (Ruehl and DeWitt, 2007; Dolbeth et al., 2008). Habitat, such as marsh creek geomorphology (Visintainer et al., 2006), can influence diet as well, at broad or smaller spatial scales.

FISHES AS PREY FOR FISHES

The role of predation in aquatic food webs (Juanes et al., 2002; Krause et al., 2002; O'Connor and Bruno, 2007) and its ecological consequences (Polunin and Pinnegar, 2002; Steneck, 2005; Heck and Valentine, 2007; Heithaus et al., 2007) is receiving increasing attention. These influences appear to apply to estuarine fishes, when they have been investigated (Crowder et al., 1997; Craig et al., 2006). Recently, the refuge value (Berryman and Hawkins, 2006) of estuarine habitats is being re-evaluated, in part because of a re-examination of "nursery" habitats in general (Beck et al., 2003; Dahlgren et al., 2006; Layman et al., 2006; Sheaves et al., 2006). As a result, it is becoming clear that piscivory by fishes is not uncommon in estuaries along the East Coast of the United States (this study), Portugal (Dolbeth et al., 2008), and Australia (Paterson and Whitfield, 2000; Sheaves, 2001; Baker and Sheaves, 2005, 2006, 2007). In our examination of the available literature for temperate estuaries along the East Coast of the United States, fishes were often important prey for other fish species (Table 8.3). More specifically, fishes were important prey for 16 (34%) species, and they occurred in the diet of 34 (72%) species. Fish eggs also were consumed frequently (38% of species studied). In a further attempt to understand the role of fishes as prey, we tabulated the occurrence of individual fish prey. Of the 34 species that consumed fish prey, the data for 24 species included sufficient information to characterize the fish species consumed. Some of the dominant (important to > 15% of species examined) fish prey were Fundulus heteroclitus (important prey for 21% of species), Menidia menidia (17%), and Anchoa mitchilli (17%). These same species were found in 29%, 29%, and 42% of the species examined, respectively. Other fish taxa that were frequently encountered (consumed by > 13%) in stomach contents were Menidia spp., Anchoa spp., family Gobiidae, Gobiosoma bosc, Gobiosoma spp., family Clupeidae, Micropogonias undulatus, Ammodytes americanus, family Atherinopsidae, and Alosa spp.

The importance of fishes as prey is somewhat surprising, given the often repeated theme that estuaries serve as refuges from predation (Joseph, 1972; Weinstein, 1979; Blaber and Blaber, 1980; Boesch and Turner, 1984; Beck et al., 2001; King et al., 2007). Most of the dominant prey species are, not unexpectedly, small species that are abundant in many estuarine habitats (Fundulus heteroclitus, Menidia menidia, and Anchoa mitchilli). Others, such as Morone saxatilis, were reported in the diet of a species that becomes piscivorous at relatively small sizes (Pomotomus saltatrix), for example, in the many studies in the Hudson River, where M. saxatilis is abundant (Mancini and Able, 2005). Thus, in all these examples, abundant prey are frequently consumed, a not surprising finding.

As a result of the same literature review for the focal species in this book (see Table 8.3) and our own examinations,

Table 8.3. Identification of fish prey ranking (important = black-filled box, minor/rare = gray-filled box, absent = open box) by young-of-the-year estuarine fish based on each study examined (modified from Mancini and Able, 2005). See Table 3.5 for species code information and details of specific studies.

Species Code (columns, left to right): 1, 2 (B), 3 (D), 3 (F), 4 (C), 5 (B), 6 (A), 6 (D), 7, 8 (G), 8 (J), 9 (H), 9 (R), 10 (G), 11 (A), 11 (C), 12 (C), 12 (D), 13 (A), 14 (C), 14 (D), 14 (E), 14 (F), 14 (I), 14 (M), 15 (A), 15 (B), 15 (C), 15 (D), 15 (E), 15 (F)

Families and species (rows):

Ammodytidae	Ammodytes americanus
Anguillidae	Anguilla rostrata
Atherinopsidae	FAMILY
	Membras martinica
	Menidia beryllina
	Menidia menidia
	Menidia spp.
Batrachoididae	Opsanus tau
Carangidae	FAMILY
	Caranx hippos
	Trachinotus carolinus
	Trachinotus falcatus
Centrarchidae	Lepomis gibbosus
Clupeidae	FAMILY
	Alosa aestivalis
	Alosa pseudoharengus
	Alosa sapidissima
	Alosa spp.
	Brevoortia tyrannus
	Clupea harengus
Cyprinidae	FAMILY
	Notropis hudsonius
Cyprinodontidae	FAMILY
	Cyprinodon variegatus
Fundulidae	Fundulus diaphanus
	Fundulus heteroclitus
	Fundulus majalis
	Fundulus spp.
	Lucania parva
Engraulidae	FAMILY
	Anchoa hepsetus
	Anchoa mitchilli
	Anchoa spp.
Gadidae	Microgadus tomcod
Gasterosteidae	Apeltes quadracus
	Gasterosteus aculeatus
Gobiidae	ORDER
	Gobiosoma bosc
	Gobiosoma spp.
	Microgobius thalassinus
Moronidae	Morone americana
	Morone saxatilis
	Morone spp.
Mugilidae	Mugil curema
Paralichthyidae	Paralichthys dentatus
Percidae	Etheostoma olmstedi
Pleuronectiformes	FAMILY
Pleuronectidae	Pseudopleuronectes americanus
Pomatomidae	Pomatomus saltatrix
Sciaenidae	Cynoscion nebulosus
	Cynoscion regalis
	Cynoscion spp.
	Leiostomus xanthurus
	Menticirrhus spp.
	Micropogonias undulatus
Soleidae	Trinectes maculatus
Sphyraenidae	Sphyraena borealis
Syngnathidae	FAMILY
	Syngnathus fuscus
	Hippocampus erectus
Tetraodontidae	Sphoeroides maculatus
	EGGS

continued

Table 8.3 continued

Family	Species Code	15 (G)	15 (H)	15 (I)	15 (I₁)	15 (I₂)	15 (I₃)	15 (J)	15 (K)	15 (L)	15 (M)	15 (N)	15 (O)	15 (P)	15 (Q)	15 (R)	15 (S)	15 (T)	16	17 (A)	17 (B)	17 (C)	18 (A)	18 (B)	18 (C)	18 (D)	18 (E)	18 (H)	18 (I)
Ammodytidae	*Ammodytes americanus*																												
Anguillidae	*Anguilla rostrata*													○															
Atherinopsidae	FAMILY																○												
	Membras martinica																												
	Menidia beryllina																												
	Menidia menidia	○	○	○	○				●	●	●			●	●	●										●			
	Menidia spp.								●	●																		○	
Batrachoididae	*Opsanus tau*																												
Carangidae	FAMILY														○														
	Caranx hippos														○														
	Trachinotus carolinus																												
	Trachinotus falcatus																												
Centrarchidae	*Lepomis gibbosus*					●																							
Clupeidae	FAMILY															○				○				○					
	Alosa aestivalis	○	○						●																●				
	Alosa pseudoharengus																												
	Alosa sapidissima								●	●	●																		
	Alosa spp.	○	○						●	●	●																	○	
	Brevoortia tyrannus									○	●					●													
	Clupea harengus																												
Cyprinidae	FAMILY																												
	Notropis hudsonius	○			○																								
Cyprinodontidae	FAMILY																												
	Cyprinodon variegatus																												
Fundulidae	*Fundulus diaphanus*																												
	Fundulus heteroclitus									○	●									○		●		●					
	Fundulus majalis									●																			
	Fundulus spp.								○			○								○									
	Lucania parva																												
Engraulidae	FAMILY					●																							
	Anchoa hepsetus																												
	Anchoa mitchilli	●	●	●	●	●		●	●	●	●	●	●	●	●	●	○	●	●	●	●	●	●	●	●	●	●	○	●
	Anchoa spp.															○													
Gadidae	*Microgadus tomcod*			●		●			○																				
Gasterosteidae	*Apeltes quadracus*																												
	Gasterosteus aculeatus																												
Gobiidae	FAMILY															○				○									
	Gobiosoma bosc																												
	Gobiosoma spp.					○										○													
	Microgobius thalassinus					●										○													
Moronidae	*Morone americana*	●	●			●					○																		
	Morone saxatilis	●	●	●				○	●	●	●																		
	Morone spp.		○	●	●			●			●																		
Mugilidae	*Mugil curema*																												
Paralichthyidae	*Paralichthys dentatus*															○													
Percidae	*Etheostoma olmstedi*			○		○																							
ORDER	Pleuronectiformes																												
Pleuronectidae	*Pseudopleuronectes americanus*																												
Pomatomidae	*Pomatomus saltatrix*	○				○		○																					
Sciaenidae	*Cynoscion nebulosus*																												
	Cynoscion regalis											○								○							○	●	
	Cynoscion spp.	○			○																								
	Leiostomus xanthurus																												
	Menticirrhus spp.																○												
	Micropogonias undulatus																○												●
Solidae	*Trinectes maculatus*																											○	
Sphyraenidae	*Sphyraena borealis*																												
Syngnathidae	FAMILY																												
	Syngnathus fuscus																												
	Hippocampus erectus																												
Tetraodontidae	*Sphoeroides maculatus*																												
	EGGS				○																								

Table 8.3 continued

Family	Species	18 (J)	18 (K)	18 (L)	18 (M)	18 (N)	19 (A)	19 (T)	20 (A)	21 (A)	21 (B)	22	23 (I)	24 (B)	24 (C)	24 (E)	24 (G)	25 (K)
	EGGS																	
Tetraodontidae	Sphoeroides maculatus																	
	Hippocampus erectus																	
	Syngnathus fuscus																	
Syngnathidae	FAMILY														▨			
Sphyraenidae	Sphyraena borealis											▨						
Soleidae	Trinectes maculatus																	
	Micropogonias undulatus					▨												
	Menticirrhus spp.													▨				
	Leiostomus xanthurus														■	▨		
	Cynoscion spp.																	
	Cynoscion regalis																	
Sciaenidae	Cynoscion nebulosus																	
Pomatomidae	Pomatomus saltatrix																	
Pleuronectidae	Pseudopleuronectes americanus																	
Pleuronectiformes	ORDER																	
Percidae	Etheostoma olmstedi																	
Paralichthyidae	Paralichthys dentatus																	
Mugilidae	Mugil curema																	
	Morone spp.																	
	Morone saxatilis																	
Moronidae	Morone americana																	
	Microgobius thalassinus																	
	Gobiosoma spp.	▨											▨	▨		▨		
	Gobiosoma bosc												▨				▨	
Gobiidae	FAMILY					▨		▨										
	Gasterosteus aculeatus																	
Gasterosteidae	Apeltes quadracus																	
Gadidae	Microgadus tomcod																	
	Anchoa spp.				■							▨						
	Anchoa mitchilli	■								▨								
	Anchoa hepsetus																▨	
Engraulidae	FAMILY																	
	Lucania parva																	
	Fundulus spp.																	
	Fundulus majalis													▨				
	Fundulus heteroclitus					▨								■				
Fundulidae	Fundulus diaphanus																	
	Cyprinodon variegatus												▨					
Cyprinodontidae	FAMILY																	
	Notropis hudsonius																	
Cyprinidae	FAMILY																	
	Clupea harengus																	
	Brevoortia tyrannus				▨													
	Alosa spp.											▨						
	Alosa sapidissima																	
	Alosa pseudoharengus																	
	Alosa aestivalis																	
Clupeidae	FAMILY		▨					■	▨									
Centrarchidae	Lepomis gibbosus																	
	Trachinotus falcatus																	
	Trachinotus carolinus																	
	Caranx hippos																	
Carangidae	FAMILY																	
Batrachoididae	Opsanus tau																	
	Menidia spp.												▨	■		▨		
	Menidia menidia														■			▨
	Menidia beryllina																	
	Membras martinica																	
Atherinopsidae	FAMILY						▨											
Anguillidae	Anguilla rostrata																	
Ammodytidae	Ammodytes americanus											▨						

it is apparent that there are several factors that cause variation in the occurrence of piscivory. Among estuarine fishes, there are a number of clearly documented instances of predators (*Pomatomus saltatrix, Caranx hippos, Prionotus evolans, P. carolinus,* and *Myoxocephalus aenaeus;* see species accounts) on fishes. In some species (*Pomatomus saltatrix, Paralichthys dentatus,* and *Cynoscion regalis*) this adoption of piscivory occurs as a marked transition in diet. For others, as in *Anchoa hepsetus, Urophycis regia,* and *Morone americana,* increasing incidence of piscivory with size is not as marked and may be a response to local abundance of fish prey. Piscivory by *Cynoscion regalis* has also been demonstrated in response to seasons (Hartman and Brandt, 1995a) and salinity (Nemerson, 2001). Interestingly, cannibalism has been frequently reported for a number of estuarine fishes, including *Opsanus tau, Anchoa mitchilli, Menidia menidia, Syngnathus fuscus, Ammodytes americanus, Cynoscion regalis,* and *Pomatomus saltatrix* (see Table 8.2). It is also useful to recognize that not all piscivores are large (> 100 mm) individuals (see Baker and Sheaves, 2006). In fact, one of the most abundant small fishes in temperate estuarine salt marshes, *Fundulus heteroclitus,* is frequently piscivorous (Able et al., 2007b).

In a further attempt to determine the role of fishes as fish predators, we examined the diet of larger, potentially piscivorous fishes, as captured by gill nets in two Middle Atlantic Bight estuaries, including marsh creeks in upper Delaware Bay and portions of the Mullica River–Great Bay. In the latter, the dominant piscivorous fishes were *Ameiurus catus, A. nebulosus, Ictalurus punctatus, Morone americana, M. saxatilis,* and *Pomatomus saltatrix* (Jones and Able, in review). Anadromous clupeids dominated the fish component of the diet across all species. Other important fish prey included fundulids, cyprinids, and moronids. The most abundant fish predator, *Morone americana,* had *Fundulus* spp. as a larger proportion of its diet than the other piscivores.

A somewhat similar pattern was evident in the Mullica River–Great Bay estuary (Fig. 8.3). For several species (*Ameiurus nebulosis, A. catus, Cynoscion regalis, Esox niger, Morone americana, M. saxatilis,* and *Pomatomus saltatrix*) the diet was dominated by fish and crabs, regardless of whether this was expressed as percent frequency of occurrence or dry weight. Other prey categories, including crustaceans and unidentified material, were ranked highly when only percent frequency of occurrence was considered. For the fish component of the diet, the important prey across all predators were *Brevoortia tyrannus* and unidentified clupeids as determined by both percent frequency of occurrence and dry weight of prey. Other important prey, by weight, were *Dorosoma cepedianum* and *Morone americana* (Fig. 8.4). Other fish prey species of some importance included *Alosa* spp., *Anchoa mitchilli, Fundulus heteroclitus, Fundulus* sp., *Menidia menidia,* and *Menidia* sp. The degree of piscivory, as measured by the same metrics, indicated that it was important along the entire salinity gradient from freshwaters at Landing Creek into the lower portion of

Great Bay, although there was a slight trend to have a higher frequency of occurrence in slightly higher salinities from Nacote Creek to upper and lower Great Bay (Fig. 8.5). Together, these analyses indicate that piscivory is common in this estuary and probably others, and that it occurs throughout all portions of the estuary (see species accounts for additional details). The important prey included those indicated in other estuaries, such as clupeids (especially *Brevoortia tyrannus*) and small prey species like engraulids, fundulids, and atherinopsids (Nemerson and Able, 2004, 2005). The best example of piscivory in temperate estuaries and estuaries worldwide is *Pomatomus saltatrix,* based on extensive analyses. It is clear that this species enters estuaries at relatively large sizes for young-of-the-year. This provides a size advantage over their fish prey throughout the first spring, summer, and fall and results in very fast growth, which is among the highest for any estuarine fish species (Juanes et al., 1994; Juanes and Conover, 1994a, b, 1995).

FISHES AS PREY FOR OTHER VERTEBRATES

The role of birds as predators on estuarine fishes is often underestimated, partly due to a lack of detailed studies of bird diets. Most birds that feed on fishes in temperate estuaries on the East Coast of the United States are seasonal visitors. During warmer months (when fish prey is most abundant; see Chapter 4), these include wading birds (herons and egrets), terns and skimmers, cormorants, brown pelicans (*Pelecanus occidentalis*), ospreys (*Pandion haliaetus*), and belted kingfishers (*Ceryle alcyon*). Fewer piscivorous birds are present during the winter in temperate estuaries, but great blue herons (*Ardea lerodias*) are often present throughout the year, belted kingfishers will remain near ice-free waters, and great cormorants (*Phalacrocorax carbo*) may occur in our study area during the colder months. Loons (two species), grebes, and several species of ducks and geese may also occur in temperate estuaries during winter. All of the foregoing examples include fishes in their diets, although some also prey on other vertebrate and invertebrate taxa as well as feed on vegetation.

Loons feed on medium-sized fishes after diving for them, often to great depths, although most prey are caught near the surface (Ehrlich et al., 1988; Sibley, 2001). Two species of loon, the common loon (*Gavia immer*) and the red-throated loon (*G. stellata*), may occur in coastal waters off the East Coast of the United States during winter. The most likely grebe to spend the winter in temperate estuaries on the East Coast of the United States is the horned grebe (*Podiceps auritus*), a short-billed species that feeds mostly on aquatic invertebrates but that also includes fishes in its diet (Ehrlich et al., 1988).

The black-crowned night heron (*Nycticorax nycticorax*) is a voracious predator that feeds on a variety of prey, including fishes (Ehrlich et al., 1988). Much of this predation occurs at night, when the heron is most active and is usually unobserved. As we pointed out in Chapter 7, the behavior

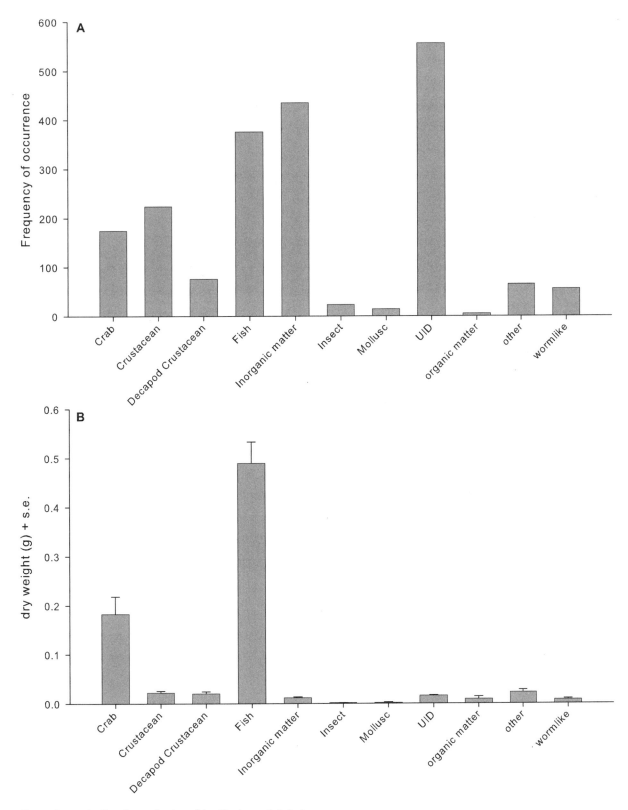

Fig. 8.3. Composite diet of seven dominant fishes (*Ameiurus nebulosis, A. catus, Cynoscion regalis, Esox niger, Morone americana, M. saxatilis,* and *Pomatomus saltatrix,* n = 665 stomachs with food) from gill net collections in the Mullica River–Great Bay estuary in southern New Jersey based on (A) frequency of occurrence and (B) weight of prey categories. UID = unidentified material.

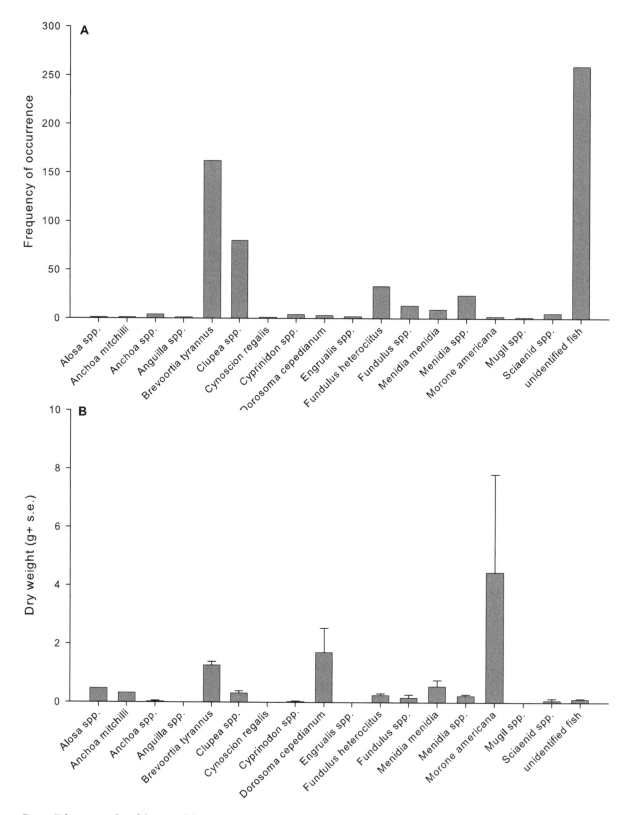

Fig. 8.4. Fish as prey in diet of dominant fishes (*Ameiurus nebulosis, A. catus, Cynoscion regalis, Esox niger, Morone americana, M. saxatilis,* and *Pomatomus saltatrix,* n = 665 stomachs with food) from gill net collections in the Mullica River–Great Bay estuary in southern New Jersey based on (A) frequency of occurrence and (B) weight of prey categories.

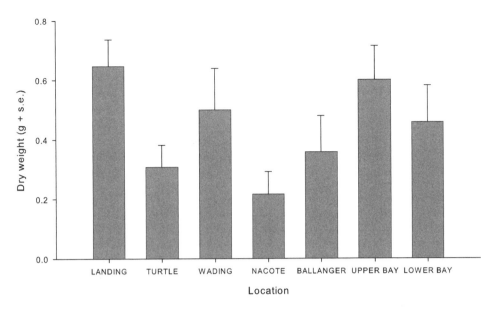

Fig. 8.5. Fish portion of diet, based on percent of diet by weight, along the salinity gradient from near the inlet and upstream to Landing Creek, in the Mullica River–Great Bay estuary in southern New Jersey.

of most estuarine fishes differs between day and night. These observations combined suggest that piscivory by the black-crowned night heron most likely results in more fish mortality than is generally estimated, although we are not aware of any studies that adequately quantify this mortality. Among some reports are those that indicate that it is particularly adept at capturing various life history stages of *Anguilla rostrata*. Other species of wading birds include the great egret (*Casmerodius albus*), the snowy egret (*Egretta thula*), the glossy ibis (*Plegadis falcinellus*), the tricolored heron (*Egretta tricolor*), and the little blue heron (*Egretta caerulea*), all of which include fishes as an important item in their diets. *Fundulus heteroclitus* constitutes up to 78% of the diet of great and snowy egrets (Maccarone and Parsons, 1994). The great blue heron (*Ardea lerodias*) occupies a wide variety of habitats (freshwater, brackish water, and saltwater) during most of the year and often overwinters in ice-free parts of temperate estuaries (Leck, 1984), concentrating on small patches of open water. They are opportunistic feeders whose diet is mainly composed of fishes (Ehrlich et al., 1988).

Waterfowl of several species occur in large numbers in the bays and estuaries of our study area throughout the winter. Although most species target vegetation or invertebrates in their diets, and only include fishes secondarily, some are primarily piscivorous. For example, both mallards (*Anas platyrhynchos*) and American black ducks (*A. rubripes*) occasionally feed on large *Fundulus heteroclitus* (M. P. Fahay, pers. observ.). The red-breasted merganser (*Mergus serrator*) and the hooded merganser (*Lophodytes cucullatus*) target fishes, which they capture by diving for them (Ehrlich et al., 1988).

Terns of the genera *Sterna* and *Rynchops* feed on small fishes found near the surface (Burger et al., in prep.). Adults provide fishes to their mates during courtship rituals, and also feed their chicks at nest sites. These activities provide the opportunity to collect dropped fishes in order to study the species composition of terns' prey (Fahay and Able, 1989).

One such study on Long Island Sound (Safina et al., 1990) involved the common tern (*Sterna hirundo*) and the roseate tern (*S. dongallii*) and found that they feed on a variety of estuarine fish species, including *Ammodytes americanus*, *Anchoa mitchilli*, *Pomatomus saltatrix*, *Syngnathus fuscus*, *Brevoortia tyrannus*, *Fundulus* spp., and flatfishes. In another specific example of avian predation on fishes from the Great Bay–Little Egg Harbor estuary (Burger et al., in prep.), three species—common and Forster's terns and black skimmers (*Rynchops niger*)—fed heavily on abundant fish species in the estuary, but prey varied between species and phases of the reproductive season. The fishes most commonly preyed upon were *Fundulus heteroclitus* (much of the salt marsh), *Menidia menidia* (pelagic in open estuary and in marsh creeks), and *M. beryllina* (most abundant in marsh pools). Other common prey species included *Anchoa* spp. and other clupeiforms (pelagic species), *Cyprinodon variegatus* (salt marsh surface pools), *Ammodytes* spp., *Pseudopleuronectes americanus,* and *Syngnathus fuscus*. The terns selected fish prey from < 25 to 74 mm in length, whereas the black skimmers fed on prey up to 100 mm. During the nesting season, the terns selected larger prey (> 50 mm) for courtship feeding, then switched to smaller prey (< 50 mm) when feeding small chicks, then switched back again to larger prey (> 50 mm) to feed larger chicks.

The brown pelican (*Pelecanus occidentalis*) commonly nests in the southern portion of our study area, between Maryland and North Carolina (Shields, 2002). Recent years have seen the brown pelican occurring farther north along the East Coast of the United States, and there have been two attempts to nest in Barnegat Bay, New Jersey (Shields, 2002). When present, it occurs along the coast, where it plunge-dives for pelagic fishes or sometimes small crustaceans. It also joins mixed-species aggregations of bird predators around salt marsh pools, where it is able to dip small fishes from the shallow waters with its large, pouch-like bill.

Both species of cormorant, the double-crested cormorant (*Phalacrocorax auritis*) (during warmer months) and the great cormorant (*P. carbo*) (during the winter), are adept divers that include fishes as their primary prey. The double-crested cormorant targets schooling fishes (Ehrlich et al., 1988) but also hunts for fishes in marsh creeks and is particularly adept at capturing and consuming eels (M. P. Fahay, pers. observ.).

The osprey (*Pandion haliaetus*) is one of the earliest piscivorous migrants to arrive in temperate estuaries of the East Coast of the United States. Its March arrival coincides with increased inshore migration activity by *Alosa* spp. and *Pseudopleuronectes americanus*. We often observe ospreys carrying these species in their talons (K. W. Able and M. P. Fahay, pers. observ.). Later in the spring, other fish species, including *Brevoortia tyrannus*, become prey of ospreys. Ospreys are incapable of diving to any appreciable depths; therefore, prey fishes must be within a few inches of the water surface so that the ospreys can take them. In order for *P. americanus* to be captured, it must occur within very shallow water (M. P. Fahay, pers. observ.).

Belted kingfishers most commonly occur in freshwater or estuarine habitats during the warmer months, but as long as there are ice-free patches of water available in the estuary, they may remain during the winter. To capture their fish prey, they dive from perches or after hovering briefly over open water. They are capable of feeding on small fishes that occur near the surface (Ehrlich et al., 1988).

Other vertebrates, such as bottlenose dolphins (*Tursiops truncatus*), are known to feed on estuarine fishes (Barros and Wells, 1998; Gubbins, 2002), including those that produce sound, such as sciaenids (Gannon et al., 2005). The same probably occurs in some temperate estuaries, where these mammals are annual visitors from spring through fall, as in the Mullica River–Great Bay estuary and adjacent nearshore waters in southern New Jersey (Toth-Brown, 2007). They are also common in Delaware Bay (K. W. Able, pers. observ.). Other marine mammals in temperate estuaries, including the same estuary, are harbor seals (*Phoca vitulina*) and gray seals (*Halichoerus grypus*). These seals are seasonally abundant in the Mullica River–Great Bay estuary from fall through spring (Slocum et al., 2005; K. W. Able, pers. observ.) and thus are some of the few piscivorous predators, besides birds, in this and, presumably, other temperate estuaries during the winter. Preliminary analysis of scat samples collected in the estuary from harbor seals included otoliths of *Clupea harengus* (most abundant), *Urophycis regia*, a scorpaenid (possibly *Myoxocephalus aenaeus*), and *Pseudopleuronectes americanus* (Slocum et al., 2005). Other less abundant species included

Marsh Surface Pools as Ecological Hotspots

Every summer, for more than two decades, we have observed large aggregations of multiple bird species gather around, over, and in individual marsh surface pools in the Sheepshead Meadows portion of the Great Bay Wildlife Management Area near the Rutgers University Marine Field Station. These feeding aggregations may be the result of local enhancement (Caldwell, 1981), social facilitation (Brzorad, 1994), or commensal feeding (Ehrlich et al., 1988). The pools in this extensive, relatively unaltered salt marsh are common features of New England-type salt marshes and often occur in temperate estuaries from Maine through southern New Jersey. The bird aggregations are most obvious in the early morning and dissipate as the day goes on. They are seldom, if ever, observed by late afternoon. The size of the aggregations is variable, reaching from a few dozen individuals to hundreds, and on at least two occasions it was estimated that more than 1000 birds were gathered together around one large pool or a set of small, adjacent pools.

Typically, great blue herons, great egrets, and snowy egrets are observed standing on the marsh surface and arrayed next to each other around the perimeter of the pool, while other species, including terns (probably both Forster's and common terns; see Burger et al., in prep.) and laughing gulls, hover above the same pool. Others, such as glossy ibises, are typically wading through the pool. Occasionally, little blue herons and double-crested cormorants were also observed in these aggregations. On one occasion, brown pelicans swam through a pool. In every instance, the birds appeared to be actively feeding, whether ambushing prey from the pool edge, diving into the pool from above, or wading through the pool.

We suspect that these birds are aggregating in response to increased availability of fish prey. Our prior studies indicate that dissolved oxygen levels decline through the night, in response to biological oxygen demand in the absence of photosynthesis, so that by dawn the oxygen levels have declined to near zero (Smith and Able, 2003). In response, many of the common marsh pool fishes (Able et al., 2005b), especially the dominant species, *Fundulus heteroclitus*, begin surface-skimming to capture the available oxygen at the water's surface. This behavior makes the fishes much more available to bird predators. The aggregation of multiple bird species probably facilitates their ability to feed on these temporarily available prey.

The impact on the fish populations in the individual marsh pools is expected to be important, given the abundance of these predators. However, it may not be deleterious to marsh fish populations in general because our observations suggest that they do not use the same pool on more than one occasion. It would be useful for future studies to more carefully examine the frequency of this behavior, determine if it varies with warmer temperatures (e.g., climate change), and evaluate how bird predation varies between fish species and life history stages.

Scophthalmus aquosus, Paralichthys dentatus, Ammodytes spp., and *Pomatomus saltatrix*. It is unknown if these species, although common in the estuary (Able and Fahay, 1998), were actually captured there or in the adjacent ocean.

FISHES AS PREY FOR INVERTEBRATES

What is seldom appreciated is that predation on fishes should be considered throughout the life history of estuarine fishes, that is, eggs, larvae, juveniles, and adults. For example, predation by an abundant ctenophore (*Mnemiopsis leidyi*) occurs on the eggs of a dominant estuarine fish, *Anchoa mitchilli* (Govoni and Olney, 1991). These interactions are likely important because of the high seasonal abundance of these predators (Burrell and Van Engel, 1976; Kremer, 1994). Predation on larvae of *Anchoa mitchilli* may also occur by ctenophores and scyphomedusae (*Chrysaora quinquecirrha*) (Burrell and Van Engel, 1976; Purcell et al., 1994). The occurrence of these early life history stages in the diet of invertebrates may be more common, but detection of the soft-bodied eggs and larvae is difficult. The abundant blue crab (*Callinectes sapidus*) may also be an important predator. Studies in the Apalachicola Estuary in the Gulf of Mexico indicated that the proportion of fish in the diet increased with crab size (> 31 mm carapace width) and these had consumed approximately 14% fish by weight, although fish occurred in all size classes examined (< 10 to > 131 mm carapace width) (Laughlin, 1982). The fish prey included *Anchoa mitchilli, Micropogonias undulatus, Microgobius* spp., *Etropus* spp., and *Trinectes* sp., all common species in temperate estuaries. *Callinectes sapidus* was also an important predator of *Fundulus heteroclitus* in a marsh pool in North Carolina, where the highest mortality (90%) occurred among the largest crabs (> 70 mm TL) (Kneib, 1982). A similar analysis of three crab species (*C. sapidus, Ovalipes ocellatus,* and *Cancer irroratus*) in the temperate Hudson River estuary found fish in the diet of all three species, but the volume was low relative to other prey types (Stehlik et al., 2004).

9

Migrations

Estuarine fishes (whether resident or transient) engage in several different kinds of migrations, depending on their life history stage, seasonal changes, or diel activity. Many of these migratory patterns are subtle and difficult to observe, and most vary temporally and spatially. Smaller-scale migrations provide connectivity between specific estuarine habitats, while certain of the larger-scale migrations do so between estuary and adjacent freshwater systems or between estuary and adjacent ocean (Metcalfe et al., 2002; Secor and Rooker, 2005). Recent findings have reviewed the increasing attention to lifecycle diversity in fishes (Secor and Kerr, 2009). In this chapter we examine the various migration patterns within three broad categories for selected estuarine fishes.

ESTUARINE MIGRATIONS BY RESIDENTS

Migrations by resident fishes within estuaries have not been well studied, but they may be more common and diverse than previously recognized. Because these migrations are smaller in scale than those between estuaries and the continental shelf or those occurring entirely in the ocean, they are sometimes difficult to detect. At these small scales, migrations allow connectivity between estuarine habitats. All of these may have cyclic (tidal, diel, seasonal) components as well as an ontogenetic one that could result in a change in habitat or in a cycle of habitat use (Fig. 9.1). Some other examples of these follow. At the upper reaches of some estuaries, such as the Delaware Bay–Delaware River system, adults of the potadromous *Acipenser brevirostrum* make spring migrations upstream to the upper limit of tide to spawn, after which they return to habitats downstream (see species account). Some populations of *Microgadus tomcod* are coastal-ocean fishes, ascending low-salinity streams for spawning, but in our study area, it occurs in the Hudson River as a strictly riverine (or estuarine) species. It spawns during the winter, after ascending from near-coastal habitats to the upstream

extent of saltwater intrusion. Hatching success is greatest in freshwater. Larvae develop as they drift downstream; transformation occurs near the river mouth; and juveniles gradually migrate back upstream during summer, when freshwater outflow is at its lowest rate.

The semianadromous species *Morone americana* migrates from deep, highly saline areas of estuaries, where it resides during winter, to shallow, brackish or freshwater habitats from early spring through summer (Beck, 1995). Spawning occurs in the lower reaches of large, coastal rivers, but there are different contingents and their success varies annually (Kerr et al., 2009). *Trinectes maculatus* makes a semicatadromous migration in the opposite direction, from mostly freshwater habitats to high-salinity areas of the Pamunkey River in Chesapeake Bay (Dovel et al., 1969). Similar patterns of migration by this species into higher salinities have also been reported from a Florida estuary (Castagna, 1955) and the Gulf of Mexico (Peterson, 1996). Within our study area, we suspect that a similar migration, associated with spawning, occurs in the Mullica River–Great Bay estuary, based on the collection of occasional small larvae near the ocean inlet and the distribution of most juveniles and adults in oligohaline portions of this estuary. Another type of migration, for feeding, occurs at a much smaller scale in *Paralichthys dentatus* (Szedlmayer and Able, 1993). This pattern involves young-of-the-year moving up and into subtidal marsh creeks on night high tides to feed, then leaving this habitat on the night ebb tide. They repeat this diel and tidal migration on nearly all consecutive night high tides during most of the summer. Ontogenetic migrations of small marsh fishes are often evident at considerably smaller scales. For example, the eggs of *Fundulus heteroclitus* are deposited intertidally, and the resulting larvae are abundant in shallow depressions on the marsh surface. As these progeny grow, however, they are found in slightly deeper intertidal creeks and marsh pools. During periods of low tide, bigger juveniles and adults are concentrated in larger creeks and at the marsh / open estuary fringe (Able et al., 2006a; Hagan et al., 2007). These small-scale migrations may occur in other, unstudied estuarine resident fishes.

MIGRATIONS INTO ESTUARIES BY TRANSIENTS

The pattern exhibited by many estuarine transient species includes spawning in the ocean, followed by a migration, or ingress, by larvae or small juveniles through inlets or bay mouths into estuarine habitats (Leggett, 1977; Baker, 1978; Northcote, 1978; McCleave et al., 1984; McKeown, 1984; Helfman et al., 1997; Dingle and Drake, 2007; Ramenofsky and Wingfield, 2007; McDowall, 2008), often at different temporal and spatial scales (see Table 5.2). In some species (e.g., *Paralichthys dentatus*) it is suspected that this ingress may be facilitated by vertical migrations by transforming individuals near the inlet, which occur in the water column during flood tides and near or on the bottom during tides ebb (Keefe

Patterns of habitat use

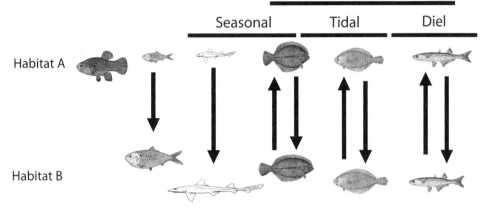

Resident
High habitat suitability
for all functions
all the time

Ontogenetic migrant
Habitat suitability
varies with size,
age or development

Cyclic migrant
High habitat suitability for one function
low suitability for another function
nekton divide among
habitats to get all functions

Seasonal   Tidal   Diel

Habitat A

Habitat B

Fig. 9.1. Patterns of temporal shifts in habitat use for temperate estuarine fishes (modified from Rountree and Able, 2007).

and Able, 1994; Hare et al., 2005a). The evidence suggests that larvae and juveniles of some other species, for example, *Pomatomus saltatrix* (Hare and Cowen, 1993), *Anguilla rostrata* and *Conger oceanicus* (Bell et al., 2003; Sullivan et al., 2006a; Wuenschel and Able, 2008), may simply actively swim into the estuary, without taking advantage of this selective tidal stream transport. For most species, mechanisms that enable larvae and juveniles to enter inlets are undescribed, but they are likely to entail some combination of those addressed in this chapter.

In at least a few cases, these ingressing larvae or juveniles are the products of reproduction that occurred during the previous summer and fall. In *Ophidion marginatum*, spawning occurs in summer on the continental shelf; the resultant larvae develop in the water column, transform, and settle to the bottom, where they overwinter until temperatures begin to increase in March. They then begin migrating into estuaries as small juveniles in the spring (Fahay, 1992; Able and Fahay, 1998). The available evidence for *Centropristis striata* suggests that spawning is limited to continental shelf waters. Larvae settle on the inner shelf at sizes of about 10 to 16 mm TL. Some small juveniles then migrate into estuaries aided by undescribed mechanisms, while others remain in coastal ocean habitats, such as beds of shell hash (Able et al., 1995a). This migration, undertaken by only part of the young-of-the-year population, suggests that this species is a facultative estuarine user.

Regardless of the methods employed by larvae and juveniles of transient species to enter temperate estuaries, young-of-the-year typically reach their greatest concentrations in estuaries during summer and early fall, after entry in the spring and before egress in the fall (Fig. 9.2). These mi-

grations are punctuated by a period of residence in the estuary in summer and in the ocean in winter. This pattern has been documented in small drowned river valley / back bay estuaries, such as Mullica River–Great Bay (Able et al., 1996b; Able and Fahay, 1998), as well as in large bay / estuaries, such as Delaware Bay–Delaware River (Able et al., 2007a). Extensive, recent sampling in the Delaware Bay system demonstrates that in all dominant species, whether resident or transient, there is usually a seasonal periodicity in abundance that is highest in spring through fall in many habitats, which corresponds closely with the period of warmest temperatures. The young-of-the-year of transient species then migrate out of estuaries (or egress) as temperatures decline in the fall. Evidence that supports this temperature-influenced egress in the fall is based on a later egress period or year-round residency in more southern estuaries, where fall cooling occurs later (e.g., *Anchoa mitchilli*; Vouglitois et al., 1987).

Several transient species, for example, anadromous species (see McDowall, 2008), migrate into temperate estuaries as adults for the purpose of reproduction. Important representatives in this category include fishes such as *Acipenser oxyrinchus*, *Osmerus mordax*, *Morone saxatilis*, and several species of *Alosa* that spawn in freshwaters. Examples of small fishes that engage in similar migrations into estuaries at the end of their first year of life include *Gasterosteus aculeatus* and *Menidia menidia* (see species accounts). However, these spawning migrations may vary among estuaries, and factors such as size of the systems may account for the selection of one system over another. The general pattern for *Osmerus mordax* includes adult residency in coastal waters until they ascend freshwater streams in the spring for spawning.

Morone saxatilis is well known to spawn in most of the

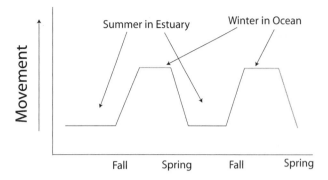

Fig. 9.2. Schematic representation of the timing of seasonal movements and duration of residence in the estuary and the ocean for representative temperate fishes.

larger systems, such as those in the Hudson River, the Delaware River, and Chesapeake Bay, but there is much less evidence that it spawns in smaller systems. In the Mullica River–Great Bay estuary, for example, there is ample evidence to suggest that adults of *M. saxatilis* migrate upstream in the spring (Able and Grothues, 2007b; Grothues et al., 2009), but there is much less proof that spawning is successful (Ng et al., 2007). Some individuals of *M. saxatilis* also migrate upstream in this system in the fall, but they return downstream very quickly, suggesting that these incursions are likely not for the purposes of feeding. It is tempting to speculate that these are exploratory migrations that may allow individuals to evaluate habitats for potential spawning sites during the following spring. Their ability to navigate and home to specific estuaries and exact locations (Ng et al., 2007; Grothues et al., 2009) would seem to make this a possibility. Although the purpose of this fall migration remains enigmatic, it does demonstrate yet another type of regularly occurring migration by a transient estuarine species.

The use of estuaries for reproduction also occurs in some elasmobranchs, although it entails the bearing of live young (pupping). Adults of *Mustelus canis* migrate into estuaries to bear their young, especially in salt marsh creeks and in adjacent portions of the estuary (Rountree and Able, 1996). Pupping also occurs by *Carcharhinus plumbeus* in Delaware Bay and the Mullica River–Great Bay estuary (McCandless et al., 2007; Merson and Pratt, 2007) as well as in Chesapeake Bay (Grubbs and Musick, 2007; Grubbs et al., 2007). See species accounts for details of these life histories.

Another pattern of estuarine migration by transient species involves adults outside of the spawning season, which enter estuaries for the purpose of feeding. Gill netting and ultrasonic tracking studies have revealed that *Alosa mediocris* engages in these feeding migrations (see species account). The diverse migratory behaviors of *Morone saxatilis* may also include long periods of estuarine residency (Ng et al., 2007). These feeding periods may be punctuated by seasonal migrations up or down the coast (Able and Grothues, 2007b; Grothues et al., 2009).

MIGRATIONS BY ESTUARINE FISHES ON THE CONTINENTAL SHELF

The seasonal egress of estuarine fishes onto the continental shelf is a common and well observed phenomenon in our study area. In the Middle Atlantic Bight, temperature appears to serve as the cue for these migrations (Musick et al., 1989). Further observations have been most synoptic on the shelf itself. Young-of-the-year and age 1+ juveniles and adults have been well sampled three times annually by the National Marine Fisheries Service (NMFS) bottom trawl survey on the continental shelf between the Gulf of Maine and Cape Hatteras, North Carolina. The methodology of this survey has been well documented elsewhere (Grosslein and Azarovitz, 1982; Able and Brown, 2005). Only the results of sampling from 1982 to 2003 are discussed here. The species we have focused on (see Table 3.4) include those that frequently use estuaries as young-of-the-year, but leave estuarine habitats and reside on the continental shelf during other stages in their life histories. We also have information on species that may only be facultative estuarine residents and whose young-of-the-year can occur either on the shelf or in the estuary (see Martino and Able, 2003), but we do not discuss these in detail.

The timing of estuarine egress is crucial to understanding the distribution of young-of-the-year and older stages on the continental shelf. Most species in the Middle Atlantic Bight leave estuaries in the fall, when temperatures begin to cool (Able and Fahay, 1998). The exact timing of this egress can vary between species as well as between different size cohorts, and may depend on the rate of temperature change (Rountree and Able, 1993). We lack data describing the magnitude of these migrations and detailing whether some species leave the estuary en masse or during a more prolonged period. A minor caveat in the trawl survey data concerns species such as *Paralichthys dentatus*; some individuals may egress as early as July in response to a storm, while others may delay their departure until December for reasons that remain undescribed (Sackett et al., 2007). Fall surveys that occur between September and November, therefore, may not detect all of these individuals. Despite this minor concern, however, it is apparent that the young-of-the-year of many (if not most) species that reside in temperate estuaries during the summer cannot survive winter estuarine temperatures, which can descend to below freezing, at least in the Middle Atlantic Bight (see Chapter 6). In contrast, temperatures well above freezing are available on the continental shelf throughout the year, and bottom temperatures on the mid- to outer shelf reach their maximums during November–December (see Fig. 2.5). Warmer temperatures are available in estuaries during the spring, and many species migrate back into estuaries; a typical pattern of alternating migration and residency for most species resembles the one illustrated in Figure 9.3 for Delaware Bay and the continental shelf.

Spatial patterns of young-of-the-year and larger juveniles

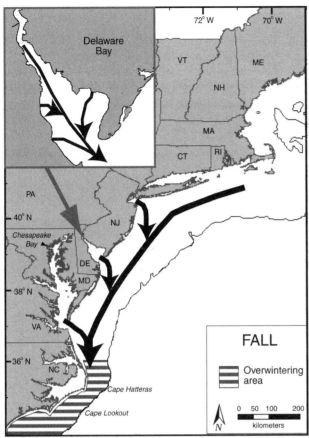

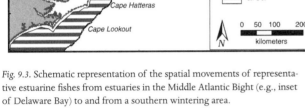

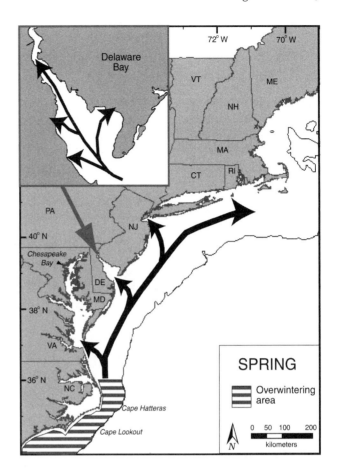

Fig. 9.3. Schematic representation of the spatial movements of representative estuarine fishes from estuaries in the Middle Atlantic Bight (e.g., inset of Delaware Bay) to and from a southern wintering area.

and adults of estuarine species on the continental shelf vary between species and season. Patterns for each species, if available, are included in the species accounts. Some species (e.g., *Pollachius virens, Clupea harengus,* and *Gasterosteus aculeatus*) reach the southern limit of their range in the northern portion of the Middle Atlantic Bight and on Georges Bank. Another group of species (e.g., *Alosa* spp., *Peprilus triacanthus, Scophthalmus aquosus,* and *Urophycis regia*) occurs over large sections of the study area during all seasons sampled. Other species (e.g., *Anchoa hepsetus, Brevoortia tyrannus, Pomatomus saltatrix,* and *Synodus foetens*) only occur in the Middle Atlantic Bight during the fall and spring, indicating that they migrate south, beyond the limits of our study area, during the winter. A few abundant species (e.g., *Morone saxatilis*) are not often collected on the continental shelf because they usually spend the first several years of their life in the estuary. A few other plentiful species are rarely detected in trawl surveys because they spend the winter in near-coastal waters, in

shallower water than the survey samples (e.g., *Tautoga onitis*), or migrate out of the Middle Atlantic Bight using very shallow water as their route (e.g., *Mugil* spp.).

The negative response of many young-of-the-year to declining estuarine temperatures may occur at relatively warm temperatures in species that originate in tropical and subtropical regions, such as *Chaetodon ocellatus,* whose young occur regularly in the Middle Atlantic Bight (McBride and Able, 1998). They apparently do not survive temperatures lower than about 10°C and therefore are unable to engage in a fall migration on the continental shelf. However, the young-of-the-year of *Caranx hippos* also derive from reproduction in subtropical waters south of our study area, and they regularly occur in estuaries in our study area (McBride and McKown, 2000). The fall migration of these young-of-the-year is well represented in trawl collections in near-coastal waters and presumably they are able to reach their southern habitats before winter (see species accounts).

10

Climate Change

EVIDENCE FOR CLIMATE CHANGE

The evidence for recent climate change is overwhelming, but this recognition has only come in the past several decades (Hughes, 2000; Levitus et al., 2000; Magnuson, 2002; Walther et al., 2002; Parmesan and Yohe, 2003; Ficke et al., 2007; Pearce and Feng, 2007; Rosenzweig et al., 2008). As a result, we still do not understand the impact of this change. This is especially true of aquatic habitats, which are less easily monitored and studied. There are fewer long-term datasets by which to measure this change and the response of the fauna (Keeling, 1998; Mountain, 2003; Nixon et al., 2004; Roessig et al., 2004; Friedland and Hare, 2007). This lack of a long-term perspective is especially problematic if it allows the "shifting baseline phenomenon" (Pauly, 1995; Pinnegar and Engelhard, 2008) to confound our detection of climate effects.

The most relevant environmental responses to climate change, at least in the North Atlantic Ocean, are reflected in the North Atlantic Oscillation (NAO) index (Fields et al., 1993; Scavia et al., 2002; Hurrell et al., 2003; Oviatt, 2004; Straile and Stenseth, 2007; Budikova, 2008). The responses correlated with this index include alterations in water temperature, precipitation, and winds and vertical mixing, and thus ocean circulation and direct habitat effects, such as sea level rise (Roessig et al., 2004). Some changes have extended to the Middle Atlantic Bight because there has been an enhanced discharge of low-salinity water from the Labrador Sea to the Middle Atlantic Bight associated with melting ice in the Arctic (Mountain, 2003; Greene et al., 2008). Of particular interest are the more pronounced and warmer temperatures during the winter (Joyce, 2002; Hare and Able, 2007; Straile and Stenseth, 2007). All of these are especially relevant to estuaries and thus fishes. An increasing focus on climate change in estuaries and estuarine ecosystems will doubtless lead to the recognition of additional environmental responses.

INFLUENCE ON FISH AND FISHERIES

The likelihood of climate effects on fishes is high, given the frequency with which these responses have been identified (see Roessig et al., 2004, for a recent review). Despite the increasing number of correlations between climate change and, particularly, rising temperatures, an understanding of the mechanisms responsible for variations in the distribution and population dynamics of fishes is often lacking. The most frequent and well-documented responses are from the North Pacific, where clear effects have been indicated for salmonids, hakes, and sardines (McFarlane et al., 2000); Pacific halibut (Clark and Hake, 2002); regime shifts for benthic versus pelagic species (Duffy-Anderson et al., 2005; Litzow et al., 2006); and, as a result, fishery management in general (King and McFarlane, 2006). In the eastern North Atlantic, climate change models have suggested shifts in the decadal-scale fluctuations in atmospheric pressure gradients and water temperatures. These are linked to an outburst of gadoids (Cushing, 1996; Sirabella et al., 2001), changes in occurrence and distribution of some species (Perry et al., 2005; Brunel and Boucher, 2007; Enghoff et al., 2007; van Damme and Couperus, 2008; Dulvy et al., 2008), growth of some juveniles but not others (Teal et al., 2008; Todd et al., 2008), and biomass of others (Tulp et al., 2008). In the western North Atlantic, several responses have been reported, including changes in recruitment patterns of several fish species (Brodziak and O'Brien, 2005; Tolan and Fisher, 2009), changes in departure and arrival times (Frank et al., 1990), and fish population dynamics (Cowen, 1985; Lenarz et al., 1995; Arcos et al., 2001; Rebstock, 2003; Smith and Moser, 2003), including those for pelagic and benthic species (Murawski, 1993) and parasites (Cook et al., 1998). Climate variability has been invoked as a contributor to the decline of *Gadus morhua*, one of the most important species in the region (Hoffman and Powell, 1998; Drinkwater et al., 2003) and has had a clear influence on the recruitment of *Limanda ferruginea* (Sullivan et al., 2005).

Few studies have considered the link between ecosystem structure, fisheries production, and climatic forcing along the East Coast of the United States (Massachusetts to Florida), but these have become more common over the past few decades. Murawski (1993) and Mountain (2002) documented links between inter-annual variability in fish distributions and water temperature. Other studies have demonstrated how warming temperatures have had effects on growth, which vary between species (Teal et al., 2008; Todd et al., 2008). An increase in tropical fish abundance and diversity on reefs on the southeast U.S. shelf (Parker and Dixon, 1998) has been associated with warmer winter water temperatures (but see Quattrini et al., 2004). At the same time, a decrease has occurred in boreal fish species in Narragansett Bay (Oviatt, 2004). In addition to changes in distribution, a link between recruitment, shelf bottom temperature, and the NAO has been demonstrated for *Limanda ferruginea* on the northeast U.S. shelf (Sullivan et al., 2005). Finally, in a study of Chesapeake Bay climate and fisheries, Austin (2002) argued that a regime shift occurred in 1977, and that this shift coincided with a change in the juvenile fish community. Another potential mechanism is the apparent correlation between pejus temperature (i.e., temperature beyond which the ability to increase aerobic metabolism is reduced) and a reduction in the scope for growth and reproduction (Pörtner and Knust, 2007). These studies indicate that climatic variation in temperature may have overarching effects on the fisheries of the East Coast of the United States.

EFFECTS ON ESTUARINE FISHES

The expected biological consequences of climate change (i.e., effects on physiology, distributions, phenology, and adaptation [Hughes, 2000]) are likely to occur for estuarine fishes as well. These influences on estuarine fishes can be expressed as assemblage-wide changes in estuarine fish ecology (Attrill and Power, 2002). Change could also occur in various life history stages outside of the estuary, based on the timing and location of spawning, as eggs and larvae of these species are being transported to or swimming to estuaries, and after they leave estuaries for overwinter refuges and feeding sites. Among those likely to be the most responsive are several species (*Centropristis striata*, *Paralichthys dentatus*, and *Pomatomus saltatrix*) that use estuaries as young-of-the-year (Murawski, 1993). Others have demonstrated change in young-of-the-year fish abundance in Chesapeake Bay (Wood, 2000; Austin, 2002; Wingate and Secor, 2008) and Narrangansett Bay and Long Island Sound (McLean, 2006; Collie et al., 2008; Wood et al., 2009a). Temperate estuaries are especially likely to be impacted by climate change because of the inherent variability in these systems (Straile and Stenseth, 2007); the common occurrence of extensive shallow areas that are likely to respond to changes in ocean and air temperature; and the potentially negative effects on productive habitats, including salt marshes and sea grasses (Stevenson et al., 2002; Day et al., 2008).

RECENT EVIDENCE FROM TEMPERATE ESTUARIES ON THE U.S. EAST COAST

There has been an increase in average temperatures in the past half century and this is evident for Chesapeake Bay

(Austin, 2002), the Hudson River (Ashizawa and Cole, 1994), and Narragansett Bay (Oviatt, 2004). This increase in temperatures has been suggested as the reason for the decline of northern species (e.g., *Osmerus mordax* and *Microgadus tomcod*) at the southern limit of their range in the Hudson River (Daniels et al., 2005). A re-examination of the hypothesis that winter temperature variability controls *Micropogonias undulatus* population dynamics was based on abundance indices that were analyzed at four life history stages from three regions along the East Coast of the United States (Hare and Able, 2007). Correlations suggested that year-class strength is decoupled from larval supply and determined by temperature-linked, overwinter survival of juveniles. Using a relation between air and water temperatures, estuarine water temperature was estimated from 1930 to 2002. Periods of high adult catch corresponded with warm winter water temperatures. Prior studies indicate that winter temperature along the East Coast is related to the NAO; variability in catch was also correlated with the NAO, thereby demonstrating a link between *Micropogonias undulatus* dynamics, thermal limited overwinter survival, and the larger climate system of the North Atlantic. We hypothesized that the environment drives the large-scale variability in abundance and distribution, but fishing and habitat loss decrease the resiliency of the population to periods of poor environmental conditions and subsequent weak year classes.

A diverse array of long-term monitoring data and selected experiments has provided new insights into the effects of climate change in the study area. A temperature time series, gathered from the mouth of the Rutgers University Marine Field Station (RUMFS) boat basin just inside Little Egg Inlet from 1976 to 2006 (see Chapter 3), has demonstrated important trends across all seasons. For example, although winter temperatures are variable from year to year, they are milder since the late 1990s, with less frequent temperatures below 0°C than in the period from the late 1970s and to the 1990s (Fig. 10.1). Over the same time period, the average annual temperatures for spring, summer, and fall have increased as well, with a pronounced and consistent rise in temperatures occurring since 2001 (Fig. 10.2).

Over the same period of increasing temperatures, there

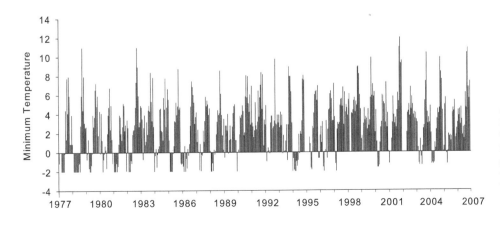

Fig. 10.1. Minimum weekly winter (weeks 47–52 and 1–12) temperatures °C in Great Bay, New Jersey, based on weekly collections at and near the mouth of the Rutgers University Marine Field Station boat basin in Great Bay, New Jersey, during 1976–2007.

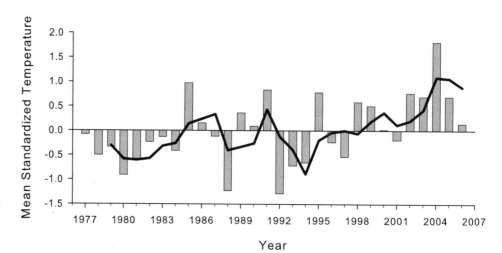

Fig. 10.2. Annual variation in monthly mean temperatures for spring, summer, and fall (weeks 12–47) during 1976–2007.

Table 10.1. Geographical sources (northern versus southern) and abundance of larval (ichthyoplankton net) and juvenile (wire mesh trap) fish collections in Great Bay, New Jersey, relative to the potential source populations. Northern species are those that spawn or have the center of their distribution north of our study area in the Middle Atlantic Bight. Southern species are those that spawn or have the center of their distribution south of our study area.

Species	Larval	Juvenile	Species	Larval	Juvenile
Northern			Lagodon rhomboides	5	1
Clupea harengus	5375	2	Chasmodes bosquianus	—	5
Tautogolabrus adspersus	119	2676	Ophichthus gomesi	5	—
Ammodytes americanus	4025	—	Chilomycterus sp.	4	—
Gasterosteus aculeatus	529	398	Membras martinica	4	—
Pollachius virens	4	218	Pogonias cromis	—	4
Pholis gunnellus	174	—	Bothus sp.	3	—
Myoxocephalus aenaeus	99	44	Caranx hippos	3	—
Enchelyopus cimbrius	117	1	Elops saurus	3	—
Ulvaria subbifurcata	31	—	Muraenidae sp.	3	—
Gadus morhua	2	—	Tetraodontidae sp.	3	—
Lumpenus maculatus	2	—	Citharichthys spilopterus	2	—
Microgadus tomcod	—	2	Epinephelus sp.	1	1
Lumpenus lumpretaeformis	1	—	Pomacentrus partitus	—	2
Southern			Acanthurus sp.	—	1
Micropogonias undulatus	18,588	5	Albula vulpes	1	—
Bairdiella chrysoura	26	1369	Aluterus schoepfii	1	—
Gobionellus boleosoma	540	2	Apogonidae sp.	—	1
Opisthonema oglinum	253	—	Bothus ocellatus	1	—
Chaetodon ocellatus	—	248	Chilomycterus schoepfii	1	—
Myrophis punctatus	153	—	Epinephelus striatus	—	1
Symphurus sp.	153	—	Etropus crossotus	1	—
Sphoeroides maculatus	132	—	Gymnothorax sp.	1	—
Leiostomus xanthurus	110	14	Haemulidae sp.	1	—
Hypsoblennius hentz	32	79	Hypleurochilus geminatus	—	1
Microgobius thalassinus	69	—	Hyporhamphus meeki	1	—
Synodus foetens	40	1	Lactophrys sp.	1	—
Mycteroperca microlepis	—	36	Microdesmus longipinnis	1	—
Serranidae sp.	1	33	Monacanthus hispidus	1	—
Gerreidae sp.	31	—	Mycteroperca phenax	—	1
Orthopristis chrysoptera	3	20	Ophidion welshi	1	—
Gobionellus oceanicus	22	—	Pomacentrus leucostictus	—	1
Gobionellus sp.	22	—	Pomacentrus sp.	—	1
Chaetodon capistratus	—	18	Rachycentron canadum	1	—
Mycteroperca sp.	—	10	Scomberomorus maculatus	1	—
Lutjanus griseus	2	6	Sphoeroides sp.	—	1
Blenniidae sp.	2	5	Sphyraena barracuda	1	—
Mugil sp.	7	—	Sphyraena borealis	1	—
Chaetodon sp.	2	4	Symphurus plagiusa	1	—
Dormitator maculatus	6	—	Synodontidae sp.	1	—

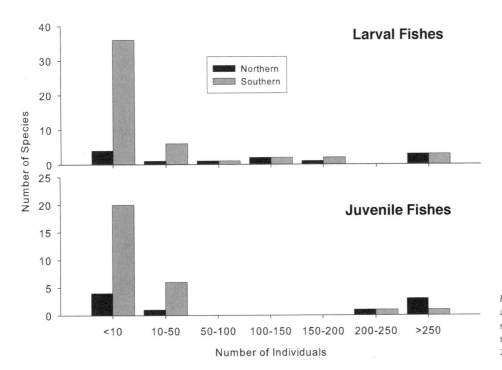

Fig. 10.3. Contribution of northern and southern larval (ichthyoplankton samples during 1987–2006) and juvenile (wire mesh traps during 1991–2006) fishes to this analysis.

has been a response by the fauna, which varies as a function of the source of the fauna. This is especially evident from the larvae of transient forms. Extensive ichthyoplankton collections from inside Little Egg Inlet from 1989 to 2006 have demonstrated that the larvae are derived from multiple sources from geographically diverse spawning sites (see Chapter 4): areas from the south, including the Sargasso Sea (Bell et al., 2003; Sullivan et al., 2006a), the South Atlantic Bight (Warlen et al., 2002), and the Gulf of Mexico (see species accounts) as well as from sites farther north, including Georges Bank and the Gulf of Maine (Witting et al., 1999; Hagan and Able, 2003) and the adjacent Middle Atlantic Bight (Able and Fahay, 1998). These overlapping ranges, from multiple sources, imply that the central portion of the Middle Atlantic Bight may represent borders for multiple species and, as such, may provide an effective location for evaluation of climate change effects on species ranges (Holt and Keitt, 2005; Dunstan and Bax, 2007).

The highest proportion of larvae is from families and species of southern origin (Table 10.1), with individual species represented by a variable number of individuals (although the occurrence of many is based on < 10 individuals; Fig. 10.3). These southern larvae typically occur during late summer and early fall with the peak in September (Fig. 10.4), when ocean water temperatures in the regions are the warmest (see Chapter 2). Given their southern source, these are presumably the result of their delivery by the northward-flowing Gulf Stream (Epifanio and Garvine, 2001; Hare et al., 2002). This type of delivery system, at least for tropical fishes, is typical in other parts of the world where oceanographic conditions can distribute larvae widely (Hutchins, 1991; Hutchins and Pearce, 1994; Booth et al., 2007). The contribution of northern species and genera of larvae is less diverse

(Table 10.1) but represented by a variable number of individuals per species (Fig. 10.3). The northern larvae occur primarily during the winter and spring (Fig. 10.4), when ocean temperatures are coldest (see Chapter 2). The source of these northern larvae is likely the result of southward-flowing water influenced by the Labrador Current.

Larval species composition and abundance varied over the period from 1989 to 2006 in Little Egg Inlet (Fig. 10.5), apparently in response to warming temperatures over the whole year (see Figs. 10.1 and 10.2). As average annual temperatures warmed during the late 1990s to 2006, the number of larvae represented by northern species declined and the number of southern species increased relative to the long-term average over that period. The pattern of decline of northern species and increase of southern species was also evident in terms of larval abundance for both of these groups. The larvae that are responsible for the composite patterns observed can be identified. For example, the larvae of several southern species, including *Opisthonema oglinum* and *Micropogonias undulatus* (see species accounts), became more abundant during the period of warming temperatures after 2000. The latter was particularly abundant during this period and our prior analysis interpreted this as enhanced juvenile survival during the first winter along with a northern expansion of the range with enhanced spawning due to milder winters in the study area (Hare and Able, 2007). Over the same period, several northern species, including the larvae of *Clupea harengus* and the adults of *Gasterosteus aculeatus* (see species accounts), became less abundant.

The species composition and abundance for juveniles were higher from southern forms as were the larvae (see Table 10.1). The contributions for juveniles were represented both by a small number of a few individuals and by large

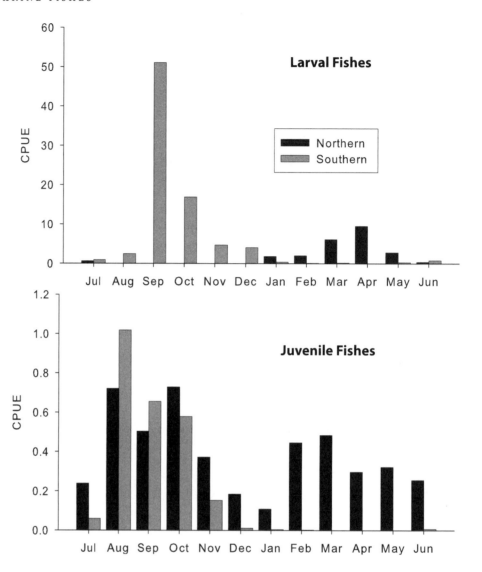

Fig. 10.4. Monthly abundance (CPUE, or catch per unit of effort) of northern and southern larval (ichthyoplankton net samples during 1987–2006) and juvenile (wire mesh traps during 1991–2006) fishes.

numbers of a few species. When we examined data from long-term sampling for juveniles of northern and southern species (see Table 10.1), the trend was not as consistent (Fig. 10.6) as that for the larvae. The juveniles of northern species were represented by fewer species and were less abundant since 2003, while those of southern origin were below or near the long-term average. The juveniles of northern individuals were represented in samples from all months, with peaks in the winter and spring (February–May) but also during summer and early fall (August–October) (see Fig. 10.4). These values are reflective of the large numbers of *Tautogolabrus adspersus* collected during the summer in this (see species accounts) and prior studies based on the same databases (Able et al., 2005c). The juveniles of southern species were represented by a relatively large numbers of species represented by < 10 individuals but also a few represented by a large number of individuals (see Fig. 10.3). The southern forms were most common in late summer and early fall during periods of highest temperatures (Fig. 10.4).

A similar pattern of increasing numbers of southern species has been suggested by studies in Narragansett Bay and Long Island Sound (Collie et al., 2008; Wood et al., 2009a). While the late summer transport of southern species to temperate waters of the East Coast has been recognized for a long time (Able and Fahay, 1998), the tempo appears to have increased over the past decade, presumably reflecting warmer estuarine temperatures. It remains to be seen whether these occurrences are translated into a permanent change in the fauna, that is, larvae and young-of-the-year surviving to reproduce.

The inconsistency between the greater contribution of southern larvae and the lack of a similar response by southern juveniles may be the result of processes beyond settlement. A likely contributor to the reduced contribution of juveniles of southern species relative to their larvae is overwinter mortality. The larvae of most southern species arrive at a time of highest water temperatures in late summer or early fall (Fig. 10.7) but shortly thereafter are exposed to

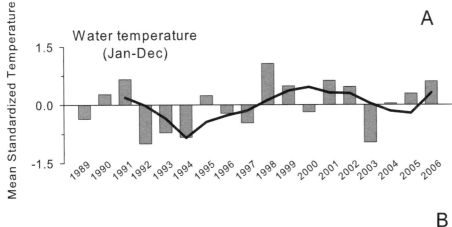

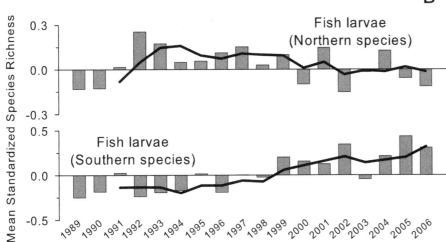

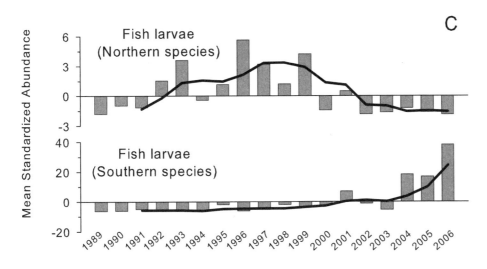

Fig. 10.5. Annual variation (1989–2006) in response of larval fish species richness (B) and abundance (C) based on ichthyoplankton collections at Little Egg Inlet relative to average annual temperatures in Great Bay (A). See Table 10.1 for designation to northern or southern species groups.

declining temperatures, with the lowest temperatures in the study area typically occurring in winter. Small individuals, with reduced swimming ability and small body sizes, may be prone to temperature-dependent mortality occurring as a result of cessation of feeding and direct response to low temperatures (Hales and Able, 2001) or predation (Barbeau, 2000; Taylor, 2005). This is evident for small individuals of

southern species that occur in the study area, such as *Chaetodon ocellatus* (McBride and Able, 1998), *Lutjanus griseus* (M. J. Wuenschel, M. E. Kimball, and K. W. Able, unpubl. data), and *Bairdiella chrysoura* (Able and Curran, 2008). Thus, while the larvae of southern species are known to occur and are even increasing in abundance, their contribution, at least in the central portion of the Middle Atlantic Bight, may be

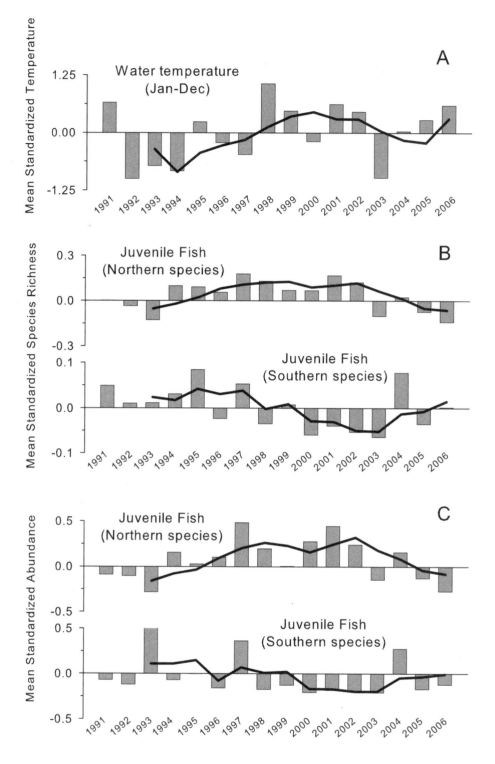

Fig. 10.6. Annual variation (1989–2006) in juvenile fish species richness (B) and abundance (C) based on wire mesh trap collections in Rutgers University Marine Field Station boat basin in Great Bay relative to average annual temperature in Great Bay (A). See Table 10.1 for designation to northern or southern species groups.

minimal or nonexistent because these individuals do not survive and thus should be considered expatriates. This might apply to a large number of southern species (see Table 5.2) and thus account for the smaller number of species collected as juveniles relative to those collected as larvae (see Table 10.1).

One of the possible mechanisms is low winter temperatures, which cause higher rates of mortality among small, relatively immobile fishes. This may occur, especially in their

first winter of life, because individuals of marine and estuarine temperate fishes may experience net energy deficits due to low temperature and food scarcity (see Sogard, 1997, for a review). Small individuals may also accrue an energy deficit due to their relatively low weight-specific standard metabolic rates, relatively low weight-specific energy reserves, and osmoregulatory stress (Calder, 1984; Thompson et al., 1991; Johnson and Evans, 1996; Hurst et al., 2000; Hales and Able, 2001). As a result, size-dependent physiological capaci-

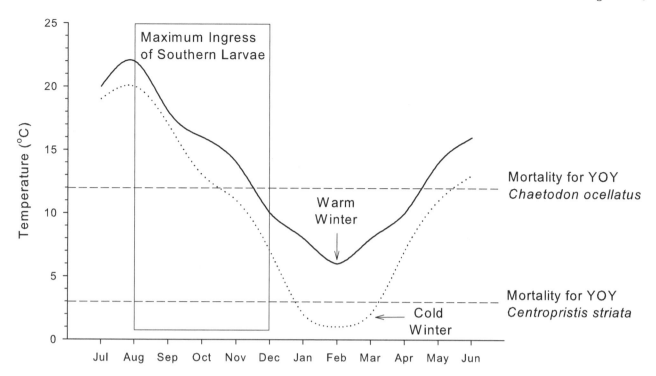

Fig. 10.7. Phenology of ingress of larvae of southern species (see Table 10.1) through a representative temperate estuary relative to representative warm and cold winters and temperatures at which mortality occurs for selected species.

ties may result in low or even negative growth rates and high and size-dependent mortality rates of young-of-the-year fishes during winter (Conover and Ross, 1982; Henderson et al., 1988; Conover and Present, 1990; Shuter and Post, 1990; Hurst and Conover, 1998; Hurst et al., 2000; Hales and Able, 2001; Hurst, 2007). In addition, size-dependent winter mortality may affect many life history traits, such as migration, growth, and reproduction (Fox and Keast, 1990, 1991; Snyder and Dingle, 1990; Conover, 1992; Lankford, 1997), and reportedly impacts species composition, trophic structure, and other aspects of poikilotherm ecology (Persson, 1986b; Hall and Ehlinger, 1989; Shuter and Post, 1990; Atkinson, 1994; Johnson and Evans, 1996).

Effects of winter mortality may be especially pronounced along the East Coast of the United States, where average winter water temperatures decline about 1°C with each degree of latitude and the annual range in sea surface temperatures exceeds 20°C (Schroeder, 1966). However, most information available on winter mortality, growth, and behavior of estuarine and marine fishes is restricted to subtropical faunas and concentrated on severe storm impacts (Günter, 1947; Hoff, 1971; Gilmore et al., 1978), although an improved understanding is developing for temperate estuarine fishes (Schwartz, 1964; Moss, 1973; Able and Fahay, 1998; Hurst and Conover, 1998; Hurst et al., 2000; Hurst, 2007).

Evidence for overwinter mortality is accumulating based on laboratory observations of several obligate and faculta-

tive estuarine users. This is most evident based on comparisons of estuarine year-round residents, seasonal summer residents, and winter residents in laboratory experiments (Table 10.2). As expected, those species that are year-round residents (*Cyprinodon variegatus, Fundulus heteroclitus, Tautoga onitis,* and *Tautogolabrus adspersus*) had the highest tolerance (0% mortality), which was similar to a winter resident (*Pseudopleuronectes americanus*). Those species that occurred in the estuary only during the summer and fall and migrated out of the estuary for a winter thermal refuge (*Bairdiella chrysoura, Centropristis striata,* and *Etropus microstomus*) had higher mortality (33 to 100%), as did the presumed expatriate (*Lutjanus griseus,* 100%). This response is consistent with the highly migratory fishes that are common in the study area (see Chapter 9).

In the conceptual model of *Micropogonias undulatus* population dynamics, winter temperature controls recruitment to the fishery and the oscillation between outburst and non-outburst population phases (Hare and Able, 2007). Winter temperature along the East Coast of the United States is related to the NAO (Joyce, 2002; Hurrell et al., 2003), and thus the dynamics of this species are linked to NAO through winter air temperature, winter estuarine water temperature, and overwinter survival of the juveniles. Although only one species was examined here, there may very well be multispecies patterns in recruitment and population dynamics that are linked to climatic forcing along the East Coast of

Table 10.2. Degree of winter mortality and associated behavior for young-of-the-year of different types of estuarine users based on laboratory experiments.

Species	Length Range Tested (mm)	Lowest Temperature Experienced (°C)	Temperature at Impaired Swimming (°C)	Temperature at Cessation of Feeding (°C)	Temperature at Mortality (°C)	Percent Mortality	Source
Year-round residents							
Cyprinodon variegatus	21–48	−1.4	—	—	—	0	Able and Curran (2008)
Fundulus heteroclitus	31–53	−1.4	—	—	—	0	Able and Curran (2008)
Opsanus tau	143–286	0	2.0	—	2.7	36	Schwartz (1964)
Tautoga onitis	45–86	2	—	< 4	3	14	Hales and Able (2001)
Tautogolabrus adspersus	39–55	2	—	2	3	3	Hales and Able (2001)
Summer residents							
Bairdiella chrysoura	52–109	−1.8	—	—	5	100	Able and Curran (2008)
Centropristis striata	45–72	2	—	4.0	2–3	100	Hales and Able (2001)
Centropristis striata	218–241	0	8.0	5.0	1.9	100	Schwartz (1964)
Chilomycterus schoepfii	107–220	0	6.0	—	4.3	100	Schwartz (1964)
Cynoscion regalis	231–295	0	10.0	—	3.3	100	Schwartz (1964)
Etropus microstomus	32–40	2	—	—	3–5	33	Hales and Able (2001)
Leiostomus xanthurus	156–258	0	4.0	4.0	3.2	100	Schwartz (1964)
Lutjanus griseus	22.5–53.7	7	—	10.3	10.2	100	Wuenschel et al. (in review)
Paralichthys dentatus	125–326	< 2	—	—	< 2	74	Szedlmayer et al. (1992)
Prionotus evolans	225–263	0	7.0	—	3.7	100	Schwartz (1964)
Sphoeroides maculatus	160–270	0	5.6	12.8	3.3–4.4	100	Schwartz (1964)
Sygnathus fuscus	<100>	1	—	—	< 4	82 (< 100 mm) 68 (> 100 mm)	See species account
Scophthalmus aquosus	15–54 (fall cohort) 83–160 (spring cohort)	−2	—	—	1	75 (fall cohort) 31 (spring cohort)	Neuman and Able (2009)
Winter residents							
Pseudopleuronectes americanus	60–126	−1.8	—	—	−1.8	25	Able and Curran (2008)
Micropogonius undulatus	15–65	1	—	—	3	98.7	Lankford and Targett (2001)

the United States. Long-term trends in abundance are known for many of the region's fisheries. These patterns are usually explained by changes in fishing mortality (Chittenden et al., 1993), but the contributing effect of climate variability is now being investigated (Sullivan et al., 2005). Further, the conceptual model for *Micropogonias undulatus* suggests that increased fishing mortality may make populations more susceptible to changes in numbers caused by climatic-scale variability in the environment by reducing the capacity of the population to sustain abundance between strong year classes. Thus, the large-scale patterns in climate described along the East Coast (Joyce, 2002; Hurrell et al., 2003) may have broad impacts on fisheries similar to those described on the West Coast of the United States for groundfish species (Hollowed et al., 2001) and in the Bay of Biscay (Poulard and Blanchard, 2005). These impacts need to be quantified and incorporated into the stock assessment process for successful long-term management of marine fish species.

While the direct effects of climate-induced temperature change are the most easily detected, the indirect effects may be just as likely. There are several ways in which this could occur in the study area. First, milder fall and early winter may provide additional time for species to migrate to a thermal refuge, either at the edge of the continental shelf or to the south (see Chapter 9). At the same time, milder temperatures may provide a refuge of greater spatial extent and thus be easier to reach. To some extent this kind of mechanism has provided increased habitat for juvenile *Micropogonias undulatus* in the Middle Atlantic Bight (Hare and Able, 2007). All of the above are most likely to have impacts on small juveniles with relatively small energy reserves. Second, indirect effects of temperature may influence adults as well. For those species that overwinter to the south, mild winters may cause a shorter retreat in the fall and winter and allow them to migrate farther north in the spring. A possible example is *Leiostomus xanthurus* (Norcross and Bodolus, 1991). This response may cause them to spawn farther north than usual and allow the estuarine-dependent larvae and juveniles to reach more northern estuaries in the summer.

While climate change effects are commonly interpreted relative to temperature change, other related impacts have been identified, including altered precipitation. On the East Coast of the United States winters characterized by high levels of precipitation typically lead to elevated levels of freshwater runoff (Schmidt and Luther, 2002). Increased freshwater discharge triggers a drop in salinity as well as an increase in inputs of chemical odorants into the water column. A number of studies have documented the strong role chemical cues play during the early life history of *Anguilla* spp. For example, the ability of *Anguilla anguilla* glass eels to detect terrestrial chemical cues increases significantly with more freshwater (Sola and Tongiorgi, 1996). Other studies have found that glass eels preferentially orient toward water treated with bacteria, algae, and other substances (Sola, 1995; McCleave and Jellyman, 2002). Discharge-related plumes carrying chemical signatures available to glass eels may extend for many kilometers outside of an estuary with a residence time of weeks. Thus, glass eels approaching a coastline may pick up this signal and orient themselves accordingly. Once near an estuary mouth, individuals may then use selective tidal stream transport to complete their long-distance journey and take up an estuarine existence as elvers. Following these arguments, there is minimal synchrony in ingress between inlets since there is minimal synchrony in precipitation. This interpretation is consistent with an analysis of time series data from two distant inlets (Sullivan et al., 2006a). In general, years with higher-than-average levels of winter precipitation (December–March) were significantly correlated with an above-average abundance of glass eels. Further, absolute peaks in precipitation for both sites (1994, Little Egg Inlet; 1998, Beaufort Inlet) mirrored the year of maximum glass eel abundance. Although precipitation was not coherent between sites on an annual basis, it appeared to be related to the El Niño-Southern Oscillation (ENSO) at

Beaufort Inlet and, to a lesser extent, the NAO at Little Egg Inlet. Levels of precipitation and freshwater discharge have been cited as important factors regulating ingress variability of glass eels for other *Anguilla* species (Chen et al., 1994; Jellyman and Lambert, 2003).

FUTURE SCENARIOS

The difficulty in determining fish response to climate change should not be surprising, given that understanding the response to environmental change, in general, is difficult (Rose, 2000). The difficulties identified include several categories, some of which are certainly relevant to climate change effects, including detectability, complex spatial heterogeneity in habitats, difficulty in obtaining regional predictions, poor understanding of community interactions, and the role of sub-lethal and cumulative effects. As an example, "complex habitat responses" may be associated with climate change, as a result of changing patterns of precipitation that could influence larval ingress to estuaries (Sullivan et al., 2006a), or rising sea levels could change the amount, location, and structure of estuarine salt marshes. This might be especially evident where salt marshes are bordered by extensive natural uplands, such as along rocky shorelines, or where armored shorelines or roads prevent the inland migration of salt marshes. Further, detectability of climate-induced changes is inherently difficult because of the high annual variation in fish populations, especially during the first year of life (Sissenwine, 1984; Rothschild, 1986; Houde, 1987; Fogarty et al., 1991; Able and Fahay, 1998). This supports the need for long-term datasets (Keeling, 1998; Holmes, 2006). All of this is further confounded by the uncertainties, both global and regional, about climate change itself (Mantua and Mote, 2002). The inability to understand community interactions, even those associated with fishes in unaltered estuaries, will increase with the added effects of climate change (Durant et al., 2007).

11

Future Directions

PERSPECTIVES ON ESTUARINE ECOSYSTEMS AND FISHES

By their very nature, estuaries are physically connected to freshwaters upstream and the ocean downstream. They are also linked to more distant waters through the life histories of fishes and their predators and prey. Because of these connections, it is difficult to define the limits of estuaries. It is clear that the area of estuarine influence is much greater than simply the physical limits of the estuary (e.g., Pritchard's semi-enclosed system) itself. At a minimum, we need to consider those portions of the inner continental shelf that are directly and physically influenced by estuaries, such as the water column that originates from nearshore buoyancy-driven flows and the ebb tidal deltas at the mouths of many estuaries. These inner shelf areas that are under the physical influence of estuaries are also used by fishes in much the same way as the semi-enclosed portion of the estuary, that is, they reproduce, develop, feed, and grow in both areas, as exemplified by several sciaenids and engraulids (see Chapter 7 and species accounts). At the other end of the estuary, the tides from the ocean extend this influence up into freshwaters, which may affect typically brackish water fishes (Kraus and Secor, 2004a). This broader perspective helps us recognize that most species of fish that use estuaries also use the ocean, as we have pointed out for all the transient species, and that many species, such as some of the ictalurids and cyprinids (see Chapter 7), which typically occur in freshwater are also found in the low-salinity portions of estuaries.

An especially important component of an improved perspective is a fuller understanding of the spatial scale of the variability in these systems. This is evident in the enhanced focus on marine metapopulations (Kritzer and Sale, 2006) as it applies to marine ecology in general (Sale et al., 2006) and conservation of marine resources (Crowder and Figueira, 2006), including fishes (Kritzer and Sale, 2006; Ciannelli et al.,

2007). The further application of these ideas to the fishes of spatially isolated estuaries provides the conceptual framework for detailed studies of population regulation (Jones, 2006; Manderson, 2008).

Just as it is important to understand the spatial (habitat) variability in estuarine ecosystems, it is also necessary to grasp the nature of the connectivity between the different components of the estuarine landscape, both within and between estuaries and adjacent ocean and freshwaters (Able, 2005; Secor and Rooker, 2005; see Chapter 9). These linkages between estuaries are becoming increasingly recognized as new techniques become available (see below). Many of these have uncovered additional complexity in the migrations of fishes (Able and Fahay, 1998; Limburg, 2001; Pittman and McAlpine, 2001; Katselis et al., 2007), including the occurrence of migration relative to individual condition (Thibault et al., 2007) and climate (Boisneau et al., 2008). For estuarine fishes these linkages may be between estuaries, as is becoming evident for *Morone saxatilis* (Able and Grothues, 2007b; Grothues et al., 2009) and *Acipenser oxyrinchus*. Within estuaries, the movements between habitats, for example, salt marshes and the rest of the estuary (Able et al., 2007a), can occur for reproduction, for feeding, and in response to abiotic (see Chapter 9) and biotic variables that we do not yet fully appreciate. In fact, these inter- and intra-estuarine connections may be more apparent in temperate estuaries, which experience pronounced seasonal temperature regimes and thus more migrations than other estuaries. This connectivity is critical to understanding how estuarine fishes use different habitats. Fortunately, the focus on habitats has been made a requirement in many systems because of the need to identify essential fish habitat in the United States (see Chapter 7).

Recent attempts to codify some of this spatial variation inherent in fish populations has led to an appreciation of the diversity of biological levels that need to be incorporated into our thinking. These include the hierarchy stretching from species to subspecies, metapopulations, populations, subpopulations, contingents, year classes, cohorts, schools, broods, and individuals (Secor, 1999). One wonders if this diversity is not more important in highly migratory fishes, such as those that dominate in temperate estuaries. Certainly, the growing recognition of the influence of metapopulations and contingents is becoming increasingly frequent for marine fishes (Kritzer and Sale, 2006), including those that use estuaries (Secor and Rooker, 2005; Secor and Kerr, 2009).

Given this broadened perspective on estuaries, it is also worth reconsidering the natural sources of variation that occur within them. In the past, much of our interpretation of the role of fishes in estuaries has been largely based on the abiotic variables that dominate these systems, including salinity, temperature, tides, and the like, perhaps because they are so dominant and easily measured. There has been much less emphasis on biotic interactions. However, it is likely that these biotic variables are central to understanding the population regulation that must occur in estuaries, at least for those species that spend much of their early life history in these systems.

We clearly lack information on the role of these biotic variables, especially predation (Moyle and Cech, 1996; Helfman et al., 1997). We earlier pointed out the diversity and abundance of predators in estuaries (see Chapter 8). Recent studies are emphasizing that broad-scale changes, including trophic cascades, have occurred in marine esosystems because of the removal of large predators through fishing (Myers and Worm, 2003; Frank et al., 2005; Myers et al., 2007; Ottersen, 2008) and this may be true of estuarine ecosystems as well (Heck and Valentine, 2007). Further, the ecological consequences of the decline in predators may be underestimated because we are seldom able to understand the behavioral modifications due to the risk of predation (risk effects) (Heithaus et al., 2007; Creel and Christianson, 2008). As a result of all of the above, we suggest that in order to grasp the ecological importance of fishes in estuaries, it may be appropriate to focus on several questions. Does the degree and/or influence of piscivory (including cannibalism) vary between estuaries and adjacent freshwaters or the ocean? Does piscivory differ within estuarine habitats? Does piscivory vary by latitude and season and predator type (fish versus birds versus mammals)? These questions may be difficult to answer because often predators are mobile or there are multiple contingents, including those that differ in terms of when and where they forage.

In addition to fishing, there are numerous other human influences, including habitat loss and alteration through eutrophication, contamination, freshwater diversions, and introduced species (Kennish, 2002; Hollister et al., 2008). These negative effects are responsible for the increasing attempts to restore estuarine watersheds (Palmer, 2009). In turn, the complexity of these systems requires a better understanding of estuarine ecosystems in general (Cowan et al., 2008; Heck et al., 2008) and ecosystem-based fishery management (Link, 2002).

As our understanding of climate change effects in the North Atlantic and elsewhere advances (see Chapter 10), our ability to understand the impact on estuaries may improve as well (Keenlyside et al., 2008). This information will help us know whether climate change might increase flooding frequency due to sea level rise (Cooper et al., 2008), cause more open water to replace marshes, as in Chesapeake Bay (Stevenson et al., 2000), or eliminate estuaries if fishes have no place to migrate inland, as caused by diked marshes in Delaware Bay or armored urbanized estuarine shorelines. Climate change may also affect fishes directly through modification of predator/prey relationships (Wells et al., 2008), physiological effects (Pörtner et al., 2008), and the phenology of migrations for fishes, as occurs for birds (Gordo, 2007; Hedenstrom et al., 2007; Coppack et al., 2008; Moller, 2008).

One of the most unknown factors affecting the population dynamics of estuarine fishes is the role of diseases and parasites. While there is an increasing number of studies of fish diseases, many of these are associated with aquaculture operations and there is a still a very big gap in our understanding of the importance of the role of disease in natural estuarine ecosystems. There is growing evidence (see Chapter 6) that parasites, including mobile temporary parasites and micropredators (Penfold et al., 2008), play an important role in the life history, ecology, and behavior of fishes (Poulin, 1995; Barber et al., 2000; Lafferty et al., 2008). Clearly, more attention to the effects of parasites on the population dynamics of their host fishes is warranted (Barber et al., 2000).

SUGGESTIONS FOR FUTURE RESEARCH

Recognizing the past and present impediments to estuarine research may help us address how and where they have created gaps in our understanding of estuarine fishes. At present, of the numerous species in temperate estuaries we only have sufficient data to summarize the basic life history and ecology for 93 species in the study area covered by this book. Some of this uncertainty or lack of information stems from the fact that fishes are still the most poorly known group of vertebrates in the world. Much of this is the result of sampling deficiencies for individual gears and other confounding factors for fishes in general (Allen et al., 1992; Johnson et al., 2008), including in estuaries (Able, 1999). It is worth noting that only recently has the importance of sound as a cue in the location of coral reef nursery habitats been emphasized (Montgomery et al., 2006). One wonders if this cue can apply to other habitats, such as surfzones on open beaches and flood tidal deltas at estuarine inlets, and thus help species spawned in the ocean to find estuaries. Additionally, a recent phenomenon is an improved understanding of the swimming ability of early life history stages and the role it may play in recruitment (Leis, 2006). Other impediments to understanding estuarine fishes include the very simple, but frequent, inability to observe estuarine fishes in their natural habitat. Typically, temperate estuaries are turbid and this often precludes even the simplest observations. The kinds of studies in coral reef systems that allow insightful observations and experiments on biotic interactions are not possible in most estuaries. Emphasis in estuaries is often on abiotic factors simply because we lack the ability to see what habitats fishes are using and how they are behaving. Of course, new techniques are being developed (see below), but these cannot replace the contribution of human vision to the study of fishes.

Further, as has often been recognized, the complex life histories of fishes make quantifying population dynamics a challenging task In addition, the connectivity between expansive ocean spawning areas and multiple estuarine nurseries makes it logistically difficult, or impossible in some instances, to adequately monitor the various life history stages in order to understand factors influencing the growth and survival of individuals and populations, especially when they are highly migratory. It is also becoming apparent that there is geographical variation in the manner in which even the same species of fishes use different estuaries (see Chapter 1). In addition, while we have focused on temperate estuaries from the East Coast of the United States, we would benefit from the broader perspective that could be gained by comparing these estuaries to other temperate estuaries on the West Coast of the United States and elsewhere in the world as well as with tropical/subtropical and boreal/arctic estuaries to determine if they are similar in structure and function.

A further factor confounding our ability to study estuarine fishes is that fisheries, both recreational and commercial, can, and in fact are designed to, have an influence on the populations of many species. For example, of the 93 species treated in this book, many are the focus of one or more fisheries, either in estuaries or in the ocean. As a result of all of the above, we still need to concentrate on the basic patterns of the life history and habitat use of estuarine fishes in much the same way that we need to understand patterns in many marine environments (Underwood et al., 2000). Simultaneously, we need to be aware of the shifting baseline that is occurring as the result of climate change (Pauly, 1995; Pinnegar and Engelhard, 2008).

New approaches can improve our understanding of estuarine fishes and ecosystems. For the fishes themselves, the methods vary with life history stage. For larvae, otolith microchemistry has proven useful in determining the timing of the larval duration and site fidelity to natal estuaries (Thorrold et al., 2001; Correia et al., 2004) and the occurrence of extreme events (Berghahn, 2000). For all life history stages, but primarily juveniles and adults, stable isotopes have helped us understand the variable contributions of different sources of primary production and how they change with movements (Currin et al., 2003; Litvin and Weinstein, 2003).

The miniaturization of coded wire tags (Wallin et al., 1997) has been especially useful for elucidating habitat use patterns, movement, and growth in small juvenile estuarine fishes (Teo and Able, 2003; Able et al., 2005a; Able et al., 2006a). For many juveniles and certainly adults of most species, acoustic techniques, such as passive acoustics to detect the presence and abundance of many species, has had many applications, although species identification continues to present some challenges. More recently, the development of Dual Frequency Identification Sonar (DIDSON) has begun to solve some of these problems through enhanced "acoustic video" that allows species identification based on shape, size, and behavior. This same approach provides for identification and mapping of structural habitat types.

For larger juveniles and adults, telemetry techniques (Able and Grothues, 2007b) have allowed the development of estuarine observatories (Grothues et al., 2005) for evaluating patterns of estuarine habitat use, residency, and movements (Metcalfe and Arnold, 1997; Able and Grothues, 2007b; Ng et al., 2007), including responses to episodic events (Anderson et al., 2008; see Chapter 7). The addition of autonomous underwater vehicles can dramatically enhance the spatial and temporal coverage of the tracking of telemetered fish while mapping habitat types with sidescan sonar and a suite of environmental sensors (Grothues et al., 2009). Archival tags (Boehlert, 1997) can detail aspects of physiology for individual fishes (Greene et al., 2009). For reproducing fishes, acoustic identification of spawning habitat is possible (Luczkovich et al., 1999). Other approaches, such as in situ video in towed camera systems (Diaz et al., 2003) and specialized lighting, have made the study of fish behavior in typically turbid estuaries easier and possible, including nighttime observations (Gibson et al., 1998; Kimball and Able, in review).

At broader scales, estuarine landscapes are being mapped and monitored with a variety of aerial and satellite mapping techniques. These same approaches allow for landscape-scale interpretation of sea surface temperatures and phytoplankton abundance. At smaller scales, data loggers are providing archived and real-time environmental monitoring, as occurs in National Estuarine Research Reserves (Kennish, 2004). At these broader scales, modeling of marine systems has proven more effective if it incorporates the major abiotic and biotic components, including the entire life history of the participants at the appropriate temporal and spatial scales as well as connections between habitats at all life history stages (Crowder and Figueira, 2006). Similar modeling approaches may enhance our ability to predict the effect of climate change on fisheries potential (Cheung et al., 2008). Obviously, a variety of the above techniques used in tandem provides a powerful toolbox for understanding the structural and functional aspects of estuarine fish ecology at diel, tidal, seasonal, and annual scales.

12

Acipenser brevirostrum Lesueur

SHORTNOSE STURGEON

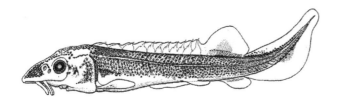

DISTRIBUTION

This species is found along the coast of North America from St. John River, New Brunswick, to the St. Johns River in northern Florida (Dadswell et al., 1984). Well-documented populations currently are found in the Delaware, Hudson, and Connecticut rivers within our study area. Other populations once occurred in major tributaries to the Chesapeake Bay, including the Potomac and Susquehanna rivers, but they are now extinct in those systems (Murdy et al., 1997). Early life history stages have been reported from tidal rivers in many estuaries in our study area.

REPRODUCTION AND DEVELOPMENT

Maturity is reached in three to five years in males, six to seven years in females. Fecundity ranges from 27,000 to 208,000 eggs per female (Dadswell et al., 1984). Spawning occurs during February (in Georgia), the middle two weeks of April (in the Delaware River), the last week in April and first week in May (in the Hudson River), the first two weeks of May (in the Connecticut River), late April (in Massachusetts), and mid-May (in Canada). Eggs are deposited on rocky substrates in freshwaters, often under falls or in rapids; temperatures are 9 to 12°C when spawning occurs (Dadswell, 1979; Taubert, 1980a, b; Kieffer and Kynard, 1996).

Early stages, from yolk sac to early juvenile, have been described, including methods to distinguish this species from *Acipenser oxyrinchus* (Snyder, 1988). Eggs are benthic and adhesive, and average 3.0 mm in diameter. They are reportedly dark brown to black, but may have a light gray polar body (Dadswell, 1979). They are separate when first spawned, but

become adhesive within 20 minutes of fertilization, effected by spoke-like protuberances of the chorion (Markov, 1978). Larvae hatch at sizes of 7.3 to 11.3 mm, and eyes are unpigmented (Taubert, 1980a, b). Barbels are formed around the mouth at a length of 12 to 14 mm SL, and yolk is absorbed at a length of 15 mm SL. Head and tail are darkly pigmented, but pigment is lacking over the ventrolateral parts of the gut and on the pectoral fin base. At a length of 20 mm, scutes (or shields) occur in dorsal, lateral, and ventral rows on the body (Pekovitch, 1979). Transformation to the juvenile stage occurs at 57 to 67 mm SL. Juveniles are best distinguished from those of *A. oxyrinchus* by pigment characters, width of mouth, and fin ray counts in the caudal, anal, and pelvic fins (Snyder, 1988).

LARVAL SUPPLY, SETTLEMENT, AND GROWTH

Larvae and juveniles derive from spawning in freshwaters adjacent to low-salinity estuaries, where they are sometimes found. They are demersal immediately after hatching. Growth is rapid in early stages. By the end of the first growing season, juveniles are between 14 and 30 cm. Juveniles in the Hudson River may reach 25 cm by the end of their first growing season, and growth rate averages 3.0 mm per 10-day period (Pekovitch, 1979). Growth rate is then very slow in individuals between 1 and 9 years old (Pottle and Dadswell, 1979).

SEASONALITY AND HABITAT USE

This species occurs in rivers, estuaries, and near-coastal waters, but they are most abundant in estuaries. Captures at sea have been within a few miles of the coast (Schaefer, 1967; Wilk and Silverman, 1976). They are anadromous, but landlocked populations occur as well (Dadswell, 1979; Taubert, 1980a, b). Ripe adults have been collected as far upstream as River Kilometer 222 in the Delaware River and River Kilometer 246 in the Hudson River (Hoff, 1965; Dovel, 1981). Larvae hatch in freshwater, well above tidal influence. Eggs and larvae have been collected at River Kilometer 190 in the Connecticut River (Taubert, 1980a, b). Spawning habitats described to date are in regions of relatively fast flow, with gravel or rubble substrates (Taubert, 1980a, b; Kieffer and Kynard, 1996), and these are generally located well upstream of summer feeding or nursery areas.

Larvae have also been found to avoid light (Buckley and Kynard, 1981). Larvae and juveniles are demersal and occupy channels with depths greater than 9 m where currents are strong (Dadswell, 1979; Pottle and Dadswell, 1979). They prefer substrates of sand or gravel and may seek cryptic habitats in order to remain hidden (Pottle and Dadswell, 1979). Juveniles may occur upstream of saline water until reaching a length of 45 cm FL at ages greater than 2 years (Dadswell, 1979). Adults usually spend the summer in areas with little or no current (McCleave et al., 1977). They generally lie motionless in deep water during the day, but move into shallower water or upstream or downstream at night (Dadswell, 1976;

Hastings et al., 1987). The possible influence of temperature on distribution has not been determined (Dadswell et al., 1984), but it has been observed that adults are seldom found in shallow water after temperatures exceed 22°C (Dadswell, 1976; Dovel, 1978). In the St. John River in Canada, they overwinter in temperatures between 0 and 13°C and in deeper waters (Dadswell et al., 1984; Li et al., 2007).

PREY AND PREDATORS

Juveniles feed mostly on benthic crustaceans, especially cladocerans (Pottle and Dadswell, 1979), and insects (Carlson and Simpson, 1987). Adults also feed on mollusks, such as *Mya arenaria* and *Crangon septemspinosa,* or small flounders (McCleave et al., 1977; Dadswell, 1979). Documentation of predation includes the report of 24 juveniles in the stomachs of perch (*Perca flavescens*) in the Androscoggin River, Maine (Squiers, 1983).

MIGRATIONS

Migrations by adults of this species are generally described as downstream in fall and upstream in spring or fall, resulting in movements from feeding to spawning to overwintering habitats (Dadswell, 1979; Dovel, 1981; Hall et al., 1991).

Some of these migrations can involve considerable distances, such as the report of 160 km between tagging and recapture sites (Dadswell, 1979). Estimates of daily movements include 20 km in Montsweag Bay, Maine (McCleave et al., 1977). In most populations, adults that will spawn the following spring migrate upstream in the fall (led by males), where they overwinter in habitats adjacent to spawning grounds (Dovel, 1981; Dadswell et al., 1984; O'Herron et al., 1993a, b). Some individuals, however, including a few ripening and all non-ripening adults, remain in deep, high-salinity areas; those that are ripening will then move upstream in the spring (Dovel, 1978). Those that do move from overwintering sites to spawning grounds do so at temperatures of 8 to 9°C (Dovel, 1978). Small portions of some populations may also occasionally move short distances to sea (Schaefer, 1967; Wilk and Silverman, 1976) or may occur near the estuary-ocean interface (Wilk and Silverman, 1976). Downstream migrations to feeding areas begin soon after spawning (Dadswell, 1979).

Juveniles, by contrast, are largely non-migratory and tend to remain in rivers just upstream from the salt wedge (Pottle and Dadswell, 1979; Brundage and Meadows, 1982a; Hall et al., 1991). With age, these juveniles tend to occupy habitats farther downstream.

13

Acipenser oxyrinchus Mitchill

ATLANTIC STURGEON

DISTRIBUTION

This anadromous species is found along the coast of North America from Labrador and Ungava Bay to northern Florida (Gruchy and Parker, 1980). Early stages are found in most large river estuary systems in our study area, but are notably absent from the Chesapeake Bay and its rivers, where the population has recently been nearly extirpated (Secor et al., 1997).

REPRODUCTION AND DEVELOPMENT

Maturation is attained in smaller and younger fish in the southern part of the species' range. In the Hudson River, in contrast, maturation is reached in males at a length of 122 cm and at ages of 11 to 21 years and in females at lengths of 183 cm and ages of 15 to 30 years (Young et al., 1988). Individuals do not spawn every year. Intervals in spawning range from one to five years for males and from two to five years for females (Vladykov and Greeley, 1963). Fecundity is related to size and ranges from 400,000 to 8 million eggs per female (Smith et al., 1982). Spawning runs begin in the spring, but actual spawning may not occur until July. Spawning occurs from February to March in northern Florida and from May to July in the St. Lawrence River.

The early development of *Acipenser oxyrinchus*, including characters useful for distinguishing the larvae and juveniles from those of *A. brevirostrum*, has been well described and summarized in Fahay (2007). Eggs are demersal and adhesive, are spherical to oval in shape, and average 2.9 mm in diameter. They are gray to black, with stellate pigment at the pole. Larvae hatch at 7 to 9 mm SL with unpigmented eyes. Yolk is absorbed at about 14 mm. Barbels form around the mouth when larvae are about 9 mm SL. The finfolds are long-lasting, not disappearing until the larvae are 60 to 70 mm

SL. Head and tail are darkly pigmented. Transformation to the juvenile stage occurs at 116 to 136 mm SL. Juveniles are identified by meristic characters, pigmentation, and relative size of the mouth parts (Bath et al., 1981; Snyder, 1988). Dorsal and ventral rows of scutes form early, before lateral rows develop.

LARVAL SUPPLY, SETTLEMENT, GROWTH, AND MORTALITY

Larvae derive from spawning in the oligohaline portions of estuarine systems. Growth rates are faster in the southern parts of the species' range. Age, length, and weight data have been summarized by Hoff (1980). Yolk sac larvae are extremely efficient at transferring yolk energy to body tissues in a wide range of temperatures, although higher temperatures increase the rates of early development (Hardy and Litvak, 2004). Laboratory-reared juveniles released into suitable nursery habitats in the Nanticoke River, Maryland, migrated downstream toward Chesapeake Bay and displayed growth rates of 1.5% per day (Secor et al., 2000). Growth rates of older stages in the Hudson River have been analyzed (Stevenson and Secor, 1999). On average, it takes about 7 years for juveniles to reach 100 cm TL. Both growth rates and survival have been shown to diminish under conditions of hypoxia (Secor and Gunderson, 1998). A conclusion reached in the latter study was that nursery habitat quality in Chesapeake Bay has deteriorated in the past century, due to the increased frequency of hypoxia, and that this degradation has contributed to the decline of Atlantic sturgeon in the bay.

SEASONALITY AND HABITAT USE

This species inhabits shallow coastal waters, usually < 20 m in depth, where it spends most of its life. Individuals also occur commonly in estuaries and rivers. They are known to move into deeper oceanic waters during the colder months and to return to freshwaters when temperatures rise in the spring. In the Delaware River–Delaware Bay system, a two-decade study of captures indicates that most larger individuals occur in the bay during the spring months, all sizes occur throughout the lower tidal river portion from spring through fall, and a few smaller individuals occur in the upper tidal river during late summer and fall (Brundage and Meadows, 1982b). Females may release as many as 2 million eggs during spawning over a rubble bottom in running water more than 3 m deep. Spawning occurs in the oligohaline zone, to and above tidal freshwaters, over solid substrates when temperatures are 13 to 20°C. No nest is built, but the adhesive eggs stick to rocks or logs on the bottom. The young sturgeon hatch a few days after spawning. No parental care is given the eggs or young. The earliest stages are photonegative and seek cover. After development, they become photopositive and begin a slow downstream migration to nursery areas, where they complete their development. This downstream migration is nocturnal and may take 12 days in the Connecticut

River (Kynard and Horgan, 2002). Juveniles have also been found to occupy nursery habitats in the upper tidal portion of the Delaware River, usually from July through December (Lazzari et al., 1986). Temperatures in this study reached as low as 0.5°C. Moser and Ross (1994) showed that swimming of juveniles in estuaries was oriented into tidal currents. Swimming against the current might ensure that they do not drift too far from their feeding habitats. Moser and Ross (1995) found that young *Acipenser oxyrinchus* move little and usually remain for long periods in the same area. Use of these discrete areas for extended periods of time seems to be common for this species, as well as other sturgeon species (Buckley and Kynard, 1985; Hall et al., 1991; Kieffer and Kynard, 1993; Foster and Clugston, 1997; Collins et al., 2000). In habitats deep enough to demonstrate thermal stratification, juveniles have been observed to remain within cooler areas or deep thermal refuges (Moser and Ross, 1995). The latter study found juveniles to prefer deep habitats (> 10 m) near the saltwater-freshwater interface.

PREY AND PREDATORS
Adults feed on benthic invertebrates, including mussels, worms, insect larvae, and shrimp. There are few data on the diets of juveniles. They have been reported to consume aquatic insects, amphipods, isopods, mollusks, and both polychaete and oligochaete worms while occupying freshwaters and brackish waters (Vladykov and Greeley, 1963; Moser and Ross, 1995), although a Chesapeake Bay study found no evidence that they preyed upon mollusks (Secor et al., 2000). Juveniles have been reported to cease feeding during summer months (Moser and Ross, 1995).

MIGRATIONS
Spawning adults migrate upriver in the spring beginning in April or May in our study area. Temperatures during this upstream migration are above 18°C. A small portion of ripe adults may make this spawning migration during the previous fall. After spawning, males may remain in river or lower estuary waters until the fall; females typically migrate out of rivers after four to six weeks. Juveniles may spend several years (one to six) in freshwater before migrating downstream; they then may remain in brackish waters for a few months. At sizes of 75 to 92 cm TL they migrate into the ocean, where they grow to adulthood. During this initial oceanic phase, juveniles are known to wander widely away from natal rivers.

14

Carcharhinus plumbeus (Nardo)

SANDBAR SHARK

DISTRIBUTION

This species occurs in most of the world's oceans, but apparently is not present in the eastern Pacific Ocean (Heist et al., 1995; Branstetter, 2002). It is found in warm temperate coastal waters on both sides of the Atlantic Ocean, and it occurs along the mid-Atlantic states of the United States as far north as Woods Hole, Massachusetts, and occasionally into the Gulf of Maine.

REPRODUCTION AND DEVELOPMENT

Males and females reach sexual maturity at lengths of about 170 cm and 180 cm, respectively (Branstetter, 2002), at ages of 12 to 15 years, although other authors have suggested that the process may take up to 30 years (Casey and Natanson, 1992). This viviparous species produces 6 to 10 young after a gestation period of 11 to 12 months. Females skip a year between pregnancies, using the intervening time as a resting period. Mating occurs in early June, and pups are born in late May or June of the following year. Details of development have been described (Springer, 1960; Baranes and Wendling, 1981), as have changes in body proportions from newly born relative to larger young-of-the-year (Schwartz, 1960).

At birth pups are 50 to 60 cm TL (Branstetter, 2002). Growth is very slow in this species; juveniles only increase in length about 10 cm per year. Older stages may only grow about 1 or 2 cm per year. Juveniles reach lengths of about 130 cm TL after 6 or 7 years (Branstetter, 2002). Maximum length is 225 to 250 cm and maximum age is more than 25 years. In recent years, this species has been reduced to approximately 20% of its former abundance by overfishing (Murdy et al., 1997).

SEASONALITY AND HABITAT USE

Adults are bottom-dwelling, coastal sharks that are found in the Middle Atlantic Bight during the summer. Pups occupy specific nursery grounds, separated from adults, and this is thought to reduce predatory mortality between cohorts (Branstetter, 1990). Pups are presumably born within estuarine habitats that serve as nurseries. South Carolina, Chesapeake Bay, the eastern shore of Virginia, and Delaware Bay presumably are some of the most important locations (Castro, 1993; Merson and Pratt, 2001; Conrath and Musick, 2007; Portnoy, 2008). The northernmost primary nursery, currently known, is in Great Bay, New Jersey (Merson and Pratt, 2007). However, some larger juveniles are common as far north as Massachusetts (Skomal, 2007). Neonates, small juveniles, and large juveniles may remain in nurseries until they reach lengths of about 130 cm TL (Conrath and Musick, 2007). In bays on the eastern shore of Virginia they typically occur in the summer at mean temperatures of 24°C and depths of 4.3 m (Conrath and Musick, 2008). Much of their behavior in these habitats appears to be in response to tidal flow (Medved and Marshall, 1983). In the winter, on the inner continental shelf, the same individuals occur at mean values of 19.9°C and 20.8 m depth. In Chesapeake Bay, most juveniles enter the bay from late May to early June and exit in late September and early October (Grubbs et al., 2007). These juveniles are most abundant in the lower bay and occur at salinities > 20.5 ppt and at depths > 5.5 m (Grubbs and Musick, 2007).

PREY AND PREDATORS

The diet of small juveniles (< 80 cm) can vary with location. Juveniles consume mostly small fishes, mantis shrimp, and blue crabs in Chesapeake Bay (Medved and Marshall, 1981; Ellis, 2003). *Brevoortia tyrannus, Centropristis striata, Symphurus* sp., and other flatfishes are important prey fishes. After they have migrated to the continental shelf, their diet consists primarily of a variety of fishes, but they also consume cephalaopds, decapod crustaceans, and human trash (Stillwell and Kohler, 1993) as well as other elasmobranchs (Collette and Klein-MacPhee, 2002).

MIGRATIONS

This species migrates seasonally along the East Coast of the United States. They move north with warming temperatures in the summer and south beginning in the fall; they can travel as far south as the Caribbean and the Bay of Campeche in Mexico (Casey and Kohler, 1990). Males and females remain in sexually segregated schools until mating season. The most detailed information is for large juveniles (121 to 142 cm TL) that were fitted with satellite tags in bays on the eastern shore of Virginia (Conrath and Musick, 2008). Most of these overwintered on the inner continental shelf off North Carolina. In Delaware Bay, other tagged individuals that were born there returned there annually for up to five years after overwintering off North Carolina (Grubbs et al., 2007; McCandless et al., 2007). Some of these same individuals migrated south to the east coast of Florida and into the Gulf of Mexico.

15

Mustelus canis canis (Mitchill)

SMOOTH DOGFISH

DISTRIBUTION

Mustelus canis canis has been reported from the Bay of Fundy, Canada, to Argentina, including islands in the Caribbean Sea. It is not found along the coast of Central America, however (Compagno, 2002). A separate subspecies, *M. canis insularis*, occurs near islands in the Caribbean Sea. It is one of the most abundant inshore sharks in the Middle Atlantic Bight, but records of early life history stages are few (see Table 4.3).

REPRODUCTION AND DEVELOPMENT

This species displays placental viviparous reproduction (Castro, 1983). Size when 50% are mature is 102 cm (females) and 86 cm (males) (Conrath and Musick, 2002). Most females reach maturation at 4 to 5 years of age, whereas most males reach maturation at 2 or 3 years of age. Mating takes place between May and September, ovulation usually occurs between late May and early June, and parturition occurs in May, after an 11- to 12-month gestation period. Females are capable of storing sperm in the terminal zone of their oviducal gland throughout the year (Conrath and Musick, 2002). Adults migrate into the central Middle Atlantic Bight from more southern areas in the spring to bear young. Litter size (3 to 18 pups) is directly related to length and age of females. Newborn closely resemble adults at birth.

GROWTH AND MORTALITY

The young are born at lengths of 28–39 cm TL. Based on modal length-frequency progression, they grow an average of 1.9 mm per day (range 1.5 to 2.1 mm) and reach 55 to 70 cm by the end of October (Fig. 15.1). Similar growth rates have been estimated from tooth width and tooth replacement data (Moss, 1972). In the portion of the Northwest Atlantic population that occurs between Cape Cod, Massachusetts, and South Carolina, length at age was determined in a study using growth bands in vertebrae (Conrath et al., 2002). Length at age 1 ranges from 50 to 85 cm TL, with smaller sizes observed in males. Length increases rapidly during the first 5 years, then tapers off for the last 5 (males) or 10 (females) years. By age 9, lengths in both sexes range from 105 to 130 cm TL. The oldest females (11 to 16 years old) are between 110 and 130 cm TL. This study also concluded that growth in age 0 and age 1 individuals occurred at a somewhat slower rate during the winter (October–April) than it did during summer (June–October) (Conrath et al., 2002). Longevity in this species has been determined to be 10 years (males) or 16 years (females).

SEASONALITY AND HABITAT USE

Adults have been reported from estuaries and bays to the edge of the continental shelf (Branstetter, 2002). The adults occur in estuaries during the spring when pupping occurs (Rountree and Able, 1996) and into the summer, but the duration of their stay and habitat use is poorly documented. In an attempt to fill these gaps, we ultrasonically tagged and tracked 10 individuals during June and July 2006 (M. Malone, T. Grothues, and K. W. Able, unpubl. data). These individuals were detected with a stationary listening array and mobile tracking in the Mullica River–Great Bay estuary (Able and Grothues, 2007b; Ng et al., 2007). With these techniques, 100% of tagged fish were detected and 80% of these were relocated with mobile tracking. All fish were found in the lower bay, typically in deep channels, and all had overlapping distributions. The exception was one individual that moved from the lower bay tagging area upstream into the lower Mullica River.

The presence of small pups with placental scars indicates that birth occurs in shallow inshore waters in New Jersey from May through early July, and young-of-the-year occur there through October (Rountree and Able, 1996). Young-of-the-year have been collected in a variety of habitats. The smallest are found in subtidal, polyhaline marsh creeks, especially those with shoals at the mouth, in the spring and early summer. They are most frequently collected in marsh creeks on flood tides, especially at night (Rountree and Able, 1992a, 1993, 1996). Larger young-of-the-year occur in deeper waters of bays later in the summer. Increased activity at night has also been observed in the laboratory (Casterlin and Reynolds, 1979). Young-of-the-year do not aggregate by sex or exhibit different emigration patterns between the sexes.

PREY AND PREDATORS

This species feeds mainly on decapod crustaceans (Gelsleichter et al., 1999; Bowman et al., 2000). Their most important prey in the ocean include crabs, lobsters, and shrimp, but they also commonly feed on squid, several bivalves, and gastropods. Teleosts are the third most important class of prey, and the list of species consumed includes *Ophichthus cruentifer, Etrumeus teres,* anchovies, *Urophycis* spp., *Ophidion* sp., *Prionotus* spp., *Stenotomus chrysops, Leiostomus xanthurus, Decapterus* sp., *Peprilus triacanthus,* and *Ammodytes* spp. (Bowman et al., 2000). It also feeds on polychaete worms and human garbage. One study examined the food habits of young-of-the-year (318 to 586 mm TL, n = 85) smooth dogfish, *Mustelus canis,* in the Barnegat Bay–Little Egg Harbor and the

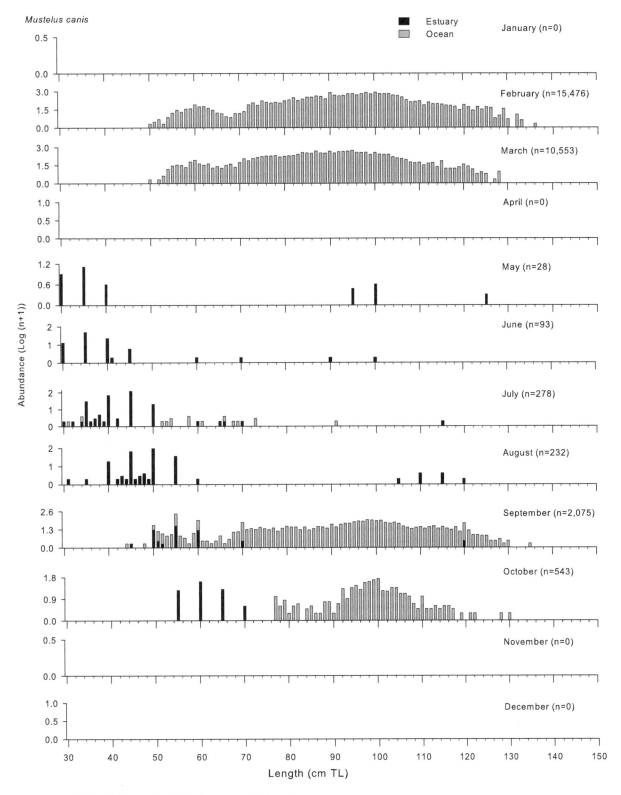

Fig. 15.1. Monthly length frequencies of *Mustelus canis canis* collected in
estuarine and coastal ocean waters of the central Middle Atlantic Bight.
Sources: National Marine Fisheries Service: otter trawl (n = 28,483);
Rutgers University Marine Field Station: seine (n = 6); 4.9 m otter trawl
(n = 25); multi-panel gill net (n = 25); and Able and Fahay 1998 (n = 739).

Age 1+ *Mustelus canis*   ## YOY *Mustelus canis*

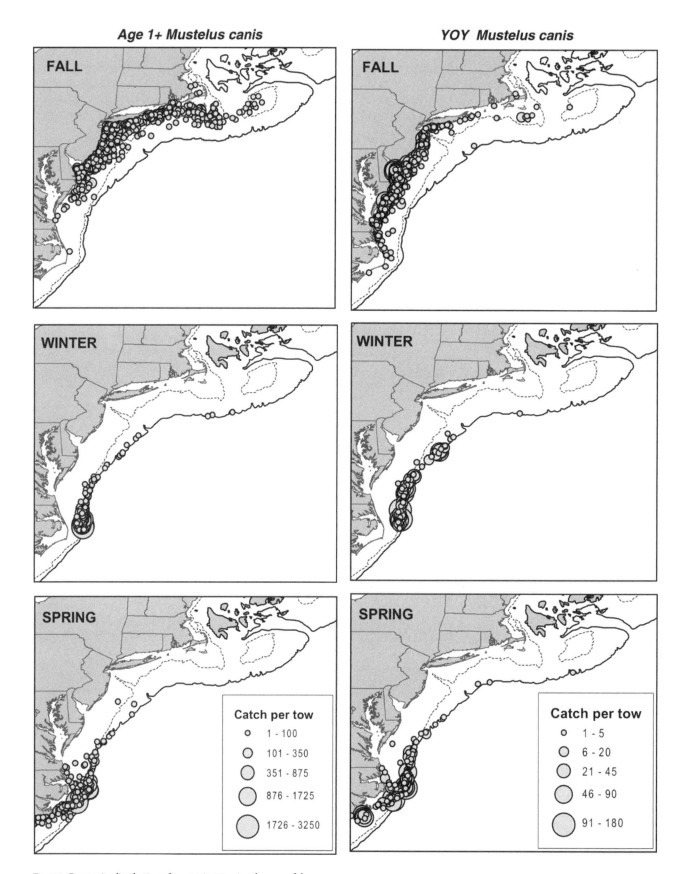

Fig. 15.2. Composite distribution of age 1+ (> 70 cm) and young-of-the-year (< 70 cm) *Mustelus canis canis* during seasonal cruises of the National Marine Fisheries Service groundfish survey. Details of sampling effort are indicated in Fig. 3.4.

Mullica River–Great Bay estuaries in New Jersey (Rountree and Able, 1996; see Table 8.1). Important prey of young-of-the-year include a variety of invertebrates (polychaetes, decapod shrimp, and crabs). Young-of-the-year also consume bivalves and small fish (*Menidia menidia* and *Fundulus heteroclitus;* see Table 8.3), although less frequently than invertebrates. Major predators are unknown, but a single occasion of predation by the sandbar shark, *Carcharhinus plumbeus,* has been reported (Rountree, 1999).

MIGRATIONS

Once the young-of-the-year (< 70 cm) leave estuaries and bays, they have a distinct seasonal distribution pattern in the ocean (Fig. 15.2). During the fall, they are found primarily in shallow waters (mean depth 173 m, range 6 to 110 m) on the inner continental shelf between Massachusetts and Cape Hatteras, North Carolina, with rare collections on Nantucket Shoals or on Georges Bank. During the winter, they migrate to the south and are found at the edge of the continental shelf (mean depth 125 m, range 43 to 388 m) between south-ern New Jersey and North Carolina. Two young-of-the-year tagged in Great Bay–Little Egg Harbor in New Jersey were collected in the winter off North Carolina (Rountree and Able, 1996). In the spring, there is evidence of movement inshore and to the north, with some individuals found on the inner shelf off North Carolina and Virginia and rare collections north of Hudson Canyon and on Georges Bank.

The age 1+ (> 70 cm) juveniles and adults have a similar seasonal distribution except in the fall, when they tend to be distributed primarily on the inner shelf (mean depth 25 m, range 8 to 79 m) in waters farther to the north between Virginia and Massachusetts and on to Georges Bank (see Fig. 15.2). In the winter, they are most abundant in deeper waters (mean depth 111 m, range 33 to 388 m) off North Carolina and rarely as far north as the Great South Channel. In the spring, they are still most abundant off North Carolina, but extend into shallow waters of the inner shelf. However, most remain near the edge of the continental shelf. These patterns account for the broad depth range (mean depth 70 m, range 9 to 143 m) at this time of year.

16

Rhinoptera bonasus (Mitchill)

COWNOSE RAY

DISTRIBUTION
This species is found from Cape Cod to Florida along the East Coast of the United States (McEachran, 2002) and throughout the Gulf of Mexico to South America (Schwartz, 1965a; Smith and Merriner, 1987).

REPRODUCTION AND DEVELOPMENT
Age at sexual maturity is estimated at 5 to 6 years for males and 7 to 8 years for females (Smith and Merriner, 1987). Males mature at approximately 90 cm disk width (DW), females at 96 cm DW (Smith and Merriner, 1986). Each female produces a single embryo that at full term averages about 40 cm DW. Pupping in Chesapeake Bay occurs in late June and July, with a second group developing by August (Smith and Merriner, 1986).

GROWTH AND MORTALITY
Neonates during their first summer in Chesapeake Bay are 33 to 52 cm DW and those collected in the fall are 47 to 59 cm DW (Smith and Merriner, 1987). The oldest individual recorded was a female of 13 years and 107 cm DW.

SEASONALITY AND HABITAT USE
These rays are seasonal occupants of Chesapeake Bay, with annual occurrences from early May through early October (Smith and Merriner, 1987) and summer occurrences in southern New Jersey in recent years (K. W. Able, pers. observ.). In Chesapeake Bay they are typically observed in shallow, nearshore waters in schools made up of several hundred (Smith and Merriner, 1987) up to several million (Blaylock, 1989) individuals. They typically occur in water temperatures of 15 to 29°C and in salinities as low as 8 ppt. Often the adults of both sexes school separately, in large schools, and show seasonal variation in sex ratio (Smith and Merriner, 1987). This differential distribution may occur among juveniles as well.

PREY AND PREDATORS
The dominant prey of juveniles and adults are estuarine bivalves. In Chesapeake Bay, the dominant prey is the soft shell clam (*Mya arenaria*) along with several other species of intertidal and subtidal bivalves (Smith and Merriner, 1985), including oysters (*Crossostrea virginica*).

MIGRATIONS
Data gathered from tagging in Chesapeake Bay indicate that *Rhinoptera bonasus* migrates out of the bay and south in the fall, with schools occurring off Cape Hatteras by mid-October and northern Florida by early December (Schwartz, 1965a). Data from Venezuela and southern Brazil suggest they arrive in mid-January and remain there until early March, when they begin a northward migration. An alternate interpretation is that the continental shelf off the South Atlantic Bight is an overwintering site (Smith and Merriner, 1987). This interpretation is supported by pop-up satellite tags on four adult females (Grusha, 2005). In the spring, schools arrive off North Carolina by April and by early May enter Chesapeake Bay, where they apparently spend the summer (Smith and Merriner, 1987).

17

Anguilla rostrata (Lesueur)

AMERICAN EEL

Estuary	Reproduction	Larvae/ Development	YOY Growth	YOY Overwinter
Ocean				
Both				
Jan				
Feb				
Mar				
Apr				
May				
Jun				
Jul				
Aug				
Sep				
Oct				
Nov				
Dec				

DISTRIBUTION

Anguilla rostrata occurs in the western North Atlantic Ocean from Greenland and Iceland to the West Indies (Bertin, 1956; Kuroki et al., 2008). It is also found in Panama and Bermuda. It is most abundant in river drainages of the United States and southern Canada. Larvae (or leptocephali) and glass eels are rarely collected at sea in the Middle Atlantic Bight (Able and Fahay, 1998), but juveniles have been reported from every estuary in the bight (see Table 4.3).

REPRODUCTION AND DEVELOPMENT

After a "yellow eel" stage lasting approximately 9 to 19 years for females and 7 to 12 years for males, both sexes undergo a fall migration downstream and eventually enter the ocean; the patterns of age and size vary with latitude and other variables (Oliveira, 1999; Jessop et al., 2004; Weeder and Hammond, 2009). In a study in Chesapeake Bay it was estimated that females left the bay at 6 or 7 years at approximately 550 mm TL, while males migrated at 5 to 7 years at approximately 300 mm TL (Hedgepeth, 1983). This spawning migration is accompanied by a metamorphosis to a "silver eel" stage, during which the eyes enlarge, the internal organs atrophy, and the pectoral fin changes its shape. Eels have been recorded leaving Chesapeake Bay in November (Wenner and Musick, 1974), but observations of adults at sea have been extremely rare (Wenner, 1973; Robins et al., 1979). It is estimated that migration to the spawning area takes about two to three months (Eales, 1968). Spawning occurs in the Sargasso Sea (Smith, 1968). Evidence of spawning is based primarily on the presence of very small (< 10 mm) larvae in the area (Miller and McCleave, 1994; Friedland et al., 2007) from late winter through early spring (McCleave, 2008). Adults presumably die after spawning.

The larval development of *Anguilla rostrata*, including characters useful for distinguishing the leptocephali from those of other eel species, has been well described and summarized (Fahay, 2007). The eggs are undescribed. Leptocephali are easily separated from those of most other species by their lack of pigmentation and relatively low meristic character counts.

Scale formation begins relatively late in development based on our examination of 32 individuals from 95 to 395 mm TL (Fig. 17.1). Scales were observed to form over the size range of 143 to 345 mm TL, but the process may continue in larger sizes. Scales are present along the lateral line from the dorsal fin origin to just before the caudal fin at 143 mm TL. Individual scales are isolated and do not overlap, as is the pattern observed in larger specimens. In general, as length increases, scale coverage becomes denser. By 180 mm TL, scales along the lateral line extend anteriorly in one or two rows to behind the pectoral fin but not posteriorly. Scales also extend dorsally and ventrally from the lateral line in several rows at the level just behind the origin of the anal fin to anterior of the caudal fin. By 190 mm TL, scales on the body are spaced closer together, but are less dense on the anterior-most and posterior-most portions of the body. Also, at this size, the dorsal and ventral extensions continue more anteriorly on the trunk. By 225 mm TL, scales extend anteriorly to the operculum and the posterior portion of the head and posteriorly farther to the base of the caudal fin. By 245 mm TL, scales extend onto the head to behind the eyes and completely cover the body. By 345 mm TL, scales have begun to form on the posterior portion of the dorsal fin; variation in scale formation on the fins is also observed. At 345 mm TL, scales cover 97.9% of the body and 73.1% of the body and fins, but scale formation may not be complete.

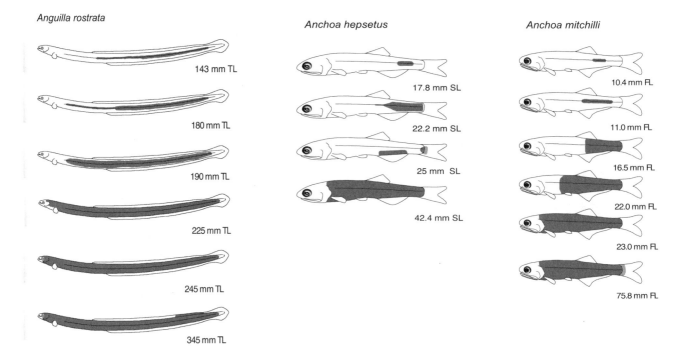

Anguilla rostrata

143 mm TL

180 mm TL

190 mm TL

225 mm TL

245 mm TL

345 mm TL

Anchoa hepsetus

17.8 mm SL

22.2 mm SL

25 mm SL

42.4 mm SL

Anchoa mitchilli

10.4 mm FL

11.0 mm FL

16.5 mm FL

22.0 mm FL

23.0 mm FL

75.8 mm FL

Fig. 17.1. Description of scale formation relative to length in *Anguilla rostrata, Anchoa hepsetus,* and *Anchoa mitchilli.*

LARVAL SUPPLY, SETTLEMENT, GROWTH, AND MORTALITY

After hatching, developing larvae (leptocephali) drift at sea for up to a year; they are transported away from the spawning area in the Sargasso Sea and northward by the Gulf Stream (Power and McCleave, 1983). Leptocephali transform into glass eels as they cross the Gulf Stream edge and approach the North American coast (Smith, 1968). Studies of their swimming ability confirm that leptocephali can swim across the continental shelf and into estuaries (Wuenschel and Able, 2008). Entrance of glass eels varies latitudinally in temperate estuaries, with the youngest individuals entering North Carolina estuaries at younger ages than those to the north (Powles and Warlen, 2002). The age and size at ingress among North Carolina, New Jersey, and New Brunswick, Canada, are 175.4, 201.2, and 209.3 days and 55.9, 60.9, and 58.1 mm TL, respectively (Powles and Warlen, 2002). In the Great Bay–Little Egg Harbor area, this ingress occurs from November through May (Figs. 17.2 and 17.3) and overlaps with collections of the largest sizes on the continental shelf. During 18 years of ichthyoplankton sampling in this study area, this ingress peaked in February in 5 years but was delayed until March in recent years (Fig. 17.3).

In a comparison between two southern New Jersey estuaries, the developmental transition between ingressing glass eels and late-stage elvers and the colonization of up-estuary sites were highly synchronized (Sullivan et al., 2009). The up-estuary movement was correlated with warming tem-

peratures reaching 10 to 12°C. However, within an estuary abundance estimates in collections varied considerably, depending on location. A number of factors may contribute to this variation, including selective tidal stream transport (McCleave and Kleckner, 1982; Edeline et al., 2007), moon phase (Sugeha et al., 2001; Jellyman and Lambert, 2003), water temperature (McKinnon and Gooley, 1998; August and Hicks, 2008), precipitation (Sullivan et al., 2006a), large-scale climate forcing (Friedland et al., 2007; Bonhommeau et al., 2008), and combinations of factors (temperature, river flow, tidal stage) (Martin, 1995).

Despite the concern over the decline in this species along the east coast of North America (Castonguay et al., 1994; Haro et al., 2000; Bonhommeau et al., 2008), the abundance of glass eels ingressing into estuaries has not declined in recent years in New Jersey or North Carolina (Sullivan et al., 2006a). The annual variation was correlated with levels of precipitation. Subsequent plankton net collections of leptocephali in Great Bay, measured over 18 years, show an increase in abundance of glass eels in the mid-1990s (Fig. 17.4).

Glass eels collected in continental shelf waters during the MARMAP study and several more recent survey cruises range from 51 to 64 mm SL (Able and Fahay, 1998). Elvers collected from several combined estuarine sources are slightly larger (see Fig. 17.2). The smallest of these correspond in length to glass eels collected offshore. Because these small elvers are the dominant size class from November through July, it is difficult to follow the progression of year classes, but

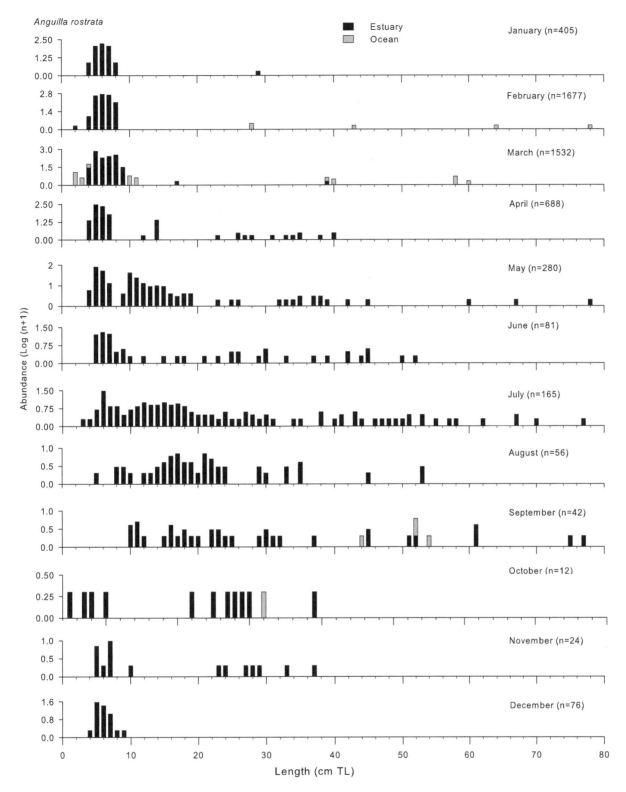

Fig. 17.2. Monthly length frequencies of *Anguilla rostrata* collected in estuarine and coastal ocean waters of the central Middle Atlantic Bight. Sources: National Marine Fisheries Service: otter trawl (n = 41); Rutgers University Marine Field Station: 1 m beam trawl (n = 13); seine (n = 12); 1 m plankton net (n = 1996); experimental trap (n = 6); killitrap (n = 64); 4.9 m otter trawl (n = 81); and Able and Fahay 1998 (n = 2825).

Seasonality for *Anguilla rostrata*

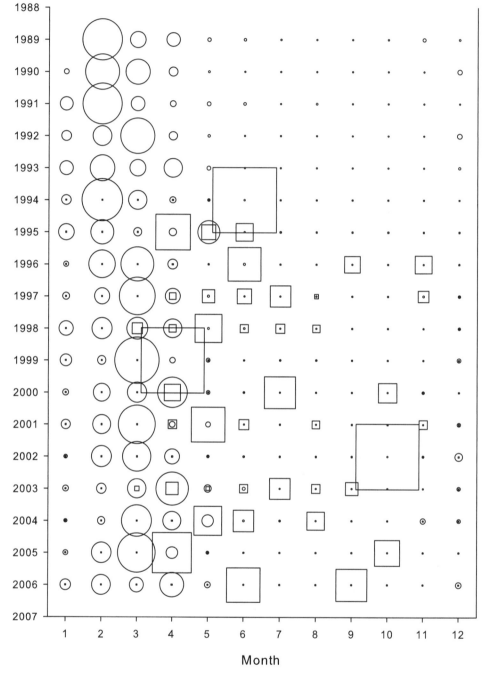

Fig. 17.3. Seasonal occurrence of larvae (based on ichthyoplankton samples collected in Little Sheepshead Creek behind Little Egg Inlet from 1989 to 2006; open circles represent the percentage of the mean number/1000 m³ of larval young-of-the-year captured by year) and juveniles of *Anguilla rostrata* (based on trap collections in Rutgers University Marine Field Station boat basin in Great Bay from 1994 to 2006; open squares represent the percentage of the mean catch per unit effort captured by year). Values range from 0 to 63% and 0 to 100% for ichthyoplankton and trap sampling, respectively. The smallest circles and squares represent when samples were taken but no individuals were collected.

individuals up to 30 cm may be young-of-the-year (see Fig. 17.2) and there are some supporting data for this interpretation from Delaware Bay (K. W. Able, unpubl. data).

Demographic characteristics in the Hudson River estuary were variable between estuarine and freshwater portions of the estuary (Morrison and Secor, 2003). Lengths were

similar among these regions (45.7 ± 0.3 cm, range approximately 30 to 70 cm), but ages differed and were lower in the estuary (8 ± 4 years) than in freshwater (17 ± 4 years) even though growth was higher in the estuary (8.0 cm/year) than in freshwater (3.4 cm/year). In several New Jersey freshwater streams, the ages ranged from 3 to 19 years with a mean of

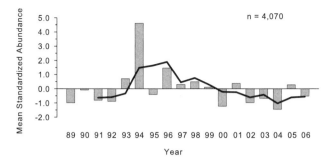

Fig. 17.4. Annual mean and three-year moving mean of standardized abundance for *Anguilla rostrata* collected in ichthyoplankton (1989–2006) sampling. Data were standardized by subtracting overall mean from annual mean abundance. Three-year moving mean was calculated by taking the mean of the standardized values from the previous two years and the year in which the data were plotted.

10 years (Ogden, 1970). In Chesapeake Bay, ages ranged from 1 through 8 years at mean sizes of 14 to 64 cm (Hedgepeth, 1983) or 1 through 13 years (Weeder and Hammond, 2009), but these values varied with location in the bay. The comparison of these data suggests that in Chesapeake Bay they grow faster and mature earlier than other populations. Silver eels are evident at ages of 6 or 7 years and approximately 55 cm (Hedgepeth, 1983).

The increasing incidence of an introduced nematode parasite (*Anguillicola crassus*), native to Asia but first reported in North America in 1995 (Fries et al., 1996), is a potential source of mortality that may contribute to its decline. During estuarine residency, juveniles are exposed to this swimbladder parasite (Barse and Secor, 1999; Barse et al., 2001; Morrison and Secor, 2003). An infection by *A. crassus* is initially acquired when an eel consumes an intermediate copepod or larger paratenic host (crustaceans, snails, fishes) containing its early-stage juveniles. Juveniles exit the gut, penetrate the swimbladder wall (where they continue to develop through several pre-adult stages), and eventually enter the swimbladder lumen as adults. In the swimbladder proper, adult *A. crassus* feed on the host's blood supply and deposit eggs. In addition to impairing the growth and survival of resident eels, infection may compromise swimbladder function during long-distance adult eel migrations to the Sargasso Sea, thus negatively impacting reproduction (Sures and Knopf, 2004).

SEASONALITY AND HABITAT USE

As glass eels enter estuaries, there is a delay in upstream migration from tidal to non-tidal habitats, and it has been suggested that this is related to a period of physiological adjustment to freshwater (Haro and Krueger, 1988). In two New Jersey estuaries, large-scale colonization of up-river sites, as well as development into late-stage elvers, were dependent on water temperature reaching approximately 10 to 12°C (Sullivan et al., 2009). During this time, glass eels undergo another transformation in which they acquire dark brown cutaneous pigmentation and begin the elver stage. Elvers occupy a wide range of coastal marine habitats, including tidal flats and marshes, harbors, barrier beach ponds, coastal rivers, creeks, and streams (Eales, 1968). They may be especially abundant in low-salinity marsh creeks. In a Delaware Bay study, relatively large numbers (n = 113, 69 to 530 mm TL, mean 158 mm TL) were collected in intertidal creeks with a small-mesh seine (Kimball and Able, 2007a, b). Elvers have been collected from temperatures as low as −0.8°C (Jeffries, 1960). Extensive analysis of otolith microchemistry suggests that individual yellow eels may be resident in one habitat, either fresh, brackish, or high salinity, for long periods of time or movements between two or more habitats may occur (Daverat et al., 2006; Jessop et al., 2008). Other studies in the Hudson River estuary suggest that dispersal is low (< 1 km) (Morrison and Secor, 2003).

During fall cooling, eels move away from coastal waters into brackish waters and freshwaters. If freshwaters are not available, wintering individuals seek out deeper parts of rivers and bays in which to spend the coldest parts of the winter (Smith and Saunders, 1955). In these areas, they are known to burrow into mud (Cox, 1916; MacKenzie, 1992) or to hibernate in burrows equipped with ventilation holes (Eales, 1968). During spring, they return to shallower areas.

Negative phototaxis becomes more pronounced with age. This response translates into nocturnal activity and results in larger individuals being found at greater depths than are smaller individuals. They are also abundant in dark habitats under and around piers in the Lower Hudson River (Able et al., 1995b), in turbid shoreline sites with a substrate composed of finely ground plant stem detritus (de Sylva et al., 1962), and in tidal creeks (Rountree et al., 1992) and tidal impoundments (Clark, 1994). Some individuals are strongly sedentary and tend to remain in a home territory. Experimentally displaced eels have returned to a home territory from as far away as 80 km (Vladykov, 1971).

PREY AND PREDATORS

Several studies (n = 3) have examined the food habits of young-of-the-year (50 to 249 mm TL, n = 187) *Anguilla rostrata* in Middle Atlantic Bight estuaries from New Jersey streams to the James River, Virginia (see Tables 8.1 and 8.3). Important prey of young-of-the-year include invertebrates (insects/arachnids, *Limulus* spp. eggs and larvae). Fish (*Menidia menidia*) are a minor dietary component. The diet shifts with ontogeny at approximately 250 to 400 mm TL, such that insects are important prey of small eels and fish and crayfish are important prey of larger eels. Ogden (1970) noted a shift in diet with ontogeny, such that eels < 400 mm TL consume primarily bottom-dwelling insects, while larger eels eat bigger prey items, almost exclusively crayfish and fish. Lookabaugh and Angermeier (1992) suggested a similar transition but at approximately 250 mm TL. Variation in diet

Oophagy in *Anguilla rostrata*

Interactions between Catadromous and Anadromous Fishes

Oophagy, or the eating of eggs, is widespread in animals (Denoel and Demars, 2008). It occurs in fishes but is less well known. This is particularly true in fishes where the determination of facultative versus obligate oophagy is lacking, and especially for estuarine fishes. These types of interactions might be especially important because eggs can be a high-quality, easily digested food source and the small size of most fish eggs and their immobility might make them especially desirable as prey for small, less mobile fishes. One possible example is that between the elvers of the catadromous *Anguilla rostrata* and the eggs of *Alosa* spp.

In preliminary observations in the Batsto River, a tributary of the Mullica River–Great Bay estuary in southern New Jersey, potential predators and their egg prey appear to overlap in space and time. In winter and early spring, the glass eels of *Anguilla rostrata* arrive in estuarine inlets, including this estuary (Sullivan et al., 2006a), from the Sargasso Sea. Subsequently, they begin transformation to pigmented elvers and a portion of the population continues moving upstream into freshwater habitats (Sullivan et al., 2009). Later in the spring, the adults of *Alosa pseudoharengus*, the most abundant of the river herrings, arrive in this and other estuaries and migrate upstream to spawn in freshwaters. Interestingly, a temperature of approximately 10°C triggers spawning in *Alosa pseudoharengus* and the same temperature prompts *Anguilla rostrata* glass eels/elvers to move upstream in estuaries (Sullivan et al., 2009). Preliminary evidence of oophagy is based on the collection of elvers below the dam in the Batsto River, where spawning adult *A. pseudoharengus* and *A. rostrata* elvers co-occur every spring. Collections of pigmented elvers have confirmed the presence of a large number of eggs in the stomachs, and sometimes these are quite evident in the distended body cavities. It would be useful to know if this phenomenon occurs in other estuaries along the East Coast of the United States, and whether this is temporally facultative or obligate to these small *Anguilla*. Also, does it play a significant role in these local food webs, or does it have broader implications for the successful recruitment of *Alosa* spp. and *Anguilla rostrata*, all of which are currently declining? One also wonders if the presence of dams might artificially increase the likelihood of these interactions by concentrating both species below the dams.

has also been noted within the James River drainage in Virginia. Stomachs of young-of-the-year (100 to 249 mm TL, n = 20) from the upper Piedmont region contain almost exclusively aquatic invertebrates, those (100 to 249 mm TL, n = 19) from the lower Piedmont contain some terrestrial invertebrates and crayfish in addition to aquatic invertebrates, while young-of-the-year (100 to 249 mm TL, n = 17) from the coastal plain contain some crayfish in addition to aquatic invertebrates (Lookabaugh and Angermeier, 1992).

The known predators of the young-of-the-year of this species include *Morone americana* and *Pomatomus saltatrix* (see Table 8.3).

MIGRATIONS

Adult eels remain in freshwater and estuarine habitats until they reach maturity and begin the spawning migration to the Sargasso Sea.

18

Myrophis punctatus Lütken

SPECKLED WORM EEL

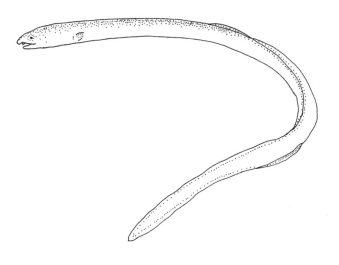

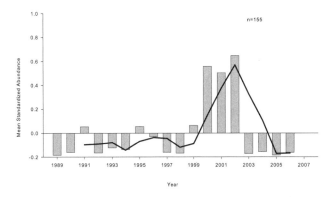

Fig. 18.1. Annual variation in abundance of *Myrophis punctatus* in Little Sheepshead Creek Bridge 1 m plankton net collections, 1989–2006. Data were standardized by subtracting the overall mean from annual abundance. Vertical bars represent standardized annual abundance; the solid line is a three-year running mean that was calculated by taking the mean of the standardized values from the previous two years and the year in which the data are plotted.

DISTRIBUTION

This species is found in the western North Atlantic Ocean from North Carolina to Brazil, including the Gulf of Mexico (Briggs, 1974; McEachran and Fechhelm, 1998; Fahay, 2007). Early stages have been collected as far north and east as the Scotian Shelf (Fahay and Obenchain, 1978; Leiby, 1989).

REPRODUCTION AND DEVELOPMENT

Spawning is undescribed. It likely occurs in the South Atlantic Bight (Fahay and Obenchain, 1978), farther south off Florida (Eldred, 1966), or in the Bahamas (Miller and Mc-Cleave, 2007) during the fall based on the collection of the smallest larvae in October and November there (Able et al., in prep.). The early development of *Myrophis punctatus*, including characters useful for distinguishing the larvae and juveniles from those of other anguilliforms, have been well described (Eldred, 1966; Fahay and Obenchain, 1978; Leiby, 1989) and summarized (Fahay, 2007). Eggs are undescribed. Larvae hatch at sizes smaller than 8.0 mm. As is typical of anguilliform fishes, the leptocephalus larvae grow to considerable lengths, then shrink as they are undergoing metamor-

phosis, and then increase in length again until they reach the adult size at a maximum size of 353 mm TL (Smith, 1989a, b). They arrive at inlets at approximately 60 to 75 mm while they are undergoing transformation from the leptocephalus to the elver stage (Able et al., in prep.).

LARVAL SUPPLY

The larvae appear to be delivered to inlets in the Middle Atlantic Bight (Powles et al., 2006) after spawning occurs in the South Atlantic Bight (Able et al., in prep.). The pattern of arrival varies between years, with a peak from 2000 to 2002 (Fig. 18.1). The seasonal timing of arrival differs as well, with some individuals arriving in November–January and others from April to July (Fig. 18.2). It is possible that these larvae are from different sources.

SEASONALITY AND HABITAT USE

Little is known of the seasonality of occurrence of most life history stages. However, the timing of larval ingress into estuarine inlets occurs during the fall and winter in Florida (Harnden et al., 1999) and at the same approximate times in South Carolina, North Carolina, and New Jersey based on collections over more than two decades of sampling (Able et al., in prep.). In New Jersey they occur in late spring and early summer as well. Once they settle and begin to bury, little is known of their biology and habitat use except that they can be observed in areas with soft mud bottom and in sand.

The adults are reported to occur in shallow waters, with greatest abundances in brackish tidal creeks and bays (Böhlke and Chaplin, 1993) as well as mangrove habitats (Barletta et al., 2000) to depths of 7 m (McEachran and Fechhelm, 1998). The juveniles and adults have only been collected with beam trawls in North Carolina creeks with soft bottoms

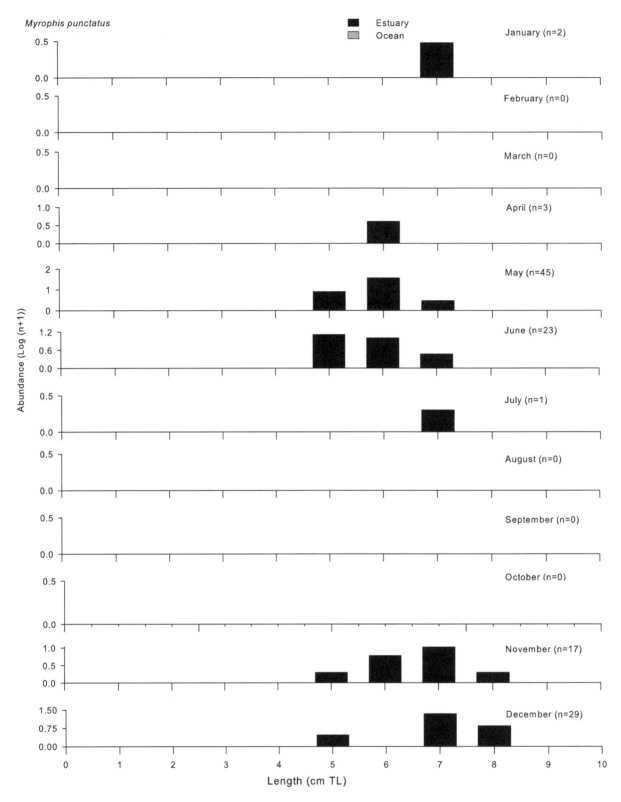

Fig. 18.2. Monthly length frequencies of *Myrophis punctatus* collected in
the Great Bay–Little Egg Harbor study area. Source: Rutgers University
Marine Field Station: 1 m plankton net (n = 120).

(Searcy, 2005), presumably because they bury in the substrate (Springer and Woodburn, 1960).

PREY AND PREDATORS
Unknown.

MIGRATIONS
Presumably, adults migrate offshore to spawn because that is where the larvae are found. The movement of larvae and glass eels into estuarine inlets appears to be directed and, at least in Florida, influenced by nighttime, moonless flood tides and strong onshore and along-shelf winds (Harnden et al., 1999).

19

Conger oceanicus (Mitchill)

CONGER EEL

	Reproduction	Larvae/ Development	YOY Growth	YOY Overwinter
Estuary				
Ocean				
Both				
Jan				
Feb				
Mar				
Apr				
May				
Jun				
Jul				
Aug				
Sep				
Oct				
Nov				
Dec				

DISTRIBUTION

Conger oceanicus occurs along the Atlantic coast of the United States and the eastern Gulf of Mexico (Bigelow and Schroeder, 1953; Smith and Tighe, 2002). It is rarely found north of Cape Cod, and is most common between Cape Cod and Delaware Bay. Although it is not usually considered an estuarine species, rare, single occurrences of young stages in estuaries have been reported (Pearson, 1941; Hauser, 1975; Moring and Moring, 1986). These isolated reports have origi-

nated from estuaries throughout the Middle Atlantic Bight (see Table 4.3; Able and Fahay, 1998).

REPRODUCTION AND DEVELOPMENT

Spawning occurs in the southwestern Sargasso Sea during the fall and winter (Schmidt, 1931; Miller and McCleave, 1994; Miller, 1995). The reproductive pattern of adult spawning migration from the continental shelf of the U.S. East Coast, spawning, and transport of the larvae (leptocephali) from the Sargasso Sea to coastal habitats is very similar to that of *Anguilla rostrata*. Previous studies have found no ripe individuals among material collected from the coast to the continental shelf/slope break in the Middle Atlantic Bight (Hood et al., 1988; Eklund and Targett, 1990). Early signs of maturation in females have been found in late spring and early summer, but these maturing females have not been observed later in the year (Hood et al., 1988). The spawning season is apparently long, perhaps extending from late summer through the winter, as evidenced by occurrences of small larvae (McCleave and Miller, 1994); however, back calculation of hatch dates from larvae successfully entering estuaries suggests spawning occurs from late October to mid-December (Correia et al., 2004). Direct observations of the oceanic spawning migration are lacking, and there is no evidence of spent adults returning to coastal or shelf areas after spawning.

The early development and characters useful for distinguishing the leptocephali and juveniles from those of other eel species have been well described and summarized (Smith, 1989a; Fahay, 2007). Eggs are undescribed. Those eggs identified by Eigenmann (1902) as *C. oceanicus* have been shown to pertain to the eggs of the ophichthid eel, *Ophichthus cruentifer* (Naplin and Obenchain, 1980). Leptocephali are characterized by a typical leaf-shaped body with a very small head. They have between 140 and 148 myomeres. Pigment along the body midline is variable, but usually features a line of melanophores paralleling a similar row along the gut tube. A crescent of pigment lies under the bottom edge of the eye.

LARVAL SUPPLY, SETTLEMENT, GROWTH, AND MORTALITY

The leptocephalus larvae are transported over five to six months from the Sargasso Sea to Middle Atlantic Bight estuaries (Correia et al., 2004); however, a clear zone in the otolith may confound these estimates. Other estimates from otoliths suggest that faster-growing leptocephali arrive at the estuary earlier than slower-growing individuals. The pattern of annual abundance at Little Egg Inlet in New Jersey is variable (Fig. 19.1). During some years, transforming leptocephali entering Great Bay have been quite numerous, although in most years between 1989 and 2006, their annual abundance was quite low. After these leptocephali transform from the glass eel stage and become elvers (juveniles) and settle, their abundance in the estuary is difficult to measure because they

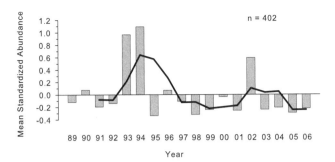

Fig. 19.1. Annual mean and three-year moving mean of standardized abundance for *Conger oceanicus* collected in ichthyoplankton (1989–2006) sampling. Data were standardized by subtracting overall mean from annual mean abundance. Three-year moving mean was calculated by taking the mean of the standardized values from the previous two years and the year in which data are plotted.

are cryptic (see below) and difficult to sample with traditional gears. Individuals captured in the plankton in the vicinity of the inlet of the same New Jersey estuary were undergoing transformation from larvae to juveniles (Bell et al., 2003). Settlement occurred at approximately this time and stage of development because no glass eels, elvers, or juveniles were collected in plankton nets and these were found only in demersal habitat traps (Bell et al., 2003).

Growth is unusual because of alternating patterns of increase in length and then shrinkage. While in its oceanic engyodontic and euryodontic (based on different dentition) leptocephalus stages, *Conger oceanicus* undergoes positive growth, reaching lengths of about 100 mm (Smith, 1989a, b, c). The leptocephalus then undergoes a transformation that includes shrinkage in length associated with a gradual change from a laterally flattened to a cylindrical body shape. These modifications conclude at the glass eel stage, the smallest size (60 to 70 mm TL) the animal will be while occupying estuarine waters. Positive growth then resumes as an elver (Able and Fahay, 1998). Describing a growth rate during the first growing season is difficult. Length frequencies derived from various collections indicate that metamorphosing larvae ingress estuaries at a modal length of about 100 mm TL

(Fig. 19.2). There are no well-defined length modes during late summer and fall, and all those recorded in Fig. 19.2 may represent young-of-the-year based on back-calculated size at age 1 from another study (Hood et al., 1988). They reach lengths of 106 cm and an age of 13 years (Hood et al., 1988). We have no information on mortality rates while this species inhabits estuarine habitats in their early stages. Cannibalism may be one source of mortality (Levy et al., 1988).

SEASONALITY AND HABITAT USE
Adults of *Conger oceanicus* are found in coastal areas to 475 m depth and are usually associated with structured habitats, such as piers, wrecks, jetties, reefs, or burrows shared with tilefish (*Lopholatilus chamaeleonticeps*) (Able et al., 1982; Grimes et al., 1986). Different habitats are used by other life history stages. Metamorphosing leptocephali are pelagic (or bentho-pelagic). They consistently enter estuaries from May through July (Fig. 19.3). It is not known if these metamorphosing leptocephali and elvers are estuarine dependent, but this is the only habitat in which they have been frequently collected. Most juveniles, as collected in traps, are in the estuary until the winter. We have few records of individuals collected in the estuary by any method from December through March, yet we also lack evidence that they leave estuaries during this period. Although young-of-the-year may burrow in estuarine substrates and remain inactive through the winter, most evidence suggests a migration into the ocean during this period (see below).

PREY AND PREDATORS
There have been no studies on the food habits of young-of-the-year (< 350 mm) *Conger oceanicus* in Middle Atlantic Bight estuaries. However, several other papers have been published on food habits of oceanic individuals (Hildebrand and Schroeder, 1928; Maurer and Bowman, 1975; Cau and Manconi, 1984; Levy et al., 1988; Bowman et al., 2000; Smith and Tighe, 2002). Predators on juvenile and adults are generally unknown, except that cannibalism is known to occur (Levy et al., 1988).

A Novel Record of Leptocephalaphagy in *Pomotomus saltatrix*

Dave Nemerson collected two *Pomatomus saltatrix* by rod and reel between approximately 18:00 and 18:30 on May 12, 2000. The fish were caught approximately 1.5 hours after a neap high tide (high tide: 16:39) in approximately 2 m of water over a sand bar just south of Shooting Thorofare and just inside Little Egg Inlet, New Jersey. Both fish were dissected for examination of stomach contents. One fish contained two adult *Anchoa mitchilli*, while the second fish contained six intact leptocephali of *Conger oceanicus* of similar size (99 to 107 mm TL). The stomach contents ranged from undigested to partially digested, suggesting that the prey had been fed on over some time. These observations are consistent with the size and timing of the occurrence of these leptocephali based on weekly collections near this inlet over 17 years. These observations are unique in that records of leptocephalophagy are rare in fish in general and this represents the first such occurrence that we are aware of in *P. saltatrix* and any other species. The detection of this type of predation may be difficult, given that digestion of these soft-bodied prey may occur quickly as for other larval fishes (Able et al., 2007a).

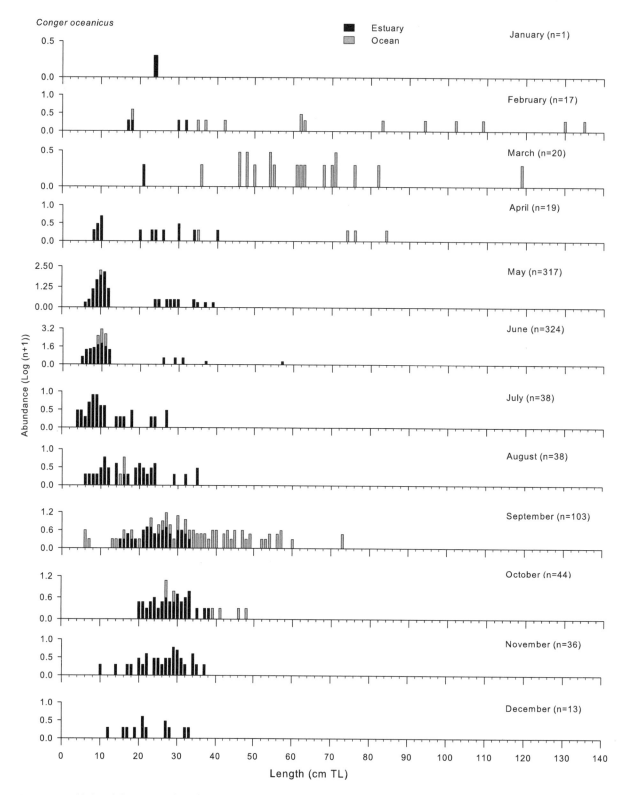

Fig. 19.2. Monthly length frequencies of *Conger oceanicus* collected in estuarine and coastal ocean waters of the central Middle Atlantic Bight. Sources: National Marine Fisheries Service: otter trawl (n = 103); Rutgers University Marine Field Station: 1 m beam trawl (n = 2); seine (n = 4); Methot trawl (n = 4); 1 m plankton net (n = 146); 2 m beam trawl (n = 1); experimental trap (n = 34); killitrap (n = 43); and Able and Fahay 1998 (n = 635).

Seasonality for *Conger oceanicus*

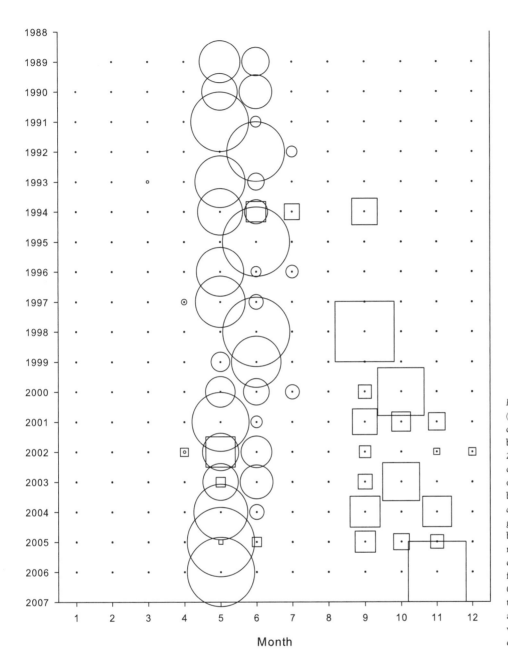

Fig. 19.3. Seasonal occurrence of larvae (based on ichthyoplankton samples collected in Little Sheepshead Creek behind Little Egg Inlet from 1989 to 2006; open circles represent the percentage of the mean number/1000 m³ of larval young-of-the-year captured by year) and juveniles of *Conger oceanicus* (based on trap collections in Rutgers University Marine Field Station boat basin in Great Bay; open squares represent the percentage of the mean catch per unit effort captured by year from 1994 to 2006). Values range from 0 to 100% for ichthyoplankton and trap sampling. The smallest circles and squares represent when samples were taken but no individuals were collected.

MIGRATIONS

Different life history stages exhibit varying patterns of migration. Early-stage leptocephali have been collected in the Sargasso Sea, north and east of the Bahamas (Schmidt, 1931; Castonguay and McCleave, 1987; Miller, 1995); larger leptocephali (up to 85 mm) have been collected from the Florida Current and Gulf Stream south of Cape Hatteras, North Carolina (McCleave and Miller, 1994). The mechanisms used by these leptocephali to exit the Gulf Stream and cross the continental shelf are undescribed. However, laboratory studies indicate that they are strong swimmers and likely capable of swimming from the Gulf Stream, across the Slope Sea and continental shelf, and into estuaries (Wuenschel and Able, 2008).

Collections of adults as by-catch in the black sea bass trap fishery on the continental shelf adjacent to our estuarine study area in New Jersey increase during November (Eklund and Targett, 1991), which is consistent with an offshore movement by the population. The conger eel by-catch in the tilefish longline fishery near the edge of the continental shelf also is at a maximum during winter and spring (Hood et al., 1988) and at its lowest during summer, again suggesting a seasonal inshore-offshore movement by the population. This pattern is supported by the known distribution from

Age 1+ *Conger oceanicus*   ## YOY *Conger oceanicus*

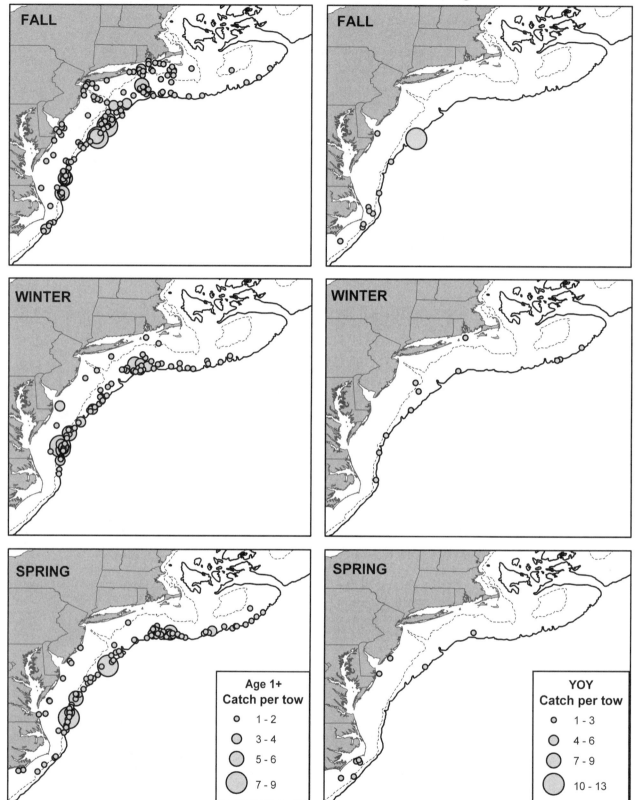

Fig. 19.4. Composite distribution of age 1+ (> 28 cm) and young-of-the-year (< 28 cm) *Conger oceanicus* during seasonal cruises of the National Marine Fisheries Service groundfish survey. Details of sampling effort are indicated in Fig. 3.4.

National Marine Fisheries Service (NMFS) bottom trawl surveys (Fig. 19.4). The young-of-the-year (< 28 cm) are infrequently encountered in these surveys. Those that have been collected occurred from southern Georges Bank to below Cape Hatteras. Regardless of season, young-of-the-year were collected at depths of 18 to 375 m (mean depth 79.3 m) with most individuals (> 80%) collected on the edge of the continental shelf. Larger juveniles and adults (age 1+) were collected at depths of 7 to 235 m (mean depth 26.7 m), but individuals were bimodally distributed, with some collected on the inner continental shelf adjacent to the coast from Nantucket Shoals to Cape Hatteras (including the Hudson River Shelf Valley) and the remainder collected on the outer edge of the continental shelf from Georges Bank to Cape Hatteras.

20

Alosa aestivalis (Mitchill)

BLUEBACK HERRING

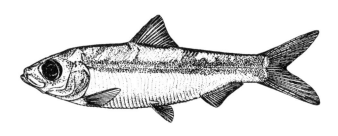

	Reproduction	Larvae/ Development	YOY Growth	YOY Overwinter
Estuary				
Ocean				
Both				
Jan				
Feb				
Mar				
Apr				
May				
Jun				
Jul				
Aug				
Sep				
Oct				
Nov				
Dec				

DISTRIBUTION

Alosa aestivalis is distributed along the east coast of North America between Nova Scotia and St. John's River, Florida (Neves, 1981). It is most abundant in the Middle Atlantic and South Atlantic bights. Juveniles occur in most estuaries throughout the study area, but eggs and larvae are found in coastal rivers (see Table 4.3). Early stages are not found in estuarine systems lacking significant freshwater input, such as Great Bay, Gardners Bay, Long Island Sound, and Nauset Marsh.

REPRODUCTION AND DEVELOPMENT

In our study area, males spawn at ages 3 to 4, females at ages 4 to 5 (Marcy, 1969; Scherer, 1972). Spawning occurs during spring or early summer (late March through July) in freshwaters or brackish waters of estuaries. Reported spawning takes place in the central part of the Middle Atlantic Bight in the Hudson, Mohawk, and Delaware rivers (Dovel, 1981; Mayo, 1982; Simonin et al., 2007); a few smaller runs occur in rivers along the New Jersey coast, including the Raritan, South, Navesink, Swimming, and Great Egg Harbor rivers (Zich, 1977; Himchak, 1983; Himchak and Allen, 1985). No evidence of spawning was found within the Mullica River–Great Bay study area (Milstein and Thomas, 1977). In another survey near the mouth of Delaware Bay, adults were abundant in gill net collections in May and June and ripe females were found as early as April (de Sylva et al., 1962). Spawning occurs mostly at night (Graham, 1956) in fast currents over a hard substrate (Loesch and Lund, 1977). Spawning can take place in temperatures as low as 14°C but optimum temperatures are 21 to 24°C (Mansueti and Hardy, 1967). *Alosa aestivalis* does not run as far upstream to spawn as does *A. pseudoharengus*, and it generally spawns three to four weeks later than its congener. After spawning, adults retreat downstream to the ocean, although some individuals spend extended periods in freshwater and others may be resident (Limburg et al., 2001).

Details of egg and larval development and methods for distinguishing the early stages of species in the family Clupeidae may be found in Fahay (2007). Larvae are 3.1 to 5.0 mm TL at hatching. Transformation from larval to juvenile morphology begins at about 20 mm TL and an age of 25 to 35 days (Watson, 1968). Fin ray development is complete and the body has begun to deepen at this time. Scales first appear at about 25 to 29 mm TL and are fully developed at 45 mm TL (Hildebrand, 1963).

LARVAL SUPPLY AND GROWTH

Larvae are supplied to the estuary from upstream freshwaters. In Delaware Bay they enter the estuary as small young-of-the-year at 20 to 45 mm (Able et al., 2007a).

Growth is slow in young-of-the-year, based on the small sizes attained by fall. Perlmutter et al. (1967) collected larvae and juveniles from the upper Hudson River in June (15 to 44 mm SL) and August (14 to 49 mm SL). Most fish collected in late summer and early fall in our study area are 50 to 100 mm TL (Fig. 20.1). This range closely approximates the sizes (36 to 71 mm TL) reported for young-of-the-year emigrating to the ocean in the fall in the Chesapeake Bay

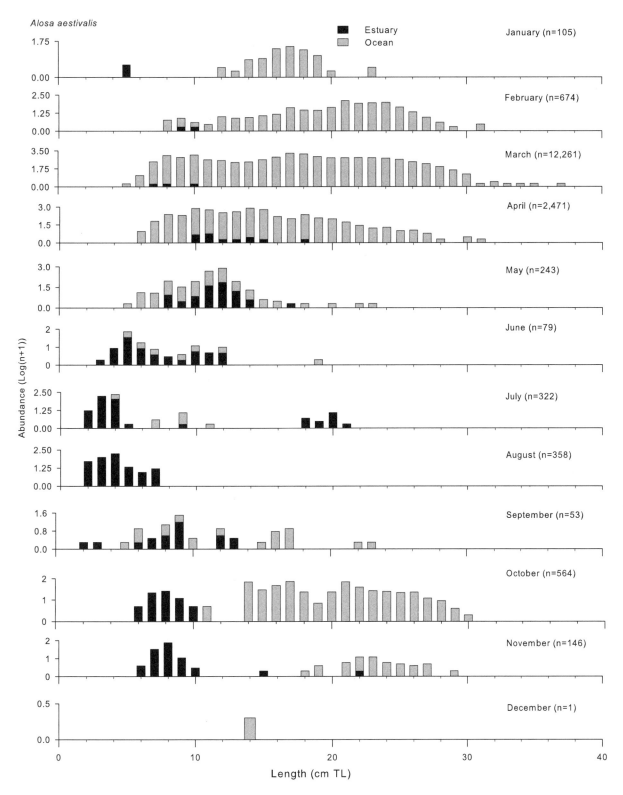

Fig. 20.1. Monthly length frequencies of *Alosa aestivalis* collected in New Jersey / New York estuarine and adjacent coastal ocean waters. Sources: National Marine Fisheries Service: otter trawl (n = 15,984); Rutgers University Marine Field Station: seine (n = 102); 1 m plankton net (n = 3); 4.9 m otter trawl (n = 6); multi-panel gill net (n = 19); and Able and Fahay 1998 (n = 1163).

region (Hildebrand and Schroeder, 1928). Most overwintering (December through March) young-of-the-year collected by Milstein (1981) were 60 to 110 mm FL. There is little, if any, growth over the first winter. Prior observations of lengths (Fig. 8.3 in Able and Fahay, 1998) indicate that juveniles from the previous year enter estuaries during March through May at lengths of about 70 to 100 mm TL. Juveniles in spring weir collections in a subtidal marsh creek in our study area ranged from 80 to 101 mm SL (Rountree and Able, 1992a) and represent the same age group of fish. This is in general agreement with Chesapeake Bay observations made by Hildebrand and Schroeder (1928), who reported age 1 during spring ranging from 64 to 119 mm TL. Through spring and summer, therefore, young-of-the-year and age 1 are only separated by about 5 to 10 cm.

SEASONALITY AND HABITAT USE

The adults occur in freshwaters or brackish waters during spring in anticipation of spawning. Eggs and larvae are found from spring through early summer in the same habitats. Juveniles can be present through the winter but also occur in the ocean at that time (see below). In Delaware Bay, larvae and/or juveniles have been observed distributed throughout the estuarine shore zone in March, moving toward the sea with the onset of warm weather in July, traveling farther downstream by August, and leaving the estuary entirely by late fall (de Sylva et al., 1962). Most *Alosa aestivalis* collected throughout the Great Bay and nearby coastal study area are young fish, apparently results of spawning in estuaries either north or south of Great Bay (Milstein and Thomas, 1977). Most of these collections were in winter (26.9%) and spring (62.7%), when they were taken in ocean (18.4%) or estuarine (81.5%) habitats. They occur in some ocean surfzone seine collections, but their abundance varies between years (K. W. Able, unpubl. data).

Juveniles, while in freshwater, remain high in the water column and avoid near-bottom depths (Warriner et al., 1969). They can tolerate abrupt rises in salinity (Chittenden, 1972a), but several studies have shown a lack of tolerance for sudden increases in temperature (Marcy, 1973; Schubel et al., 1977). Emigration from freshwater habitats occurs, as a single peak (Iafrate and Oliveira, 2008), in response to heavy rainfall, high water, a decline in temperature (Pardue, 1983), and other abiotic variables (Kosa and Mather, 2001), and is delayed by low water levels (Yako et al., 2002). In extensive sampling across 12 habitat types in the Delaware Bay estuary, most young-of-the-year were collected in river-shoreline habitats; smaller numbers occurred in bay shoreline, although they were detected in most habitats (Able et al., 2007a). The smallest young-of-the-year (20 to 45 mm) appear in intertidal and subtidal marsh habitats in April–June, while larger young-of-the-year (20 to 95 mm) are collected along river and bay shorelines and the river water column from July to November. After leaving estuaries in the fall, some young-of-the-year

overwinter in areas near their estuarine nurseries (Milstein, 1981), which are segregated from overwintering areas described for older year classes (Neves, 1981). These overwintering fish are collected within 7.4 km of shore in an area consistently influenced or affected by waters from the estuary, and they are evident in National Marine Fisheries Service (NMFS) surveys at that time (Fig. 20.2). Winter bottom temperatures in this nearshore area are generally warmer than those in the adjacent estuary and range from a high of 10.0°C in December to a low of 2.0°C in February. As temperatures warm in March (4.4 to 6.5°C), the inner continental shelf occurrences decrease, presumably coincident with a spring migration back into estuaries. A similar phenomenon has been reported from Chesapeake Bay, where some young-of-the-year remain in areas near their nurseries in deeper, more saline parts of the bay through the winter (Hildebrand and Schroeder, 1928).

Large numbers of age 1+ and older individuals are evident from collections by NMFS on the continental shelf from February to April and again in October and November (Fig. 20.2).

PREY AND PREDATORS

Several studies (n = 4) have examined the food habits of young-of-the-year (5 to 106 mm, n = 1610) *Alosa aestivalis* in Middle Atlantic Bight estuaries (Juanes and Conover, 1994) from the Connecticut River to James River, Virginia (see Tables 8.1 and 8.3). Important prey of young-of-the-year include a variety of invertebrates (copepods, cladocerans, insects/arachnids, rotifers). These studies do not show a shift in diet with ontogeny or over the geographical range. Juveniles in Connecticut and Virginia feed in the water column and concentrate on copepods or cladocerans (Burbridge, 1974; Domermuth and Reed, 1980). Very small amounts of benthic prey are consumed. Feeding takes place during daylight and ceases at night. Some feeding on zooplankton and benthic invertebrates also occurs during the spawning migration (Simonin et al., 2007). Young-of-the-year are important prey for a number of fish species, including *Anguilla rostrata*, *Perca flavescens*, and *Morone americana*. Juveniles have also been found to be important prey for young *Pomatomus saltatrix*, in the Hudson River (Juanes et al., 1993). In our summary of the available literature, this species was evident in the diet of two species, but *Pomatomus saltatrix* was the most important predator (Mancini and Able, 2005). *Alosa aestivalis* was an important prey in one study and a minor component of the diet in seven other studies.

MIGRATIONS

After leaving estuaries in the fall, the young-of-the-year (< 15 cm) have distinct seasonal distributions on the continental shelf (see Fig. 20.2). In the fall, they are found primarily on the inner continental shelf. However, some are collected farther offshore (mean depth 31 m, range 11 to 260 m) from

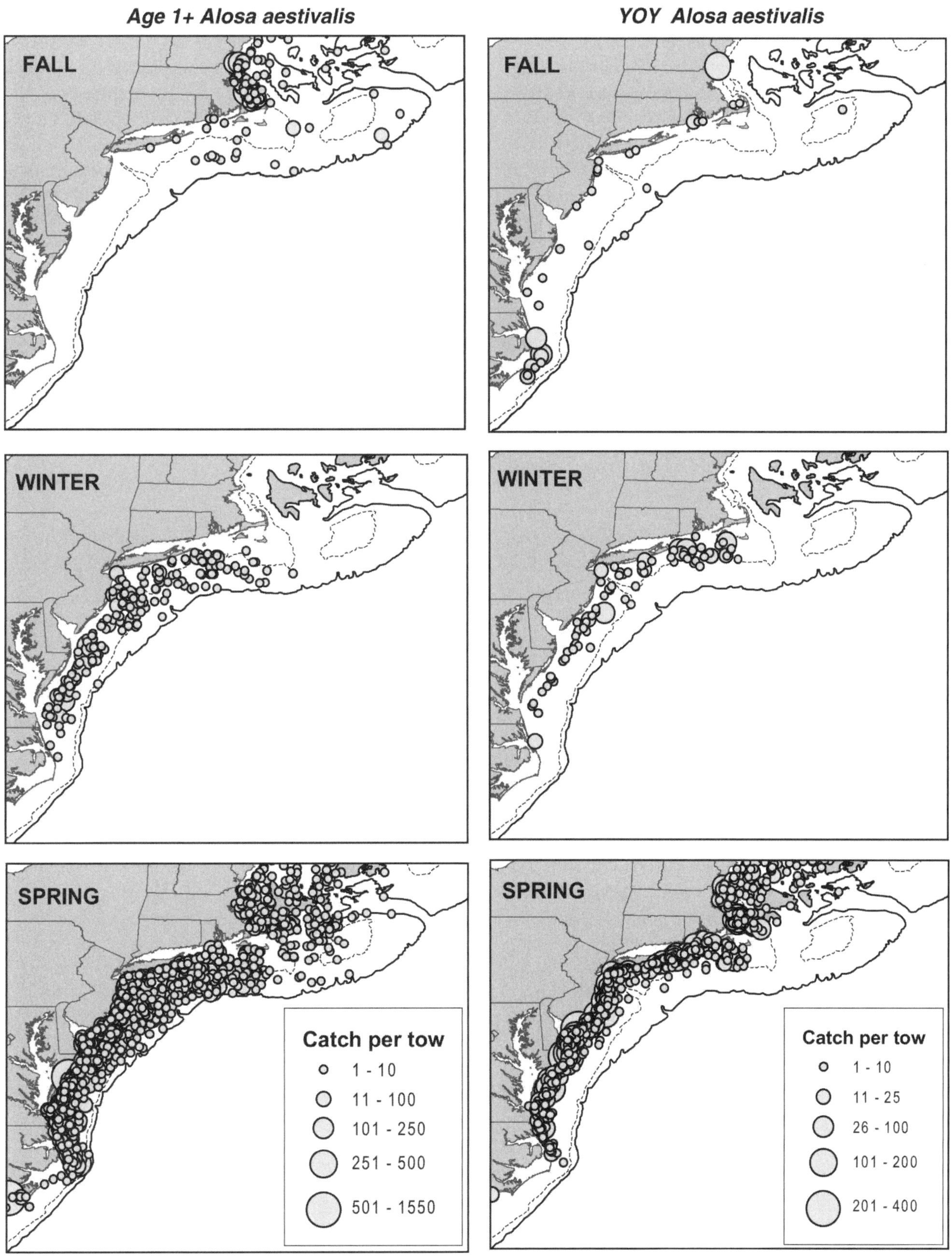

Age 1+ *Alosa aestivalis*

YOY *Alosa aestivalis*

Fig. 20.2. Composite distribution of age 1+ (> 15 cm) and young-of-the-year
(< 15 cm) *Alosa aestivalis* during seasonal cruises of the National Marine
Fisheries Service groundfish survey. Details of sampling effort are indicated
in Fig. 3.4.

Massachusetts to Cape Hatteras, with perhaps the greatest abundances near the latter location. In the winter, they are limited to the inner and mid-shelf (mean depth 37 m, range 13 to 17 m) between Nantucket Shoals and North Carolina. In the spring, they are widely distributed between the Gulf of Maine and Cape Hatteras. In the Middle Atlantic Bight, they are found on the inner shelf to the mid-shelf (mean depth 23 m, range 5 to 207 m).

The age 1+ (> 15 cm) juveniles and adults have a different distribution pattern relative to the young-of-the-year (see Fig. 20.2). In the fall, they are widely distributed on the inner portion of the Gulf of Maine, the deeper portions of Georges Bank, and south to waters off Rhode Island and Long Island (mean depth 47 m, range 15 to 194 m). In the winter, they are found over a large portion of the continental shelf (mean depth 44 m, range 13 to 162 m) from south of Nantucket Shoals to North Carolina. In the spring, they have their widest distribution, occurring from the Gulf of Maine and the western portion of Georges Bank to south of Cape Hatteras (mean depth 38 m, range 5 to 223 m).

21

Alosa mediocris
(Mitchill)

HICKORY SHAD

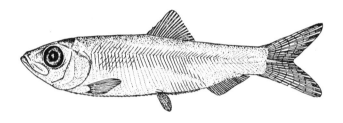

DISTRIBUTION

Alosa mediocris is found along the coast of the United States between the Bay of Fundy and northern Florida (Mansueti, 1962), but it is rare north of Cape Cod (Collette and Hartel, 1988). It is most abundant between Chesapeake Bay and North Carolina. Detailed examination of literature accounts in the study area is summarized in Able and Fahay (1998). Within the Middle Atlantic Bight, eggs and larvae have only been reported from the Chesapeake Bay, and juveniles are rarely reported from estuaries (see Table 4.3).

REPRODUCTION AND DEVELOPMENT

Spawning occurs from late April through early June. This is an anadromous species that enters bays and tidal freshwaters during spring to spawn, but there is latitudinal variation in the timing of spawning, with populations spawning later in the study area than those to the south (Harris et al., 2007). In certain Chesapeake Bay tributaries, spawning occurs in tidal freshwater, apparently between dusk and midnight (Mansueti, 1962; Mansueti and Hardy, 1967). No evidence has been found of spawning in Delaware Bay waters, although ripe adults have been collected in the Chesapeake and Delaware (C&D) Canal (Wang and Kernehan, 1979). There is some evidence of spawning in the Delaware River based on the presence of larvae and "young" (Himchak, 1981), but no evidence of spawning in Great Bay and the adjacent ocean area near Little Egg Inlet (Milstein and Thomas, 1977). These authors considered the few specimens they collected as "migrant adults." After spawning, adults leave estuaries and return to the ocean. The collections of adults made in July in Little Egg Harbor (Rountree and Able, 1997) may represent spent fish returning to the ocean from freshwater spawning areas. The ontogeny of this species, and methods for distinguishing it from the larvae of congeners and other clupeids, are moderately well described (Fahay, 2007).

LARVAL SUPPLY AND GROWTH

Little is known about the early life history of this species. The larvae and small juveniles enter estuaries from upstream freshwaters. Limited data suggest they attain lengths of 150 to 200 mm when they reach age 1 (Mansueti, 1962).

SEASONALITY AND HABITAT USE

Young-of-the-year apparently leave nursery areas in tidal freshwaters and migrate to the ocean early in the summer, although some evidence from the Chesapeake Bay suggests age 1 individuals occur in bays irregularly throughout the year (Mansueti, 1962).

The adults occur in estuaries during the summer and fall based on gill net collections (241 to 350 mm TL) in the Little Egg Harbor estuary, where they are taken in marsh creek and adjacent shallow bay habitats (Rountree and Able, 1997). In addition, they have frequently been sampled by hook and line in the Rutgers University Marine Field Station (RUMFS) boat basin (K. W. Able, pers. observ.) and by gill nets (Able et al., 2009b) and hook and line in Delaware Bay during the summer (C. Epifanio and K. W. Able, pers. observ.). In an attempt to learn more about their habitat use and behavior, we gastrically implanted ultrasonic tags in adults (n = 10, 284 to 343 mm FL) in July 2006. Fish were detected with a series of stationary hydrophones distributed throughout the estuary (see Chapter 3), including in the RUMFS boat basin, the original site of capture and release (T. M. Grothues, K. W. Able, and G. Henkes, pers. observ.). These fish remained in the lower portion of the estuary during the entire period that they were detected. One individual moved extensively over a 20-day period and was detected in marsh creeks and main channels of the estuary. Of these fish, 30% (n = 666 contacts) returned to the boat basin over a 20-day period, typically from dusk until dawn at around the period of highest tides. Average duration of residency in the basin was approximately 46.7 + 80.5 minutes.

PREY AND PREDATORS

This species is the most highly piscivorous in the genus *Alosa*. It consumes the juveniles of many species, including *Ammodytes*, clupeids, engraulids, atherinopsids, *Tautogolabrus adspersus*, and *Stenotomus chrysops*, and also eats squid, fish eggs, and small crabs (Collette and Klein-MacPhee, 2002).

MIGRATIONS

Although it is well known that members of this species spend most of their lives at sea and are anadromous, migrating into brackish and freshwaters to spawn, very little is known regarding their distribution in the ocean. They have been reported to overwinter in North and South Carolina (Smith, 1907; Hildebrand, 1963).

22

Alosa pseudoharengus (Wilson)

ALEWIFE

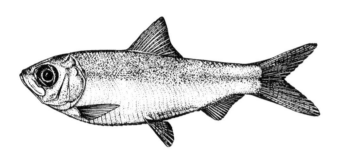

Estuary	Reproduction	Larvae/ Development	YOY Growth	YOY Overwinter
Ocean				
Both				
Jan				
Feb				
Mar				
Apr				
May				
Jun				
Jul				
Aug				
Sep				
Oct				
Nov				
Dec				

DISTRIBUTION

Alosa pseudoharengus occurs along the coast of eastern North America from the Gulf of St. Lawrence to South Carolina, but is most abundant between the Gulf of Maine and Chesapeake Bay (Berry, 1964; Winters et al., 1973). There are landlocked populations in New York (Smith, 1985). A non-migratory dwarf population has been reported to occur in the mouth of the Susquehanna River (Foerster and Goodbred, 1978) but

has not been observed recently. Eggs, larvae and juveniles occur in estuaries throughout the Middle Atlantic Bight, except in those with limited freshwater input, such as Great South Bay, Gardner's Bay, and Nauset Marsh (see Table 4.3).

REPRODUCTION AND DEVELOPMENT

Adults migrate into coastal rivers to spawn during the spring. Spawning runs in the central part of the Middle Atlantic Bight occur in the Hudson, Raritan, South, Navesink, Swimming, Great Egg Harbor, and Delaware rivers (Perlmutter et al., 1967; Zich, 1977; Dovel, 1981; Himchak, 1982a, 1983; Himchak and Allen, 1985). Evidence of spawning has also been found in the Batsto River (K. W. Able, pers. observ.), Mullica River, Nacote Creek, the Brigantine Wildlife Refuge (Milstein and Thomas, 1977), and several small tributaries of the Delaware River estuary (Wang and Kernehan, 1979). Males apparently mature at an earlier age (3–4 years) than females (4–5 years), but do not live as long (Joseph and Davis, 1965). Mean size at maturity ranges from 265 to 278 mm for males and 284 to 308 mm for females (Mayo, 1974). In studies in Connecticut and Chesapeake Bay, spawners were dominated by fish ages 3 to 8; about two-thirds of these had spawned in more than one year (Joseph and Davis, 1965; Marcy, 1969). There is some evidence that adults return to their natal rivers to spawn (Thunberg, 1971; Havey, 1973). *Alosa pseudoharengus* runs farther upstream to spawn than does *A. aestivalis*, and it generally spawns three to four weeks earlier than its congener (Jones et al., 1978). Spawning sites are typically shallow, with sluggish water flow, ranging from oxbows in large rivers to small streams; there are also records of spawning in ponds located on barrier beaches (Bigelow and Welsh, 1925; Kissil, 1974; Wang and Kernehan, 1979; Mullen et al., 1986). Spawning begins at temperatures between 13 and 15°C, and ends when waters are warmer than 27°C (Loesch, 1969). An upper lethal temperature of 29.7°C has been reported for eggs in the Hudson River (Kellogg, 1982).

Details of egg and larval development and methods for distinguishing the early stages of species in the family Clupeidae may be found in Fahay (2007). Incubation time varies with temperature: 2.1 days at 28.9°C to 15 days at 7.2°C (Jones et al., 1978). Larvae hatch at 3.1 to 5.0 mm. Transformation from larval to juvenile morphology begins at about 25 mm TL, when fin ray development is complete and the body has begun to deepen. Scales first appear at about 25 to 29 mm TL and are fully developed at 45 mm TL (Hildebrand, 1963).

LARVAL SUPPLY AND GROWTH

The larvae are supplied to estuaries from upstream freshwaters. In Delaware Bay the young-of-the-year first appear at approximately 5 to 30 mm in April through June (Able et al., 2007a).

Our estimates of growth are based on collections made in the central part of the Middle Atlantic Bight (Fig. 22.1). In

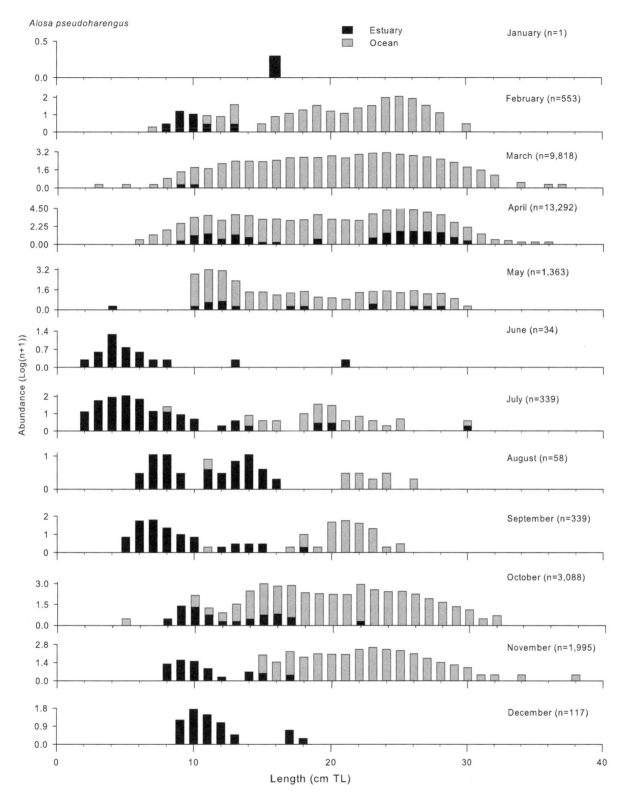

Alosa pseudoharengus

Fig. 22.1. Monthly length frequencies of *Alosa pseudoharengus* juveniles and adults collected in New Jersey / New York estuarine and adjacent coastal ocean waters. Sources: National Marine Fisheries Service: otter trawl (n = 28,908); Rutgers University Marine Field Station: seine (n = 33); 4.9 m otter trawl (n = 362); multi-panel gill net (n = 10); and Able and Fahay 1998 (n = 1784).

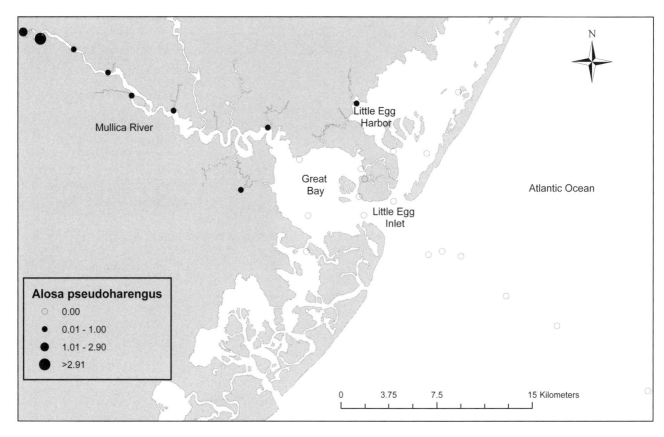

Fig. 22.2. Habitat use and average catch per tow (see key) for juvenile and adult *Alosa pseudoharengus* in the Ocean–Great Bay–Mullica River corridor in southern New Jersey based on small otter trawl collections during July and September from 1988 to 2006. See Table 3.2 for characteristics at each sampling collection.

the bight, young-of-the-year exhibit a mode at about 5 cm during July and reach about 10 cm by the end of summer; this is in agreement with our earlier interpretation (Able and Fahay, 1998). Growth appears to be arrested during winter and resumes the following spring. At age 1, the average size is between 10 and 15 cm. These rates are in general agreement with studies made elsewhere: (1) small larvae from the upper Hudson River were 10 to 59 and 45 to 74 mm SL in June and August, respectively (Perlmutter et al., 1967); (2) emigrants moving downstream during fall from a reservoir in Rhode Island ranged from 30 to 105 mm TL (Richkus, 1975); (3) wintering young-of-the-year ranged from 60 to 170 mm FL on the inner continental shelf (Milstein, 1981); (4) age 1+ males at the end of their second summer in the Connecticut River averaged 147 mm TL (Marcy, 1969); (5) fish of age group 2 were about 180 mm TL by the end of the third summer (Netzel and Stanek, 1966); and (6) spawning fish migrating up the Delaware River were 220 to 320 mm TL in late March and early April (Rohde and Schuler, 1974). The growth rates of juveniles in two spawning ponds in Massachusetts ranged from 0.2 to 0.5 mm per day (Cole et al., 1980). Collections from Delaware Bay show similar patterns, with spring through summer growth, followed by a cessation of size in-

crease during winter and resumption of growth as young-of-the-year reach age 1 in the spring (Able and Fahay, 1998).

SEASONALITY AND HABITAT USE

This species uses a full range of habitats—from the edge of the continental shelf into freshwater during the first year. Eggs occur exclusively in freshwater; larvae, in freshwater or the upper estuary. Juveniles are found in upper levels of the water column until October, when they spend more time near the bottom (Mullen et al., 1986). This behavior may change from night to day based on extensive sampling in an embayment in Great Bay, New Jersey (Hagan and Able, 2008). Juveniles have been shown to prefer temperatures of 20 to 22°C and salinities of 4 to 6 ppt in laboratory tests (Meldrim and Gift, 1971), although they are collected from a much wider range in natural habitats. In sampling across the tidal freshwater to ocean transects in the Mullica River–Great Bay estuary, this species was most frequently collected at the lowest but not at the highest salinities in the estuary and in the ocean (Fig. 22.2).

During extensive sampling across 12 habitat types in the Delaware Bay estuary, most larvae and small juveniles (5 to 30 mm) occurred in the bay and river water column in

Age 1+ Alosa pseudoharengus

YOY Alosa pseudoharengus

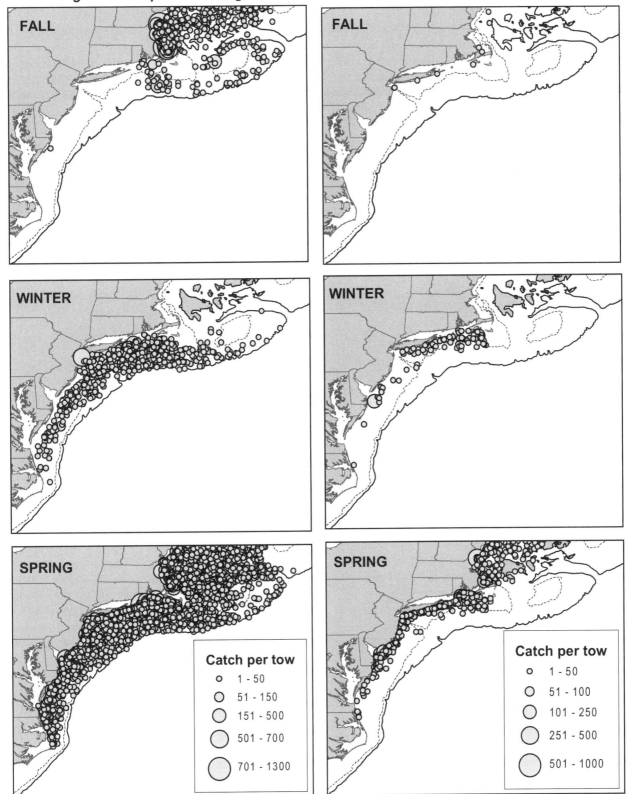

Fig. 22.3. Composite distribution of age 1+ (>12 cm) and young-of-the-year (< 12 cm) *Alosa pseudoharengus* during seasonal cruises of the National Marine Fisheries Service groundfish survey. Details of sampling effort are indicated in Fig. 3.4.

April–June (Able et al., 2007a). At larger sizes (20 to 80 mm), young-of-the-year were most abundant in river and bay shorelines at low salinities (< 10 ppt). By October–November, they had migrated out of the river and upper bay and into the deeper portions of the bay. Smaller numbers of individuals were found in all other bay and river habitats sampled (Able et al., 2007a). Some older individuals (age 1+, 80 to 110 mm) also occurred in subtidal marsh creeks and the deeper water of the bay.

Young-of-the-year leave estuaries and presumably overwinter in areas near their estuarine nurseries (Milstein, 1981), segregated from major wintering areas described for older year classes (Neves, 1981). As temperatures warm in March (to 4.4 to 6.5°C) the continental shelf occurrences of young-of-the-year decrease, presumably coinciding with a spring re-entry into freshwaters and brackish waters at age 1 (Milstein, 1981). Young-of-the-year have been collected in the Indian River Inlet during February and May at lengths between 66 and 94 mm FL (Pacheco and Grant, 1973) and have also overwintered in the upper part of the Delaware River estuary (Wang and Kernehan, 1979), further attesting to the proclivity of some populations to remain near the nursery area through the winter.

PREY AND PREDATORS

One study examined the food habits of young-of-the-year (mean 64.8 mm TL, n = 48) *Alosa pseudoharengus* in the Hudson River estuary (Grabe, 1996) (see Tables 8.1 and 8.3). Young-of-the-year fed primarily on chironomid insects and amphipods. Juveniles are inclined to eat mostly pelagic zooplankton until they are about 120 mm TL, but larger juveniles consume increasing amounts of benthic amphipods (Morsell and Norden, 1968). Young-of-the-year are important prey for a number of fish species, including *Anguilla rostrata*, *Perca flavescens*, and *Morone americana* (Moring and Mink, 2002). Larvae and juveniles have also been found to be important prey for young *Pomatomus saltatrix* in the Marsh River estuary, Maine (Creaser and Perkins, 1994).

MIGRATIONS

The spawning migration commences when river temperatures reach 10.5°C (Cianci, 1969; Kissil, 1969). During this time, adults are highly tolerant of salinity changes (Chittenden, 1972b). After spawning, an adult downstream movement is apparently triggered by an increase in water flow (Huber, 1978).

Juveniles emigrate from freshwaters and brackish waters during both late summer and fall (Yako et al., 2002; Iafrate and Oliveira, 2008). Waves of this downstream movement are apparently triggered by heavy rainfall (Cooper, 1961), high water levels and dropping temperatures (Richkus, 1975), and other abiotic variables (Kosa and Mather, 2001), but may be inhibited somewhat by bright, sunny days (Richkus, 1974) and low water levels (Yako et al., 2002).

Young-of-the-year (< 12 cm) have distinct seasonal differences in distribution on the continental shelf (Fig. 22.3). In the fall, the young-of-the-year are rare, with only occasional occurrences inshore (mean depth 23 m, range 13 to 35 m) in the Gulf of Maine and south to Long Island. Many young-of-the-year are presumably still found in estuaries at this time of the year. In the winter, when sampling is spatially restricted (see Chapter 3), they are limited to areas south and west of Nantucket Shoals to off Delaware Bay with rare occurrences to the south. At this time, they also are found at relatively shallow depths (mean depth 39 m, range 15 to 70 m) on the inner shelf. In the spring, they are more abundant, with large numbers in the Gulf of Maine and on the inner shelf of the Middle Atlantic Bight (mean depth 26 m, range 7 to 217 m) from south and west of Nantucket Shoals to North Carolina.

The age 1+ (> 12 cm) juveniles and adults are more widely distributed than young-of-the-year (Fig. 22.3). They are most abundant in the Gulf of Maine and on the shallow portions of Georges Bank (mean depth 64 m, range 15 to 220 m) to off southern Massachusetts and Rhode Island in the fall. By the winter, they are distributed much more to the south, from the southern edge of Georges Bank and over much of the continental shelf of the Middle Atlantic Bight in varying depths (mean depth 61 m, range 12 to 275 m). In the spring, they have their widest distribution, from the Gulf of Maine to all but the shallowest portions of Georges Bank and over most depths sampled (mean depth 52 m, range 5 to 295 m) in the Middle Atlantic Bight. These patterns are similar to those reported by Neves (1981).

23

Alosa sapidissima (Wilson)

AMERICAN SHAD

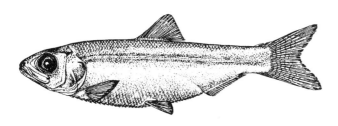

Estuary	Reproduction	Larvae/ Development	YOY Growth	YOY Overwinter
Ocean				
Both				
Jan				
Feb				
Mar				
Apr				
May				
Jun				
Jul				
Aug				
Sep				
Oct				
Nov				
Dec				

DISTRIBUTION

Alosa sapidissima is an anadromous species found along the coast of North America from Labrador to Florida (Robins and Ray, 1986). The center of abundance is between Connecticut and North Carolina. Early life history stages are found in several estuaries in the central part of the Middle Atlantic Bight, but eggs, larvae, and juveniles are absent in systems with limited freshwater input, such as Nauset Marsh, Great Bay, and estuaries on the eastern shores of Virginia and Maryland (see Table 4.3).

REPRODUCTION AND DEVELOPMENT

This species spends several years in the ocean before maturing and homing to natal rivers for spawning during the spring. In the York River in Virginia, estimates of homing were 94%, but homing to subtributaries was more variable (Walther and Thorrold, 2008). Males mature at 3 to 5 years, females at 4 to 6 years (Leim, 1924; Leggett, 1976), but their reproductive characteristics vary with latitude (Leggett and Carscadden, 1978) and between tributaries (Carscadden and Leggett, 1975). The timing of spawning migrations, as well as spawning itself, is regulated by water temperatures (Chittenden, 1969). Spawning begins when temperatures reach 12°C and ends when they exceed 20°C. Spawning occurs at night, generally in shallow waters with moderate currents (Marcy, 1972). Important spawning runs occur in the Hudson and Delaware rivers as well as smaller runs in tributaries to these and other rivers (Chittenden, 1969; Smith, 1985; Miller, 1995, Bilkovic et al., 2002). In the Hudson River, perhaps half of the spawning adults survive the upstream/downstream spawning migration and spawn again the following year, but this can be mediated by migration distance (Leonard and McCormick, 1999). In the Delaware River, fewer than 5% do so (Talbot and Sykes, 1958; Chittenden, 1975; Leggett and Carscadden, 1978), although this number may increase in response to recent improved dissolved oxygen conditions (Weisberg and Burton, 1993).

The numbers of spawning *Alosa sapidissima* in the Delaware River and Virginia (Maki et al., 2006) fell sharply at the beginning of this century. The decline in the Delaware River has been attributed to reductions in suitable spawning and nursery habitat caused by pollution (Chittenden, 1976). Recent improvements in water quality in the lower Delaware River have resulted in expansions in this habitat and increases in the size of the spawning population (Weisberg and Burton, 1993; Weisberg et al., 1996).

Details of egg and larval development and methods for distinguishing the early stages of species in the family Clupeidae may be found in Fahay (2007). The embryos incubate for 2 days (at 27°C) to 17 days (at 12°C) (Ryder, 1887; Marcy and Jacobson, 1976). After hatching at 5 to 10 mm TL, larvae are pelagic for 2 to 3 weeks, and reach a length of 25 to 28 mm before transforming to the juvenile stage (Jones et al., 1978). The relative body depth increases through the larval and juvenile stages. Fin ray development is complete at about 30 mm. Scales are first evident above the mid-lateral area at about 34.5 mm and are complete at 52 mm (Jones et al., 1978).

LARVAL SUPPLY, GROWTH, AND MORTALITY

Small juveniles first occur in estuarine samples in May and June at sizes of 3 to 8 cm and are present in samples through November (Fig. 23.1). Some young-of-the-year are apparent on the continental shelf in February, March, and April, and by that time they have reached approximately < 20 cm. It is difficult to assess growth rates of young-of-the-year because

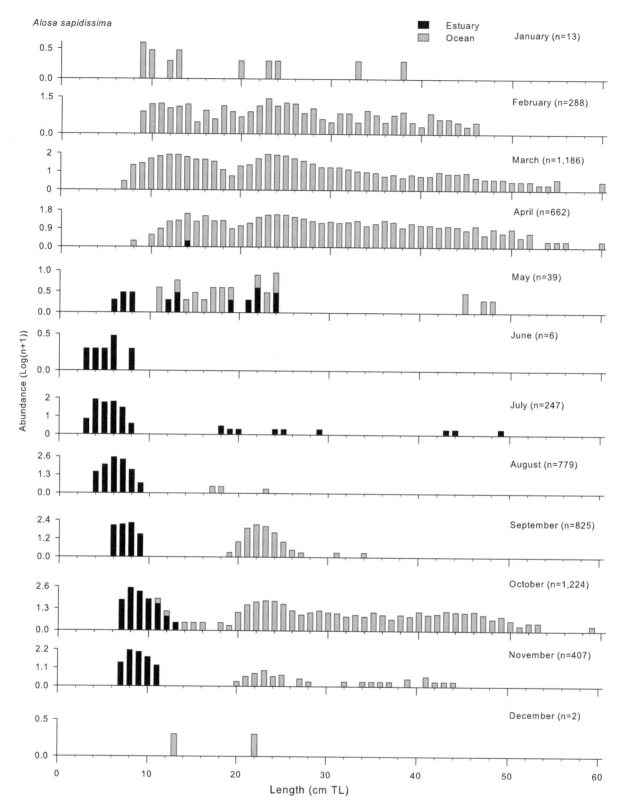

Fig. 23.1. Monthly length frequencies of *Alosa sapidissima* juveniles and adults collected in New Jersey / New York estuarine and adjacent coastal ocean waters. Sources: National Marine Fisheries Service: otter trawl (n = 3032); Rutgers University Marine Field Station: seine (n = 6); and Able and Fahay 1998 (n = 2640).

Age 1+ *Alosa sapidissima*　　　　**YOY *Alosa sapidissima***

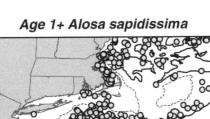
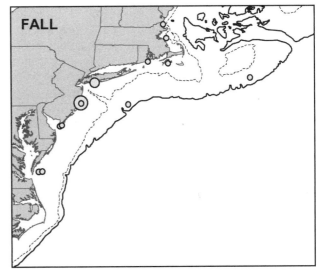
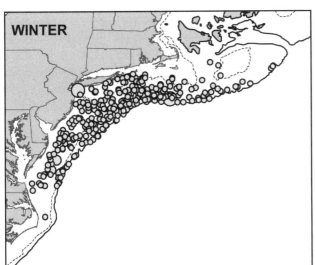
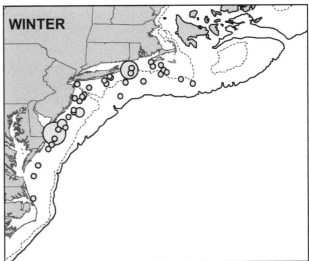
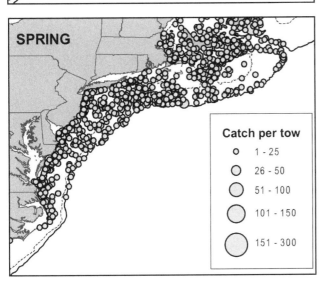
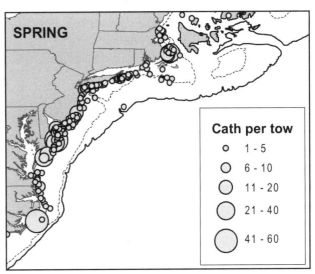

Fig. 23.2. Composite distribution of age 1+ (> 15 cm) and young-of-the-year
(< 15 cm) *Alosa sapidissima* during seasonal cruises of the National Marine
Fisheries Service groundfish survey. Details of sampling effort are indicated
in Fig. 3.4.

there is much within-river variation in size, with the smallest individuals always being found farther upstream, as has been demonstrated for the Delaware River (Chittenden, 1969). A recent study in the Hudson River suggests the possibility that individuals are capable of enhanced growth under certain conditions, and that these individuals may migrate to marine waters as early as late June and at ages of only 6 to 9 weeks (Limburg, 1995). Young-of-the-year reach approximately 120 mm TL after a full year and 240 mm TL after 2 years, based on data derived primarily from the Hudson River population and from the Bay of Fundy (Leim, 1924).

The range of acceptable environmental conditions is fairly well known for young-of-the-year. Prolonged exposure to temperatures below 4 to 6°C produces sub-lethal effects or death, especially during the fall and winter before young-of-the-year have acclimated (Chittenden, 1972b). Young-of-the-year tolerate salinity increases (5 to 30 ppt) much better than salinity decreases (30 to 0 ppt); mortality results if the decrease is abrupt (Chittenden, 1973a). However, a late migration and entry into the ocean can be physiologically disadvantageous (Zydlewski et al., 2003). In a laboratory study, young-of-the-year began to die when dissolved oxygen levels were below 1.92 mg/l, and total mortality occurred at levels below 0.64 mg/l (Chittenden, 1973b). This study also found that minimum daily levels of 2.5 to 3.0 mg/l were sufficient to permit migration through polluted areas.

SEASONALITY AND HABITAT USE

Young-of-the-year remain in freshwaters of the York River for 15 to 45 days and in the lower estuary for 32 to 66 days through the first summer based on stable isotope analysis (Hoffman et al., 2007a). Then they begin to gradually disperse downstream by fall. This appears to occur for other estuaries, such as the Delaware River, as well (Chittenden, 1969). After reaching lengths of 75 to 125 mm in the fall, young-of-the-year pass through estuarine habitats and move out to the ocean. This migration down the estuary by young-of-the-year is apparently triggered by decreasing water temperatures and occurs earlier in upstream locations. A peak in this downstream migration occurs in late October in rivers draining into Delaware Bay (Chittenden, 1972b; Leggett, 1977), but could occur as early as June (Limburg, 1995). This bimodality is reflected in adults as well (Limburg, 2001). In Chesapeake Bay young-of-the-year migrate into the ocean during February and March (Hoffman et al., 2008).

PREY AND PREDATORS

Several studies (n = 4) have examined the food habits of young-of-the-year (10 to 95 mm FL, n = 2861) in estuaries from the Connecticut River to Neuse River, North Carolina (see Tables 8.1 and 8.3). Important prey of young-of-the-year include a variety of invertebrates (insects/arachnids, copepods, ostracods, and nematodes) and fishes (cyprinids and *Anchoa* spp.). An analysis of tissue stable isotopes for young-of-the-year indicates that diet switched to terrestrial sources of carbon and zooplankton availability in response to changes in river discharge (Hoffman et al., 2007b). Juveniles in the lower Hudson River are preyed upon by young bluefish, *Pomatomus saltatrix* (Juanes et al., 1993).

MIGRATIONS

After leaving their natal rivers in the fall, juveniles remain at sea until reaching maturity. There are pronounced seasonal differences in the distribution of the young-of-the-year (< 15 cm) on the continental shelf (Fig. 23.2). In the fall, the collections are relatively rare, presumably because many young-of-the-year are still in estuaries. The captures that do occur at this time of the year are primarily on the inner continental shelf, with a few at the edge of the continental shelf (mean depth 31 m, range 9 to 110 m). By the winter, the occurrences are more frequent and range slightly deeper (mean depth 37 m, range 12 to 81 m), with most collections between southern Massachusetts and North Carolina on the inner continental shelf (Milstein, 1981). These fish range in length from less than 100 mm to about 150 mm through the winter, consistent with our current findings for lengths during those months (see Fig. 23.1). By the spring, there are more frequent, abundant collections on the shallow portions (mean depth 21 m, range 5 to 119 m) of the inner continental shelf from north of Cape Cod to south of Cape Hatteras.

The age 1+ (> 15 cm) juveniles and adults are distributed differently than the young-of-the-year and their locations vary with season as well (see Fig. 23.2). In the fall, most occur in the Gulf of Maine and in the deeper waters of Georges Bank and southern New England, with smaller collections on the inner continental shelf from Long Island to Virginia (mean depth 76 m, range 9 to 296 m). In the winter, most collections are farther south, between southern Georges Bank and waters off the mouth of Chesapeake Bay, and are largely limited to the north by the northern boundaries of the survey. At this time, they occur across the entire breadth of the continental shelf, with many extending to the deeper depths of the survey (mean depth 81 m, range 12 to 360 m). By the spring, they have the widest distribution of any season, with adults extending from the Gulf of Maine, around the perimeter of Georges Bank, and throughout the Middle Atlantic Bight to waters off North Carolina over all depths of the survey (mean depth 64 m, range 7 to 326 m). These patterns may be confounded by different groups of fish that migrate through the study area at different times (Leggett, 1977).

24

Brevoortia tyrannus (Latrobe)

ATLANTIC MENHADEN

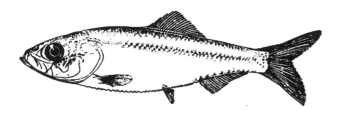

Estuary	Reproduction	Larvae/ Development	YOY Growth	YOY Overwinter
Ocean				
Both				
Jan				
Feb				
Mar				
Apr				
May				
Jun				
Jul				
Aug				
Sep				
Oct				
Nov				
Dec				

DISTRIBUTION

Brevoortia tyrannus occurs along the Atlantic coast of North America from the Gulf of St. Lawrence to northern Florida (Munroe, 2002). Large schools migrate long distances within this range. Spawning during these migrations results in early life history stages being found in all estuaries in the Middle Atlantic Bight (see Table 4.3).

REPRODUCTION AND DEVELOPMENT

Spawning may occur during every month of the year, depending on location, with peaks during spring and fall (McHugh et al., 1959). Most spawning activity in the Middle Atlantic Bight occurs in fall into early winter, but there may be limited spawning in coastal waters during spring. Egg collections in the area show that most spawning occurs over the inner continental shelf (Berrien and Sibunka, 1999), and some may extend into the lower regions of major bays and estuaries (Pearson, 1941; Dovel, 1971; Able et al., 2007a). Based on extensive analyses in past decades, there is limited spawning activity during the northward spring migration and in early summer as far north as Cape Cod (and into the Gulf of Maine), then increased spawning activity during the southward fall migration (Berrien and Sibunka, 1999). This pattern is followed by intense spawning in the South Atlantic Bight during winter (Higham and Nicholson, 1964).

The early development of *Brevoortia tyrannus*, including characters useful for distinguishing the eggs, larvae, and juveniles from those of other species of clupeids, has been well described and summarized (Fahay, 2007). Eggs are pelagic, spherical, and 1.30 to 1.95 mm in diameter. They have a single oil globule. The elongate larvae hatch at 2.4 to 4.5 mm with pigmented eyes and fully functional mouthparts. The swimbladder develops at approximately 8 mm SL (Govoni and Hoss, 2001). Visual and auditory responses begin as early as 8 to 19 mm TL (Higgs and Fuiman, 1996). Fin ray counts and the relative positions of the dorsal and anal fins are important diagnostic characters. Transformation to the juvenile stage occurs at about 30 mm, during which slender, scaleless, and largely unpigmented larvae become deep-bodied, large-headed juveniles and develop pigmentation and full complements of scales and ventral scutes.

LARVAL SUPPLY, GROWTH, AND MORTALITY

Collections of larvae in the Middle Atlantic Bight during the mid-1960s substantiate the pattern of spawning in the spring and fall, with peaks in mid-May to early June and again in September and October (Kendall and Reintjes, 1975). This pattern has also been confirmed for Peconic Bays, New York (Ferraro, 1980). In that study, spawning was most intense at temperatures between 15 and 18°C. Our analysis (Able and Fahay, 1998) has confirmed that larvae in the Middle Atlantic Bight are abundant over the inshore part of the continental shelf in the fall (September–November) and are also common during the winter (December–March). Eggs, however, did not occur in our study area between December and May, at least in 1977 through 1987 (Berrien and Sibunka, 1999).

An extensive analysis of the age, size, and hatching dates for larvae collected in the Great Bay–Little Egg Harbor estuary, based on formation of daily increments (Ahrenholz et al., 2000), indicates that these larvae are from different spawning areas (Warlen et al., 2002). Larvae collected in the fall are likely from local spawning off New Jersey during the adult fall

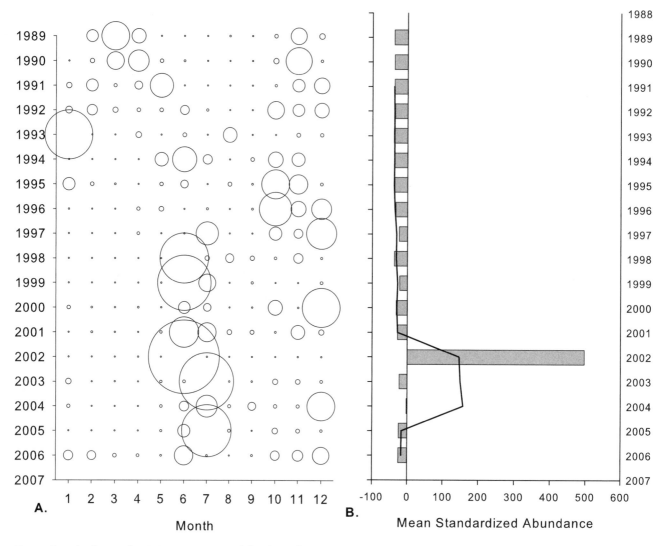

Fig. 24.1. Annual and seasonal variation in occurrence and abundance of *Brevoortia tyrannus*. (A) Seasonal variation of the larvae is based on ichthyoplankton samples collected in Little Sheepshead Creek behind Little Egg Inlet from 1989 to 2006. Open circles represent the percentage of the mean number/1000 m³ captured within each year. Values range from 0 to 100%. The smallest circles represent when samples were taken but no larvae were collected. (B) Annual variation in abundance for the same data was standardized by subtracting overall mean from annual abundance. Horizontal bars represent standardized annual abundance; the solid line is a three-year running mean that was calculated by taking the mean of the standardized values from the previous two years and the year in which the data are plotted.

migration to the south. Larvae collected during the winter are probably the products of spawning in the winter south of Cape Hatteras (Govoni and Pietrafesa, 1994). The contribution of larvae from south of Cape Hatteras varied widely for the three years analyzed (1989–1990, 1990–1991, 1992–1993) and ranged from 10 to 87% (Warlen et al., 2002), with the remainder contributed by local spawning off New Jersey. More recent examination of the seasonal occurrence of larvae inside Little Egg Inlet, New Jersey, indicates a distinct change in availability (Fig. 24.1). From 1998 to 2006, the relative contribution of larvae during June and July has been greater than any other time of the year, although larvae have continued to be collected fairly consistently in the fall. The pattern may correspond to the increasing abundance of larvae/older adults farther north, such as off New Jersey, during the summer (J. Smith, NOAA-Beaufort Laboratory, pers. comm.).

During a 15-year ichthyoplankton sampling program in Great Bay–Little Egg Harbor, the inter-annual abundance of larvae was remarkably constant, with the exception of a single year in which levels were extremely high (see Fig. 24.1), but the contamination of different spawning cohorts confounds the interpretation (Fitzhugh et al., 1997; Warlen et al., 2002).

Growth of juveniles has been estimated at 1 mm per day in some areas based on the progression of length modes (Reintjes, 1969). Growth is not overtly affected by hypoxia until very low levels of oxygen (1.5 mg O_2/l^{-1}) (McNatt and Rice, 2004). Overall growth is similar for juveniles in Great Bay–Little Egg Harbor through the spring and summer, and these young-of-the-year are approximately 8 to 17 cm TL by September (Fig. 24.2); this is true in Delaware Bay as well (K. W. Able, unpubl. data). Larvae that enter Great Bay–

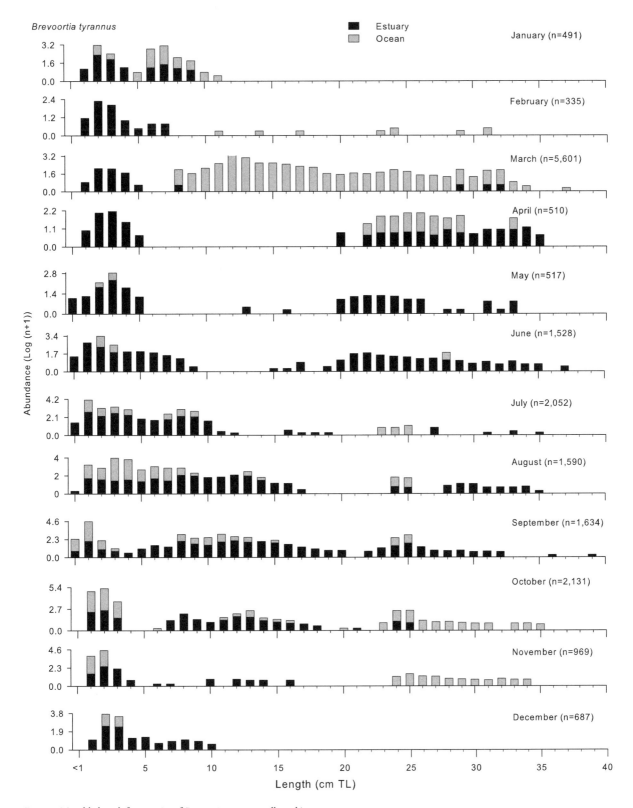

Fig. 24.2. Monthly length frequencies of *Brevoortia tyrannus* collected in estuarine and coastal ocean waters of the central Middle Atlantic Bight. Sources: National Marine Fisheries Service: otter trawl (n = 5311); Rutgers University Marine Field Station: seine (n = 1023); Methot trawl (n = 1147); 1 m plankton net (n = 4113); 4.9 m otter trawl (n = 822); multi-panel gill net (n = 1141); and Able and Fahay 1998 (n = 4488).

Little Egg Harbor during September and October increase in size from 1 to 2 cm TL to a modal size of 7 cm TL in January. Both of these cohorts are evident in October. The fate of the late summer to fall-spawned progeny is unknown, since collections are lacking that would complete the progression of length modes through the remainder of the winter and into the following spring. The large collection in the ocean in March, at a modal size of approximately 8 to 18 mm, probably represents young-of-the-year spawned during the previous year.

Mortality has been reported when temperatures drop below 3°C for several days or when temperatures decrease rapidly to 4.5°C (Lewis, 1965; Reintjes and Pacheco, 1966). Therefore, as previously mentioned, those larvae spawned during the fall may be particularly susceptible to cold-induced mortality, while those resulting from spring spawning are not. Mortality can also occur in response to low dissolved oxygen conditions (Smith, 1999) even though they are capable of detecting and avoiding 1 mg O_2/l^{-1} levels (Wannamaker and Rice, 2000) and thus potentially avoiding size-related mortality that occurs at 0.6 ppm O_2 (Shimps et al., 2005).

SEASONALITY AND HABITAT USE

Adults occur most abundantly in large schools in coastal areas, and these schools are especially plentiful in and adjacent to major estuaries and bays. The habitat of eggs and larvae is the water column in coastal areas, including the surfzone and estuaries (Able et al., in review). Larvae enter estuaries, where they transform into juveniles (Pacheco and Grant, 1965; Reintjes, 1969). In Great Bay–Little Egg Harbor, they are primarily in the postflexion stage when they make this entrance (Witting et al., 1999). In Delaware Bay estuaries, larvae are typically 10 to 20 mm TL when they ingress (Wang and Kernehan, 1979), and the height of this activity is December through May. In the Indian River in Delaware, young occur from September to June and are most abundant from December through May (Reintjes and Pacheco, 1966). In New Jersey, during some years, the locally spawned larvae enter Great Bay–Little Egg Harbor in the fall, and this peak is often followed by a period of few or no collections (Warlen et al., 2002); this pattern has been reported elsewhere (Lewis, 1965; Able et al., 2007a). These periods of reduced catches during the winter are often associated with low water temperatures (1 to 5°C), followed by generally warming temperatures and larger numbers of larvae from south of Cape Hatteras, although there is some annual variation in this pattern. Thus, the larval contributions to this estuary, and presumably others in the region, are from different spawning areas at different times. The relative percentage of these that survive to the juvenile stage is not known.

After transformation, young-of-the-year reside in estuaries through the summer (Ahrenholz et al., 1989), where they have been reported to occur in salinities of less than 1 to 36 ppt. Important documented habitats in the central part of the

Middle Atlantic Bight include the Chesapeake and Delaware (C&D) Canal and tidal tributaries of the lower Delaware River (Smith, 1971; Wang and Kernehan, 1979). As juveniles grow, they move upstream into lower-salinity waters (Lewis et al., 1972) and areas of maximum phytoplankton production (Friedland et al., 1996). Differences in the distribution of young-of-the-year of this species can be observed between day and night. They occur in an embayment during the night but not during the day, likely because of variations in vertical distribution and dispersal (Hagan and Able, 2003). Extensive sampling across 12 habitat types in the Delaware Bay estuary showed that most larvae and small juveniles (5 to 30 mm) occur in the bay and river water column in April–June (Able et al., 2007a). At larger sizes (20 to 80 mm), young-of-the-year are most abundant in river and bay shorelines at low salinities (< 10 ppt). By October–November, they have migrated out of the river and upper bay and into the deeper portion of the bay. Other, smaller numbers of individuals are found in all other bay and river habitats sampled (Able et al., 2007a). Some older individuals (age 1+, 80 to 100 mm) also occur in subtidal marsh creeks and the deeper water of the bay.

Overwintering by juveniles in estuarine areas has been recorded between Chesapeake Bay and Florida (Reintjes, 1969), but some have managed to overwinter in elevated temperatures of power-plant discharge plumes as far north as Delaware (Wang and Kernehan, 1979). Larger and older individuals also use estuaries outside the primary spawning season. In detailed sampling with gill nets in Little Egg Harbor, New Jersey, individuals 200 to 300 mm SL (age 2+) occurred from spring through fall in marsh creeks and particularly in adjacent bay habitats (Rountree and Able, 1997). Larger individuals have also been reported from other gill net sampling programs in Delaware Bay (Able et al., 2009b).

PREY AND PREDATORS

Several studies (n = 6) have examined the food habits of young-of-the-year (10 to 120.1 mm, n > 719) *Brevoortia tyrannus* in Middle Atlantic Bight estuaries from Narragansett, Rhode Island, to Newport River, North Carolina (see Table 8.1). Important prey of young-of-the-year include a variety of small invertebrates (copepods, zooplankton, mysids, and cirripedes) and phytoplankton (diatoms, flagellates, and algae). Fishes (*Anchoa mitchilli*) are a minor dietary component. The diet of young-of-the-year shifts with ontogeny, from one dominated by zooplankton at smaller sizes to one dominated by phytoplankton at larger sizes. Small young-of-the-year (19 to 34 mm FL, n = 243) consume zooplankton (copepods) almost exclusively, whereas gut contents of progressively larger young-of-the-year (28 to 94 mm FL, n = 117) shift to a mixture of detritus, diatoms, and flagellates (June and Carlson, 1971). However, there is no change in diet with size of 10 to 50 mm young-of-the-year (n = 300) (Mulkana, 1966). Geographical variation in diet has also been noted within Narragansett Bay, Rhode Island (Jeffries, 1975), with some

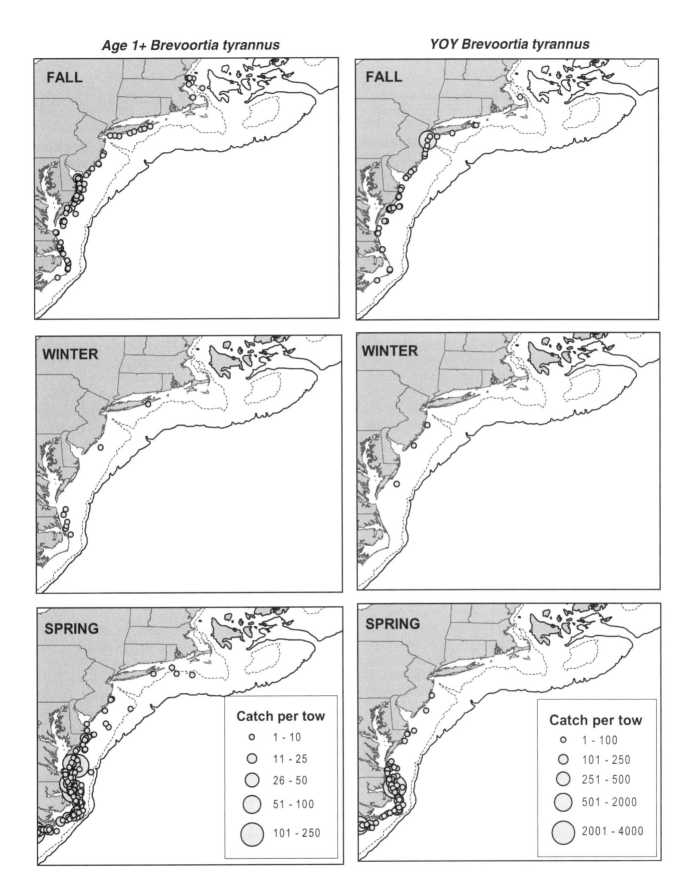

Age 1+ Brevoortia tyrannus

FALL

WINTER

SPRING

YOY Brevoortia tyrannus

FALL

WINTER

SPRING

Catch per tow
- 1 - 10
- 11 - 25
- 26 - 50
- 51 - 100
- 101 - 250

Catch per tow
- 1 - 100
- 101 - 250
- 251 - 500
- 501 - 2000
- 2001 - 4000

Fig. 24.3. Composite distribution of age 1+ (> 17 cm) and young-of-the-year (< 17 cm) *Brevoortia tyrannus* during seasonal cruises of the National Marine Fisheries Service groundfish survey. Details of sampling effort are indicated in Fig. 3.4.

young-of-the-year (40 to 60 mm, n = 30 to 40) relying less on zooplankton and more on suspended organic matter from bay through river to marsh.

Brevoortia tyrannus is important prey for many piscivorous estuarine fishes. In our review of the literature for young-of-the-year fishes, several fish species preyed on this species (see Table 8.3; Mancini and Able, 2005). Of the available studies, the species was important in three and minor in four studies, with most of these included in the diet of *Pomatomus saltatrix*. This species is also important prey for larger predators in estuaries based on percent frequency of occurrence and dry weight (see Fig. 8.4). The predators include *Ameiurus nebulosus* and *A. catus*.

MIGRATIONS

Most juveniles remain in estuarine habitats until September or October (Reintjes, 1969), when declining temperatures initiate emigration to the ocean (Friedland and Haas, 1988). Most young-of-the-year migrate south and overwinter in offshore waters, although it is clear that numerous young-of-the-year are found in Great Bay during some winters (Hagan and Able, 2003). Emigration to the ocean begins earliest in the north and progressively later to the south.

The distribution of young-of-the-year (< 17 cm) on the continental shelf varies markedly with the seasons (Fig. 24.3). In the fall, young-of-the-year have been collected primarily from Long Island to North Carolina on the inner con-tinental shelf (mean depth 12 m, range 9 to 23 m). In the winter, they are rare in the Middle Atlantic Bight, but the few individuals collected remain on the inner shelf (mean depth 23 m, range 18 to 31 m). By the spring, many young-of-the-year have migrated back to the Middle Atlantic Bight inner continental shelf (mean depth 16 m, range 5 to 38 m), where they have been collected in greatest numbers from New Jersey south to North Carolina.

The age 1+ (> 17 cm) juveniles and adults have a generally similar pattern (see Fig 24.3). In the fall, they are found as far north as the Gulf of Maine but more consistently from Long Island to North Carolina on the inner shelf (mean depth 14, range 6 to 32 m). In the winter, collections are rare, with most individuals found off North Carolina in slightly deeper depths (mean depth 21 m, range 13 to 27 m) although some may also occur in Chesapeake Bay tributaries (June and Chamberlain, 1959). In the spring, they are widely distributed in shallow waters (mean depth 19 m, range 8 to 61 m) off southern Massachusetts to Virginia and North Carolina, but they are most abundant in the southern part of the Middle Atlantic Bight. The latter concentrations are presumably the result of migrations from farther south into the Middle Atlantic Bight. Seasonal migrations during spring and fall reportedly coincide with shifts in position of the 10°C isotherm (Reintjes, 1969). They migrate north through the central part of the Middle Atlantic Bight during spring and return south during a fall migration (Nicholson, 1971).

25

Clupea harengus
Linnaeus

ATLANTIC HERRING

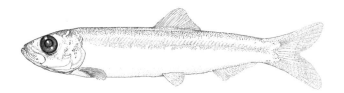

Estuary	Reproduction	Larvae/ Development	YOY Growth	YOY Overwinter
Ocean				
Both				
Jan				
Feb				
Mar				
Apr				
May				
Jun				
Jul				
Aug				
Sep				
Oct				
Nov				
Dec				

DISTRIBUTION

Clupea harengus is found on both sides of the North Atlantic Ocean. In the western North Atlantic it occurs from Greenland and Labrador to Cape Hatteras (Scott and Scott, 1988). Within this range, separate stocks occur in the Gulf of St. Lawrence, Banquereau Bank, the Scotian Shelf, the Gulf of Maine, and Georges Bank. Each of these stocks has preferred spawning, feeding, and wintering areas (Iles and Sinclair, 1982; Munroe, 2002; Stevenson and Scott, 2005). These populations also have characteristic migration patterns between

these areas, sometimes resulting in mingling of separate groups. Larvae seldom occur south of Hudson Canyon or off New Jersey (Morse et al., 1987). However, juveniles have been recorded from most estuaries in our study area from Chesapeake Bay and north (see Table 4.3).

REPRODUCTION AND DEVELOPMENT

Clupea harengus spawns somewhere within its range throughout the year, with peaks of activity in spring and fall. Spawning fish sometimes form huge schools before depositing eggs in shallow waters during spring or in deeper waters during fall. Based on collections of larvae, spawning occurs during every month between the northern part of the Middle Atlantic Bight and the Gulf of Maine but peaks in the fall (Able and Fahay, 1998). Fish in the stock closest to our study area (on Georges Bank and Nantucket Shoals) spawn in September and October (Anthony, 1982). Spawning probably does not occur in continental shelf waters in the central part of the Middle Atlantic Bight because larvae were almost totally nonexistent there between 1977 and 1987, although a very few were collected in the fall and winter (Morse et al., 1987).

The early development of *Clupea harengus*, including characters useful for distinguishing the eggs, larvae, and juveniles from those of other species of Clupeidae, has been well described and summarized (Fahay, 2007). Eggs are demersal and adhesive, forming clumps. Their diameter ranges from 1.0 to 1.4 mm and oil globules are lacking. Eggs remain attached to the bottom through the incubation period, which lasts 10 to 30 days, depending on temperature. Larvae hatch at 4 to 10 mm TL and are extremely elongate. Pigment characters and relative fin positions are important for identification. Larvae attain lengths of about 30 mm before beginning to transform into juveniles (Fahay, 2007) but may attain a maximum larval size of 42.5 mm before this transformation (Saila and Lough, 1981). Development from larval to juvenile stages requires another five to seven months, depending on temperature. During this process, the gut length decreases relative to body length, the dorsal fin migrates forward on the body, the body deepens, the air bladder becomes prominent, and all fin rays are ossified. In the Georges Bank/Nantucket Shoals area, estimated duration of larval life is 210 days (Saila and Lough, 1981).

LARVAL SUPPLY, GROWTH, AND MORTALITY

The commercially exploitable stock nearest our study area occurs on Georges Bank. It has gone through several wide fluctuations in abundance, having been decimated by overfishing in the 1960s and 1970s but experiencing a resurgence in the 1990s (Overholtz and Link, 2006). Evidence for this resurgence is provided by increased spawning activity, especially in the Nantucket Shoals/western Georges Bank region (Smith and Morse, 1990; Munroe, 2002; Overholtz, 2002; Overholtz and Friedland, 2002). Larvae presumably result-

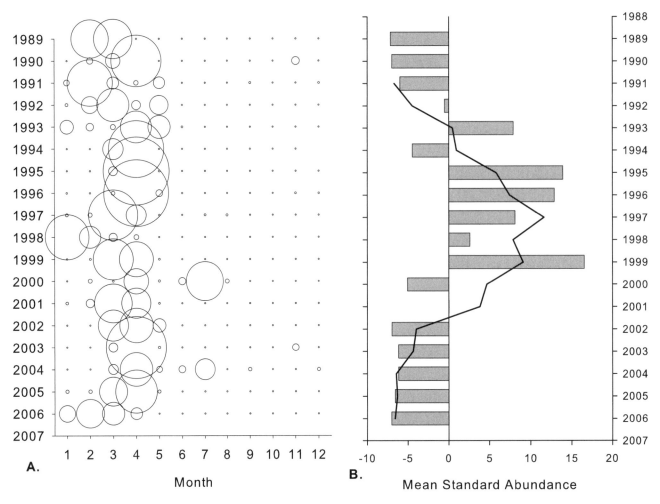

A.   Month

B.   Mean Standard Abundance

Fig. 25.1. Annual and seasonal variation in occurrence and abundance of *Clupea harengus*. (A) Seasonal variation of the larvae is based on ichthyoplankton samples collected in Little Sheepshead Creek behind Little Egg Inlet from 1989 to 2006. Open circles represent the percentage of the mean number/1000 m³ captured within each year. Values range from 0 to 90%. The smallest circles represent when samples were taken but no larvae were collected. (B) Annual variation in abundance for the same data was standardized by subtracting overall mean from annual abundance. Horizontal bars represent standardized annual abundance; the solid line is a three-year running mean that was calculated by taking the mean of the standardized values from the previous two years and the year in which the data are plotted.

ing from spawning in this region occur regularly in plankton net sampling in Little Egg Inlet (Able and Fahay, 1998). Fluctuations in the abundance of these larvae have occurred at the same frequencies as the spawning stock, with low levels prior to 1993, a period of increased numbers between the years 1993 and 2001, and a subsequent decline (Fig. 25.1). The smallest size cohorts ingress into the Great Bay–Little Egg Inlet Harbor estuary in southern New Jersey between November and August, and these individuals reach sizes of about 10 cm in May and June (Fig. 25.2).

The origin of the few larvae that appear in New Jersey inshore waters is equivocal. They may result from late-season, residual spawning by a Georges Bank group that overwinters in the New York Bight, although these fish have not been reported to spawn until their return to Georges Bank in the spring (Anthony, 1982). Larvae (< 5 cm) have also been collected from the ocean during many of the same months that they have been collected in Great Bay–Little Egg Harbor (see

Fig. 25.2). Larger young-of-the-year may be represented at approximately 9 to 17 cm in the fall in the ocean (see Fig. 25.2). Age 1+ individuals are represented in sampling in the ocean in the spring, but they are difficult to separate into year classes.

It is difficult to describe a growth rate for the Great Bay–Little Egg Harbor study area, since almost all available data on young-of-the-year end in early summer when they apparently leave the system at about 10 cm and are no longer available to collecting gear. Predation by a variety of vertebrates may be an important source of mortality (see below).

SEASONALITY AND HABITAT USE

Larvae enter the Great Bay–Little Egg Harbor estuary during the winter and spring. Ingressing larvae (19 to 63 mm) have been taken in plankton net sampling inside Little Egg Inlet in nearby Little Sheepshead Creek as early as November and December, but most occur between January and

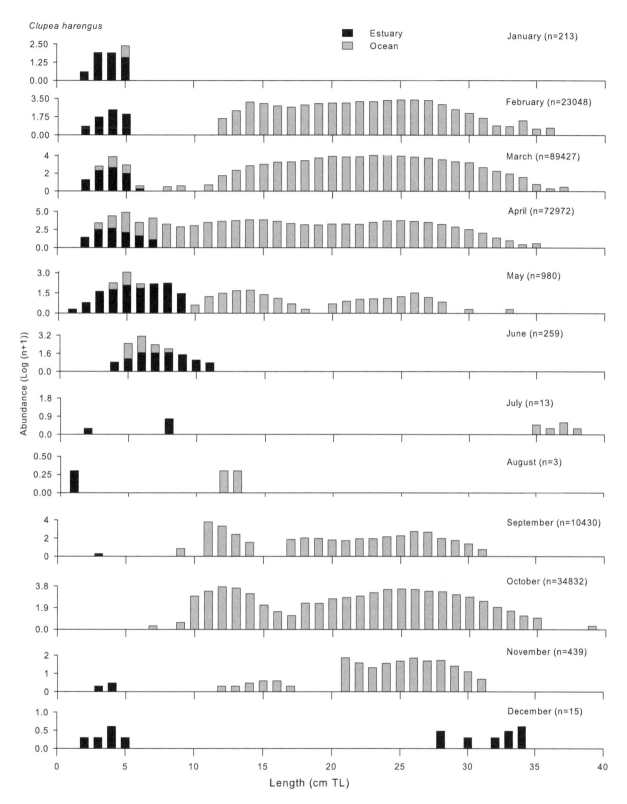

Fig. 25.2. Monthly length frequencies of *Clupea harengus* collected in estuarine and coastal ocean waters of the central Middle Atlantic Bight. Sources: National Marine Fisheries Service: otter trawl (n = 22,9156); Rutgers University Marine Field Station: seine (n = 79); 1 m plankton net (n = 1646); 4.9 m otter trawl (n = 2); and Able and Fahay 1998 (n = 1748).

Age 1+ Clupea harengus   **YOY Clupea harengus**

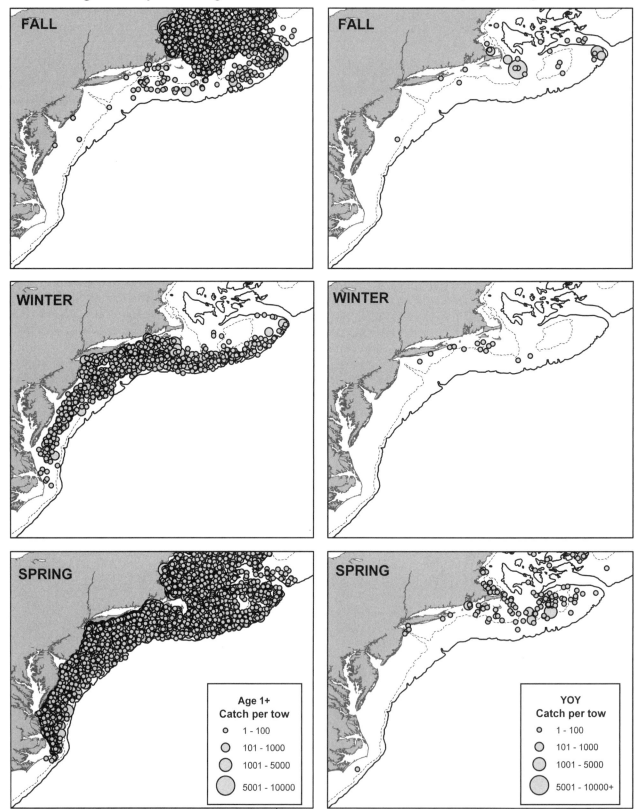

Fig. 25.3. Composite distribution of age 1+ (> 12 cm) and young-of-the-year
(< 12 cm) *Clupea harengus* during seasonal cruises of the National Marine
Fisheries Service groundfish survey. Details of sampling effort are indicated
in Fig. 3.4.

May (see Fig. 25.2). Larvae (23 to 45 mm) have also been recorded entering Indian River Inlet in Delaware from January through April (Pacheco and Grant, 1973). There are also a few records of juveniles as far as 68 km up the Hudson River (Smith, 1985).

The source of transforming larvae entering New Jersey inlets remains enigmatic. It is possible that the resurgence of spawning by this species over Georges Bank and Nantucket Shoals (Smith and Morse, 1990) has also included increased reproductive activity in the northern part of the Middle Atlantic Bight, but we presently lack the sampling effort to demonstrate this. Larval occurrences in Great Bay–Little Egg Harbor in the 1990s also include juveniles in pelagic fish sampling in 1995–1996 (Hagan and Able, 2003), and this may be another signal of the resurgence at that time. At a finer temporal scale, the young-of-the-year occurred more frequently and were more abundant at night in the same sampling program, presumably as the result of a vertical movement that allows them to move into shallow water at that time (Hagan and Able, 2008).

PREY AND PREDATORS

Our examination of food habits of this species (32.3 to 85.2 mm FL, n = 121) from Great Bay, New Jersey, found that the diet of young-of-the-year was dominated by zooplankton (37.4 to 84.3% by weight) with smaller amounts of unidentified crustaceans and fish. These preferences were similar to those of other young-of-the-year individuals (33.9 to 51.1 mm, n = 18) in Long Island Sound, where copepods were the bulk of the diet, but mysids, diatoms, and dinoflagellates were also consumed (Richards, 1963). Juveniles and adults are fed heavily on by a variety of predators, including squid, fishes, marine mammals, and birds (Hildebrand and Schroeder, 1928; Overholtz and Link, 2006).

MIGRATIONS

After they leave estuaries, young-of-the-year (<12 cm) are primarily distributed in the shallow waters of the inner continental shelf, on Georges Bank, and into the Gulf of Maine (Fig. 25.3). In the fall, they are collected from shallow waters (mean depth 70.0 m, range 10 to 180 m) in the southern Gulf of Maine, on Georges Bank, near Nantucket Shoals, and occasionally farther south. In winter, the few collections of the young-of-the-year are off southern Long Island, on Nantucket Shoals, and in the Great South Channel in depths from 27 to 77 m (mean depth 45.2 m). However, large numbers are also collected in Great Bay at this time (Hagan and Able, 2003). In the spring, they are more frequently collected, especially in the central portion of Georges Bank. At this time, they are distributed across 11 to 202 m (mean depth 48.4 m).

The age 1+ and older and larger individuals (> 12 cm) are much more frequently represented in bottom trawl surveys, and are found much farther south than young-of-the-year, especially in the winter and spring (see Fig. 25.3). In the fall, they are common throughout the Gulf of Maine and on the eastern and southern portions of Georges Bank, with fewer collections off Massachusetts and Rhode Island and fewer still farther south. At this time the mean depth of capture was 96.3 m but ranged from 10 to 220 m. In the winter, they are distributed much farther south, with collections across much of the continental shelf (mean depth 57.7, range 13 to 240 m) from the southern and eastern tips of Georges Bank continuously to off the mouth of Chesapeake Bay and, occasionally, to Cape Hatteras. In the spring, they once again moved back north but were found almost continuously from throughout the Gulf of Maine to Cape Hatteras, at least based on these cumulative collections. The depths ranged from 5 to 281 m (mean depth 45.9 m) at this time.

26

Dorosoma cepedianum (Lesueur)

GIZZARD SHAD

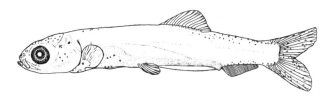

Estuary	Reproduction	Larvae/ Development	YOY Growth	YOY Overwinter
Ocean				
Both				
Jan				
Feb				
Mar				
Apr				
May				
Jun				
Jul				
Aug				
Sep				
Oct				
Nov				
Dec				

DISTRIBUTION

Dorosoma cepedianum occurs along the east coast of North America from the Hudson River estuary to the Gulf of Mexico (Lee et al., 1980). Larvae and juveniles occur in freshwater tributaries that empty into large estuaries such as Delaware and Chesapeake bays and the Hudson River.

REPRODUCTION AND DEVELOPMENT

Age 1 females are capable of spawning (Smith, 1985). This occurs when temperatures reach 10 to 21°C (stimulated at

16°C) in the spring in shallow waters (15 to 30 cm) at rising water levels (Murdy et al., 1997). Details of egg and larval development and methods for distinguishing the early stages from other clupeids can be found in Fahay (2007). The eggs are demersal and adhesive, with a smooth chorion and one large and one to five smaller oil globules. The egg diameter is 0.75 mm. Hatching occurs after 36 hours at 27°C or 95 hours at 17°C (Smith, 1985). Larvae hatch at 3.2 mm. The body is elongate with a straight gut. The anus is always posterior to the dorsal fin. Caudal fin flexion occurs between 11 and 17 mm. Pigmentation includes large melanophores near the anus and a series of spots on the dorsal surface of the anterior gut. Transformation occurs at sizes larger than 24 mm. The body begins to deepen at about 35 mm. Scales are complete at about 50 mm.

LARVAL SUPPLY AND GROWTH

The larvae and small juveniles are supplied to the estuary from upstream freshwaters. A few young-of-the-year become evident in Delaware Bay as early as June, but most are found in August through October at sizes of < 20 cm (Fig. 26.1). By the winter they reach a modal size of 12 to 20 cm. In Chesapeake Bay the young-of-the-year attain lengths of 109 to 160 mm by October (Hildebrand and Schroeder, 1928). They apparently do not grow over the winter, as similar size fish are detected in the estuary the following spring. The combined length frequencies during August suggest that three length modes, representing three year classes, are evident in the estuary. This species attains a maximum length of nearly 50 cm and ages of 6 or 7 (Smith, 1985).

SEASONALITY AND HABITAT USE

Spawning occurs in quiet, fresh waters in the spring and late summer in Chesapeake Bay (Murdy et al., 1997). The larvae and juveniles develop in freshwaters, where they often occur around submerged aquatic vegetation. At larger sizes they move downstream into brackish waters. They can tolerate salinities up to 22 ppt (Murdy et al., 1997) or 33 to 34 ppt (Smith, 1985). During the fall they move back up into freshwaters for the winter (Murdy et al., 1997).

PREY AND PREDATORS

Newly hatched larvae feed on protozoans and other plankton (Smith, 1985). After a few weeks the diet changes to include phytoplankton and algae as well as mud and detritus from the bottom (Hildebrand and Schroeder, 1928; Murdy et al., 1997).

MIGRATIONS

An upstream run from brackish waters into freshwater has been reported to occur in Chesapeake Bay in the fall (Hildebrand and Schroeder, 1928).

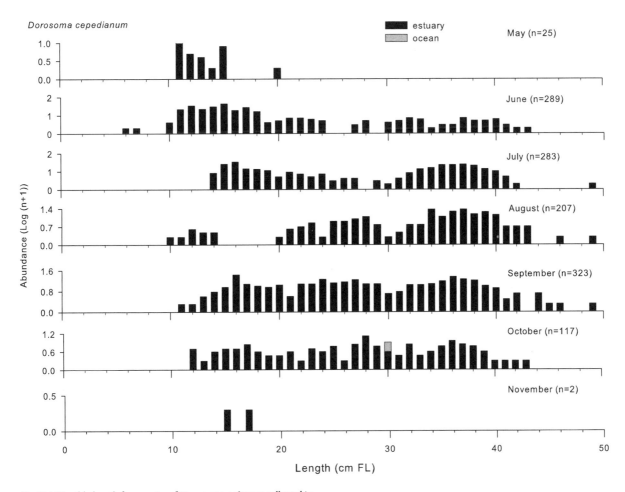

Fig. 26.1. Monthly length frequencies of *Dorosoma cepedianum* collected in
estuarine waters of the Mullica River–Great Bay estuary and Delaware
Bay. Source: Rutgers University Marine Field Station: multi-panel gill net
(n = 1234).

27

Opisthonema oglinum (Lesueur)

ATLANTIC THREAD HERRING

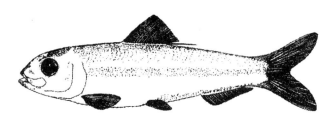

DISTRIBUTION

This species is found in the western North Atlantic Ocean from the Gulf of Maine and Bermuda, throughout the Gulf of Mexico, to the Caribbean Sea and West Indies (Munroe and Nizinsky, 2002). Early stages were once considered rare in our study area, but in the past decade have become increasingly common.

REPRODUCTION AND DEVELOPMENT

Spawning occurs from May to June off North Carolina (Munroe and Nizinsky, 2002). The early development of *Opisthonema oglinum*, including characters useful for distinguishing the larvae and juveniles from those of other clupeids, has

been well described and summarized in Fahay (2007). Eggs are pelagic, spherical, and 1.08 to 1.31 mm in diameter. Larvae hatch at 3.0 mm with pigmented eyes. Transformation to the juvenile stage occurs at 15 to 25 mm SL. Juveniles are diagnosed by the prolonged posteriormost dorsal fin ray. Transformation occurs between 15 and 25 mm SL (Fahay, 2007).

LARVAL SUPPLY AND GROWTH

Larvae were rarely collected in the Middle Atlantic Bight in synoptic sampling prior to 1987 (M. P. Fahay, pers. observ.). Early stages of this species have appeared in our study area annually since 1995 and during two years (2001 and 2006) their numbers increased (Fig. 27.1). This species represents another example of a southern taxon that has become increasingly common in recent years. Larvae are 10 to 30 mm TL when they initially appear in our study area (Fig. 27.2), indicating that spawning probably occurs here. These larvae grow to a modal size of 60 mm TL by October, after which time they are no longer present.

SEASONALITY AND HABITAT USE

Larvae have been collected almost annually in our plankton nets since 1995, and they have consistently arrived in July or August (see Fig. 27.2). During the fall, small juveniles occur on open ocean beaches.

PREY AND PREDATORS

Adults feed on plankton, especially copepods, by filter feeding, but they also include small fishes, crabs, and shrimp in their diets (Munroe, 2002). Juveniles emigrating from ocean inlets in North Carolina during the fall are heavily preyed upon by *Pomatomus saltatrix* and *Scomberomorus maculatus* (Smith, 1994).

MIGRATIONS

We have no data describing an emigration by juveniles or adults of this species from New Jersey estuaries.

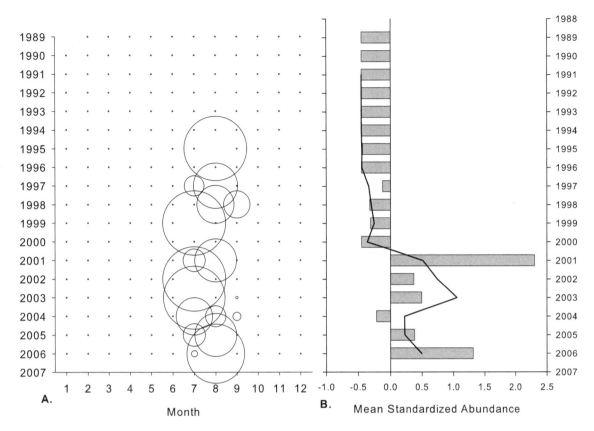

Fig. 27.1. Annual and seasonal variation in occurrence and abundance of *Opisthonema oglinum*. (A) Seasonal variation of the larvae is based on ichthyoplankton samples collected in Little Sheepshead Creek behind Little Egg Inlet from 1989 to 2006. Open circles represent the percentage of the mean number/1000 m^3 captured within each year. Values range from 0 to 100%. The smallest circles represent when samples were taken but no larvae were collected. (B) Annual variation in abundance for the same data was standardized by subtracting overall mean from annual abundance. Horizontal bars represent standardized annual abundance; the solid line is a three-year running mean that was calculated by taking the mean of the standardized values from the previous two years and the year in which the data are plotted.

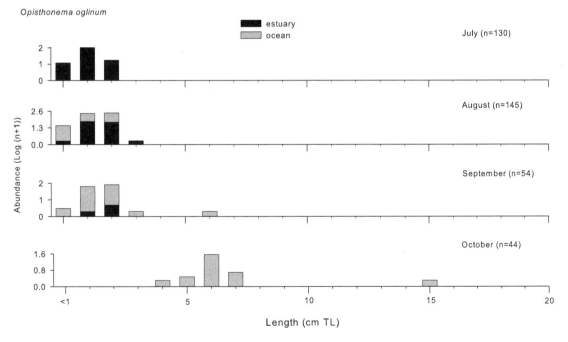

Fig. 27.2. Monthly length frequencies of *Opisthonema oglinum* collected in coastal ocean waters of the central Middle Atlantic Bight. Source: Rutgers University Marine Field Station: seine (n = 44); Methot trawl (n = 69); 1 m plankton net (n = 259); 4.9 m otter trawl (n = 1).

28

Anchoa hepsetus (Linnaeus)
STRIPED ANCHOVY

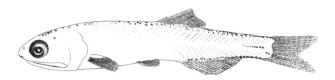

Estuary	Reproduction	Larvae/Development	YOY Growth	YOY Overwinter
Ocean				
Both				
Jan				
Feb				
Mar				
Apr				
May				
Jun				
Jul				
Aug				
Sep				
Oct				
Nov				
Dec				

DISTRIBUTION

Anchoa hepsetus is found along the Atlantic coasts of North America and South America from Nova Scotia to Uruguay, but is most abundant from Chesapeake Bay to the West Indies (Hildebrand, 1963). In the Middle Atlantic Bight the early life history stages have been reported from most estuaries as far north as Great South Bay and Narragansett Bay but seldom farther north (see Table 4.3).

REPRODUCTION AND DEVELOPMENT

Spawning in the ocean occurs during the spring and summer, with the greatest activity occurring from southern New Jersey southward, based on the distribution and abundance of eggs (Berrien and Sibunka, 1999). Spawning begins as early as April near Cape Hatteras, then extends northward to Delaware Bay by May. In lower Delaware Bay, it occurs from May through August (Stevenson, 1958), and continues into September off southern New England. Eggs have been collected in May in Great South Bay, Long Island (Monteleone, 1992); however, the smallest larvae have not been found before September in Great Bay–Little Egg Harbor. Spawning in the southern part of the Middle Atlantic Bight reportedly takes place within harbors, estuaries, and sounds (Hildebrand and Cable, 1930), but eggs also occur over much of the inner continental shelf.

Details of early development are summarized in Fahay (2007). The pelagic eggs are strongly elliptical and lack oil globules. Larvae hatch at 3.6 to 4.0 mm, characterized by unpigmented eyes and a yolk sac that tapers posteriorly. Larvae are lightly pigmented, and the relative positions of the fins are important for identification.

Individuals between 10 and 66.5 mm FL (n = 33) were examined for scale formation (see Fig. 17.1). Scales were often missing in preserved individuals; thus the patterns illustrated are based on composite observations from both sides of individuals as well as the occurrence of scale pockets. Scale formation occurred between 19 and 46 mm FL and began after the completion of fin rays in the dorsal, anal, and caudal fins. The sole point of origin appeared to be on the caudal peduncle. This occurred at approximately 19 mm FL, the smallest specimen in our material observed with scales. By 24 mm FL, scales had expanded anteriorly to the level of the posterior portion of the dorsal fin and laterally to cover the surfaces of the caudal peduncle. By at least 46 mm FL, the scales covered all surfaces of the body from the base of the caudal fin to the operculum. Only the fins and the head lacked scales. At this size, scales covered 77.8% of the body.

LARVAL SUPPLY AND GROWTH

The sole estimate of annual variation in larval abundance indicates larger than average abundance in 1997 and 1999 (Fig. 28.1). The extended period of reproduction accounts for the prolonged duration of larvae (< 2 cm) from June through November in both the estuary and the ocean (Fig. 28.2). After that time they have only been collected in the ocean and at much lower numbers. Two length modes of presumed young-of-the-year are evident in July, August, and October. By October, the largest young-of-the-year have a modal size of approximately 9 cm but some individuals are up to 14 cm.

SEASONALITY AND HABITAT USE

Eggs and larvae are most frequently collected in the ocean, mostly over the inner continental shelf (Able and Fahay,

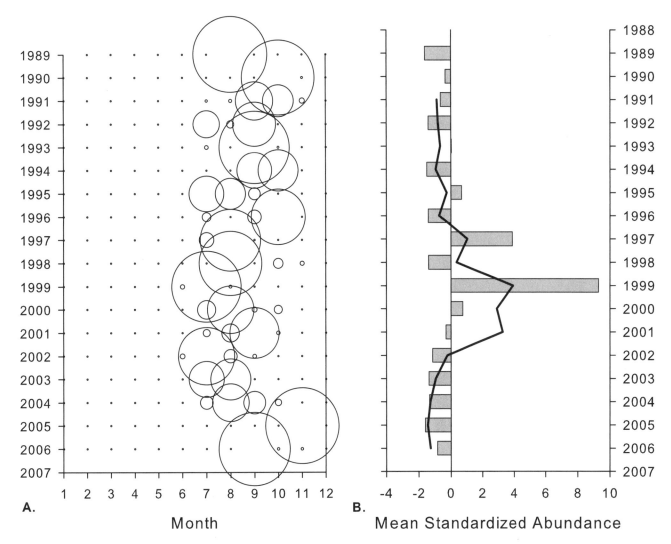

Fig. 28.1. Annual and seasonal variation in occurrence and abundance of *Anchoa hepsetus*. (A) Seasonal variation of the larvae is based on ichthyoplankton samples collected in Little Sheepshead Creek behind Little Egg Inlet from 1989 to 2006. Open circles represent the percentage of the mean number/1000 m³ captured within each year. Values range from 0 to 100%. The smallest circles represent when samples were taken but no larvae were collected. (B) Annual variation in abundance for the same data was standardized by subtracting overall mean from annual abundance. Horizontal bars represent standardized annual abundance; the solid line is a three-year running mean that was calculated by taking the mean of the standardized values from the previous two years and the year in which the data are plotted.

1998; Berrien and Sibunka, 1999), but they also occur in estuaries, bays, and harbors in the southern part of the Middle Atlantic Bight (Hildebrand and Cable, 1930). Juveniles and adults have been collected primarily from the polyhaline portions of many estuaries, such as Delaware Bay (de Sylva et al., 1962) and Great Bay–Little Egg Harbor (Able et al., 1996b; Hagan and Able, 2003). In Great Bay–Little Egg Harbor young-of-the-year (1 to 10 cm TL) have been collected from July through November from subtidal creeks and an embayment (Rountree et al., 1992; Able et al., 1996b; Hagan and Able, 2003) (Fig. 28.2). Collections from other estuaries, including Hereford Inlet in southern New Jersey (Allen et al., 1978) and Indian River Inlet (Wang and Kernehan, 1979), indicate temporarily overlapping patterns. Young-

of-the-year and adults are much more abundant at night, perhaps as the result of diel vertical migration and dispersal (Hagan and Able, 2008).

PREY AND PREDATORS

Our examination of food habits of this species (27.3 to 125.5 mm, n = 145 stomachs) found a fairly consistent diet of zooplankton (22.9 to 52.1% by weight of stomach contents) and other crustaceans (5.9 to 24.4% by weight) at sizes up to 100 mm. Fishes were a relatively low but consistent component of the diet (0.2 to 12.8%) at sizes < 79 mm but made up 39.4% at sizes > 100 mm FL. Other prey included mysids (8.2 to 31.4%) at sizes between 40 and > 100 mm.

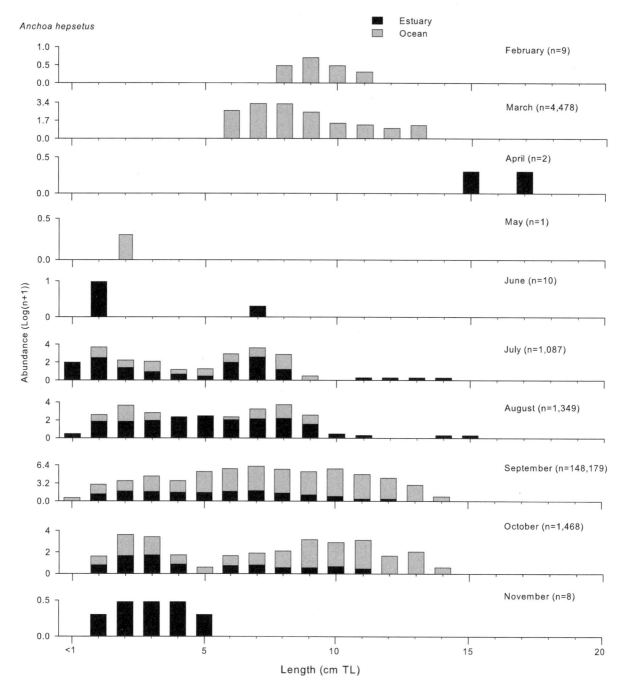

Fig. 28.2. Monthly length frequencies of *Anchoa hepsetus* collected in estuarine and coastal ocean waters of the central Middle Atlantic Bight. Sources: National Marine Fisheries Service: otter trawl (n = 153,181); Rutgers University Marine Field Station: seine (n = 350); Methot trawl (n = 239); 1 m plankton net (n = 686); 2 m beam trawl (n = 3); 4.9 m otter trawl (n = 271); and Able and Fahay 1998 (n = 1861).

SEASONAL MIGRATIONS

The young-of-the-year (< 14 cm) migrate seasonally in and out of our study area (Fig. 28.3). In the fall, when they leave estuaries, they are concentrated in shallow waters (mean depth 18.4 m, range 5 to 286 m) from the inner shelf off North Carolina and Virginia. In the winter, they are absent, presumably because they have migrated south of the study

area. In the spring, only a few individuals have been collected.

The age 1+ (> 7 cm) juveniles and adults also occur seasonally (see Fig. 28.3). In the spring, the age 1+ individuals were most abundant off North Carolina at Cape Hatteras but extended north as far as Massachusetts in shallow depths (mean depth 19.3 m, range 10 to 47 m). As waters warm, they

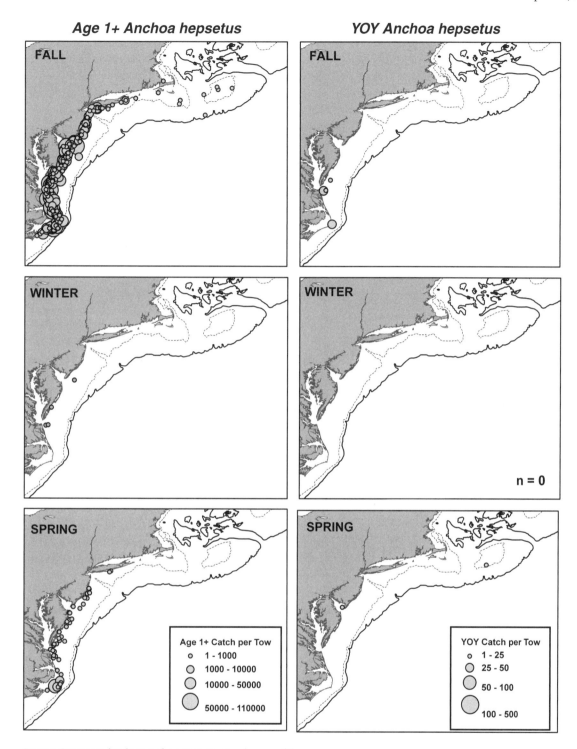

Age 1+ *Anchoa hepsetus*   **YOY *Anchoa hepsetus***

Fig. 28.3. Composite distribution of age 1+ (> 7 cm) and young-of-the year (< 7 cm) *Anchoa hepsetus* during seasonal cruises of the National Marine Fisheries Service groundfish survey. Details of sampling effort are indicated in Fig. 3.4.

presumably migrate north and into estuaries or at estuarine inlets for spawning. In the fall, after migrating out of estuaries, they are concentrated in shallow waters (mean depth 17.1 m, range 5 to 149 m) from the inner shelf off North Carolina to the eastern end of Long Island, with some extending to off Massachusetts and on to Georges Bank. This migration

behavior is also apparent in Chesapeake Bay, where juveniles and adults migrate to deeper waters in fall and winter (Musick, 1972) and then leave the bay (Hildebrand and Schroeder, 1928). In the winter, none have been collected on the inner shelf.

29

Anchoa mitchilli (Valenciennes)

BAY ANCHOVY

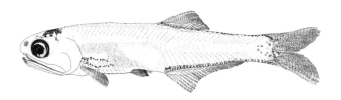

Estuary	Reproduction	Larvae/Development	YOY Growth	YOY Overwinter
Ocean				
Both				
Jan				
Feb				
Mar				
Apr				
May				
Jun				
Jul				
Aug				
Sep				
Oct				
Nov				
Dec				

DISTRIBUTION

Anchoa mitchilli is found from Maine to Florida and throughout the Gulf of Mexico to the Yucatan Peninsula (Hildebrand, 1943; Bigelow and Schroeder, 1953). It may be the most abundant and productive fish species in the western North Atlantic Ocean (McHugh, 1967; Morton, 1989; Houde and Zastrow, 1991; Jung and Houde, 2004a, b). Eggs, larvae, juveniles, and adults occur in almost every estuary in the Middle Atlantic Bight (see Table 4.3).

REPRODUCTION AND DEVELOPMENT

Fish of both sexes mature at 40 to 45 mm FL, at approximately 10 months of age (Zastrow et al., 1991), but they are also reported to mature at sizes as small as 31 mm FL and at ages of 2.5 to 3 months (Luo and Musick, 1991). However, individuals from the northern part of the range (Hudson River) are older and larger (Bassista and Hartman, 2005). Individuals in Chesapeake Bay spawn every 1.3 to 4 days, with up to 54 total spawnings per female (Luo and Musick, 1991). Egg densities on the inner continental shelf (Berrien and Sibunka, 1999) may be of the same order of magnitude as in the estuary (Vouglitois et al., 1987). In the latter study, egg abundance was high as far as 6 km offshore but declined at locations 15 to 18 km offshore. Spawning in the coastal ocean begins as early as April off Cape Hatteras, and expands rapidly northward over inner continental shelf waters during May and June (Berrien and Sibunka, 1999). It occurs as far north and east as Long Island and Narragansett Bay in July, then tapers off between September and October. Most spawning in Middle Atlantic Bight estuaries occurs between April and November, as indicated by gonad development and the occurrence of eggs and larvae (Stevenson, 1958; Croker, 1965; Dovel, 1981; Vouglitois et al., 1987; Castro and Cowen, 1991; Luo and Musick, 1991; Zastrow et al., 1991; Monteleone, 1992; Wang and Houde, 1995), including in Great Bay, New Jersey. The spawning season in northern populations was more abbreviated, as in the Hudson River (Bassista and Hartman, 2005) and Narragansett Bay (Lapolla, 2001a, b). Based on a rigorous sampling in Chesapeake Bay, a correlation analysis suggests that spawning is most intense in areas of high zooplankton abundance and where the ctenophore *Mnemiopsis leidyi*, a potential predator on eggs and larvae, is least abundant (Dorsey et al., 1996). Most spawning occurs in the evening (Ferraro, 1980; Luo and Musick, 1991; Zastrow et al., 1991).

Eggs are off-round, but less elliptical than those of *Anchoa hepsetus*. Oil globules are lacking and the long axis measures 0.84 to 1.11 mm. Hatching occurs at a length of 1.8 to 2.7 mm. Larvae are characterized by an elongate body with yolk sac tapering posteriorly. Pigment is very light. Relative fin positions are very important in identification. Transformation to the juvenile stage occurs at about 20 mm. Hatching occurs in approximately 24 hours at summer temperatures (Mansueti and Hardy, 1967; Monteleone, 1992). Further details of ontogeny may be found in Fahay (2007).

A total of 29 specimens between 10.0 and 75.8 mm FL were examined for patterns of scale formation (see Fig. 17.1). Scales were often missing; thus illustrations were based on observation of both sides of specimens and represent a combination of actual scales and scale pockets. Scale formation occurred over a size range of 10 to 76 mm FL, although most scales formed at sizes between 10 and 20 mm FL; thus most scales are well developed at about the size when fin ray formation is complete. Scales begin to appear on the lateral line anterior to the caudal peduncle at approximately 10 mm FL.

By 16 mm FL, scales cover the body posterior to the middle of the dorsal fin. By 20 to 23 mm FL, scales cover the body anteriorly as far as the operculum in some specimens. In a 75.8 mm FL specimen, scales were observed on the base of the caudal fin, which is consistent with the presumed adult condition. No scales were observed on the head region. Thus, scales covered 84.2% of the body and 63.6% of the body and fins at this size and in adults.

LARVAL SUPPLY, GROWTH, AND MORTALITY

In Great Bay–Little Egg Harbor, there is considerable annual variation in abundance and size of larvae. Years of peak abundance (e.g., 1989, 1995, 2004) frequently may be followed by years with very low levels; there does not seem to be a correlation with high abundances of juveniles (e.g., 2001) (Fig. 29.1). Similar extreme variation has been suggested for the Hudson River (Dovel, 1981), Barnegat Bay (Vouglitois et al., 1987), Delaware Bay (Derickson and Price, 1973), and Chesapeake Bay (Newberger and Houde, 1995). These fluctuations are likely to result in recruitment variation because the populations are dominated by young-of-the-year and age 1 individuals (Vouglitois et al., 1987; Newberger and Houde, 1995; Wang and Houde, 1995). In upper and mid-Chesapeake Bay, annual production of young-of-the-year was 856 g/100 m³, 87.9% of which occurred in the first 3 months of life (Wang and Houde, 1995). Young-of-the-year (92.6%) accounted for nearly all of the annual contribution for this species. Production of *A. mitchilli* young-of-the-year is of such large magnitude that it could influence the total fish production of many estuaries.

The supply of larvae can originate from a couple of sources as a result of spawning in coastal oceanic waters and in adjacent estuaries. The degree of exchange is unknown, but tidal flushing in both directions and upwelling may con-

tribute. Presumably these factors would have more impact on the smaller, more weakly swimming larvae. The occurrence of larvae in surfzone and nearshore habitats varied annually over the period from 1995 to 1999 (Able et al., 2010).

In estuaries, dispersal after spawning appears to occur in an up-estuary direction, but this may vary between estuaries (Kimura et al., 2000; Schultz et al., 2000; North and Houde, 2004). Several factors affected the general pattern of distribution and survival, including spawning stock biomass, spring-summer dissolved oxygen levels, wind-forced circulation patterns, and this species' copepod prey (Taylor and Rand, 2003; Jung and Houde, 2004a, b; North and Houde, 2004).

Several estimates of growth are available for young-of-the-year. In Great South Bay, larval growth rates averaged 0.53 to 0.55 mm per day (Castro and Cowen, 1991). In Chesapeake Bay, these varied from 0.32 to 0.49 mm per day (Zastrow et al., 1991; Newberger and Houde, 1995). Growth appeared to be faster for young-of-the-year in a northern population in Narragansett Bay, with rates up to 0.70 mm per day (Lapolla, 2001a, b). In Great Bay–Little Egg Harbor, the young-of-the-year are approximately 2 to 10 cm TL by October (Fig. 29.2); the same pattern is evident for Delaware Bay (Able and Fahay, 1998). In Barnegat Bay, the relative growth and survival varied between years, with the number of larvae and juveniles smaller than 10 mm FL ranging from 31 to 70% of the total population in three successive years (Vouglitois et al., 1987). For Great Bay–Little Egg Harbor, estimates of the size composition and growth of young-of-the-year in October or later may be confounded by the likelihood that larger individuals leave the estuary first. This would explain the decrease and eventual disappearance of larger individuals (6 to 9 cm TL) during November and December. A similar pattern of size composition has been suggested for Chesapeake Bay (Wang and Houde, 1995).

Mortality during embryonic and larval development is high, with annual mortality rates estimated at 89 to 95% per year in Chesapeake Bay (Newberger and Houde, 1995) and > 90% with no fish surviving beyond year 2 in Narragansett Bay (Lapolla, 2001a, b). In a Chesapeake Bay field study, 73% of the eggs died before hatching. For recently hatched larvae (24 hours after hatching), mortality can be as high as 64% (Dorsey et al., 1996). Mesocosm experiments with eggs and yolk sac larvae in Chesapeake Bay indicated that 95% of a cohort died within 2 days of hatching (Houde et al., 1994). In Great South Bay, the average seasonal mortality rates for eggs varied annually, but ranged from 69 to 98.2% per day, while larval (3- to 15-day-old) rates were 34.6 to 37.0% per day (Castro and Cowen, 1991). These high mortality rates provide evidence for the importance of survival during the egg and larval stages in recruitment processes for this species.

SEASONALITY AND HABITAT USE

The larvae and small juveniles are most abundant in Great Bay–Little Egg Harbor from July through December, with

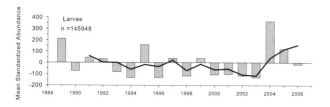

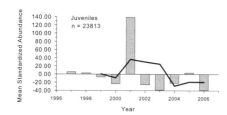

Fig. 29.1. Annual mean (bars) and three-year moving mean (line) of standardized abundance for *Anchoa mitchilli* collected in larval sampling with plankton net (*upper*) and juvenile sampling with otter trawl (*lower*). Data were standardized by subtracting overall mean from annual abundance. Three-year moving mean was calculated by taking the mean of the standardized values from the previous two years and the year in which data are plotted.

peaks occurring in August through October in most years (Fig. 29.3). However, the young-of-the-year also occur in the ocean and estuary from June through October.

The young-of-the-year tolerate a wide range of estuarine habitats. Eggs have been collected over a wide salinity gradient in both Delaware Bay (Wang and Kernehan, 1979) and Chesapeake Bay (Dovel, 1971), but egg viability is highest when salinities are greater than 8.0 ppt (Wang and Kernehan, 1979). In a Great South Bay study, egg density was highest near an inlet, but larvae were homogeneous throughout the bay (Monteleone, 1992). In Great Bay–Little Egg Harbor, young-of-the-year are uniformly distributed from the deeper portions of estuaries (Szedlmayer and Able, 1996) to the shorelines (Able et al., 1996b). On a broader scale, based on samples collected during July and September across the freshwater-estuarine inner continental shelf ecocline, the young-of-the-year were widely distributed from freshwater to inner shelf (Fig. 29.4). Interestingly, the highest abundances encountered were on the inner shelf in the vicinity of a sand ridge.

The movements of young-of-the-year in estuaries appear to have a regular pattern. In the Hudson River, many newly hatched larvae move upstream from high-salinity (> 10 ppt) to lower-salinity (< 10 ppt) areas as they grow. In early fall, larvae and juveniles then begin to move downstream into higher-salinity areas (Wang and Kernehan, 1979; Dovel, 1981; Loos and Perry, 1991). In Great Bay–Little Egg Harbor, young-of-the-year are abundant from June to January, although most leave by December (see Fig. 29.2). While in the estuary, the young-of-the-year were much more available to collections in an embayment in Great Bay, New Jersey, at night than during the day, presumably as a result of diel vertical migrations (Hagan and Able, 2008).

PREY AND PREDATORS

Several studies (n = 6) have examined the food habits of young-of-the-year (2.5 to 60 mm, n = 6179) *Anchoa mitchilli* in Middle Atlantic Bight estuaries from Delaware Bay to Newport River, North Carolina (see Table 8.1). Important prey of young-of-the-year include a variety of invertebrates (copepods, mysids, crabs, and ostracods). Fishes (*Anchoa mitchilli* and *Micropogonias undulatus*) are a minor dietary component with instances of cannibalism (Nemerson, 2001). The diet of young-of-the-year shifts from one that includes various small invertebrates (copepods, crabs, and ostracods) to one dominated by mysids supplemented by other invertebrates and fish with increasing size. The diet of young-of-the-year (2.5 to 16.4 mm, n = 1485) larvae from Chesapeake Bay also changes to include greater amounts of copepods with increasing size, such that young-of-the-year 2.5 to 7.4 mm feed on copepod eggs and nauplii, rotifers, tintinnids, and diatoms, while young-of-the-year 13.5 to 16.4 mm consume primarily copepodites and adult copepods (78%) (Auth, 2003). In Delaware Bay, the diet of young-of-the-year (10 to 60 mm, n = 3227) shifts at approximately 50 mm from zooplankton

to mysids, crustaceans, and fish (Nemerson, 2001). The diet changes earlier (approximately 20 mm) in York River, Virginia, such that the dominant prey of young-of-the-year 16 to 20 mm are crabs, ostracods, and copepods, while the dominant prey of young-of-the-year 21 to 80 mm are mysids supplemented by various invertebrates and detritus (Smith et al., 1984). The diet suggested by these studies is consistent over the geographic range. Relative importance of individual prey items also varies intra-annually in Delaware Bay (Nemerson, 2001) and inter-annually in Chesapeake Bay (Auth, 2003).

Anchoa mitchilli are important prey for many piscivorous estuarine fishes, including *Pomatomous saltatrix, Morone saxatilis, Cynoscion regalis,* and *Paralichthys dentatus* (see Scharf et al., 2002, for relevant literature). In our review of the available literature, *Anchoa mitchilli* occurs in the diet of 7 species and is an important component in 26 studies and a minor component in 13 (see Table 8.3; Mancini and Able, 2005). This species has also been found in the stomach contents of piscivorous predators in the Mullica River–Great Bay estuary (see Table 8.3). In Chesapeake Bay, *Anchoa mitchilli* may make up 60 to 90% of the diets of piscivorous fishes (Baird and Ulanowicz, 1989; Hartman and Brandt, 1995b). It is noteworthy that unidentified *Anchoa* make up a consistent component of the diet as well (see Table 8.3). These are likely *Anchoa mitchilli,* given the dominance of this species. Together, these observations demonstrate the importance of this species in estuarine predator-prey interactions (see Scharf et al., 2002).

MIGRATIONS

At the end of the first summer, most individuals migrate out of Middle Atlantic Bight estuaries and are then very abundant on the inner continental shelf (Vouglitois et al., 1987). The young-of-the-year (< 10 cm) have distinct seasonal migrations in and out of inner continental shelf waters of the Middle Atlantic Bight (Fig. 29.5). In the fall, they are concentrated in shallow waters (mean depth 17 m, range 6 to 87 m) from the inner shelf off North Carolina to the eastern end of Long Island, with a few individuals captured as far east as Massachusetts and Georges Bank. In the winter, they are virtually absent from the Middle Atlantic Bight, with only a few collections off North Carolina and Virginia (in depths of 21 to 25 m). In the spring, they appear to begin migrating back into the Middle Atlantic Bight because they are abundant on the inner shelf (mean depth 16 m, range 5 to 39 m) from North Carolina to Delaware, with a few collections off New Jersey.

The age 1+ (> 10 cm) juveniles and adults are not as abundant over the continental shelf, but the seasonal pattern is the same (see Fig. 29.5). In the fall, they are concentrated on the inner shelf (mean depth 19 m, range 10 to 36 m) from North Carolina to Long Island. They are not collected during the winter, but some are found in the spring on the inner shelf (mean depth 16 m, range 10 to 25 m) from North Carolina to off Delaware Bay.

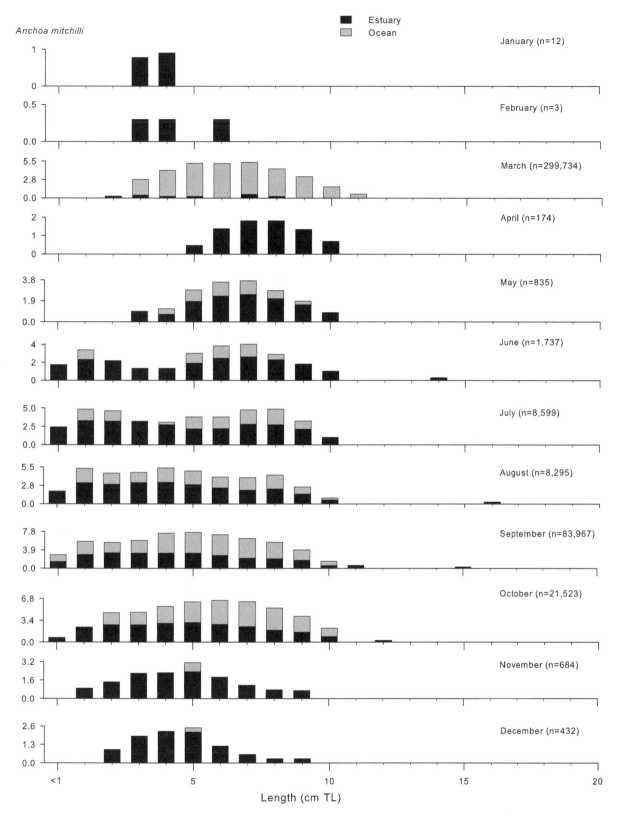

Anchoa mitchilli

Estuary
Ocean

January (n=12)

February (n=3)

March (n=299,734)

April (n=174)

May (n=835)

June (n=1,737)

July (n=8,599)

August (n=8,295)

September (n=83,967)

October (n=21,523)

November (n=684)

December (n=432)

Length (cm TL)

Fig. 29.2. Monthly length frequencies of *Anchoa mitchilli* collected in estuarine and coastal ocean waters of the central Middle Atlantic Bight. Sources: National Marine Fisheries Service: otter trawl (n =386,361); Rutgers University Marine Field Station: 1 m beam trawl (n = 4); seine (n = 3336); Methot trawl (n = 888); 1 m plankton net (n = 11,377); 2 m beam trawl (n = 32); 4.9 m otter trawl (n = 9631); and Able and Fahay 1998 (n = 14,366).

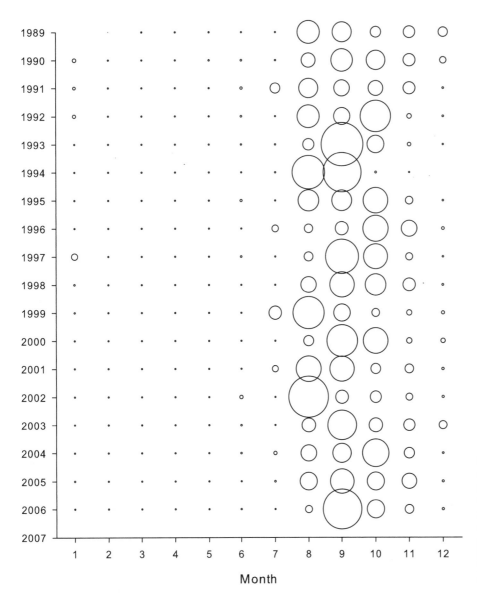

Fig. 29.3. Seasonal occurrence of the larvae of *Anchoa mitchilli* based on ichthyoplankton samples collected in Little Sheepshead Creek behind Little Egg Inlet from 1989 to 2006. Full circles represent the percentage of the mean number/1000 m³ of larval young-of-the-year *Anchoa mitchilli* captured by given year. Values range from 0 to 61%. The smallest circles represent when samples were taken but no larvae were collected.

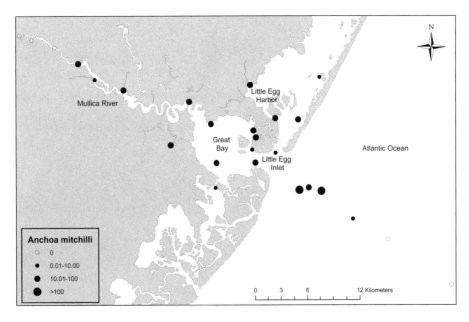

Fig. 29.4. Habitat use and average catch per tow (see key) for juvenile and adult *Anchoa mitchilli* in the Ocean–Great Bay–Mullica River corridor in southern New Jersey based on small otter trawl collections during July and September from 1988 to 2006. See Table 3.2 for characteristics at each sampling collection.

Age 1+ *Anchoa mitchilli*   **YOY *Anchoa mitchilli***

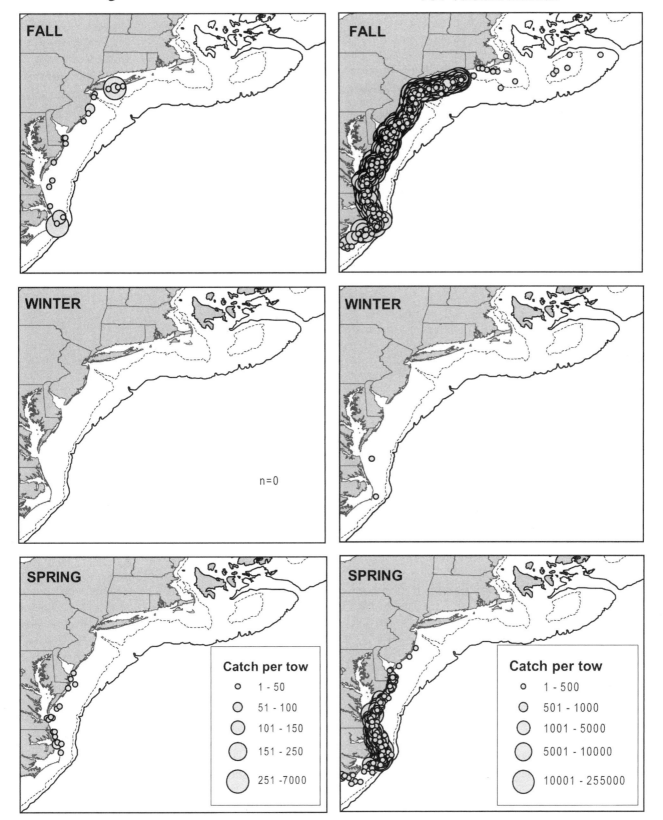

Fig. 29.5. Composite distribution of age 1+ (> 10 cm) and young-of-the-year (< 10 cm) *Anchoa mitchilli* during seasonal cruises of the National Marine Fisheries Service groundfish survey. Details of sampling effort are indicated in Fig. 3.4.

30

Engraulis eurystole
(Swain and Meek)

SILVER ANCHOVY

DISTRIBUTION

An early synopsis of data concerning this species suggested the range as Woods Hole, Massachusetts, to Beaufort, North Carolina (Hildebrand, 1963). More recent collections in otter trawls and under nightlights have extended the range to the northern Gulf of Mexico (Daly, 1970; Klima, 1971) and ichthyoplankton collections have further expanded the range northward to the continental shelf off Nova Scotia (Markle et al., 1980). As presently understood, adults of this species occur off the Atlantic coast of North America from Massachusetts to the northern Gulf of Mexico, and a disjunct population apparently occurs from Venezuela to northern Brazil (Nizinsky and Munroe, 2002). It has not been reported from Chesapeake Bay (Murdy et al., 1997), but has been recorded from coastal New Jersey (Nichols and Breder, 1927; Hildebrand, 1963). Larvae and early stages occur over the continental shelf of the United States from Cape Fear, North Carolina, to as far north and east as the Scotian Shelf (Fahay, 1975; Markle et al., 1980).

REPRODUCTION AND DEVELOPMENT

Spawning is not well described, but probably extends from spring into fall. There are several anecdotal records of spawning location, but comprehensive studies are lacking. One such early record indicated that larvae occur within the Gulf Stream (Oscar E. Sette, pers. comm., in Jones et al., 1978) leading to subsequent speculation that spawning occurred there. The collection of early stages over the continental slope off Nova Scotia (Markle et al., 1980) reinforced this notion somewhat, and also implied that reproduction may occur well to the south of the Middle Atlantic Bight, with subsequent transport of young stages via the Gulf Stream.

Larvae have been collected in Onslow Bay, North Carolina, during June and October (Powell and Robbins, 1998) and a single early juvenile, 35.9 mm FL, was reported during the winter southeast of Cape Fear, North Carolina (Fahay, 1975). Both of these observations indicate a more southern limit to spawning. However, see below for evidence of spawning within the Middle Atlantic Bight.

The early development of *Engraulis eurystole*, including characters useful for distinguishing the eggs, larvae, and juveniles from those of other clupeiform larvae, has been well described and summarized (Fahay, 2007). Eggs are pelagic and elliptical, with the long axis diameter between 1.02 and 1.25 mm and the short axis diameter between 0.50 and 0.80 mm, have a segmented yolk, and lack oil globules and pigment (Berrien and Sibunka, 2006). Egg dimensions may be intermediate between, and overlap slightly with, those of *Anchoa mitchilli* and *A. hepsetus* (P. L. Berrien, pers. comm.). Larvae at hatching are about 2.0 to 3.0 mm long, and have unpigmented eyes and unformed mouth parts. Pigment is sparse throughout ontogeny. Larvae and juveniles are diagnosed by positions of their dorsal and anal fins. These overlap by varying degrees in *Anchoa* spp., but in the present species the origin of the anal fin is directly under, or posterior to, the insertion of the dorsal fin. We have no information on the formation of scales in this species.

LARVAL SUPPLY AND GROWTH

Larvae have been reported from the continental shelf and in continental slope waters as far offshore as the Gulf Stream (Fahay, 2007). Collections from the Museum of Comparative Zoology (Harvard University) include larvae collected during August over water depths of 40 to 457 m within the Middle Atlantic Bight (M. P. Fahay, pers. observ.). Larvae were also collected in the southern part of the Middle Atlantic Bight during the 11-year MARMAP ichthyoplankton survey (1977–1987). They occurred from May to January, with a peak in August and September (M. P. Fahay, pers. observ.). In June, larvae were concentrated in coastal waters from the mouth of Delaware Bay to the border of Virginia. In July, large numbers of larvae were collected off North Carolina and over deep waters adjacent to the Gulf Stream. There were also scattered occurrences as far north as the New York Bight off Sandy Hook, New Jersey. Larval occurrences were most widespread in August and September, when collections were made across the breadth of the continental shelf from Delaware Bay to Cape Hatteras. Off Cape Hatteras, these collections were made in waters affected by the Gulf Stream, and Gulf Stream transport probably accounts for the larvae reported in August off Nova Scotia (Markle et al., 1980). From October through January there are only scattered occurrences of a few larvae throughout the Middle Atlantic Bight as far north as Georges Bank (M. P. Fahay, pers. observ.). Maximum sizes in all larvae collected range from 7.9 mm in June to 38.0 mm in November. This size range is comparable

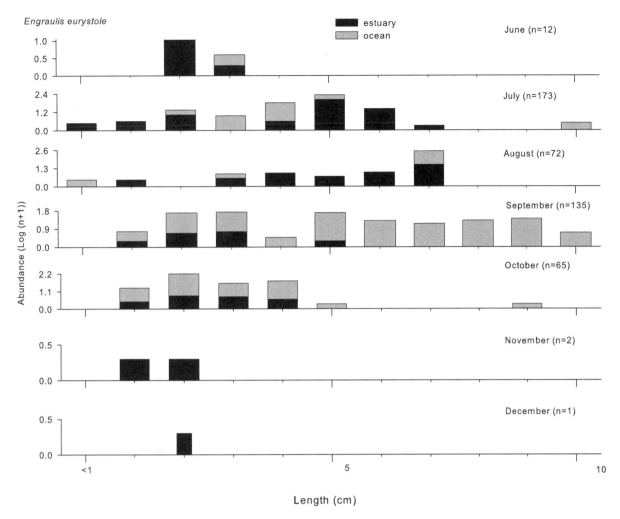

Fig. 30.1. Monthly length frequencies of *Engraulis eurystole* collected in
estuarine and coastal ocean waters of the central Middle Atlantic Bight.
Length measurements include both fork length (FL) and standard length
(SL). Source: Rutgers University Marine Field Station: seine (n = 230);
Methot trawl (n = 48); 1 m plankton net (n = 22); 4.9 m otter trawl (n = 15);
pop net (n = 145).

to that reported from Nova Scotia, where a range of 16.4 to
30.7 mm was observed (Markle et al., 1980). Larvae of this
species have been collected in the Mullica River–Great Bay at
a minimum size of 20 mm FL in June and small individuals
continue to occur through November and December (Fig.
30.1). These larvae may originate from the intense spawning event that occurs in the southernmost Middle Atlantic
Bight (July through September), in which case larvae may be
transported northward by the Gulf Stream and then across
the continental shelf by some undescribed mechanism. Alternately, they might simply result from less intense local
spawning off the coast of New Jersey.

The smallest size classes we have observed in our collections are about 10 mm FL (see Fig. 30.1) during June through
November. These may increase to about 100 mm FL by September. Our observations may accurately represent growth
during these months in our study area, or they may be the
result of continuous recruitment of early stages from a prolonged spawning event. Certain of these recruits may enter
our study area when 20 mm in length, whereas others may
arrive at lengths of 50 mm or more. Until accurate details of
spawning in this species are determined, we can only surmise
that the latter scenario is more likely.

SEASONALITY AND HABITAT USE
Adults are reported to be schooling and pelagic in coastal
waters, especially in sheltered harbors, but they are also observed over depths of 20 to 65 m (Nizinsky and Munroe,
2002). Earlier studies have suggested that this species occurs
well offshore, perhaps in close proximity to the Gulf Stream
(Hildebrand, 1963). The uncertainty associated with the usual
range and habitats of this species may be due to misidentifi-
cations with other anchovies. Our observations of young-
of-the-year have been made both within the water column in

the estuary during August through the following December (Hagan and Able, 2003) and in nearshore ocean habitats in less than 1 m water depth. In the estuary they are more abundant at night than during the day (Hagan and Able, 2008).

PREY AND PREDATORS
Adults feed on plankton, but we have no data on food habits of early stages. Predators are also unknown.

MIGRATIONS
Unknown.

31

Osmerus mordax (Mitchill)

RAINBOW SMELT

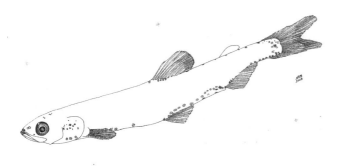

DISTRIBUTION

Osmerus mordax occurs along the Atlantic coast of North America from southern Labrador to New Jersey (Scott and Crossman, 1973; Smith, 1985; Hartel et al., 2002). It also is found in the Arctic and North Pacific oceans, where it is recognized as a separate subspecies by some authors. The Atlantic population is most numerous in the southern Maritime Provinces of Canada and the state of Maine. The species is mostly anadromous, but there are several naturally occurring landlocked populations in lakes throughout its range. Two spatially segregated populations occur in the St. Lawrence River estuary, one along the north shore and the other along the south shore of this system (Lecomte and Dodson, 2004). The present southern range limit of this species is in northern New Jersey, although it once occurred in Delaware Bay (Robins and Ray, 1986) and Virginia (Bigelow and Schroeder, 1953) and was considered common in the New York area (Nichols and Breder, 1927). More recent records indicate that anadromous populations occur in the Hudson River and numerous streams on Long Island, as well as several landlocked populations in New York (Smith, 1985), but these appear to be declining (Waldman et al., 2006). A recent report suggests that spawning populations are limited to areas north of Cape Cod (Fuda et al., 2007). However, early life history stages have been reported from Middle Atlantic Bight estuaries from the Hudson River and north (see Table 4.3).

REPRODUCTION AND DEVELOPMENT

Spawning occurs from late winter through early spring (Klein-MacPhee, 2002a). Reproduction in the two St. Lawrence River estuary populations is temporally and spatially segregated (Lecomte and Dodson, 2004). The north shore population spawns two weeks earlier than the south shore population, and deposits eggs directly on shallow shoals within fluvial, estuarine waters. The south shore population spawns in small tributaries to the middle estuary. Most populations ascend freshwater streams from coastal waters, bays, or estuaries, where age 2 and older fish spawn. In the central part of the Middle Atlantic Bight, most information on spawning concerns landlocked populations; we know little about those that utilize estuarine waters. Historically, spawning presumably occurred in the Hudson River and in certain small, Long Island streams in February or March, when water temperatures reached about 9°C (Smith, 1985). Mature adults have been collected in Newark Bay, New Jersey, from January through April 1994 (S. J. Wilk, pers. comm.).

The early development of *Osmerus mordax*, and characters for distinguishing the eggs, larvae, and juveniles from those of their more northerly occurring relative *Mallotus villosus*, have been well described and summarized (Fahay, 2007). Eggs are demersal and adhesive with a diameter of about 1.0 mm. The outer membrane of the egg turns "inside-out," forming a stalk, and the eggs are then deposited in clusters. Larvae hatch at 5.0 to 6.0 mm and are very elongate with an extremely long pre-anus length of 65 to 75% TL. The mouth is small and terminal. Pigment is light and includes a line of melanophores along the ventral edge. Scale formation begins on the caudal peduncle, then proceeds anteriorly. At 36 mm, scales cover the posterior half of the area between the vent and caudal fin (Cooper, 1978).

LARVAL SUPPLY AND GROWTH

Larvae are derived from local spawning, even in the St. Lawrence River estuary, where two populations occur (Lecomte and Dodson, 2004). Larvae may grow to 20 to 40 mm in a few months and may reach 50 mm by August (Scott and Crossman, 1973). In the Woods Hole, Massachusetts region, they may reach 63 mm by the end of the first summer (Klein-MacPhee, 2002a).

The combined effects of turbidity and food density on growth rates in larvae have been evaluated in laboratory studies and in the St. Lawrence River estuary (Sirois and Dodson, 2000). Larvae benefit from feeding within the estuarine turbidity maximum (ETM) zone, because they expend less energy finding and ingesting food than they do in less turbid areas. However, these authors also found that up to 38% of larvae collected in the ETM were parasitized by cestodes; these parasitized larvae fed at only half the rate of non-parasitized larvae, ultimately resulting in substantially lower growth rates.

SEASONALITY AND HABITAT USE

Adults inhabit bays, estuaries, coastal waters, and freshwater lakes. They generally seek deeper, cooler waters during summer. Spawning occurs after an anadromous migration up freshwater streams in the spring (Smith, 1985; Hartel et al., 2002). After the eggs hatch, larvae drift downstream to brackish water or to larger lakes in landlocked populations. There is no information on habitat use by young-of-the-year in the central part of the Middle Atlantic Bight; however, a single young-of-the-year cohort with a mode at about 30 mm was collected in Newark Bay during April 1994 (S. J. Wilk, pers. comm.). In the St. Lawrence River, developing larvae are retained in turbid estuarine waters by using "selective tidal stream transport," by which they migrate to near-surface layers during flood tides and descend to the bottom during the ebb (Laprise and Dodson, 1989), thereby minimizing net downstream displacement. Larvae of the two St. Lawrence River estuary populations are also segregated among discrete nursery habitats (Lecomte and Dodson, 2004). Larvae of the north shore population occur in all channel habitats within the ETM zone and to a lesser degree adjacent shoals and small embayments. Larvae of the south shore population are confined to shallow shoals and large, shallow bays at the downstream limit of the ETM. Young-of-the-year in the northern Middle Atlantic Bight most likely migrate downstream and into the ocean by early summer (Bigelow and Schroeder, 1953). Beds of Zostera marina provide nursery habitat for young-of-the-year in shallow, nearshore habitats in Maine (Lazzari et al., 2003) and Great Bay, New Hampshire, where juveniles of this species are the most abundant species of fish collected (Ganger, 1999). During winter, this species is capable of producing an antifreeze protein that enables the fish to survive subfreezing temperatures (Ewart and Fletcher, 1990). It also accumulates glycerol in high concentrations, which also depresses the freezing point during winter (Raymond, 1992). Therefore, at least a portion of the population is able to survive winter conditions without the necessity of migrating to deeper, warmer oceanic habitats.

Several environmental factors have been studied in relation to hatching success and larval survival (Fuda et al., 2007). Hatching was impaired in conditions of low dissolved oxygen, acidic pH levels, and high salinities (> 30 ppt). Larval survival was adversely affected by pH levels < 5.0 ppt. Other variables (nitrates and phosphates) had little or no effects on hatching success or survival. These authors concluded that early life history survival was most seriously impacted by acidification of the water, sediment covering in silty conditions, and fungal growth.

PREY AND PREDATORS

Osmerus mordax is a voracious carnivore (Klein-MacPhee, 2002a). Small fishes and a long list of invertebrate taxa make up the list of prey. Important fish prey species include Clupea harengus, Anchoa spp., Tautogolabrus adspersus, Ammodytes americanus, Menidia menidia, sticklebacks, and Alosa spp. There have been no studies of the food habits of young-of-the-year (< 75 mm) in Middle Atlantic Bight estuaries. This species is an important prey item for larger fishes, including Gadus morhua, Salmo salar, Morone saxatilis, and Pomatomus saltatrix (Clayton et al., 1978), and many aquatic birds, including mergansers, cormorants, gulls, and terns (Klein-MacPhee, 2002a). The eggs are eaten by sticklebacks and trout (Baird, 1967).

MIGRATIONS

This anadromous species ascends freshwater streams in the spring for spawning, and seeks deeper waters in summer to avoid warming waters (Smith, 1985).

32

Synodus foetens (Linnaeus)

INSHORE LIZARDFISH

DISTRIBUTION

Synodus foetens is found from Massachusetts and Bermuda to Brazil, including the Gulf of Mexico and the Caribbean Sea (Robins and Ray, 1986). It is most common south of South Carolina (Breder, 1948). Pelagic larvae are frequently collected over the continental shelf, but are only rarely reported in bays or estuaries. Juveniles occur regularly in estuaries in the southern part of the Middle Atlantic Bight (see Table 4.3).

REPRODUCTION AND DEVELOPMENT

Details of reproduction are poorly known. Distribution of synodontid larvae near the edge of the continental shelf indicates that they are probably transported into the Middle Atlantic Bight via the Gulf Stream from south of Cape Hatteras, where the prolonged spawning season presumably begins in spring and continues through summer and fall (Able and Fahay, 1998; Fahay, 2007). There is no evidence to suggest that widespread adult occurrences in the Middle Atlantic Bight during some years are associated with a spawning migration. In the Tampa Bay, Florida area the smallest larvae occurred during November and December (Springer and Woodburn, 1960), although larvae less than 40 mm have been collected during the entire year (Gibbs, 1959). In Middle Atlantic Bight continental shelf waters, small larvae have been collected as early as January and continuously from May through October. This, coupled with a prolonged period of ingress into estuaries, provides further evidence of a prolonged spawning season.

Details of early development have been recently summarized (Fahay, 2007). Eggs are undescribed. Larvae are characterized by peritoneal pigment patches along the long gut. As larvae transform into juveniles, the snout elongates; the prominent melanophores along the gut internalize and, with further growth, become overlain with dermal pigment and

disappear. Juveniles have distinctive cross-shaped pigment accumulations along the lateral line. Individuals from 33.5 to 152.9 mm TL (n = 14) were examined for scale formation (Fig. 32.1). Scales are first present at a size of 35.2 mm TL, when they occur above and below the lateral line as well as on the base of the caudal fin. In larger specimens, scales also begin forming in two loci on the side of the head. Scales spread across the entire head (excluding the branchiostegal rays) and also extend posteriorly across the proximal caudal fin rays. At about 125 mm, scales cover 89.9% of the body and 71.5% of the body and fins; this represents the adult condition (see also Boschung, 1957; Thomson et al., 1971).

LARVAL SUPPLY, SETTLEMENT, AND GROWTH

The annual variation in abundance of early life history stages has not been measured. However, several studies have remarked on inter-annual variation in abundance of adults in New Jersey waters (Allen et al., 1978). There does not appear to be any relationship between years when adults are abundant in the study area and abundance of early stages. Larvae and transforming juveniles begin to settle to the bottom at sizes of about 30 to 40 mm (Hoese, 1965).

Determining growth rates from available length data (Fig. 32.2) is equivocal without attendant age data, because it is not possible to establish the progression of modes of discrete length groups. Confounding this situation is the fact that older year classes, as well as young-of-the-year, may occasionally migrate into the Middle Atlantic Bight for the summer, and that the smallest size categories continue to be collected as late as October. Lacking evidence to the contrary, we suggest that individuals up to 25 cm collected in late summer and fall in our study area are all young-of-the-year.

SEASONALITY AND HABITAT USE

Adults occur in bays, in coastal ocean habitats, and on the continental shelf to a maximum depth of 200 m, but they favor shallow waters with sand bottoms, where they are able to bury in the substrate (Robins and Ray, 1986). Sampling directed at different habitat types in the Great Bay–Little Egg Harbor estuary indicates the species is most abundant in sea lettuce (*Ulva lactuca*) beds over sandy substrates, although they are also frequently collected over silty bottoms with structure ranging from shell or peat to flat and featureless. In North Carolina, Hettler and Barker (1993) collected juveniles entering estuaries with mean lengths of 26.8 mm SL at Oregon Inlet and 33.5 mm SL at Okracoke Inlet during June, July, and August. In the central portion of the Middle Atlantic Bight larvae and older young-of-the-year occur during the summer and into the fall. Recently settled individuals (30 to 35 mm) have been collected in epibenthic sled hauls on the sea bottom near Hereford Inlet, New Jersey (Allen et al., 1978) during the summer. These authors also report seine collections of pre-transformation larvae (about 35 mm) in coastal ocean and estuarine habitats. It is unknown

Synodus foetens   *Morone americana*   *Centropristis striata*

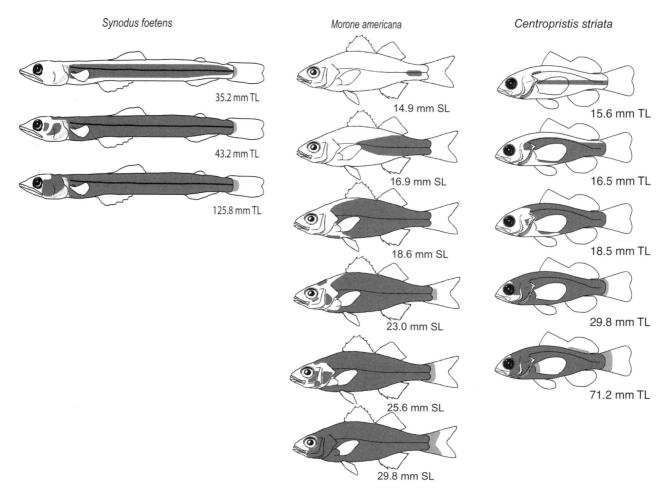

Fig. 32.1. Description of scale formation relative to total length (TL) in
Synodus foetens, Morone americana, and *Centropristis striata.*

whether settlement occurs in one habitat more so than in others, although there are reports of early settlers burying in muddy substrates (Breder, 1962; Hoese, 1965). It is also unknown whether all young-of-the-year enter estuaries, either before or after settlement. Collections of small juveniles in the Little Egg Inlet–Great Bay area begin in June and young-of-the-year can be found there through the summer and fall, but juveniles of similar lengths continue to occur in coastal ocean habitats at that time (see Fig. 32.2).

PREY AND PREDATORS

This species is a voracious predator that burrows into the substrate and darts out to capture prey (Allen et al., 1978; Russell, 2002). Major diet items are small invertebrates and fishes, including young *Cynoscion regalis* (Hildebrand and Schroeder, 1928). Crabs, shrimp, worms, and other animals are also sometimes eaten. There have been no studies of the food habits of young-of-the-year in Middle Atlantic Bight estuaries.

MIGRATIONS

During the summer, young-of-the-year (< 25 cm) are usually present in estuaries in the study area (Able and Fahay, 1998).

In the fall, they presumably move out of estuaries, and then occur in shallow waters (mean depth 23.7 m, range 5 to 171 m) on the inner continental shelf from Long Island to Cape Hatteras. They are most frequently collected from the mouth of Delaware Bay to south of Cape Hatteras (Fig. 32.3). Young-of-the-year are infrequently collected in the winter, presumably because they migrate south and out of the sampling area. However, occasional large collections are made near the continental shelf edge in the southern part of the Middle Atlantic Bight. In the spring, most occurrences are restricted to south of Cape Hatteras.

Age 1+ individuals (> 25 cm) are only seasonal visitors to much of the study area (see Fig. 32.3), and their abundance fluctuates from year to year. In the fall, they occur in shallow waters (mean depth 19.6 m, range 5 to 61 m) of the inner continental shelf from western Long Island to south of Cape Hatteras, but they are most abundant from Virginia and south. They are infrequently collected during winter and spring on the mid- to outer shelf, presumably because most migrate south of Cape Hatteras during the coldest time of the year.

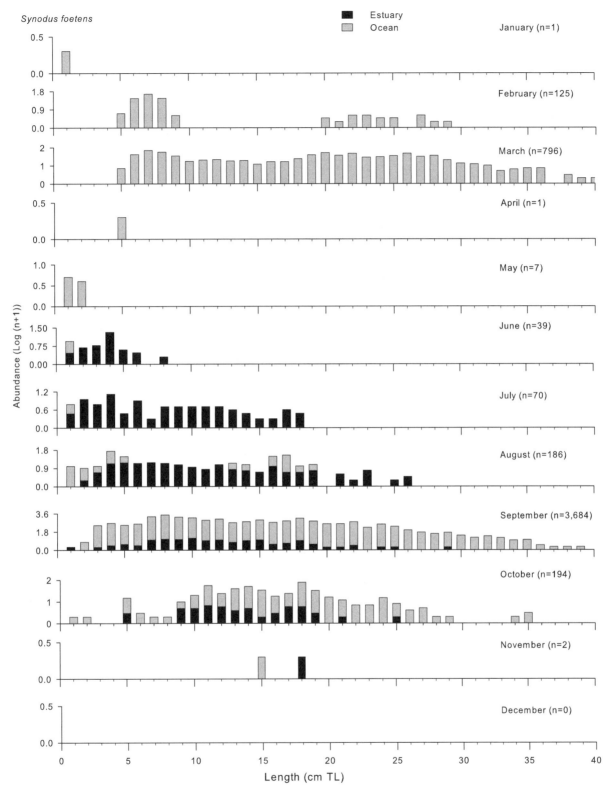

Fig. 32.2. Monthly length frequencies of *Synodus foetens* collected in estuarine and coastal ocean waters of the central Middle Atlantic Bight. Sources: National Marine Fisheries Service: otter trawl (n = 4631); Rutgers University Marine Field Station: seine (n = 233); Methot trawl (n = 1); 1 m plankton net (n = 14); 2 m beam trawl (n = 3); 4.9 m otter trawl (n = 9); multipanel gill net (n = 1); and Able and Fahay 1998 (n = 231).

Age 1+ Synodus foetens　　　　**YOY Synodus foetens**

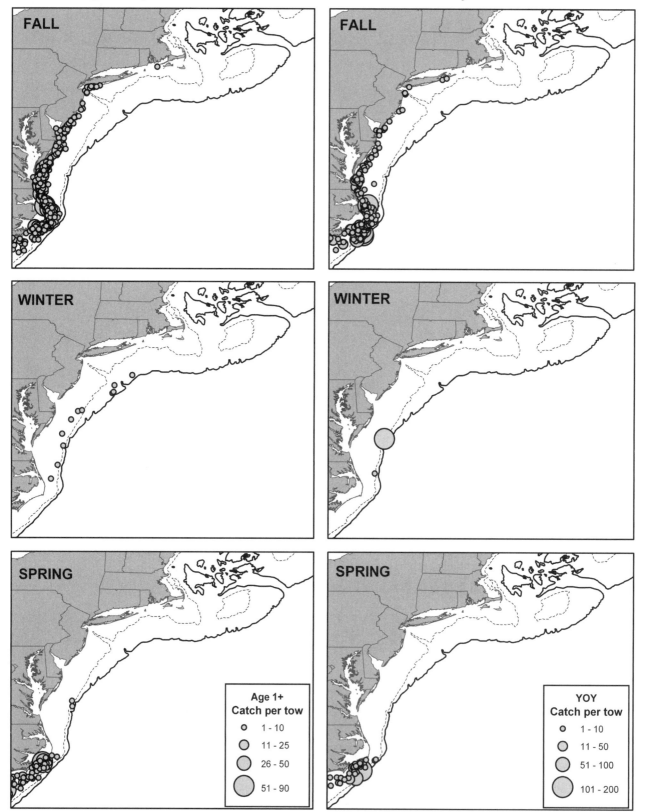

Fig. 32.3. Composite distribution of age 1+ (> 25 cm) and young-of-the-year (< 25 cm) *Synodus foetens* during seasonal cruises of the National Marine Fisheries Service groundfish survey. Details of sampling effort are indicated in Fig. 3.4.

33

Enchelyopus cimbrius (Linnaeus)

FOURBEARD ROCKLING

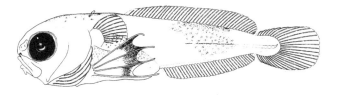

DISTRIBUTION
This species occurs along the coast of North America from Newfoundland to the Gulf of Mexico (Klein-MacPhee, 2002a). Eggs, larvae, and juveniles are sometimes found in larger bays and estuaries in the northern part of our study area, such as Long Island Sound (Fahay, 1983).

REPRODUCTION AND DEVELOPMENT
Spawning occurs year-round, with most taking place between April and September, with a strong peak in June (Berrien and Sibunka, 1999). Most spawning occurs in the Gulf of Maine, but eggs are also found in Long Island Sound (Fahay, 1983). Prior studies on the inner continental shelf off southern New Jersey have reported eggs and larvae (Able and Hagan, 1995). Some small larvae (< 10 mm) have been collected in the estuary behind Little Egg Inlet, also indicating that spawning may occur off New Jersey in some years (see Fig. 33.1).

The early development, including characters useful for distinguishing the eggs, larvae, and juveniles from those of other gadiforms, has been well described (Fahay, 2007). Eggs are pelagic and spherical and measure 0.66 to 0.98 mm in diameter. The usually single oil globule is 0.13 to 0.27 mm in diameter. Oil globules are sometimes multiple, in which case they soon coalesce into one. Larvae at hatching are about 1.6 to 2.4 mm long and have lightly pigmented eyes and unformed mouth parts. Pigment is concentrated into a band encircling the tail midway between the anus and the notochord tip. Early-forming, enlarged pelvic fin rays are heavily pigmented. Larvae have four pelvic fin rays, adults have five to seven. Pelagic juveniles are diagnosed by their elongated bodies that are counter-colored with silvery flanks and blue dorsums. After settlement the development of the unique first dorsal fin occurs. This consists of a single long ray, followed by a series of very short filaments lying in a groove.

LARVAL SUPPLY, SETTLEMENT, GROWTH, AND MORTALITY
The highest densities of eggs collected between 1978 and 1987 were from the Gulf of Maine and the continental shelf south of Rhode Island (Berrien and Sibunka, 1999), and any larvae collected in our study area may be derived from spawning in the latter area. Transformation and settlement to demersal habitats occur at sizes of about 20 mm (Fahay, 1983). We have no information on growth rates in early stages, nor can we estimate sources or rates of mortality.

SEASONALITY AND HABITAT USE
Adults occur primarily on the continental shelf and slope, in depths between 20 and 650 m, with most occurrences between 20 and 50 m (Cohen and Russo, 1979). They inhabit muddy sand or soft silt areas, often located between harder substrates. Pelagic juveniles occur neustonically over the continental shelf, until settlement (see above). Most larvae collected behind Little Egg Inlet occur in May and June (Fig. 33.2)

PREY AND PREDATORS
The diet of this species in the ocean is well described, but there are no studies of their food habits while occupying estuarine waters. In age 1 fish in the Gulf of Maine, copepods constitute more than 70% of the diet, whereas bivalves make up more than 20%. This ratio changes in older age classes, where copepods' importance is reduced to 28% and bivalves' increased to 56% (Deree, 1999).

MIGRATIONS
Unknown.

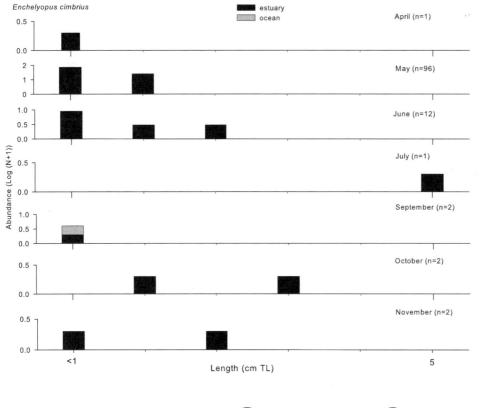

Enchelyopus cimbrius

estuary
ocean

April (n=1)

May (n=96)

June (n=12)

July (n=1)

September (n=2)

October (n=2)

November (n=2)

Abundance (Log (N+1))

Length (cm TL)

Fig. 33.1. Monthly length frequencies of *Enchelyopus cimbrius* collected in estuarine and coastal ocean waters of the Mullica River–Great Bay Estuary. Source: Rutgers University Marine Field Station: Methot trawl (n = 1); 1 m plankton net (n = 114); killitrap (n = 1).

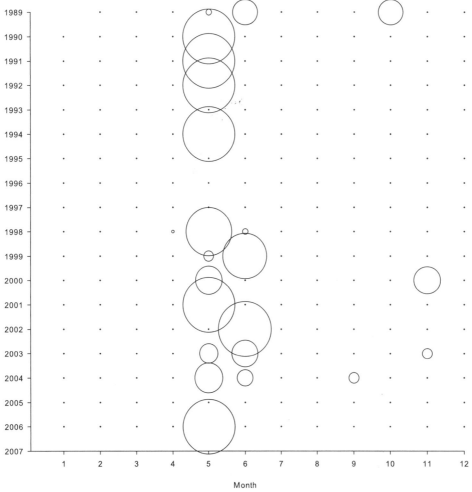

Month

Fig. 33.2. Seasonal occurrence of the larvae of *Enchelyopus cimbrius* based on ichthyoplankton samples collected in Little Sheepshead Creek behind Little Egg Inlet from 1989 to 2006. Full circles represent the percentage of the mean number/1000 m³ of larval young-of-the year *Enchelyopus cimbrius* captured by given year. Values range from 0 to 100%. The smallest circles represent when samples were taken but no larvae were collected.

34

Urophycis chuss (Walbaum)

RED HAKE

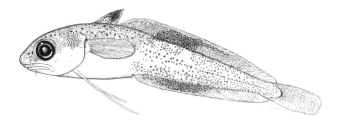

Estuary	Reproduction	Larvae/ Development	YOY Growth	YOY Overwinter
Ocean				
Both				
Jan				
Feb				
Mar				
Apr				
May				
Jun				
Jul				
Aug				
Sep				
Oct				
Nov				
Dec				

DISTRIBUTION

Urophycis chuss occurs along the North American Atlantic coast from the Gulf of St. Lawrence to North Carolina, with a center of abundance between Georges Bank and Hudson Canyon (Anderson, 1982). Juveniles have been collected in most estuaries throughout the Middle Atlantic Bight, but eggs and larvae are more restricted to southern New England estuaries and larger embayments east of the Hudson River (see Table 4.3). Comparisons of seasonal distributions

for various life history stages in three co-occurring species of *Urophycis* can be found in Able and Fahay (1998).

REPRODUCTION AND DEVELOPMENT

Urophycis chuss reaches sexual maturity during the second year at about 300 mm or larger (Markle et al., 1982). Spawning occurs in the Middle Atlantic Bight as early as April and may extend into October (Musick, 1969; Wilk et al., 1990). There is a strong peak in spawning activity during late June and July off Maryland and northern Virginia (Eklund and Targett, 1990). Ichthyoplankton collections indicate spawning in the central part of the Middle Atlantic Bight begins in June at the earliest and continues through September, with sporadic larval occurrences continuing into November (M. P. Fahay, pers. observ.; Comyns and Grant, 1993). During this period, spawning individuals are most concentrated on the continental shelf in waters less than 110 m deep between Martha's Vineyard and Long Island (Anderson, 1982), where larvae are also heavily concentrated (Able and Fahay, 1998).

Aspects of the ontogeny have been previously summarized, including methods to distinguish early stages from those of congeners in our study area (Able and Fahay, 1998; Fahay, 2007). Morphological changes, as larvae develop into pelagic juveniles and then demersal juveniles, take place during the first few months of life and before young-of-the-year occur in estuaries or near-coastal waters.

LARVAL SUPPLY, SETTLEMENT, AND GROWTH

Larvae derive from reproduction that occurs during the summer over the continental shelf of the Middle Atlantic Bight (Able and Fahay, 1998). Larvae were most abundant during August and September (when the water column was most strongly stratified) in a study conducted over the central part of the Middle Atlantic Bight continental shelf (Comyns and Grant, 1993). Some larvae may also occur in shallow, nearshore waters, in the adjacent surfzone (Able et al., 2010), and in estuaries or coastal waters. Settlement occurs after approximately two months in larval and pelagic juvenile stages. Descent to the bottom occurs during the summer at sizes of 5 to 10 cm TL. Laboratory observations suggest that pelagic juveniles descending through a thermocline require an acclimation period to adjust to the sharp, negative change in temperature (Steiner and Olla, 1985). This acclimation is apparently accomplished while remaining within the thermocline for a period. Lacking acclimation, descending fish become moribund on encountering colder bottom waters. With this descent, fish initially retain pelagic juvenile coloration and filamentous pelvic fin rays (Musick, 1969; Steiner and Olla, 1985). After this descent, and within a matter of hours, the body shape becomes terete, neustonic coloration changes to an adult-like pattern, and the pelvic fin rays begin to be deployed in a forward orientation (Musick, 1969).

Evidence of a prolonged spawning season, variable growth rates, and successive cohorts of early demersal-stage juveniles all contribute to a large amount of size overlap within year classes of *Urophycis chuss*. Some estimates of growth in early demersal stages include 15.7 mm per month in our study area (Steiner et al., 1982) and a minimum average size of 100 mm in the first year off Nova Scotia (Markle et al., 1982). Laboratory studies offering large volumes of food found juveniles (92 to 133 mm TL) capable of growing 1.0 to 1.5 mm per day in a 24-day experiment (Luczkovich and Olla, 1983). The histograms derived from sampling in our study area (Fig. 34.1) show a modal length of about 1 to 15 cm at the end of fall; indications of nearly no growth through the first winter; a group of age 1 individuals in June at about 10+ cm; and growth through their second summer and fall to lengths more than 250 mm. The largest of these may attain lengths of 65 to 70 cm by November. Relatively few juveniles, between 10 and 20 cm, occur during the spring.

SEASONALITY AND HABITAT USE

Adult *Urophycis chuss* are typically found on soft mud, silt, or sand bottoms (Fritz, 1965; Musick, 1974), but they also can occur over rocky bottoms (Eklund and Targett, 1991). Depths of occurrence range from bays to the outer continental shelf (as deep as 550 m) and vary with age and season. They are most common in depths of 35 to 130 m, although younger stages occur in shallower habitats. During the fall, young-of-the-year occur pelagically in waters of the Gulf of Maine, Georges Bank, and the northern part of the Middle Atlantic Bight. After they descend to the bottom during this season, they find shelter in structured habitats, especially scallop beds (Able and Fahay, 1998). During winter, most have departed from the Gulf of Maine and are found in sheltered habitats in the Middle Atlantic Bight (Sullivan et al., 2006) and along the southern flank of Georges Bank.

The available evidence suggests that structure of some kind is critical to survival of just-settled *Urophycis chuss* (Steiner et al., 1982). During their initial demersal stage they have been observed lying in the troughs between sand waves with their bodies curled into a C-shape (M. P. Fahay and K. W. Able, unpubl. observ. from submersible). Although these fish will utilize nonliving objects, it is clear that the most common form of shelter use involves an inquiline association with the sea scallop, *Placopecten magellanicus*. This relationship, in which the initial demersal stages live within the mantles of sea scallops during the fall and winter, has been well described (Goode and Bean, 1896; Welsh, 1915; Musick, 1969; Markle et al., 1982). Settlement to sea scallop beds begins in September, about two months after peak spawning in our study area, and continues through December (Steiner et al., 1982). During September, these juveniles range from 23 to 100 mm; in November, they are as large as 100 mm. Overall, juveniles from 23 to 140 mm occur in scallops, where they are most commonly found during the

day, leaving this shelter to forage at night (Musick, 1974). There is evidence that the largest sea scallops are sought out by juveniles (Markle et al., 1982); other studies suggest that large fish seek out large scallops, while smaller fish inhabit all sizes of scallops (Steiner et al., 1982). Young-of-the-year apparently emigrate from scallop (or other) shelters between January and May, either due to their large sizes or because temperatures drop below 4°C (Musick, 1974). When many large scallops are available, hakes may remain inquiline for a longer period of time. Conversely, when only small scallops are present (e.g., after overfishing), hakes must leave this relationship earlier and at smaller sizes (Musick, 1974). Some of these émigrés occupy bay or estuarine habitats for a brief period after leaving the shelter of scallop beds. In New Jersey waters, juveniles 7 to 20 cm (presumably following the scallop stage) have been found in Sandy Hook Bay and Great Bay during March and April (Thomas et al., 1974; Pacheco, 1983), when temperatures are typically slightly warmer than 4°C. Collection records from these studies also include juveniles lingering through May and June, but in much reduced numbers. Not all of these juveniles occur in bays, however. Studies on the inner continental shelf in our study area have also collected large numbers of these post-scallop stages from March through May (Wilk et al., 1992).

Hypoxia (low dissolved oxygen) is a common, but unpredictable, occurrence in coastal and estuarine habitats in our study area (Garlo et al., 1979; Swanson and Sindermann, 1979; Officer et al., 1984). In studies among different age groups of *Urophycis chuss*, young-of-the-year (more so than 1- to 3-year-olds) become behaviorally active in avoiding hypoxic waters. This activity takes the form of increased swimming and orientation to near-surface layers (Bejda et al., 1987).

PREY AND PREDATORS

The most important items in the diet of this species are crustaceans (63.3% by weight) and small fishes (21.4%) (Bowman et al., 2000). Juveniles are known to feed almost exclusively on amphipods on the outer continental shelf during the fall (Sedberry, 1983), but also utilize decapod crustaceans and polychaetes. There is a shift in diet with increasing size. Fish about 5 cm eat mostly chaetognaths, along with smaller amounts of copepods, amphipods, and mysids (Bowman, 1981). Fish 6 to 10 cm eat mostly decapod shrimp, including *Crangon*, but amphipods, polychaetes, and chaetognaths are also important. Diets are similar in fish 11 to 20 cm, but they then include quantities of euphausiids as well (Bowman et al., 1987). One study examined the food habits of young-of-the-year (70 to 203.8 mm, n = 48) in Long Island Sound (Richards, 1963). Decapod shrimp were the basis of the diet of young-of-the-year, supplemented by small proportions of other invertebrates (see Table 8.1). Young stages of *Urophycis chuss* on the continental shelf are preyed upon by many fishes, but predator studies in estuaries are lacking.

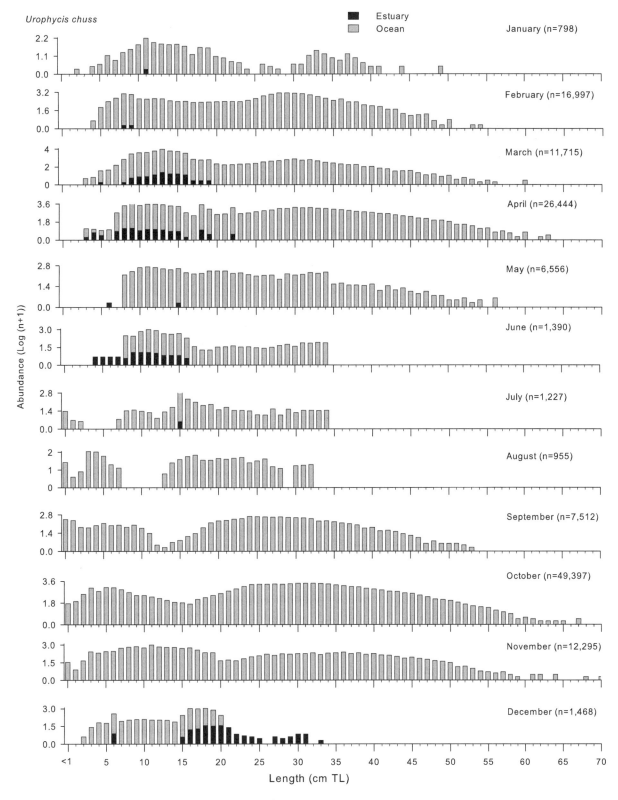

Fig. 34.1. Monthly length frequencies of *Urophycis chuss* collected in estuarine and coastal ocean waters of the central Middle Atlantic Bight. Sources: National Marine Fisheries Service: otter trawl (n = 106,554); Rutgers University Marine Field Station: Methot trawl (n = 449); 1 m plankton net (n = 4); experimental trap (n = 1); killitrap (n = 2); 4.9 m otter trawl (n = 9); and Able and Fahay 1998 (n = 29,735).

Age 1+ *Urophycis chuss*   ## YOY *Urophycis chuss*

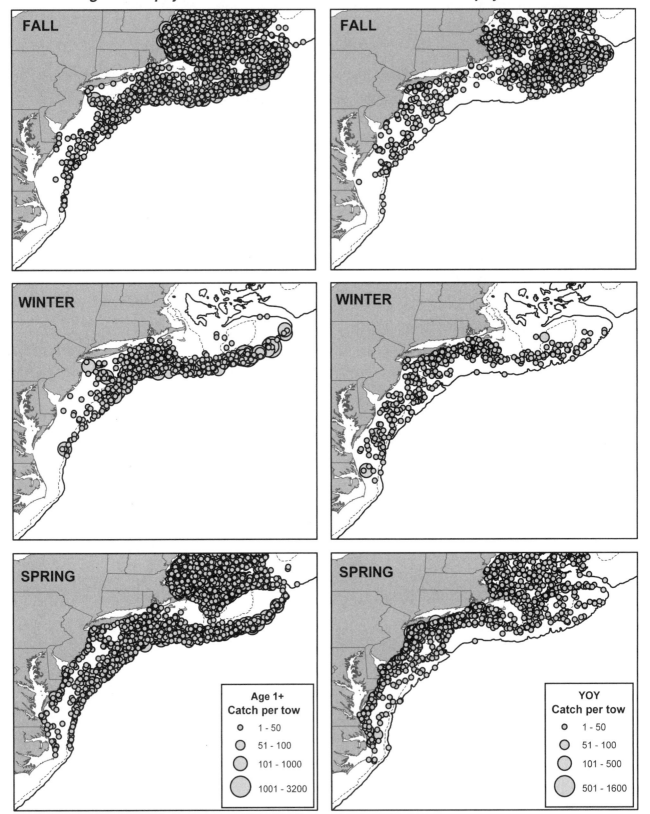

Fig. 34.2. Composite distribution of age 1+ (> 17 cm) and young-of-the-year
(< 17 cm) *Urophycis chuss* during seasonal cruises of the National Marine
Fisheries Service groundfish survey. Details of sampling effort are indicated
in Fig. 3.4.

MIGRATIONS

Young-of-the-year (< 17 cm) are most abundant in the Gulf of Maine, Georges Bank, and the northern part of the Middle Atlantic Bight during the fall, and many are associated with scallop beds (Scott and Scott, 1988) (Fig. 34.2). Most spend the winter between Georges Bank and Cape Hatteras. Many of these then move closer to shore during the following spring. The young-of-the-year then join older year classes (> 17 cm) in an offshore migration during their second winter. In general, adults inhabit very deep water during winter, primarily along the southern flank of Georges Bank and the Middle Atlantic Bight. Some then migrate inshore during the spring, and then move offshore in summer associated with spawning (Musick, 1974). These generalizations are evident in the large numbers collected by trawl in the Middle Atlantic Bight.

35

Urophycis regia (Walbaum)

SPOTTED HAKE

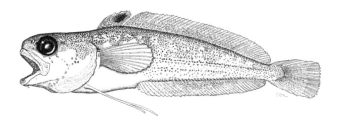

Estuary	Reproduction	Larvae/ Development	YOY Growth	YOY Overwinter
Ocean				
Both				
Jan				
Feb				
Mar				
Apr				
May				
Jun				
Jul				
Aug				
Sep				
Oct				
Nov				
Dec				

DISTRIBUTION

Urophycis regia occurs most commonly in continental shelf waters between Massachusetts and the northeastern Gulf of Mexico (Hoese and Moore, 1977), with a center of abundance in the Chesapeake Bay area (Bigelow and Schroeder, 1953). It is rarely collected as far north as Nova Scotia (Hildebrand and Schroeder, 1928; Scott and Scott, 1988) and reports from as far south as Brownsville, Texas (Springer and Bullis, 1956), may be based on confusion with a congener, *U. flori-*

dana (Hoese and Moore, 1977). It is a very common component of the estuarine fauna in North Carolina (Burr and Schwartz, 1986), but was once thought to be uncommon in the remainder of our study area (Nichols and Breder, 1927). In the Great Bay–Little Egg Harbor estuary, however, it has been the most common *Urophycis* collected in recent years (Szedlmayer and Able, 1996). Juveniles occur in most estuaries in the Middle Atlantic Bight (see Table 4.3), especially those in the southern part of the bight. Comparisons of seasonal distributions for various life history stages in three co-occurring species of *Urophycis* may be found in Able and Fahay (1998).

REPRODUCTION AND DEVELOPMENT

Spawning occurs from summer into winter. Different size groups may spawn at different times of the year, resulting in a bimodal pattern (Able and Fahay, 1998). In our study area, there are two temporal modes of spawning activity. Gonadosomatic studies (Wilk et al., 1990) indicate that ripe females occur from August to October (at lengths of 191 to 360 mm) and again from February through April (at lengths of 283 to 396 mm). A similar August–October observation was made off Maryland and northern Virginia (Eklund and Targett, 1990). In the Chesapeake Bay region, ripe females (> 225 mm) were collected from August to November (Barans, 1969), and have also been found in December off North Carolina (Bigelow and Schroeder, 1953). South of Cape Hatteras, the presence of larvae indicates most spawning activity occurs during the winter (Fahay, 1975). The pattern emerges, therefore, of spawning commencing in late summer in the central part of the Middle Atlantic Bight and occurring progressively later with distance to the south. A second discrete spawning event then occurs in our study area during early spring in waters near the continental shelf edge. Wilk et al. (1990) demonstrated that the spawning cohort present during spring is composed of older individuals than the age 1 spawners present during late summer and fall. Therefore, the apparent bimodality in reproduction may reflect different distribution, migration, and maturation patterns of different age groups of spawners.

Aspects of the ontogeny have been previously summarized, including methods to distinguish early stages from those of congeners in our study area (Able and Fahay, 1998; Fahay, 2007). The diagnostic black spot surrounded by a white margin on the first dorsal fin is formed in early demersal juveniles between 35 and 50 mm, but the series of pale spots occurring along the length of the lateral line apparently does not become visible until juveniles are about 60 mm (Hildebrand and Cable, 1938).

LARVAL SUPPLY, SETTLEMENT, AND GROWTH

The occurrences of larvae and pelagic juveniles off New Jersey and Virginia (Comyns and Grant, 1993) support the bimodal pattern of reproduction described above. In the latter

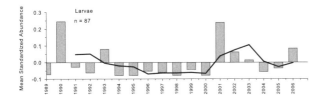

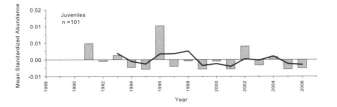

Fig. 35.1. Annual mean (bars) and three-year moving mean (line) of standardized abundance for *Urophycis regia* collected in larval sampling with plankton net (*upper*) and juvenile sampling with traps (*lower*). Data were standardized by subtracting overall mean from annual abundance. Three-year moving mean was calculated by taking the mean of the standardized values from the previous two years and the year in which data are plotted.

study, larvae were most abundant in October and November in mid-shelf regions but also occurred during February, March, and May on stations near the edge of the shelf. Pelagic juveniles were also reported between November and May in Middle Atlantic Bight continental shelf waters (Fahay, 1987). Collections during May at stations beyond the edge of the continental shelf off the Delmarva Peninsula (Hare et al., 2001) yielded large numbers of neustonic larvae in an area dominated by Slope Sea water. The patterns of circulation in this area suggest that these fish were produced locally or perhaps slightly to the north (e.g., in Slope Sea waters off New Jersey) (Cowen et al., 1993). Patterns of spring occurrences of adults presented by Barans (1969) are consistent with these larval occurrences and anecdotal observations reported therein indicate the presence of larvae at least as late as April off Chesapeake Bay (Massmann et al., 1961, 1962). Inter-annual variation in abundance of larvae at Little Egg Inlet, from both cohorts combined, indicates slight increases at approximately decadal intervals (Fig. 35.1). Juveniles collected in traps were not abundant, and their inter-annual variation was marked only by occasional increases.

Settlement occurs at about 25 to 30 mm, based on maximum sizes of the neustonic stage in the Middle Atlantic Bight, although rare specimens may remain in the neuston until about 75 mm in the South Atlantic Bight (Fahay, 1975, 1987). Data on specific habitats where this descent occurs are lacking, however. Early demersal juveniles (n = 43, collected by bottom trawl) were found in various depths on the continental shelf during March between New Jersey and Cape Hatteras and ranged from 34.1 to 64.6 mm SL (Fahay, 1987).

The various sources of larvae summarized under "Reproduction" help explain the seemingly multiple size cohorts observed in estuaries during spring in our study area (Fig. 35.2). Whatever the source for these modes, it is apparent

in neuston sampling on the inner continental shelf that two size classes of young-of-the-year are present in our study area (Able and Fahay, 1998). The smallest sizes are collected in May–June and again in September–October, and these represent the two cohorts spawned in spring and fall. Any evidence of a winter spawning would probably not be apparent in collections north of Cape Hatteras. The smallest size mode observed from late winter into spring probably incorporates progeny from both previously described cohorts, and together these attain lengths of 15 to 20 cm by the following fall. Size at age 1 must be similar in our study area and in Chesapeake Bay. On the first anniversary of the fall spawning event in our study area (September), the mode (about 20 cm) is precisely the estimate provided by Barans (1972) for age 1 fish in his Chesapeake Bay study. Estimates of sizes of this cohort after a season of estuarine growth south of Cape Hatteras are also remarkably similar (Burr and Schwartz, 1986). The size at age 1 for progeny from the spring-spawned cohort is less clear. Presumably, their sizes merge with either the fall cohort after a summer of slow growth or with age 1+ fish during early summer. Our observations and those of Barans (1972) indicate that not all young-of-the-year enter estuaries and that estuarine fish are larger than their continental shelf counterparts. The estuarine components of the year class account for the largest part of the monthly histograms, while fish apparently remaining offshore lag behind in size. There are also growth differences between young-of-the-year in Great Bay–Little Egg Harbor (data included in Fig. 35.2) and those in the Chesapeake Bay study (Barans, 1972). The modes in Chesapeake Bay increase from about 100 mm in March to about 200 mm in June. Comparable figures from the Great Bay–Little Egg Harbor data are about 50 mm in March and only about 75 mm in June (although a few individuals reach 200 mm and the lack of larger fish may reflect a sampling bias).

SEASONALITY AND HABITAT USE

A survey of an inner continental shelf site at the apex of the New York Bight (Wilk et al., 1992) indicated *Urophycis regia* adults were most abundant during August through October and absent from January to April. In a study of the fauna of the upper continental slope (Haedrich et al., 1980), *Urophycis regia* ranked high in the upper depth zone (40 to 264 m) by weight, but not by number, indicating the presence of larger individuals in these deep habitats. Collections of larvae at Little Egg Inlet indicate an estuarine ingress occurring as early as October or November as well as a second one in February–April (Fig. 35.3). Most of these collections were composed of individuals in the process of descending to the early demersal stage (Witting et al., 1999). Most previously available data concerning estuarine occurrences of young-of-the-year *Urophycis regia* derive from studies in Chesapeake Bay (Barans, 1972). Young-of-the-year enter that bay during March, although a lower bottom temperature limit of about 6.5°C seems to

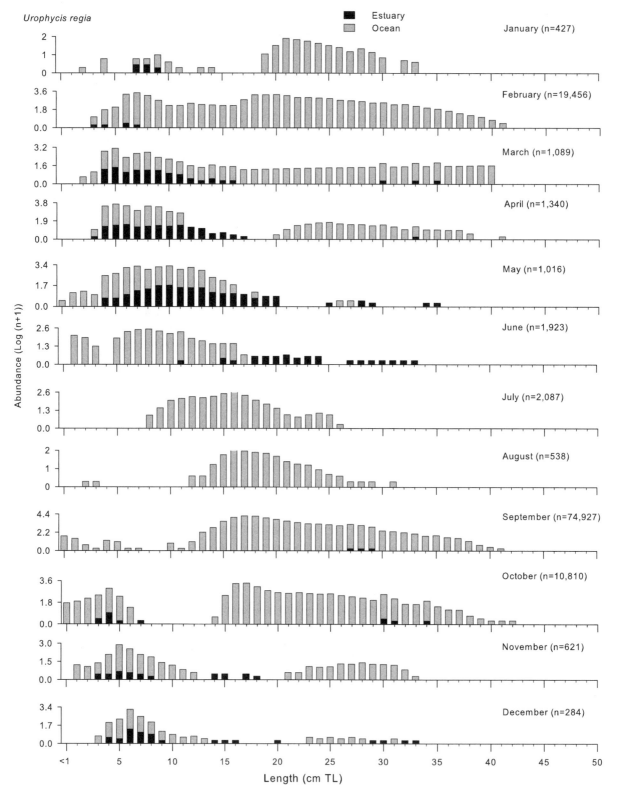

Fig. 35.2. Monthly length frequencies of *Urophycis regia* collected in estuarine and coastal ocean waters of the central Middle Atlantic Bight. Sources: National Marine Fisheries Service: otter trawl (n = 105,631); Rutgers University Marine Field Station: 1 m beam trawl (n = 1); seine (n = 23); Methot trawl (n = 614); 1 m plankton net (n = 31); 2 m beam trawl (n = 4); experimental trap (n = 10); killitrap (n = 49); 4.9 m otter trawl (n = 158); and Able and Fahay 1998 (n = 7997).

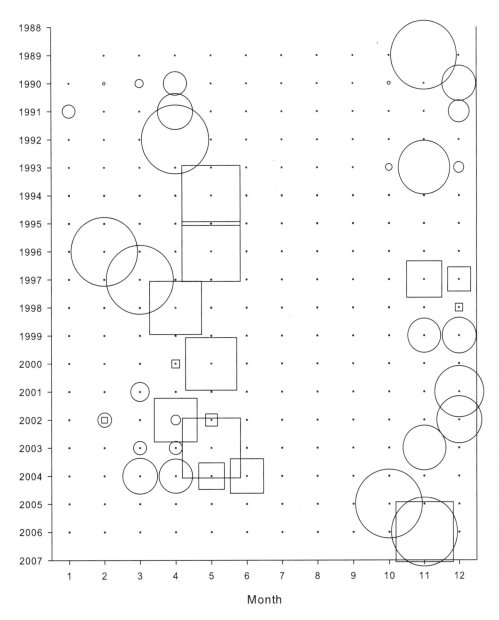

Month

Fig. 35.3. Seasonal occurrence of larvae based on ichthyoplankton samples collected in Little Sheepshead Creek behind Little Egg Inlet from 1989 to 2006 (open circles represent the percentage of the mean number/1000 m³ of larval young-of-the-year captured by year) and juveniles of *Urophycis regia* based on trap collections in Rutgers University Marine Field Station boat basin in Great Bay (open squares represent the percentage of the mean catch per unit effort captured by year) from 1994 to 2006. Values range from 0 to 100% for ichthyoplankton and trap sampling. The smallest circles and squares represent when samples were taken but no individuals were collected.

impede this ingress. After entering the bay, these young-of-the-year penetrate lower-salinity habitats upriver, reaching a maximum upstream limit in April and May (at a minimum salinity of 7 ppt). Growth accompanies this upstream penetration in Chesapeake Bay, where young-of-the-year found farther upstream are larger than those remaining in lower bay habitats. In our focused study area, juveniles from the previous year occur on the shelf or in estuaries at least until June, when they reach a maximum length of about 15 cm. From July onward, these occur in ocean habitats only. Exceptions to this generalization occur in the fall, when a few small juveniles from the fall cohort may inhabit estuarine habitats.

There is little information on specific habitats occupied by *Urophycis regia* young-of-the-year while in their estuarine phase, although Barans (1972) suggested they were somewhat more abundant in channels than in adjacent shallows. Observations of small *Urophycis regia* (107 to 135 mm) indicate a pro-

pensity for burying in sand until the entire body is hidden except for the eyes and snout (Barans, 1969). Juveniles in a New Jersey estuary appear to be associated with a yellow sponge-peat habitat (Szedlmayer and Able, 1996). However, others report small *Urophycis regia* are not associated with structures on bottom as much as *U. chuss* (Edwards and Emory, 1968). In our study area, young-of-the-year from the spring cohort are more likely to be collected in traps than those from the fall cohort. These traps were deployed in the Rutgers University Marine Field Station (RUMFS) boat basin, and the differences from spring and fall may reflect a difference in habitat preference between juveniles of the two cohorts.

PREY AND PREDATORS

Two studies examined the food habits of young-of-the-year (34 to 199 mm, n = 78) in Long Island Sound and Magothy Bay, Virginia (see Tables 8.1 and 8.3). Important prey include

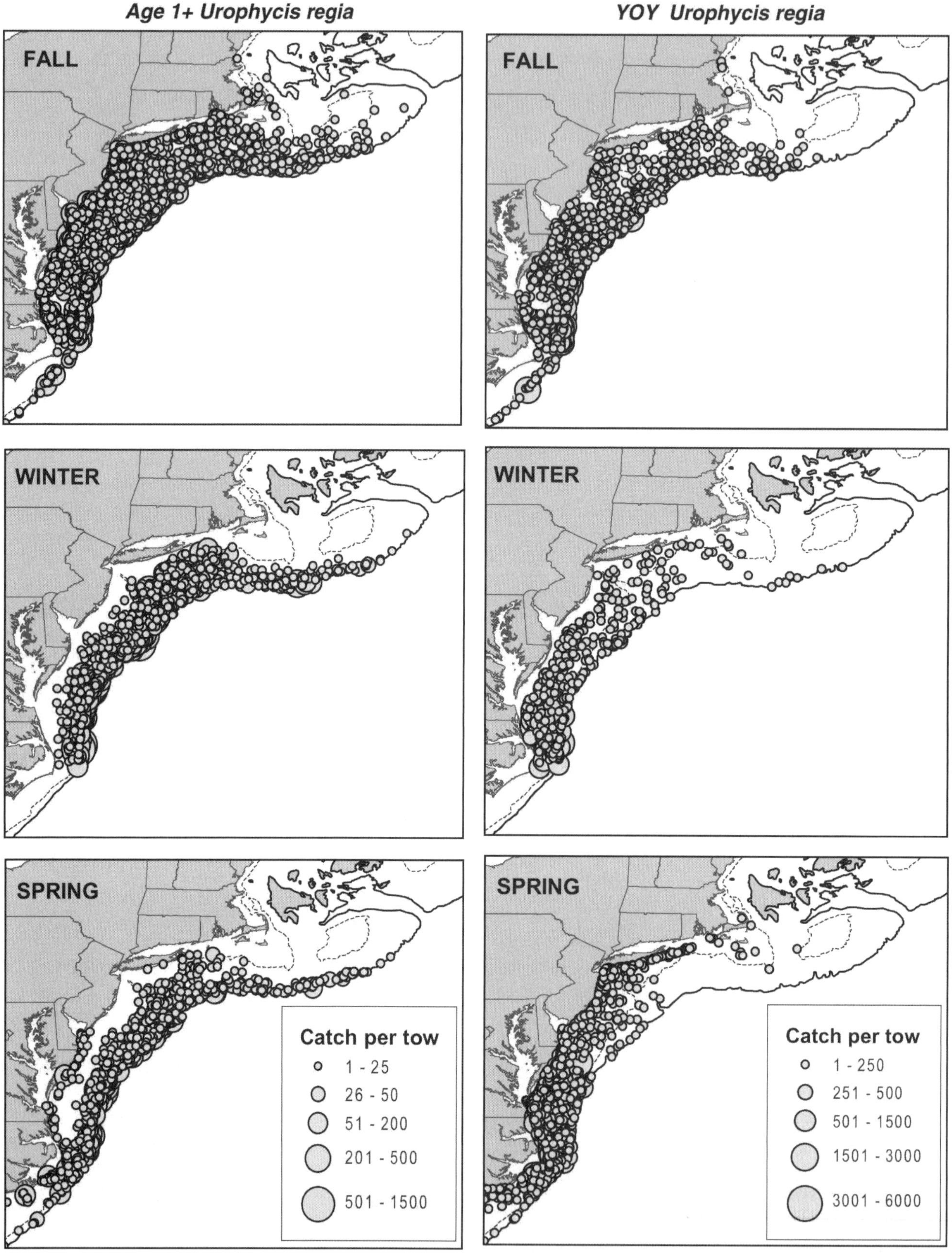

Age 1+ *Urophycis regia*   ## YOY *Urophycis regia*

Fig. 35.4. Composite distribution of age 1+ (> 20 cm) and young-of-the-year (< 20 cm) *Urophyciss regia* during seasonal cruises of the National Marine Fisheries Service groundfish survey. Details of sampling effort are indicated in Fig. 3.4.

a variety of invertebrates (decapod shrimps, amphipods, and mysids). Fishes are a minor dietary component—primarily *Ammodytes americanus* in Long Island Sound (Richards, 1963) and *Anchoa mitchilli* in Magothy Bay (Kimmel, 1973). In the latter study, smaller young-of-the-year (34 to 80 mm, n = 26) consumed large percentages of smaller food items (amphipods, mysids, copepods, and ostracods), while larger individuals (81 to 199 mm, n = 19) consumed larger items including decapod shrimp (*Crangon septemspinosa*) and fish (*Anchoa mitchilli*) (Kimmel, 1973).

MIGRATIONS

Some young-of-the-year *Urophycis regia* occur in estuaries during the winter and spring, but we have found no records of any remaining there past May. These fish apparently spend the rest of their first year in continental shelf habitats. In the Chesapeake Bay, most migrate downstream and leave the bay by June, rarely as late as July (Barans, 1972). Similar results have been reported for estuaries south of Cape Hatteras (Burr and Schwartz, 1986).

The young-of-the-year (< 20 cm) consistently occur over the continental shelf but the northern limits of their distribution vary somewhat with the seasons (Fig. 35.4). In the fall, they are found over all depths (mean depth 77 m, range 9 to 457 m) of the continental shelf between Cape Hatteras and Nantucket Shoals and the Great South Channel. South of Cape Hatteras they are only collected at the edge of the shelf. In the winter, they are found more concentrated to the south of Long Island and are less abundant off Rhode Island and southern Massachusetts and southern Georges Bank. Over these locations, they occur over a wide range of depths (mean depth 65 m, range 12 to 388 m). In the spring, they have a similar distribution and occur over similar depths (mean depth 37 m, range 5 to 367 m) to that in the fall, except they are distributed more in the southern part of the Middle Atlantic Bight. Note, however, that some occurrences shown in Figure 35.4 for young-of-the-year in spring, include fish approaching the cut-off size of 20 cm for this species. Therefore, these include some age 1 fish, and are not exclusively young-of-the-year.

The age 1+ (> 20 cm) juveniles and adults have a similar geographical and depth distribution to the young-of-the-year in fall (mean depth 75 m, range 6 to 457 m) and winter (mean depth 95 m, range 20 to 388 m) except that older fish are found farther east on the southern edge of Georges Bank (see Fig. 35.4). Also, they are less abundant in shallow water in the spring (mean depth 102 m, range 7 to 413 m). During the coldest time of the year (spring) there appears to be a somewhat bimodal habitat distribution with most individuals on the mid- to outer continental shelf, while some can be found on the inner shelf from North Carolina to Delaware Bay. Some of these "inshore" occurrences may presumably be augmented by the largest size class of "young-of-the-year" portrayed in Figure 35.4. This inshore distribution may also reflect the movement of some individuals inshore as part of a spring migration.

36

Urophycis tenuis (Mitchill)

WHITE HAKE

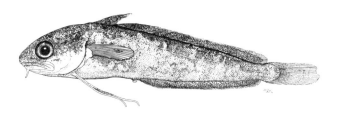

Estuary	Reproduction	Larvae/ Development	YOY Growth	YOY Overwinter
Ocean				
Both				
Jan				
Feb				
Mar				
Apr				
May				
Jun				
Jul				
Aug				
Sep				
Oct				
Nov				
Dec				

DISTRIBUTION

The overall range of *Urophycis tenuis* in the North Atlantic Ocean is Iceland to North Carolina (Scott and Scott, 1988) with occurrences in very deep water as far south as Florida (Musick, 1974). The present account pertains to that portion of the population that occurs in the Gulf of Maine, Georges Bank, and the Middle Atlantic Bight. Many details concerning occurrences, life history, and biology are obscured by the close resemblance between this species and *U. chuss* in all

life history stages. For this reason, the reported occurrences of juveniles in only a few Middle Atlantic Bight estuaries between Massachusetts and southern New Jersey may be underestimates (see Table 4.3). Comparisons of seasonal distributions for various life history stages in three co-occurring species of *Urophycis* may be found in Fahay and Able (1989) and Able and Fahay (1998).

REPRODUCTION AND DEVELOPMENT

Spawning occurs in late winter or early spring in waters over the continental slope off Georges Bank and the Middle Atlantic Bight (Fahay and Able, 1989; Lang et al., 1996). Attempts to describe the reproductive biology of fish in the Middle Atlantic Bight or Gulf of Maine have been frustrated by the inability of researchers to locate ripe females (Musick, 1969; Burnett et al., 1984). Recent studies (Hare et al., 2001) have collected very small larvae (< 5.0 mm NL) at a number of locations in Slope Sea waters off the Middle Atlantic Bight during May, providing further evidence of spawning there. These larvae were part of a ubiquitous slope assemblage, and the water masses with which they were associated suggested an origin within the Slope Sea rather than transport from other areas. Spawning on the Middle Atlantic Bight continental slope, therefore, has been inferred by the presence of the smallest larvae there, and the seasonality of reproduction in this area (early spring) has been verified by studies of otolith ages (Lang et al., 1996).

Details of development have been previously summarized, including methods to distinguish early stages from those of congeners in our study area (Able and Fahay, 1998; Fahay, 2007).

LARVAL SUPPLY, SETTLEMENT, GROWTH, AND MORTALITY

A recent study conducted between the continental shelf and the Gulf Stream off the southern part of the Middle Atlantic Bight during the spring found that the smallest larvae occur in warm Slope Sea waters, but as they develop, they are increasingly found in continental shelf waters, which are cooler during this time of year (Hare et al., 2001). Larvae and pelagic juveniles have also been collected along two transects over the continental shelf in the central part of the Middle Atlantic Bight (Comyns and Grant, 1993). In this study, larvae as small as 3 to 4 mm were collected near the shelf/slope break, and the largest sizes (40 to 50 mm SL) were found closest to shore, suggesting an offshore-inshore migration with growth. Available evidence indicates young-of-the-year remain in a pelagic juvenile stage until they enter certain New England and Canadian estuaries, at which time they begin to occupy bottom habitats and their coloration and morphology change, as described for *U. chuss* (Markle et al., 1982; Fahay and Able, 1989). Individuals (67 to 76 mm TL) going through this morphological change have also been reported from coastal ocean water near Shark River, New Jersey, during May (Nichols and Breder, 1927). In view of these results, and considering those of other studies (Fahay and Able, 1989; Comyns and Grant, 1993), we suggest that while a portion of the year class can certainly be traced from an offshore spawning across the shelf to eventual settlement in estuarine nursery areas, another portion might simply descend to undescribed continental shelf habitats during the early demersal stage. Although an influx of juveniles occurs annually in various estuaries in our study area, we lack sufficient data to infer how that influx might vary annually, either in total or between estuaries.

Estimates of larval (Markle et al., 1982) and pelagic juvenile (Fahay and Able, 1989) growth rates range from 10 to 22 to 35 mm per month. Estimates of the size at which the habitat shift from pelagic juvenile stage to the initial demersal stage occurs range from 50 to 60 mm TL (Markle et al., 1982) to more than 63.8 mm SL (Fahay and Able, 1989), thereby implying that the early pelagic stages occupy about 2 months. In an earlier study, we based the rate of growth on the progression of modes and estimated that young-of-the-year in Nauset Marsh, Massachusetts, grew at the rate of about 30 mm per month (Fahay and Able, 1989). A more recent study, based on otolith analyses and covering various areas in the Middle Atlantic Bight, found a juvenile growth rate of 0.99 mm per day in June through September (Lang et al., 1996). We have too few specimens from our New Jersey study area to make equivalent monthly growth rate estimates, but the length frequencies of available material suggest that young-of-the-year reach about 30 to 35 cm as they begin their first winter (Fig. 36.1). This large size at age 1 is a demonstration of the "get big quick" strategy employed by this species (Markle et al., 1982).

SEASONALITY AND HABITAT USE

In the Middle Atlantic Bight, early stages of *Urophycis tenuis* spend much of their first year in the ocean, after an early spring spawning near the continental shelf edge. After settlement in spring and early summer, most juveniles occur in bottom habitats on the continental shelf, but some enter estuaries, especially those in southern New England (Fahay and Able, 1989). Juveniles, less than a year old, retreat to deeper oceanic waters in fall and winter. When young-of-the-year have been detected entering New England estuaries, they have arrived there during June and July (Fahay and Able, 1989). In our central Middle Atlantic Bight study area, a small number of young-of-the-year have been collected in estuaries as early as April and May (see Fig. 36.1). Recent intensive sampling with a variety of gears in a variety of Great Bay–Little Egg Inlet estuarine habitats, however, has failed to find young-of-the-year (Able and Fahay, 1998). It appears from available evidence, therefore, that young-of-the-year *Urophy-cis tenuis*, at sizes comparable to those of New England estuarine inhabitants, also settle in inner continental shelf sites, such as the Beach Haven Ridge near Little Egg Inlet (Able and Hagan, 1995).

Several sources have mentioned the importance of eelgrass beds as habitat for young-of-the-year in estuaries (Bigelow and Schroeder, 1953; Fahay and Able, 1989; Lazzari et al., 2003). In our study area, however, as well as in the Bay of Fundy (Markle et al., 1982), it is apparent that while young-of-the-year are spatially segregated from older year classes, occurring generally in shallower depths, they are not necessarily tied to eelgrass, other vegetation, or structured bottom habitats. Because of their relative scarcity in our study area, however, it is difficult to ascertain habitat associations. A singular observation of burying in sand substrates (McAllister, 1960) indicates a behavior similar to that described for young-of-the-year *U. regia*.

Most adults occur deeper than 200 m (Scott and Scott, 1988), where they are found over the continental shelf edge, in basins in the Gulf of Maine, and in submarine canyons along the continental slope (Bigelow and Schroeder, 1953; Cooper et al., 1987). In the central part of the Middle Atlantic Bight, they are regularly collected in submarine canyon habitats, 400 to 800 m deep, near the edge of the continental shelf (Markle and Musick, 1974; Haedrich et al., 1980).

PREY AND PREDATORS

This species feeds mostly on a wide variety of fishes (57%), cephalopods (21%), and crustaceans (20%) (Bowman et al., 2000). Crustaceans dominated the diet in fish up to 40 cm TL; fishes and squids were the dominant prey in larger fish. There have been no studies of the food habits of young-of-the-year (< 350 mm) in Middle Atlantic Bight estuaries. However, at sea, fish < age 1 feed mainly on *Crangon, Neomysis,* amphipods, other small crustaceans, and polychaetes (Bowman, 1981). Predators of young-of-the-year in the Gulf of Maine include Atlantic puffins and arctic terns (Fahay and Able, 1989). Larger fish are eaten by *Carcharhinus plumbeus,* bigger *U. tenuis,* and *Gadus morhua,* and several other piscine predators (Bowman et al., 2000).

MIGRATIONS

During fall and winter, both young-of-the-year and older stages retreat to deeper waters in the Middle Atlantic Bight and Gulf of Maine (Fig. 36.2). During the spring, young-of-the-year at sizes up to 35 cm (the cut-off size for determining year class) occur throughout the Gulf of Maine and along deeper parts of the continental shelf off southern New England and Georges Bank. Adults are also found in slightly deeper waters at this time and along the shelf edge as far south as Cape Hatteras.

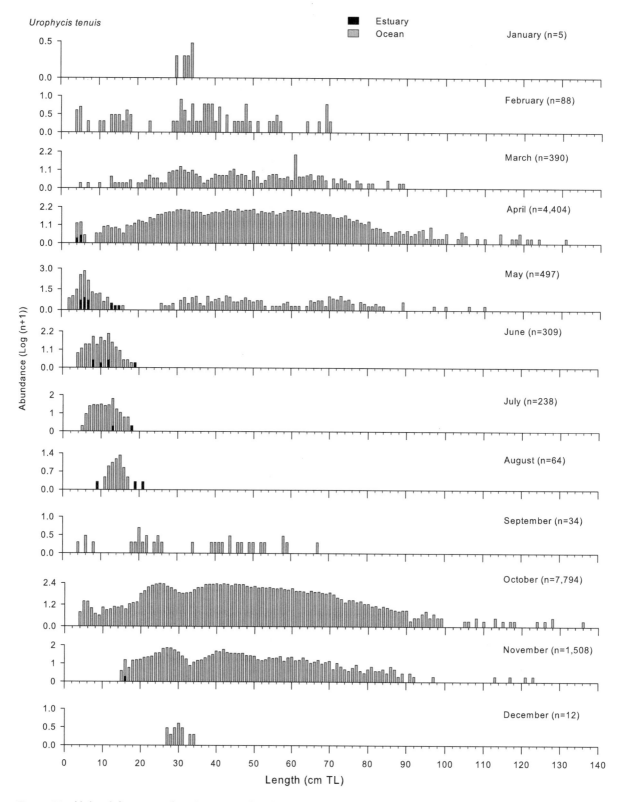

Urophycis tenuis

Estuary
Ocean

January (n=5)

February (n=88)

March (n=390)

April (n=4,404)

May (n=497)

June (n=309)

July (n=238)

August (n=64)

September (n=34)

October (n=7,794)

November (n=1,508)

December (n=12)

Abundance (Log (n+1))

Length (cm TL)

Fig. 36.1. Monthly length frequencies of *Urophycis tenuis* collected in estuarine and coastal ocean waters of the central Middle Atlantic Bight. Sources: National Marine Fisheries Service: otter trawl (n = 14,398); and Able and Fahay 1998 (n = 945).

Age 1+ *Urophycis tenuis* ## YOY *Urophycis tenuis*

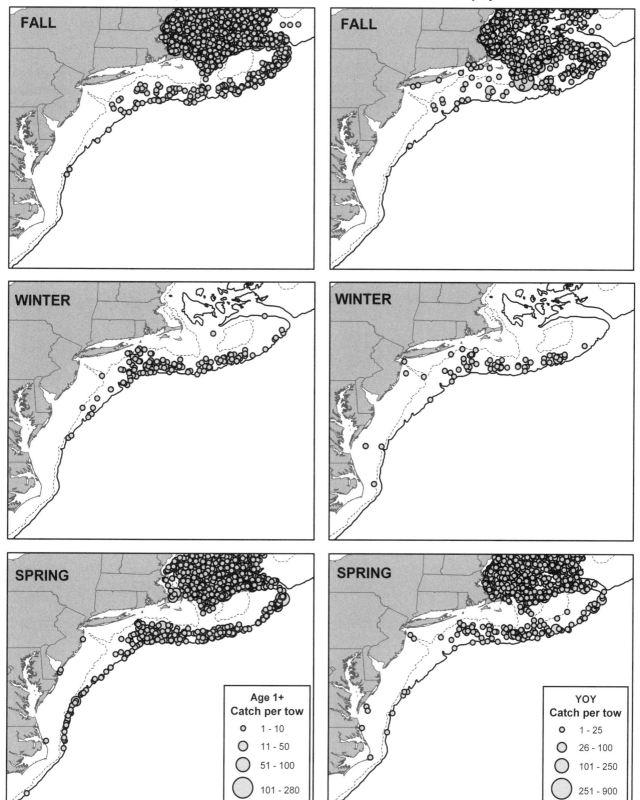

Fig. 36.2. Composite distribution of age 1+ (> 35 cm) and young-of-the-year (< 35 cm) *Urophycis tenuis* during seasonal cruises of the National Marine Fisheries Service groundfish survey. Details of sampling effort are indicated in Fig. 3.4.

37

Microgadus tomcod (Walbaum)

ATLANTIC TOMCOD

Estuary	Reproduction	Larvae/Development	YOY Growth	YOY Overwinter
Ocean				
Both				
Jan				
Feb				
Mar				
Apr				
May				
Jun				
Jul				
Aug				
Sep				
Oct				
Nov				
Dec				

DISTRIBUTION

Microgadus tomcod occurs along the coast of North America from Labrador to Virginia. It has been found as far south as North Carolina (F. J. Schwartz, pers. comm., in Scott and Scott, 1988), but it is rare anywhere south of the Hudson River, New York. In the past 30 years, the southernmost stock (that which occurs in our study area) has undergone a decrease in life span, abundance, and geographical range (Dew, 1991). It was once considered abundant, especially from fall through early spring, in the New York–Sandy

Hook Bay area (Nichols and Breder, 1927) and through winter in the Sea Isle City and Asbury Park areas of the New Jersey coast (Fowler, 1906). In a comparison of relative abundances of species collected in our study area during 1929–1933 and 1972–1973 (Thomas and Milstein, 1974), *Microgadus tomcod* was ranked the fifth most abundant species trawled in the ocean and bays off southern New Jersey in the earlier period; in the more recent time period, it was not collected at all. It was once especially common during spring in the Barnegat Bay and Barnegat Inlet region, but has not been collected there recently (Tatham et al., 1984). Its present status in the central part of the Middle Atlantic Bight is uncertain (Heintzelman, 1971; Miller, 1972). Recently, it has been collected (but only rarely) in Sandy Hook Bay (Pacheco, 1983, 1984) and more frequently in the Hudson River estuary (Dew and Hecht, 1994a, b; Able et al., 1995b; Able and Fahay, 1998) and Newark Bay (S. J. Wilk, pers. comm.). The reduced abundance in the southern part of its range is likely due to increasingly warm temperatures as the result of changing climate trends (Daniels et al., 2005). Within the Middle Atlantic Bight, early life history stages have been reported from estuaries east and north of the Hudson River (see Table 4.3), but they are not found in systems such as Nauset Marsh, Massachusetts, where tidal freshwater zones are absent.

REPRODUCTION AND DEVELOPMENT

The Hudson River marks the current southern limit to spawning (Grabe, 1978). Spawning throughout this species' range occurs during the winter by both young-of-the-year (just less than age 1) and older year classes (if present) and involves an elaborate courtship behavior involving small groups (Howe, 1971). Most spawning takes place in freshwater, near the upstream extent of saltwater intrusion. In the Hudson River, almost all of the spawners (93 to 99%) are age 0 resident fish approaching their first birthday (McLaren et al., 1988), and most of these do not survive beyond the first year.

The early development of *Microgadus tomcod*, including characters useful for distinguishing eggs, larvae, and juveniles from closely related species, has been well described and summarized (Fahay, 2007). Eggs are demersal, slightly adhesive, and either spherical or slightly oval. Their diameters are 1.39 to 1.70 mm. Oil globules are absent in most eggs, although some have 3 to 12 tiny ones. Larvae hatch at 5.0 to 7.0 mm, and the anus opens laterally on the finfold, not on its edge. Pigment patterns and meristic characters are important for accurate identification. Juveniles develop a mottled, blotchy pigment pattern.

LARVAL SUPPLY, SETTLEMENT, AND GROWTH

Larvae are capable swimmers and are positively phototactic immediately after hatching. They may swim to the surface in order to fill their air bladders (Peterson et al., 1980), although studies in the Hudson River found that they stay near the bottom and are only moved passively upward into the water column (Dew and Hecht, 1994a). Larvae remain semipelagic (Dew and Hecht, 1994a) until reaching lengths of about 12 mm TL, when they settle and begin a demersal habit (Booth, 1967).

First-year growth is characterized by an initial fast rate in spring and early summer, followed by a slow rate in midsummer and resumed faster rates during fall (Howe, 1971; Able and Fahay, 1998). Early growth is enhanced by higher spring temperatures, but is then impeded in late May by temperatures above 13°C (Dew and Hecht, 1994b). Modes of 40 to 49 and 60 to 69 mm TL have been reported for May and June, respectively (Young et al., 1991). A second fast growth rate resumes in the fall (Grabe, 1978). They reach 9 cm TL by September (Howe, 1971); upper size limits of 15 to 16 cm for young-of-the-year during November and December have been reported (Grabe, 1978). This growth rate pattern (fast in the spring, slow during summer, fast in the fall) has also been observed from 1993 to 1994 in Newark Bay (Able and Fahay, 1998), where young-of-the-year reached modes of 10 and 18 cm during September and November, respectively (Fig. 37.1). Multiple year classes are also evident in collections from Newark Bay, suggesting that they live beyond the first year, in contrast to the situation described for the Hudson River.

Growth rates may be affected by habitats. This species is able to occupy dark, under-pier areas in the lower Hudson River, partly because of its mode of feeding (see below), where other species (*Tautoga onitis* and *Pseudopleuronectes americanus*) exhibit suboptimal or negative growth (Metzger et al., 2001). However, growth rates for *Microgadus tomcod* are slower under these piers (despite higher concentrations of benthic prey), when compared to rates in pier-edge or open water habitats.

SEASONALITY AND HABITAT USE

This is a demersal species that occurs in coastal, estuarine, and freshwaters. Some populations are landlocked in lakes. Other populations are found in coastal ocean waters and ascending low-salinity streams. Some are strictly riverine. The evidence suggests that a local population of *Microgadus tomcod* spends its entire life history in the Hudson River estuary, rather than migrating seasonally to sea as it does in some other parts of its range (Lawler et al., 1975, cited in Grabe, 1978). This behavior may have been different in earlier decades, when the species was more abundant and potentially achieved older ages and larger sizes.

Most eggs are deposited in December and January (Booth, 1967), and incubation occupies 24 to 60 days (Scott and Crossman, 1973) or 61 to 70 days (Dew and Hecht, 1994a, b), depending on temperature. The habitat where spawning and egg deposition occurs is typically freshwater near the upriver extent of saltwater intrusion during high tides. In the Hudson

River, this intrusion can occur as far upriver as km 243, near Albany (Dew and Hecht, 1994b), but during the reproductive season it is typically located between kms 18 (George Washington Bridge) and 80 (Con Hook). Bottom temperatures are likely to be below 3°C in spawning areas (Grabe, 1978). Most hatching in the Hudson River occurs between mid-February and mid-March (Dew and Hecht, 1994a). Hatching success is greatest in freshwater and declines with increasing salinity. Larval size at hatching is related to the duration of the incubation period, which in turn is related to temperature (see Pepin, 1991; Dew and Hecht, 1994b).

Some studies have found that larvae migrate into saline water immediately after hatching (Peterson et al., 1980). In the Hudson River estuary, larvae are gradually displaced downstream as they develop (Dew, 1995), but the bulk of post-yolk sac larvae remain concentrated just downstream from the salt front, the position of which is determined by the volume of spring freshwater flows. Therefore, as this flow varies, the highest concentrations of larvae are likely to be shifted both upstream and downstream as development proceeds. Larvae and juveniles, combined, are most abundant in salinities of 4.5 to 8.7 ppt (Dew and Hecht, 1994a). In the lower Hudson River, juveniles often occur in darkened areas under large piers (Able and Duffy-Anderson, 2006).

PREY AND PREDATORS

Several studies (n = 2) have examined the food habits of young-of-the-year (28.9 to 143.8 mm, n > 557) *Microgadus tomcod* in the Hudson River estuary of the Middle Atlantic Bight (see Table 8.1). Important prey consist of a variety of small invertebrates, including amphipods, copepods, mysids, and isopods. The diet of young-of-the-year (28.9 to 143.8 mm, n = 577) shifts with ontogeny in Haverstraw Bay, New York, such that the importance of copepods decreases and that of amphipods increases as length increases, with amphipods becoming a dominant item by 90 mm (Grabe, 1978). Relative importance of individual prey items also varies intraannually. Copepods dominate the diet during May–June; amphipods, mysids, and isopods are abundant prey from July to December (Grabe, 1978). The benthic prey that support young-of-the-year under piers in the lower Hudson River include harpacticoid copepods and amphipods (Metzger et al., 2001). Adults also feed mostly on small crustaceans, including shrimp and amphipods, but they also consume worms, small mollusks, squid, and larvae of a variety of fishes. Adults use their sense of smell to locate prey in the substrate. Chin barbels and the tips of their pelvic fin rays are also used for this purpose (Collette and Klein-MacPhee, 2002).

Predators include *Morone saxatilis* and *Pomatomus saltatrix* in the Hudson River (Dew and Hecht, 1994b; Wilk, 1977). However, other studies have found that *Microgadus tomcod* is not included in the diet of *Morone saxatilis* in the lower Hudson River, at least not during winter (Dunning et al., 1997).

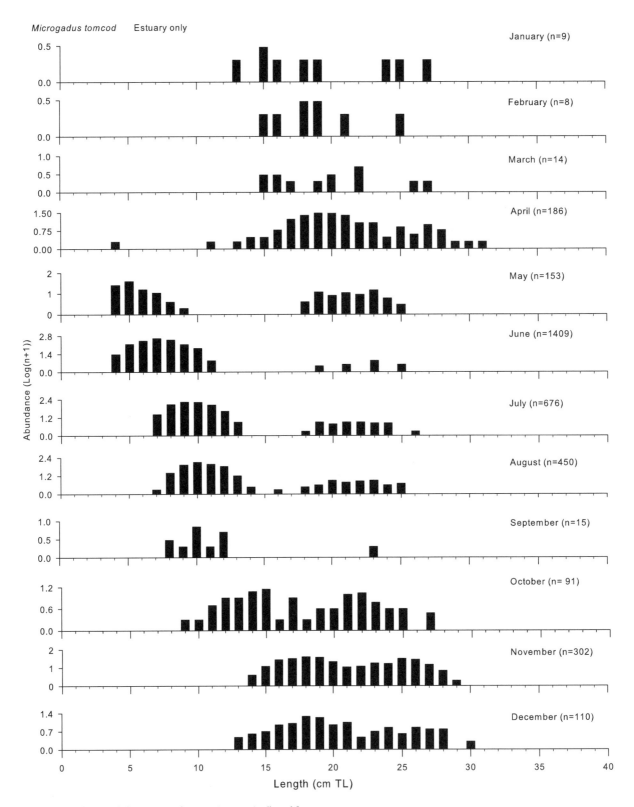

Microgadus tomcod Estuary only

Fig. 37.1. Monthly length frequencies of *Microgadus tomcod* collected from Newark Bay, May 1993–April 1994, with 8.5 m otter trawl (n = 3423). Data courtesy S. J. Wilk, NOAA, NMFS, Sandy Hook Marine Laboratory.

MIGRATIONS

A combination of river flow and temperatures may determine to a large extent the distribution of young-of-the-year within estuarine habitats of the Hudson River. Juveniles are first observed in the Hudson River estuary in March, are concentrated in the downstream portions during April and May, and begin to move back upstream between late April (Dew, 1995) and July, possibly associated with the upstream encroachment of the salt front that occurs during summer when freshwater flow is lowest. Consistent with this pattern, young-of-the-year reach a peak in June under piers and among pilings in the lower Hudson River Estuary (Able et al., 1995b).

38

Pollachius virens (Linnaeus)

POLLOCK

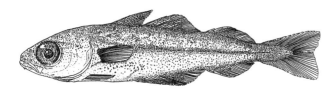

Estuary	Reproduction	Larvae/ Development	YOY Growth	YOY Overwinter
Ocean				
Both				
Jan				
Feb				
Mar				
Apr				
May				
Jun				
Jul				
Aug				
Sep				
Oct				
Nov				
Dec				

DISTRIBUTION

Pollachius virens is primarily a boreal species that occurs on both sides of the North Atlantic. In the western North Atlantic its distribution extends from western Greenland, Hudson Strait, and Labrador to Cape Hatteras, North Carolina (Scott and Scott, 1988), but it is uncommon south of New Jersey. Historical (as well as contemporary) collections reported on here indicate that the range of juveniles extends at least as far south as New Jersey and, less commonly, to Chesapeake Bay (see Table 4.3). The biology and behavior

of this species, described below, may differ in more northern parts of its range, where it is more abundant.

REPRODUCTION AND DEVELOPMENT

Both sexes reach maturity during their third year, at lengths of 50.5 cm in males and 47.9 cm in females (Mayo et al., 1989). Some of these large adults form huge spawning aggregations during the winter and spring. One such aggregation was described as a giant school more than 6.4 km in diameter south of Block Island, Rhode Island, during May (Wilk et al., 1979). Most reported landings in the central part of the Middle Atlantic Bight occur between March and June and are made up of these larger individuals. Spawning occurs between September and April, with a peak from early November to mid-January. Important spawning locations include the western Gulf of Maine, Massachusetts Bay, Georges Bank, and Browns Bank. Occurrences of eggs (Berrien and Sibunka, 1999) and larvae (Able and Fahay, 1998) indicate that spawning begins in the fall in the northern part of the Middle Atlantic Bight when temperatures are below 9.4°C. During the winter, spawning is most intense around the periphery of the Gulf of Maine and on Georges Bank. The occurrence of eggs and very small larvae indicates reproduction also takes place, although less intensely, during spring months in the central part of the Middle Atlantic Bight. Eggs have been found as far south as Delaware Bay (Berrien and Sibunka, 1999), and larvae have been collected over the continental shelf as far south as the Virginia Capes area (Able and Fahay, 1998). The relatively small sizes and lack of fin ray development in these larvae indicate that they were spawned locally.

The early development of *Pollachius virens*, including characters useful for distinguishing eggs and larvae from those of closely related species, has been well described and summarized (Fahay, 2007). Eggs are pelagic, buoyant, and spherical, but lack oil globules. Diameters range from 1.0 to 1.2 mm. They are very similar to eggs of *Gadus morhua*, but are smaller. Larvae hatch at lengths of 3.0 to 4.0 mm and have unpigmented eyes. Pigment patterns and caudal fin ray counts are important for accurate identification. Transformation occurs at sizes greater than 25 mm, when all fin rays are formed. Pelagic juveniles acquire a uniform olive-green coloration.

Individuals from 34.1 to 95.8 mm TL (n = 22) were examined for scale formation (Fig. 38.1). Scales originate along the posterior portion of the lateral line at about 37 mm. They then spread dorsally and ventrally to cover a larger proportion of the posterior body by 42.7 mm, and reach the dorsal and ventral surfaces of the body by 53.1 mm. Scale coverage includes all surfaces of the body posterior to the operculum, the bases of caudal fin rays, and the gular region on the venter under the operculum by 71.4 mm. A patch of scales develops on the operculum behind the eye at 86.0 mm. Variation was observed in head scale development. In some specimens a second patch of scales forms on the operculum, and in

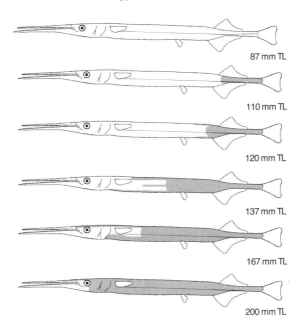

Stongylura marina

87 mm TL

110 mm TL

120 mm TL

137 mm TL

167 mm TL

200 mm TL

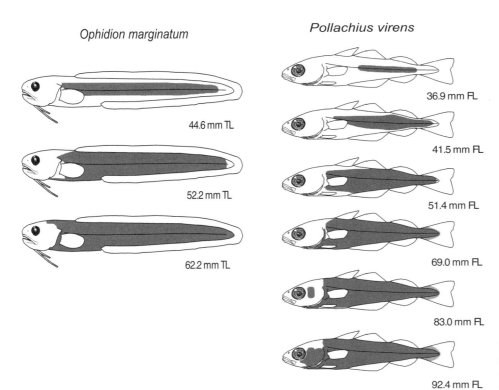

Ophidion marginatum

44.6 mm TL

52.2 mm TL

62.2 mm TL

Pollachius virens

36.9 mm FL

41.5 mm FL

51.4 mm FL

69.0 mm FL

83.0 mm FL

92.4 mm FL

Fig. 38.1. Description of scale formation relative to length in *Pollachius virens, Ophidion marginatum,* and *Strongylura marina.*

others these patches merge and spread to include the top of the head. Scales cover 93.9% of the body and 88.4% of the body and fins when the fish reaches a size of about 95 mm.

LARVAL SUPPLY, SETTLEMENT, GROWTH, AND MORTALITY

Spawning over the continental shelf produces larvae that recruit to estuarine habitats in our study area (Able and Fahay,

1998). Transformation occurs at about 25 mm, when all fin rays are formed (Fahay, 2007). However, post-transformation juveniles remain strongly pelagic and do not settle to bottom habitats as readily as other gadiform species. Nevertheless, these stages are vulnerable to traps set on the substrate, indicating that some near-bottom foraging occurs. Young-of-the-year have declined in numbers in recent years based on these trap collections in the Great Bay estuary (Fig. 38.2). The most

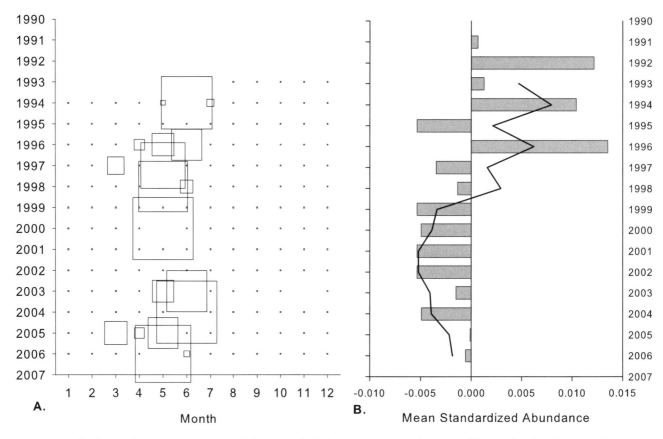

Fig. 38.2. Annual and seasonal variation in occurrence and abundance of *Pollachius virens*. (A) Seasonal variation of the juveniles is based on trap collections in Rutgers University Marine Field Station boat basin in Great Bay from 1994 to 2006. Open squares represent the percentage of the mean catch per unit effort captured by year. Values range from 0 to 100%. The smallest squares represent when samples were taken but no juveniles were collected. (B) Annual variation in abundance for the same data from 1991 to 2006 was standardized by subtracting overall mean from annual abundance. Horizontal bars represent standardized annual abundance; the solid line is a three-year running mean that was calculated by taking the mean of the standardized values from the previous two years and the year in which the data are plotted.

abundant collections occurred in the early to mid-1990s, but have been consistently low since that time, perhaps due to increasingly warm winter and spring temperatures that have occurred here (see Chapter 10).

Growth rates estimated from the progression of length modes in Delaware Bay and coastal New Jersey fish (Fig. 38.3) are consistent with those described for Nova Scotian young-of-the-year (Clay et al., 1989) and indicate that they reach lengths of 4 to 13 cm by midsummer, when they are at a presumed age of 4 or 5 months (Able and Fahay, 1998). Lengths of older year classes collected over the continental shelf, primarily during spring and fall, indicate that young-of-the-year join the adult population during fall, at lengths approaching 20 cm. We have no estimates of mortality rates in young-of-the-year.

SEASONALITY AND HABITAT USE

In the central part of the Middle Atlantic Bight, young-of-the-year are only found from late winter (in the ocean) to early summer (in estuaries or ocean) (see Fig. 38.3). Larvae are distributed over the entire continental shelf from February to May (Able and Fahay, 1998). Small pelagic juveniles (< 50 mm)

approach the coast and begin to enter inlets in February and March. During this stage in their development, these individuals are highly pelagic and capable swimmers. Therefore, passive collecting gear often fails to detect them, and other collection methods probably underestimate their abundance. Nevertheless, seines and trawls collect them often enough that a pattern emerges indicating an ingress from oceanic to estuarine habitats, similar to the pattern reported for Nova Scotia (Clay et al., 1989) and other localities to the north (e.g., Nauset Marsh on Cape Cod; Able et al., 2002). Juveniles are susceptible to capture in traps in the Great Bay–Little Egg Harbor estuary as early as March, but primarily in May or June (see Figs. 38.2 and 38.3). They are seldom collected after June, possibly because they leave estuaries as a result of higher water temperatures, which average 23°C by July (Able et al., 1992).

Although patterns of estuarine occupation in young-of-the-year in the central part of the Middle Atlantic Bight are similar to those from more northern locations, there are apparent differences in the extent to which succeeding cohorts utilize inshore areas. In Nova Scotia, both young-of-the-year and age 1 occur in bays and estuaries, where they

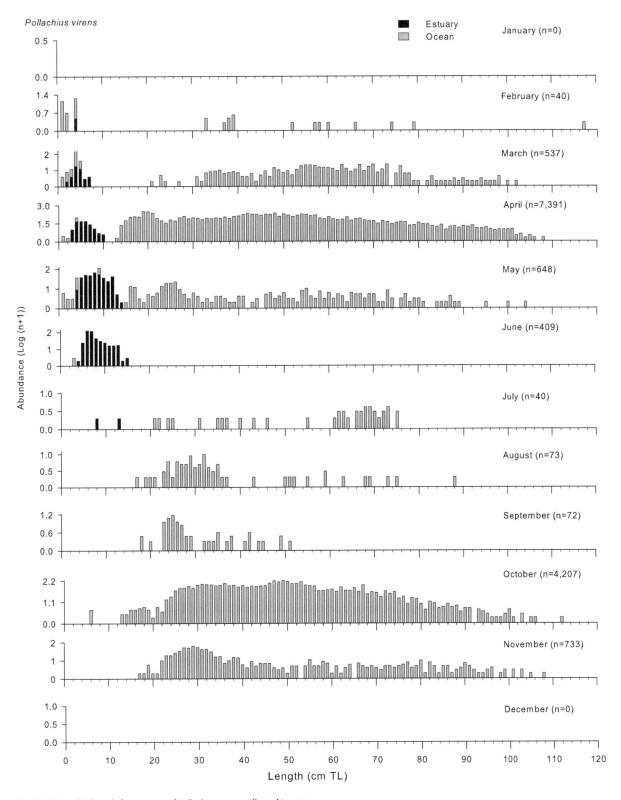

Fig. 38.3. Monthly length frequencies of *Pollachius virens* collected in estuarine and coastal ocean waters of the central Middle Atlantic Bight. Sources: National Marine Fisheries Service: otter trawl (n = 13,084); Rutgers University Marine Field Station: seine (n = 5); Methot trawl (n = 2); 1 m plankton net (n = 3); experimental trap (n = 36); killitrap (n = 57); and Able and Fahay 1998 (n = 963).

are referred to as "harbor pollock" (Clay et al., 1989). In central Middle Atlantic Bight estuaries, length-frequency distributions only indicate the presence of young-of-the-year (see Fig. 38.3). Age 1+ occur over the continental shelf. Another difference between the two populations concerns the temporal pattern of estuarine occupation by young-of-the-year. In Nova Scotian waters, young-of-the-year spend the first six months at sea (after spawning arbitrarily centered on January 1) and the second six months (July through December) within estuarine waters (Clay et al., 1989).

Correlations with specific estuarine habitat types are difficult to demonstrate, because young-of-the-year are highly pelagic. However, they have been shown to favor shallow, nearshore habitats in Maine, especially those with beds of vegetation (Lazzari et al., 2003). Collections elsewhere have revealed that young-of-the-year feed heavily on amphipods that inhabit hydroid beds (Richards, 1963). The rocky intertidal zone has been shown to be an important habitat for juveniles in Canadian waters (Rangely and Kramer, 1995a, b). On rising tides, juveniles move (as individuals or in small schools) from subtidal zones into intertidal zones characterized by dense fucoid macroalgae beds. On falling tides, juveniles school in more open subtidal zones. As the first growing season progresses, there is a distributional shift toward greater depths in the intertidal zone or into subtidal zones exclusively.

On the Scotian Shelf, studies indicate that adults prefer a depth range between 110 and 181 m (Scott, 1982), although this range can vary with season and food supply. They demonstrate little selectivity for substrate type. Adult fish occur in waters as cold as 0°C and apparently do not tolerate temperatures greater than 11°C (Bigelow and Schroeder, 1953). Young stages are found in bays and estuaries throughout the species' range, although many reports are largely anecdotal (Nichols and Breder, 1927; Bigelow and Schroeder, 1953). In some parts of this species' range, coastal and estuarine occurrences have been largely overlooked, in part due to difficulties in sampling this fast-swimming fish.

PREY AND PREDATORS

Adults primarily feed on pelagic crustaceans (especially large euphausiids), small fishes, and squid (Collette and Klein-MacPhee, 2002). Another study showed that the diet was made up of fishes (54% by weight), cephalopods (23.5%), and euphausiids (20.9%) (Bowman et al., 2000). Juveniles feed mostly on euphausiids, but also on smaller quantities of amphipods, decapod larvae, isopods, copepods, polychaetes, and small fishes (Bowman et al., 1987). First-feeding larvae consume phytoplankton and copepod nauplii, then shift to copepods, amphipods, cumaceans, isopods, and larval fishes (Marak, 1960). A study of food habits of young-of-the-year (23.1 to 66.2 mm, n = 22) in Long Island Sound (Richards, 1963) reported that mysids were most frequently eaten by fish 38 to 66 mm, while larger fish (55 to 66 mm) also consumed amphipods found over hydroid beds (see Table 8.1).

MIGRATIONS

Details of processes associated with emigration from estuaries in our study area are lacking, but collections end abruptly in early July at the latest, when maximum lengths of juveniles are about 11 to 14 cm TL (Able and Fahay, 1998). Presumably this coincides with an offshore migration, possibly associated with rising estuarine temperatures. In general, young-of-the-year (< 20 cm) during the fall are most common in shallow, inshore waters (mean depth 40 m, range 21 to 150 m) of southern New England and the Gulf of Maine (Fig. 38.4). No young-of-the-year are collected in the winter, which probably reflects, to some degree, the reduced sampling in the Gulf of Maine and northern Georges Bank at that time. During the spring, the geographical and depth distribution of young-of-the-year is similar to that in the fall (mean depth 47, range 17 to 101 m). Age 1+ (> 20 cm) juveniles and adults have a distribution generally similar to that of the young-of-the-year except that a few individuals are collected in the winter in areas that are well-sampled. In the fall, they are concentrated over a wide depth range (mean depth 99 m, range 14 to 314 m) in the Gulf of Maine and on northern Georges Bank, with a few individuals off southern New England. In the winter, they are infrequently collected from deeper waters (mean depth 101 m, range 51 to 217 m) from the eastern edge of Georges Bank to off New Jersey. In the spring, they are most abundant over a wider area (mean depth 96 m, range 15 to 200 m), primarily in the Gulf of Maine and on northern Georges Bank, with some individuals present as far south as New Jersey and North Carolina.

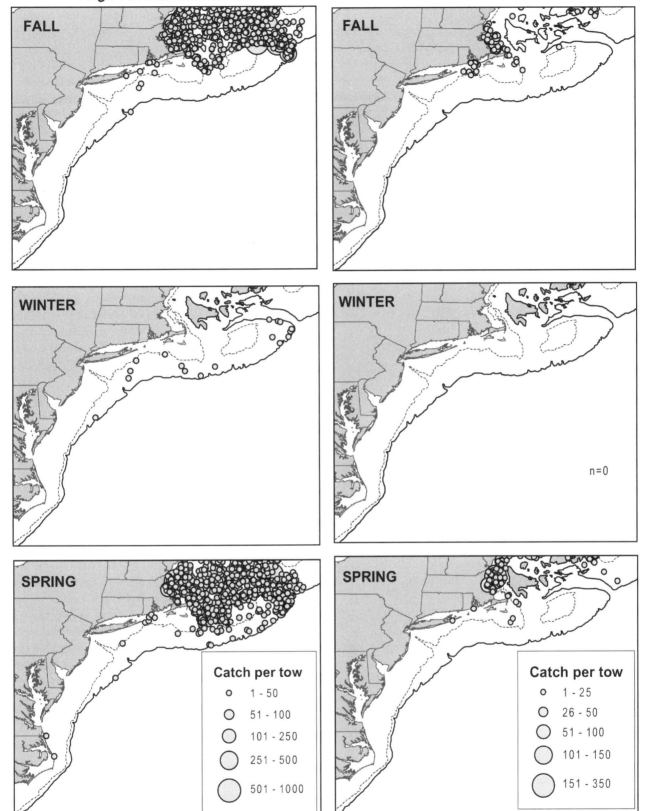

Fig. 38.4. Composite distribution of age 1+ (> 20 cm) and young-of-the-year (< 20 cm) *Pollachius virens* during seasonal cruises of the National Marine Fisheries Service groundfish survey. Details of sampling effort are indicated in Fig. 3.4.

39

Ophidion marginatum (DeKay)

STRIPED CUSK EEL

	Reproduction	Larvae/Development	YOY Growth	YOY Overwinter
Estuary				
Ocean				
Both				
Jan				
Feb				
Mar				
Apr				
May				
Jun				
Jul				
Aug				
Sep				
Oct				
Nov				
Dec				

DISTRIBUTION

This species is distributed along the coast of the United States from southern New England (rarely as far north as New Hampshire) to northeastern Florida (Fahay, 2007). Larvae occur over the continental shelf of the United States (Fahay, 1992); larvae and juveniles have also been reported from estuaries between Barnegat and Chesapeake bays (see Table 4.3), with most reports resulting from studies that focus on inlets.

REPRODUCTION AND DEVELOPMENT

This species spawns in the summer (Fahay, 1992; Able and Fahay, 1998). Observations in the laboratory indicate that courtship and spawning involve the production of sound (Mann et al., 1997) and close tandem swimming by a spawning pair (Fahay, 1992; Bowers-Altman, 1993). Individual females may release a small batch of eggs, contained in a mucilaginous "raft," nightly for a period up to two months. Centers of reproductive activity are found off Delaware Bay and the Delmarva Peninsula and correspond latitudinally to estuarine occurrences cited above. Larval occurrences indicate that spawning extends from June to November, with a peak off the New Jersey coast in August and September (Fahay, 1992).

Morphological development has been described and summarized (Fahay, 1992, 2007). Eggs are slightly off-round and encased in a buoyant, gelatinous veil. Larvae hatch at about 2.0 mm after an incubation period of 36 hours at 24 to 26°C. The elongate larvae have relatively short pre-anus lengths and large heads with big mouths. The gut forms a coil midway along its length at about 4.0 mm TL. The pelvic fin rays arise from a point near the cleithral symphysis in early larvae, but in later stages they originate near the lower jaw tip. This forward migration is due to lengthening of the pelvic fin basipterygium. Pigment is light in early stages, but includes a series of pale spots along the lateral line on the posterior part of the tail.

Scale formation occurs after larvae settle to the bottom at about 22 mm SL (Fahay, 1992). Individuals from 19 to 120 mm TL (n = 9) were examined for scale formation (see Fig. 38.1); those from 19 to 21 mm TL lacked scales. In samples collected from our estuarine study area, the smallest specimen observed with scales was 44.6 mm TL; scales occurred along the lateral line from behind the operculum to near the caudal fin. By 52.2 mm TL, scales had expanded laterally to the dorsal and ventral surfaces of the body. By 62.2 mm TL, the scales had reached the adult condition, covering the body from the dorsum above the operculum to the base of the caudal fin. At this size, scales covered 81.9% of the body. There were no scales on the head or fins.

LARVAL SUPPLY, SETTLEMENT, AND GROWTH

Reproduction takes place over the continental shelf of the Middle Atlantic Bight (Fahay, 1992); larvae collected at Little Egg Inlet are the products of spawning that occurred there the previous summer (Able and Fahay, 1998). Settlement occurs on the continental shelf at sizes of about 22.0 mm SL, an estimate based on sizes of the largest larvae collected pelagically. All available evidence suggests that most settlement occurs at depths between 20 and 40 m (Fahay, 1992). During the past two decades, annual occurrences of larvae collected at Little Egg Inlet have been remarkably consistent; only one year (1990) provided much higher abundance (Fig. 39.1). The earliest juveniles collected in the spring (March and April) average 65.8 mm SL as they enter Little Egg Inlet (Witting et al., 1999), and many of these are approximately the same sizes as larvae settling to bottom the previous fall.

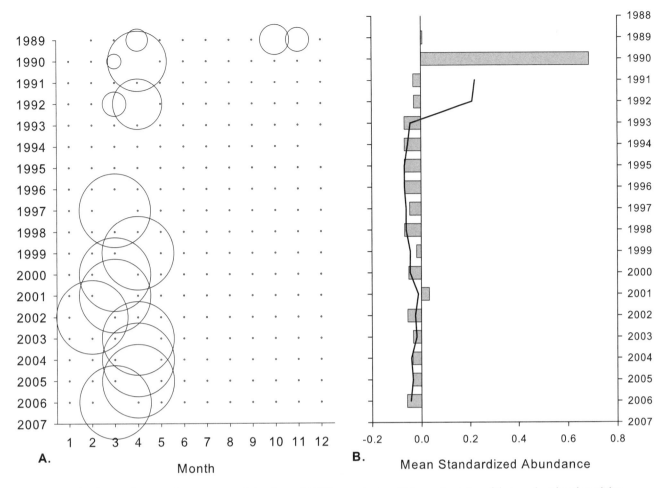

A.

Month

B.

Mean Standardized Abundance

Fig. 39.1. Annual and seasonal variation in occurrence and abundance of *Ophidion marginatum*. (A) Seasonal variation of the juveniles is based on ichthyo-plankton samples collected in Little Sheepshead Creek behind Little Egg Inlet from 1989 to 2006. Open circles represent the percentage of the mean number/1000 m³ captured within each year. Values range from 0 to 100%. The smallest circles represent when samples were taken but no larvae were collected. (B) Annual variation in abundance for the same data was standardized by subtracting overall mean from annual abundance. Horizontal bars represent standardized annual abundance; the solid line is a three-year running mean that was calculated by taking the mean of the standardized values from the previous two years and the year in which the data are plotted.

Thus, inferences concerning growth rates derived from accumulated length frequencies (Fig. 39.2) suggest that there is essentially no growth in just-settled individuals over their first winter, but they grow at an estimated rate of about 1.0 mm per day during the following summer. One-year-olds apparently reach sizes of 100 to 200 mm by the end of the summer, with a mode of about 160 mm. The group we interpret as 2-year-olds also apparently grow little over their second winter, as they approximate this same size range during their second spring.

SEASONALITY AND HABITAT USE

Habitats used by this species vary with life history stage. Larvae are pelagic and are distributed in nearshore waters from the coast of Long Island to the Cape Hatteras area from July through September (Fahay, 1992; Able and Fahay 1998). There is no evidence that they also occur in estuarine waters. Preflexion and flexion stages are more common in inner continental shelf waters < 30 m deep; postflexion stages occur

more commonly over depths between 20 and 40 m. Settlement to bottom habitats on the continental shelf presumably occurs during the fall, and young-of-the-year spend their first winter in these habitats. Several studies sampling in the vicinity of major inlets have documented the appearance of juveniles (20 to 70 mm TL) from mid-March to mid-April (Pacheco and Grant, 1973; Able and Fahay, 1998), and this is also true in the vicinity of Little Egg Inlet. In rare years this immigration may occur as early as late February (see Fig. 39.2). They continue to occupy estuarine habitats between March and October as 1-year-olds (see Fig. 39.2). There is evidence that 2-year-olds are also present in estuaries in the spring, and this largest size class then presumably migrates offshore to spawn over the continental shelf during the summer. This burrowing species is usually found on sandy bottom habitats in relatively shallow bays, estuaries, and inner continental shelf waters. They have also commonly been collected in marsh creeks near submerged blocks of marsh sod (Allen et al., 1978). They are also well known to be strongly noc-

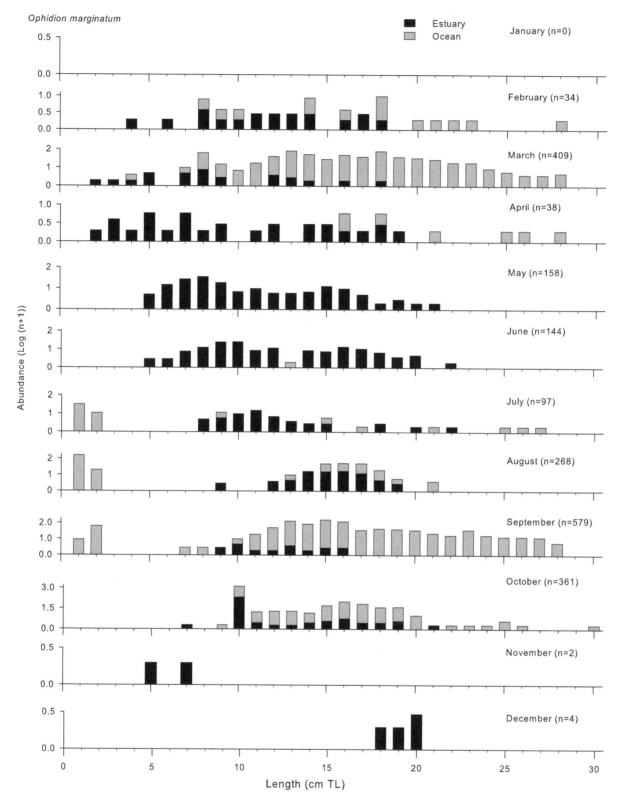

Fig. 39.2. Monthly length frequencies of *Ophidion marginatum* collected in estuarine and coastal ocean waters of the central Middle Atlantic Bight. Sources: National Marine Fisheries Service: otter trawl (n = 974); Rutgers University Marine Field Station: seine (n = 380); Methot trawl (n = 2); 1 m plankton net (n = 45); 2 m beam trawl (n = 7); 4.9 m otter trawl (n = 27); and Able and Fahay 1998 (n = 659).

Daytime Burial in *Ophidion marginatum* Reduces Availability to Estuarine Sampling Gears

Ophidion marginatum is a common cusk eel that reproduces on the continental shelf in the Middle Atlantic Bight. However, it may also reproduce in estuaries based on the frequency with which its courtship sounds are recorded in North Carolina (Luczkovich et al., 1999) and New Jersey (Bowers-Altman, 1993; Rountree, 1999; T. M. Grothues, C. Kennedy, and K. W. Able, unpubl. data). Despite this evidence, these cusk eels are seldom collected in temperate estuaries, although they are often observed on power plant intake screens, especially at night (Tatham et al., 1984). Our own sampling has largely failed to detect them. For example, based on numerous otter trawl tows regularly conducted from the ocean through the estuary and into freshwater in the Mullica River–Great Bay estuary in southern New Jersey over nearly 20 years, this species is seldom collected during daytime sampling. However, in the few instances when sampling was conducted at night, they were common in a channel in the high-salinity portion of the estuary.

Behavioral observations in the laboratory have helped to provide further insights into this seeming anomaly. Small individuals (50 to 75 mm TL) collected in plankton nets quickly buried when they were exposed to sandy substrates in a beaker. Other larger individuals (> 250 mm TL) were exposed to sandy substrates in specially made, narrow aquaria that allow us to see them under the substrate surface. Initially, they assumed an S-shape to the entire body (see Bowers-Altman, 1993, for an accurate rendition of this behavior) and then swam backward into the substrate. As they did, they displaced deeper substrates to the surface, as noted by the movement of different colored, deeper sediments to the surface. Under the substrate, the body was buried more or less vertically in the substrate. They remained buried there for long periods, with the snout slightly exposed or not at all. In this condition, it is highly unlikely that an otter trawl or seine, which rides on the surface of the substrate, would ever collect these buried individuals. Of further interest would be to determine whether this behavior is limited to certain substrate types. In addition, does burial occur diurnally, or do the cusk eels remain buried for long periods of time when they are not feeding or reproducing? Further, once they leave the substrate, do they return to the same location to rebury? This may be a model system by which we could determine the importance of burial behavior for other temperate estuarine species.

turnal (Greenfield, 1968; Matallanas and Riba, 1980). They burrow tail first into the substrate during daylight hours and actively forage for food at night (Allen et al., 1978; Bowers-Altman, 1993).

PREY AND PREDATORS

Our examination of food habits of this species (56.0 to 201 mm TL, n = 26) found that the diet was made up almost exclusively of crustaceans (89.2% by weight), with much smaller amounts of zooplankton, including mysids. Females in the Cape Fear River, North Carolina, were found to consume quantities of fishes, especially anchovies, gobies, and *Symphurus* spp. (Schwartz, 1997); a Chesapeake Bay study found that crustaceans and gobies were important prey (Hildebrand and Schroeder, 1928). Several fish species (notably spiny dogfish and skates) prey on *Ophidion marginatum* on the continental shelf, but studies of predation in estuarine habitats are lacking.

MIGRATIONS

Winter collections of any life history stage are rare in any of the sources available to us. This lack of records indicates that this species probably burrows into the substrate during the winter, but whether this occurs in estuaries, on the ocean bottom, or both, is unknown, because any seasonal migrations that may occur have gone mostly undetected. Young-of-the-year (< 10 cm) were infrequently collected in National Marine Fisheries Service (NMFS) bottom trawl surveys (Fig. 39.3), either because they were not retained in the nets due to their slender shape and small size or because they burrow into the substrate (Able and Fahay, 1998). Those few that were collected ranged from off Long Island to the mouth of Chesapeake Bay in depths from 13 to 66 m (mean depth 23.2 m) and most frequently off Maryland and southern New Jersey. Age 1+ individuals (> 10 cm) were consistently collected between Long Island to south of Cape Hatteras in relatively shallow depths (mean depth 19.5 m, range 12 to 32 m) on the inner continental shelf, with the largest concentrations observed during fall near the mouths of Delaware and Chesapeake bays. Winter records of older life history stages remain rare. Spring records of age 1+ are concentrated along the inner continental shelf.

Age 1+ *Ophidion marginatum*

YOY *Ophidion marginatum*

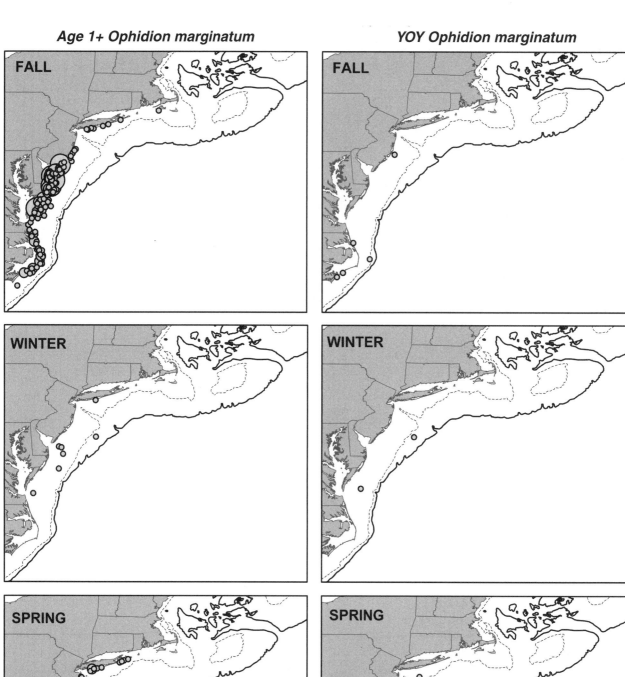

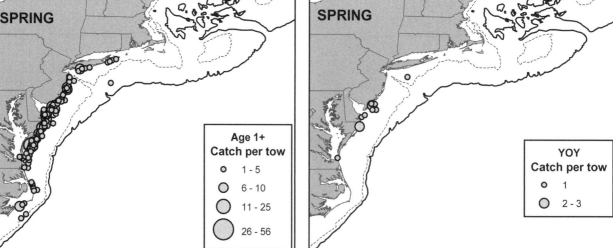

Fig. 39.3. Composite distribution of age 1+ (> 10 cm) and young-of-the-year (< 10 cm) *Ophidion marginatum* during seasonal cruises of the National Marine Fisheries Service groundfish survey. Details of sampling effort are indicated in Fig. 3.4.

40

Opsanus tau
(Linnaeus)

OYSTER TOADFISH

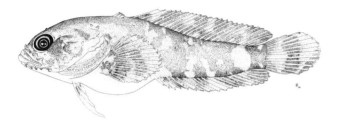

Estuary	Reproduction	Larvae/ Development	YOY Growth	YOY Overwinter
Ocean				
Both				
Jan				
Feb				
Mar				
Apr				
May				
Jun				
Jul				
Aug				
Sep				
Oct				
Nov				
Dec				

DISTRIBUTION

Opsanus tau occurs along the East Coast of the United States from Massachusetts to Florida, with a few records north of Cape Cod (Collette and Klein-MacPhee, 2002). In the Middle Atlantic Bight, early life history stages are common in most estuaries (see Table 4.3).

REPRODUCTION AND DEVELOPMENT

Spawning has been reported to occur from June through August in Woods Hole, Massachusetts (Gudger, 1910), and in June and July in the vicinity of New York City (Nichols and

Breder, 1927). In Chesapeake Bay, the spawning season extends from April to early August or October (Hildebrand and Schroeder, 1928; Dovel, 1960; Schwartz, 1965b) and occurs at temperatures of 17.5 to 27°C (Gray and Winn, 1961). Sound production, especially by the male, occurs during courtship and while defending the spawning site, eggs, and larvae (Gray and Winn, 1961). The production of mating calls decreases toward the end of July in the York River (Fine, 1978), coinciding with gonadal atrophy. Nests have been observed in June, July, and August in North Carolina (Gudger, 1910). In our observations in Great Bay–Little Egg Harbor, the smallest individuals are collected in July and August, presumably from spawning during the summer.

The early development of *Opsanus tau* has been well described and summarized (Fahay, 2007). The embryos develop while attached to the substrate in a nest (Tracy, 1959; Dovel, 1960). Eggs are large and spherical, but flattened where attached to the substrate. Diameters are 5.0 to 5.5 mm and the chorion is yellow. Depending on temperature, larvae hatch in 5 to 12 days at lengths of 6.0 to 7.4 mm. The young remain attached to the nest site for another 6 to 19 days (Gray and Winn, 1961) until yolk absorption at about 16 to 18 mm TL, when they become free swimming. Development is gradual and adult characters are complete approximately 20 days after hatching (Dovel, 1960). This species lacks scales.

LARVAL SUPPLY, SETTLEMENT, AND GROWTH

This species spawns in the estuary. Young leave the nest and seek shelter on the bottom. They have no major dispersal stage; therefore, the collections we have made all derive from local reproduction. The production of young, as represented by sampling with a plankton net, fluctuates little, with occasional years of increased abundance spaced at approximate decadal intervals (Fig. 40.1). Older stages occur at relatively

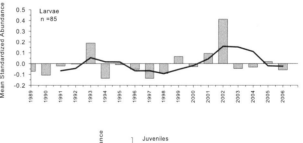

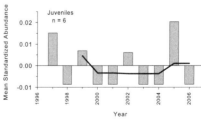

Fig. 40.1. Annual mean (bars) and three-year moving mean (line) of standardized abundance for juvenile *Opsanus tau* collected in sampling with plankton net (*upper*) and sampling with otter trawl (*lower*). Data were standardized by subtracting overall mean from annual abundance. Three-year moving mean was calculated by taking the mean of the standardized values from the previous two years and the year in which data are plotted.

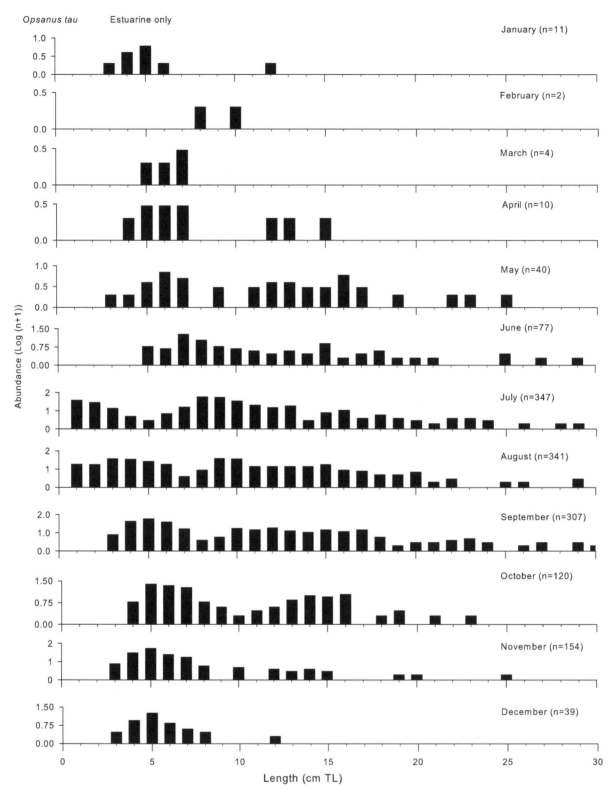

Opsanus tau   Estuarine only

Abundance (Log (n+1))

Length (cm TL)

Fig. 40.2. Monthly length frequencies of *Opsanus tau* collected in the Great
Bay–Little Egg Harbor study area and adjacent coastal ocean waters.
Sources: Rutgers University Marine Field Station: 1 m beam trawl (n = 31);
seine (n = 203); 1 m plankton net (n = 54); experimental trap (n = 53); kil-
litrap (n = 124); 4.9 m otter trawl (n = 195); multi-panel gill net (n = 1); and
Able and Fahay 1998 (n = 791).

constant, annual levels of abundance, as measured by otter trawling.

Young-of-the-year begin to appear in collections in July of most years (Fig. 40.2). Their modal sizes increase from 1 to 4 cm TL in July to about 4 to 9 cm TL in October. There is very little growth during winter. Age 1 individuals grow from a modal size of 6 cm TL in May to 14 to 16 cm TL in October. The size at age 1 in New Jersey overlaps that for Maryland (Schwartz and Dutcher, 1963) and Virginia (Radtke et al., 1985). In a study in South Carolina (Wilson et al., 1982), the size at age 1 was 14 cm. This estimate is consistent with our estimate for size at age 2 (during spring and early summer). Wilson et al. (1982) admitted that their size at age 1 may have been an underestimate because the first opaque zone of the otolith may have obscured an earlier annulus. Alternately, their samples may have been collected in the fall, when many fish in this cohort are observed to be approximately this length.

SEASONALITY AND HABITAT USE

All life history stages are found in estuarine habitats. This species typically occurs on sandy and muddy bottom and often in eelgrass (Bigelow and Schroeder, 1953) as well as in bays and estuaries, sometimes near oyster reefs or rocky outcrops. They are typically found in shallow waters. They are rarely collected during the winter, presumably because during that season they are in deeper channels, where they bury in muddy substrates (Gudger, 1910). In the Great Bay–Little Egg Harbor study area, young-of-the-year occur in a variety of habitats, but they are clearly more abundant in eelgrass (Able and Fahay, 1998). Along a transect extending from freshwater to oceanic waters, this species was limited to sampling stations exhibiting mesohaline conditions (Fig. 40.3). Young-of-the-year are most abundant in July and August (in plankton nets) and May–October (in otter trawls). They may be collected during any month of the year, but numbers decline during winter. Older individuals (age 1+) occur in collections from April through November, with most individuals collected from June through November (see Fig. 40.2).

PREY AND PREDATORS

This species is omnivorous, but feeds most frequently on small decapods (see Tables 8.1 and 8.3). A Chesapeake Bay study examined food habits of young-of-the-year (20 to 100 mm TL, n = 56) (Chrobot, 1951). Principal food items are arthropods, crabs (*Eurypanopeus depressus* and *Callinectes sapi-*

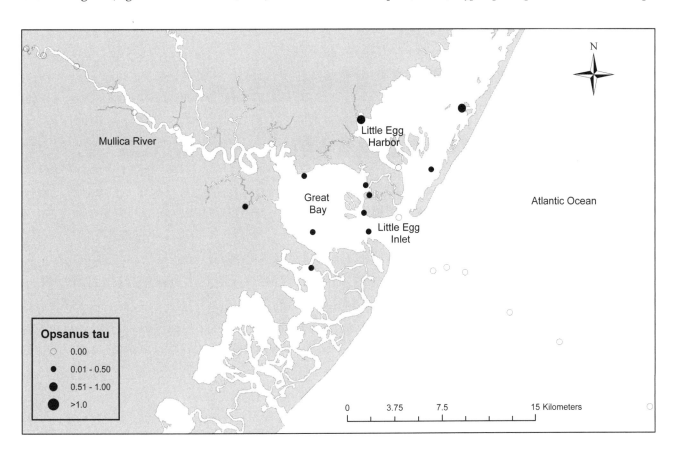

Fig. 40.3. Habitat use and average catch per tow (see key) for juvenile and adult *Opsanus tau* in the Ocean–Great Bay–Mullica River corridor in southern New Jersey based on small otter trawl collections during July and September from 1988 to 2006. See Table 3.2 for characteristics at each sampling collection.

dus), decapod shrimp (*Palaemonetes vulgaris*), and amphipods. Fishes (*Apeltes quadracus, Menidia menidia, M. beryllina, Fundulus majalis, Lucania parva,* and *Gobiosoma bosc*) are a minor part of the diet. Young-of-the-year also exhibit cannibalistic behavior. The diet of young-of-the-year shifts with ontogeny at approximately 100 mm TL, from a diet of invertebrates (crabs, arthropods, amphipods, and decapod shrimp) to a diet of crabs and fishes. Relative importance of individual prey items also varies intra-annually (Chrobot, 1951). In May, young-of-the-year consume mostly fishes and a small percentage of crabs. From July to August there is an abundance of both fishes and crabs, as well as numerous arthropods, amphipods, and shrimp. Crabs are the only abundant food item in September.

MIGRATIONS
We have no evidence that this species migrates away from or within estuaries during any of its life history stages.

41

Strongylura marina (Walbaum)

ATLANTIC NEEDLEFISH

Estuary Ocean Both	Reproduction	Larvae/ Development	YOY Growth	YOY Overwinter
Jan				
Feb				■
Mar				■
Apr				■
May				■
Jun				
Jul				
Aug				
Sep				
Oct				
Nov				■
Dec				■

DISTRIBUTION

Strongylura marina occurs in the western North Atlantic Ocean from Maine to Florida, throughout the Gulf of Mexico, and south to Brazil (Kendall, 1914; Collette, 1968; Dahlberg, 1975). It is absent from the Bahamas and West Indies. In the Middle Atlantic Bight, the larvae and juveniles are reported in estuaries as far north as Great South Bay, New York (see Table 4.3).

REPRODUCTION AND DEVELOPMENT

Relatively little is known about reproduction. Spawning takes place during spring and summer in shallow estuaries and freshwaters. Eggs are deposited within masses of submerged algae (Foster, 1974). In Great Bay–Little Egg Harbor, small individuals are found as early as May. Spawning and hatching probably occur over a relatively long period because the smallest size classes have been collected from May through August. The early development of *Strongylura marina*, and characters useful for distinguishing the eggs, larvae, and juveniles from those of other belonids and hemiramphids, have been well described and summarized (Fahay, 2007). Eggs are demersal and spherical and range from 3.5 to 3.6 mm in diameter. The chorion is equipped with numerous filaments that enable the eggs to be attached to vegetation or mats of algae. Larvae are 9.2 to 14.0 mm long at hatching. They are elongate and have a prominently bulging gut. The dorsal and anal fins form at the extreme posterior part of the body. Pigment is very dense. Early juveniles pass through a "half beak stage" wherein the lower jaw elongates before the upper. Both jaws are about the same length when they reach about 100 mm in length.

Individuals from 41.3 to 200 mm TL (n = 25) were examined for scale formation (see Fig. 38.1; Able et al., 2009a). Those from 41.3 to 80 mm TL lacked scales. Scales form at relatively large sizes (87 to 200 mm TL). The only obvious origin of scales is on the caudal peduncle on the lateral line at approximately 87 mm TL. By 110 mm TL, scales have expanded anteriorly on the lateral surface to near the origin of the dorsal and anal fins. By 120 mm TL, scales have extended anteriorly on the lateral surface and onto the ventral surface. By 137 mm TL, scales on the ventral and dorsal surfaces have expanded to the operculum and level of the terminus of the pectoral fin, respectively. At the same size, scales continue to form anteriorly on the lateral line. By 167 mm TL, scales on the dorsal surface extend to the snout at the base of the upper jaw. Scales on the ventral surface have expanded more to the lateral surface, while those on the midline have extended anteriorly behind the pectoral fin. By 200 mm TL, scales have reached the presumed adult condition, covering the entire surface of the body behind the eye and on the dorsal surface of the snout over the posterior edge of the gape. In this condition, scales cover 89% of the body; the only nonscaled portions are anterior to the eye, on most of the jaws, and on the fins.

LARVAL SUPPLY AND GROWTH

Larvae originate from local spawning, and there is no evidence that they derive from spawning on the continental shelf. The smallest size class appears as early as May (Fig. 41.1). These slender fish increase rapidly in length during the summer so that by September they range from 16 to 55 cm TL and from 27–64 cm TL during October. The increase in modal length frequency shows that estimated growth is relatively fast, approximately 2.3 mm per day.

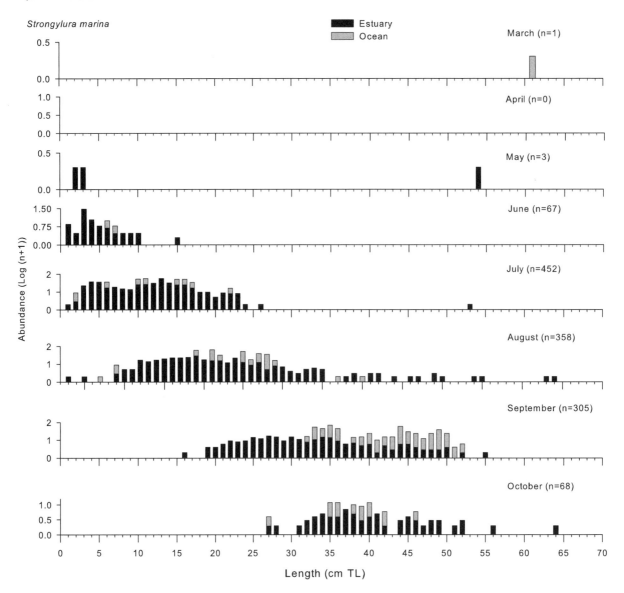

Fig. 41.1. Monthly length frequencies of *Strongylura marina* collected in estuarine and coastal ocean waters of the central Middle Atlantic Bight. Sources: National Marine Fisheries Service: otter trawl (n = 1); Rutgers University Marine Field Station: 1 m beam trawl (n = 3); seine (n = 394); Methot trawl (n = 2); 1 m plankton net (n = 13); killitrap (n = 1); 4.9 m otter trawl (n = 3); multi-panel gill net (n = 5); and Able and Fahay 1998 (n = 832).

SEASONALITY AND HABITAT USE

This species is common in marine and estuarine waters and also extends up into freshwaters (Massmann, 1954; Tagatz and Dudley, 1961). The adults enter estuaries in the spring and remain there until the fall. The young-of-the-year are present in estuaries from May through October (see Fig. 41.1). All life history stages are epipelagic. Our knowledge of the habitat of the young-of-the-year is limited because of their habit of swimming very close to the surface; their speed probably makes them inaccessible to many sampling gears. Visual observation in the Rutgers University Marine Field Station (RUMFS) boat basin and frequent collection with small dip

nets (Able et al., 1997) indicate that the habitat of the smallest individuals is open surface waters. Throughout Great Bay, they have been found in subtidal creeks (Rountree and Able, 1992a) and along subtidal shorelines but not in intertidal creeks or in marsh surface pools (Able et al., 1996b; K. W. Able and M. P. Fahay, unpubl. data). In Delaware Bay, young-of-the-year have been collected in the vicinity of the Chesapeake and Delaware (C&D) Canal (Smith, 1971) and reported as far upstream as the head of tide at Trenton (Fowler, 1906). In Chesapeake Bay, juveniles are found throughout the bay, including up into freshwater (Musick, 1972). This same general pattern has been observed in the Hudson River, where presumed young-of-the-year are found in marshes and weed beds from just above the Hudson Highlands to just below the Tappen Zee Bridge (Lake, 1983).

PREY AND PREDATORS

Adults prey mostly on small, surface-dwelling fishes. Species of *Fundulus* and *Menidia* are high on the list of diet items (Collette and Klein-MacPhee, 2002). Juveniles in the "halfbeak stage" feed on small crustaceans, such as shrimp, mysids, and copepods, but after the transition to the stage where both jaws are elongate, they shift to an almost completely piscivorous stage (Carr and Adams, 1973). Our examination of food habits of this species (28.0 to 206.0 mm, n = 69) found that the diet of young-of-the-year was dominated by fishes (42.1% by weight) and unidentified material (54.1%) with traces of annelids, crustaceans, and zooplankton.

MIGRATIONS

Adults apparently move north in the spring with rising temperature. *Strongylura marina* is seldom collected in our study area after October, and is not represented in winter collections, presumably because individuals leave the estuary and retreat to areas south of Cape Hatteras in the fall.

42

Cyprinodon variegatus Lacepède

SHEEPSHEAD MINNOW

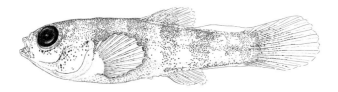

Estuary	Reproduction	Larvae/ Development	YOY Growth	YOY Overwinter
Ocean				
Both				
Jan				
Feb				
Mar				
Apr				
May				
Jun				
Jul				
Aug				
Sep				
Oct				
Nov				
Dec				

DISTRIBUTION

Cyprinodon variegatus occurs along the coast of the United States from Cape Cod to Florida, the Bahamas, northern Mexico, and the Caribbean (Haney et al., 2007). Several investigations have suggested that there are two subspecies, *C. v. ovinus* north of Cape Hatteras and *C. v. variegatus* to the south (Hubbs, 1936; Finne, 2001; Haney et al., 2007). Early life history stages and adults occur in all of the estuaries in the Middle Atlantic Bight (see Table 4.3).

REPRODUCTION AND DEVELOPMENT

Reproductive maturity is reached at a minimum length of 27 mm SL for females and 24 mm for males (Warlen, 1964). Ovarian development and the occurrence of small larvae from May through September indicate that spawning occurs at age 1+ in New Jersey (Talbot and Able, 1984; Able, 1990; Smith, 1995) and Delaware Bay (Wang and Kernehan, 1979; Able et al., 2007a). The primary spawning sites are shallow areas, particularly marsh surface pools with vegetation, where the smallest larvae are collected (Talbot and Able, 1984; Smith, 1995; Chitty and Able, 2004).

The early development of *Cyprinodon variegatus* is fairly well described and summarized (Fahay, 2007), but it is not known whether the early stages differ substantially from those of the more southern subspecies. Eggs are demersal, spherical, and equipped with many adhesive filaments on the chorion. The diameter is 1.2 to 1.4 mm, and there is a single, large oil globule surrounded by many smaller ones. Larvae hatch at about 4.2 to 5.2 mm TL with a large yolk mass. The head is small, with a short, blunt snout and small, terminal mouth. All fin rays are complete by 12.0 mm. Pigment includes six or more bars crossing the body, although it may be lacking in some specimens. Juveniles are characterized by a short, plump body and vague pigment bars.

A total of 74 lab-reared and wild-caught individuals between 3.4 and 30.0 mm TL were examined for scale formation (Fig. 42.1). Individuals from 3.4 to 8.4 mm TL had no scales. Scales formed between the lengths of 8.5 and 18.5 mm TL. Loci where scales originate occur on the lateral surface of the caudal peduncle, at two locations on the operculum, and on the ventral surface. Scale formation begins on the caudal peduncle at the location of the lateral line at about 8 mm TL. Subsequently, they spread anteriorly and laterally to cover the body behind the dorsal and anal fins. By approximately 13 mm TL, scales develop anteriorly until they reached the gill opening. They appear next on the dorso-lateral surface of the head, and under the eye at about 15 mm TL. Scales do not form on the dorsal surface of the head early as in the fundulids (*Fundulus* spp., *Lucania*) (Able et al., 2009a). Later, at about 17 mm TL, scales originate on the operculum and on the ventral surface of the body. By approximately 18 mm TL, the scales reach the adult condition, covering 86% of the body. Scales are lacking on the snout, portions of the operculum, and anterior and ventral to the pectoral fin and on the fins. The rate at which the adult scale coverage is attained is compared among six cyprinodontiform species in Figures 42.1 and 46.1.

LARVAL SUPPLY, GROWTH, AND MORTALITY

The variation in the abundance of this species reaches distinct peaks in some years (Fig. 42.2). Between 1989 and 2006, they were lacking in the collections in an embayment at the

Taxonomy of *Cyprinodon variegatus*
The Need for Further Resolution

One might expect that there would be few taxonomic issues still extant in the heavily studied estuarine fish fauna of the northeastern United States. Further, it would be surprising if these issues persisted for a small, easily collected (nothing more complicated than a dip net is sufficient), widely distributed, and locally abundant form such as *Cyprinodon variegatus*. This is decidedly not the case based on two recent studies (Finne, 2001; Haney et al., 2007). Both studies arrived at the conclusion that this species diverges, based on genetic differences, along the East Coast of the United States in the vicinity of a major biogeographic barrier at Cape Hatteras, North Carolina. The earlier study found little evidence for a hybrid zone (Finne, 2001), suggesting these populations to the north and south might be considered separate species. The other study considered the differences observed to reflect subspecific status, that is, *C. variegatus ovinus* to the north and *C. variegatus variegatus* to the south following an earlier suggestion (Hubbs, 1936). It is interesting that the divergence is reported to occur in the vicinity of Cape Hatteras. While this is a clearly recognized biogeographic barrier (see Chapter 1), it is usually with reference to species occurring in the ocean, where the thermal difference north and south of the cape is likely to have an important influence (Briggs, 1974). It would be unusual that such a shallow water, estuarine form would be influenced by oceanographic barriers.

   Given the uncertainty on the status of this species complex, it might be productive to re-evaluate this issue, especially given the ability to compare the distribution and characteristics across this potential barrier based on the geographically isolated mainland and Outer Banks populations. Further, it might be useful to apply, besides genetics, other characters that have been helpful in resolving the taxonomic status of a closely related form, *Fundulus heteroclitus*. For this form, geographical variation in egg morphology (egg size, number of oil droplets, chorionic filaments), larval morphology (size at hatching), adult morphology (pigmentation), and behavior (spawning site) have proven useful (Able and Felley, 1988) and might be instructive and important given the abundance of the *C. variegatus* complex in estuaries along the East Coast.

Rutgers University Marine Field Station (RUMFS) during the first two years, but they were very abundant in the following year (1991). In the years hence, their relative numbers have been remarkably consistent. Larval supply is almost certainly local because the eggs are attached to the substrate and the recently hatched larvae are bottom oriented and weak swimmers (Sakowicz, 2003).

   On the basis of monthly increases in modal length frequencies in Great Bay, New Jersey salt marshes, young-of-the-year appear to grow at the rate of approximately 0.3 mm per day. Tagged individuals grew at lower rates for both adults (0.15 mm per day) and young-of-the-year (0.07 mm per day) in a study in Delaware Bay (Chitty and Able, 2004). Individuals reach 19 to 49 mm TL (Able, 1990) or perhaps up to 60 mm (Fig. 42.3) by the end of the first summer (Chitty and Able, 2004). Growth may be faster in vegetated pools than in pools without vegetation in southern New Jersey marshes (Smith, 1995). Most of the population in the fall is composed of young-of-the-year, as has been reported for other New Jersey populations (Pyle, 1964), including in Delaware Bay (Able et al., 2007a). There is essentially no growth over the winter so that at the end of the first year individuals are 3 to 6 cm TL. It has been suggested that some members of a Delaware population may live up to 3 years (Warlen, 1964). Individuals that spend the winter on the marsh surface in ditches or pools may be susceptible to mortality during especially cold winters, at least in New Jersey (K. W. Able, pers. observ.).

SEASONALITY AND HABITAT USE
The habitat is broad and variable with all life history stages occurring in the same areas. In general, they are found in very shallow estuarine habitats, in depths < 0.5 m. They are found in protected coves, bays, ponds, and intertidal marsh creeks, but mostly in marsh surface pools (Talbot and Able, 1984; Rountree and Able, 1992a; Smith, 1995; Able et al., 1996b) and less so in subtidal marsh creeks (Chitty and Able, 2004). Demersal larvae of *Cyprinodon variegatus* (Seligman, 1951; Sakowicz, 2003) are less likely to be collected in pit traps on the marsh surface than *Fundulus* spp. (Talbot and Able, 1984), apparently because *Cyprinodon variegatus* larvae move less with flooding tides and prefer to stay in deeper pools and ditches. However, some do leave marsh pools and move onto the marsh surface during high tides (Smith, 1995). In direct comparisons between vegetated (*Ruppia maritima*) and unvegetated pools on the marsh surface, young-of-the-year are more abundant in vegetated pools (Smith, 1995).

   This species is extremely tolerant of a range of environmental conditions (Bennett and Beitinger, 1997). When young-of-the-year were exposed to low levels of dissolved oxygen similar to that observed in natural pools, they survived at levels as low as 0.5 ppm (Smith and Able, 2003). Individuals initially respond to a lowering of dissolved oxygen by resting on the bottom, then changing to aquatic surface respiration. Similar broad ranges are reported for salinity (e.g., 4 to 35 ppt; de Sylva et al., 1962) and temperature (Hildebrand and Schroeder, 1928) in other study areas.

Cyprinodon variegatus   Fundulus heteroclitus   Fundulus confluentus

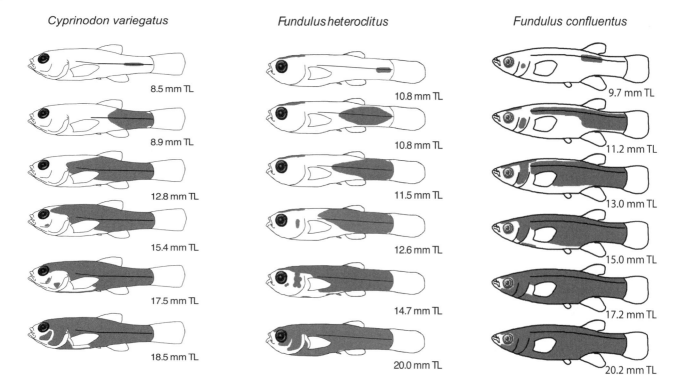

Fig. 42.1. Description of scale formation relative to total length (TL) in
Cyprinodon variegatus, Fundulus confluentus, and *Fundulus heteroclitus.*

Burial in the substrate can confound the interpretation of habitat use if it is based on abundance in net catches. This species buries in the substrate in sand and silt. The smallest individuals tested (15.0 to 19.9 mm TL size class) buried in sand and silt. Overall, among fish tested (15.0 to > 25 mm TL), 50% (n = 6 tested) buried in sand while 92% of those exposed buried in silt.

PREY AND PREDATORS

Several studies (n = 3) have examined the food habits of young-of-the-year (10 to 57 mm SL, n = 1147) in Middle Atlantic Bight estuaries from Hereford Inlet, New Jersey, to Indian River Bay, Delaware (see Table 8.1). Important prey included phytoplankton (algae, diatoms, foraminiferans, and flagellates) and a variety of invertebrates (nematodes, polychaetes, amphipods, and copepods), as well as detritus and sediment. Fishes are a minor dietary component (see Table 8.3). At Hereford Inlet, New Jersey, the diet of young-of-the-year (10 to 35 mm SL, n = 50) shifts with ontogeny at approximately 20 mm, from a diet of meiobenthic animals to a more herbivorous diet (Clymer, 1978). Individuals < 20 mm consume primarily invertebrates (copepods, polychaetes, and nematodes), whereas larger young-of-the-year consume primarily phytoplankton (algae, diatoms, and foraminiferans) and detritus.

MIGRATIONS

During the summer, this species is largely limited to intertidal and shallow subtidal habitats based on tag/recapture (Chitty and Able, 2004). Some movements of tagged fish occur between intertidal creeks and the creeks into a pond (Chitty and Able, 2004). Some have been collected in a deeper embayment (1 to 2 m), adjacent to a marsh creek in November and December in Great Bay, suggesting a movement into deeper water for the winter (Able and Fahay, 1998; Chitty and Able, 2004).

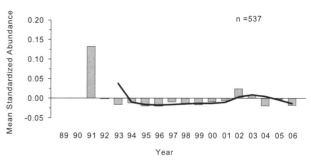

Fig. 42.2. Annual variation in abundance of *Cyprinodon variegatus* collected in trap sampling, 1991–2006. Data were standardized by subtracting the overall mean from annual abundance. Vertical bars represent standardized annual abundance; the solid line is a three-year running mean that was calculated by taking the mean of the standardized values from the previous two years and the year in which the data are plotted.

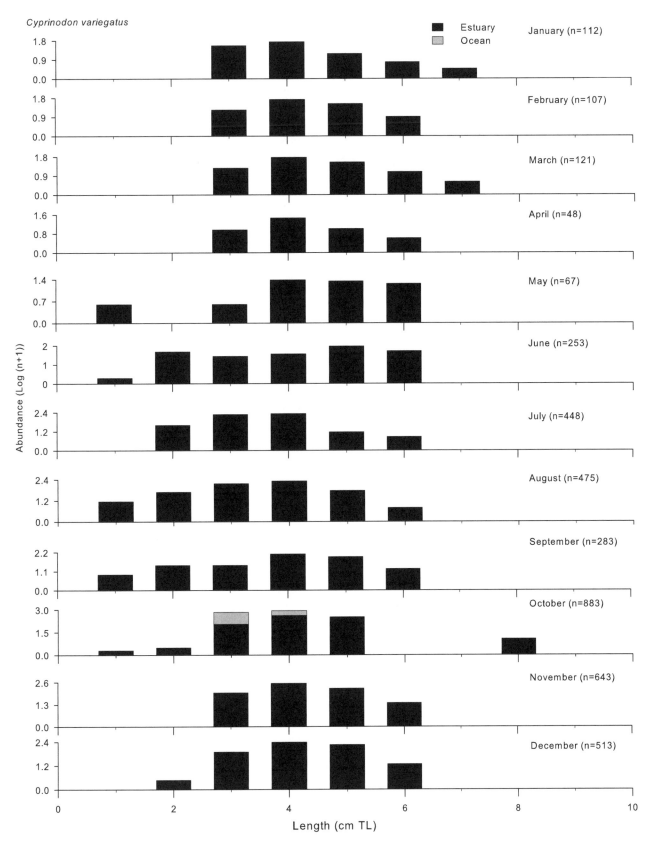

Fig. 42.3. Monthly length frequencies of *Cyprinodon variegatus* collected in estuarine and coastal ocean waters of the central Middle Atlantic Bight. Sources: Rutgers University Marine Field Station: seine (n = 60); 1 m plankton net (n = 4); experimental trap (n = 1); killitrap (n = 265); and Able and Fahay 1998 (n = 3665).

43

Fundulus confluentus Goode and Bean

MARSH KILLIFISH

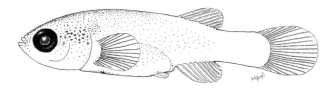

DISTRIBUTION

Most authors indicate that the subspecies *Fundulus confluentus confluentus* occurs from Chesapeake Bay to southern Florida (including the Florida Keys) and the northeastern Gulf of Mexico (to Alabama). Although this subspecies has been recorded from New Jersey, Delaware, Maryland, and Virginia, the northernmost verifiable records are from Lynnhaven Inlet in southern Chesapeake Bay (Relyea, 1965; Hardy and Hudson, 1975). Recently, it has been suggested that it may have been extirpated from that area (Murdy et al., 1997) and our attempts to collect this species from southern Chesapeake Bay have been unsuccessful. The early reported collections from North Carolina have frequently been repeated (Hildebrand, 1916, 1941; Hubbs, 1931; Lee et al., 1980; Menhinick, 1991). More recently, extensive collections in freshwater ponds on North Carolina's Outer Banks have recorded this species from 11 scattered localities from the northern banks to Shackleford Banks (Schwartz, 1992). We also collected specimens from near Corolla, just south of Nags Head, and from Salvo, just south of Hatteras Inlet, as well as on Shackleford Banks (K. W. Able, J. Merriner, and J. Govoni, unpubl. data). Our attempts to clarify the northern limit of the range include collections from MacKay National Wildlife Refuge in North Landing River south of the Virginia–North Carolina border.

REPRODUCTION AND DEVELOPMENT

There has been relatively little added to our understanding of this species since early accounts based on North Carolina observations (Hildebrand and Schroeder, 1928). Sexual maturity is reached at approximate lengths of 35 mm. Spawning may occur within masses of vegetation (including algae) in fresh-

water or brackish water. Sites for reproduction include rain-filled ditches, small pools, or edges of brackish tidal marshes. Members of the Florida population may spawn during any month of the year. Reports of spawning are from April to May in Chesapeake Bay and from April to October near Beaufort, North Carolina (Hildebrand and Schroeder, 1928). The eggs of this species from a Vero Beach, Florida population are capable of hatching up to three months after deposition (Harrington and Haeger, 1958; Harrington, 1959a, b). In aquaria, a North River, North Carolina population reproduced over several months, with eggs often deposited singly or occasionally in a cluster of three to five.

The early development of *Fundulus confluentus* has not been completely described. The eggs are demersal and spherical or slightly off-round. Their diameter has been described as 1.6 to 1.8 mm after fertilization. The chorion surface is equipped with attachment filaments as well as "chorionic bristles" or "minute punctae" (small spherical knobs) (Hardy, 1978a). There are 10 to 15 oil globules in the yolk. The eggs are typically attached to plants or masses of algae. Incubation takes as many as 28 days, although when eggs are held out of water, hatching may be delayed for up to 95 days. Larvae hatch at 4.0 to 5.6 mm in length (K. W. Able, pers. observ.) Yolk-sac larvae are equipped with 12 well-formed caudal fin rays. Pigment is heavy along the dorsum of the head and extends the length of the body and tail along the mid-dorsal edge. The yolk sac and lower half of the tail are heavily pigmented as well. There is a dark bar on the side of the head posterior to the eye, and a few melanophores occur on the developing caudal and pectoral fin rays. Arrays of yellow, orange, or white chromatophores are also present, but these do not survive preservation. Larger larvae have not been described. We have recently preserved a series of developing larvae, hatched from eggs collected from North River, near Beaufort, North Carolina, and fertilized in the laboratory. An early juvenile from this series (see above) exhibits full complements of fin rays. Pigment is faint or absent over most of the body, but a weak scattering of melanophores is most dense on the head behind the eye. A ventral series of melanophores on the venter of the belly includes one prominent spot, surrounded by smaller ones. Most of the caudal peduncle is unpigmented.

Juveniles (< 30 mm) are usually similar to adult females in pigmentation. However, juvenile males may have an ocellus in the dorsal fin, a character otherwise found only in adult females (Relyea, 1965). Scale formation in this species is generally similar to that of other *Fundulus* spp. examined (Able et al., 2009a). The formation of scales on the operculum at a small size is a unique pattern found in all the *Fundulus* spp. examined, although it forms earliest in this species (Fig. 42.1). Individuals from 5.6 to 8.7 mm TL lack scales. Scales originate in several loci, including on the lateral surface under the dorsal fin, on the dorsal surface of the head, on the operculum behind the eye, and on the ventral surface at the level of the pectoral fin. Onset of scale formation occurs at 9.7 mm TL on the caudal peduncle and on the operculum. Scales form on the dorsal surface of the head at approximately 11 mm TL. By this same size the scales expand laterally to cover much of the caudal peduncle and also extend anteriorly along the lateral line to the level of the gill opening. By 13 mm TL, the scales have continued to expand laterally on the trunk but not to the dorsal or ventral surfaces. At this same size, scales on the operculum and dorsal surface of the head have also expanded and merge. By 15 mm TL, scales have extended farther to cover the dorsal surface of the trunk. By 17 mm TL, scales cover almost all of the trunk and head, as in the adult condition, but are absent just anterior to the pectoral fin base. By 20 mm TL, the areal extent of scale coverage has reached the adult condition, with scales covering 98.0% of the trunk and head and 72.2% if fins are included. In the adult condition, scales do not cover the snout, lower jaw, or any of the fins.

LARVAL SUPPLY AND GROWTH
The larval supply is likely local because the eggs are attached to spawning sites. Maximum size is 65 mm for females and 60 mm for males (*F. c. pulvereus*; Relyea, 1965).

SEASONALITY AND HABITAT USE
This species occurs mainly in grassy, muddy backwaters and brackish bays, and not along open beaches (Hildebrand and Schroeder, 1928). It may also be found in freshwater. Our own observations during the spring in the vicinity of Beaufort and on the Outer Banks of North Carolina indicate that juveniles and adults can be found in oligohaline salinities. In both localities we have been most successful collecting them in shallow, small (< 1 m wide) roadside ditches with natural marsh vegetation (*Spartina* and *Juncus*) in non-tidal pools. At the North River site, *Fundulus confluentus* was most commonly collected with a dip net over soft, detrital substrates and under vegetation at the edge of roadside ditches (12 to 48 cm wide) that were probably only flooded with saltwater on storm tides. In these instances it often co-occurred with *F. heteroclitus*, *F. luciae*, and *Gambusia holbrooki*.

Small individuals of this species (approximately 7 to 10 mm TL) reared in aquaria displayed some of the behaviors of adults with regard to feeding, seeking refuge, and schooling. These individuals typically swim singly and never school. The only time these small individuals and adults aggregated was when food was introduced. When disturbed, all fish of these sizes hide in detritus and remain inactive.

PREY AND PREDATORS
The food of this species includes algae, vascular-plant detritus and fresh tissue, protozoans, rotifers, a bryozoan, gastropod mollusks, pelecypods; polychaete annelids, oligochaetes, crustaceans, arachnids, diplopodans, insects, and three species of fishes, mostly larval *Gambusia affinis* (Harrington and Harrington, 1972).

MIGRATIONS
We have no evidence that this species engages in migrations within the limited habitat where it resides.

44

Fundulus diaphanus
(Lesueur)

BANDED KILLIFISH

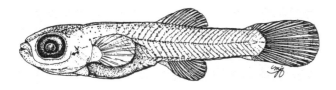

Estuary	Reproduction	Larvae/ Development	YOY Growth	YOY Overwinter
Ocean				
Both				
Jan				
Feb				
Mar				
Apr				
May				
Jun				
Jul				
Aug				
Sep				
Oct				
Nov				
Dec				

DISTRIBUTION

Fundulus diaphanus occurs along the Atlantic coast of the United States from Newfoundland to South Carolina (Lee et al., 1980). The subspecies in the study area, *F. d. diaphanus*, is the common form along the East Coast, while *F. d. minona* is an inland form. These intergrade where they come together in some of the Great Lakes and the St. Lawrence River (April and Turgeon, 2006). The early life history stages are likely found in all estuarine tributaries in the study area.

REPRODUCTION AND DEVELOPMENT

Spawning first occurs at age 1, at sizes of approximately 55 mm (Hildebrand and Schroeder, 1928), and takes place from May through July or August based on gonadosomatic indices and the occurrence of larvae in New Jersey (Able, 1990). Spawning extends from April until September in Chesapeake Bay (Hildebrand and Schroeder, 1928). During spawning the males of this subspecies defend territories and court females (Richardson, 1939). The eggs are extruded and suspended below the female by chorionic filaments; the male clasps the female against aquatic vegetation, where the eggs are released. Similar behavior was observed at high tide in tidal freshwaters of the Mullica River, New Jersey, where males also defended territories (K.W. Able, pers. observ.). In this instance, when a female was ready to spawn she nipped the substrate and the male immediately clasped the female (a behavior also seen in *F. heteroclitus*); the eggs were released into the sandy substrate. Sneaker males also clasped females that had been courted by a primary male.

Early development of this species is fairly well described and summarized (Fahay, 2007), but it is not known if development varies between subspecies. Eggs are demersal and spherical and have many filaments, some of them quite long, on the surface of the chorion. The diameter is 1.5 to 2.0 mm and there are many oil globules. Larvae hatch at 5.0 to 5.5 mm with a large yolk mass. The head is small with a short, blunt snout and a small, terminal mouth. Fin rays are formed by at least 12 mm TL. Pigment is light in larvae; juveniles initially have approximately 7 vertical bars (Jones and Tabery, 1980) but eventually develop 16 to 20 vertical bars on the lateral surface of the body.

LARVAL SUPPLY AND GROWTH

Relatively little is known about the larval ecology of this species except that the supply is likely to be local because of the attached eggs. Based on monthly increases in modal length frequencies in a New Jersey impoundment, young-of-the-year appear to grow slowly (Fig. 44.1). Individuals reach 24 to 58 mm TL at the end of the first summer and 35 to 63 mm TL by 12 months. A maximum size of 98 mm TL is reached when individuals are likely 2 to 3 years old (Able, 1990).

SEASONALITY AND HABITAT USE

The adults are evident year round but larvae and small juveniles (< 20 mm TL) are only evident from June through September in a New Jersey study site (Able, 1990). They typically occur along shallow shorelines in freshwaters and tidal brackish waters and are rarely found at salinities > 5 ppt (Murdy et al., 1997). They appear, in some instances, to select vegetation for spawning (Richardson, 1939) and juvenile and adult habitat (Serafy and Harrell, 1993). Some populations are capable of burying in the substrate (Colgan and Costeloe, 1980).

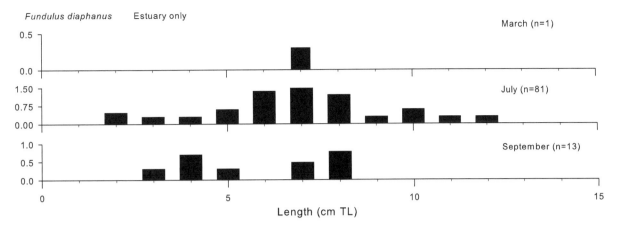

Fig. 44.1. Monthly length frequencies of *Fundulus diaphanus* collected in estuarine waters of the Great Bay–Little Egg Harbor study area. Source: Rutgers University Marine Field Station: killitrap (n = 1); 4.9 m otter trawl (n = 94).

PREY AND PREDATORS

The diet is diverse and consists of small crustaceans, insects, mollusks, and annelid worms (Hildebrand and Schroeder, 1928).

MIGRATIONS

Unknown.

45

Fundulus heteroclitus (Linnaeus)

MUMMICHOG

Estuary / Ocean / Both	Reproduction	Larvae/ Development	YOY Growth	YOY Overwinter
Jan				
Feb				
Mar				
Apr				
May				
Jun				
Jul				
Aug				
Sep				
Oct				
Nov				
Dec				

DISTRIBUTION

Fundulus heteroclitus occurs in shallow estuarine waters from southwestern Newfoundland and Prince Edward Island to northern Florida (Relyea, 1983; Scott and Scott, 1988). Freshwater populations also occur (Hastings and Good, 1977; Denoncourt et al., 1978; Samaritan and Schmidt, 1982). Some authors recognize two subspecies (Able and Felley, 1986). In this interpretation, *F. h. heteroclitus* occurs from northern New Jersey south to Florida, including lower Chesapeake and Delaware bays, while *F. h. macrolepidotus* is distributed from Connecticut north to Newfoundland with disjunct popula-

tions in upper Chesapeake and Delaware bays. Intergrade zones occur where the two subspecies come together. In the Middle Atlantic Bight, these occur in northern New Jersey, on Long Island, and in upper Chesapeake and Delaware bays (Morin and Able, 1983; Marteinsdottir and Able, 1988; Gonzalez-Villasenor and Powers, 1990; Smith et al., 1992; Powers et al., 1993). This interpretation is supported by more recent genetic analysis (Smith et al., 1998; Adams et al., 2006). Early life history stages and adults of this species occur in every estuarine system in our study area (see Table 4.3).

REPRODUCTION AND DEVELOPMENT

Spawning occurs for many populations at age 1, in early spring through the summer, but the season varies in response to local temperatures. Reproduction occurs in Massachusetts from May to July (Wallace and Selman, 1981), in New Jersey from late April through August (Talbot and Able, 1984; Able, 1990; Able et al., 2007a), and in North Carolina from March to August (Kneib and Stiven, 1978). Some populations, primarily those identified as *F. h. heteroclitus*, exhibit pronounced lunar periodicity in their spawning. This has been demonstrated most clearly for a population in lower Delaware Bay (Taylor et al., 1977, 1979; DiMichele and Taylor, 1980; Taylor, 1999), where spawning occurs during the three to five days around lunar spring tides. The same lunar periodicity is evident in Great Bay–Little Egg Harbor (K. W. Able, pers. observ.), North Carolina (Kneib and Stiven, 1978), South Carolina (Radtke and Dean, 1982), and Georgia (Kneib, 1986). Populations of *F. h. macrolepidotus*, although not as well documented, may spawn daily during the reproductive season in Maine (Petersen et al., in review), Massachusetts (Taylor et al., 1982), and Long Island (Conover and Kynard, 1984). As further evidence of subspecific differences in spawning periodicity, in New Jersey a *F. h. heteroclitus* population from lower Delaware Bay had a pronounced peak in spawning on spring tides, while a *F. h. macrolepidotus* population from the lower Delaware River spawned almost daily (Marteinsdottir, 1991). On the other hand, presumed *F. h. macrolepidotus* in upper Chesapeake Bay had distinct semimonthly cycles, which were usually in phase with the lunar cycle (Hines et al., 1985).

A review of the literature on development during ontogeny is summarized in Able and Fahay (1998) and Fahay (2007). The characteristics of eggs differ between subspecies. In *F. h. heteroclitus*, eggs are 2.0 to 2.2 mm in diameter, have numerous small-diameter and short attachment filaments on the chorion, and have numerous oil globules. In *F. h. macrolepidotus*, eggs are 1.6 to 1.9 mm in diameter, have a few large-diameter and long attachment filaments on the chorion, and have few oil globules (Morin and Able, 1983). Hatching normally occurs in 9 to 18 days but is a function of temperature, dissolved oxygen level, and elevation in the intertidal spawning site, because hatching is cued to emersion after the eggs are exposed at low tide (DiMichele and Taylor, 1980). In addition, the intertidal site may cause hatching to be delayed for longer periods if emersion does not occur because of lower than normal tides (Taylor et al., 1977; DiMichele and Powers, 1982). Therefore, larvae from delayed hatching are often more advanced developmentally (fin ray formation, pigmentation; K. W. Able, pers. observ.). This variation may be especially relevant to cohort success (Kneib, 1986), as apparently occurs for other fishes (Porter and Bailey, 2007). Larvae are elongate, typically have a short, bulging gut, and have large eyes and a small, terminal mouth. Pigment is initially distributed on the dorsum of the head and body, but then becomes scattered over the entire body. The short-based dorsal and anal fins are situated on the posterior part of the body. Pelvic fin rays are late forming.

The sequence of scale formation was assessed from examination of 173 lab-reared and wild-caught individuals between 5.4 and 30.8 mm TL (see Fig. 42.1; Able et al., 2009a). Individuals from 5.4 to 9.7 mm TL have no scales. Scales originate on several loci, including the lateral surfaces of the caudal peduncle, the dorsal surface of the head, the operculum behind the eye, and the venter of the body at the level of the pectoral fin. Onset of scale formation occurs at approximately 9 mm TL on the caudal peduncle, and on the dorsal surface of the head at 10.2 mm TL. Later, the scales on the caudal peduncle expand in all directions until they cover the caudal peduncle, and then begin to spread out anteriorly along the midline of the body at about 11 mm TL. By approximately 12 mm TL, scales extend on to the dorso-lateral surface to above the pectoral fin. At this length the head scales have not changed appreciably in spatial coverage but scales begin to appear on the operculum. At slightly larger sizes (approximately 14 mm TL), scales are originating behind the eye, and those on the body have expanded to cover the dorsal surface of the body and parts of the head. At this length, they are also beginning to appear on the ventral surface anterior and posterior to the pectoral fin. By approximately 20 mm TL, the scale pattern has reached the adult condition, with scales covering 89.2% of the body. There are no scales on the snout, the ventro-lateral surface of the head, the posterior portion of the operculum, anterior to the pectoral fin, or the fins. See Figures 42.1 and 46.1 for comparisons of scale formation with other cyprinodontiforms.

LARVAL SUPPLY, GROWTH, AND MORTALITY

There are no long-term measures of larval supply except from our ichthyoplankton sampling from a thoroughfare behind Little Egg Inlet. The occurrence in this time series is sporadic (Fig. 45.1), but we suspect these larvae are the result of episodic transport off the marsh surface, where they typically occur (see below). Annual variation in the abundance of an adjacent population in Great Bay, New Jersey, as estimated by trap collections in a boat basin during a 15-year period, indicates considerable variation from multiple years of low abundance to those with relatively high abundance, when both are compared to average conditions.

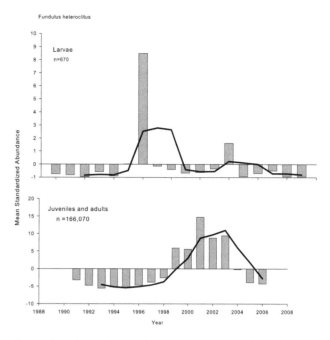

Fig. 45.1. Annual mean (bars) and three-year moving mean (line) of standardized abundance for *Fundulus heteroclitus* collected in larval sampling with plankton net (*upper*) and juvenile sampling with traps (*lower*). Data were standardized by subtracting overall mean from annual abundance. Three-year moving mean was calculated by taking the mean of the standardized values from the previous two years and the year in which data are plotted.

Growth of young-of-the-year is typically slow; in Great Bay, it averages 0.26 mm per day based on the slope of length regressions (Rountree, 1992). Young-of-the-year attain lengths of about 3 to 7 cm TL by late fall (Fig. 45.2; Able, 1990). Growth of young-of-the-year varies latitudinally, in a countergradient manner, with the fastest growth in the northern part of the range (Schultz et al., 1996). There is little evidence of growth during the winter, and they are 4 to 8 cm TL by the end of the first 12 months. These sizes are in approximate agreement with the values for age 1 fish in North Carolina (Kneib and Stiven, 1978). Young-of-the-year generate at least 70.8% (New Jersey salt marsh, Hagan et al., 2007) to 78% (Delaware salt marsh, Meredith and Lotrich, 1979) of the annual production for all year classes. Given the abundance of this species in many salt marsh systems, these parameters indicate the ecological significance of the population dynamics of this species to the trophodynamics of salt marsh ecosystems. During exceptionally cold winters, at least in New Jersey, individuals in marsh surface pools and ditches are susceptible to increased mortality (K. W. Able, pers. observ.).

SEASONALITY AND HABITAT USE

This species occurs in a diversity of shallow habitats, particularly in salt marshes (Clymer, 1978; Talbot and Able, 1984; Rountree and Able, 1992a; Yozzo and Smith, 1998; Halpin, 2000; MacKenzie and Dionne, 2008), but they can be collected in almost any shallow estuarine habitat—from eelgrass

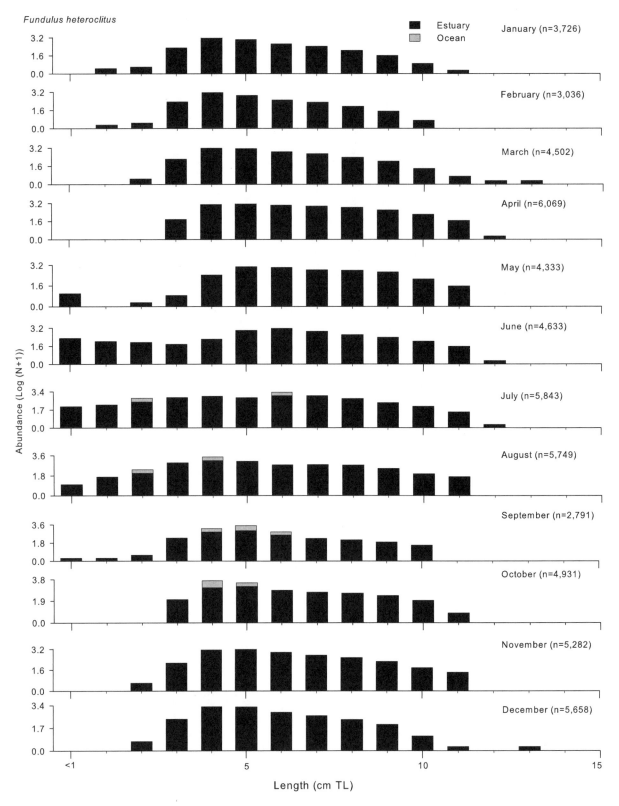

Fig. 45.2. Monthly length frequencies of *Fundulus heteroclitus* collected
in estuarine and coastal ocean waters of the central Middle Atlantic
Bight. Sources: Rutgers University Marine Field Station: 1 m beam trawl
(n = 204); seine (n = 1017); 1 m plankton net (n = 353); experimental trap
(n = 583); killitrap (n = 36,977); 4.9 m otter trawl (n = 155); multi-panel gill
net (n = 1); and Able and Fahay 1998 (n = 17,263).

beds (Sogard and Able, 1991) to open shorelines over substrates that range from sand to silt (Able et al., 1996b). Habitat use varies with life history stage. The adult habitat is slightly broader than that of the young-of-the-year. Movements of adults in salt marshes appear to be restricted during the summer. Intensive mark/recapture studies in a Delaware marsh found a home range of approximately 18 m (Lotrich, 1975), while movements up to 1 km occurred over the winter (Fritz et al., 1975). This species, at least in the laboratory, is selective about the habitat in which it buries. Burial in silt occurred for 33% (n = 12 tested) of individuals less than 10 mm TL; thus they are capable of burying before scale formation begins. Over all sizes tested (< 10 to > 25 mm TL), 41% buried in silt while none (n = 11 tested, 15.0 to > 25 mm) buried in sand. Burial only in silty substrates would be consistent with the types of marsh habitats in which this species typically is found. This behavior is different from that of *F. majalis*, which typically occurs over and buries in sandy substrates. Burying of *F. heteroclitus* to a depth of 15 to 20 cm in marsh pools has been reported during the winter (Chidester, 1916).

All populations appear to spawn intertidally and primarily in salt marshes (Able and Hata, 1984; Able and Hagan, 2003). The eggs are deposited intertidally and, as a result, are often exposed to the air. Subspecific differences may be responsible for the site of egg deposition as well (Able, 1984). Females, regardless of subspecies, deposit eggs in cracks, crevices, and small interstices, a behavior that is facilitated by the anal sheath (ovipositor) of the female (Able and Castagna, 1975). Populations of *F. h. heteroclitus* are known to deposit eggs between the empty valves of the ribbed mussel, *Geukeasia demissa* (Able, 1984), and the basal leaves of *Spartina alterniflora* (Taylor and DiMichele, 1983) or in stems of *Phragmites australis* (Able et al., 2003a). Populations of *F. h. macrolepidotus* deposit their eggs in the substrate, in mats of vegetation, or over mud or rocky substrate (Able, 1984; Petersen et al., in review). Spawning over mud substrates has been reported for Delaware populations, but it is not certain which subspecies demonstrated this behavior (Wang and Kernehan, 1979).

After hatching on flooding tides, the larvae are typically carried into the high marsh, where they are found in shallow pools, ponds, ditches, and depressions on the marsh surface (Kneib and Stiven, 1978; Wang and Kernehan, 1979; Smith, 1995). In New Jersey marshes, the larvae and juveniles were the most abundant forms found across a variety of high marsh habitat types (Talbot and Able, 1984). On rare occasions, the larvae have been collected in plankton tows in deeper estuarine waters (Croker, 1965; Himchak, 1982b; Witting, 1995; Witting et al., 1999). A recent analysis of ichthyoplankton samples in a deep (3 to 4 m) thoroughfare through the Sheepshead Meadows in southern New Jersey collected an average of 40 individuals per year, with a high of 234 in one year. This occurrence is annually quite variable (see Fig. 45.1). Transforming larvae and small juveniles have been captured from mid-May through September in high marsh

pools, ponds, and ditches, but were most plentiful in these habitats during June and July (Talbot and Able, 1984; Able, 1990). During the same period, other young-of-the-year are abundant along a variety of intertidal and subtidal creeks and shorelines (Able et al., 1996b; Kimball and Able, 2007b; Kimball et al., in press). Many of the larger juveniles apparently remain in marsh pools during the first summer (Smith, 1995). In an extensive study of fish habitat use in Delaware Bay, this species was largely restricted to intertidal and shallow subtidal habitats (Able et al., 2007a). They also occurred in other shallow river and bay habitats, but those in the river may be *F. h. macrolepidotus* while those in the bay were likely *F. h. heteroclitus*, based on earlier studies (Able and Felley, 1986).

The periodicity with which the larvae and juveniles occur is quite regular but differs with sampling gear in Great Bay–Little Egg Harbor. The larvae from ichthyoplankton sampling are most abundant during June and July (Fig. 45.3). The larger juveniles have a somewhat complementary seasonal distribution in an embayment when they are typically most abundant in the fall and spring. Some patterns of microhabitat use may vary between sexes, with males relying more on the marsh surface and females more on pools (Hunter et al., 2007; MacKenzie and Dionne, 2008).

This species is highly tolerant to low dissolved oxygen conditions in estuaries (Layman et al., 2000; Wannamaker and Rice, 2000; Smith and Able, 2003). There is little evidence of negative effects on growth and mortality, presumably because of its ability to use aquatic surface respiration (Smith and Able, 2003; Stierhoff et al., 2003).

PREY AND PREDATORS

All life history stages of *Fundulus heteroclitus* feed opportunistically on small crustaceans, annelids, and gastropods. Several studies (n = 6) have examined the food habits of young-of-the-year (6.6 to 41 mm SL, n = 828) in Middle Atlantic Bight estuaries from Hereford Inlet, New Jersey, to Newport River, North Carolina (see Table 8.1). Important prey of young-of-the-year include a variety of invertebrates (copepods, amphipods, polychaetes, tanaids, insects, and oligochaetes) and phytoplankton (algae and diatoms). Fishes (*Fundulus* spp.) and eggs are minor dietary components. The diet of young-of-the-year shifts with ontogeny at approximately 30 mm, from largely carnivorous at smallest sizes (amphipods, copepods, and tanaids) to more omnivorous (detritus, algae, crabs, insects, and polychaetes) at larger sizes (Schmelz, 1964; Kneib and Stiven, 1978; Morgan, 1990; Smith et al., 2000). However, in Hereford Inlet, New Jersey, young-of-the-year remain carnivorous at larger sizes (16 to 40 mm, n = 50) (Clymer, 1978). Relative importance of individual prey items also varies intra-annually in North Carolina salt marshes; algae, insects, crabs, and detritus are consumed more frequently in summer and fall, while small crustaceans and polychaetes are often eaten during the entire year (Kneib and Stiven, 1978). Food habits can also vary in

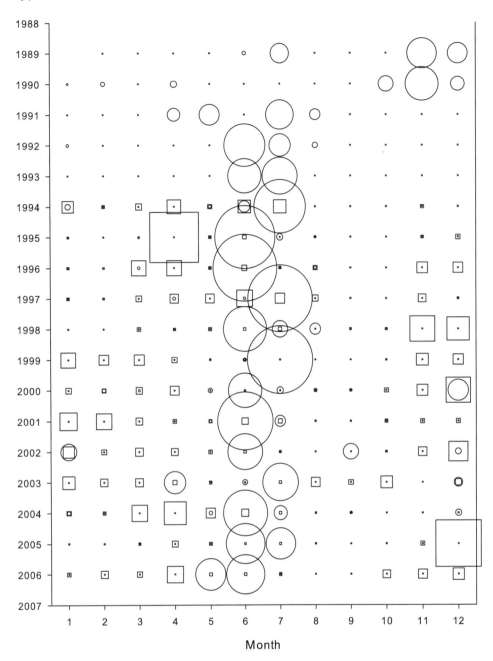

Fig. 45.3. Seasonal occurrence of larvae based on ichthyoplankton samples collected in Little Sheepshead Creek behind Little Egg Inlet from 1989 to 2006 (open circles represent the percentage of the mean number/1000 m³ of larval young-of-the-year captured by year) and juveniles of *Fundulus heteroclitus* based on trap collections in Rutgers University Marine Field Station boat basin in Great Bay (open squares represent the percentage of the mean catch per unit effort captured by year) from 1994 to 2006. Values range from 0 to 97% and 0 to 84% for ichthyoplankton and trap sampling, respectively. The smallest circles and squares represent when samples were taken but no individuals were collected.

response to novel food sources, such as when juveniles of an invasive crab become an important component of the diet (Brousseau et al., 2008). The trophic interactions for this species include cannibalism and scavenging fish prey (Able et al., 2007b). This occurs for fish from 12 to 106 mm TL with a frequency ranging from 0.2 to 9.1% for cannibalism and 0.5 to 9.9% for scavenging.

Fundulus heteroclitus is an important prey for many young-of-the-year estuarine piscivorous fishes (see Table 8.3). In our review of the literature for temperate estuarine fishes, *F. heteroclitus* was preyed upon by five different species and was an important prey (primarily for *Pomatomus saltatrix*) in five studies and present in the diet of four others. This species, and *Fundulus* spp., was also important in the diet of larger

predators, including *Esox niger*, in the Mullica River–Great Bay estuary (see Fig. 8.3).

MIGRATIONS

The movements of the young-of-the-year in the Mullica River–Great Bay estuary are restricted, with most tagged individuals recaptured at or near (0 to 5 m) the release site (Able et al., 2006b). However, a few were recaptured up to 299 m away up to 160 days later. Another study indicated that tagged individuals had high fidelity to individual marsh creeks over one year (Hagan et al., 2007). At least some portion of the population in Great Bay, including young-of-the-year, move into pools on the marsh surface during the fall and remain through the winter (Smith and Able, 1994).

Fundulus heteroclitus Attack Terrestrial Insects on High Tide

Terrestrial insects are often detected in the diets of salt marsh *Fundulus heteroclitus* (see Table 8.1; Haas et al., 2009). The high-tide behavior provides clues as to how this might occur based on observations in a southern New Jersey marsh.

The long, elevated (3 m above marsh surface) causeway leading to the Rutgers University Marine Field Station provides an effective vantage point for observing a portion of the Sheepshead Meadow marshes. As is characteristic of *Fundulus heteroclitus*, the adults move on high tide onto the marsh surface, where they typically feed on benthic invertebrates (Weisberg, 1986). However, on several occasions we observed fish jumping out of the water on higher than normal spring tides. This behavior occurred several times over different years even though there was no directed study. On at least one occasion, when we attempted to quantify its occurrence, we observed this jumping out of the water hundreds of times in a 15-minute period.

On closer observation, with binoculars, it became obvious that the fish jumping was directed at the taller, above-water portions of the stems of *Spartina alterniflora* (short form) that occurred 10s of meters from the creek edge. On occasion, after the fish attacked a stem, small insects were observed leaving the stem and eventually falling into the water. In each instance, the fish appeared to visually follow the flying insect until it hit the water and was subsequently ingested. On occasion, other *Fundulus heteroclitus* joined in the chase for the insect. At other times, it appeared that the fish bumped the base of the stem with their heads to knock insects off the stem and then capture them. While it was difficult to consistently observe the insects on the stems, other extensive studies in the same marsh system stem documented the abundance of large numbers of terrestrial insects living on *Spartina* leaves (Denno et al., 2003; Gratton and Denno, 2003). These observations, with further documentation, would not only explain the occurrence of these terrestrial insects in the diet of *Fundulus heteroclitus* but may also elucidate an important component of the marsh food web that has been previously underappreciated. This may be especially important given the energetic importance of insects in salt marshes (Bertness, 1999).

46

Fundulus luciae (Baird)

SPOTFIN KILLIFISH

Estuary	Reproduction	Larvae/ Development	YOY Growth	YOY Overwinter
Ocean				
Both				
Jan				
Feb				
Mar				
Apr				
May				
Jun				
Jul				
Aug				
Sep				
Oct				
Nov				
Dec				

DISTRIBUTION

Fundulus luciae is found along the East Coast of the United States from Massachusetts to Georgia (Lee et al., 1980; Hartel et al., 2002). It is probably much more common than is recognized because it is likely to be confused with *F. heteroclitus* (Byrne, 1978). When special efforts have been made to collect this species, it has proven to be common but never abundant (Able et al., 1983; Osgood et al., 2003; Yozzo and Ottman, 2003). Early life history stages are most commonly collected in the southern part of the Middle Atlantic Bight (see Table 4.3).

REPRODUCTION AND DEVELOPMENT

Maturation occurs at age 1 when males are 24 to 27 mm TL and females are 28 to 30 mm TL (Byrne, 1978). Spawning occurs from late May to August in New Jersey (Talbot and Able, 1984; Able, 1990), April to August in the Chesapeake Bay (Byrne, 1978), and April to October in North Carolina (Hildebrand and Schroeder, 1928).

The early development of *Fundulus luciae*, and characters useful for distinguishing the eggs, larvae, and juveniles from those of other fundulid species, have been summarized in Able and Fahay (1998) and Fahay (2007). Eggs are demersal, spherical, and 2.0 to 3.0 mm in diameter. There are about 50 small oil globules and many large attachment filaments on the chorion. Micro-filaments occur on the larger attachment filaments and on the chorion itself. Incubation time determines length at hatching. Those hatching in 12 to 16 days are 5.3 to 6.0 mm TL and have large yolk sacs and undeveloped fin rays. When hatching is delayed, larvae are 6.0 to 6.3 mm long and have small yolk sacs with partially formed anal fin rays (Byrne, 1978). Remnants of finfolds are evident at 11 to 12 mm TL, but all larger individuals have completely formed fin rays.

The sequence of scale formation is based on the examination of 29 lab-reared and wild-caught individuals between 6.8 and 34.0 mm TL (see Fig. 46.1; Able et al., 2009a). Individuals from 6.8 to 10.0 mm TL have no scales. Scales form at sizes of 10.1 to 11.5 to 18.3 mm TL. Scales originate at several loci, including along the lateral line anterior to the caudal peduncle, on the dorsal surface of the head, on the operculum, and behind the eye. The scales originating on the lateral line begin to expand anteriorly and posteriorly at approximately 11 mm TL in some specimens. By approximately 13 mm TL, scales form on the head. Eventually scales expand laterally and then anteriorly to above the pectoral fin by approximately 15 mm TL. At this same size, scales originate on the operculum, and the scales on the head have expanded. By approximately 18 mm TL, the scale pattern has reached the adult condition, with scales covering most of the body (94.7%) except the snout and just anterior to the base of the pectoral fin. None of the fins are scaled. See Figures 42.1 and 46.1 for comparisons of scale formation with other cyprinodontiforms.

LARVAL SUPPLY, SETTLEMENT, GROWTH, AND MORTALITY

The larvae originate from adhesive eggs and probably remain on the marsh surface during the entire first summer. They have never been collected in extensive ichthyoplankton sampling in a marsh thoroughfare in Great Bay–Little Egg Harbor, New Jersey, over 17 years of sampling even though the larvae of co-occurring *F. heteroclitus* are occasionally abundant. Settlement per se does not occur because the larvae are never pelagic (Byrne, 1978).

Based on modal length-frequency progression, young-of-the-year grow an average of 0.2 mm TL per day between June and October and reach 2 to 5 cm TL by the fall (Fig.

Fundulus luciae   Fundulus majalis   Lucania parva

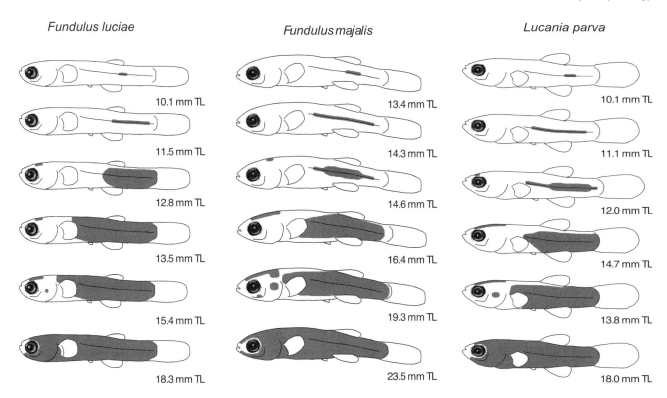

Fig. 46.1. Description of scale formation relative to total length (TL) in *Fundulus luciae, Fundulus majalis,* and *Lucania parva.*

46.2). There appears to be little growth over the winter, and they are the same size at the end of their first year.

Laboratory observations suggest that cannibalism may be a source of mortality for this species. When larvae are exposed to potential predation by conspecifics, in a manner similar to that presented in Able et al. (2007), larvae (6.5 to 11.7 mm TL) were preyed upon by predators (34.6 to 44.4 mm TL) 55% of the time in 20 trials (K. W. Able, unpubl. data).

SEASONALITY AND HABITAT USE
Larvae, juveniles, and adults share habitat. They occur in shallow bodies of water on the marsh surface, where they experience a broad range of temperatures and salinities. They have been collected in New Jersey estuaries at 0 to 46 ppt and 0.6 to 34.5°C (Able et al., 1983; Talbot and Able, 1984), and similar broad ranges have been reported from Virginia (Byrne, 1978). They have also been collected in intertidal creeks (Able et al., 1996b). They tolerate a variety of marsh alteration for mosquito control, which creates shallow pools on the marsh surface, but they do not occur in large expanses of shallow, standing water such as impoundments (Talbot et al., 1986). Young-of-the-year may be more abundant in pools without vegetation (*Ruppia maritima*) or where ammonia levels are high (Smith, 1995). Although little is known about winter habitat, it is assumed to be the marsh surface because there is no evidence of immigration to deeper habitats (Byrne, 1978).

This species can bury in silt substrates, at least in the laboratory. Individuals from 15.0 to 19.9 mm TL, did not bury in

silt (n = 2 tested), but larger individuals (> 20 mm TL) did so 61% of the time (n = 13 tested). This difference in a potential predation avoidance behavior may contribute to their preference for habitats with silty substrates.

PREY AND PREDATORS
Two studies have examined the food habits of young-of-the-year (11 to 47 mm, n = 370) *Fundulus luciae* in Middle Atlantic Bight estuaries at York River, Virginia, and Bogue Sound, North Carolina (see Table 8.1). Gut contents of young-of-the-year included detritus, phytoplankton (diatoms), and a variety of invertebrates (ostracods, insects, copepods, and tanaids). At Fox Creek Marsh, Virginia, fish eggs are a minor dietary component (Byrne, 1978). Relative importance of individual prey items varies intra-annually at this same site (Byrne, 1978) as well as in Tar Landing Bay, North Carolina, where summer and fall gut contents are dominated by organic matter (detritus, diatoms, and sand), while winter and spring diet is dominated by small crustaceans (copepods, tanaids, and amphipods) and insects (Kneib, 1978).

MIGRATIONS
We have no data to suggest that this species engages in seasonal migrations within or from the estuarine habitats where it occurs. In fact, tagged young-of-the-year (17 to 40 mm TL) had a high degree of fidelity in marsh habitats, with 99% of recaptured fish resident at the tagging site for up to 66 days (Able et al., 2006b).

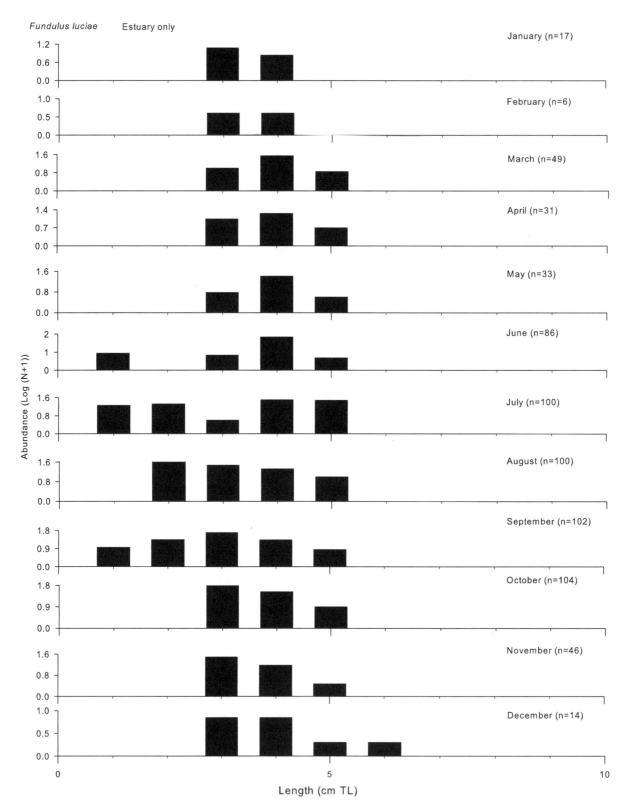

Fig. 46.2. Monthly length frequencies of *Fundulus luciae* collected in estuarine and coastal ocean waters of the central Middle Atlantic Bight. Source: Able and Fahay 1998 (n = 688).

47

Fundulus majalis (Walbaum)

STRIPED KILLIFISH

Estuary	Reproduction	Larvae/ Development	YOY Growth	YOY Overwinter
Ocean				
Both				
Jan				
Feb				
Mar				
Apr				
May				
Jun				
Jul				
Aug				
Sep				
Oct				
Nov				
Dec				

DISTRIBUTION

Fundulus majalis is found along the coast of the United States from New Hampshire to the Gulf of Mexico (Jackson, 1953; Relyea, 1983). In the study area, early life history stages and adults occur in most estuaries but are less frequently reported in the northern part of the bight (see Table 4.3).

REPRODUCTION AND DEVELOPMENT

Spawning, by adults as small as 75 and 63 mm for females and males, respectively, occurs from late May to July in New Jersey and from April to September in Chesapeake Bay (Hilde-

brand and Schroeder, 1928; Able and Fahay, 1998). The presence of multiple young-of-the-year cohorts in June, July, and August among recently hatched individuals at Corson's Inlet, New Jersey, suggests two or three spawning peaks, perhaps associated with spring tides (Able and Fahay, 1998).

The early development of *Fundulus majalis*, and characters useful for distinguishing the eggs, larvae, and juveniles from those of other fundulids, have been well described. Most aspects of ontogeny have been summarized in Able and Fahay (1998). Hatching occurs after an incubation period of 22 to 23 days (Newman, 1908, 1914). Eggs are demersal, spherical, and yellowish to amber in color. Diameter is between 2.0 and 3.0 mm and there are about 50 small oil globules. Large attachment filaments, each equipped with micro-filaments, occur on the chorion. Temperature and degree of crowding of the eggs influence the meristic characters of embryos and the subsequent larvae (Fahy, 1978, 1979, 1980, 1982). Larvae are elongate, with short pre-anus lengths and bulging guts. The mouth is small and terminal. Scattered pigment soon becomes aggregated into vague bars on the body. The larvae develop into juveniles with typical vertical bars (as in adult males) within a few weeks of hatching. They have six branchiostegal rays, whereas juveniles of the similar *F. heteroclitus* have five (notwithstanding the values given in Able and Fahay [1998], now corrected).

The sequence of scale formation is based on the examination of wild-caught individuals from 7.5 to 28.5 mm TL (n = 56) (see Fig 46.1; Able et al., 2009a). Scales begin to form at about 13.4 mm TL; specimens from 7.5 to 13.1 mm TL lack scales (Able et al., 2009a). Scales originate at several loci on the body, including on the developing lateral line anterior to the caudal peduncle, on the dorsal surface of the head, on the operculum, below the eye, and on the ventral surface at the level of the pectoral fin. Scales begin forming on the lateral line at the level of the posterior parts of the dorsal and anal fins at about 13 mm TL. By approximately 14 mm TL, scales have extended along the lateral line both anteriorly and posteriorly. At slightly larger sizes (14.6 mm TL), scales along the lateral line begin extending laterally, and scales appear on the head between the eye and the gill opening. By approximately 16 mm TL, scales on the body have extended posteriorly and anteriorly to the level of the pectoral fin. At this size, scales on the head have also extended anteriorly to the level of the front of the eye. By 19 mm TL, the scales on the body have continued to develop posteriorly to near the base of the caudal fin and anteriorly on the lateral surface to the gill opening. At this size, other scales originate on the operculum and below the eye. By 22 mm TL, scales are present on the ventral surface of the body. By approximately 23 mm TL, scales have reached the adult condition, where 92.5% of the body is covered with scales. The snout and fins are the only regions that lack scales. See Figures 42.1 and 46.1 for comparisons of scale formation with other cyprinodontiforms.

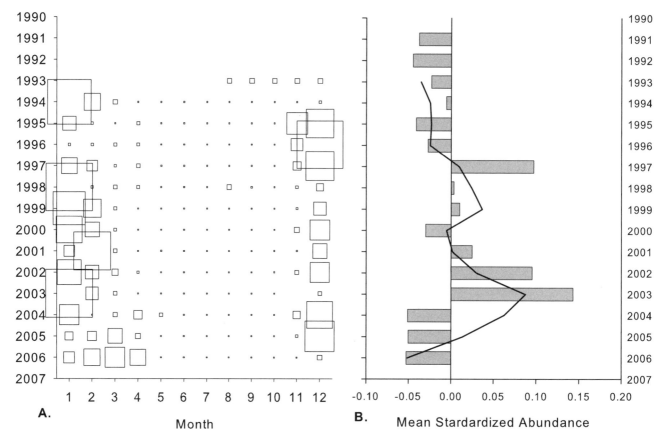

Fig. 47.1. Annual and seasonal variation in occurrence and abundance of *Fundulus majalis.* (A) Seasonal variation of the juveniles and adults is based on trap collections in Rutgers University Marine Field Station boat basin in Great Bay from 1994 to 2006. Open squares represent the percentage of the mean catch per unit effort captured by year. Values range from 0 to 98%. The smallest squares represent when samples were taken but no juveniles and adults were collected. (B) Annual variation in abundance for the same data from 1991 to 2006 was standardized by subtracting overall mean from annual abundance. Horizontal bars represent standardized annual abundance; the solid line is a three-year running mean that was calculated by taking the mean of the standardized values from the previous two years and the year in which the data are plotted.

LARVAL SUPPLY, SETTLEMENT, AND GROWTH

The larvae originate locally from adhesive eggs deposited in the substrate of sandy beaches. The larvae have been found in intertidal pools and along beach edges (Able and Fahay, 1998). The larvae have never been collected in extensive ichthyoplankton sampling in a deep thoroughfare through a marsh in Great Bay–Little Egg Harbor, New Jersey, over 17 years of sampling even though *Fundulus heteroclitus* is occasionally abundant. The annual variation in abundance, based on occurrence in traps in Great Bay, New Jersey, is variable, with several years of greater catches during 2001–2003 (Fig. 47.1). Settlement does not occur because the larvae are never pelagic.

Multiple cohorts of young-of-the-year are often present (Able and Fahay, 1998). Larvae are evident during late June, July, and August (Fig. 47.2). The modal size is 60 mm TL by the end of September in New Jersey. In the following May, the majority of individuals, presumably representing the same year class (from the previous year), are 50 to 100 mm TL. This suggests that very little or no growth occurs during the winter. By June, when these individuals are age 1, they

have grown to approximately 50 to 100 mm TL, although it is difficult to separate the larger individuals from those one year older. Scales have been used to age a Chesapeake Bay population (Clemmer and Schwartz, 1964), with the resulting interpretation that males reach sizes of approximately 65 mm and females 61 mm at the end of the first year.

SEASONALITY AND HABITAT USE

Fundulus majalis occurs in a variety of shallow habitats— from open beaches in the ocean to coves, bays, and occasionally marsh creeks in estuaries (Rountree and Able, 1992a) but typically over sandy substrates (Briggs and O'Connor, 1971; Harvey, 1998). Individuals of this species are found in salinities ranging from 1.0 to 37.0 ppt, but are usually most abundant in the more saline waters (Foster, 1967; Weinstein, 1979; Weisberg, 1986). They have not been captured at greater depths (> 8 m) in open waters (Szedlmayer and Able, 1996). Spawning occurs in shallow, intertidal pools over sand substrates. Some authors have reported that eggs are buried in substrates as much as 7 to 10 cm deep near the water's edge (Newman, 1909; Sumner et al., 1913). Habitat use varies with

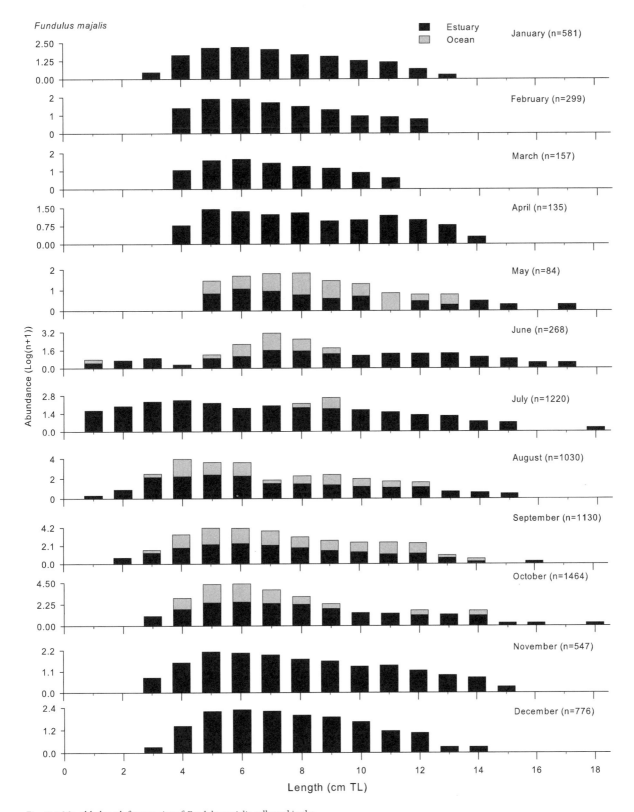

Fig. 47.2. Monthly length frequencies of *Fundulus majalis* collected in the
Great Bay–Little Egg Harbor and the coastal ocean study area. Sources:
Rutgers University Marine Field Station: 1 m beam trawl (n = 1); seine
(n = 2641); 1 m plankton net (n = 19); experimental trap (n = 78); killitrap
(n = 1419); and Able and Fahay 1998 (n = 3533).

life history stage and season during the first year. Larvae typically occur in shallow (< 3 cm at low tide), intertidal, sand bottom pools (Able and Fahay, 1998). Generally, young-of-the-year are found in a variety of shallow habitats along estuarine shores (Able et al., 1996b)—from open beaches to coves and bays. They can be abundant in marsh creeks (Werme, 1981; Rountree and Able, 1992a, 1993) and, from there, move onto the marsh surface (Werme, 1981), although they are never resident in the latter habitat (Weisberg, 1986; Smith, 1995). Temperature-salinity experiments for young-of-the-year (8 to 69 mm TL) indicate that survival time and salinity tolerance vary primarily with temperature, with the greatest survival at 20°C (Schmelz, 1970). Larvae are less resistant to temperature and salinity extremes (Schmelz, 1970). The seasonal change from shallow shoreline habitats to deeper water in the winter, as reflected in greater abundance in traps in a deep (to 3 m) embayment (see Fig. 47.1), probably reflects the movement into a more stable temperature regime.

To determine the movement patterns of adults (n = 80) and young-of-the-year (n = 129), individuals were tagged using coded wire tags at a salt marsh in Dennis Township, New Jersey (Chitty, 2000). Fish (40 to 119 mm TL) were tagged during May, June, and October 1998 in two small intertidal creeks approximately 53 m apart. The overall recapture rate was 14.4% (30/209). Adult fish were more frequently recaptured (34.9%, 29/83) than were young-of-the-year (< 1%, 1/126). Time at liberty for recaptured fish ranged from 1 to 153 days. The return rate of fish tagged in one creek was 10.3% (18/175), with the majority of fish (17/18) recaptured there. The return rate of fish tagged in a second creek was 32.4% (11/34); however, the majority of fish (10/11) were recaptured outside of this creek. In general, the tagged fish stayed within the same marsh but some individuals moved between creeks.

The ability to bury may develop with ontogeny. In the laboratory, individuals < 14.9 mm TL (n = 2) do not bury in sand or silt. At sizes between 15.0 and 19.9 mm TL (n = 19), 42% bury in sand. At this size, scales have formed on the lateral surface of the body and all of the lateral line is covered with scales. At sizes > 15.0 to 19.9 mm TL (n = 41), the percent burying in sand is 42% and those burying in silt is 55%. This burial behavior is commonly observed in the field, especially when attempting to collect individuals over sandy substrate. In the laboratory and the field, burial consists of head-first "swimming" into the substrate with exaggerated lateral movements of the posterior half of the body.

PREY AND PREDATORS

Prey items include a variety of mollusks, crustaceans, insects, and fishes. This species is generally more carnivorous than *F. heteroclitus* (Jeffries, 1972). *Fundulus majalis* is apparently capable of feeding on prey buried in sand more than *F. heteroclitus* (Werme, 1981). Two studies have examined the food habits of young-of-the-year (26 to 70 mm SL, n = 62) at Hereford Inlet, New Jersey, and the Delaware River estuary (see Table 8.1). Important prey of young-of-the-year included a variety of invertebrates (copepods, nematodes, and *Limulus* spp. eggs and larvae). Fishes are a minor dietary component in the Delaware River estuary (de Sylva et al., 1962).

MIGRATIONS

We lack sufficient data to infer whether this species engages in seasonal movements within the estuary or in adjacent coastal ocean habitats.

48

Lucania parva
(Baird and Girard)

RAINWATER KILLIFISH

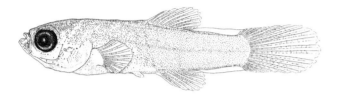

Estuary	Reproduction	Larvae/Development	YOY Growth	YOY Overwinter
Ocean				
Both				
Jan				
Feb				
Mar				
Apr				
May				
Jun				
Jul				
Aug				
Sep				
Oct				
Nov				
Dec				

DISTRIBUTION

Lucania parva occurs along the Atlantic coast of North America from Cape Cod to Mexico (Hubbs and Miller, 1965). It is a euryhaline species, found in shallow, vegetated habitats or along open shores in coves, bays, and creeks (Hubbs and Miller, 1965; Duggins et al., 1983). Early life history stages and adults are commonly found in shallow estuaries in the southern portion of the Middle Atlantic Bight but are less frequently encountered north and east of Long Island (see Table 4.3).

REPRODUCTION AND DEVELOPMENT

This species matures at about 25 mm TL (Hildebrand and Schroeder, 1928; Able, 1990), presumably at age 1. Spawning occurs during the summer, with the same female spawning more than one time (Hildebrand and Schroeder, 1928). On the basis of gonadosomatic indices (Able, 1990) and the occurrence of the smallest larvae, spawning lasts from June through August or September in New Jersey (Fig. 48.1; Talbot and Able, 1984), but it may occur from May through July in Delaware (Wang and Kernehan, 1979). Spawning has been reported to occur in freshwater in Delaware (Wang and Kernehan, 1979) and the Gulf of Mexico (Günter, 1945, 1950). It also takes place at higher salinities based on the occurrence of recently hatched larvae in polyhaline marshes in New Jersey (Able et al., 2005a).

The early development of *Lucania parva* is only moderately well described. Eggs are demersal and 1.0 to 1.3 mm in diameter, and have 8 to 12 large oil globules. Attachment filaments are present in a tuft on the chorion. Larvae are 4.0 to 5.0 mm when they hatch, and they have a large, early-forming caudal fin. Fin rays are completely formed by 15 to 20 mm TL. Other details of development are summarized in Fahay (2007). Hatching occurs 11 days after fertilization at 25°C in the laboratory (Wang and Kernehan, 1979) but has also been reported to occur in as little as 5 to 6 days at "summer" temperatures (Hildebrand and Schroeder, 1928).

The sequence of scale formation is based on the examination of wild-caught individuals from 6.5 to 30.5 mm TL (n = 65) (see Fig. 46.1; Able et al., 2009a). Individuals < 10 mm lacked scales entirely. Scales originate at several loci, including the posterior part of the lateral line at about 10 mm, on the dorsum of the head above the eye at about 12.0 mm, on the operculum at about 13.8 mm, below the eye, and on the ventral surfaces at about 18.0 mm. Scales that first appear on the lateral line spread anteriorly and posteriorly in larger individuals, then spread dorsally and ventrally on the body. By 12 mm TL, scales first appear on the head above the eye. By approximately 13 to 15 mm TL, the scales extend laterally, especially below the lateral line and anteriorly to the pectoral fin, and become more developed on the head. In some individuals scales originate on the operculum. Between 15 and 18 mm TL, scales appear on the venter of the body and behind the eye. By 18 mm TL, the scales cover 99.2% of the body, which is the adult condition. The snout and fins are the only regions that lack scales throughout development. See Figures 42.1 and 46.1 for comparisons of scale formation with that of other cyprinodontiforms.

LARVAL SUPPLY, SETTLEMENT, AND GROWTH

The source of larvae is likely local since the adhesive eggs are deposited in vegetation. One definite source of larvae is marsh pools and especially the algal mats around the perimeter of the pools, at least in New Jersey (Able et al., 2005b). Extensive ichthyoplankton collections in a deep marsh thor-

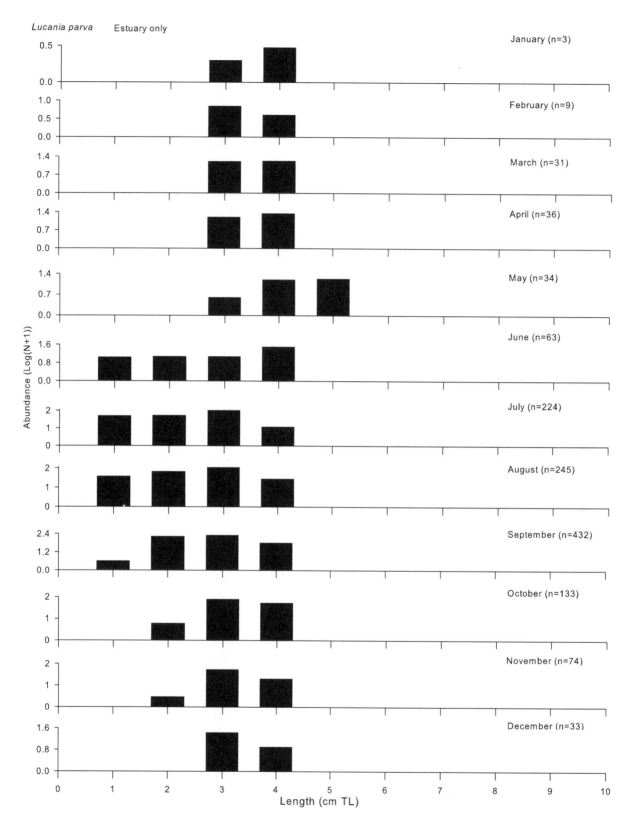

Lucania parva Estuary only

Fig. 48.1. Monthly length frequencies of *Lucania parva* collected in the
Great Bay–Little Egg Harbor study area and adjacent coastal ocean waters.
Sources: Rutgers University Marine Field Station: 1 m beam trawl (n = 6);
seine (n = 243); experimental trap (n = 1); killitrap (n = 5); 4.9 m otter
trawl (n = 3); and Able and Fahay 1998 (n = 1059).

oughfare over 17 years have never encountered them (K. W. Able, unpubl. data).

A comparison of the modal lengths between June and September suggests that larvae grow approximately 0.3 mm per day. The young-of-the-year are 1 to 4 cm TL by the end of the first summer (Able, 1990; see Fig. 48.1). There is little apparent growth during the winter, and the length after 12 months is 30 to 40 mm TL.

SEASONALITY AND HABITAT USE

This species spends its entire life on the marsh surface or in shallow-estuarine submerged aquatic vegetation (Able and Fahay, 1998). The larvae are found primarily in shallow (0.3 to 0.6 m) marsh surface pools with submerged vegetation *(Ruppia maritima)* (Talbot and Able, 1984; Smith, 1995). Larger juveniles are found in the same habitats (Able et al., 2005a). Through the winter, juveniles do not appear to use the intertidal marsh surface during most high tides but remain in pools or subtidal ditches (Talbot and Able, 1984; Smith, 1995).

When exposed to low dissolved oxygen, as occurs naturally in marsh pools, laboratory populations of young-of-the-year survive levels as low as 0.5 ppm by switching from midwater swimming to aquatic surface respiration (Smith and Able, 2003).

This species buries in silt and sand substrates in the laboratory (K. W. Able, unpubl. data). They did not bury in sand (n = 3 tested) but 33% (n = 6) buried in silt, which is a typical substrate for the habitats in which they commonly occur.

PREY AND PREDATORS

This species is known to feed on small crustaceans (Hildebrand and Schroeder, 1928).

MIGRATIONS

A mass downstream migration of juveniles and adults (> 270,000 individuals in a few hours) was observed during October in the York River, Virginia (Beck and Massmann, 1951). The cause of, or purpose for, this migration is unknown.

49

Gambusia holbrooki Girard

EASTERN MOSQUITOFISH

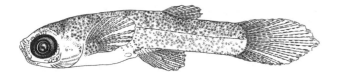

Estuary	Reproduction	Larvae/ Development	YOY Growth	YOY Overwinter
Ocean				
Both				
Jan				
Feb				
Mar				
Apr				
May				
Jun				
Jul				
Aug				
Sep				
Oct				
Nov				
Dec				

DISTRIBUTION

Gambusia holbrooki is reported from central Alabama east into Florida and along Atlantic coast drainages as far north as New Jersey (Rivas, 1963; Wang and Kernehan, 1979; Wooten et al., 1988). Early life history stages are found in the same habitats as the adults. The natural occurrence of this species in New Jersey has been questioned. The first records, from the turn of the century, coincide with the introduction of *Gambusia* spp. for mosquito control (Gooley and Lesser, 1977). Some of the introductions have been from other

populations in the central United States that are now recognized as *G. affinis* (Wooten et al., 1988). Thus, the taxonomy of *Gambusia* in the study area is confounded by potential introductions (Duryea et al., 1996). Available data suggest the presence of two separate species (i.e., *G. affinis* and *G. holbrooki*) along the Atlantic coast of the United States (Wooten et al., 1988).

REPRODUCTION AND DEVELOPMENT

Maturation is reached 4 to 6 weeks after birth (Krumholz, 1948) at sizes of approximately 20 mm. Fertilization and development of eggs are internal. Birth occurs during August and September in New Jersey (based on the occurrence of the smallest size classes) and May to September in Chesapeake Bay (Hildebrand and Schroeder, 1928).

The early development of *Gambusia holbrooki* has been described and summarized (Fahay, 2007), but methods for distinguishing the eggs and larvae from those of *G. affinis* are lacking. The young of these livebearers are born at an advanced stage of development (Kuntz, 1914) at sizes of 8 to 10 mm (Hildebrand and Schroeder, 1928). Embryos from about 6.6 to 8.0 mm TL and post-birth larvae from 9.1 to 29.7 mm TL (n = 27) were examined for sequence of scale formation (Fig. 49.1). Scales begin forming before birth, with individuals as small as 7.2 mm TL having scales along the lateral line and extending laterally from before the origin of the dorsal fin to near the base of the caudal fin. Some of these small individuals also have scales more widely distributed over most of the lateral surface of the body from behind the pectoral fin to the base of the caudal fin. After birth, at approximately 9.1 mm TL, the scales extend over much of the body behind the pectoral fin, including the dorsal and ventral surfaces of the body, and begin to form on the operculum. By 12 mm TL, scales extend onto the dorsal surface of the head and those on the body stretch anteriorly to in front of the pectoral fin, leaving a small scaleless area on the fin base. By 12.5 mm TL, scales cover more of the head and by 15.7 mm TL, only the snout and the fins remain unscaled. This represents the adult condition in which 96.0% of the body and 68.0% of the body and fins are covered with scales.

LARVAL SUPPLY, GROWTH, AND MORTALITY

The supply of young from these livebearers is local but dispersal can be high (> 800 m/day) (Alemadi and Jenkins, 2008). The young-of-the-year occur in some New Jersey locations from June through September (Talbot et al., 1980). During this time, modal length frequency progression indicates that they grow approximately 0.1 mm per day. They reach 20 to 40 mm TL by October (Fig. 49.2), but there is no evidence of growth during the winter. Many may succumb to low temperatures in the winter (Talbot et al., 1980, 1986). Thus, we expected that more southern populations would

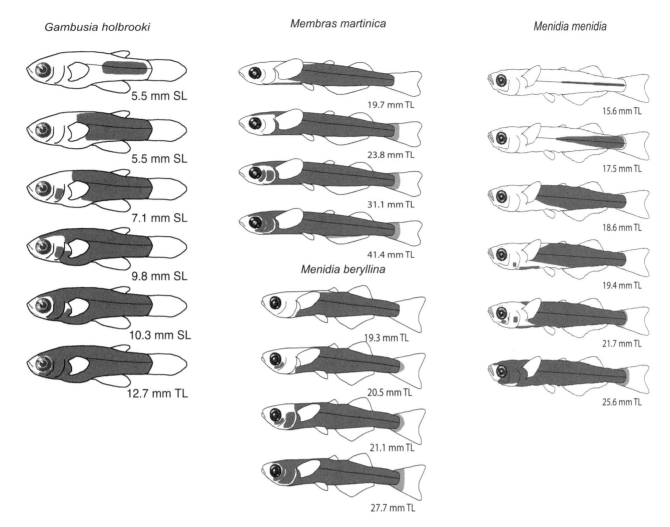

Fig. 49.1. Description of scale formation relative to length in *Gambusia holbrooki, Membras martinica, Menidia beryllina,* and *Menidia menidia.*

be even less tolerant. In an attempt to test this idea we exposed the young-of-the-year (17 to 35 mm TL) from a North Carolina population (reared from adults collected from The Straits near Harkers Island, North Carolina) to winter temperatures in an aquarium in a small freshwater pond in New Jersey during the winter of 2006–2007. At least some of these (approximately 50%) survived very low temperatures (< or = 3°C) on at least 20 nights over the same winter. In addition, there were even colder temperatures (0°C) for periods ranging from 6 to 13 days under ice cover. Another group of young-of-the-year was held in an aquarium at more moderate temperatures (5 to 20°C) and these had 100% survival. In fact, some females became sexually mature and appeared to be pregnant over the same time period.

SEASONALITY AND HABITAT USE

Young-of-the-year and adults typically occur in quiet, shallow, brackish water and freshwater marshes (Yozzo and Smith, 1998) and can be very abundant in man-made impoundments (Talbot et al., 1980), especially where there is aquatic vegetation (Wang and Kernehan, 1979). They are found in a wide range of temperatures and salinities.

PREY AND PREDATORS

There have been no studies of the food habits of young-of-the-year (< 40 mm TL) in Middle Atlantic Bight estuaries.

MIGRATIONS

Unknown.

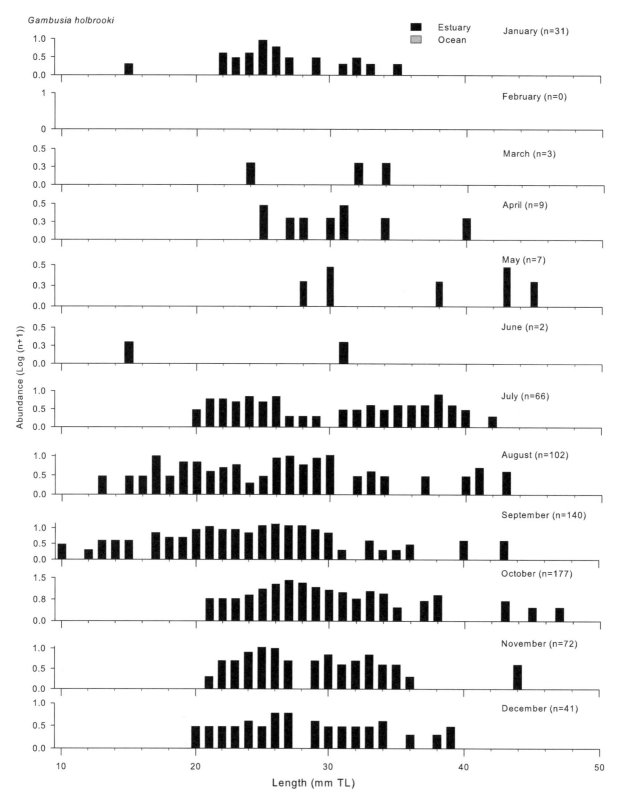

Fig. 49.2. Monthly length frequencies of *Gambusia holbrooki* collected from Manahawkin Bay between 1978 and 1979. Source: Able and Fahay 1998 (n = 650).

50

Membras martinica (Valenciennes)

ROUGH SILVERSIDE

Estuary	Reproduction	Larvae/ Development	YOY Growth	YOY Overwinter
Ocean				
Both				
Jan				
Feb				
Mar				
Apr				
May				
Jun				
Jul				
Aug				
Sep				
Oct				
Nov				
Dec				

DISTRIBUTION

Membras martinica occurs along the Atlantic coast of North America from New York to the Yucatan Peninsula, including the Gulf of Mexico (Fahay, 2007). It is more common in the southern part of its range. In the Middle Atlantic Bight, early life history stages have been reported from estuaries in New Jersey and south (see Table 4.3).

REPRODUCTION AND DEVELOPMENT

Little is known about reproduction in this species. Spawning occurs from late spring through most of the summer in polyhaline to oligohaline waters of Delaware Bay, in temperatures of 20 to 30°C (Wang and Kernehan, 1979). A single, large collection of eggs in Delaware Bay suggests that many individuals may spawn simultaneously (Wang and Kernehan, 1979), but this observation has not been made by subsequent authors. The early development, and characters useful for distinguishing the eggs, larvae, and juveniles from those of two co-occurring atherinopsid species (*M. beryllina* and *M. menidia*), have been described and summarized (Fahay, 2007). Eggs are demersal and spherical, and range from 0.7 to 0.8 mm in diameter. Eight to fifteen oil globules in early eggs coalesce to a smaller number in later egg stages. One to three attachment filaments originate from a single point on the chorion. Larvae hatch at 3.0 to 5.0 mm TL. The body is elongate with a very short pre-anus length. The snout becomes short and pointed, with a very small, terminal mouth. Juveniles have a few large, widely spaced melanophores on the dorsal midline of the body; the upper sides are usually clear of pigment.

Specimens over the size range of 19.7 to 99.1 mm TL (n = 16) were examined for scale formation (see Fig. 49.1). Scales originate in several loci, including behind the eye and on the operculum. Scales on the body may originate along the posterior part of the lateral line (as in *M. menidia*), but there are no small specimens that demonstrate this. The smallest specimen examined (19.7 mm TL) had scales covering most of the body behind the pectoral fin and on the ventral surface. At 23.8 mm TL, scales cover all of the body behind the operculum as well as the dorsal surface of the head and the base of the caudal fin. At 31.1 mm TL, scales occur behind the eye and on the operculum as well. At 41.4 mm TL, scales on the head have coalesced and cover most surfaces behind the midpoint of the eye; scales are also present on the base of the anal fin. At this size, the presumed adult condition is achieved, where scales cover 94.4% of the body and 66.6% of the body and fins. Lacinations, or rough edges, occur on the margins of scales in fish larger than 30 mm (Hildebrand and Schroeder, 1928). Small projections along the margins may be visible in smaller fish under microscopic magnification.

LARVAL SUPPLY AND GROWTH

The precise origin of larvae that we have collected is unknown. At Sandy Hook, New Jersey, the occurrence of young-of-the-year has been irregular, perhaps exhibiting a pattern that is typical for a species that is near the northern limit of its range. In regular collections in the fall of several years, they were abundant in one year (1979) but absent in most ensuing years (1980–1982). The presence of larvae in coastal oceanic waters may indicate that some reproduction takes place there, or that progeny are subject to being flushed out of estuarine systems during ebbing tides. When synoptic collections occurred in Sandy Hook Bay and the adjacent ocean during 1979, they appeared to represent both young-

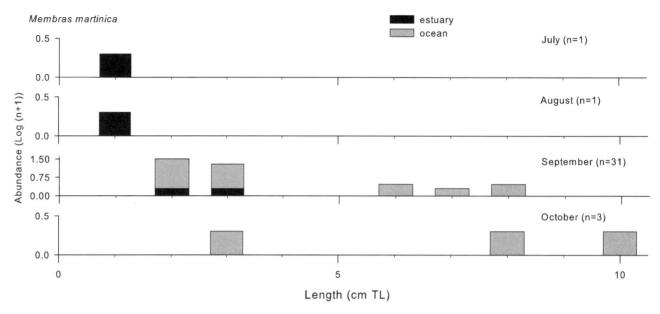

Fig. 50.1. Monthly length frequencies of *Membras martinica* collected in estuarine and coastal ocean waters of the central Middle Atlantic Bight. Sources: National Marine Fisheries Service: otter trawl (n = 4); Rutgers University Marine Field Station: Methot trawl (n = 28); 1 m plankton net (n = 4).

of-the-year (up to 30 mm TL) and a second year class at 60 to 80 mm TL during October (Able and Fahay, 1998). Lengths of specimens collected more recently in our study area are indicated in Figure 50.1. The smallest sizes have occurred during July and August, the largest in October. Maximum size in this species is about 110 mm (Hildebrand and Schroeder, 1928).

SEASONALITY AND HABITAT USE

Adults are usually found in open, shallow waters of bays, estuaries, and coastal ocean, along exposed shorelines and sandy beaches with little or no vegetation (Able and Fahay, 1998). Collections in New Jersey, in both the estuary and ocean, are over sand substrates. This species is apparently most abundant at higher salinities of > 25 ppt (Shuster, 1959). In Chesapeake Bay, it has been reported as mesohaline in its distribution and common over grass flats and channel edges (Musick, 1972). In Delaware Bay it is reported along beaches with hard sand or mud substrate in polyhaline to oligohaline shallow waters (Wang and Kernehan, 1979). Eggs are attached to submerged vegetation, algae, and other de-

bris in shallow waters. In Delaware Bay, larvae and juveniles are most abundant throughout the summer and early fall along beaches with either sand or mud substrates (Wang and Kernehan, 1979).

PREY AND PREDATORS

Adults feed mostly on copepods and other planktonic crustaceans (Murdy et al., 1997), but insect fragments and unidentified eggs are also included in the diet (Hildebrand and Schroeder, 1928). There have been no studies of the food habits of young-of-the-year *Membras martinica* in Middle Atlantic Bight estuaries. They are an important food item for larger predatory fishes, species not specified (Hildebrand and Schroeder, 1928).

MIGRATIONS

During the winter this species becomes less abundant in estuarine and coastal waters, indicating a movement into deeper waters offshore (Wang and Kernehan, 1979), but nothing else is known of its migrations.

51

Menidia beryllina (Cope)
INLAND SILVERSIDE

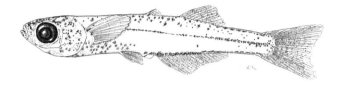

Estuary	Reproduction	Larvae/ Development	YOY Growth	YOY Overwinter
Ocean				
Both				
Jan				
Feb				
Mar				
Apr				
May				
Jun				
Jul				
Aug				
Sep				
Oct				
Nov				
Dec				

DISTRIBUTION

Menidia beryllina occurs along the Atlantic coast of North America from Cape Cod, Massachusetts, to Vera Cruz, Mexico, including the Gulf of Mexico (Gosline, 1948; Robbins, 1969; Johnson, 1974). Early life history stages are distributed in estuaries in the Middle Atlantic Bight as far north as Long Island, less commonly farther north (see Table 4.3).

REPRODUCTION AND DEVELOPMENT

Spawning occurs during spring and summer. Duration of the spawning season varies from north to south: it is rela-

tively short (June–July) in Rhode Island (Bengtson, 1984) and longer (May–August) in Delaware Bay (Wang and Kernehan, 1979). On the basis of a female gonadosomatic index, variation in ova diameters, and the occurrence of early life history stages, spawning takes place from May to August in a New Jersey salt marsh in Barnegat Bay (Coorey et al., 1985). Spawning occurs in adults as small as 42 mm TL. Most spawning is reported from oligohaline waters, which is consistent with reports from Delaware Bay (Wang and Kernehan, 1979) and Chesapeake Bay (Dovel, 1971).

The early development and characters useful for distinguishing the eggs, larvae, and juveniles from those of two co-occurring atherinopsid species have been described and summarized (Fahay, 2007). Eggs are demersal, spherical, and 0.8 to 0.9 mm in diameter. There are three to five oil globules, variously sized. One large and five smaller, attachment filaments originate from a single point on the chorion. Hatching occurs after 8 days at 17 to 25°C (Wang and Kernehan, 1979). Larvae are 3.5 to 4.0 mm TL at hatching. The body is elongate with a very short pre-anus length. The snout becomes short and pointed, with a very small, terminal mouth. Fin rays are completely formed by 20.0 mm. Juveniles have a scattering of small spots along the dorsal midline of the body, and these extend down onto the upper sides.

Specimens from field collections over the size range of 19.3 to 49.5 mm TL (n = 17) were examined to determine patterns of scale formation (see Fig. 49.1). Scales originate on several loci, including one on the head below the eye and one on the operculum. Scales on the body may originate along the posterior part of the lateral line (as in *Menidia menidia*), but there are no small specimens that demonstrate this. The smallest specimen examined (19.3 mm TL) has scales on the lateral surface of the body behind the pectoral fin but not on the dorsal surface of the body anterior to the dorsal fin. By 20.5 mm TL, scales are present on the dorsal surface of the body, on the anterior portion of the caudal fin, and as a discrete patch on the operculum. By 27.7 mm TL, scales on the side of the head cover a larger area, including the top of the head, and they also expand onto the caudal fin. At this size the presumed adult condition is achieved, where scales cover 93.0% of the body and 63.2% of the body and fins.

LARVAL SUPPLY AND GROWTH

All available evidence suggests that spawning occurs within estuarine waters; therefore, the larvae that we collect are all the products of local populations. Transformation occurs between 16.0 and 24.0 mm TL (Fahay, 2007). Young-of-the-year reach lengths of approximately 4 to 6 cm TL by the end of the first summer in New Jersey (Fig. 51.1) and Rhode Island and show little growth over the winter (Bengston, 1984; Coorey et al., 1985). Growth of juveniles in New Jersey marsh systems, based on modal length-frequency progression,

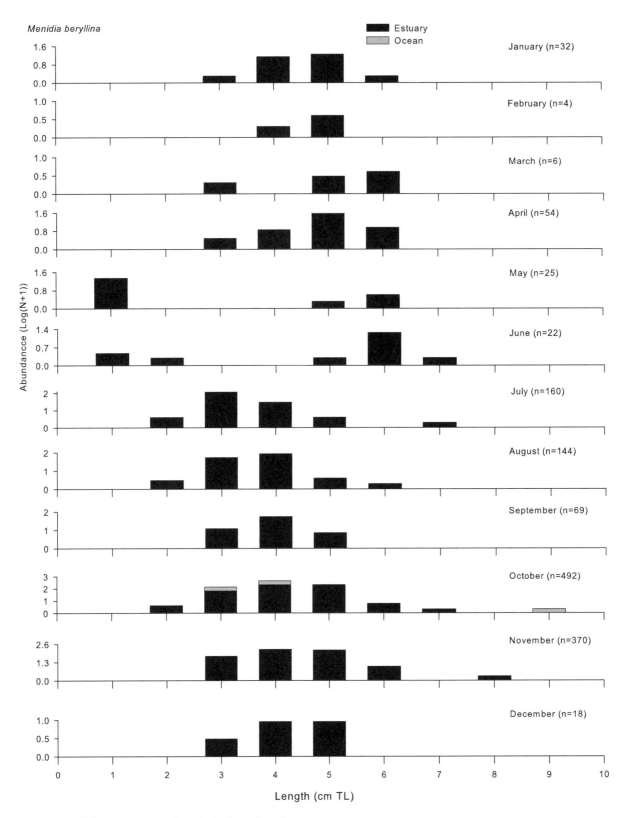

Fig. 51.1. Monthly length frequencies of *Menidia beryllina* collected in estuarine and coastal ocean waters of the central Middle Atlantic Bight. Sources:
Rutgers University Marine Field Station: 1 m beam trawl (n = 2); seine
(n = 222); Methot trawl (n = 1); 1 m plankton net (n = 23); 4.9 m otter trawl
(n = 12); and Able and Fahay 1998 (n = 1136).

averages 0.2 mm per day (Marcellus, 1972; Coorey et al., 1985). Mean instantaneous growth of larvae (7 to 14 days post-hatch) held in mesocosms in a Rhode Island estuary was 0.12 to 0.13 mm per day (Gleason and Bengston, 1996).

SEASONALITY AND HABITAT USE

Adults inhabit shallow estuarine and freshwater marshes, where they are commonly found in tidal channels, particularly in the vicinity of submerged vegetation. In New Jersey, they enter marshes during the spring, spend the summer, and then move out of the marshes into deeper water during the fall (Coorey et al., 1985). Adults also occur in shallow, coastal ocean waters (K. W. Able, unpubl. data).

Eggs are typically attached to vegetation when they are spawned (Wang and Kernehan, 1979; Coorey et al., 1985). Young-of-the-year (4.0 to 20.0 mm TL) are present in Barnegat Bay, New Jersey, from mid-May through early August but are most abundant during May and June (Coorey et al., 1985). Plankton sampling in Little Sheepshead Creek for all life history stages of this summer resident has revealed peaks in abundance during most months from April to December (Able and Fahay, 1998). Larvae are often absent during May, however, presumably because they inhabit marsh surface pools during that month (Coorey et al., 1985; Able et al., 2005b). Young-of-the-year are also abundant on inter- and subtidal beaches during fall and into the following spring (Able and Fahay, 1998). Recent intensive sampling on estuarine shores in the Great Bay–Little Egg Harbor estuary during summer found all life history stages across the gradient from marsh surface pools to intertidal creeks to shallow subtidal creeks (Able et al., 1996b) but not in the deeper portions (< 1 to 8 m) of the estuary (Szedlmayer and Able, 1996). Larvae are most abundant at low salinities (< 10 ppt) (Dovel, 1971; Wang and Kernehan, 1979; Coorey et al., 1985). Dissolved oxygen levels appear to influence habitat use and behavior of young-of-the-year. When exposed to low levels of dissolved oxygen that occur naturally in marsh pool habitats, young-of-the-year survive at levels as low as 0.5 ppm by using aquatic surface respiration (Smith and Able, 2003). This ability to tolerate low dissolved oxygen allows them to use marsh surface pools, where larvae and juveniles are often collected (Coorey et al., 1985; Able et al., 2005b).

PREY AND PREDATORS

Adults feed mostly on small crustaceans, small mollusks, insects, and worms (Hildebrand and Schroeder, 1928). Several studies (n = 13) have examined the food habits of young-of-the-year (15 to 67 mm TL, n = 522) in Middle Atlantic Bight estuaries from Great South Bay, New York, to Barnegat Bay, New Jersey (see Table 8.1). Important prey of young-of-the-year include a variety of invertebrates (copepods, rotifers, insects, and crabs), phytoplankton (algae), detritus, fishes, and fish eggs. These studies show variation in diet between estuaries, such that young-of-the-year in Great South Bay, New York, rely on decapod larvae, fish eggs, and larvae (Grover, 1982), while insects, rotifers, and detritus are important prey items of young-of-the-year in New Jersey salt marshes (Coorey, 1981; Coorey et al., 1985). Copepods are important prey items in both estuaries. Geographical variation in diet has also been noted between salt marshes within the same New Jersey estuary (Coorey, 1981). At oligo-polyhaline Cattus Island, copepods (46.7% composition) are most important to the diet of young-of-the-year (15 to 40 mm TL, n = 218), supplemented by rotifers, detritus, and algae (19.1, 20.8, and 12.4%, respectively), while insects (60.9%) are most important to young-of-the-year (15 to 40 mm TL, n = 6) at mesohaline Popular Point, supplemented by copepods and rotifers (16.8 and 5.8%, respectively). However, detritus and algae are not consumed. Relative importance of individual prey items also varies intra-annually in Great South Bay, New York, such that copepods decrease in percent composition of the diet from spring to summer, and then are the only prey consumed in fall, while decapod larvae and fish eggs and larvae become more important in summer (Grover, 1982). Similar seasonal trends were noted at Cattus Island and Cedar Run, New Jersey, where copepods dominate the diet during the entire year and algae and rotifers are utilized during summer and fall months (Coorey, 1981). These studies do not show an ontogenetic shift in diet during the first year. This species provides food for a variety of larger predatory fishes (Hildebrand and Schroeder, 1928).

MIGRATIONS

Seasonal variation in abundance in Barnegat Bay indicates that the young-of-the-year move out of salt marshes during late fall and winter, presumably into deeper waters of the estuary, and re-enter during the spring (Coorey et al., 1985). The same pattern is evident in Great Bay–Little Egg Harbor, where seasonal sampling along shallow polyhaline subtidal areas indicates that they are rare in the summer, when they are in marshes, but become increasingly abundant in the fall and can be found in low numbers through the winter and early spring (Able and Fahay, 1998). In Delaware Bay, young-of-the-year and adults reportedly migrate to higher-salinity waters in the lower bay during late fall when temperatures are declining (Wang and Kernehan, 1979).

52

Menidia menidia (Linnaeus)

ATLANTIC SILVERSIDE

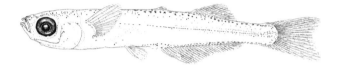

	Reproduction	Larvae/ Development	YOY Growth	YOY Overwinter
Estuary				
Ocean				
Both				
Jan				▓
Feb				▓
Mar				▓
Apr	░			
May	░	▓		
Jun	░	▓		
Jul	░	▓		
Aug			▓	
Sep			▓	
Oct				▓
Nov				▓
Dec				▓

DISTRIBUTION

Menidia menidia occurs along the Atlantic coast of North America from the Gulf of St. Lawrence to Florida (Chernoff, 2002). It is most abundant between Cape Cod and South Carolina. This is also one of the most abundant fishes throughout the Middle Atlantic Bight, and early life history stages have been collected in all estuaries examined (see Table 4.3).

REPRODUCTION AND DEVELOPMENT

Spawning occurs from spring through the summer, with slight geographic variations in timing and duration. Throughout the range of this species, spawning begins at approximately equivalent temperatures, but at later dates with increasing latitudes (Conover and Present, 1990). Spawning takes place from April to June in Massachusetts (Conover and Kynard, 1984) and from late April to early July in Rhode Island (Bengston, 1984). In New Jersey, the smallest size classes, as an indication of recent spawning, continue to be available from May to September. This species spawns during daytime high tides in intertidal areas, including marsh creeks (Middaugh and Takita, 1983). This is an annual species and most spawners are 1 year old. Fewer than 1% of breeding adults were 2 years old in one New England study (Conover and Ross, 1982).

The early development of *Menidia menidia*, and characters useful for distinguishing the eggs, larvae, and juveniles from those of two co-occurring atherinopsid species, have been described and summarized (Fahay, 2007). Eggs are demersal, spherical, and 1.0 to 1.5 mm in diameter. There are 5 to 12 large oil globules. Numerous long attachment filaments form a tuft on the chorion. Larvae are 3.8 to 5.0 mm TL at hatching. The body is elongate with a very short pre-anus length. The snout becomes short and pointed, with a very small, terminal mouth. Juveniles have a scattering of small spots on the upper sides of the body, and a silvery stripe extends along the side.

A total of 77 specimens between 12.7 and 40.9 mm TL were examined for patterns of scale formation (Fig. 49.1; Able et al., 2009a). Scales originate at several loci, including the posterior portion of the lateral line, the venter of the body under the pectoral fin origin, on the operculum, and under and behind the eye. Although specimens were treated with care during collection and preservation, some specimens had lost scales. As a result, illustrations represent a combination of actual scales and scale pockets from both sides of the same individuals. Specimens from 12.7 to 15.6 mm TL lack scales. Scale formation occurs over the size range of 15.6 to 25.6 mm TL. The onset of scale formation is initially observed at approximately 15 mm TL along the lateral line on the caudal peduncle. By 18 mm TL, scales expand anteriorly along the lateral line to the level of the origin of the first dorsal fin. Over the same size range the scales expand laterally to the dorsal surface of the body but do not cover the ventral surface anterior to the origin of the anal fin. By 19 to 20 mm TL, scales also originate on the ventral surface under the pectoral fin and on the operculum. Between 19 and 22 mm TL, scales nearly cover the body posterior to the operculum and are further advanced on the head region. By 21 mm TL, scales extend onto the base of the caudal fin and originate under and behind the eye. By 25 mm TL, scale formation is complete and scales cover the body and head from the caudal peduncle to the tip of the snout. At this size scales cover 95.8% of the body and 67.5% of the body and fins, approximating the adult condition.

LARVAL SUPPLY, GROWTH, AND MORTALITY

There is no evidence that this species spawns in oceanic waters; therefore, larvae in our immediate study site are products of local spawning, primarily in intertidal habitats. The pattern of annual variation, based on the occurrence and abundance of larvae captured in weekly, nighttime plankton samples in Great Bay–Little Egg Harbor or juveniles collected in a variety of gears, changes from year to year (Fig. 52.1).

Juveniles grow quickly during the summer. Estimates from otolith daily growth rings suggest rates of 0.84 mm per day in Rhode Island (Barkman et al., 1981), and these growth rates appear to be influenced by temperature (Bengtson and Barkman, 1981). In our study site, the overall length range of young-of-the-year is 1 to 9 cm by September (Fig. 52.2). There is little evidence of growth during the winter in New Jersey or Rhode Island (Bengston, 1984). In New Jersey, during May, at approximately age 1, these individuals are 5 to 15 cm. This general pattern of length distribution by month is consistent for other areas of the Middle Atlantic Bight, including the Hudson River (Perlmutter et al., 1967) and Delaware Bay (Schuler, 1974; Able et al., 2007a), and has been observed in earlier studies in Great Bay (Thomas et al., 1972). In Delaware Bay, collections made in a variety of habitats exhibit similar length modes, although adults (1+) are rarely collected there after May (K. W. Able, pers. observ.). By November, most young-of-the-year are between 27 and 99 mm FL. The size distributions we observe are also consistent with the model of countergradient growth relative to latitude (Conover and Present, 1990; Conover, 1992). This model predicts that fish at high latitudes (where spawning begins later and growing seasons are relatively short) grow at a faster rate than do those at lower latitudes (where growing seasons are longer), although both reach approximately the same maximum size after a season's growth. Fish in our study area exhibit intermediate growth rates between those from northern populations (Gulf of St. Lawrence) and southern populations (northern Florida).

Mortality during the winter is size-selective in high-latitude populations (Schultz et al., 1998). Severely low winter temperatures may result in mortality in all sizes. The shorter growing season at higher latitudes causes fish to accumulate lipids at a faster rate than those at lower latitudes, and this energy accumulation enables these fish to withstand and survive lower temperatures (Schultz et al., 1998). Habitat-specific mortality occurs in small young-of-the-year in New Jersey marsh pools during the summer (Smith and Able, 2003) and in Virginia (Layman et al., 2000). Other common sources of mortality include predation on young-of-the-year by *Pomatomus saltatrix* (Juanes et al., 1993), and at least in Rhode Island, this size-selective predation focuses on larger individuals (Gleason and Bengtson, 1996).

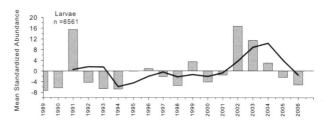
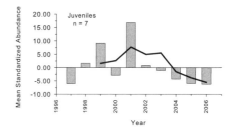

Fig. 52.1. Annual mean (bars) and three-year moving mean (line) of standardized abundance for *Menidia menidia* collected in larval sampling with plankton net (*upper*) and juvenile sampling with otter trawl (*lower*). Data were standardized by subtracting overall mean from annual abundance. Three-year moving mean was calculated by taking the mean of the standardized values from the previous two years and the year in which data are plotted.

SEASONALITY AND HABITAT USE

Adults are found in a variety of shallow estuarine and marine habitats. They are particularly common on open beaches over sandy and gravelly bottoms and in tidal creeks. Spawning adults move from the lower estuary into intertidal zones during daytime high tides, synchronized by new and full moon phases. Eggs are deposited 1.5 to 1.8 m above mean low water on the stems or roots of *Spartina* or on mats of detritus (Bengston, 1984). In Great Bay–Little Egg Harbor, New Jersey, small, recently hatched individuals occur from May through July (see Fig. 52.2), but in Delaware Bay (New Jersey side), they can occur as early as April (Able et al., 2007a). In Great Bay–Little Egg Harbor, larvae are dominant components of the ichthyoplankton and reach a peak in abundance in June (Fig. 52.3; Able and Fahay, 1998). Plankton sampling in Little Sheepshead Creek collects young-of-the-year during two phases of their life history (see Fig. 52.2). During most years, small larvae are well represented during June, shortly after the peak in spawning activity (Able and Fahay, 1998). During several years when the youngest larvae were not well sampled, the largest percentage of young-of-the-year were collected during November or December as they were migrating out of the estuary, or selecting different habitats, for the winter. Juveniles reach their peak in July, but they are abundant through the summer and fall and then decline until December. Thereafter, they are less abundant in estuarine waters, especially from January to March. This seasonal pattern is similar to that reported for Delaware Bay (Able et al., 2007a) and Massachusetts (Conover and Kynard, 1984).

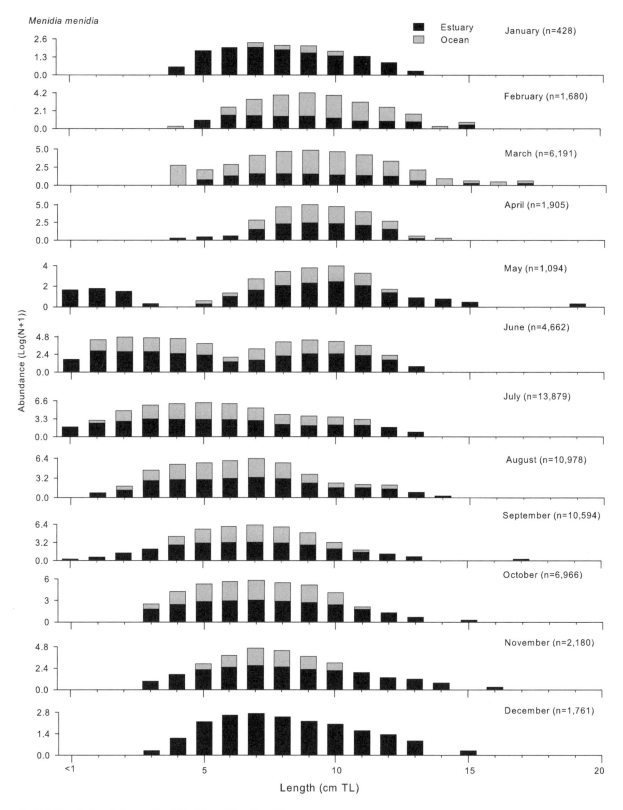

Fig. 52.2. Monthly length frequencies of *Menidia menidia* collected in estuarine and coastal ocean waters of the central Middle Atlantic Bight. Sources: National Marine Fisheries Service: otter trawl (n = 8382); Rutgers University Marine Field Station: 1 m beam trawl (n = 453); seine (n = 20,852); Methot trawl (n = 3); 1 m plankton net (n = 2549); experimental trap (n = 39); killitrap (n = 1517); 4.9 m otter trawl (n = 4359); and Able and Fahay 1998 (n = 24,164).

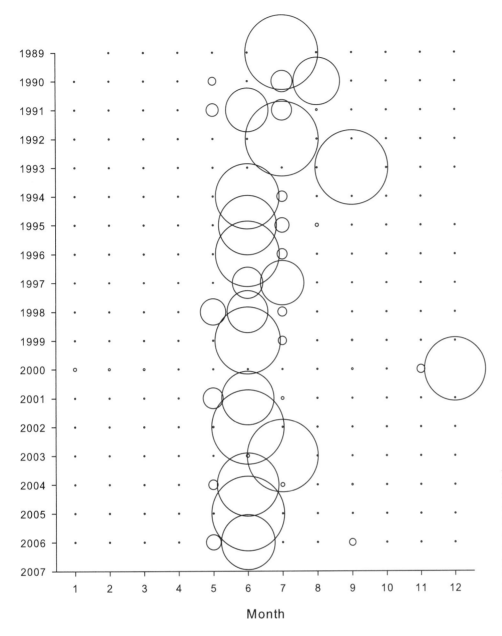

Fig. 52.3. Seasonal occurrence of the larvae of *Menidia menidia* based on ichthyoplankton samples collected in Little Sheepshead Creek behind Little Egg Inlet from 1989 to 2006. Full circles represent the percentage of the mean number/1000 m³ of larval young-of-the year *Menidia menidia* captured by given year. Values range from 0 to 87%. The smallest circles represent when samples were taken but no larvae were collected.

Young-of-the-year are abundant in a variety of intertidal and subtidal habitats on estuarine shores and in deeper waters and this is well documented for Great Bay (Able et al., 1996b; Szedlmayer and Able, 1996). An examination of collections, primarily from otter trawls, indicates that they are generally more numerous in the lower portions of the estuary (Fig. 52.4) at shallower depths (1 to 3 m), over sandier sediments (< 20% silt/clay), and where the structural portions of the habitat includes eelgrass, amphipod tubes, shell, and sea lettuce (Able and Fahay, 1998). Intensive sampling in the intertidal and subtidal portions of marsh creeks indicates ontogenetic patterns in habitat use that are largely influenced by diel differences between young-of-the-year cohorts (Rountree and Able, 1993). The smaller size cohorts make limited use of intertidal creeks during the day. However, the larger size cohorts are abundant in intertidal and subtidal creeks at night, presumably as the result of nocturnal movements into marsh creeks (Rountree and Able, 1997). Ontogenetic differences in habitat use have also been reported for Great South Bay, New York, where smaller individuals appear to prefer eelgrass (Briggs and O'Connor, 1971). In Delaware Bay, this species is abundant in the shallow waters of intertidal marshes and bay shoreline, but it is also found in subtidal marsh creeks over a wide salinity range (Able et al., 2007a).

When young-of-the-year are exposed to low dissolved oxygen, they first show signs of stress (demonstrated by aquatic surface respiration) at approximately 2.6 ppm. Thus they are much less tolerant than the co-occurring congener *M. beryllina* or several species of fundulids or cyprinodontids that also occur in marsh surface pools (Smith and Able, 2003). This relative intolerance may explain why larvae and juveniles are only present in marsh pools until July. Thereafter,

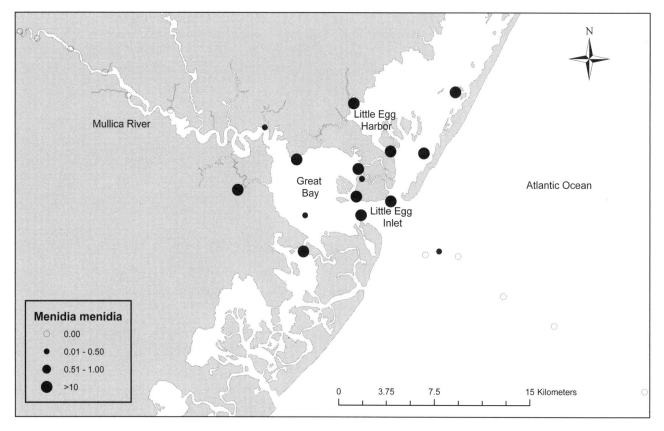

Fig. 52.4. Habitat use and average catch per tow (see key) for juvenile and adult *Menidia menidia* in the Ocean–Great Bay–Mullica River corridor in southern New Jersey based on small otter trawl collections during July and September from 1988 to 2006. See Table 3.2 for characteristics at each sampling collection.

the nocturnal periods of low dissolved oxygen are more consistent and of longer duration, and may cause emigration or mortality (Smith and Able, 2003).

PREY AND PREDATORS

This omnivorous species feeds on a wide variety of prey, including copepods, barnacle nauplii, horseshoe crab larvae, mysids, shrimp, small decapods, amphipods, Cladocera, fish eggs (including their own), young squid, annelid worms, molluskan larvae, insects, algae, and diatoms (various sources, summarized in Chernoff, 2002). They are diurnal feeders, although some feeding continues through the night (Spraker and Austin, 1997).

Several studies (n = 7) have examined the food habits of young-of-the-year (10 to 120 mm, n = 1439) in Middle Atlantic Bight estuaries from Lower Pettaquamscutt River, Rhode Island, to Newport River, North Carolina (see Table 8.1). Important prey of young-of-the-year include a variety of invertebrates (rotifers, polychaetes, insects, mysids, amphipods, decapod shrimp, crabs, copepods, cirripedes, and *Limulus* spp. eggs), fishes *(Menidia menidia* and *Fundulus* spp.) eggs, phytoplankton (diatoms), and detritus. Young-of-the-year may exhibit cannibalistic behavior (Cadigan and Fell,

1985). The diet of young-of-the-year shifts with ontogeny (between 30 and 80 mm), from a diet of small invertebrates (copepods, cirripedes, and insects), invertebrate eggs, and phytoplankton (diatoms) to a diet of larger invertebrates (decapod shrimp, mysids, amphipods, and isopods) and fishes with increasing size (Mulkana, 1966; Kimmel, 1973; Cadigan and Fell, 1985; Morgan, 1990). In the Pataguanset estuary in Connecticut, the diet of young-of-the-year (64 to 103 mm, n = 147) shifts at approximately 80 mm, such that small fish most frequently eat copepods, barnacle nauplii, and insects, while larger individuals prefer shrimp, fishes, and amphipods (Cadigan and Fell, 1985). Morgan (1990) found results similar to these in the Newport River estuary in North Carolina. In Magothy Bay, Virginia, small young-of-the-year (16 to 69 mm, n = 19) consume higher percentages of copepods, while larger fish (71 to 118 mm, n = 28) eat higher percentages of mysids (Kimmel, 1973). However, in the Lower Pettaquamscutt River in Rhode Island, the diet shifts at approximately 30 mm, where important food components for small young-of-the-year (10 to 30 mm, n = 69) include diatoms, rotifers, barnacle nauplii, and invertebrate eggs; important prey items of larger young-of-the-year (31 to 80 mm, n = 81) include amphipods, isopods, and insects, while copepods

Age 1+ *Menidia menidia*

YOY *Menidia menidia*

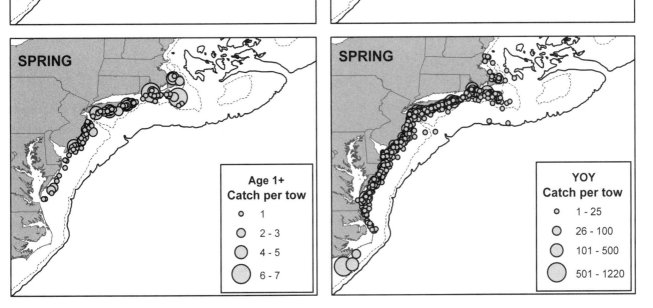

Fig. 52.5. Composite distribution of age 1+ (> 12 cm) and young-of-the-year (< 12 cm) *Menidia menidia* during seasonal cruises of the National Marine Fisheries Service groundfish survey. Details of sampling effort are indicated in Fig. 3.4.

of various stages were important prey at all sizes examined (Mulkana, 1966). Spatial variation in diet has also been noted in Pataguanset estuary in Connecticut, such that shrimp, copepods, and eggs are dominant prey in the lower estuary, while plant material, fishes, and copepods are most abundant in the upper estuary (Cadigan and Fell, 1985), and in Narragansett, Rhode Island, where young-of-the-year from lower Pettaquamscutt River consume a mixture of plant and animal food, while young-of-the-year from lower Point Judith Pond are largely dependent upon phytoplankton (Mulkana, 1966). Relative importance of individual prey items also varies intra-annually in Magothy Bay, Virginia, such that during winter (December–March) mysids and polychaetes are the principal prey, while amphipods and calanoid copepods are the main diet during spring and summer. Seasonal diet variation was also noted in the Pataguanset River estuary in Connecticut (Cadigan and Fell, 1985) and Great South Bay, New York (Grover, 1982).

There are diverse predators on this species that attest to its importance as a prey species (see Table 8.3). The best-documented and most important are *Pomatomus saltatrix*, *Cynoscion regalis,* and other *Menidia menidia*. Other frequent predators are *Mustelus canis*, *Anguilla rostrata*, *Opsanus tau*, and *Paralichthys dentatus*. Other known predators include *Morone saxatilis* and *Prionotus evolans*, as well as bottlenose dolphins, aquatic birds, and blue crabs (see Chernoff, 2002).

MIGRATIONS

When young-of-the-year reach 60 to 80 mm TL in early August, they move into deeper waters of the estuary (Rountree and Able, 1992a). Many young-of-the-year leave the estuary during the late summer and early fall to move into offshore ocean habitats for the winter (Fig. 52.5), and then return the following spring to spawn (Conover and Murawski, 1982; Able et al., 2007a). This movement out of the estuary may occur as late as December because several small peaks in collections have occurred at that time. Fish from more northern latitudes engage in a pronounced shift to oceanic habitats during the winter (Conover and Present, 1990), while those from more temperate latitudes (including our study area) may survive in deeper estuarine habitats. Overwintering fish are most commonly collected over the inner continental shelf from Cape Cod to the mouth of Delaware Bay (Fig. 52.5).

53

Apeltes quadracus (Mitchill)

FOURSPINE STICKLEBACK

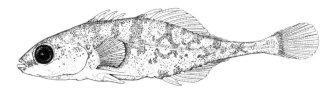

Estuary	Reproduction	Larvae/ Development	YOY Growth	YOY Overwinter
Ocean				
Both				
Jan				
Feb				
Mar				
Apr				
May				
Jun				
Jul				
Aug				
Sep				
Oct				
Nov				
Dec				

DISTRIBUTION

Apeltes quadracus is found in coastal waters of the western North Atlantic Ocean from Labrador and Newfoundland to Virginia (Scott and Crossman, 1973). Early life history stages and adults are common in most estuaries in the Middle Atlantic Bight (see Table 4.3).

REPRODUCTION AND DEVELOPMENT

Courtship and spawning occur in the spring, later in northern portions of their range. Nests are constructed of vegeta-

tion glued together by kidney secretions (Breder, 1936). The eggs and larvae are defended by the male (Reisman, 1963; Scott and Crossman, 1973). Based on the occurrence of the smallest individuals, most spawning occurs during April and May in Delaware Bay (Wang and Kernehan, 1979) and Chesapeake Bay (Dovel, 1971) and during April–July in Great Bay–Little Egg Harbor.

The early development of *Apeltes quadracus*, including characters useful for distinguishing eggs, larvae, and juveniles from those of other sticklebacks, has been well described and summarized (Fahay, 2007). Eggs (1.3 to 1.6 mm) are spherical and adhesive. They are deposited in nests built of vegetation fragments that are guarded by the males. The larvae hatch in 7 to 8 days at 21°C (Wang and Kernehan, 1979) at sizes of 4.2 to 4.5 mm TL and they continue to be protected by the male parent. All early stages are densely pigmented. The very slender caudal peduncle is apparent at sizes of < 10 mm TL. All fin rays (except pelvic) are complete by 16.0 mm TL. Juveniles display a mottled pigment pattern, unlike other juvenile sticklebacks.

LARVAL SUPPLY AND GROWTH

The larvae are restricted to the immediate nest area; thus they are generally absent in plankton collections. The young-of-the-year grow quickly after hatching and by 8 days old are 7 mm (Kuntz and Radcliffe, 1917). Young-of-the-year first appear in May at sizes of approximately 6 to 30 mm TL (Fig. 53.1). By October, young-of-the-year are approximately 27 to 60 mm TL. Estimates based on the increases in modal length frequency over this period indicate a growth rate of approximately 0.1 mm TL per day. There appears to be little growth during the winter so that by May of the following year this same year class is 33 to 63 mm TL. At this time, the bimodal length frequencies representing both the young-of-the-year and age 1 individuals are apparent. These estimates of size at age are greater than those from Chesapeake Bay (Schwartz, 1965a), where males, based on examination of annuli in vertebrae, were estimated to be 36 to 41 mm at age 1, while females were on average 41 mm. Some females may reach 2 to 3 years of age.

SEASONALITY AND HABITAT USE

In Great Bay–Little Egg Harbor, the young-of-the-year of this estuarine resident have been collected in every month (see Fig. 53.1), making them one of the most consistent components of the estuarine fauna. They are most abundant in late summer and fall with a peak in September based on collections with otter trawl (Able and Fahay, 1998). During the winter they are consistently collected in the deeper, quieter waters of an embayment (Fig. 53.2). In a comparison across a variety of habitat types at different depths across the ocean-estuary-tidal freshwater gradient (< 1 to 8 m), juveniles were sporadic in their spatial distribution. They were consistently most abundant in the shallow (< 1 m) *Zostera marina* habitat (Able et al., 1989; Szedlmayer and Able, 1996; Able and Fahay,

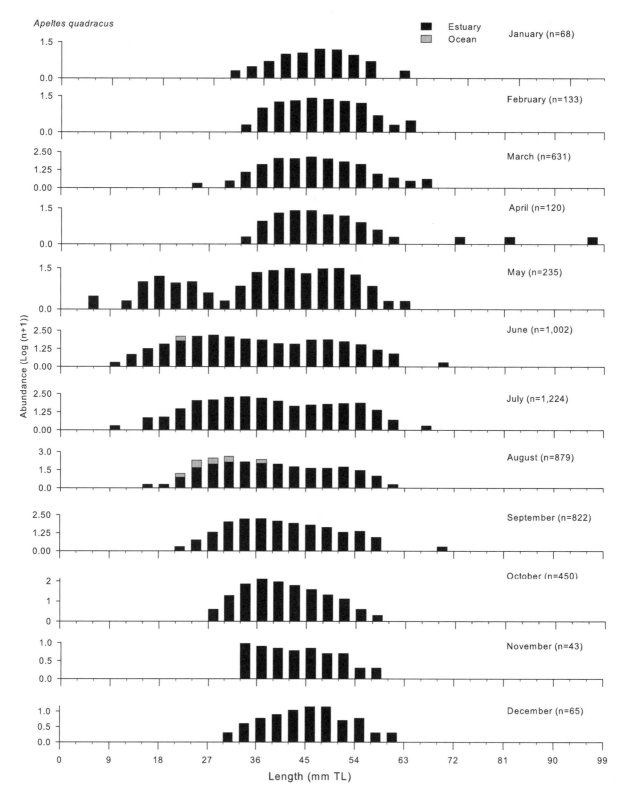

Fig. 53.1. Monthly length frequencies of *Apeltes quadracus* collected in
the Great Bay–Little Egg harbor and adjacent coastal ocean study area.
Sources: Rutgers University Marine Field Station: 1 m beam trawl (n = 981);
seine (n = 20); 1 m plankton net (n = 44); experimental trap (n = 18); killitrap (n = 847); 4.9 m otter trawl (n = 760); and Able and Fahay 1998
(n = 3002).

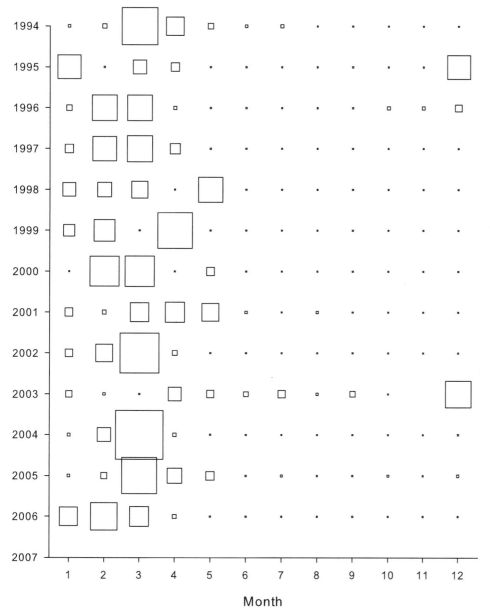

Fig. 53.2. Seasonal occurrence of the juveniles of *Apeltes quadracus* based on trap collections in Great Bay (Rutgers University Marine Field Station boat basin) from 1994 to 2006. Full squares represent the percentage of the mean catch per unit effort of *Apeltes quadracus* captured in a given year. Values range from 0 to 100%. The smallest squares represent when samples were taken but no individuals were collected.

1998). This same pattern has been observed in more comprehensive collections during July and September (Fig. 53.3). Other collections in this sampling occur near the freshwater-saltwater interface, as well as in some tributaries of the estuary that include quantities of macroalgae (primarily *Ulva lactuca*). This species also has been reported in macroalgae or marsh creeks or pools on the marsh surface and along subtidal shorelines in the same system (Sogard and Able, 1991; Rountree and Able, 1992a; Able et al., 1996b). Elsewhere, they have been collected in low-salinity and freshwater habitats (Scott and Scott, 1988).

PREY AND PREDATORS

Our examination of food habits of this species (19.0 to 43.0 mm TL, n = 18) found that the diet was dominated by mysids (33.7% by weight), followed by other crustaceans (23.0%) and meiofauna (20.8%), with smaller amounts of other zooplankton (4.8%).

MIGRATIONS

All life history stages may move into the deep water of estuaries during the winter; they have been found as deep as 31 m in Chesapeake Bay (Hildebrand and Schroeder, 1928).

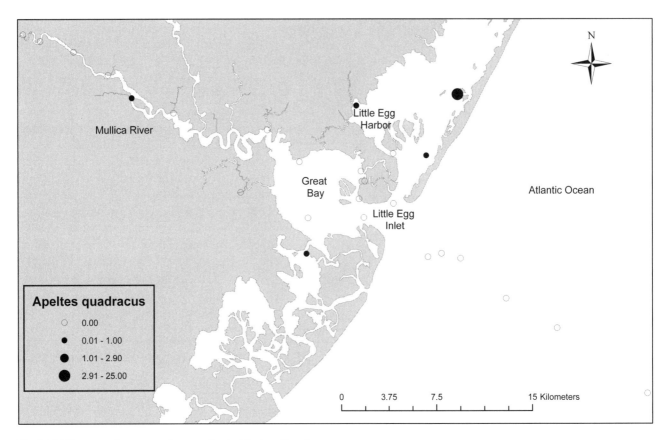

Fig. 53.3. Habitat use and average catch per tow (see key) for juvenile and adult *Apeltes quadracus* in the Ocean–Great Bay–Mullica River corridor in southern New Jersey based on small otter trawl collections during July and September from 1988 to 2006. See Table 3.2 for characteristics at each sampling collection.

54

Gasterosteus aculeatus Linnaeus

THREESPINE STICKLEBACK

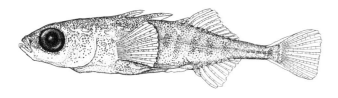

Estuary	Reproduction	Larvae/Development	YOY Growth	YOY Overwinter
Ocean				
Both				

Jan				
Feb				
Mar				
Apr				
May				
Jun				
Jul				
Aug				
Sep				
Oct				
Nov				
Dec				

DISTRIBUTION

Gasterosteus aculeatus occurs in nearly all coastal habitats in the northern hemisphere. It is found along the east coast of North America from Baffin Island to North Carolina (Musick, 1972; Scott and Scott, 1988). Early life history stages are recorded from estuaries throughout the Middle Atlantic Bight (see Table 4.3).

REPRODUCTION AND DEVELOPMENT

An anadromous pattern of spawning occurs in winter through early spring, when adults move from oceanic waters into estuaries or freshwater to spawn. Reproduction in our Great Bay–Little Egg Harbor, New Jersey study site occurs from March through May. Based on the examination of numerous females (56 to 71 mm TL, n = 748), the gonadosomatic index increases during the winter and spring (Jan = 8.9, Feb = 13.2, Mar = 20.6, Apr = 26.5, May = 34.7). Also, otolith increment counts from juveniles collected in the ocean indicate they are 29 to 56 days old and thus had a birth date in mid-April to mid-May (Cowen et al., 1991). This agrees with the estimated spawning and hatching period based on the gonadosomatic index from Great Bay–Little Egg Harbor and other observations on Long Island (Perlmutter, 1963; Monteleone, 1992). In other coastal areas of the western North Atlantic, spawning in the littoral zone has been reported for Virginia (Richards and Castagna, 1970), Long Island (Perlmutter, 1963), and the St. Lawrence estuary (Picard et al., 1990). Reproduction involves establishment of territories and nest-building by males. Fertilization of the eggs occurs after an elaborate courtship behavior. Fertilized eggs are held in the nest until after the larvae hatch, after which time the male guards the young. By May, most females have died or left the estuary.

The early development of *Gasterosteus aculeatus*, including comparisons with eggs, larvae, and juveniles of closely related species, has been well described and summarized (Fahay, 2007). Embryonic development occurs over 6 to 10 days while eggs are tended by the male in the nest (Wooton, 1976). Eggs are demersal and spherical, and adhere to each other. Their diameters range from 1.0 to 2.0 mm and they have numerous oil globules of varying sizes. Larvae are 3.0 to 7.0 mm (most from 4.2 to 4.5 mm TL) when they hatch and have fully functioning eyes and mouth parts. The head and body are densely pigmented. The demersal larvae are guarded by the parent until they become free swimming. All fin rays are completely formed by 15.0 mm TL.

Individuals from 17.4 to 58.4 mm TL (n = 19) were examined for scale formation (Fig. 54.1). Scutes, or modified scales, are observed over this entire size range. Scales begin forming at the time fin rays are completing their development. Scutes originate along the lateral line under the pectoral fin. In larger sizes, scutes spread posteriorly to include the entire length of the body, and then expand dorsally and ventrally until they reach the proportions exhibited by adults, covering about 35% of the body.

LARVAL SUPPLY AND GROWTH

The larvae originate from individuals that hatch from nests and this accounts for the occurrence of some larvae and small juveniles in marsh pools in Great Bay, New Jersey (Able et al., 2005b). Juveniles (< 27 mm) have been collected in the estuary (Fig. 54.2) as early as April on into July and August. By June through August, other individuals have been collected in the ocean. These captures are the results of movement out of the estuary, perhaps as temperatures warm there. Based

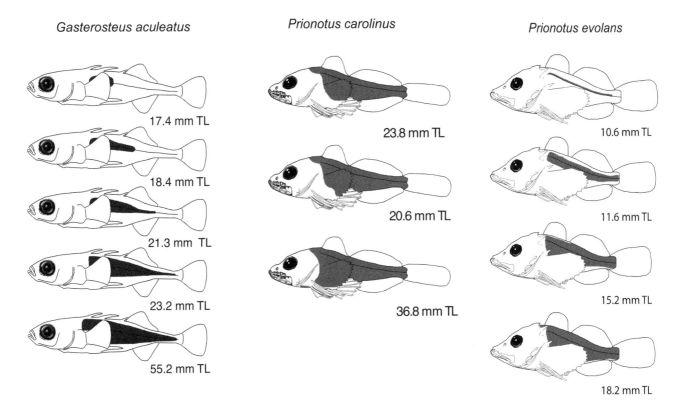

Fig. 54.1. Description of scale formation relative to total length (TL) in *Gasterosteus aculeatus, Prionotus carolinus,* and *Prionotus evolans.*

on the collection of year-of-the-young in ichthyoplankton net sampling, numbers of pelagic juveniles captured as they are leaving the estuary appear to fluctuate from year to year, and there has been a long-term trend for a decrease in abundance measured over the prior 17 years (Fig. 54.3).

Considering the rate of growth estimated from collections in our Great Bay study area, wherein young-of-the-year may measure 39 to 63 mm by the fall, some individuals may reach reproductive size by the next spring, at age 1, and these individuals are evident in the estuary in the spring at 39 to 72 mm (see Fig. 54.2).

SEASONALITY AND HABITAT USE

The seasonal occurrence of adults in the estuary is reflected in their capture in plankton nets from December through March, in pop nets during the same period (Hagan and Able, 2003), and in otter trawls in March and April (Fig. 54.4). We suspect that nest-building, courtship, and spawning occur in shallow marsh pools in the polyhaline portion of Great Bay–Little Egg Harbor estuary because adult females with high gonadosomatic indices are abundant in these pools during the spring and small young-of-the-year are collected later from these pools. The young-of-the-year peaked in abundance in May in plankton net collections, but continued to be collected well into June.

While in the estuary, the various life history stages use a variety of habitats, including pools (Able et al., 2005b), shallow depressions on the marsh surface, and intertidal and subtidal creeks (Talbot and Able, 1984; Rountree and Able, 1992a). During 1992, when larval fish traps were deployed in marsh pools from May 6 to July 29, large numbers of small juveniles were collected in May (n = 106, mean 24.8 mm TL, range 17 to 29 mm TL) and June (n = 28, mean 22.4 mm TL, range 19 to 27 mm TL). On a finer temporal scale, this species did not occur in water column samples during the day but were frequently and abundantly collected at night (Hagan and Able, 2008). The adults migrating into the estuary and juveniles migrating out of the estuary have also been collected in larger, deeper thoroughfares through the marsh, such as Little Sheepshead Creek (Able and Fahay, 1998; Witting et al., 1999).

PREY AND PREDATORS

Adults consume a wide variety of invertebrate prey, including copepods, gammarids, oligochaetes, hemipterans, and chironomids (Delbeek and Williams, 1987). They also prey on the eggs of other sticklebacks. Young-of-the-year reportedly eat copepods, ostracods, branchiurans, rotifers, and dipterans (Poulin and FitzGerald, 1989). Our own examination of food habits found that the diet of young-of-the-year (17.0 to 25.0 mm TL, n = 14) was dominated by zooplankton (39.3% by weight) and fishes (20.4%), with the balance made up of unidentified material, detritus, phytoplankton, and unidentified crustaceans.

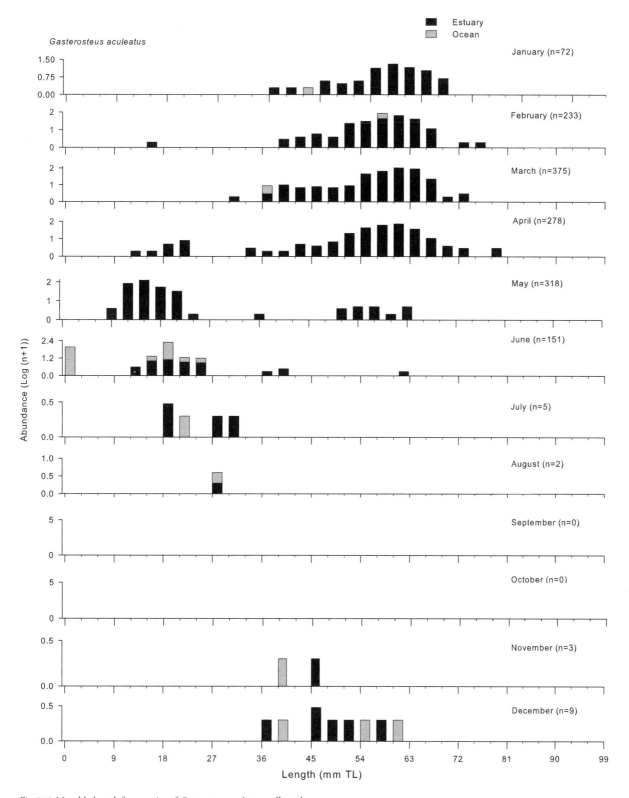

Fig. 54.2. Monthly length frequencies of *Gasterosteus aculeatus* collected
in the Great Bay–Little Egg Harbor and adjacent coastal ocean waters.
Sources: Rutgers University Marine Field Station: 1 m plankton net
(n = 181); experimental trap (n = 4); killitrap (n = 327); 4.9 m otter trawl
(n = 2); and Able and Fahay 1998 (n = 842).

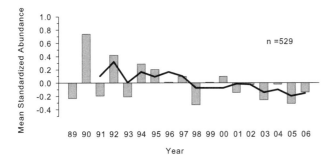

Fig. 54.3. Annual variation in abundance of *Gasterosteus aculeatus* in Little
Sheepshead Creek Bridge 1 m plankton net collections, 1989–2006. Data
were standardized by subtracting the overall mean from annual abundance.
Vertical bars represent standardized annual abundance; the solid line is a
three-year running mean that was calculated by taking the mean of the
standardized values from the previous two years and the year in which the
data are plotted.

MIGRATIONS

After leaving estuaries in the summer following spawning, the surviving adults and eventually the young-of-the-
year occur in near-surface layers of the ocean (Cowen et al.,
1991, 1998). Nevertheless, a seasonal pattern is evident for
both young-of-the-year and adults (Fig. 54.5). In the fall,

the young-of-the-year (< 6 cm) are collected primarily in
the Gulf of Maine and on Georges Bank over a wide range
of depths (mean depth 37.0 m, range 23 to 57 m). In the
spring, they are collected almost exclusively on the inner
continental shelf from Nantucket Shoals to northern Virginia, at shallow depths (mean depth 15.4 m, range 9 to 26
m). These individuals may have been captured as they left
estuarine spawning areas. Their absence during the winter
may be due to the lack of sampling in the Gulf of Maine at
that time.

The distribution of age 1+ (> 6 cm) is similar to that of
the young-of-the-year in fall and spring (Fig. 54.5). In the fall,
all collections occurred in the Gulf of Maine, with a few
off Massachusetts over a wide range of depths (mean depth
43.5 m, range 30 to 57 m). In the spring, almost all collections
occurred on the inner continental shelf between the Gulf of
Maine and Virginia, in shallow waters (mean depth 15.6 m,
range 10 to 37 m). One interpretation of these seasonal patterns for young-of-the-year and age 1+ individuals is that they
are distributed in the fall and winter in the Gulf of Maine
but by the spring some have migrated onto the inner shelf in
the Middle Atlantic Bight prior to spawning in estuarine and
freshwaters (Able and Fahay, 1998).

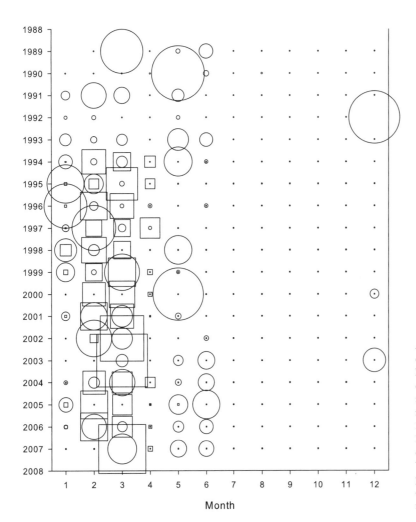

Fig. 54.4. Seasonal occurrence of pelagic juveniles based
on ichthyoplankton samples collected in Little Sheepshead Creek behind Little Egg Inlet from 1989 to 2006 (open
circles represent the percentage of the mean number/1000
m³ of larval young-of-the-year captured by year) and juveniles of *Gasterosteus aculeatus* based on trap collections in
Rutgers University Marine Field Station boat basin in Great
Bay (open squares represent the percentage of the mean
catch per unit effort captured by year) from 1994 to 2006.
Values range from 0 to 100% and 0 to 92% for ichthyoplankton and trap sampling, respectively. The smallest
circles and squares represent when samples were taken but
no individuals were collected.

Age 1+ *Gasterosteus aculeatus*

YOY *Gasterosteus aculeatus*

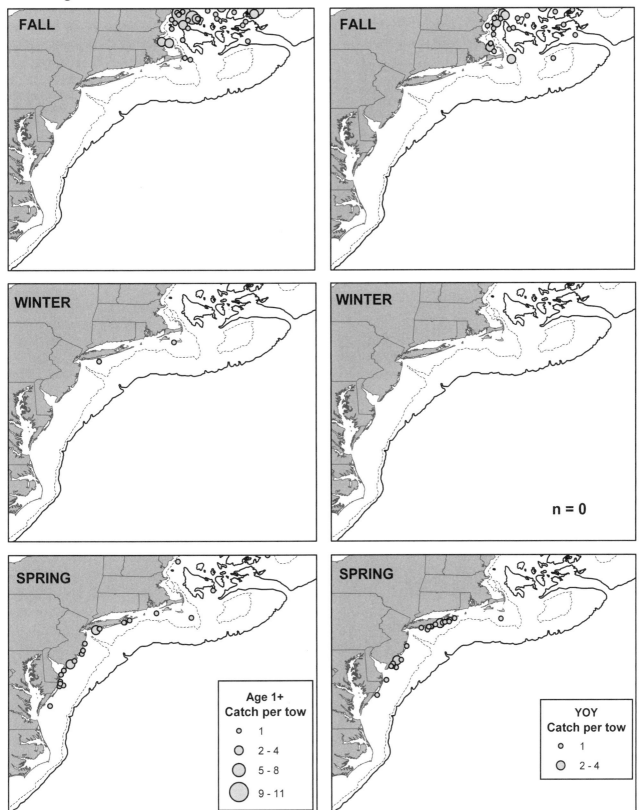

Fig. 54.5. Composite distribution of age 1+ (> 6 cm) and young-of-the-year (< 6 cm) *Gasterosteus aculeatus* during seasonal cruises of the National Marine Fisheries Service groundfish survey. Details of sampling effort are indicated in Fig. 3.4.

55

Hippocampus erectus Perry

LINED SEAHORSE

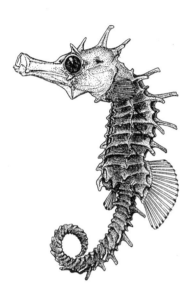

Estuary	Reproduction	Larvae/ Development	YOY Growth	YOY Overwinter
Ocean				
Both				
Jan				
Feb				
Mar				
Apr				
May				
Jun				
Jul				
Aug				
Sep				
Oct				
Nov				
Dec				

DISTRIBUTION

Hippocampus erectus occurs in estuaries and on the inner continental shelf from Nova Scotia to Uruguay (Vari, 1982). The early life history stages and adults are sporadically reported in Middle Atlantic Bight estuaries (see Table 4.3).

REPRODUCTION AND DEVELOPMENT

This species, like many other syngnathids, has reversed sex roles (Vincent et al., 1992), with the male bearing the fertilized embryos and larvae in a "placenta-like" brood pouch (Linton and Soloff, 1964). Mating and birth occur over a protracted period in the Middle Atlantic Bight. In Chesapeake Bay, the gonadosomatic index for adult females (> 70 mm TL) is highest from May through August or October (Teixeira, 1995). In southern New Jersey, gonadosomatic index values for females (n = 37) begin to increase in May, reach a peak in July, and stay relatively high through September (Able and Fahay, 1998). The males examined (63 to 127 mm TL, n = 39) have as many as 1515 embryos in the pouch; brood size increases with the size of the male. Aspects of the ontogeny have been previously summarized (Fahay, 2007).

LARVAL SUPPLY AND GROWTH

The recently born larvae are found in both the estuary and the ocean in most months (Fig. 55.1). Thus, it is unclear if they are derived from both locations or if those in the ocean occur as a result of outwelling of estuarine larvae. Alternatively, they reside in both areas, at least as young-of-the-year. The pattern of growth based on length-frequency distribution is difficult to estimate. Small, recently born individuals (< 1 to 2 cm TL) were present in the vicinity of Great Bay–Little Egg Harbor in the estuary and the ocean from June through October. These may reach 6 to 7 cm TL in July. By October, most are 1 to 10 cm TL. On the basis of the sizes collected during March and April, they do not appear to grow over the winter. By approximately 12 months, they are 7 to 10 cm TL and clearly differentiated from the recently born young-of-the-year.

SEASONALITY AND HABITAT USE

In estuaries, this species is typically found in eelgrass or other types of vegetation (Sogard and Able, 1991), oyster beds, salt marshes, and deep channels. It occasionally occurs at the surface, even over deep water. Either the pattern of habitat use is variable or we simply lack sufficient data to interpret it for our study area. In early summer (June and July), most young-of-the-year are in the Great Bay–Little Egg Harbor estuary. During August and September, when collections are dominated by recently born individuals, many are collected from the inner continental shelf. The abundance of recently born individuals in the ocean during August and September is inconsistent with seasonal inshore-offshore migrations, but perhaps occurrences in the ocean at this time are the result of estuarine outwelling

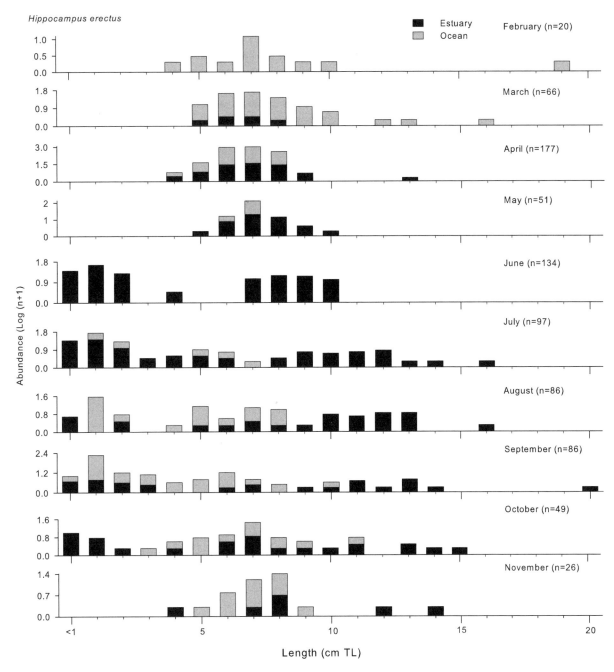

Fig. 55.1. Monthly length frequencies of *Hippocampus erectus* collected in estuarine and coastal ocean waters of the central Middle Atlantic Bight. Sources: National Marine Fisheries Service: otter trawl (n = 78); Rutgers University Marine Field Station: 1 m beam trawl (n = 3); seine (n = 9); Methot trawl (n = 19); 1 m plankton net (n = 135); 2 m beam trawl (n = 2); killitrap (n = 1); 4.9 m otter trawl (n = 19); and Able and Fahay 1998 (n = 526).

of these small individuals. From November through March, most are collected from the ocean. The largest individuals collected in the summer (> 10 cm) are typically from the estuary (see Fig. 55.1). In deeper, more southern estuaries, such as Chesapeake Bay, they may overwinter in channels (Musick, 1972).

PREY AND PREDATORS

One study examined the food habits of young-of-the-year (23 to 126 mm TL, n = 136) in lower York River, Virginia (Teixeira, 1995) (see Table 8.1). Amphipods are the most important prey, followed by copepods. The diet shifts with ontogeny at approximately 60 mm, such that individuals < 60 mm TL feed mainly on amphipods and secondarily on copepods,

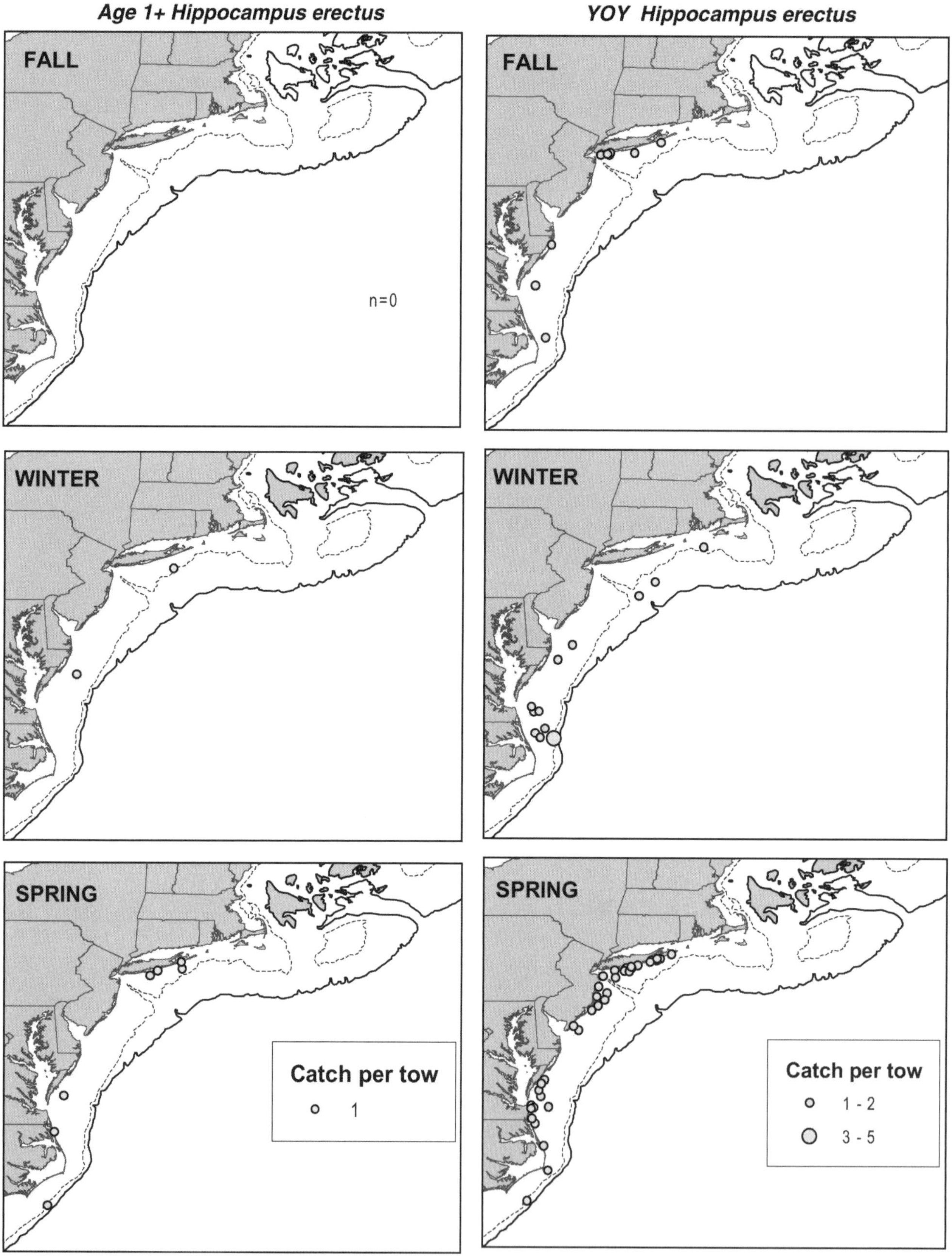

Age 1+ *Hippocampus erectus*

YOY *Hippocampus erectus*

Fig. 55.2. Composite distribution of age 1+ (> 10 cm) and young-of-the-year (< 10 cm) *Hippocampus erectus* during seasonal cruises of the National Marine Fisheries Service groundfish survey. Details of sampling effort are indicated in Fig. 3.4.

whereas those > 60 mm feed almost exclusively on amphipods. Several individuals, size unreported, have also been found in the guts of *Urophycis chuss* and *Merluccius bilinearis* outside Delaware Bay (46 m) in the early spring (Allen et al., 1978).

MIGRATIONS
While this species is relatively infrequently collected in continental shelf surveys, the distribution of young-of-the-year (< 10 cm) suggests that they undergo seasonal inshore-offshore migrations (Fig. 55.2). In the fall, a few individuals have been collected from relatively shallow waters (mean depth 18 m, range 8 to 32 m) on the inner and mid-shelf between North Carolina and Long Island. In the winter, they have been collected in deeper waters (mean depth 37 m, range 21 to 68 m) over the same approximate range. In the spring, they are primarily found back on the inner shelf (mean depth 19 m, range 8 to 79 m) in the same general locations. The age 1+ (> 10 cm) juveniles and adults are less well represented in these collections, but where collected have distributions similar to those of the young-of-the-year. During the winter, at a temperature of 10.6°C, individuals have been observed by divers on the inner continental shelf off Long Island, where they lay motionless on the substrate (Wicklund et al., 1968).

56

Syngnathus fuscus
Storer

NORTHERN PIPEFISH

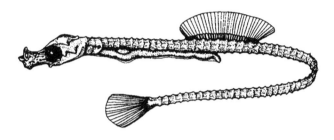

Estuary	Reproduction	Larvae/ Development	YOY Growth	YOY Overwinter
Ocean				
Both				
Jan				
Feb				
Mar				
Apr				
May				
Jun				
Jul				
Aug				
Sep				
Oct				
Nov				
Dec				

DISTRIBUTION

Syngnathus fuscus is distributed along the coast of North America from the Gulf of St. Lawrence to Florida (Dawson, 1982). In the Middle Atlantic Bight, early life history stages occur in every estuary for which there are data (see Table 4.3).

REPRODUCTION AND DEVELOPMENT

The mode of reproduction for this species, as in other syngnathids, is one with reversed sex roles, in which the female deposits the eggs in the male brood pouch and the male carries the fertilized eggs through the gestation period (Vincent et al., 1992; Teixeira, 1995). Most spawning is done by age 1 fish, at lengths of 8 to 24 cm TL, as they enter estuaries from the ocean. Males as small as 83 mm may carry a brood, but most brooding males are at least 90 to 100 mm (Dawson, 1982). In Great Bay–Little Egg Harbor, reproduction occurs from spring through late summer as indicated by the timing of the maturation of the gonads, the presence of embryos in the male pouch, and the occurrence of small, recently born larvae (Campbell and Able, 1998). Egg size is 0.75 to 1.0 mm in diameter and there are multiple oil globules. Brood size is variable, ranging from 45 to 1380 embryos in males from 119 to 222 mm TL in size. Each brood appears to be from a single mating because all of the embryos are in the same developmental stage (Campbell and Able, 1998). A similar pattern occurs in Chesapeake Bay (Teixeira, 1995).

Aspects of the ontogeny are summarized in Fahay (2007). Incubation lasts approximately 10 days (Bigelow and Schroeder, 1953), after which the larvae are released from the pouch. Early larvae are extremely elongate and have a prominent, large dorsal fin midway along the body length. The mouth is very small and situated at the tip of the elongate snout. Bony rings form on the body at small sizes. The earliest stages lack pigment, but after about 8.5 mm, the entire body becomes very dark.

LARVAL SUPPLY, SETTLEMENT, GROWTH, AND MORTALITY

Larvae produced within the study area are relatively common during every year, but occasional peaks in the local brood size occur at intervals of five years or more (Fig. 56.1). Larvae remain planktonic until they attain lengths of approximately 40 mm TL; they then settle into shallow, vegetated habitats (Lazzari and Able, 1990; Campbell and Able, 1998). Growth of the earliest larvae to be released from brood pouches proceeds until about September or October, when the mode reaches about 10 to 15 cm TL (Fig. 56.2). The young-of-the-year are extremely variable in size by the end of the first growing season (November), ranging from approximately < 5 to > 20 cm TL, which makes growth rates difficult to estimate. This broad range is due in part to the long period of reproduction and release of larvae (Campbell and Able, 1998). There is probably little growth during the winter. By the following spring, at approximately 12 months of age, the biggest young-of-the-year are 7 to 21 cm TL and are large enough to reproduce (Bigelow and Welsh, 1925). There is no evidence that reproduction by age 2 or older occurs. Thus, the population may be dominated by young-of-the-year individuals. Others have suggested that some individuals may reach 2 years of age and 28 to 30 cm (Warfel and Merriman,

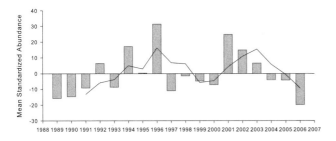

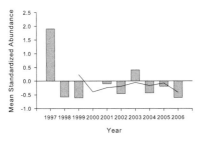

Fig. 56.1. Annual mean (bars) and three-year moving mean (line) of standardized abundance for *Syngnathus fuscus* collected in larval sampling with plankton net (*upper*) and juvenile sampling with otter trawl (*lower*). Data were standardized by subtracting overall mean from annual abundance. Three-year moving mean was calculated by taking the mean of the standardized values from the previous two years and the year in which data are plotted.

1944; Mercer, 1973; Dawson, 1982; Tatham et al., 1984) but definitive studies of the age and growth need to be conducted.

A significant source of mortality occurs during the winter. Comparison of length frequencies between the fall (November) and spring (April) indicate that individuals smaller than 5 cm do not appear to survive the winter (Able and Fahay, 1998) and this is supported by laboratory experiments. Mortality of small (< 100 mm TL) and large (> 100 mm TL) *Syngnathus fuscus* occurs during a period of lower temperatures (< 4°C) during December and January based on laboratory observations at ambient temperatures (Fig. 56.3). This drop results in a marked increase in mortality relative to the period of milder temperatures (generally > 8°C) during November through mid-December. This mortality reaches 82% for small and 68% for large individuals. This difference may be the result of size-selective mortality, with greater mortality for small individuals as occurs for other temperate estuarine fishes (see Chapter 6). After this period, little mortality occurs, probably in response to the milder temperature (generally > 4°C) during the rest of the winter. These observations suggest that low temperatures cause the seasonal migrations into the ocean, where milder temperatures are available through the season (see below).

SEASONALITY AND HABITAT USE
This species is found in shallow bays, harbors, rivers, creeks, and salt marshes and is usually associated with submerged vegetation. The abundance of small planktonic individuals peaks in midsummer in the estuary near Little Egg Inlet (Fig. 56.4) and in adjacent surfzone and nearshore waters (Able

et al., 2009c). Young-of-the-year are most abundant during the summer in shallow, vegetated habitats, including eelgrass (*Zostera marina*) and sea lettuce (*Ulva lactuca*) (Hildebrand and Schroeder, 1928; Briggs and O'Connor, 1971; Tatham et al., 1984). They also occur in unvegetated areas adjacent to these habitats, in marsh creeks (Sogard and Able, 1991; Rountree and Able, 1992b), and in a variety of shallow, estuarine shoreline habitats (Able et al., 1996b). In sampling in slightly deeper (< 1 to 8 m) waters they have also been collected across a variety of habitat types but are most abundant in the shallowest, vegetated habitats (Szedlmayer and Able, 1996; Able and Fahay, 1998), primarily in the lower estuary and immediately offshore (Fig. 56.5). In experiments with artificial eelgrass in the same estuary (Sogard, 1989; Sogard and Able, 1994), young-of-the-year readily colonized these habitats, even if they had to cross broad expanses of sand to do so. Colonization occurred during both day and night, but most small individuals (< 70 mm TL) colonized at night. Some evidence suggests movements between habitats may be gender-based, with males carrying embryos less likely to leave protective habitats than females (Roelke and Sogard, 1993).

PREY AND PREDATORS
Adults feed during the day and concentrate on minute copepods, amphipods, fish eggs, and small fish larvae (Hildebrand and Schroeder, 1928). Several studies (n = 4) have examined the food habits of young-of-the-year (28 to 209 mm TL, n = 5542) in Middle Atlantic Bight estuaries from Long Island Sound, New York, to lower York River, Virginia (see Table 8.1). Important prey of young-of-the-year include a variety of invertebrates (amphipods, copepods, isopods, mysids, cladocerans, and gastropods). Fishes (*Syngnathus fuscus* and *Hippocampus erectus*) are a minor dietary component (see Table 8.3). Young-of-the-year exhibit cannibalistic behavior on a minor scale (Teixeira, 1995). The diet of young-of-the-year shifts with ontogeny at approximately 100 mm in Chesapeake Bay, from a diet of predominantly copepods at smaller sizes to amphipods and isopods when larger (Ryer and Orth, 1987; Teixeira, 1995). Relative importance of individual prey items also varies intra-annually in the lower Chesapeake Bay, such that gammarid and caprellid amphipods and calanoid copepods dominate the diet in spring and summer, isopods during fall, and mysids and calanoid copepods in winter (Ryer and Orth, 1987).

Predators include *Mustelus canis, Gadus morhua, Hemitripterus americanus, Centropristis striata,* and *Cynoscion regalis* in the ocean (Rountree, 1999). *Opsanus tau* and *Pomatomus saltatrix* are also known predators in the estuary (Roelke and Sogard, 1993; Buckel et al., 1999a).

MIGRATIONS
Seasonal migrations in the northern Middle Atlantic Bight are common, with most individuals leaving the estuaries to spend the winter on the inner continental shelf

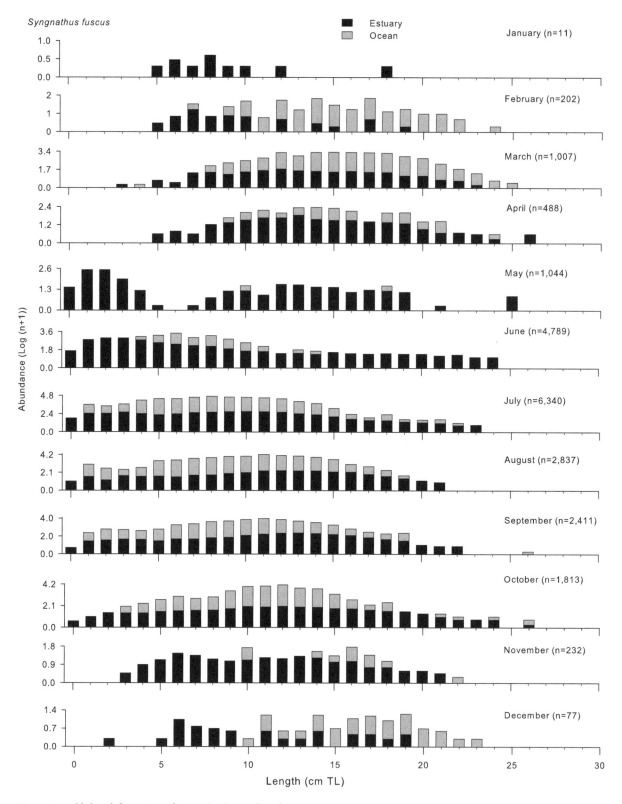

Fig. 56.2. Monthly length frequencies of *Syngnathus fuscus* collected in estuarine and coastal ocean waters of the central Middle Atlantic Bight. Sources: National Marine Fisheries Service: otter trawl (n = 1235); Rutgers University Marine Field Station: 1 m beam trawl (n = 660); seine (n = 4156); Methot trawl (n = 89); 1 m plankton net (n = 4622); 2 m beam trawl (n = 2); experimental trap (n = 8); killitrap (n = 58); 4.9 m otter trawl (n = 2005); and Able and Fahay 1998 (n = 8416).

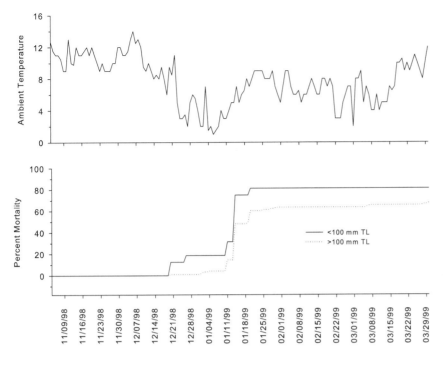

Fig. 56.3. Winter mortality of small (< 100 mm TL) and large (> 100 mm TL) *Syngnathus fuscus* relative to ambient water temperatures based on laboratory observations in flow through seawater at the Rutgers University Marine Field Station during the winter of 1998–1999.

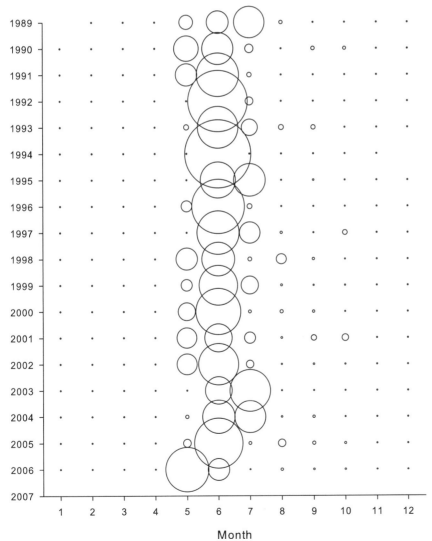

Fig. 56.4. Seasonal occurrence of the larvae of *Syngnathus fuscus* based on ichthyoplankton samples collected in Little Sheepshead Creek behind Little Egg Inlet from 1989 to 2006. Full circles represent the percentage of the mean number / 1000 m³ of larval young-of-the year *Syngnathus fuscus* captured by given year. Values range from 0 to 100%. The smallest circles represent when samples were taken but no larvae were collected.

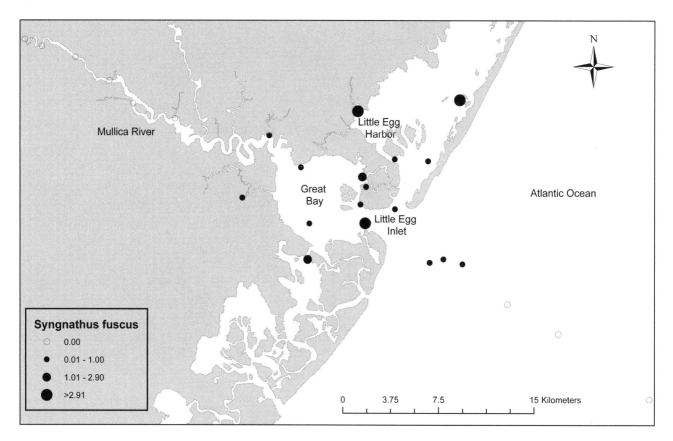

Fig. 56.5. Habitat use and average catch per tow (see key) for juvenile and adult *Syngnathus fuscus* in the Ocean–Great Bay–Mullica River corridor in southern New Jersey based on small otter trawl collections during July and September from 1988 to 2006. See Table 3.2 for characteristics at each sampling collection.

(Lazzari and Able, 1990). They move into the ocean during November and December and return to the estuary in March and April. During the winter, they are found in the ocean within 10 km of shore in depths of 2 to 19 m (Lazzari and Able, 1990). Other observations in the winter off Long Island in depths of 14 to 17 m found torpid individuals partially buried in the sand with only the head and caudal fin exposed (Wicklund et al., 1968). The observations off New Jersey and Long Island suggest these movements into the ocean are in response to low winter temperatures.

The distribution of young-of-the-year (< 19 cm), as reflected by National Marine Fisheries Service (NMFS) collections (see Fig. 56.6), indicates seasonal inshore-offshore migrations as suggested previously based on earlier sampling programs during 1963–1986. In the current analysis, young-of-the-year are most frequently collected in the fall in shallow waters (mean depth 34 m, range 9 to 111 m) between Massachusetts and Virginia and onto the shallow portion of

Georges Bank, where they are most abundant. In the winter they are collected in deeper waters on the inner to mid-shelf (mean depth 46 m, range 15 to 119 m) over the same approximate geographical range. In the spring, they are again most abundant on the inner shelf and along the shallow portions of Georges Bank (mean depth 24 m, range 6 to 96 m). The age 1+ (> 19 cm) juveniles and adults had a similar distribution to the young-of-the-year (see Fig. 56.6). In the fall, the few adults collected occurred in shallow inshore waters (mean depth 22 m, range 9 to 49 m) from North Carolina to Massachusetts and on Georges Bank. In the winter they were found in deeper waters (mean depth 48 m, range 17 to 105 m) on inner and mid-shelf waters over the same range. By the spring, they were once again primarily collected on the inner shelf (mean depth 18 m, range 7 to 69 m) between Virginia and Long Island but, in a few instances, extending from North Carolina to Massachusetts to the Great South Channel.

Age 1+ *Syngnathus fuscus*

YOY *Syngnathus fuscus*

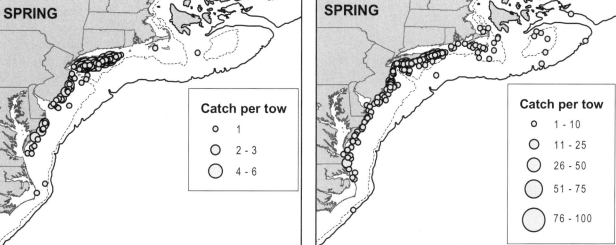

Fig. 56.6. Composite distribution of age 1+ (> 19 cm) and young-of-the-year (< 19 cm) *Syngnathus fuscus* during seasonal cruises of the National Marine Fisheries Service groundfish survey. Details of sampling effort are indicated in Fig. 3.4.

57

Prionotus carolinus (Linnaeus)

NORTHERN SEAROBIN

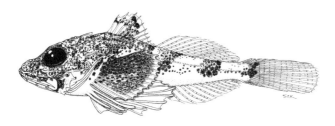

Estuary	Reproduction	Larvae/ Development	YOY Growth	YOY Overwinter
Ocean				
Both				
Jan				
Feb				
Mar				
Apr				
May				
Jun				
Jul				
Aug				
Sep				
Oct				
Nov				
Dec				

DISTRIBUTION

Prionotus carolinus occurs along the coast of North America from the Gulf of Maine to Florida. This species is more abundant than its congener *P. evolans* in the Middle Atlantic Bight (McBride and Able, 1994; McBride et al., 2002). Juveniles and adults are generally more common in estuaries in the northern part of the Middle Atlantic Bight than in the southern part (Stone et al., 1994). It is among the four most abundant species in Sandy Hook Bay during the summer (Wilk and Sil-

verman, 1976), but it is less common in the shallower Great Bay–Little Egg Harbor (Szedlmayer and Able, 1996). Eggs, larvae, and juveniles have been reported from most estuaries in the Middle Atlantic Bight (see Table 4.3).

REPRODUCTION AND DEVELOPMENT

Spawning occurs during the summer in continental shelf (mostly) and estuarine waters throughout the Middle Atlantic Bight, based on back-calculated hatching dates determined from otoliths (McBride et al., 2002). Gonadosomatic index levels indicate females are ripe from May through September and peak in July in the New York Bight (Wilk et al., 1990). A study in Peconic Bay, New York, found most spawning occurred in the evening or at night (Ferraro, 1980). On the Middle Atlantic Bight continental shelf, eggs (identified only to the genus *Prionotus*) occurred during every month of the year, except February and December, with a strong peak in August and September (Berrien and Sibunka, 1999). Eggs are found in the southern part of the bight as early as March or April, then typically spread north as far as Georges Bank in July though September (in three of nine years surveyed). In May, eggs were typically found in two clusters, one along the shore and another near the shelf edge. At an inner continental shelf site off the coast of New Jersey, eggs (identified only to the genus *Prionotus*) were present from May to October, and their abundance peaked during July or August (McBride and Able, 1994). Eggs also occurred in Great Bay, New Jersey, with an early peak in June, although these also were not identified to species (McBride and Able, 1994). In Sandy Hook Bay, eggs of *Prionotus* sp. were the most abundant taxon collected in May and June (Croker, 1965), but larvae were not collected. Eggs occurred between June and August in Narragansett Bay, where they were fifth or sixth in abundance, although larvae were rare (Herman, 1963; Bourne and Govoni, 1988). A study in the Hereford Inlet, New Jersey area used immunochemical methods to discriminate the eggs of *P. carolinus* from those of *P. evolans* (Keirans et al., 1986). Keirans et al. detected an initial peak in spawning intensity of *P. carolinus* in June or July, followed by a second peak in August of one year and September in another. Larvae are most abundant in Middle Atlantic Bight continental shelf waters from July through October, particularly south of the drowned Hudson River Valley to Cape Hatteras (Able and Fahay, 1998). At the Beach Haven Ridge study site off southern New Jersey, larvae were found only during September; a few also occurred then in plankton sampling near Little Egg Inlet (McBride and Able, 1994).

Details of early development, and methods to discriminate the early stages of congeneric *Prionotus* in the Middle Atlantic Bight, have been summarized (Able and Fahay, 1998; Fahay, 2007). Eggs cannot confidently be distinguished from those of *P. evolans*. Their diameters range from 0.86 to 1.5 mm. Oil globules number 11 to 37 and occur in one hemisphere. Incubation takes 60 hours at 15 to 22°C; the larvae are about 3.0 mm NL at hatching. They undergo flexion at 6 to 7 mm SL, 13 days after hatching (McBride et al., 2002). Larvae have early-forming pectoral fin rays and several series of head spines. The lower three rays of the pectoral fin become separated from the rest of the fin at about 12 mm SL, shortly after settlement, in specimens we have examined. The pectoral fins are long in postflexion larvae, but not as long as they are in *P. evolans* larvae.

Relatively few specimens (9.0 to 36.8 mm TL, n = 12) were available to examine for scale formation (see Fig. 54.1). The smallest specimen with scales was 23.8 mm TL. In this specimen, scales were present from the base of the caudal fin on the lateral and dorsal surfaces to the origin of the pectoral fin. By 36.8 mm TL, these scales extended anteriorly on the body to just behind the operculum and covered the ventral surface, but not the area under the pectoral fin.

LARVAL SUPPLY, SETTLEMENT, AND GROWTH

The larval dispersal period for this species is relatively short (about three weeks); therefore, larval distributions are good indicators of spawning locations (McBride et al., 2002). In a recent study, the largest planktonic larvae collected were 9.8 mm SL, and settlement occurred at sizes of 8 to 9 mm SL and ages of 18 to 19 days (with a maximum of 25 days) after hatching (McBride et al., 2002). Those settled individuals that are encountered in estuarine habitats occur in July through October. The generally late peak in spawning activity (as late as August) combined with a slow growth rate of about 0.5 mm per day combine to produce relatively small juveniles (2 to 10 cm TL) as the winter begins (Fig. 57.1). They are also found during the following spring, when they are a few months younger than 1 year old. Others settle in the ocean. Growth rates vary latitudinally. Northern members of year classes are larger than those from the southern part of the Middle Atlantic Bight (McBride et al., 2002). Thus age 2 fish from the Chesapeake Bay area average 163 mm SL, whereas those from New England waters average 195 mm SL. Age 4 fish in the Chesapeake Bay area average 197 mm SL, whereas those from New England waters average 245 mm SL.

Winter mortality among young-of-the-year on the continental shelf is estimated to occur at a low rate, possibly due to this species' ability to bury into the substrate (an anti-predator tactic) (McBride et al., 2002). While in some other species that overwinter on the shelf size-selective mortality is indicated by truncated length distributions between fall and the following spring, such is not the case in *Prionotus carolinus*.

SEASONALITY AND HABITAT USE

This species is found from estuaries to the edge of the continental shelf, depending on the time of year. It prefers sand-bottomed substrates in coastal waters between May and October and mid- to outer continental shelf depths during winter months (Roberts, 1978). Early life history stages are present on the continental shelf. Eggs and larvae occur in the water column there and recently settled juveniles are found on the bottom in maximum densities of $> 10/100m^2$ (McBride et al., 2002). In coastal waters off New Jersey, eggs can be found in the surfzone and shallow inshore waters in some years (Able et al., 2009c, 2010). The larvae are present from July to October and peak in September, when they greatly outnumber those of *P. evolans* (McBride et al., 2002). Larvae collected at Little Egg Inlet occur as early as April or May but are also present from July through October (Fig. 57.2). Young-of-the-year are distributed on the continental shelf according to size and age. Older and larger young-of-the-year occur on substrates in deeper water (McBride et al., 2002). Any use of estuaries (or inner continental shelf sites) begins during the spring, when young-of-the-year are just less than 12 months old and at sizes between 4 and 15 cm (see Fig. 57.1). Members of this year class are also observed as they emigrate offshore during late summer and fall, possibly in response to rising temperatures.

These patterns of estuarine use vary throughout the Middle Atlantic Bight. In the central part of the Middle Atlantic Bight, any young-of-the-year or early age 1 fish that might occur in estuaries leave those habitats during the warmer parts of summer and move offshore to cooler temperatures. However, in southern New England, they are found in coastal areas throughout the summer and then emigrate in October or November in response to dropping temperatures (Lux and Nichy, 1971). Young-of-the-year generally emigrate from Long Island Sound during the fall, but occasional specimens remain through the winter (Richards et al., 1979). Young-of-the-year between 3 and 8 cm TL have been collected in the deeper portions of Chesapeake Bay between December and May, indicating they sometimes spend the winter there (Hildebrand and Schroeder, 1928).

PREY AND PREDATORS

This species feeds on the bottom, facilitated by the chemosensory tips of their free pectoral fin rays (Bardach and Case, 1965). Three major components of this species' diet are decapod crustaceans (51.5% by weight), polychaetes (15.2%), and fishes (14.7%) (Bowman et al., 2000). The most important crustaceans in this diet include *Crangon septemspinosa* and *Cancer irroratus*. Other items consumed less frequently include bivalves, cumaceans, isopods, mysids, amphipods, annelids, and gastropods. Juveniles have been reported to eat copepods (Richards et al., 1979).

Several studies (n = 4) have examined the food habits of young-of-the-year (20 to 159.7 mm TL, n = 295) in Middle Atlantic Bight estuaries from Woods Hole, Massachusetts, to Magothy Bay, Virginia (see Tables 8.1 and 8.3). Important prey of young-of-the-year include a variety of invertebrates (mysids, copepods, amphipods, decapod shrimp, polychaetes, cumaceans, isopods, and crabs) and fishes (*Anchoa mitchilli*).

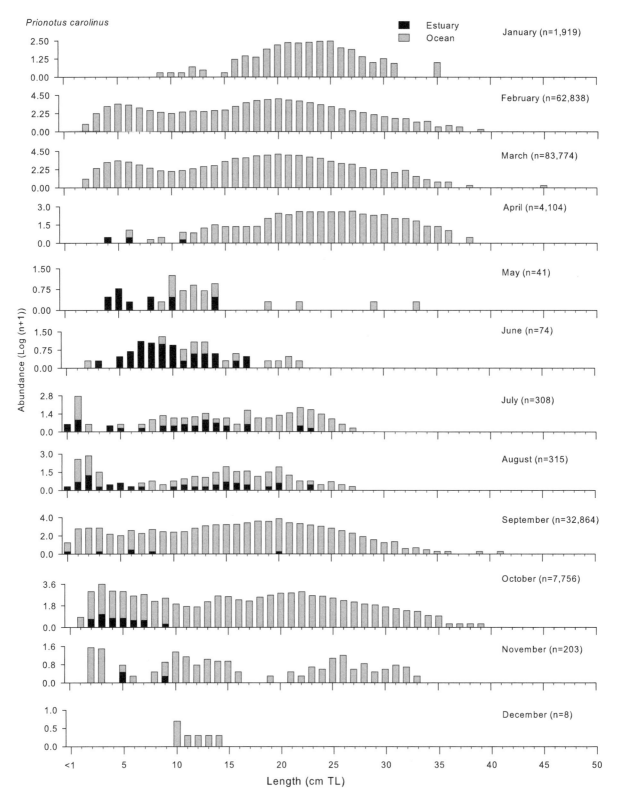

Fig. 57.1. Monthly length frequencies of *Prionotus carolinus* collected in
estuarine and coastal ocean waters of the central Middle Atlantic Bight.
Sources: National Marine Fisheries Service: otter trawl (n = 191,914); Rutgers University Marine Field Station: seine (n = 56); Methot trawl (n = 7);
1 m plankton net (n = 26); 2 m beam trawl (n = 142); experimental trap
(n = 1); killitrap (n = 1); 4.9 m otter trawl (n = 324); multi-panel gill net
(n = 1); and Able and Fahay 1998 (n = 1732).

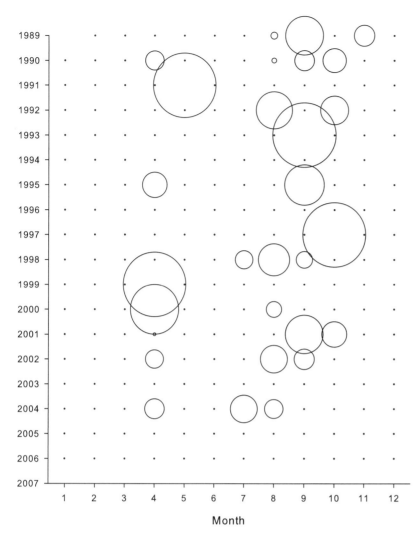

Fig. 57.2. Seasonal occurrence of the larvae of *Prionotus carolinus* based on ichthyoplankton samples collected in Little Sheepshead Creek behind Little Egg Inlet from 1989 to 2006. Full circles represent the percentage of the mean number/1000 m³ of larval young-of-the year *Prionotus carolinus* captured by given year. Values range from 0 to 100%. The smallest circles represent when samples were taken but no larvae were collected.

Sediment was also prominent in stomachs. In Long Island Sound, the diet of young-of-the-year (37 to 119 mm SL, n = 120) shifts with ontogeny. Small young-of-the-year consume copepods, and as they grow through the fall, their diet changes to amphipods, isopods, cumaceans, and small crabs (Richards et al., 1979). Similarly, in Magothy Bay, Virginia, smaller (22 to 54 mm, n = 24) young-of-the-year feed mainly on *Neomysis americana* (82.2% volume), supplemented by amphipods (10.1%) and small *Crangon septemspinosa* (4.7%), while larger (57 to 115 mm, n = 24) young-of-the-year do not feed as heavily on *Neomysis americana* (44.5%), but increase consumption of amphipods (16.3%), decapod shrimp (12.5%), isopods (8.7%), crabs (4.0%), and plant material. Predators include a wide variety of sharks and finfish, and juveniles are sometimes preyed upon by adult *P. carolinus* or *P. evolans* (Richards et al. 1979). However, reports of young-of-the-year included in the gut contents of larger predatory fishes are rare (Maurer and Bowman, 1975).

MIGRATIONS

The young-of-the-year (< 8 cm) are distributed differently with the seasons, based on National Marine Fisheries Ser-

vice (NMFS) bottom trawl surveys (Fig. 57.3). In the fall, they are found on the inner continental shelf (mean depth 29.6 m, range 9 to 240 m) from the central portion of Georges Bank and Nantucket Shoals to south of Cape Hatteras. In the winter they are restricted to waters between Long Island and Cape Hatteras. In this area they are most abundant off Delaware Bay and farther south, where they are found in deeper waters (mean depth 40.7 m, range 15 to 244 m) than in the fall. In the spring, they have a similar distribution but are found in shallower waters (mean depth 27.6 m, range 7 to 82 m). Age 1+ (> 8 cm) individuals move inshore to offshore with the seasons. In the fall, they are distributed in shallow inner continental shelf depths (mean depth 28.6 m, range 6 to 240 m) from the coastal Gulf of Maine and central Georges Bank to south of Cape Hatteras, but from New Jersey and south they are found in the deeper portions of this range. In the winter, they are concentrated at mid- and outer shelf depths (mean depth 80.1 m, range 15 to 429 m) from Great South Channel to Cape Hatteras. In the spring, some individuals can be found on the inner continental shelf, especially south of Chesapeake Bay, but most are concentrated on the outer continental shelf (mean depth 85.5 m, range 8 to 413 m).

Age 1+ *Prionotus carolinus*

YOY *Prionotus carolinus*

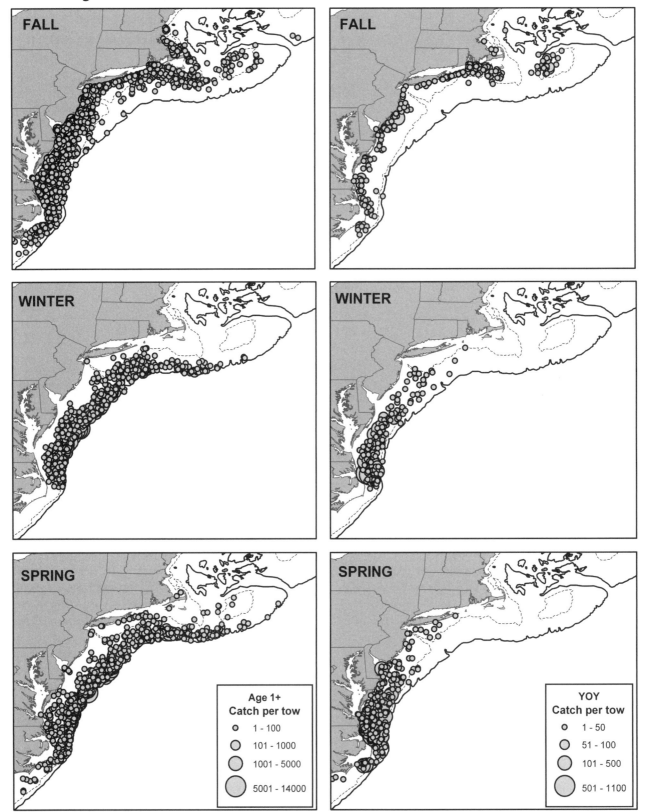

Age 1+
Catch per tow

○ 1 - 100
○ 101 - 1000
○ 1001 - 5000
○ 5001 - 14000

YOY
Catch per tow

○ 1 - 50
○ 51 - 100
○ 101 - 500
○ 501 - 1100

Fig. 57.3. Composite distribution of age 1+ (> 8 cm) and young-of-the-year (< 8 cm) *Prionotus carolinus* during seasonal cruises of the National Marine Fisheries Service groundfish survey. Details of sampling effort are indicated in Fig. 3.4.

58

Prionotus evolans (Linnaeus)

STRIPED SEAROBIN

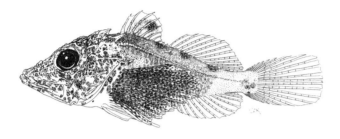

Estuary				
Ocean	Reproduction	Larvae/ Development	YOY Growth	YOY Overwinter
Both				
Jan				
Feb				
Mar				
Apr				
May				
Jun				
Jul				
Aug				
Sep				
Oct				
Nov				
Dec				

DISTRIBUTION

Prionotus evolans occurs from the Bay of Fundy to the coast of northeast Florida (Gilmore, 1977; Scott and Scott, 1988) but is most common between Cape Cod and South Carolina. Juveniles have been reported from several estuaries in the Middle Atlantic Bight, but eggs and larvae have been reported from only a few (see Table 4.3). This oversight might be due to the inability of investigators to discriminate *Prionotus* species in early stages.

REPRODUCTION AND DEVELOPMENT

This species begins spawning at age 2 (McEachran and Davis, 1970). Spawning occurs at night over sandy substrates, and is associated with sound production. Spawning takes place from May to October with a peak in July and August (Yuschak, 1985; McBride and Able, 1994; McBride et al., 2002). In the central part of the Middle Atlantic Bight, gonadosomatic evidence indicates that spawning occurs between May and August with a peak in June and July (Wilk et al., 1990), and the occurrence of eggs and larvae indicates that spawning climaxes in July or August (McBride and Able, 1994). Most spawning takes place in inner continental shelf waters, although eggs and larvae have also been collected near inlets and in estuaries. Eggs (identified by immunochemical techniques) have been collected near Hereford Inlet, New Jersey, between May and September, with the strongest peak in the latter month (Keirans et al., 1986). During 4 years of MAR-MAP sampling (1984–1987), only 24 larvae were collected in the entire Middle Atlantic Bight, and these were found between March and September and were scattered between Long Island and Cape Hatteras (Able and Fahay, 1998). A study in the New York Bight collected larvae between July and August; these occurred mostly within 50 km of shore and southwest of the Hudson Shelf Valley (Able and Fahay, 1998).

Details of early development and methods to discriminate the early stages of congeneric *Prionotus* in the Middle Atlantic Bight have recently been summarized (Able and Fahay, 1998; Fahay, 2007). Eggs of *Prionotus evolans* are 1.05 to 1.25 mm in diameter and have 16 to 37 small oil globules located in one hemisphere. Incubation takes 80 to 90 hours at 19 to 20°C; larvae are about 2.7 mm NL at hatching (Yuschak and Lund, 1984; Fahay, 2007). The lower three rays of the pectoral fin become separated from the rest of the fin at about 10 to 15 mm (Morrill, 1895). Individuals from 7.4 to 20.1 mm TL (n = 13) were examined for scale formation (Fig. 54.1). The smallest individual examined (10.6 mm TL) had scales forming along the lateral line from the level of the dorsal fin origin to the caudal peduncle. By 11.6 mm TL, the area covered with scales had spread dorsally and ventrally over the sides of the body. By 15.2 mm TL, the area of coverage extended farther anteriorly and laterally, approximating the adult condition.

LARVAL SUPPLY, SETTLEMENT, AND GROWTH

Larvae originate from spawning located over the inner continental shelf, near inlets, or in estuaries. The larvae can be common in surfzone and nearshore collections, but not in every year (Able et al., 2009c, 2010). Pelagic larvae settle to the bottom at lengths of 8 to 9 mm and ages of 18 to 19 days

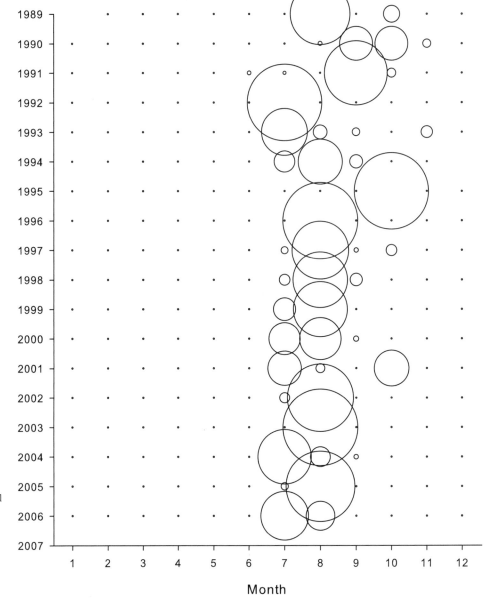

Fig. 58.1. Seasonal occurrence of the larvae of *Prionotus evolans* based on ichthyoplankton samples collected in Little Sheepshead Creek behind Little Egg Inlet from 1989 to 2006. Full circles represent the percentage of the mean number/1000 m³ of larval young-of-the year *Prionotus evolans* captured by given year. Values range from 0 to 100%. The smallest circles represent when samples were taken but no larvae were collected.

(McBride et al., 2002). This settlement occurs during August and September at the Beach Haven Ridge study site off Little Egg Inlet. Plankton net collections at nearby Little Egg Inlet indicate that early stages occur most frequently between July and October (Fig. 58.1).

The growth rate is slow in young-of-the-year, only about 0.5 mm per day during the first year (McBride et al., 2002). Young-of-the-year 23 to 88 mm SL have been reported from New Haven Harbor, Connecticut, in August and September (Warfel and Merriman, 1944), and juveniles have been reported from Sandy Hook Bay, where they average 55 mm SL in August and 70 mm SL by October (Nichols and Breder, 1927). During the fall (October–December), we suggest that the young-of-the-year are 1 to 10 cm in the estuary and ocean (Fig. 58.2). There is apparently little or no growth over the winter, and juveniles are observed the following spring (May)

in the same size range. Some studies have suggested the size at age 1 as 151 or 173 mm FL (in 2 different years) (McEachran and Davis, 1970) or 173 mm SL (Richards et al., 1979).

SEASONALITY AND HABITAT USE

Prionotus evolans prefers sand bottoms in coastal waters between May and November and spends the winter months at outer continental shelf depths (Roberts-Goodwin, 1981; McBride and Able, 2002). An analysis of seasonal distributions found that while *Prionotus carolinus* and *Prionotus evolans* often co-occur during summer, *P. evolans* tends to occur in warmer, less oxygenated, and more turbid habitats than *P. carolinus*; it also arrives at, and departs from, coastal areas later than its congener (McBride and Able, 1994). It is among the four most abundant species in Sandy Hook Bay during the summer (Wilk and Silverman, 1976), but is less common

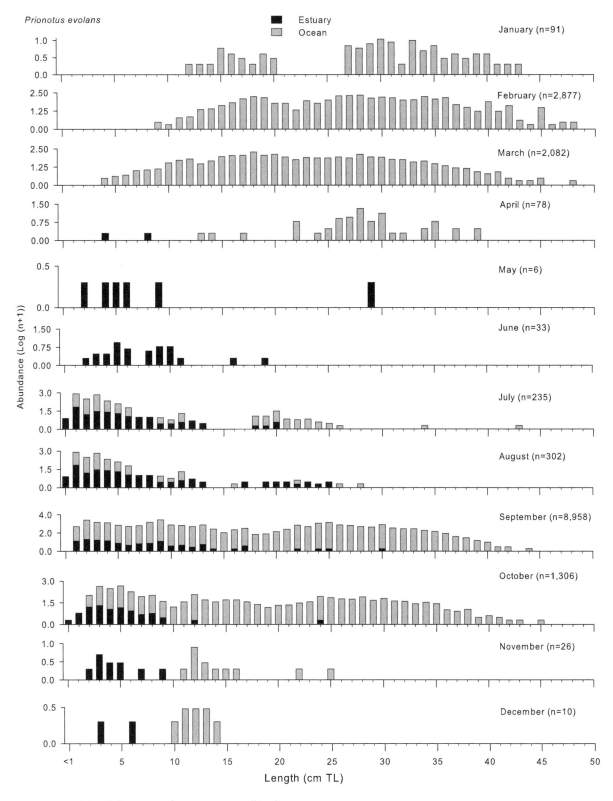

Fig. 58.2. Monthly length frequencies of *Prionotus evolans* collected in estuarine and coastal ocean waters of the central Middle Atlantic Bight. Sources: National Marine Fisheries Service: otter trawl (n = 14,979); Rutgers University Marine Field Station: 1 m beam trawl (n = 1); seine (n = 201); Methot trawl (n = 16); 1 m plankton net (n = 168); 2 m beam trawl (n = 8); experimental trap (n = 90); killitrap (n = 1); 4.9 m otter trawl (n = 116); and Able and Fahay 1998 (n = 425).

in Great Bay–Little Egg Harbor (Szedlmayer and Able, 1996). Although both species winter on the mid- or outer continental shelf, *P. carolinus* is then found at latitudes farther north than *P. evolans*, which is restricted to offshore areas south of Delaware (McBride and Able, 1994; McBride et al., 2002). Young-of-the-year of both species spend their first winter with adults, judging from bottom trawl survey results (Edwards et al., 1962).

The young-of-the-year of this species occurs in late summer and early fall in the Great Bay–Little Egg Harbor estuary (Fig. 58.2). Habitat use varies with life history stage. Larvae are found most frequently over the inner continental shelf, including the surfzone, although a few have been collected in plankton sampling inside Little Egg Inlet (Morse et al., 1987; Witting et al., 1999). Most recently settled juveniles also occur on the shelf. In the few months after settlement, some young-of-the-year are also found in estuaries over a wide range of substrates in depths greater than 2 m. Unlike its congener, *P. carolinus*, *P. evolans* young-of-the-year are more likely to be collected in estuaries between spawning and their first winter. In the Hudson River estuary, juveniles enter the system from Raritan Bay at sizes of 16 to 27 mm SL in June, and this is followed by growth accompanied by upstream migration. These fish may reach 67 km upriver and attain sizes of about 60 mm SL by October (Schmidt, 2007).

PREY AND PREDATORS

This species feeds on the bottom on small crustaceans (amphipods, mysids, and *Crangon*), annelids, cumaceans, crabs, mollusks, fishes, and fish eggs (Marshall, 1946). Fishes in their diet include *Anchoa mitchilli*, *Syngnathus* sp., *Menidia menidia*, *Stenotomus chrysops*, *Pseudopleuronectes americanus*, *Scophthalmus aquosus,* and smaller sizes of both *Prionotus* sp. (Richards et al., 1979). Studies in Sandy Hook Bay and the nearby Navesink River estuary showed that this species primarily ate mysids and *Crangon septemspinosa* (Manderson et al., 1999). However, young-of-the-year *Pseudopleuronectes americanus* were also important components of their diets. These young-of-the-year were found in the diets of 69% of *P. evolans* stomachs examined during the month of June in Sandy Hook Bay. Observations in the laboratory revealed that *P. evolans* use their modified pectoral fin rays to detect, flush, and occasionally excavate buried *P. americanus*. When *P. evolans* 212 to 319 mm TL were provided with *P. americanus* 30 to 114 mm TL, they selected prey < 70 mm TL; maximum prey size

appeared to be limited by the width of the predator's esophagus (Manderson et al., 1999). One conclusion from this study was that when these two species co-occur in estuarine habitats, *P. evolans* consumes large quantities of *P. americanus* in the size range of 15 to 70 mm TL. Young-of-the-year *P. americanus* did not suffer this predation in the nearby Navesink River estuary, primarily because *P. evolans* did not extend into that habitat.

Two studies examined the food habits of young-of-the-year (33 to 150 mm SL, n = 309) *Prionotus evolans* in Woods Hole, Massachusetts, New Haven, and the Connecticut shore of Long Island Sound (see Table 8.1). Important prey of young-of-the-year included a variety of invertebrates (decapod shrimp, copepods, gastropods, bivalves, mysid shrimp, and eggs), fishes, and fish eggs. In Long Island Sound, diet of young-of-the-year (40 to 150 mm SL, n = 271) shifts with ontogeny, such that small young-of-the-year consume copepods and larger young-of-the-year in the fall consume more *Neomysis americana*, *Crangon septemspinosa*, a few mollusks, and fishes (Richards et al., 1979). Important predators of young-of-the-year include adult *P. evolans* (Richards et al., 1979) and spiny dogfish (Rountree, 1999).

MIGRATIONS

The young-of-the-year (< 9 cm) are seasonal visitors to the continental shelf (Fig. 58.3). In the summer they usually occur in estuaries (Able and Fahay, 1998). In the fall, when they are most abundant, they are distributed on the shallow (mean depth 16.6 m, range 6 to 37 m) inner continental shelf from Massachusetts to Cape Hatteras. Occasional young-of-the-year specimens have been found in deep water in Long Island Sound as late as February (Richards et al., 1979). In the winter, they have never been collected, presumably because they have migrated out of the study area. In the spring, only a few individuals have been collected off Virginia and North Carolina.

The age 1+ (> 9 cm) individuals undergo seasonal inshore-offshore migrations (see Fig. 58.3). In the fall, they are concentrated in shallow waters (mean depth 19.1 m, range 5 to 118 m) on the inner continental shelf from north of Cape Cod to south of Cape Hatteras. By the winter, they migrate south and offshore to the edge of the continental shelf and occur in deeper waters (mean depth 77.8 m, range 21 to 200 m). In the spring, they are still found primarily at the edge of the continental shelf, although a few individuals move to the inner continental shelf (mean depth 64.6 m, range 8 to 300 m).

Age 1+ *Prionotus evolans*   ## YOY *Prionotus evolans*

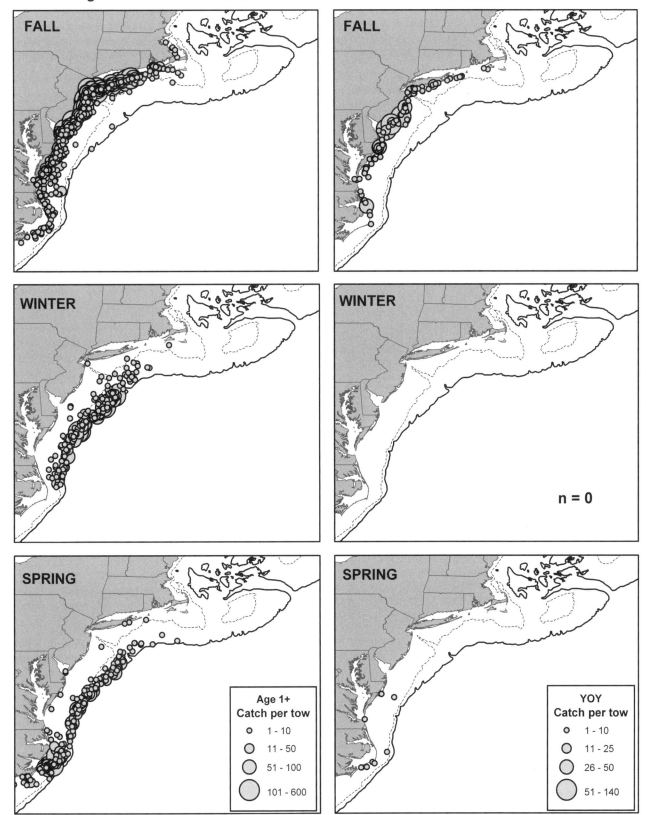

Fig. 58.3. Composite distribution of age 1+ (> 9 cm) and young-of-the-year
(< 9 cm) *Prionotus evolans* during seasonal cruises of the National Marine
Fisheries Service groundfish survey. Details of sampling effort are indicated
in Fig. 3.4.

59

Myoxocephalus aenaeus (Mitchill)

GRUBBY

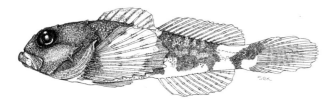

Estuary	Reproduction	Larvae/ Development	YOY Growth	YOY Overwinter
Ocean				
Both				
Jan				▓
Feb	▓			
Mar	▓			
Apr	▓			
May	▓		▓	
Jun			▓	
Jul			▓	
Aug			▓	
Sep			▓	
Oct				▓
Nov				▓
Dec				▓

DISTRIBUTION

Myoxocephalus aenaeus is found in the western North Atlantic Ocean from the Gulf of St. Lawrence and Newfoundland to New Jersey (Bigelow and Schroeder, 1953; Ennis, 1969). Early life history stages have been reported from several estuaries between Cape Cod and Delaware Bay (see Table 4.3).

REPRODUCTION AND DEVELOPMENT

Spawning occurs from winter into spring. On the basis of literature accounts and ichthyoplankton sampling through-

out the Middle Atlantic Bight, it appears that spawning begins in coastal waters and takes place later in offshore oceanic waters (Able and Fahay, 1998). In coastal waters spawning occurs in winter (Sumner et al., 1913; Morrow, 1951; Bigelow and Schroeder, 1953) or early spring (Richards, 1959; Smith, 1985). Winter spawning in a Cape Cod estuary has been verified by gonadosomatic analysis, presence of egg masses, and a peak in reproductive coloration of males (Lazzari et al., 1989). In Newfoundland, spawning may begin as early as fall and continue into winter (Ennis, 1969). Based on larval occurrences in the Middle Atlantic Bight between 1977 and 1987, spawning takes place between Georges Bank and New Jersey (Able and Fahay, 1998). This spawning occurs between March and June, but in the central part of the Middle Atlantic Bight it is limited to February through May (Able and Fahay, 1998; Fahay, 2007).

The early development of *Myoxocephalus aenaeus*, including characters useful for distinguishing the eggs and larvae from those of other members of the family Cottidae, has been well described and summarized (Fahay, 2007). Eggs are demersal, adhesive, and usually colored yellow, green, or red. They are deposited in clumps and adhere to algae or other objects. Larvae hatch at an average size of 5.4 mm, after an incubation period of 40 to 57 days (Lund and Marcy, 1975). Characteristic pigment patterns and an array of head spines form during the larval stage. Larvae transform into juveniles at an age of about 55 days and at lengths of about 9 to 10 mm. Most adult characters are developed at this size, although the urostyle tip is not yet resorbed and scales have not yet formed (Lund and Marcy, 1975). A characteristic juvenile pigment pattern develops at about 12 mm.

LARVAL SUPPLY, SETTLEMENT, AND GROWTH

Lengths of larvae occurring over the continental shelf range from about 5 mm (corresponding to the size at hatching) to about 15 mm, but most are smaller than 10 mm (Able and Fahay, 1998). The size at settlement can be estimated at about 10 to 15 mm, since this represents the largest size reached in planktonic individuals and the smallest collected in bottom-sampling in estuaries. The abundance of early stages in a New Jersey estuary is relatively constant annually, but numbers of larvae were high in the mid-1990s. The collections of juveniles in otter trawls demonstrate similar trends (Fig. 59.1).

Otolith analysis shows that juveniles grow to 60 to 65 mm SL in the first year (Lazzari et al., 1989). Observations on growth in the Great Bay–Little Egg Harbor study area (Fig. 59.2) indicate that young-of-the-year reach lengths of 3 to 10 cm by their first fall. The similarity in sizes the following spring indicates that there is little or no growth through the winter. A summer growth rate, calculated on the progression of the observed modes, would therefore be about 0.3 mm per day between April and October. Although our records are few, it appears that the maximum size of about

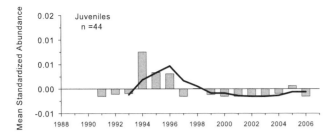

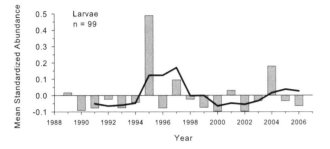

Fig. 59.1. Annual mean (bars) and three-year moving mean (line) of standardized abundance for *Myoxocephalus aenaeus* collected in juvenile sampling with otter trawls (*upper*) and larval sampling with plankton net (*lower*). Data were standardized by subtracting overall mean from annual abundance. Three-year moving mean was calculated by taking the mean of the standardized values from the previous two years and the year in which data are plotted.

15 cm (Scott and Scott, 1988) is reached by fish beginning their second winter.

SEASONALITY AND HABITAT USE

This species inhabits sand, mud, or gravel substrates in coastal waters from the low-tide mark to depths of about 27 m (Sumner et al., 1913), but it has also been reported from as deep as 130 m (Robins and Ray, 1986). Planktonic larvae have been collected from oceanic and estuarine waters (Lazzari et al., 1989). Although young-of-the-year of this species are never abundant in estuaries of the Middle Atlantic Bight, an influx of settlement-size individuals occurs in southern New Jersey estuaries in April (Witting et al., 1999). An analysis of a larger time series from the same locations found them from February through April (Fig. 59.3). Young-of-the-year were frequently collected in traps from April through July and occasionally in August and October in the Mullica River–Great Bay estuary. In Nauset Marsh, Massachusetts, larvae were abundant in April, and juveniles appeared in summer collections (Lazzari et al., 1989). Both estuarine and continental shelf habitats are used during the first year. Young-of-the-year, as well as older stages, occur most commonly in eelgrass beds in Nauset Marsh (Lazzari et al., 1989), and this habitat apparently is also most commonly used in the area

around Woods Hole, Massachusetts (Bigelow and Schroeder, 1953). All sizes were collected in similar habitats in the Bay of Fundy (Huntsman, 1922). Young-of-the-year may also undertake diel movements between habitats. In a Great Bay–Little Egg Harbor study, these movements (into artificial seagrass beds) were greatest at night (Sogard and Able, 1994). Their numbers in some estuaries may be underestimated as they have been observed by divers to be common in channels among blocks of peat, sloughed off the marsh surface, and this habitat is not readily sampled by traditional gears (Able et al., 1988; Lazzari et al., 1989).

PREY AND PREDATORS

This species feeds on a wide variety of bottom-dwelling animals, including annelid worms, shrimp, crabs, copepods, snails, nudibranch mollusks, and ascidians (Collette and Klein-MacPhee, 2002). Small fishes may be a minor component of their diet (Lazzari et al., 1989); the list of prey includes *Alosa pseudoharengus, Tautogolabrus adspersus, Anguilla rostrata, Fundulus heteroclitus, Ammodytes* spp., *Menidia menidia, Gasterosteus aculeatus,* and *Microgadus tomcod* (Collette and Klein-MacPhee, 2002).

Two studies have examined the food habits of young-of-the-year (37 to 134.9 mm, n = 196) in Cape Cod, Massachusetts, and on the Connecticut shore of Long Island Sound (Lazzari et al., 1989; see Table 8.1). Important prey of young-of-the-year include a variety of invertebrates (decapod shrimp, mysids, amphipods, and isopods) and fishes (*Fundulus heteroclitus*). In Nauset Marsh, Massachusetts, the diet of young-of-the-year (37 to 109 mm SL, n = 168) shifts with ontogeny at approximately 60 mm. The diet of smaller size fish consists primarily of crustaceans (*Crangon septemspinosa,* isopods, and amphipods). Larger size fish shift to a diet consisting of crustaceans (*C. septemspinosa* and *Carcinus maenas*) and fishes (*F. heteroclitus* and *Apeltes quadracus*) (see Table 8.3).

MIGRATIONS

Young-of-the-year (< 6 cm) and age 1+ (> 6 cm) individuals are consistently distributed in shallow waters (mean depth 46.8 m, range 12 to 147 m) on the continental shelf from northern Georges Bank to as far south as the mouth of Delaware Bay, with a center of abundance on Nantucket Shoals (Fig. 59.4). However, they occur in all months in the estuary (see Fig. 59.2), indicating that some individuals may be resident there. They are less available to the sampling gear on the continental shelf in the winter possibly because most are in estuarine habitats at that time for spawning (Able and Fahay, 1998).

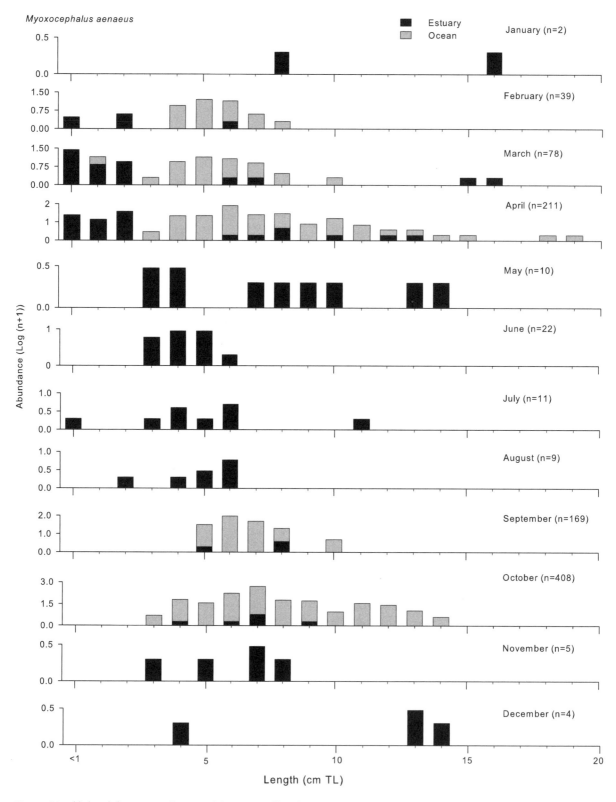

Fig. 59.2. Monthly length frequencies of *Myoxocephalus aenaeus* collected
in estuarine and coastal ocean waters of the central Middle Atlantic Bight.
Sources: National Marine Fisheries Service: otter trawl (n = 759); Rutgers
University Marine Field Station: 1 m beam trawl (n = 1); 1 m plankton net
(n = 57); experimental trap (n = 6); killitrap (n = 20); 4.9 m otter trawl
(n = 2); and Able and Fahay 1998 (n = 123).

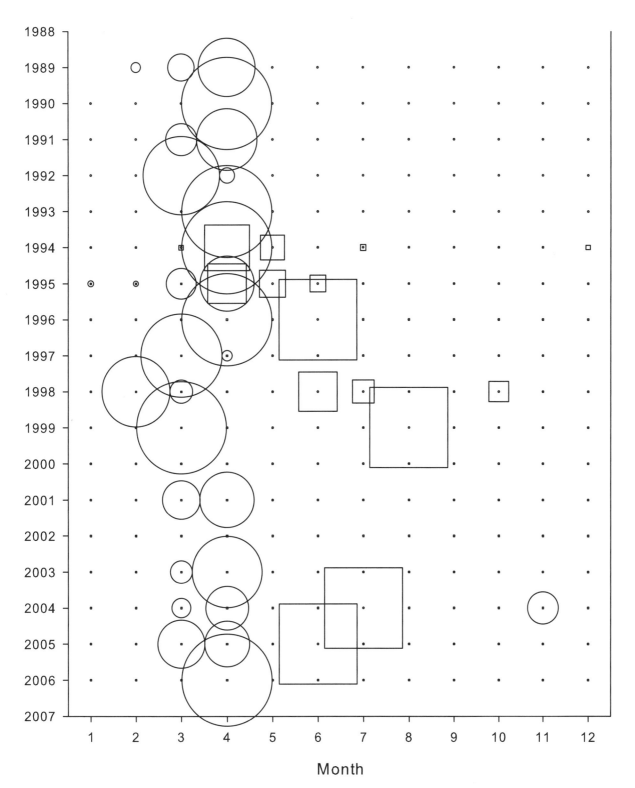

Fig. 59.3. Seasonal occurrence of larvae based on ichthyoplankton samples collected in Little Sheepshead Creek behind Little Egg Inlet from 1989 to 2006 (open circles represent the percentage of the mean number/1000 m³ of larval young-of-the-year captured by year) and juveniles of *Myoxocephalus aenaeus* based on trap collections in Rutgers University Marine Field Station boat basin in Great Bay (open squares represent the percentage of the mean catch per unit effort captured by year) from 1994 to 2006. Values range from 0 to 100% for ichthyoplankton and trap sampling. The smallest circles and squares represent when samples were taken but no individuals were collected.

Age 1+ *Myoxocephalus aenaeus*

YOY *Myoxocephalus aenaeus*

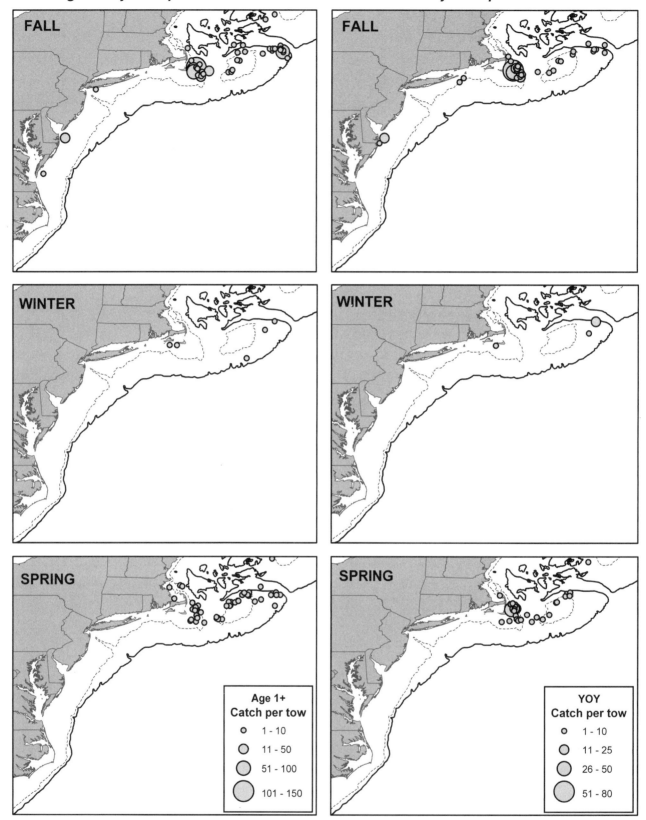

Fig. 59.4. Composite distribution of age 1+ (> 6 cm) and young-of-the-year (< 6 cm) *Myoxocephalus aenaeus* during seasonal cruises of the National Marine Fisheries Service groundfish survey. Details of sampling effort are indicated in Fig. 3.4.

60

Morone americana (Gmelin)

WHITE PERCH

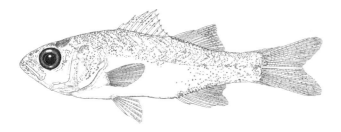

Estuary	Reproduction	Larvae/ Development	YOY Growth	YOY Overwinter
Ocean				
Both				
Jan				
Feb				
Mar				
Apr				
May				
Jun				
Jul				
Aug				
Sep				
Oct				
Nov				
Dec				

DISTRIBUTION

Morone americana occurs in eastern North America from New Brunswick and Nova Scotia to South Carolina (Murdy et al., 1997). It is endemic to estuaries, but landlocked populations are present in Lake Erie and Lake Ontario. It is one of the most abundant resident species in parts of the Hudson River estuary (Bath and O'Connor, 1982), Delaware Bay (O'Herron et al., 1994), and Chesapeake Bay (Setzler-Hamilton, 1991).

Its range in the Delaware River extends from Marcus Hook, Pennsylvania (river km 0), to Long Eddy, New York (river km 364) (Ashton et al., 1975), but it is most common in the tidal portion of the river south of Trenton. In the Hudson River, its 250 km range extends from Manhattan to Albany (Bath and O'Connor, 1982). Genetically distinct subpopulations may occur in certain larger tributaries to the lower Delaware Bay, as they do in tributaries to Chesapeake Bay (Mulligan and Chapman, 1989). Early life history stages occur in most Middle Atlantic Bight estuaries, although eggs and larvae are not found in systems lacking tidal freshwaters, such as Nauset Marsh, Massachusetts, or Great South Bay, New York (see Table 4.3).

REPRODUCTION AND DEVELOPMENT

Most spawning takes place in the lower reaches of large coastal rivers. Spawning occurs in brackish water (< 4 ppt) and freshwater areas during the early spring (late March through early June) after a migration from deeper, wintering habitats (Hardy, 1978b; Setzler-Hamilton, 1991; Beck, 1995; Jackson and Sullivan, 1995; Able et al., 2006a). In Delaware Bay, most activity takes place when temperatures are between 14 and 18°C (Wang and Kernehan, 1979). Spawning also occurs in the Chesapeake and Delaware (C&D) Canal, and in most larger tributaries to Delaware Bay (Wang and Kernehan, 1979; Weisberg and Burton, 1993). In Chesapeake Bay, spawning reaches a peak at temperatures of 10 to 16°C and usually occurs in freshwater or brackish water over beds of fine gravel or sand (Setzler-Hamilton, 1991). Spawning in the Hudson River takes place in shallow flats, embayments, and tidal creeks, primarily between river km 138 and 198 (Klauda et al., 1988). Spawning begins in this river in late April when temperatures are 10 to 12°C and reaches a peak in late May when they are 16 to 20°C.

The early development of *Morone americana*, and characters useful for distinguishing eggs, larvae, and juveniles from those of *Morone saxatilis*, have been well described and summarized (Fahay, 2007). Eggs are spherical and 0.65 to 1.09 mm in diameter. An attachment disk adheres the developing egg to the substrate. Hatching occurs at lengths of 1.7 to 3.0 mm, after 2 to 6 days, depending on temperature (Hardy, 1978b). Larvae are elongate, and the pre-anus length is > 50% of the SL. Pigment is very light. Larvae complete transformation to the juvenile stage at about 20 to 30 mm after about 6 weeks of passive, planktonic existence in near-surface waters (Wang and Kernehan, 1979; Public Service Electric and Gas Co., 1984). Body depth relative to length increases during this size interval. Fin rays are complete by 30 mm TL.

Individuals from 14.4 to 45.3 mm FL (n = 42) were examined for scale formation (see Fig. 32.1; Able et al., 2009a). Individuals from 14.4 to 15.5 mm FL lacked scales. Scales formed over the size range of 16.2 to 34.0 mm FL, as reported elsewhere (Marcy and Richards, 1974). Scales originate at several loci, including on the caudal peduncle, on the ventral surface below the pectoral fins, on the operculum, on the dorsal surface of the head, and behind the eye. Scales first appear on the caudal peduncle on the lateral line. By approximately 20 mm FL, scales have spread laterally and anteriorly so that they cover the caudal peduncle and up to the posterior edge of the pectoral fin. In some specimens of approximately the same size, scales cover much of the lateral surface of the body to behind the operculum but not the dorsal or ventral surfaces. By 22 mm FL, scales have spread dorsally and ventrally to cover much of the lateral surface of the body. At this size scales also appear on the ventral surface anterior to the origin of the pectoral fin. By 23 mm FL, scales also originate on the operculum and below the eye. At 25 mm FL, they are present on the dorsal surface of the head and spread to the base of the caudal fin. Scales continue to expand, and by 34 mm FL individuals attain the adult condition wherein scales extend farther on to the caudal fin and the only portions not covered are the snout and fins. By this size scales cover 96.5% of the body and 69.4% of the body and fins. Scales were complete over most of the body in 100% of those young-of-the-year larger than 23 mm FL in a Delaware Bay study (Wallace, 1971).

LARVAL SUPPLY, SETTLEMENT, GROWTH, AND MORTALITY

Larvae are often concentrated in "estuarine turbidity maximum" areas of upper estuaries where salinities are low (Shoji et al., 2005; North and Houde, 2006). This explains why ichthyoplankton sampling at Little Egg Inlet does not typically collect these stages. Settlement in more saline habitats can occur soon after the larvae transform into the juvenile stage (at approximately 45 days), although a portion of the cohort may remain in tidal freshwater habitats throughout its first year (Kraus and Secor, 2004b; Kerr et al., 2009). We do not understand why, but the abundance of juveniles and adults has increased in the Mullica River–Great Bay estuary in recent years (Fig. 60.1).

In general, growth is fastest during the first summer; 37 to 40% of the ultimate maximum size is reached after the first year (Wallace, 1971; St. Pierre and Davis, 1972; Bath and O'Connor, 1982). Growth rates of young-of-the-year are influenced by environmental and habitat factors and differ strongly between study areas (Setzler-Hamilton, 1991), in different temperatures (Klauda et al., 1988), and in varying salinities (Kraus and Secor, 2004b). For example, populations in freshwater impoundments grow faster than those in rivers (Bath and O'Connor, 1982). Sizes after the first year in several study areas range from 72 to 93 mm TL (Able and Fahay, 1998). Young-of-the-year grow through the summer and fall and reach these sizes in November, while older year classes stop growing in September of each year (Wallace, 1971). A positive correlation between mean river temperatures and annual variation in growth rate has been demonstrated for larvae and early juveniles (< 25 mm TL) in the Hudson River

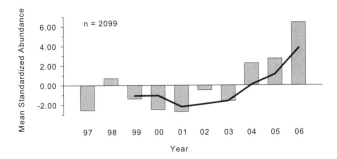

Fig. 60.1. Annual variation in abundance of *Morone americana* collected in otter trawl sampling, 1997–2006. Data were standardized by subtracting the overall mean from annual abundance. Vertical bars represent standardized annual abundance; the solid line is a three-year running mean that was calculated by taking the mean of the standardized values from the previous two years and the year in which the data are plotted.

(Able and Fahay, 1998), but in another study, rates during older juvenile stages (25 to 60 mm TL) were not correlated with temperatures, freshwater flow rate, or juvenile abundance (Klauda et al., 1988). Results of all these cited studies indicate that growth in all year classes is nearly nil through the winter. The lengths presented here from the Great Bay–Little Egg Harbor study area (Fig. 60.2) demonstrate some of these trends. Young-of-the-year exhibit a modal length of 7 cm FL as they enter their first winter. Modes for this cohort are about the same the following spring, and at age 1 they have grown to about 10 to 15 cm. Growth in the Hudson River is comparable to that of young-of-the-year from the Delaware region, where a mode of 8 cm TL is apparent at the end of the first summer (Able and Fahay, 1998). Adults reach ages of 7 to 10 years and sizes of 200 to 270 mm in Middle Atlantic estuaries (Wallace, 1971; St. Pierre and Davis, 1972). Eggs generally incubate at temperatures between 10 and 20°C, but higher temperatures or suddenly dropping temperatures can have lethal effects on the eggs (Setzler-Hamilton, 1991). Optimum survival and growth of larvae is attained at temperatures between 15 and 20°C (Marguiles, 1988).

SEASONALITY AND HABITAT USE

All life history stages are semianadromous, with seasonal migrations occurring between deeper, highly saline areas of estuaries during winter to shallow, brackish, or freshwater areas from early spring through summer (Beck, 1995). Adults are often found near structured habitats in bays, estuaries, and brackish streams. Eggs incubate in temperatures between 10 and 20°C, with the optimum at about 14°C (Morgan and Rasin, 1982). All early stages have a high tolerance to salinity differences, ranging from 0 to 13 ppt (Dovel, 1971). Turbidity affects the distribution and feeding ecology of larvae. Many coastal plain estuaries include "estuarine turbidity maximum" zones that are characterized by relatively high turbidity and concentrations of suspended sediments, mesozooplankton, and ichthyoplankton (Schubel, 1968; Shoji

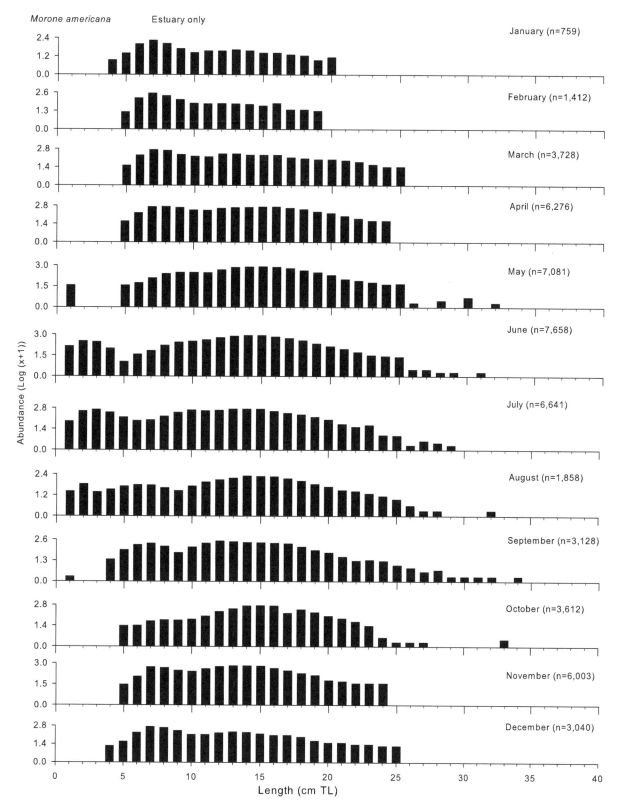

Fig. 60.2. Monthly length frequencies of *Morone americana* collected in
the Great Bay–Little Egg Harbor study area. Sources: Rutgers University
Marine Field Station: seine (n = 4); 4.9 m otter trawl (n = 3045); multi-
panel gill net (n = 1397); and Able and Fahay 1998 (n = 37,707).

et al., 2005). Studies have demonstrated that these areas are important nurseries for fishes in the Chesapeake Bay and other systems in the world (Dodson et al., 1989; North and Houde, 2001, 2006; Bennett et al., 2002). Improvements in water quality in the Delaware River, especially between Philadelphia and Wilmington, have resulted in enhanced survival of a congener, *M. saxatilis*, but may have had less effect on *M. americana* production, presumably because young-of-the-year of the latter species complete their development upstream from the area formerly impacted (Weisberg and Burton, 1993). The latter study, however, did find increased abundances of larvae in the impacted part of the river compared to studies undertaken before the improvements.

Juvenile habitat during the first summer tends to be located downstream from spawning areas, although a portion of the cohort may remain in tidal, freshwater habitats (Kraus and Secor, 2004b). Young-of-the-year occur most frequently in relatively shallow, brackish to freshwater areas, including tidal creeks. They appear to prefer level bottoms of compact silt, mud, sand, or clay with little or no cover and to avoid soft muck, highly organic substrate, and gravelly or rocky areas (Stanley and Danie, 1983; Setzler-Hamilton, 1991). Vegetation or other structural components are not essential habi-

tats (Hardy, 1978b; Stanley and Danie, 1983). In an extensive study of the Delaware Bay and River, small young-of-the-year (30 to 50 mm) were present in marsh shallow habitats by June (Able et al., 2007a). Larger young-of-the-year (30 to 60 mm) can be found in most habitats by July and remain there through the summer and early fall. Typically they were most abundant at 0 to 12 ppt; however, some move downstream and into deeper water as they increase in length—a pattern consistent with that in many other estuaries (Mansueti, 1964; Klauda et al., 1988; Setzler-Hamilton, 1991). In general, age 1+ individuals (50 to 290 mm) were most abundant in the river during May to October and in the bay from November through April. In a comparison of the distribution from the ocean to tidal freshwater in the Mullica River–Great Bay estuary, most individuals were captured in the low-salinity portion of the estuary (Fig. 60.3). In an examination of habitat use by acoustically tagged fish (208 to 255 mm FL) in the tributaries of the York River, Virginia, the pattern of use varied over the average of 21 days of tracking (McGrath and Austin, 2009). During the summer, individuals in marshes used both the marsh surface and shallow creeks at high tide and deeper channels at low tide. Individuals associated with submerged structures remained in the same location. In both habitats,

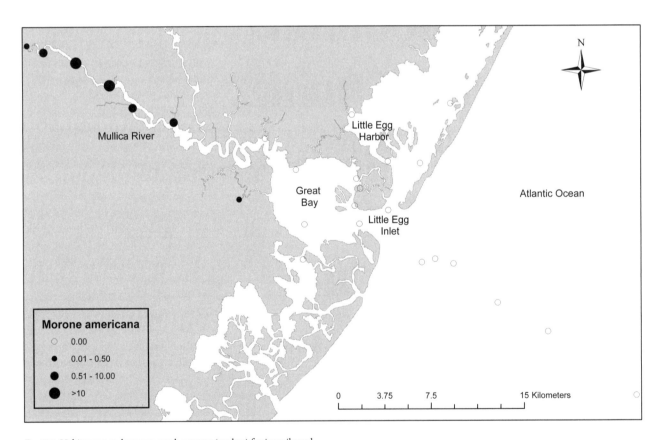

Fig. 60.3. Habitat use and average catch per tow (see key) for juvenile and adult *Morone americana* in the Ocean–Great Bay–Mullica River corridor in southern New Jersey based on small otter trawl collections during July and September, 1988 to 2006. See Table 3.2 for characteristics at each sampling collection.

the home range was small (0.11 km²) and there was a high degree of site fidelity.

PREY AND PREDATORS

Adults are carnivorous predators with diets focusing on crabs, shrimp, and small fishes (Murdy et al., 1997). A study in the Hackensack Meadowlands of northern New Jersey found that the dominant diet items were gammarid amphipods, followed by shrimp and small fishes (Weis, 2005). Larvae feed on copepodite and adult stages of the copepod *Eurytemora affinis,* which they find in high concentrations within estuarine turbidity maximum zones (Shoji et al., 2005). Studies have demonstrated that annual variability in freshwater runoff may affect the structure of the turbidity maximum. In years when river flow rates result in higher concentrations of this copepod, larval feeding success and survival are enhanced (Shoji et al., 2005). Several other studies have examined the food habits of young-of-the-year (21 to 139 mm, n = 2971) in Middle Atlantic Bight estuaries, from the Connecticut River to the York River, Virginia (see Tables 8.1 and 8.3). Important prey items of young-of-the-year include a variety of invertebrates (zooplankton, nemertines, polychaetes, oligochaetes, insects/arachnids, mysids, cumaceans, isopods, amphipods, decapod shrimp, crabs, ostracods, copepods, cladocerans, invertebrate eggs, and detritus) and fishes (*Anguilla rostrata* and *Notropis hudsonius*), including incidences of cannibalism (Marcy, 1976). The diet of young-of-the-year shifts with ontogeny at approximately 60 to 110 mm, becoming more diverse with increased size (Miller, 1963; Marcy, 1976; Smith et al., 1984; Bath and O'Connor, 1985). Marcy (1976) found young-of-the-year (< 100 mm TL, n = 759) fed less on plankton (copepods, cladocerans, and amphipods) and more on benthic organisms (oligochaetes and polychaetes) and insects as size increases. Diet also varies with salinity, such that insects and zooplankton are important to young-of-the-year (60 to 139 mm, n = 611) in the limnetic-mesohaline Mill Creek area of upper Delaware Bay, while mysids are more important to young-of-the-year (60 to 139 mm, n = 97) in the mesohaline-polyhaline Dennis/Moores areas of lower Delaware Bay salt marshes (Nemerson, 2001). Bath and O'Connor (1985) found young-of-the-year (< 110 mm) in the Hudson River ate amphipods, insect larvae, and cladocerans in the spring, while amphipods, isopods, and decapod shrimp were important food items in the summer. Relative importance of individual prey items also varies intra-annually in the Delaware River, the Hudson River, and the lower Connecticut River estuaries (Miller, 1963; Marcy, 1976; Bath and O'Connor, 1985).

Morone americana is a frequent prey for estuarine piscivores. A review of the available literature of diets of young-of-the-year fishes found this species was preyed upon by three different species, was an important prey for one of these species, and occurred in the diet in another 18 studies, primarily *Pomatomus saltatrix* (see Table 8.3; Mancini and Able, 2005). *Morone americana* also occurred in the diet of other large predators in the Mullica River–Great Bay estuary, including *Esox niger* and *Ameiurus catus* (see Chapter 8).

MIGRATIONS

In the Delaware Bay area, young-of-the-year leave shallow nursery areas during the fall and spend their first winter in deeper parts of the lower river and upper bay (Beck, 1995). In the Hudson River, older juveniles begin moving from shore and shoal areas into deep midriver areas in October, and by mid-December a majority of this age class is found in deep areas (Klauda et al., 1988). In the Chesapeake Bay area, they overwinter in deep channel areas with depths between 12 and 18 m (maximum 42 m) and in temperatures between 2 and 5°C (Setzler-Hamilton, 1991). In Chesapeake Bay, a portion of the Patuxant River population resides in freshwater natal habitats, while another portion migrates down the estuary into brackish waters (Kerr et al., 2009).

61

Morone saxatilis (Walbaum)

STRIPED BASS

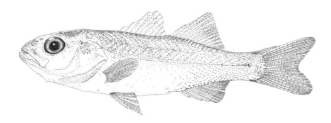

Estuary	Reproduction	Larvae/Development	YOY Growth	YOY Overwinter
Ocean				
Both				
Jan				
Feb				
Mar				
Apr				
May				
Jun				
Jul				
Aug				
Sep				
Oct				
Nov				
Dec				

DISTRIBUTION

Morone saxatilis occurs along the Atlantic coast of North America between the St. Lawrence River, Canada, and St. Johns River, Florida, and in the northern Gulf of Mexico from Florida to Louisiana (Heemstra et al., 2002). Within this range, different populations are found, many migratory, some non-migratory, and some in freshwater lakes. Several principal anadromous stocks have been recognized in the Middle Atlantic Bight, centered in Roanoke River (North Carolina); Chesapeake Bay and tributaries; Delaware River; and Hudson River (Fabrizio, 1987; Waldman et al., 1990). Juveniles occur in most estuaries throughout the Middle Atlantic Bight, but eggs and larvae have primarily been reported from the Hudson River and the Chesapeake and Delaware bays (see Table 4.3).

REPRODUCTION AND DEVELOPMENT

Males mature at 2 years of age and sizes of 17 cm TL. Females mature at 4 to 6 years of age and sizes of 45 to 55 cm TL (Heemstra et al., 2002). Throughout its range, spawning may occur between February and July, primarily in rivers or heads of estuaries. This species spawns from the limit of brackish water to freshwater in small tributaries of the Chesapeake Bay as well as certain tributaries to, and mainstems of, the Delaware and Hudson rivers from early April through the end of May (Fabrizio, 1987; Waldman et al., 1990). Major spawning areas include tributaries of the Chesapeake Bay—Potomac, James, and York rivers, including the Mattaponi and Pamunkey rivers in Virginia (Bilkovic et al., 2002)—and many smaller rivers along the eastern shore of Maryland, but significant reproduction also occurs in the Roanoke River. We have found very limited evidence that reproduction occurs in the Mullica River–Great Bay system (see "Seasonality and Habitat Use" below).

The early development of this species has been well described and summarized (Fahay, 2007). Eggs, larvae, and juveniles are similar to those of *Morone americana*. Patterns of interdigitation of fin ray pterygiophores and adjacent neural and haemal spines are critical for accurate identification of early stages of these two species (Olney et al., 1983). Eggs are spherical and semi-buoyant. Their diameters range from 1.3 to 4.6 mm, and vary widely with different salinities. The single oil globule ranges from 0.40 to 0.85 mm. Hatching occurs after 48 hours of incubation at temperatures of about 17 to 18°C but ranges between 29 hours (at 24°C) and 80 hours (at 12°C) (Hardy, 1978b). Larvae are 2.0 to 3.7 mm long at hatching and are slender, with a pre-anus length > 50% of the SL. Individuals from 16 to 88.2 mm FL (n = 28) were examined for scale formation (Fig. 61.1; Able et al., 2009a). Scales form over the size range of 16 to 56.9 mm FL. Scales originate at several loci, including on the caudal peduncle, on the ventral surface of the body below the pectoral fin, on the operculum, and behind the eye. Scale formation begins on the caudal peduncle on the lateral line at approximately 13.9 mm SL. These scales spread anteriorly and laterally such that by 15.9 mm SL they reach the posterior margin of the pectoral fin and the dorsal and ventral surfaces of the body next to the second dorsal and anal fins. By 18.2 mm SL, the scales on the lateral surface of the body extend to the operculum, begin forming on the operculum and behind the eye, and also start to form on the ventral surface of the body below the pectoral fin. By 20.6 mm SL, scales cover almost all of the body behind the operculum except anterior to the pectoral

---

fin. By 25.5 mm SL, the body behind the operculum is completely covered, the lateral surfaces of the head have more scales, and scales are present on the base of the caudal fin. By 38 mm SL, scales cover most of the body except the snout. By 50 mm SL, scales have also developed on the base of the second dorsal and anal fins. At this size, scales cover 93.5% of the body and 73.2% of the body and fins.

LARVAL SUPPLY, SETTLEMENT, GROWTH, AND MORTALITY

Eggs and larvae are supplied from the major spawning areas described above. Successful transport of these larvae into the estuarine turbidity maximum appears critical to survival at these stages (Martino, 2008).

Growth rates of young-of-the-year appear to be uniform across a number of locations (Able and Fahay, 1998). In the Delaware River population, growth continues from spring through summer and the year class reaches a mode of about 10 cm by late fall (Fig. 61.2). This same pattern was evident for the Hudson River (Kahnle and Hattala, 1988; McKown,

1991; Able and Fahay, 1998). However, lengths of the age 1 size class in the lower Hudson River and in western Long Island embayments differ; the latter were between 39 and 57 mm larger than the former (McKown, 1991). Growth apparently ceases during the winter.

When growth rates of young-of-the-year were compared to growth rates of young-of-the-year *Pomatomus saltatrix* in the Hudson River, no evidence was found to suggest that young-of-the-year of the two species compete for resources. Growth rates in both species were found to be independent of the other species' densities (Buckel and McKown, 2002). The combined effects of temperature and salinity may affect growth rates. A two-year study in the Pamunkey River, Virginia, showed that temperature can affect the timing of hatching and larval development (McGovern and Olney, 1996). When temperatures are lower, food densities are also lower, and the duration of development is extended, thus exposing larvae to elevated numbers of predators for a longer period. When temperatures are higher, development takes a shorter period of time and coincides with an increase in food

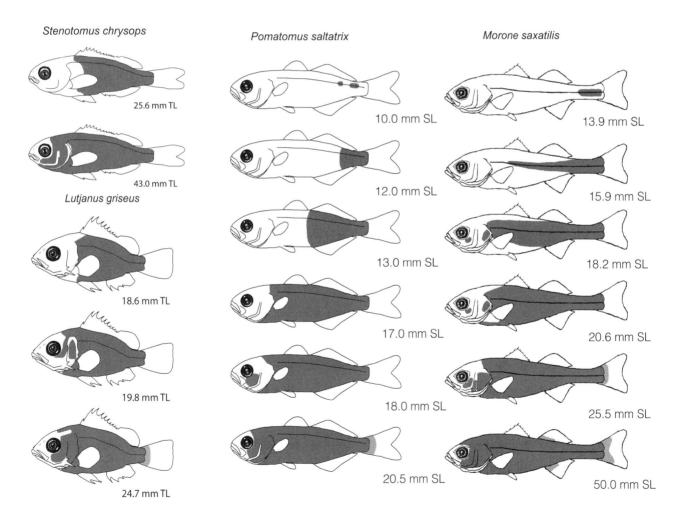

Fig. 61.1. Description of scale formation relative to length in *Morone saxatilis, Pomatomus saltatrix, Lutjanus griseus,* and *Stenotomus chrysops.*

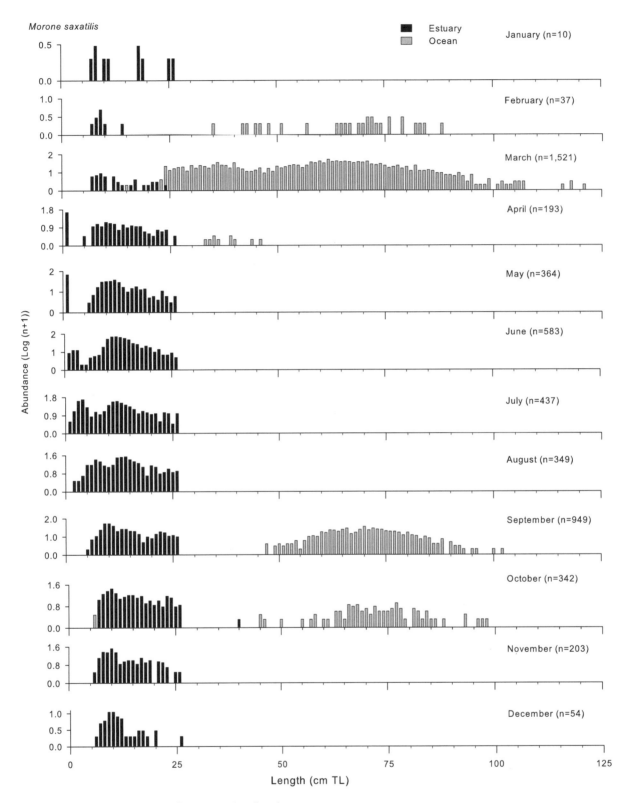

Fig. 61.2. Monthly length frequencies of *Morone saxatilis* collected in estuarine and coastal ocean waters of the central Middle Atlantic Bight, including the Upper Delaware Bay and tidal portions of the Delaware River. Sources: National Marine Fisheries Service: otter trawl (n = 2156); Rutgers University Marine Field Station: seine (n = 1); 4.9 m otter trawl (n = 1); and Able and Fahay 1998 (n = 2884).

items in a field of reduced predator density. Year-class success is greater in the year with higher temperatures during the time of peak egg production. Salinity was found to have an effect on Chesapeake Bay fishes, such that those at 7 ppt experienced 40% higher growth than those at 0.5 or 15 ppt. In addition, populations from higher latitudes experience faster growth rates than populations from lower latitudes (Conover, 1990). In Chesapeake Bay, the faster growth of larvae is associated with wetter springs and improved retention within the estuarine turbidity maximum (Martino, 2008).

Winter mortality is size-selective in this species (Hurst and Conover, 1998). This study found that the abundance of age 0 fish was not correlated with the numbers of age 1+ fish. Furthermore, age 1+ fish were negatively correlated with the severity of the first winter. Therefore, winter mortality has a great effect on recruitment. Winter mortality of young-of-the-year may be affected by the amount of energy they expend in finding prey (Hurst and Conover, 2001). When prey fishes are available in sufficient quantity, energetic stress may be lowered because the metabolic cost of swimming between suitable habitats is reduced. When the combined effects of temperature and salinity were related to winter mortality in young-of-the-year, survival at low temperatures was highest in intermediate salinities (between 0 and 35 ppt) (Hurst and Conover, 2001). These authors suggest that the winter distribution of young-of-the-year within their study estuary is largely determined by osmoregulatory abilities of young-of-the-year at low temperatures. Size-selective mortality for young-of-the-year was also detected in Chesapeake Bay (Martino, 2008). This effect was magnified in years in which density-dependent growth was evident.

SEASONALITY AND HABITAT USE

A general model of reproduction and early life history in anadromous populations, as it relates to habitats, suggests that during the spring, adults migrate from coastal areas and spawn in freshwater. In some systems, fish that are resident in estuaries will spawn in upstream freshwaters. Eggs are deposited just upstream from the salt front (Secor and Houde, 1995), generally in the first 40 km upstream (Rathjen and Miller, 1957). In the Delaware River, spawning activity occurs between river km 94 and 201 (Murawski, 1969), where salinity is less than 3.0 ppt (Wang and Kernehan, 1979). Eggs are also abundant in the C&D Canal, between Chesapeake and Delaware bays, where concentrations as high as 110 eggs and 95 larvae/m³ have been collected (Kernehan et al., 1977). Most spawning in the Hudson River occurs between West Point and Kingston (Smith, 1985), and recent collections of eggs have spanned the 90 to 153 km and 37 to 164 km areas above saltwater (Texas Instruments, 1973, cited in Dovel, 1981). Limited spawning activity has also been reported in the Mullica River in southern New Jersey (Hoff, 1976), but it is not clear that it is successful because very few young-of-the-year have been collected (Able and Grothues, 2007b).

In the Chesapeake and Hudson systems, spawning begins when temperatures reach 11°C (Dovel, 1981). In the Delaware Bay region, spawning starts when temperatures reach 14°C (Wang and Kernehan, 1979; Fay et al., 1983).

Habitat use during the first year varies with life history stage. After spawning, there is a net movement of young-of-the-year from upstream locations to those in lower tidal reaches; this dispersal may be related to areas with increased abundance of prey (Markle and Grant, 1970). The control of recruitment in Chesapeake Bay appears to vary with hydrological regimes as related to the development of the estuarine turbidity maximum (Martino, 2008). In wet years, it is well developed and eggs and larvae are retained there, as well as more prey, thus resulting in better feeding and enhanced survival. Subsequent downstream dispersal also helps juveniles find improved overwintering habitats in the lower estuary (Dovel, 1992). Adults inhabit coastal waters, including bays, estuaries, and rivers. They are found off sandy beaches and in rocky areas. They winter in depths up to 37 m. Late larvae and early juveniles favor shallow waters (including tidal creeks) with sluggish currents and sand or gravel bottoms (Smith, 1971; Wang and Kernehan, 1979; Boynton et al., 1987). Young-of-the-year of the Delaware River population migrate downstream from spawning locations and spend their first summer within the tidal portions of the river and other habitats (Nemerson and Able, 2003; Able et al., 2007a). Most young-of-the-year from the Hudson River also move downstream from spawning areas and spend their first summer and winter in the lower Hudson River (McKown, 1991). In this river, young-of-the-year have been found in and around pier pilings (Stoecker et al., 1992; Able et al., 1998); in the lower Hudson River estuary, they appear to be more abundant in deep inter-pier habitats than in shallow ones during late summer (Cantelmo and Wahtola, 1992). After young-of-the-year overwinter in the lower Hudson River estuary, age 1 individuals are found the following spring both in the lower Hudson and in various bays on the north and south shores of western Long Island (McKown, 1991), especially Jamaica Bay, Little Neck Bay, Manhasset Bay, and, in some years, Bellport Bay. In years with strong year classes, young-of-the-year are found beyond the lower Hudson River and are included in the catches of age 1 (and older) individuals in certain of these embayments (McKown, 1991).

The range of acceptable environmental conditions is relatively well known for this species. An important requirement for the spawning area is a current strong enough to keep the eggs suspended in the water column, lest they settle to the bottom and be smothered by silt (Bigelow and Schroeder, 1953). The upper lethal temperature for developing embryos is 27°C (Morgan and Rasin, 1982). Eggs and larvae are less tolerant of salinity fluctuations than adults, and their survival is enhanced in low salinities (2 to 10 ppt) (Fay et al., 1983). Turbidity also adversely affects larvaes' ability to capture prey (Fay et al., 1983), although there may be more food concentrated in the estuarine turbidity maximum. Optimum temperatures for larval growth and survival are 15 to 22°C (Fay et al., 1983; Funderburk et al., 1991); rapidly changing temperatures can be detrimental to this life history stage (Hollis, 1967). The optimum range for juveniles is 18 to 23°C (Fay et al., 1983).

Acoustically tagged large juveniles and adults in the Mullica River–Great Bay estuary preferred warmer than average temperatures in the spring and cooler temperatures in the summer (Ng et al., 2007). For dissolved oxygen, the optimum levels for all life history stages is 6 to 12 mg/l (Fay et al., 1983). Levels below 2.4 mg/l are lethal for larvae, and those below 3 mg/l are lethal for juveniles (Fay et al., 1983). Some males, with elevated levels of PCB (polychlorinated biphenyl) contamination, tend to remain in freshwater habitats, rarely exceeding 5 ppt (Zlokovitz and Secor, 1999). Another study demonstrated that oyster reefs have been shown to be more important to the distribution of *Morone saxatilis* than sand substrates in the Chesapeake Bay (Harding and Mann, 2003).

Pollution and resulting low oxygen levels historically prevented successful reproduction in the lower Delaware River (Chittenden, 1971), but improved water quality in the region over the period 1980 to 1993 has resulted in increased reproduction and enhanced production and survival of juveniles (Weisberg and Burton, 1993). The latter authors cite a thousandfold increase in the abundance of juveniles during the decade studied and correlate this increase with improved water quality, not with an overall increase in the population. In particular, this improvement has resulted in reductions of anoxic conditions that once formed a block to migratory species, especially in that part of the river downstream from Philadelphia during late summer. Thus, historically important spawning and nursery areas that had been lost to the population (Chittenden, 1971) have been restored to a condition allowing for its use. The effects of hypoxia are manifold. Juvenile *Morone saxatilis* have been shown to decrease predation on *Gobiosoma bosc* larvae at low dissolved oxygen concentrations (< 2 m/l⁻¹) (Breitburg et al., 1994). On the contrary, in Chesapeake Bay, growth rates in age 2 and age 4 fish were enhanced in high-quality habitats affected by hypoxia, primarily because prey fishes were forced into warmer, well-oxygenated layers where *Morone saxatilis* typically forages (Costantini et al., 2008). However, this study also found that this hypoxia-enhanced predation, combined with increased numbers of *Morone saxatilis* during a recent recovery of most populations, has led to overconsumption of available prey and has resulted in poor growth and health in these fish in recent years.

In the Delaware River system, young-of-the-year and ensuing year classes may spend two years or more within the estuary before migrating offshore and joining the migratory population (Miller, 1995). Most overwinter in deeper portions of the estuary, but some also overwinter in tidal creeks (Smith, 1971). Studies in western Long Island embayments

indicate that Hudson River progeny may spend up to three years in these estuaries (McKown, 1991) before beginning coastal migrations with the adults.

The results of extensive acoustic tagging studies in the Mullica River–Great Bay estuary clearly identified spring and fall arrivals and departures of larger (508 to 978 mm TL) coastal migrants but also a diverse array of individual movement patterns in the estuary that varied among seasons and years (Able and Grothues, 2007b). Upstream movements in the spring, a relatively common pattern, may have been associated with attempts to follow anadromous *Alosa* spp. upstream as prey (Walter and Austin, 2003) or attempts, apparently unsuccessful, to spawn (Ng et al., 2007). Other studies in the same system have shown that fish (483 to 953 mm TL) that are seasonal residents can spend long periods of time during the summer with relatively little movement (Ng et al., 2007). During these periods, they often prefer structural habitats such as marsh banks, sandbars, and other structures. These findings are consistent for other estuaries; habitats include oyster reefs (Harding and Mann, 2003) and man-made structures, such as bridge supports, channel markers, and pound net stakes (Haeseker et al., 1996). In addition, fish tracked in the Mullica River–Great Bay estuary preferred deeper than average sites, such as channels and drop-offs, especially during the summer (Ng et al., 2007). These patterns were affected to some degree by diel patterns in which there was slightly greater activity at night than during the day, with increased activity around sunset.

The adults of *Morone saxatilis* are known to be periodic residents and/or visitors to the Mullica River–Great Bay estuary (Able and Grothues, 2007b; Ng et al., 2007; Grothues et al., 2009) but less is known about the seasonality and habitat use of juveniles in this and other estuaries. The available data from other temperate estuaries indicate they may be resident for two or three years in the natal stream or leave the natal stream to spread onto adjacent estuaries during the first summer (Waldman et al., 1990). In an attempt to resolve these differences in the literature we ultrasonically tagged and tracked juveniles with a combination of stationary listening arrays and regular mobile tracking in the Mullica River–Great Bay estuary during the summer of 2006 (P. Clerkin, T. M. Grothues, and K. W. Able, unpubl. data). These fish were tagged in the upper portion of the estuary and were resident at the same location during most of the summer.

PREY AND PREDATORS

Adults feed mostly on fishes and crustaceans (Heemstra et al., 2002). Juveniles concentrate on crustaceans. A recent compilation has surveyed the feeding habits of young-of-the-year and adults along the Atlantic coast (northern New Jersey to Nova Scotia), Chesapeake and Delaware bays and the adjacent coastline of the Middle Atlantic Bight, and North Carolina (Walter et al., 2003). Adults in all studies made fishes the most important component of their diet. *Brevoortia tyrannus* was often cited as the most important prey fish, but various species of sciaenids or clupeids were also primary targets. In certain isolated studies, blue crabs or fundulids were the most important prey (by weight). In the several studies surveyed, young-of-the-year concentrated on mysids, amphipods, or *Crangon septemspinosa* in most areas, but in some studies in the Chesapeake Bay or its major tributaries, small fishes were most important.

The diets of Massachusetts large juvenile and adult coastal fish have been analyzed (Nelson et al., 2003). Fishes (Clupeidae, *Menidia* sp., and *Ammodytes* sp.) and crustaceans (*Crangon septemspinosa, Cancer irroratus,* and *Homarus americanus*) dominated the diet, but there were differences between seasons, habitats, and size of the predator. *Brevoortia tyrannus* dominated the diet items in estuarine and rocky shorelines during August and September, but was less important during June and July. Further, the diets of fish from rocky shorelines included more *Cancer irroratus* than the diets of estuarine fish. Larger fish (> 675 mm TL) in rocky habitats were more likely to feed on *Homarus americanus* than smaller fish in similar habitats.

Larger juveniles and adults captured by gill nets in the Mullica River–Great Bay estuary feed on large amounts of fishes and crabs as well as other crustaceans based on the frequency of occurrence and weight of prey in their stomachs (see Figs. 8.3 and 8.4). Other prey include worms with a high frequency of occurrence of unidentified prey. Among the fish prey, abundant species includes *Brevoortia tyrannus* and other clupeids as well as *Fundulus heteroclitus*. Engraulids and other unidentified fishes were also included. Similar food habits were observed in Delaware Bay marsh creeks (Tupper and Able, 2000).

Several studies have examined the food habits of young-of-the-year (25 to 160 mm, n = 2657) *Morone saxatilis* in Middle Atlantic Bight estuaries from the Hudson River, New York, to Albemarle Sound, North Carolina (see Tables 8.1 and 8.3). Important prey items of young-of-the-year include a variety of invertebrates (polychaetes, insects, mysids, decapod shrimp, copepods, cladocerans, and *Limulus spp.* eggs and larvae) and fishes (*Fundulus diaphanus, F. heteroclitus, Anchoa mitchilli, Gobiosoma bosc,* and *Micropogonias undulatus*). The diet of young-of-the-year shifts with ontogeny, at approximately 70 to 100 mm, from small invertebrates (mysids, insects, amphipods, copepods, cladocerans, polychaetes, and isopods) at smaller sizes to larger invertebrates (decapod shrimp and crabs) and fishes (*Microgadus tomcod, Fundulus diaphanus, F. heteroclitus, Micropogonias undulatus,* and *Anchoa mitchilli*) at larger sizes (Markle and Grant, 1970; Gardinier and Hoff, 1982; Hartman and Brandt, 1995c; Cooper et al., 1998; Nemerson and Able, 2003). The diet also varies with salinity. When salinity decreases from mesohaline to limnetic/oligohaline portions of the estuaries, piscivory increases and mysid consumption decreases (Markle and Grant, 1970; Boynton et al., 1981; Coo-

per et al., 1998; Nemerson and Able, 2003). Insects are also more common food items in limnetic/oligohaline areas than in mesohaline areas (Markle and Grant, 1970; Boynton et al., 1981). Geographical variation in diet has also been noted within the Potomac, Albemarle Sound, and Delaware Bay estuaries (Boynton et al., 1981; Cooper et al., 1998; Nemerson and Able, 2003) and between Virginia rivers (Markle and Grant, 1970). Relative importance of individual prey items also varies both intra-annually in the Hudson River and Chesapeake Bay estuaries (Gardinier and Hoff, 1982; Hartman and Brandt, 1995b) and inter-annually in the Chesapeake Bay (Hurst and Conover, 2001). Young-of-the-year from Albemarle Sound preferred mysid shrimp as the most important prey, and fishes were minor components of the diet (Cooper et al., 1998). Furthermore, the importance of mysid shrimp increased with increasing lengths of the young-of-the-year. These authors also found that higher percentages of mysids were consumed in habitats with higher salinities.

In the Hudson River estuary, young-of-the-year *Pomatomus saltatrix* impose significant mortality on young-of-the-year *Morone saxatilis*. In this system, when densities of *M. saxatilis* young-of-the-year are high, predation by young-of-the-year *P. saltatrix* can also be very high (Buckel and Stoner, 2000). Between 50 and 100% of *M. saxatilis* young-of-the-year losses may be attributable to predation by young-of-the-year *P. saltatrix*. As a result, in this system, when *P. saltatrix* abundance is high, *M. saxatilis* year-class strength may be severely impacted (Buckel et al., 1999b). The feeding response by young-of-the-year *P. saltatrix* is density dependent and increases when young-of-the-year *M. saxatilis* are most abundant.

MIGRATIONS

Young-of-the-year (< 10 cm) are usually not represented in trawl collections on the continental shelf. An unusual exception is a single individual reported on the southern edge of Georges Bank (Fig. 61.3). Age 1+ (> 10 cm) juveniles and adults, however, are frequently encountered on the inner and middle portions of the continental shelf, depending on season. During the fall, these catches occur on the inner continental shelf (mean depth 20 m, range 8 to 55 m) from north of Cape Cod on to Nantucket Shoals and off Long Island, with a few individuals off New Jersey. During the winter, catches are reduced and most individuals average slightly deeper (mean depth 27 m, range 13 to 37 m) from off Long Island to North Carolina. By spring, they occur abundantly on the inner shelf (mean depth 34 m, range 13 to 37 m) from North Carolina to Long Island, with a few individuals north of Cape Cod. The larger juveniles and adults of both sexes of many populations undertake seasonal migrations, moving north in late winter to early spring and south in the fall (Waldman et al., 1990; Heemstra et al., 2002). Females gradually shift to increased use of ocean habitats at about ages 5 to 8. The percentage of females using oceanic habitats increases from 50% at age 7 to 75% at age 13. A smaller percentage

Age 1+ *Morone saxatilis*

FALL

WINTER

SPRING

YOY *Morone saxatilis*

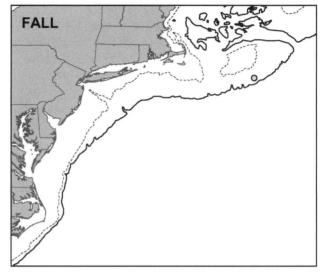

FALL

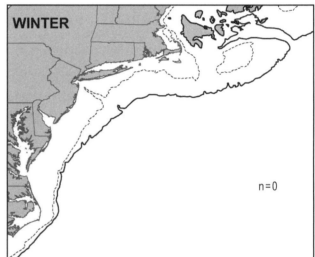

WINTER

n=0

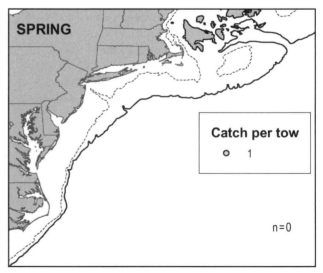

SPRING

n=0

Catch per tow

○   1 - 5
○   6 - 15
○   16 - 30
○   31 - 60
○   61 - 120

Catch per tow

○   1

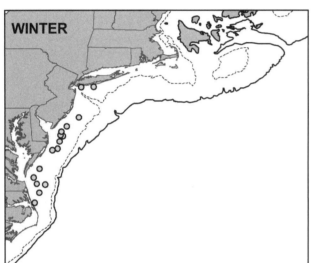

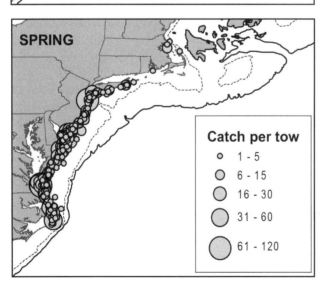

Fig. 61.3. Composite distribution of age 1+ (> 10 cm) and young-of-the-year (< 10 cm) *Morone saxatilis* during seasonal cruises of the National Marine Fisheries Service groundfish survey. Details of sampling effort are indicated in Fig. 3.4.

of males shift to an oceanic existence, with an increase from 20% at age 4 to 50% at age 13 (Secor and Piccoli, 2007). The degree to which temperate populations migrate varies between contingents (Secor, 1999; Secor et al., 2001; Zloko-vitz et al., 2003) and individuals (Able and Grothues, 2007b; Grothues et al., 2009). Detailed studies of acoustically tagged large juveniles and adults demonstrated seasonal movements, including those between southern New Jersey and Maine es-tuaries, and that repeated seasonal movements could be de-tected, including details of estuarine site fidelity (Ng et al., 2007).

62

Centropristis striata (Linnaeus)

BLACK SEA BASS

Estuary	Reproduction	Larvae/ Development	YOY Growth	YOY Overwinter
Ocean				
Both				
Jan				
Feb				
Mar				
Apr				
May				
Jun				
Jul				
Aug				
Sep				
Oct				
Nov				
Dec				

DISTRIBUTION

Centropristis striata is found from the Gulf of Maine to Florida (Miller, 1959), but it is most abundant in the Middle Atlantic Bight (Kendall and Mercer, 1982). The Middle Atlantic Bight population is considered to be a separate stock from populations south of Cape Hatteras (Mercer, 1978; but see Bowen and Avise, 1990). This interpretation is supported by studies that show seasonal inshore-offshore and north-south move-

ments in the Middle Atlantic Bight but only localized movements in the South Atlantic Bight (see Able et al., 1995a). Morphological studies have suggested that stock structure may not be homogeneous in the Middle Atlantic Bight (Shepherd, 1991). Juveniles and adults are found in most estuaries in the Middle Atlantic Bight (see Table 4.3).

REPRODUCTION AND DEVELOPMENT

This species is a sequential hermaphrodite that matures first as a female, then changes to a male at ages of 1 to 8 years (Lavenda, 1949; Mercer, 1978; Wenner et al., 1986). Collections of eggs over several years (Berrien and Sibunka, 1999) indicate that spawning occurs on the Middle Atlantic Bight continental shelf from April through October. Spawning begins between Cape Hatteras and Chesapeake Bay in April or May and expands northward, especially along the inner continental shelf, and reaches a peak in August, when eggs are most concentrated between Cape Hatteras and northern New Jersey (Able and Fahay, 1998). Subsequently, egg abundance declines, and the distribution of eggs becomes patchy. Off the central portion of the study area, spawning occurs from June through October (Able et al., 1995a) as indicated by small larvae on the inner and middle continental shelf (Able and Fahay, 1998). There is no evidence that spawning takes place in bays or estuaries, although larvae are occasionally collected in these habitats.

Details of early development in this species have been summarized (Fahay, 2007). More specifically, the pelagic eggs hatch in 35 to 75 hours, depending on temperature (Wilson, 1891; Tucker, 1989). Larvae are about 2.0 mm long at hatching and are characterized by a deep body, light pigment pattern, and early development of small spines on various bones of the head. Fin spines and rays are completely formed at about 9.0 mm.

Individuals from 13.3 to 71.2 mm TL (n = 71) were examined for scale formation (see Fig. 32.1; Able et al., 2009a). Scales formed over the size range of 15.6 to 71.2 mm TL and thus occurred later than the size at completion of fin rays (9 mm). Those individuals from 13.3 to 14.6 mm TL lacked scales. The onset of scale formation occurred in several locations simultaneously, including the midline of the body, at the anterior portion of the lateral line, on the edge of the opercle, and behind the eye. In our samples, scales are first evident on the midline of the body from behind the opercle to the base of the caudal fin at approximately 16 mm TL, but smaller specimens were not available. At this size, scales are also forming on the anterior portion of the lateral line just above and behind the operculum. By 18 mm TL, scales occur on the edge of the opercle and have spread laterally to cover much of the surface of the body with the exception of the ventral surface below the pectoral fin. By approximately 23 mm TL, scales first appear on the base of the caudal fin and cover all of the body, including a large portion of the operculum. By 25 mm TL, scales have formed on the pre-opercle behind the eye as well. By 36 mm TL, most of the operculum is covered with scales. By 71 mm TL, the scales have reached the adult condition, including scales on the bases of the caudal fin, pectoral fin, and second dorsal fin. Only the snout and small portions of the operculum are unscaled. At this size, 94.0% of the body is covered with scales and 59.3% of the body and fins is covered.

LARVAL SUPPLY, SETTLEMENT, AND GROWTH

Development of pelagic larvae takes place in continental shelf waters in the Middle Atlantic Bight (Kendall, 1972). Developing larvae (3 to 13 mm SL) are most abundant in the southern portion of the bight, partly because spawning occurs over a longer period in that region (Able et al., 1995a). At Beach Haven Ridge on the inner continental shelf off southern New Jersey, small larvae (3 to 10 mm NL; Able et al., 2006) are collected in July and August. Others occur in surfzone and nearshore habitats off northern New Jersey (Able et al., 2009c, 2010). Settlement occurs at sizes of 10 to 16 mm TL (Kendall, 1972; Able et al., 1995a) on the inner continental shelf. Movement of recently settled individuals from the continental shelf spawning area into estuaries (Able et al., 1995a) is implied because no larvae have been collected in the Great Bay estuary in weekly ichthyoplankton collections (Witting et al., 1999), yet small juveniles occur there. Small individuals (< 20 mm TL) are collected from the Great Bay–Little Egg Harbor study area July through October (Fig. 62.1).

Calculations of growth, based on the progression of length modes, indicate rates of approximately 0.6 mm per day. A separate study, examining the growth of marked and recaptured fish (age 0+) arrived at slightly lower estimates of 0.42 to 0.45 mm per day (Able and Hales, 1997). Young-of-the-year from the estuary and the inner continental shelf reach lengths of 2 to 11 cm by October (Able et al., 1995a). There appears to be little growth during the winter, but it resumes at a relatively fast rate (0.74 mm per day) (Able and Hales, 1997) the following spring and summer so that by midsummer, approximately one year after hatching, individuals are 8 to 21 cm. By fall, at an approximate age of 14 to 17 months, they are 13 to 23 cm TL (Able et al., 1995a). This growth pattern was consistent with data from the early 1970s and the early 1990s in southern New Jersey in both the estuary and ocean (Able et al., 1995a). Laboratory studies have indicated that low dissolved oxygen levels may negatively influence the somatic growth rate of young-of-the-year (Hales and Able, 1995). In multiple experiments between 2.2 and 5.8 ppm dissolved oxygen, somatic growth was positively related to oxygen level, with growth rates varying from 0 to 0.30% per day.

SEASONALITY AND HABITAT USE

Seasonal peaks in abundance of young-of-the-year in the estuary occur during two different periods based on trap collections (Fig. 62.2). The peak in the spring is due to fish (just less than 1 year old) returning to the estuary after wintering on the continental shelf; the second peak (August to October) represents recently settled individuals from the current year class. Collections of older year classes become reduced in July as these larger individuals become less available to our sampling methods.

Young-of-the-year and adults have been collected and observed in a variety of structured habitats both in the estuary and at most depths on the continental shelf, including shallow and deep drowned river valleys and in rocky areas along the coast. Structured habitats in all these areas include reefs, wrecks, and rock piles (Bigelow and Schroeder, 1953; Able et al., 1995a). Summer densities may be similar in the ocean and the estuary, suggesting that both areas have comparable habitat value (Able et al., 1995a). A sampling transect from the inner shelf to tidal freshwater in southern New Jersey found individuals distributed across much of the inner continental shelf and at most stations in the Great Bay portion of the estuary (Fig. 62.3). In the ocean, young-of-the-year are more frequently collected in catches with large amounts of shell hash, particularly valves of the surf clam, *Spisula solidissima* (Able et al., 1995a). In the estuary, they occur at a variety of sites, including those with shell, those with amphipod tubes (*Ampelisca abdita*), and deep channels with rubble (Able and Fahay, 1998). They are especially abundant in the Rutgers University Marine Field Station (RUMFS) boat basin, which is an embayment off a marsh creek (Able and Hales, 1997), and in deep holes in that and other marsh creeks (see below). The young-of-the-year have also been captured around pier pilings and in open water in the Hudson River (Able et al., 1995b). The importance of structured habitats for young-of-the-year is further substantiated by the increase in the catch rate of juveniles after mollusk shell was experimentally added to featureless estuarine substrate in Chesapeake Bay (Arve, 1960).

Laboratory observations of individually tagged young-of-the-year (38 to 64 mm TL, n = 12) indicate a preference for oyster shells over barren sand substrate (2 cm deep) in large tanks (250 cm × 115 cm) (Able and Fahay, 1998). Under these conditions, young-of-the-year spend most of their time resting during the day under or against shells. They seldom share these habitats and are often distributed evenly, with one individual per shell. When aggression occurs, it is during the day and associated with defense or an attempt to take over the shell from a prior occupant. At night, most of these individuals left their shells, schooled together, and continuously circled the tank. This behavior might be interpreted as migratory because, during the period of the experiments (September 22–October 3, 1991), the juveniles are leaving the estuary and moving offshore (Able et al., 1995a).

There seems to be a high degree of habitat fidelity during the summer and fall in the estuary. In preliminary experiments in the summer of 1991, juveniles (n = 34) were captured in deep (2 to 3 m) holes in polyhaline marsh creeks in Great Bay

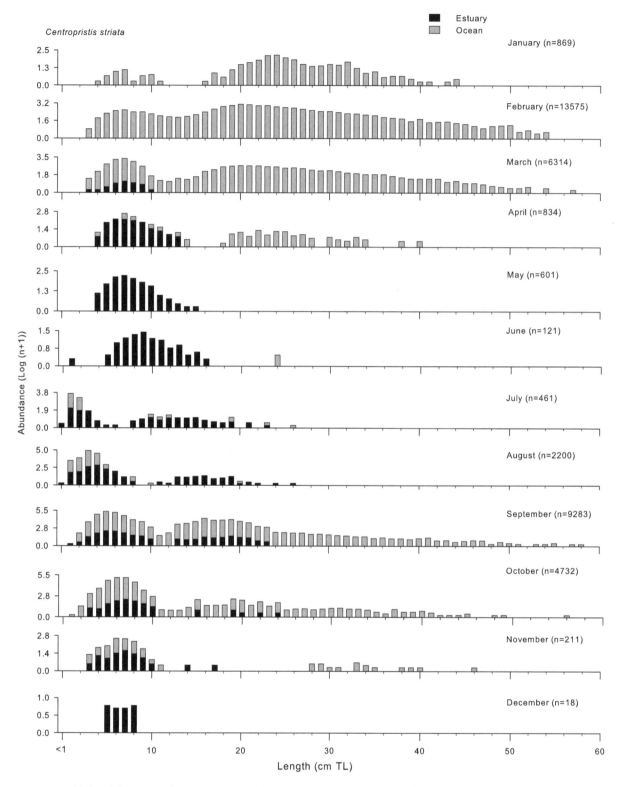

Fig. 62.1. Monthly length frequencies of *Centropristis striata* collected in estuarine and coastal ocean waters of the central Middle Atlantic Bight. Sources: National Marine Fisheries Service: otter trawl (n = 32,862); Rutgers University Marine Field Station: 1 m beam trawl (n = 2); seine (n = 101); 1 m plankton net (n = 235); 2 m beam trawl (n = 128); experimental trap (n = 192); killitrap (n = 992); 4.9 m otter trawl (n = 125); multipanel gill net (n = 4); and Able and Fahay 1998 (n = 4578).

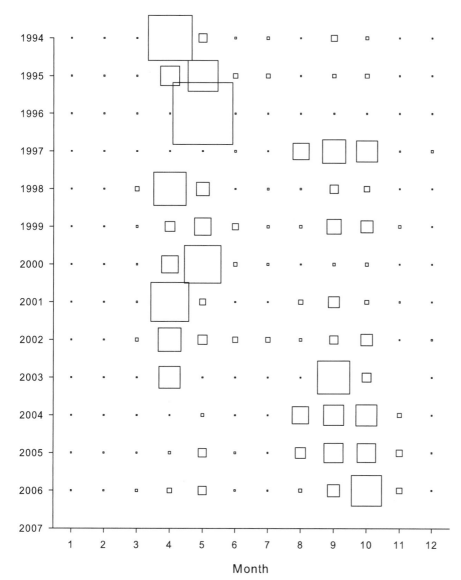

Fig. 62.2. Seasonal occurrence of the juveniles of *Centropristis striata* based on trap collections in Great Bay (Rutgers University Marine Field Station boat basin) from 1994 to 2006. Full squares represent the percentage of the mean catch per unit effort captured in a given year. Values range from 0 to 100%. The smallest squares represent when samples were taken but no individuals were collected.

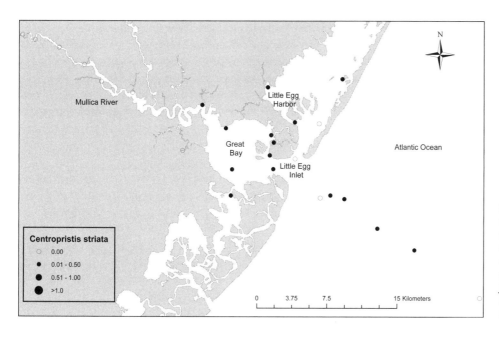

Fig. 62.3. Habitat use and average catch per tow (see key) for juvenile and adult *Centropristis striata* in the Ocean–Great Bay–Mullica River corridor in southern New Jersey based on small otter trawl collections during July and September from 1988 to 2006. See Table 3.2 for characteristics at each sampling collection.

and the RUMFS boat basin. These individuals were marked with injections of acrylic dyes and released at the same locations (Able and Fahay, 1998). In several instances (n = 11), they were collected in the same locations after 4 to 90 days. In 1992, this same pattern of habitat fidelity occurred in a much larger tagging study in the same area (Able and Hales, 1997). In that study, age 0+ (n = 337, 34 to 106 mm TL) and age 1+ (n = 367, 43 to 111 mm TL) individuals were tagged, with 225 recaptures of 180 fish up to 121 days after initial release (Able and Hales, 1997). Recapture frequency was 20% for age 0+ fish released in summer and fall and 30% for age 1+ fish released in spring and summer. Almost all recaptures (99%) were made within 30 m of the release site. Of 35 fish that were recaptured twice, 46% were recaptured at the original site.

Temperature and oxygen seem to be especially important components of the habitat. In intensive laboratory observations of age 0+ individuals under ambient estuarine seasonal temperatures, there was a strong response to low winter temperatures (Hales and Able, 1995). When temperature fell to 6°C, they occasionally buried in sand substrate. Below 4°C, they ceased feeding. At 2 to 3°C, mortality increased sharply, and no individuals survived the entire winter.

Average summer densities of young-of-the-year (0.33 individuals/m²) are often similar in the ocean and the estuary. This similarity and the correspondences in growth rates suggest that both areas have comparable habitat value as nurseries, and thus this species is not strictly estuarine dependent (Able et al., 1995a).

PREY AND PREDATORS

Two studies examined the food habits of young-of-the-year (19.1 to 146 mm, n = 76) in Long Island Sound and Magothy Bay, Virginia (see Tables 8.1 and 8.3). Important prey of young-of-the-year include a variety of invertebrates (mysids, amphipods, and crabs). Fishes are a minor dietary component in Magothy Bay, Virginia (Kimmel, 1973). At the same site, more than half of the diet of younger fish (30 to 91 mm, n = 28) consists of mysids, supplemented by amphipods, sediment, polychaetes, shrimp, isopods, cumaceans, and plant remains, while larger fish (92 to 146 mm, n = 20) feed primarily on mud crab and commensal crabs, supplemented by mysids, amphipods, polychaetes, mollusks, shrimp, and

sediment. Relative importance of individual prey items also varies intra-annually in Magothy Bay, Virginia, such that during March and April more than half of the diet by volume is composed of mysids, while the diet during spring and summer is more diverse, with mud crabs and amphipods consumed more frequently and mysids less frequently (Kimmel, 1973). A variety of fishes have been shown to be important predators on adults, but studies of predators on young-of-the-year are lacking.

MIGRATIONS

The young-of-the-year (< 17 cm) are migratory during some portions of the first year. They migrate out of the estuary and away from inner continental shelf nursery areas during the fall as water temperatures drop (Able et al., 1995a; Able and Hales, 1997). They move into deeper water on the continental shelf for the winter and then back into estuaries and the inner continental shelf the following spring and summer (Fig. 62.4; Able et al., 1995a). In the fall, young-of-the-year occur over a broad depth range (mean depth 26 m, range 5 to 105 m), but the distribution varies within the Middle Atlantic Bight. Individuals south of Delaware Bay are captured over most portions of the continental shelf, but north, from New Jersey to Massachusetts, they occur primarily on the inner shelf. In the winter, most are found from the mid- to outer shelf in deeper depths (mean depth 74 m, range 19 to 244 m) between southern New England and North Carolina. In the spring, they are distributed over most portions of the continental shelf (mean depth 59 m, range 5 to 300 m) and show evidence of a return to inshore habitats, especially south of Delaware Bay. North of Delaware Bay, most fish are in deeper waters during the spring.

Age 1+ (> 17 cm) juveniles and adults have a similar seasonal distribution to that of the young-of-the-year in most seasons (see Fig. 62.4). In the fall they have an inner to mid-shelf distribution pattern (mean depth 25 m, range 5 to 177 m) between southern New England and North Carolina. In winter and spring, most of these fish are found in deeper parts of the continental shelf, with depth ranges of 21 to 370 m for juveniles and 11 to 413 m for adults. Note that young-of-the-year are more likely to be found in inner continental shelf depths during the spring than older year classes.

Age 1+ Centropristis striata   **YOY Centropristis striata**

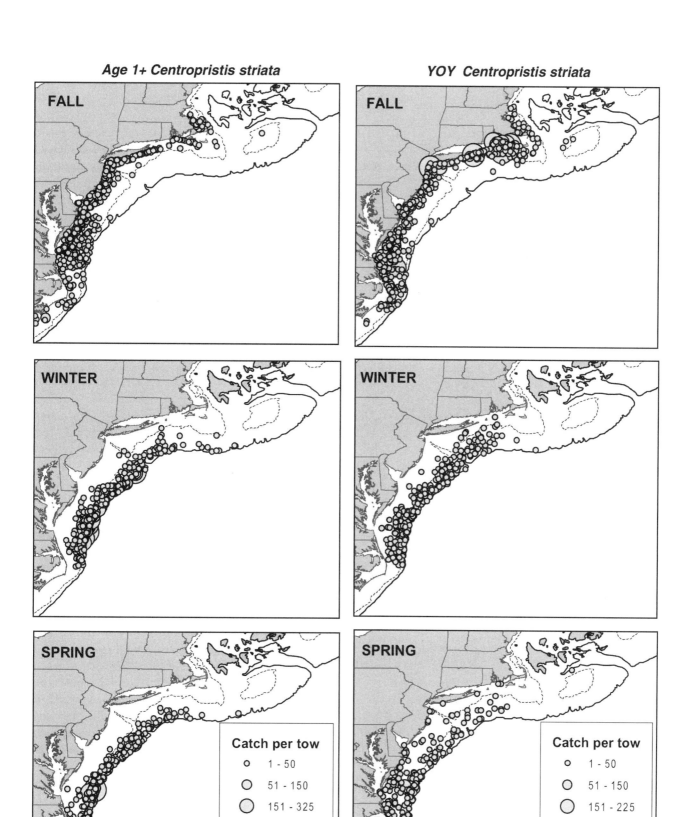

Fig. 62.4. Composite distribution of age 1+ (> 17 cm) and young-of-the-year (< 17 cm) *Centropristis striata* during seasonal cruises of the National Marine Fisheries Service groundfish survey. Details of sampling effort are indicated in Fig. 3.4.

63

Mycteroperca microlepis (Goode and Bean)

GAG GROUPER

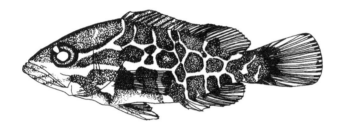

Estuary	Reproduction	Larvae/ Development	YOY Growth	YOY Overwinter
Ocean				
Both				
Jan	■			
Feb	■			
Mar	■			
Apr	■			
May				
Jun				
Jul				
Aug				
Sep				■
Oct				■
Nov				■
Dec				■

DISTRIBUTION

Mycteroperca microlepis occurs in the western Atlantic Ocean from North Carolina and Bermuda (where it is rare), through the Gulf of Mexico to the Yucatan Peninsula (Heemstra et al., 2002). There is a single record from Cuba. Juveniles are found as far north as Massachusetts and as far south as Cape Kennedy, Florida.

REPRODUCTION AND DEVELOPMENT

Sexual maturity is attained at age 5 or 6. These fish are protogynous hermaphrodites, and change from males to females at about 10 to 11 years of age. Little is known about eggs or their production, except that spawners gather in large aggregations. Spawning occurs between December and April in the Gulf of Mexico (Heemstra et al., 2002), January to March off the west coast of Florida (McErlean, 1963), and winter to early spring, with a peak in late March and early April off North Carolina (Collins et al., 1987). There is no evidence that this species spawns north of Cape Hatteras. Spawning aggregations composed of transient individuals form over depths of 50 to 120 m in the eastern Gulf of Mexico (Heemstra et al., 2002).

The early development of *Mycteroperca microlepis*, and characters useful for distinguishing the larvae and juveniles from those of other species in the family Serranidae, have been well described and summarized (Fahay, 2007). The bodies of early larvae are elongate, with large heads and early-forming pelvic and second dorsal fin spines. Pigment is sparse, but usually includes a prominent spot on the caudal peduncle. Juveniles have an array of spines on several bones of the head, including a prominent, serrated one at the preopercle angle. Pigmentation in juveniles features a light reticulated pattern against a dark background.

LARVAL SUPPLY, SETTLEMENT, AND GROWTH

Studies south of the Middle Atlantic Bight indicate that after spawning in offshore waters, larvae ingress into estuarine habitats in late spring and early summer at a mean age of 44 days (Rutten, 1998), followed by residence there (mean duration 114 days). This is a regularly occurring phase of their early life history (Keener et al., 1988; Ross and Moser, 1995). In those studies, however, the size of ingressing larvae is about 15 mm, which is considerably smaller than those that are detected as they enter Little Egg Inlet annually. We conclude that larval growth occurs during transport via the Gulf Stream into our study area; therefore juveniles that enter our study area in mid-summer are larger than members of the same cohort that ingress in the South Atlantic Bight. Early stages of this species occur irregularly in our study area. Juveniles collected in traps are found periodically in widely separated years (Fig. 63.1). This species first becomes evident in New Jersey when small juveniles (< 10 cm) are collected in traps during July and August (Fig. 63.2).

Growth can be estimated from the distribution of sizes in juveniles collected in our study site (see Fig. 63.2). Fish as small as 3 cm TL have been collected in July, when the apparent size mode is 6 cm TL. By September, this mode has increased to 11 cm TL. Reports from off Virginia (Hoese et al., 1961) indicate the appearance of similar sizes (6 cm) there in July, and these seemingly increase to lengths of 19 cm by November. These individuals were observed in Chincoteague

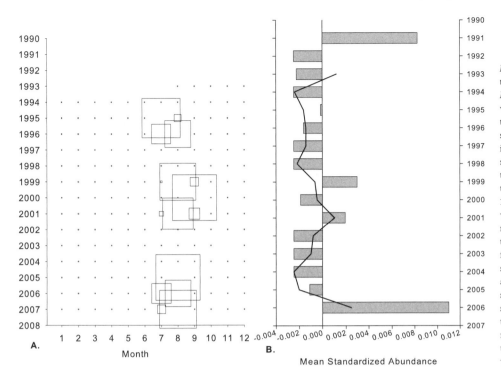

Fig. 63.1. Annual and seasonal variation in occurrence and abundance of *Mycteroperca microlepis.* (A) Seasonal variation of the juveniles is based on trap collections in Rutgers University Marine Field Station boat basin in Great Bay from 1994 to 2006. Open squares represent the percentage of the mean catch per unit effort captured by year. Values range from 0 to 100%. The smallest squares represent when samples were taken but no juveniles were collected. (B) Annual variation in abundance for the same data from 1991 to 2006 was standardized by subtracting overall mean from annual abundance. Horizontal bars represent standardized annual abundance; the solid line is a three-year running mean that was calculated by taking the mean of the standardized values from the previous two years and the year in which the data are plotted.

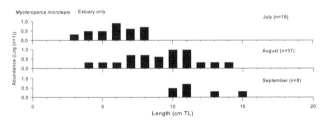

Fig. 63.2. Monthly length frequencies of *Mycteroperca microlepis* collected in the Great Bay–Little Egg Harbor study area. Source: Rutgers University Marine Field Station: seine (n = 2); experimental trap (n = 1); killitrap (n = 26); 4.9 m otter trawl (n = 1).

Bay, Virginia, and it is not known whether they remained there during the ensuing winter or migrated offshore after this date.

SEASONALITY AND HABITAT USE

Adults are usually found over rocky substrates, in depths of 40 to 80 m, but they may also occur closer to shore over rocky or grassy substrates (Heemstra et al., 2002). Spawning aggregations are found in the eastern Gulf of Mexico in depths of 50 to 120 m. Most, if not all, spawning takes place south of Cape Hatteras; the larvae occur at sea, possibly in close prox-

imity to the Gulf Stream. Juveniles are present in estuaries as far north as Massachusetts (Heemstra et al., 2002), where they are often found in seagrass beds. Those that have been collected in our study area have been taken in traps, indicative of their demersal, possibly cryptic habits. They are found regularly in the late summer and fall.

PREY AND PREDATORS

Adults are highly piscivorous, but they may also prey on crabs, shrimp, or squid. Juveniles feed mostly on crustaceans that live in grass beds (Heemstra et al., 2002).

MIGRATIONS

There is some evidence of a spawning migration away from South Carolina waters, but the trend is for small reproducing females to remain there, whereas larger spawners may migrate south, as far as southern Florida, to spawn (Van Sant et al., 1994). Juveniles leave estuaries and return to oceanic habitats off North Carolina in August and September, at sizes of 120 to 240 mm TL (Rutten, 1998). Their age at ingress in this study was 44 days (mean) and at egress was 158 days (mean). Their residence time in the estuary, therefore, was a mean of 114 days, or a range of 2.5 to 5 months.

64

Pomatomus saltatrix (Linnaeus)

BLUEFISH

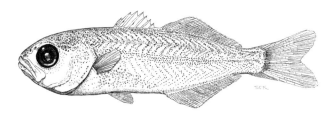

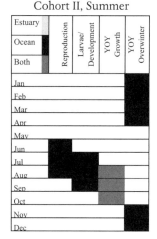

Cohort I, Spring					
Estuary	Reproduction	Larvae/Development	YOY Growth	YOY Overwinter	
Ocean					
Both					
Jan					
Feb					
Mar					
Apr					
May					
Jun					
Jul					
Aug					
Sep					
Oct					
Nov					
Dec					

Cohort II, Summer					
Estuary	Reproduction	Larvae/Development	YOY Growth	YOY Overwinter	
Ocean					
Both					
Jan					
Feb					
Mar					
Apr					
May					
Jun					
Jul					
Aug					
Sep					
Oct					
Nov					
Dec					

DISTRIBUTION

Pomatomus saltatrix occurs in temperate and semi-tropical ocean waters worldwide (Briggs, 1960; Juanes et al., 1996), but it is absent from the eastern Pacific and Indo-Pacific north of the equator. In the western North Atlantic Ocean, it ranges from Nova Scotia and Bermuda to the Gulf of Mexico. It is rare between southern Florida and the northern coast of South America. A disjunct population occurs off Argentina. Juveniles have been reported from all estuaries surveyed within the Middle Atlantic Bight, but eggs are rare and larvae have been recorded from only a few systems (see Table 4.3).

REPRODUCTION AND DEVELOPMENT

Mean age and size for females at first maturity are 1.9 years and 480 mm TL, respectively. Oocyte development is asyn-

chronous, and females are multiple spawners. Mean batch fecundity averages 402,247 eggs and ranges from 114,513 to 920,746 eggs, depending on the age and weight of the female (Robillard et al., 2008). Another study compared males and females and found the median length at maturity to be 33.9 cm for males and 33.4 cm for females (Salerno et al., 2001). The median age at maturity was 1.2 years for males and 1.1 years for females.

An early study, based largely on the distribution of eggs and larvae, concluded that there were two discrete spawning events in the western Atlantic population (Kendall and Walford, 1979). The first occurs from March to May near the edge of the continental shelf of the South Atlantic Bight. The second occurs between June and August in the Middle Atlantic Bight. Recent studies have re-examined this conclusion and refined our knowledge of a complex reproductive pattern, supporting the concept of a single, migrating spawning stock (Hare and Cowen, 1993; Smith et al., 1994) that begins spawning south of Cape Hatteras as early as March and finishes spawning in our study area during the late summer (Robillard et al., 2008). Sexual maturity and gonad ripening occur in early spring off Florida, in early summer off North Carolina, and in late summer off New York (Hare and Cowen, 1993). In the New York Bight, gonadosomatic studies indicate both sexes are ripe or ripening between June and September, with a strong peak in July (Chiarella and Conover, 1990). A fall (September–January) spawning event is also suggested (McBride et al., 1993) because gonads reach a second peak in ripeness in fish off Florida in September and larvae reoccur in the South Atlantic Bight in the fall (Collins and Stender, 1987). Since the latter progeny presumably have no impact on the Middle Atlantic Bight, we will only consider the spring-spawned and summer-spawned cohorts in this chapter.

Evidence from an intensive study of the distribution of eggs and larvae in the Middle Atlantic Bight supports the suggestion that the spawning season is a single, protracted one that begins in May off North Carolina (Smith et al., 1994; Berrien and Sibunka, 1999). Egg and larval occurrences then progress northward as far as Cape Cod, where they peak in July, then become less dense in August. During some years, they are concentrated over the inner continental shelf; in other years, they are more widely distributed across the entire shelf (Morse et al., 1987; Smith et al., 1994; Berrien and Sibunka, 1999).

The early development of *Pomatomus saltatrix*, and characters useful for distinguishing the eggs, larvae, and juveniles from those of similar species, have been well described and summarized (Fahay, 2007). Eggs are pelagic and spherical and measure 0.95 to 1.00 mm in diameter. The single oil globule is 0.26 to 0.29 mm in diameter. Larvae hatch at 2.0 to 2.4 mm, with unpigmented eyes and unformed mouth parts. Pigment in larvae includes a streak along the posterior part of the lateral line. Juveniles exhibit a characteristic pigment pattern where melanophores trace the outlines of the myomeres. After completion of fin ray development, they go through a pelagic juvenile stage characterized by a silvery, laterally compressed body (Deuel et al., 1966; Norcross et al., 1974). This transition occurs at ages of 18 to 25 days and at sizes between 10 and 12 mm SL (Hare and Cowen, 1994).

A total of 70 individuals between 5.6 and 30.0 mm TL were examined for scale formation (see Fig. 61.1). As is typical for many pelagic species like bluefish, scales are frequently lost. Patterns of scale formation presented here were based on examination of both sides of specimens and represent a combination of actual scales and scale pockets. This may account for some of the differences in the characters we observed compared to those reported by Silverman (1975). Scale formation occurred over sizes of 10.0 to 20.5 mm SL. Two loci for scale origin are located on the caudal peduncle and behind and below the eye. Onset of scale formation occurs at 10.0 mm SL with the development of a single row of scales along the lateral line on the caudal peduncle. By 12.0 mm SL, scales have spread laterally and cover the caudal peduncle. By approximately 13.0 mm SL, scales have spread anteriorly and laterally on the body to the origin of the second dorsal and anal fins. By 17.0 mm SL, scales cover the body posterior to the operculum, and by approximately 18.0 mm SL, scales are observed on the head region, below and behind the eye. Completion of scale formation was observed on a 20.5 mm SL specimen. This represents the adult condition, where scales cover the body from the base of the caudal fin to the head. At this size there are no scales on the tip of snout. In this condition scales covered 91.6% of the body and 68.4% of the body and fins (see Fig. 61.1). However, on some specimens of slightly larger sizes (28.1 to 29.7 mm TL) scale formation is not yet complete (i.e., no scales observed on the caudal fin).

LARVAL SUPPLY, SETTLEMENT, GROWTH, AND MORTALITY

There are two sources for young-of-the-year that we sample in our study area. Some of the progeny resulting from early spawning in the South Atlantic Bight in March and April are entrained in the Gulf Stream and are transported northward into the Middle Atlantic Bight (Hare and Cowen, 1993; Able and Fahay, 1998; Hare et al., 2001). After developing into pelagic juveniles, these fish either actively swim or are transported by a variety of oceanographic features across the Slope Sea and continental shelf. They then enter estuaries in our study area at a mean size of 60 mm and at ages of 60 to 76 days (McBride and Conover, 1991; Able and Fahay, 1998). Local spawning during the summer yields eggs and larvae beginning off North Carolina in May and peaking in July, when they occur over the continental shelf between Rhode Island and North Carolina (Berrien and Sibunka, 1999). These "summer-spawned" young-of-the-year enter estuaries or remain in coastal ocean or surfzone habitats (Able et al., 2009c, 2010), in mid- to late August at a mean size of 47 mm

and at ages of 33 to 47 days. The lack of a long-term dataset for either of the two cohorts prevents us from estimating annual variation in abundance for these recruits.

Growth rates vary between life history stages and cohorts. Before larvae enter estuaries, growth ranges from 0.3 to 0.8 mm per day (Deuel et al., 1966; Hare and Cowen, 1994). In comparisons between spring- and summer-spawned cohorts from several years, a summer (1988) group was found to be the slowest growing through the juvenile stage (J. Hare, pers. comm.). A separate study, comparing different groups, found a summer cohort to be the fastest growing among a different set of cohorts (McBride and Conover, 1991). Juveniles entering estuaries during the first (spring) recruitment range from about 30 to 70 mm in length (Juanes et al., 1996), but this size varies somewhat between years and study sites throughout the Middle Atlantic Bight (Able and Fahay, 1998). Sizes of fish from the summer spawning are somewhat smaller when they appear in estuaries.

Length-frequency histograms resulting from sampling in the Great Bay–Little Egg Harbor and Hudson River study areas clearly demonstrate the ingress of the two cohorts in May and July through September (Fig. 64.1). Modal sizes of the earliest cohort increase from 2 cm in May to about 18 cm in October. Individuals in the summer-spawned cohort only rarely exceed 10 cm by October. Another well-marked size mode of 30 to 40 cm occurs in our study area between June and October. We interpret these to represent age 1 individuals.

Juveniles grow at the rate of 0.9 to 2.1 mm per day (McBride and Conover, 1991; Juanes et al., 1993, 1996; McBride et al., 1995). Members of the spring-spawned cohort are much larger than those of the summer-spawned cohort at the onset of fall migration (McBride and Conover, 1991) and continue to be larger at ages 1 through 4 (Lassiter, 1962). Size modes of the first (spring) cohort reach about 13 to 14 cm by late August in Long Island waters (Nyman and Conover, 1988) or 15 to 20 cm in New England (McBride et al., 1995). In the late 1950s, the contribution of these two cohorts to the overall population was observed to be approximately equal. However, studies in the New York Bight during the 1980s demonstrated that the spring-spawned cohort strongly dominates (Chiarella and Conover, 1990). More recent studies have shown that the cohort structure is more complicated than previously recognized, at least based on sampling in 1998 (Able et al., 2003b; Taylor and Able, 2006). This species grows rapidly until the age of 5 to 7 years, after which the growth rate slows considerably (Robillard et al., 2009). They can reach 13 years and 800 mm TL.

Studies of the effects of winter are relevant because in recent years recruits to the adult population have been dominated by the spring-spawned cohort (Munch and Conover, 2000; Conover et al., 2003) and there have been questions as to the ability of the summer-spawned cohort to survive the winter. The overwinter mortality of summer-spawned

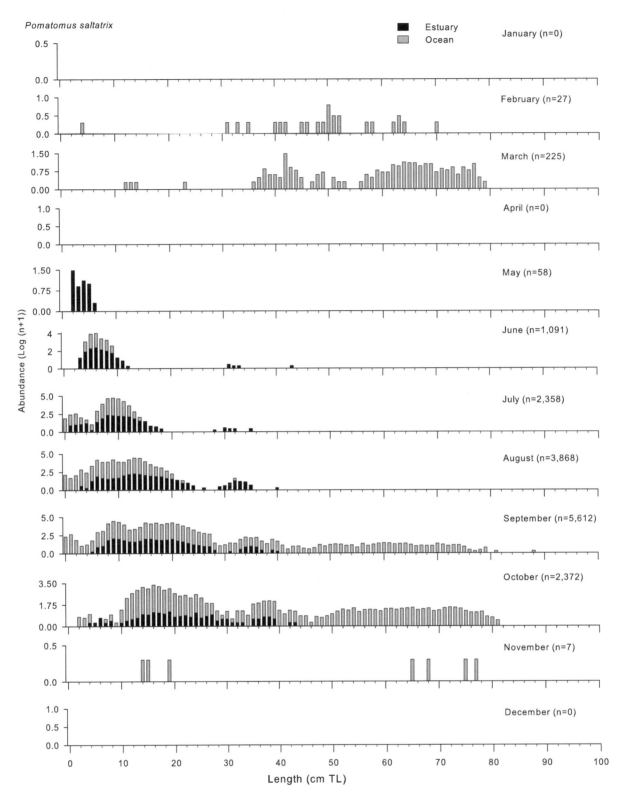

Fig. 64.1. Monthly length frequencies of *Pomatomus saltatrix* collected in estuarine and coastal ocean waters of the central Middle Atlantic Bight. Sources: National Marine Fisheries Service: otter trawl (n = 4399); Rutgers University Marine Field Station: seine (n = 5858); Methot trawl (n = 1329); 1 m plankton net (n = 1); 4.9 m otter trawl (n = 81); multi-panel gill net (n = 342); and Able and Fahay 1998 (n = 3608).

Pomatomus saltatrix has been studied in Onslow Bay, North Carolina (Morley et al., 2007). While this study demonstrated that members of this cohort are able to survive ambient winter temperatures in this embayment, their relative abundance decreased with declining temperatures and higher numbers were observed in milder winter conditions. In these fish, gut fullness decreased with decreasing temperatures. Energy reserves in larger fish began at higher levels than in smaller fish, but then were depleted more, such that in the middle of winter, all cohorts tested had approximately equal energy reserve levels. Possibly compounding this issue, prey densities off the coast of North Carolina have been shown to decline during winter (Morley, 2004).

The effects of reduced winter temperatures and prolonged starvation were further investigated in a mesocosm study in North Carolina (Slater et al., 2007). When spring-spawned (larger fish) and summer-spawned (smaller fish) cohorts were compared, the smaller fish were found to be more susceptible to winter starvation than the larger fish. When provided with food, however, fish in both cohorts were observed to store lipids in late fall, deplete those reserves during winter, and ultimately survive until the following spring. Furthermore, fish in the summer-spawned cohort also demonstrated high "starvation endurance," whereby > 90% survived 120 days without food. The spring-spawned cohort suffered increased mortality during a short-lived cold event where temperatures dropped below 6°C, suggesting that these larger individuals are less able to withstand acute cold stress. In summary, the summer-spawned cohort were observed to store adequate amounts of lipids in the fall, deplete these reserves at reduced rates in cold temperatures, endure long periods of starvation, and withstand sudden drops in temperatures. Therefore, these investigators found no evidence of size-selective winter mortality that might lead to recruitment failure in either of the cohorts.

SEASONALITY AND HABITAT USE

Adults occur in fast-swimming schools (Grothues and Able, 2007; Stehlik, 2009) or small groups in large bays and estuaries as well as across the entire continental shelf. Younger fish on the shelf are usually found closer to shore than older fish (Salerno et al., 2001). The eggs and larvae occur in the upper levels (primarily between the surface and 15 m) of the water column in oceanic waters (Kendall and Naplin, 1981). After spawning on the outer shelf of the South Atlantic Bight from March though April (Collins and Stender, 1987), some larvae are retained there and enter estuaries south of Cape Hatteras (McBride et al., 1993), while others are entrained northeastward by the Gulf Stream (Hare and Cowen, 1996; Hare et al., 2001). The most developed of these, after entering slope waters off the Middle Atlantic Bight (Hare et al., 2001), actively swim across the Slope Sea until they reach continental shelf waters. Less developed larvae are entrained in warm water filaments or streamers associated with warm-core

rings, and this also results in their introduction into continental shelf waters. Evidence accrued from neuston sampling indicates that pelagic juveniles gather in outer continental shelf waters in the central part of the Middle Atlantic Bight before actively crossing the shelf toward nursery areas in bays and estuaries of the region (Shima, 1989; Cowen et al., 1993) after the shelf/slope temperature front dissipates in late spring or early summer. This initial ingress apparently occurs abruptly, with most of these pelagic juveniles appearing in estuaries simultaneously (Nyman and Conover, 1988). The second pulse of pelagic juveniles results from summer spawning in the New York Bight and occurs during mid- to late August (McBride and Conover, 1991). Some authors have suggested that this cohort (or a major part of it) undergoes juvenile development in inner continental shelf waters rather than entering estuaries (Kendall and Walford, 1979; Able et al., 2003b; Taylor et al., 2006) and this appears to be the case based on our recent sampling (see Fig. 64.1). Additional studies suggest that choices between estuarine and ocean habitats may be facultative (Taylor et al., 2007).

Little information is available on specific habitats where *Pomatomus saltatrix* young-of-the-year occur. Most studies of growth and feeding habits, however, have made collections with beach seines, indicating they occur in relatively shallow estuarine or coastal ocean habitats. Other studies have used small otter trawls in slightly deeper water to advantage. There are some indications that young-of-the-year undertake diel and tidal movements between marsh creeks and open bay habitats (Rountree and Able, 1993). Acoustically tagged individuals (286 to 622 mm FL) preferred the shallow, polyhaline portion of the Mullica River–Great Bay estuary, but several individuals moved upstream where salinities were as low as 12 ppt (Grothues and Able, 2007).

PREY AND PREDATORS

Adults are voracious and accomplished piscivorous carnivores. Contrary to most piscivores, adults are capable of severing their prey into manageable pieces. The list of favored prey is long, and includes more than 28 species of bony fishes and 10 species of invertebrates (Buckel et al., 1999a). High on the list are schooling taxa, such as *Peprilus tricanthus*, clupeids, *Anchoa mitchilli,* and squid. Growth rates accelerate after juveniles shift their diet from zooplankton to young-of-the-year of several fish species, with much of the increase expressed in weight (Friedland and Haas, 1988; Juanes et al., 1993, 1994). Age 1+ *Pomatomus saltatrix* may prey on age 0 *P. saltatrix*, but given a choice in experiments, these older stages preferred to prey on *Menidia menidia* (Bell et al., 1999). In habitat-related diet studies in Chesapeake Bay, age 1+ *P. saltatrix* consumed a wider variety of fish prey over oyster reefs than they did over oyster bars or sandy substrates (Harding and Mann, 2001). These prey were mostly members of the Clupeidae, Engraulidae, Atherinidae, Gobiidae, Sciaenidae, and several unidentified taxa.

Several studies (n = 13) have examined the food habits of young-of-the-year (20 to 290 mm FL, n = 6853) in Middle Atlantic Bight estuaries from Great South Bay, New York, to lower Chesapeake Bay (see Tables 8.1 and 8.3). Important prey of young-of-the-year include a variety of fishes (*Menidia beryllina*, *M. menidia*, *Brevoortia tyrannus*, *Alosa aestivalis*, *A. sapidissima*, *Alosa* spp., *Fundulus heteroclitus*, *F. majalis*, *Anchoa mitchilli*, *Microgadus tomcod*, *Morone saxatilis*, and *Morone* spp.) and a few invertebrates (decapod shrimp and mysids). Young-of-the-year also exhibit cannibalistic behavior on a minor scale (Juanes et al., 1993; Buckel and Conover, 1997; Gartland, 2002). Summer-spawned young-of-the-year exhibit a less diverse diet (almost exclusively fishes, *Anchoa mitchilli*, with invertebrates rare) than spring-spawned young-of-the-year (*Morone saxatilis*, *Anchoa mitchilli*, *Alosa* spp., and *Menidia menidia*, supplemented by invertebrates, crab zoea, and decapod shrimp) (Juanes et al., 1993; Juanes and Conover, 1995; Buckel and Conover, 1997; Buckel et al., 1999a, b). Geographical variation in diet has also been noted between estuaries (Juanes et al., 1994) and between estuarine and oceanic sites (Gartland, 2002; Able et al., 2003b). Relative importance of individual prey items also varies intra-annually (Juanes and Conover, 1994a, 1995; Hartman and Brandt, 1995 a, b, c; Buckel and Conover, 1997; Buckel et al., 1999a, b; Juanes et al., 2001; Gartland, 2002). Gartland (2002) found the dominant prey of young-of-the-year (33 to 290 mm, n = 406) in Chesapeake Bay was *Menidia menidia* from late spring into early summer and then *Anchoa mitchilli* from mid-summer to fall. Relative importance of individual prey items also varies inter-annually (Friedland et al., 1988; Juanes et al., 1994; Juanes and Conover, 1995; Buckel and Conover, 1997; Buckel et al., 1999a, b). Juanes and Conover (1995) found decapod shrimp and fishes of near equal importance (49.53% and 48.92% by weight, respectively) to spring-spawned young-of-the-year (90 to 180 mm TL, n = 176) in 1988, whereas in 1989 fishes were much more important than decapod shrimp (67.95% and 21.73% by weight, respectively) to the diet of young-of-the-year (90 to 180 mm TL, n = 534) in Great South Bay, New York. Finally, a study of diets in the Hudson River estuary found that young-of-the-year *Pomatomus saltatrix* may have a considerable impact on year-class strength of young-of-the-year *Morone saxatilis* (Buckel et al., 1999b). In this system, young-of-the-year *Pomatomus saltatrix* preyed on *Anchoa mitchilli*, *Menidia menidia*, *Alosa* spp., and young-of-the-year *Morone saxatilis*. The *M. saxatilis* were most heavily preyed upon when their densities were high, suggesting a density-dependent feeding response on the part of the young-of-the-year *P. saltatrix*. Furthermore, this study found that during one year (1993) *P. saltatrix* predation accounted for 50 to 100% of all losses in the *M. saxatilis* year class, and concluded that predation by young-of-the-year of this species can have a substantial impact on recruitment of several fish species.

Diets in young-of-the-year can affect both growth rates and elemental composition in the otoliths (Buckel et al., 2004). This study demonstrated that concentrations of certain elements in growth zones of otoliths can be attributed to diet type (fishes versus shrimp), and need not necessarily be attributed to physical or chemical properties of the water the fish lives in. While the predators of *Pomatomus saltatrix* have not been extensively studied, the adults are an important component of the diet of *Isurus oxyrhinchus* in temperate, oceanic waters (Wood et al., 2009a).

MIGRATIONS

Our data indicate that some young-of-the-year remain in coastal ocean habitats during summer and fall and some stay in estuarine habitats through October (see Fig. 64.1). After emigration from estuaries in the fall, the evidence suggests they move south in coastal waters and do not occur in the Middle Atlantic Bight during winter (Fig. 64.2). During most years, the spring-spawned cohort dominates in these emigrating young-of-the-year. Recent evidence suggests that energetically efficient glide-upswimming is used during these migrations (Stehlik, 2009). The following spring, as they approach age 1, they occur abundantly in the vicinity of Cape Hatteras, and then begin to reoccupy habitats in the Middle Atlantic Bight. Older stages exhibit the same pattern except that there is evidence that some overwinter near the shelf-edge off Virginia and North Carolina (Hamer, 1959; Miller, 1969; Wilk, 1982) and possibly farther north, off southern New England, which implies that a portion of the population engages in an inshore-offshore seasonal movement (Shepherd et al., 2006). The return in spring also includes a few individuals collected near the continental shelf edge, near Hudson Canyon. Tagging studies also support a north-south coastal migration (Miller, 1969; Shepherd et al., 2006). These studies demonstrate that adults may travel 2.6 km per day during much of the year, and that this rate may increase to 5 km per day during migratory periods (spring and fall). Year-round occurrences of all life history stages have been reported from the South Atlantic Bight (Anderson, 1968), possibly indicating that a portion of the population is non-migratory.

Age 1+ *Pomatomus saltatrix*

YOY *Pomatomus saltatrix*

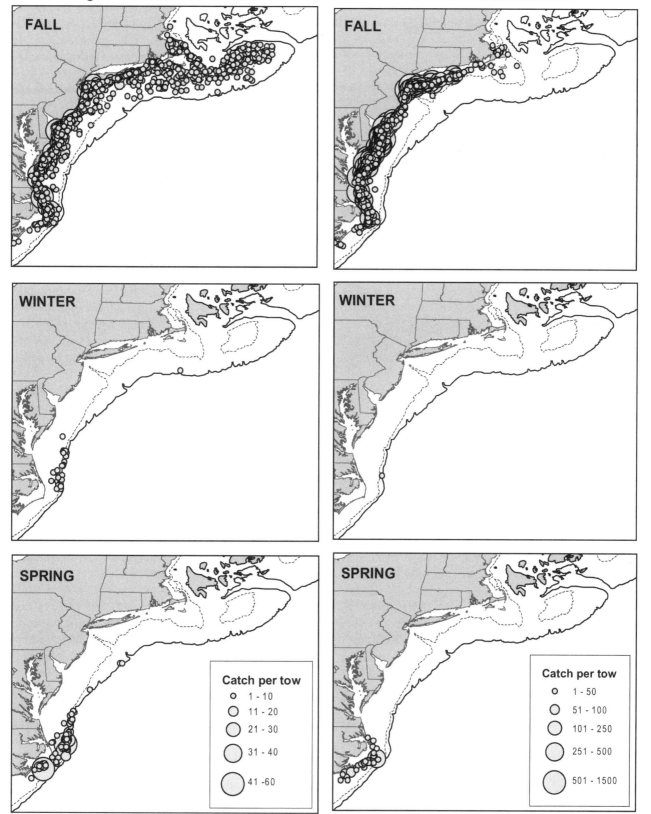

Fig. 64.2. Composite distribution of age 1+ (> 26 cm) and young-of-the-year (< 26 cm) *Pomatomus saltatrix* during seasonal cruises of the National Marine Fisheries Service groundfish survey. Details of sampling effort are indicated in Fig. 3.4.

65

Caranx hippos (Linnaeus)

CREVALLE JACK

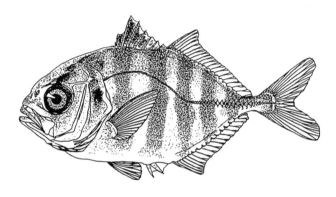

DISTRIBUTION

Caranx hippos is found from Nova Scotia to Florida, throughout the Gulf of Mexico, and south to Uruguay (Berry, 1959; Scott and Scott, 1988). It is also present in the eastern Atlantic Ocean and the Mediterranean Sea. The young are common in inshore and estuarine waters and up into freshwater. In the Middle Atlantic Bight, juveniles have been found in a number of estuaries (see Table 4.3). McBride (1995) provides a complete listing of study area occurrences based on original sampling effort and museum material.

REPRODUCTION AND DEVELOPMENT

Spawning occurs in offshore subtropical and tropical waters based on the occurrence of young stages (18 to 21 mm), but actual locations are not completely described (Berry, 1959). Some locations where spawning has been documented include the Straits of Florida and the tropical Caribbean Sea (Berry, 1959; Fahay, 1975; Montolio, 1978). Eggs are undescribed. Egg characters cited in Able and Fahay (1998) were based on Indian Ocean material (Chacko, 1949; Subrahmanyam, 1964) where this species, unfortunately, does not occur. Early larvae are not distinguishable from those of a congener, *Caranx latus*. Development of later stages is summarized in Fahay (2007).

Only seven specimens (26.2 to 40.8 mm FL) were avail-

able for examination of scales (Fig. 65.1). Of these, scale formation was complete on most of the body but this extended more anteriorly on the dorsal and ventral surfaces for the larger specimens. Over this range of sizes, scutes had already formed on the caudal peduncle. The development of scales and scutes in smaller individuals may be similar to that described for *Caranx crysos* (Berry, 1960).

LARVAL SUPPLY, SETTLEMENT, GROWTH, AND MORTALITY

Larvae in the study area are presumably derived from spawning in subtropical regions along the East Coast or in the Gulf of Mexico. Pelagic juveniles (12.2 to 22.5 mm) have been collected from surface waters of the South Atlantic Bight from May through August (Fahay, 1975). Juveniles are subsequently dispersed to temperate estuaries from mid-summer into fall in New York and New Jersey bays and rivers in most years (see McBride and McKown, 2000). The smallest individuals encountered in the Middle Atlantic Bight are approximately 3 to 4 cm, as they are in more southern estuaries (McBride and McKown, 2000). These small individuals occur from June through September, suggesting either a protracted spawning period or long delivery time to temperate estuaries (Fig. 65.2).

Based on modal length-frequency progression in the study area, the largest juveniles are capable of growing approximately 1.3 mm per day; similar growth is reported from other estuaries in the region (McBride and McKown, 2000). Some of these grow to lengths of 120 to 200 mm TL by September, when they begin to emigrate out of the estuary. Fish larger than 230 mm TL collected in September and October on the continental shelf (see Fig. 65.2; Fig. 7 in McBride and McKown, 2000) are presumably age 1. Overwinter mortality due to estuarine temperatures near 9°C have been reported from temperate (Hoff, 1971) and subtropical (see McBride and McKown, 2000) estuaries.

SEASONALITY AND HABITAT USE

Young stages are common in inshore and estuarine waters and up into freshwater. Juveniles are reported in Delaware estuaries from June to October at sizes greater than 20 mm (de Sylva et al., 1962; Wang and Kernehan, 1979). In the Hudson River, Jamaica Bay, and Haverstraw Bay, New York, as well as in Delaware Bay, small individuals occurred in every year sampled from 1986 to 1993 from July to November (McBride, 1995; McBride and McKown, 2000). In the Hudson River and Jamaica Bay, their sizes were approximately 25 to 75 mm FL in July and ranged from 35 to 175 mm FL by September.

Young-of-the-year seem to occupy a wide range of estuarine environmental conditions. In the Hudson River and Jamaica Bay, they were collected across a wide range of bottom temperatures (9 to 30°C), salinities (1.3 to 32.0 ppt), and dissolved oxygen (2.0 to 13.6 ppm) (McBride, 1995). They have

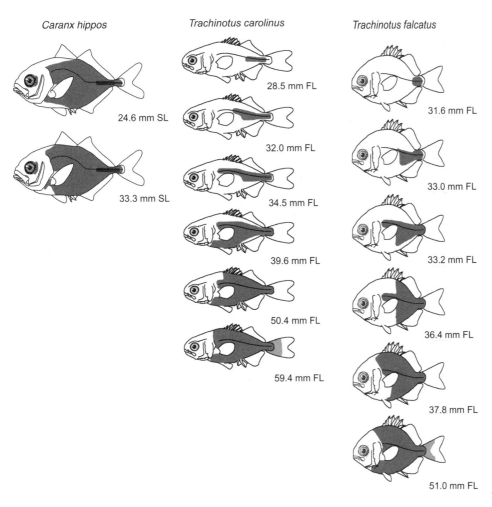

Fig. 65.1. Description of scale formation relative to length in *Caranx hippos*, *Trachinotus carolinus*, and *Trachinotus falcatus*.

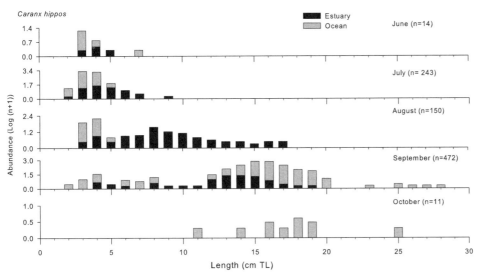

Fig. 65.2. Monthly length frequencies of *Caranx hippos* collected in estuarine and coastal ocean waters of the central Middle Atlantic Bight. Sources: National Marine Fisheries Service: otter trawl (n = 367); Rutgers University Marine Field Station: seine (n = 396); 1 m plankton net (n = 3); 4.9 m otter trawl (n = 19); multi-panel gill net (n = 8); and Able and Fahay 1998 (n = 97).

occurred as far upstream as the freshwater interface (Mc-Bride and McKown, 2000). In Delaware Bay, young-of-the-year have been found across a wide range of salinities extending up into brackish waters (Wang and Kernehan, 1979). Young-of-the-year are also found in estuarine habitats south of Cape Hatteras, where their monthly length frequencies are similar to those from the present temperate study area (McBride and McKown, 2000), suggesting that the entire East Coast of the United States is used as juvenile habitat. Details of their distribution in specific habitat types are poorly

Age 1+ Caranx hippos **YOY Caranx hippos**

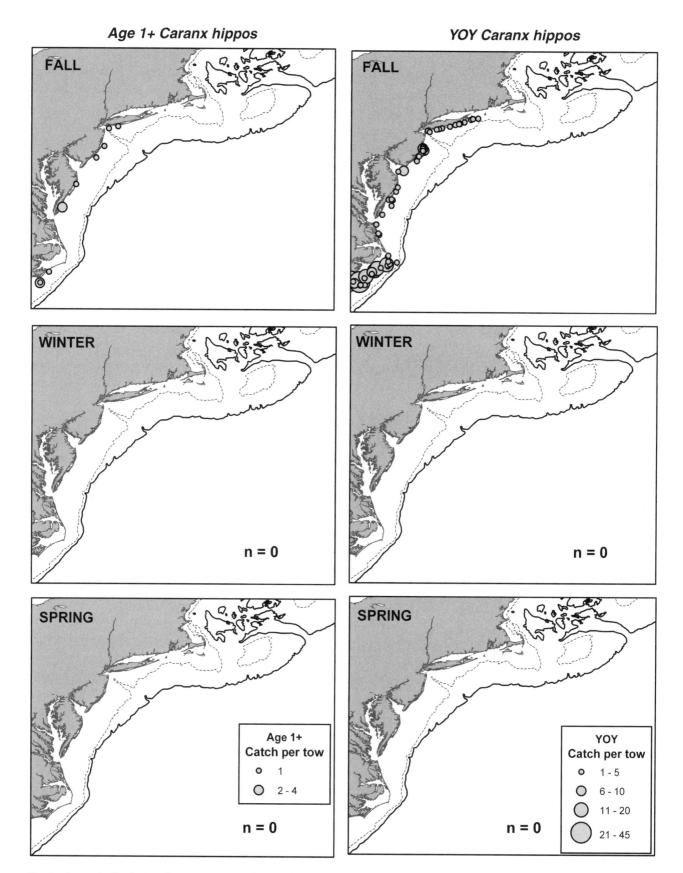

Fig. 65.3. Composite distribution of age 1+ (> 20 cm) and young-of-the-year
(< 20 cm) *Caranx hippos* during seasonal cruises of the National Marine
Fisheries Service groundfish survey. Details of sampling effort are indicated
in Fig. 3.4.

known, although they have been collected in marsh creeks (Rountree and Able, 1993) and along shallow estuarine shorelines (Able et al., 1996b) in Great Bay–Little Egg Harbor, New Jersey.

PREY AND PREDATORS

One study examined the food habits of young-of-the-year (30 to 160 mm TL, n = 40) *Caranx hippos* in the Delaware River estuary (de Sylva et al., 1962), where they frequently consumed mysid shrimp (*Neomysis americana*), decapod shrimp (*Palaemonetes* spp. and *Crangon* spp.), and fishes (*Gobiidae, Anchoa* spp., and *Atherinidae*) (see Table 8.1). Important predators of *C. hippos*, at any age or size, are largely unknown.

MIGRATIONS

Young-of-the-year leave estuaries, at least in the northern Middle Atlantic Bight, in the fall based on their declining availability by October. In Jamaica Bay and Haverstraw Bay, both the number and the average size of fish declined by October and November. In the study area, young-of-the-year (< 20 cm) and age 1+ (> 20 cm) are collected during trawl surveys on the continental shelf in the fall (Fig. 65.3). At this time the young-of-the-year are distributed on the inner continental shelf (mean depth 14.5 m, range 7 to 33 m) from North Carolina to the eastern end of Long Island, with greatest abundance south of Cape Hatteras. The few larger individuals (age 1+) occur annually but are less frequent in the surveys. They have a similar distribution and depth (mean depth 15.3 m, range 9 to 25 m) but they have not been collected as far east along the shore of Long Island. The absence of young-of-the-year and age 1+ individuals in the winter and spring suggests that they have retreated south of the study area during these seasons (McBride and McKown, 2000). In the South Atlantic Bight, adults occur in inshore waters during the summer and fall and then move to offshore waters near the Gulf Stream in the winter.

66

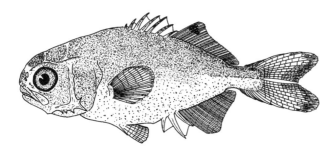

Trachinotus carolinus (Linnaeus)

FLORIDA POMPANO

DISTRIBUTION

Trachinotus carolinus occurs in the western Atlantic Ocean from Massachusetts to Brazil, including Bermuda and the Gulf of Mexico (Manooch, 1984; Gilbert, 1986). It is rare in the Caribbean Sea. The distribution of juveniles is not well described.

REPRODUCTION AND DEVELOPMENT

Spawning occurs between March and September (possibly more protracted) off the southeastern part of the United States (Finucane, 1969) and presumably in deep water (Gilbert, 1986). The early development of *Trachinotus carolinus*, and characters useful for distinguishing the eggs, larvae, and juveniles from those of other species in *Trachinotus*, have been well described and summarized (Fahay, 2007). Eggs are pelagic and spherical and measure 0.7 mm in diameter (unfertilized). There is a single oil globule. The bodies of larvae deepen soon after hatching, which occurs at an undescribed length. Much of the body is heavily pigmented in larvae, and there is usually a darkly pigmented lateral line that is displayed against a background of less intense pigment. Juveniles have an array of spines on several bones of the head, including a prominent parietal ridge, not present in other genera of the family Carangidae.

LARVAL SUPPLY, SETTLEMENT, GROWTH, AND MORTALITY

The source of the individuals that occur in the Middle Atlantic Bight is presumably the South Atlantic Bight. The smallest individuals (1 to 2 cm FL) are collected during June–August on ocean beaches and only a few in the estuary. By September they reach a modal size of 12 to 13 cm (Fig. 66.1). Individuals from 16.2 to 78.0 mm FL (n = 25) were examined for scale formation (Fig. 65.1). Scales originate on the caudal peduncle at approximately 28 mm FL. The scales extend anteriorly along the lateral line and by 32 mm FL occur at the level of the first dorsal fin and extend ventrally to the level of the second dorsal fin. By 34 mm FL, the scales extend along the lateral line to the level of the base of the pectoral fin. By

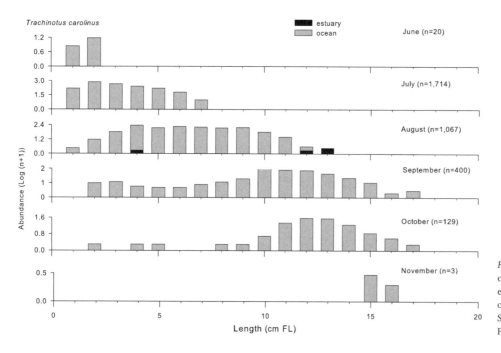

Fig. 66.1. Monthly length frequencies of *Trachinotus carolinus* collected in estuarine and coastal ocean waters of the central Middle Atlantic Bight. Source: Rutgers University Marine Field Station: seine (n = 3333).

40 mm FL, the scales extend to cover a larger area below the lateral line and behind the pectoral fin. In some individuals of the same length the scales occur to the level of the operculum below the lateral line and extend onto the ventral surface. By 59 mm FL, scales cover the entire body behind the operculum except for a small area anterior to the base of the pectoral fin. They also extend onto the caudal fin. At this size, scales cover 69.4% of the body and 52.7% of the body and fins and this represents the adult condition (see Johnson, 1978).

SEASONALITY AND HABITAT USE
Adults occur along sandy beaches and near inlets to bays and estuaries but most spawning occurs well offshore in oceanic waters. They are often associated with turbid water conditions. Juveniles often occur in the surfzone (Fields, 1962). In New Jersey, the young-of-the-year were on ocean and, occasionally, estuarine beaches.

PREY AND PREDATORS
Unknown.

MIGRATIONS
We suspect that the juveniles collected in New Jersey are capable of retreating south for the winter. Very few were collected after November (see Fig. 66.1).

67

Trachinotus falcatus (Linnaeus)

PERMIT

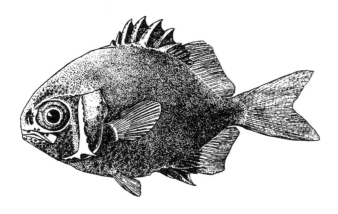

DISTRIBUTION

Trachinotus falcatus occurs in the western Atlantic Ocean from Massachusetts to Brazil, including Bermuda, the Gulf of Mexico, and the Caribbean Sea (Robins and Ray, 1986). The distribution of juveniles is not well described.

REPRODUCTION AND DEVELOPMENT

Maturation for 50% of females occurs by 547 mm and at 3.1 years (Crabtree et al., 2002) and for males at 486 mm at 2.3

years. Spawning off Florida takes place May–July in depths of 10 to 30 m (Crabtree et al., 2002). However, some authors believe spawning might occur farther offshore in proximity to the Gulf Stream (Fields, 1962). The early development of *Trachinotus falcatus*, and characters useful for distinguishing the larvae and juveniles from those of other species in *Trachinotus*, have been well described and summarized (Fahay, 2007). Eggs are undescribed. The bodies of larvae deepen soon after hatching, which occurs at an undescribed length. Much of the larval body is heavily pigmented. Late larvae and juveniles have an array of spines on several bones of the head, including a prominent parietal ridge, not present in other genera of the family Carangidae.

LARVAL SUPPLY, SETTLEMENT, GROWTH, AND MORTALITY

The source of larvae to the Middle Atlantic Bight is presumably from the South Atlantic Bight. The young-of-the-year first appear in the New Jersey surfzone and estuarine beaches in June and July at sizes of 1 cm FL (Fig. 67.1). These small individuals continue to occur through October. Individuals from 11.0 to 58.2 mm FL (n = 25) were examined for scale formation (see Fig. 65.1). Scale formation starts on the caudal peduncle by approximately 32 mm FL. In some individuals of similar size, the scales extended anteriorly along the lateral line. By 37 mm FL, scales occur farther anteriorly along the lateral line to the level of the base of the pectoral fin and ventrally below the lateral line behind the posterior edge of the same fin. In some individuals of similar size (36 mm FL), scales extend to the base of the dorsal fin. By 48 mm FL, scales cover almost the entire body to behind the operculum except for a small area anterior to the base of the pectoral fin. By 51 mm FL, the scales extend onto the caudal fin. At this size, scales cover 69.5% of the body and 50.9% of the body and fins; this represents the adult condition (see Ginsburg, 1952).

By September, the young-of-the-year have reached a

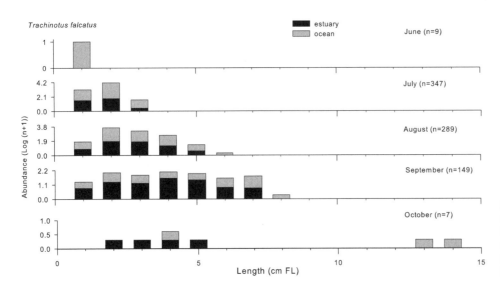

Fig. 67.1. Monthly length frequencies of *Trachinotus falcatus* collected in estuarine and coastal ocean waters of the central Middle Atlantic Bight. Source: Rutgers University Marine Field Station: seine (n = 801).

modal size of approximately 4 cm (see Fig. 67.1). By October, the average size is smaller, suggesting that the larger young-of-the-year (> 6 cm) have left the area. The total numbers collected are much reduced by October.

SEASONALITY AND HABITAT USE

Adults occur over sand flats, over muddy bottoms, in the surfzone, or over reefs to a maximum depth of 30 m (Fields, 1962). Spawning usually occurs well offshore, and young stages grow to 12 to 50 mm before moving toward coastal waters. Juveniles often occur in small schools on sandy beaches (Fields, 1962; Finucane, 1969).

PREY AND PREDATORS

Unknown.

MIGRATIONS

We suspect that this species leaves New Jersey beaches in the fall with cooling temperatures and migrates to the south.

68

Lutjanus griseus
(Linnaeus)

GRAY SNAPPER

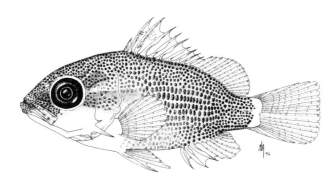

Estuary	Reproduction	Larvae/Development	YOY Growth	YOY Overwinter
Ocean				
Both				
Jan				■
Feb				■
Mar				■
Apr				■
May	■	■		
Jun	■	■		
Jul	■		▨	
Aug	■	■	▨	
Sep	■		▨	
Oct			▨	■
Nov				■
Dec				■

DISTRIBUTION

Lutjanus griseus occurs in the western North Atlantic Ocean from Massachusetts, Bermuda, and the northern Gulf of Mexico to southeastern Brazil (Hildebrand and Schroeder, 1928; Hoese and Moore, 1977). It is also found in the eastern Atlantic off West Africa. In the Middle Atlantic Bight,

juveniles are occasionally reported from estuaries from Buzzard Bay and south (see Table 4.3).

REPRODUCTION AND DEVELOPMENT

Spawning occurs from May to September, especially during new and full moon phases. Eggs are produced in multiple batches in offshore waters over deep reefs, with the northernmost spawning occurring off Florida on the east (Starck, 1971; Rutherford et al., 1989; Domeir and Colin, 1997) and west (Allman and Grimes, 2002) coasts.

The early development has been well described and summarized (Fahay, 2007). The pelagic eggs are spherical and 0.70 to 0.85 mm in diameter. Hatching occurs after 17 to 20 hours at 27 to 30°C. Larvae hatch at about 2.5 mm throughout the lunar cycle, with a peak near the new moon (Tzeng et al., 2003) The larvae are very elongate, but the body soon deepens. Dorsal fin spines and pelvic fin rays develop early, and all fin spines and rays are complete at sizes < 10 mm. An array of spines develops on several bones of the head, with prominent ones forming at the angle of the pre-opercle. Pigment is generally heavy over the head and body, with unpigmented areas on the caudal peduncle and lower cheek area.

Only individuals relatively late in development (18.6 to 28.6 mm TL, n = 9) were examined for scale formation (see Fig. 61.1). Scales originate on the body as well as on the operculum and behind the eye. By approximately 18 mm TL, scales cover the body on the lateral, ventral, and most of the dorsal surfaces, from the base of the caudal fin anteriorly to behind the operculum. By 19 mm TL, scales have formed on the opercle and upper pre-opercle and have extended to the dorsal surface of the head and anterior to the pectoral fin base. By 24 to 28 mm TL, scales have expanded to the lateral surfaces of the head and to the bases of the caudal fin rays. At 28 mm, scales cover 86.6% of the body and 56.2% of the body and fins, and this approximates the adult condition.

LARVAL SUPPLY, SETTLEMENT, GROWTH, AND MORTALITY

The supply of larvae of this species to temperate estuaries is probably from distant sources. The only presumed spawning sites on the East Coast of the United States are from Florida. It has been speculated that larvae could be transported northward along the shelf by wind-driven flows or be entrained into the Gulf Stream and then advected across the shelf into North Carolina estuaries (Tzeng et al., 2003); the same might apply to those occurring farther to the north in New Jersey estuaries. In North Carolina, transforming larvae ingress during the new and full moons and are 11.3 to 15 mm SL at 21 to 34 days old (Tzeng et al., 2003). Settlement occurs after the larvae have been pelagic for 22 to 32 days at sizes > 20 mm (Richards and Saksena, 1980).

Little is known about rates of growth in temperate estuaries. Transforming larvae have growth rates of approximately 0.50 mm per day (Tzeng, 2000). The smallest individ-

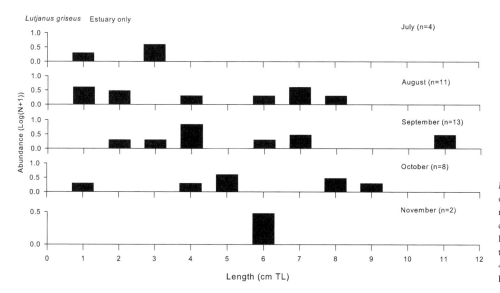

Fig. 68.1. Monthly length frequencies of *Lutjanus griseus* collected in estuarine and coastal ocean waters of the central Middle Atlantic Bight. Sources: Rutgers University Marine Field Station: seine (n = 14); killitrap (n = 1); 4.9 m otter trawl (n = 3); and Able and Fahay 1998 (n = 20).

uals arrive in New Jersey in the summer as settled juveniles at lengths between 1 and 3 cm (Fig. 68.1). By late summer and fall, presumed young-of-the-year are as large as 10 to 110 mm. Size at egress from a North Carolina estuary is 70 to 130 mm SL (mean = 88.3 mm SL) and an age of 85 to 103 days (mean = 94 days) (Tzeng, 2000). The largest individuals are 760 mm TL and 24 years old (Burton, 2001).

One source of mortality, especially for individuals that settle in temperate estuaries, is that which occurs during the winter. The minimum temperature for survival is reported as 11°C (Starck, 1971); however, some young-of-the-year have survived in temperatures as low as 5.7°C (Tzeng, 2000). A further analysis indicated that young-of-the-year (22.5 to 53.7 mm TL) captured in Great Bay, New Jersey, suffered 100% mortality when ambient estuarine temperatures reached 10.2°C (see Table 10.2). Lower temperatures are common in every winter in Middle Atlantic Bight estuaries (see Chapter 2). Thus, if young-of-the-year cannot successfully migrate south or offshore to reach a thermal refuge, they will not survive.

SEASONALITY AND HABITAT USE
Relatively little is known of this species' distribution or ecology in the Middle Atlantic Bight, probably because it is an uncommon visitor. Presumed young-of-the-year (105 to 111 mm)

are known from Chesapeake Bay as strays (Hildebrand and Schroeder, 1928), and individuals 14 to 72 mm have been collected in Delaware Bay (de Sylva et al., 1962; Smith, 1971; Wang and Kernehan, 1979). A similar size range (14 to 39 mm) has been reported in New Jersey (Milstein and Thomas, 1976b). In all these studies, as well as recent collections in the Great Bay–Little Egg Harbor area, occurrences ranged from July to November (see Fig. 68.1) and young-of-the-year have been collected from several different habitats. In North Carolina, the young-of-the-year are often found in eelgrass beds (Adams, 1976). In New Jersey, they have been collected at shallow beaches and appear to be associated with macroalgae at polyhaline salinities (K. W. Able, pers. observ.).

PREY AND PREDATORS
Unknown in the study area.

MIGRATIONS
Little is known concerning the fate of young-of-the-year that emigrate from Middle Atlantic Bight estuaries in the fall. There is no evidence to suggest that they successfully return to habitats south of Cape Hatteras (as do young-of-the-year *Caranx hippos*, for example), especially given their slow feeding and growth rates at moderately low temperatures (18°C) (Wuenschel et al., 2004).

69

Lagodon rhomboides (Linnaeus)

PINFISH

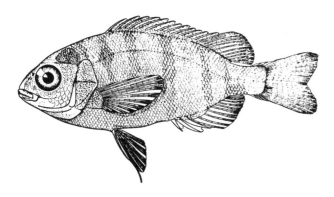

Estuary	Reproduction	Larvae/ Development	YOY Growth	YOY Overwinter
Ocean				
Both				
Jan	■	■		
Feb	■	■		■
Mar	■	■		■
Apr				■
May				■
Jun			▨	
Jul			▨	
Aug			▨	
Sep			▨	
Oct	■			■
Nov	■			■
Dec	■	■		■

DISTRIBUTION

Lagodon rhomboides is distributed from Cape Cod (rarely) and Bermuda to the Gulf of Mexico and the northern coast of Cuba (Carpenter, 2002). Larvae, juveniles, and adults are reported from Pamlico and Albemarle sounds, juveniles from

Great Bay, New Jersey, and only a few adults are known to occur in large embayments in the northern part of our study area (see Table 4.3). Most of the biological data reported for this species derives from research south of our study area. We have summarized the available life history data from the Atlantic coast, but advise that additional data are available from the Gulf of Mexico.

REPRODUCTION AND DEVELOPMENT

Spawning occurs offshore (Hildebrand and Schroeder, 1928; Hildebrand and Cable, 1938) from October to March from North Carolina to Florida (Murdy et al., 1997). The early development of this species has been well described (Powell and Greene, 2002; Fahay, 2007). Eggs are pelagic and spherical with a diameter of 0.99 to 1.05 mm. Larvae hatch at an unknown size. They have an elongate body and small head. A row of pigment spots defines the ventral edge of the body. Juveniles are diagnosed by the presence of five or six pigment bars crossing the body.

LARVAL SUPPLY, SETTLEMENT, GROWTH, AND MORTALITY

Eggs are fertilized in offshore waters, and the smallest developing larvae (5.0 to 10.0 mm TL) have been collected between 19.2 and 20.1 km off the coast of North Carolina (Hildebrand and Cable, 1938), among other locations. Larvae migrate inshore at sizes of about 10 to 12 mm SL before completing the transformation to the juvenile stage. They enter Chesapeake Bay during the spring. At selected sites south of Cape Hatteras, they enter inlets between October and April, with a peak in December (Beaufort, North Carolina); between November and April, with a peak in February (Newport River estuary, North Carolina); or peak in January, February, and March (North Inlet estuary, South Carolina) (Darcy 1985). At all these sites, the sizes of ingressing larvae are between 10.0 and 15.5 mm SL (Shenker and Dean, 1979). Subsequent studies have further characterized ingress dynamics in North Carolina (see Warlen and Burke, 1990; Hettler et al., 1997).

There are many studies of growth from the Gulf of Mexico (Darcy, 1985). If the monthly length-frequency distribution reported from the west coast of Florida (Stoner, 1980) is typical for the study area, then two year classes (young-of-the-year < 10 cm and age 1+ > 10 cm) may be represented in the study area (Fig. 69.1).

Causes of mortality are unknown, but the absence of age 1+ individuals in spring and summer suggests that the young-of-the-year do not survive the winter in our study area.

SEASONALITY AND HABITAT USE

After larvae move into estuaries from the continental shelf, the juveniles occupy a variety of habitats, but usually over shallow, vegetated areas such as grass beds (Darcy, 1985) and infrequently over rocks or in mangroves. They may enter

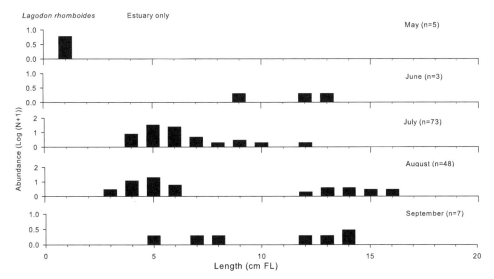

Fig. 69.1. Monthly length frequencies of *Lagodon rhomboides* collected in the Great Bay–Little Egg Harbor study area. Source: Rutgers University Marine Field Station: seine (n = 102); 1 m plankton net (n = 5); killitrap (n = 1); 4.9 m otter trawl (n = 14); multi-panel gill net (n = 14).

brackish waters or freshwaters. Once settled in estuaries, they establish strong fidelity for sites in marsh creeks, with an average home range of 9.4 m (Potthoff and Allen, 2003). Despite this, they also exhibit tidal periodicity in both movements and feeding. Some of these include movements onto the marsh surface for feeding (Hettler, 1989b).

PREY AND PREDATORS

This species' diet varies through five ontogenetic stages, ranging from planktivory to omnivory, strict carnivory, and strict herbivory (Stoner, 1980). Larvae and juveniles feed in the water column, whereas adults are primarily benthic feeders (Darcy, 1985). The species is basically omnivorous as an adult, but earlier life history stages go though transitions where diet items change. Larvae feed primarily during the day and locate their animal prey by sight. Larvae between 11 and 15 mm SL are planktivorous and feed mostly on copepods (Carr and Adams, 1973). At about 16 to 20 mm SL, larvae switch to amphipods, oligochaetes, and polychaetes. At sizes between 20 and 35 mm SL, they become more carnivorous, focusing on shrimp, amphipods, mysids, harpactocoid copepods, and invertebrate eggs. Mysids become important items when small juveniles are about 31 to 35 mm SL (Kinch, 1979). Larger juveniles (36 to 80 mm SL) are basically omnivorous, incorporating more diet items than at any other life history stage (Darcy, 1985). Epiphytes, shrimp, fishes, and algae are all utilized at these sizes. Depending on location, larger juveniles (up to 80 mm SL) feed on a wide variety of benthic organisms, including epiphytes, algae, polychaetes, larval fishes, and crustaceans (summarized in Darcy, 1985). Adults feeding in eelgrass beds near Beaufort, North Carolina, were found to contain mostly detritus, picked up while foraging on polychaetes, calanoid copepods, caprellid amphipods, gammarid amphipods, and eelgrass, along with smaller amounts of several other constituents (Adams, 1976).

Predators include many species of larger fishes, spotted dolphin (*Stenella plagiodon*), brown pelican (*Pelecanus occidentalis*), magnificent frigatebird (*Fregata magnificens*), and double-crested cormorant (*Phalacrocorax auritus*) (see Darcy, 1985). Fishes known to prey on *Lagodon rhomboides* include *Carcharhinus plumbeus* (Hildebrand and Schroeder, 1928), *Elops saurus* (Darnell, 1958), *Opsanus beta* (Reid, 1954), *Urophycis floridana* (Reid, 1954), *Sciaenops ocellatus* (Peterson and Peterson, 1979), *Cynoscion nebulosus* (Moody, 1950), *C. regalis* (Merriner, 1975), *Istiophorus platypterus* (Voss, 1953), and *Paralichthys lethostigma* (Darnell, 1958).

MIGRATIONS

At the end of their first summer, young-of-the-year have been reported to migrate to deeper waters on the continental shelf (Grimes and Mountain, 1971; Weinstein et al., 1977). However, another report describes young-of-the-year (12 to 16 mm TL) schooling with two other species during the winter, near jetties in Chesapeake Bay (Hildebrand and Cable, 1938). The larger juveniles and adults are also believed to move offshore during the winter in the South Atlantic Bight (Darcy, 1985).

70

Stenotomus chrysops (Linnaeus)

SCUP

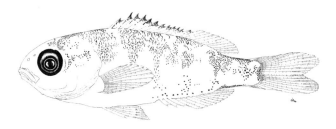

Estuary	Reproduction	Larvae/ Development	YOY Growth	YOY Overwinter
Ocean				
Both				
Jan				
Feb				
Mar				
Apr				
May				
Jun				
Jul				
Aug				
Sep				
Oct				
Nov				
Dec				

DISTRIBUTION

Stenotomus chrysops is found along the coast of North America from Nova Scotia to South Carolina, but is most common between Cape Cod and Cape Hatteras (Hildebrand and Schroeder, 1928; Bigelow and Schroeder, 1953; Scott and Scott, 1988). Juveniles have been reported from all estuaries between Buzzards Bay and Chesapeake Bay (see Table 4.3), but eggs and larvae have only been found in those north and east of the Hudson River estuary.

REPRODUCTION AND DEVELOPMENT

Sexual maturity is reached at age 2 (Nichols and Breder, 1927; Finkelstein, 1969a; Morse, 1982). Spawning occurs from May to August, with a peak in June, as shown in several studies between Vineyard Sound, Massachusetts, and Sandy Hook, New Jersey (Nichols and Breder, 1927; Perlmutter, 1939; Wheatland, 1956; Finkelstein, 1969b; Sisson, 1974; O'Brien et al., 1993). All of these studies indicate a single peak in abundance of eggs and larvae; thus, there is apparently only one spawning per season. The precise location of spawning, and of concentrations of eggs and larvae, has remained enigmatic. Spawning appears to take place in larger bodies of water, such as Long Island Sound (especially the more saline eastern part; Richards, 1959), Sandy Hook Bay (possibly near the bay-ocean interface), Narragansett Bay, and coastal ocean waters between those systems (Morse, 1982). Gardiner's and Peconic bays have also been characterized as important spawning areas (Perlmutter, 1939). However, eggs and larvae are rare or absent in other well-studied systems, such as Block Island Sound, Great South Bay, the Hudson River estuary, and Great Bay (Merriman and Sclar, 1952; Swiecicki and Tatham, 1977; Kahnle and Hattala, 1988; Monteleone, 1992; Witting et al., 1999). It has been suggested that spawning occurs over sandy and weed-covered bottoms (Morse, 1978). Spawning presumably does not take place over the continental shelf, because during 11 years (1977–1987), the MARMAP surveys collected only 14 larvae, and all of these were from 3 sampling stations near the mouth of Narragansett Bay.

The early development of *Stenotomus chrysops* is well described and summarized (Fahay, 2007). Eggs are pelagic and spherical and measure 0.8 to 1.0 mm in diameter. The embryos incubate for 70 to 75 hours at 18°C and 44 to 54 hours at 21°C. Larvae hatch at about 2.0 mm, and their yolk is absorbed after 2 or 3 days, when active feeding begins (Griswold and McKenney, 1984). Weak spines form along the preopercle edge. Full complements of fin rays are formed by about 25 mm. The deep-bodied, sharp-snouted adult morphology is acquired by 40 to 60 mm (Nichols and Breder, 1927).

Scale formation was examined in 9 individuals, 25.6 to 55.1 mm TL (see Fig. 61.1). By approximately 25 mm TL, scales cover the lateral surface of the body from behind the pectoral fin to the base of the caudal fin. By 43 mm TL, most scales are completely formed on the body and head, including the operculum, an area below and behind the eye, and on the ventral surface in the gular region. The adult condition, where scales cover 92.5% of the body and head, is reached at a length of about 50 mm.

LARVAL SUPPLY, SETTLEMENT, AND GROWTH

Larvae have only rarely been collected in our study area, presumably the result of sporadic spawning along the adjacent coastal zone. When they have been collected, they have appeared from May through August (Fig. 70.1). Otter trawl

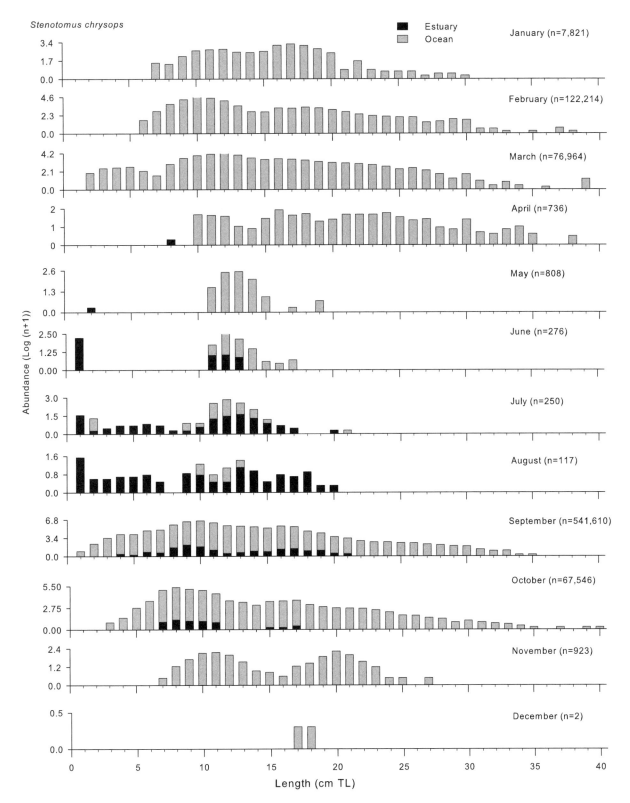

Fig. 70.1. Monthly length frequencies of *Stenotomus chrysops* collected in estuarine and coastal ocean waters of the central Middle Atlantic Bight. Sources: National Marine Fisheries Service: otter trawl (n = 816,020); Rutgers University Marine Field Station: seine (n = 16); 1 m plankton net (n = 4); 2 m beam trawl (n = 1); killitrap (n = 2); 4.9 m otter trawl (n = 280); multi-panel gill net (n = 10); and Able and Fahay 1998 (n = 2834).

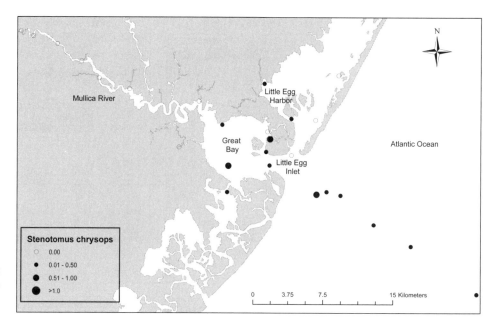

Fig. 70.2. Habitat use and average catch per tow (see key) for juvenile and adult *Stenotomus chrysops* in the Ocean–Great Bay–Mullica River corridor in southern New Jersey based on small otter trawl collections during July and September from 1988 to 2006. See Table 3.2 for characteristics at each sampling collection.

sampling between 1997 and 2006 indicates that there may be two or three more abundant years per decade in our area (K. W. Able, pers. observ.). Settlement to the bottom occurs when larvae are about 15 to 30 mm (Lux and Nichy, 1971). Growth is relatively slow during the first year, based on combined collections made in the Great Bay–Little Egg Harbor estuary, Long Island embayments, and waters within 24 km of shore in the central part of the Middle Atlantic Bight. By October, most young-of-the-year are 3 to 15 cm TL. These sizes are somewhat larger than those reported for New England young-of-the-year (Lux and Nichy, 1971). Growth over the winter continues to be slow, and age 0+ fish returning in the spring are as small as 2 cm but as large as 15 cm TL (Wilk et al., 1992). The average length of New England fish at age 1 is 10.6 cm (Morse, 1982).

SEASONALITY AND HABITAT USE

This species is most abundant on the inner continental shelf and in larger and deeper estuaries, such as Delaware Bay and Narragansett Bay, than in smaller ones, such as the Lower Hudson–Raritan estuary or Great Bay (Berg and Levinton, 1985). In the northern Middle Atlantic Bight, schools move inshore during April and May and spend the summer in bays or coastal waters within 10 km of the coast (Perlmutter, 1939; Bigelow and Schroeder, 1953; Wheatland, 1956). Fish in the central part of the Middle Atlantic Bight move inshore during early May (Morse, 1982) and June (Eklund and Targett, 1991), beginning with the larger size classes. In this part of the bight, large individuals 4 years old and older tend to remain in coastal ocean waters or near mouths of larger bays and do not penetrate estuaries. Younger fish are found inside bays and sounds and on the inner continental shelf (Fig. 70.2). Most *Stenotomus chrysops* occupying Sandy Hook Bay during summer are age 1 (Wilk and Silverman, 1976) and presum-

ably are too young to spawn. Similar observations have been made near the mouth of Delaware Bay (de Sylva et al., 1962).

This species typically occurs over hard substrates and structured habitats. The smallest young-of-the-year appear in estuaries during May or June, and they are only found there through October (Able and Fahay, 1998). We have only rarely collected larvae or juveniles in ichthyoplankton sampling. There is little available information regarding specific habitats favored by juveniles. In Great Bay–Little Egg Harbor our sampling has covered a variety of habitats; juveniles occur over a variety of substrates, but are most abundant over a bottom with no structure and in depths ranging from 3 to 5 m. They do not penetrate low-salinity areas (Morse, 1982). They have been reported as "abundant" in the highest-salinity portions (> 25.0 ppt) of the Hudson River–Raritan Bay estuary and Long Island Sound, but are considered "rare" in the mixed-salinity zone (0.5 to 25.0 ppt) of these systems (Stone et al., 1994). The same trend is apparent in the Mullica River–Great Bay estuary, with most individuals collected on the adjacent inner continental shelf and in the lower, high-salinity portion of the estuary (see Fig. 70.2).

PREY AND PREDATORS

This species feeds directly over the bottom and consumes mostly cnidarians, squid, polychaetes, crustaceans, and small fishes (Bowman and Michaels, 1984; Bowman et al., 2000). One study examined the food habits of young-of-the-year (20 to 146 mm, n = 181) *Stenotomus chrysops* in Long Island Sound (Richards, 1963). Young-of-the-year concentrate on polychaetes, amphipods, other crustaceans, and mollusks (see Table 8.1). The diet of young-of-the-year shifts with ontogeny at approximately 70 to 100 mm, such that copepods and mysids are more prevalent in fish < 100 mm, while larger fish consume more mollusks, decapod shrimp, and crabs.

Age 1+ *Stenotomus chrysops*   ## YOY *Stenotomus chrysops*

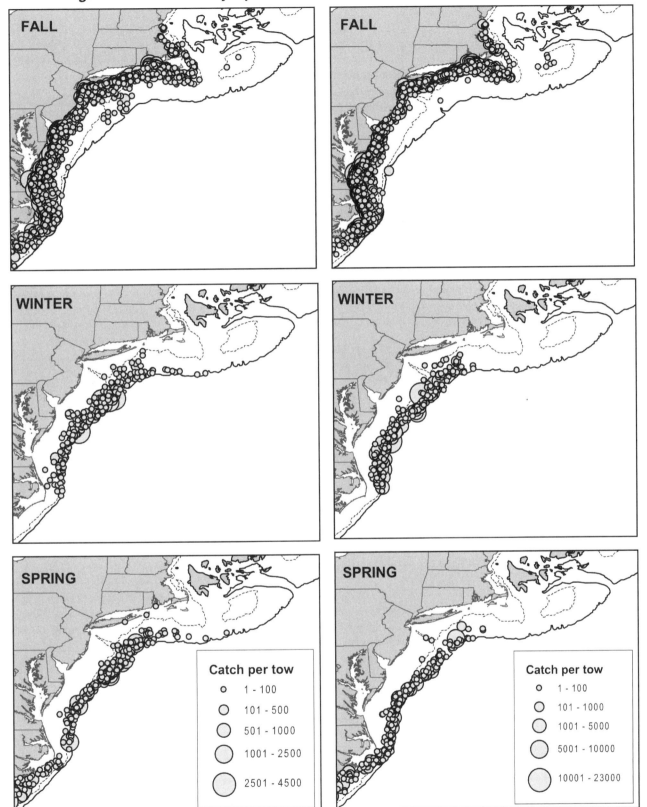

Fig. 70.3. Composite distribution of age 1+ (> 13 cm) and young-of-the-
year (< 13 cm) *Stenotomus chrysops* during seasonal cruises of the National
Marine Fisheries Service groundfish survey. Details of sampling effort are
indicated in Fig. 3.4.

Important predators are several species of elasmobranchs and bony fishes, including *Squalus acanthias, Mustelus canis,* and *Pomatomus saltatrix* (Rountree, 1999).

MIGRATIONS

The emigration of young-of-the-year from estuaries and bays into the ocean begins in September and lasts through November. During this period, young-of-the-year 6 to 13 cm TL (along with larger size classes) are frequently collected by trawl in coastal ocean waters in our study area (Fig. 70.3; Wilk et al., 1992). The fall migration proceeds southward inside the 10 m isobath and is then followed by an offshore movement (Hamer, 1970). The population then spends the winter in depths between 37 and 146 m, mostly on the outer continental shelf between New Jersey and Cape Hatteras (Neville and Talbot, 1964; Morse, 1982; Shepherd and Ter-

ceiro, 1994). There is no evidence suggesting that young-of-the-year overwinter in areas separate from older year classes. Fish from New England also spend the winter on the outer continental shelf, but in areas slightly farther to the north and east than those from the southern Middle Atlantic Bight (Hamer, 1970). Winter distributions appear to be determined largely by temperature because the species occurs mostly in bottom waters at temperatures above 7.3°C (the lower temperature tolerance limit), and the location of that isotherm can vary annually (Neville and Talbot, 1964). These patterns are generally supported by our more recent observations. In the spring, they have a similar geographical and depth (mean depth 78 m, range 8 to 300 m) distribution as in the winter with the exception that the available spring sampling below Cape Hatteras found them on the inner continental shelf.

71

Bairdiella chrysoura (Lacepède)

SILVER PERCH

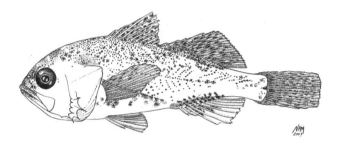

Estuary	Reproduction	Larvae/ Development	YOY Growth	YOY Overwinter
Ocean				
Both				
Jan				
Feb				
Mar				
Apr				
May				
Jun				
Jul				
Aug				
Sep				
Oct				
Nov				
Dec				

DISTRIBUTION

Bairdiella chrysoura occurs in the western North Atlantic Ocean from Cape Cod to Florida and the northern Gulf of Mexico (Robins and Ray, 1986). It reaches its northern limit in the Middle Atlantic Bight, where early life history stages are common in estuaries from as far north as Great South Bay and the Connecticut River to the southern part of our study area (see Table 4.3).

REPRODUCTION AND DEVELOPMENT

Individuals mature in their third year (Welsh and Breder, 1923). Spawning occurs from late spring to early summer. Spawning in northern New Jersey has been reported from June to August, with a peak in June (Welsh and Breder, 1923). In Great Bay, New Jersey, we have collected running ripe individuals (204 to 245 cm TL) in late June. In Delaware Bay, most spawning takes place in shallow areas from June through August, with peaks in June and July (Thomas, 1971). At Beaufort, North Carolina, spawning is reported to occur within the harbor, in estuarine marshes, in sounds, and also at some distance at sea (Hildebrand and Cable, 1930). Most spawning may take place primarily in estuaries because larvae were almost never collected (n = 12) in the Middle Atlantic Bight during the MARMAP surveys (1977–1987) (Able and Fahay, 1998).

The early development of *Bairdiella chrysoura*, including comparisons with the eggs and larvae of other sciaenids, has been well described and summarized (Fahay, 2007). Eggs are pelagic and spherical and range from 0.66 to 0.88 mm in diameter. There is a single oil globule, measuring 0.2 to 0.6 mm. Larvae at hatching are deep-headed and taper to a slimmer caudal end. Pigmentation throughout development includes a wide swath of pigment through the nape and opercular region. All fin spines and rays are formed at a length of 8.8 mm SL. Transformation to the juvenile stage is gradual and begins with the initiation of scale formation at about 19 mm. All scales and fin rays are formed at a size of about 56 mm TL.

Scale development in a total of 22 wild-caught individuals between 6.4 and 69.6 mm TL was examined (Fig. 71.1). No scales were observed on specimens between 6.4 and 19.6 mm TL. Scale formation occurred over the size range of 19.7 to 56.7 mm TL, and originated on the lateral line and three points on the head (operculum, dorso-lateral surface of the head, behind the eye). At about 19.7 mm, scales first appeared as a single row, which occurred along the lateral line on the posterior portion of the body. Scales spread anteriorly from this point in larger sizes until most of the body was covered, with the exception of the dorsum under the first dorsal fin. At about 32 mm, scales began forming on three portions of the head: on the operculum, in the temporal area, and behind the eye. Scales spread onto the bases of the caudal fin rays at about 36 mm, and onto the bases of the second dorsal and anal fins at about 46 mm. By 57 mm TL, scale formation was complete and had reached the adult condition, with scales covering 99% of the body and 65.5% of the body and fins.

LARVAL SUPPLY, SETTLEMENT, AND GROWTH

During most years, the larvae are collected in low numbers. However, their occurrences have spiked during some years and this is reflected differently, depending on the sampling gear used (Fig. 71.2). Their larvae were plentiful during 1999,

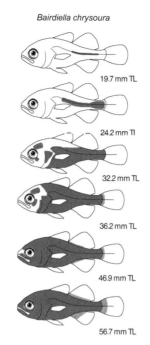

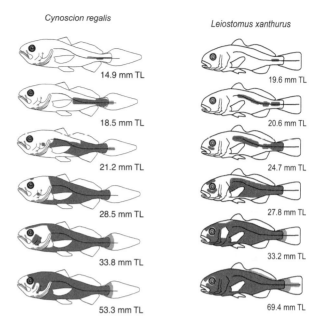

Bairdiella chrysoura

19.7 mm TL
24.2 mm Tl
32.2 mm TL
36.2 mm TL
46.9 mm TL
56.7 mm TL

Cynoscion regalis

14.9 mm TL
18.5 mm TL
21.2 mm TL
28.5 mm TL
33.8 mm TL
53.3 mm TL

Leiostomus xanthurus

19.6 mm TL
20.6 mm TL
24.7 mm TL
27.8 mm TL
33.2 mm TL
69.4 mm TL

Fig. 71.1. Description of scale formation relative to total length (TL) in *Bairdiella chrysoura, Cynoscion regalis,* and *Leiostomus xanthurus.*

as reflected by collections with a plankton net. Early-settled individuals were abundant during 1993, based on trap collections. Juveniles were numerous during 2006, based on otter trawl collections. It is not well understood why these years of relative high abundance do not coincide between the gear types.

The young first appear in the estuary as larvae and small juveniles in June and July at sizes of approximately 1 to 2 cm (Fig. 71.3). A single size mode of young-of-the-year is evident through the summer. In most years, they reach greatest abundance in the estuary in August and decline thereafter when they are captured in the ocean. By October, the young-of-the-year range from 3 to 15 cm TL or 6 to 14 cm as found by Welsh and Breder (1923). The modal progression in lengths in Great Bay–Little Egg Harbor samples indicates they grow an average of 0.6 mm per day during the summer and early fall. By the following March, the mode of the presumed age-1 individuals is 8 to 20 cm, which overlaps with the size (13 to 14 cm) at the end of the first year, as determined by analysis of scales (Thomas, 1971).

SEASONALITY AND HABITAT USE

Throughout its range, this species occurs in a variety of habitats, including sandy or muddy substrates in coastal ocean waters, and it is also often associated with marshes and seagrass beds in estuaries during the summer and fall (Fig. 71.4). The range of habitats occupied may vary as the result of seasonal migrations. In Great Bay–Little Egg Harbor, most young-of-the-year collected during the summer in otter trawls were found in eelgrass beds, but they have also been reported from marsh creeks (Rountree and Able, 1992a) and estuarine shorelines (Able et al., 1996b), primarily in the lower, polyhaline portions of the estuary and on the in-

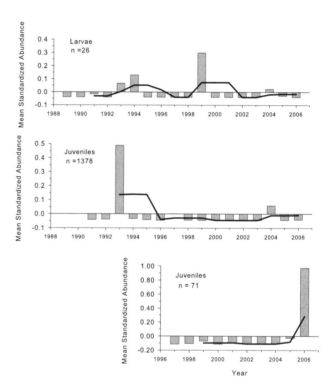

Fig. 71.2. Annual mean (bars) and three-year moving mean (line) of standardized abundance for *Bairdiella chrysoura* collected in larval sampling with plankton net (*upper*), juvenile sampling with traps (*middle*), and juvenile sampling with otter trawls (*lower*). Data were standardized by subtracting overall mean from annual abundance. Three-year moving mean was calculated by taking the mean of the standardized values from the previous two years and the year in which data are plotted.

nermost portion of the continental shelf (Fig. 71.5). In Delaware Bay, young-of-the-year are most abundant in shallow waters, typically in the lower portions of creeks and ditches and along the bay in protected areas, where they are found

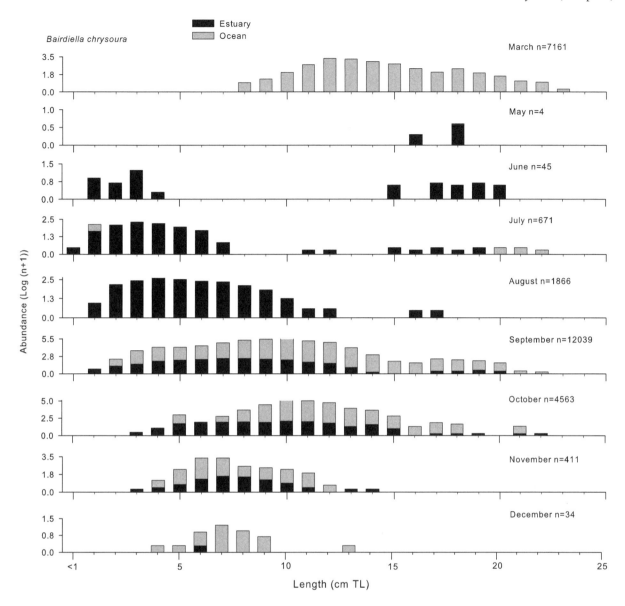

Fig. 71.3. Monthly length frequencies of *Bairdiella chrysoura* collected in estuarine and coastal ocean waters of the central Middle Atlantic Bight. Sources: National Marine Fisheries Service: otter trawl (n =20,997); Rutgers University Marine Field Station: 1 m beam trawl (n = 3); seine (n = 226); 1 m plankton net (n = 18); killitrap (n = 240); 4.9 m otter trawl (n = 366); and Able and Fahay 1998 (n = 4954).

over a bottom of mud or mud and sand where detritus is abundant (Thomas, 1971). Young stages occur up the bay as far as the Chesapeake and Delaware (C&D) Canal. Some are collected in the ocean in September and October, and by November and December, most of the collections are from the ocean (Beach Haven Ridge; see Fig. 71.3). The average size of young-of-the-year decreases in November and December, perhaps because larger individuals are moving out of the estuary into the ocean.

PREY AND PREDATORS

Several studies (n = 5) have examined the food habits of young-of-the-year (7 to 157 mm TL, n = 630) in Middle Atlantic Bight estuaries from the Delaware River to Beaufort, North Carolina (see Table 8.1). Important prey of young-of-the-year include a variety of invertebrates (mysids, amphipods, decapod shrimp, polychaetes, oligochaetes, crabs, and copepods) and fishes (*Anchoa mitchilli*). The diet of young-of-the-year shifts with ontogeny at approximately 40 to 70 mm, from a diet of small invertebrates (copepods and ostracods) to relatively larger invertebrates (decapod shrimp and crabs) and fishes. In the York River estuary in Virginia

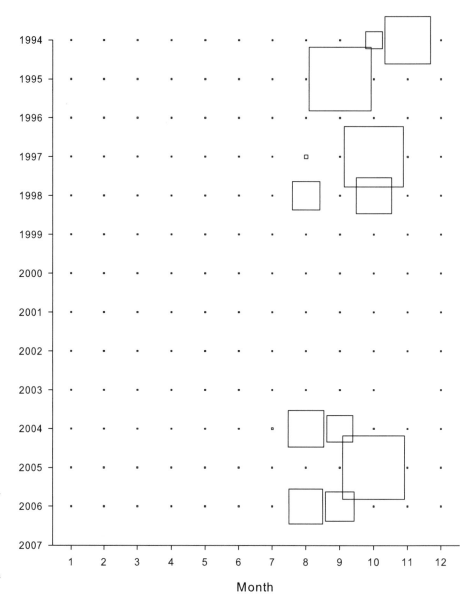

Fig. 71.4. Seasonal occurrence of the juveniles of *Bairdiella chrysoura* based on trap collections in Great Bay (Rutgers University Marine Field Station boat basin) from 1994 to 2006. Full squares represent the percentage of the mean catch per unit effort of *Bairdiella chrysoura* captured in a given year. Values range from 0 to 100%. The smallest squares represent when samples were taken but no individuals were collected.

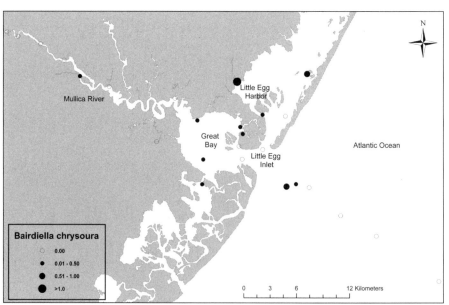

Fig. 71.5. Habitat use and average catch per tow (see key) for juvenile and adult *Bairdiella chrysoura* in the Ocean–Great Bay–Mullica River corridor in southern New Jersey based on small otter trawl collections during July and September from 1988 to 2006. See Table 3.2 for characteristics at each sampling collection.

Age 1+ *Bairdiella chrysoura*

YOY *Bairdiella chrysoura*

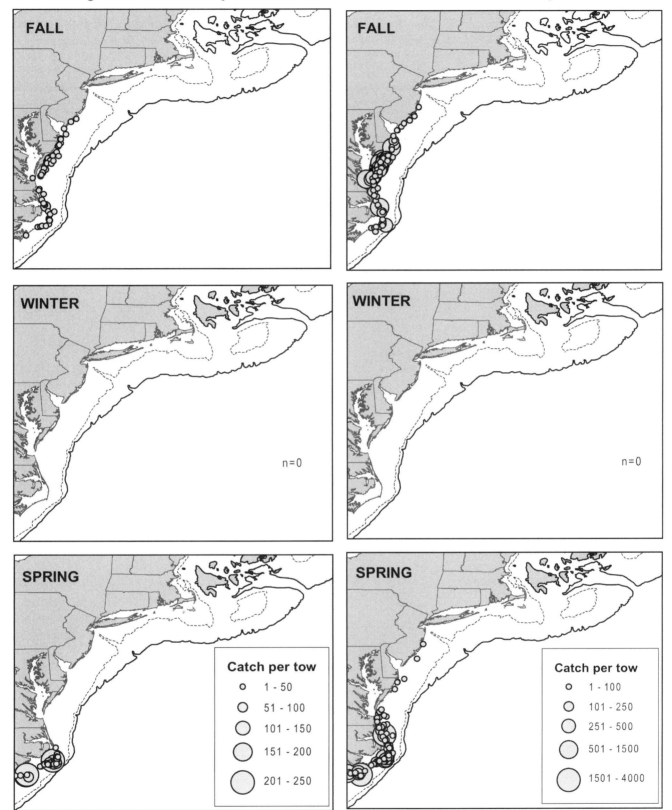

Fig. 71.6. Composite distribution of age 1+ (> 15 cm) and young-of-the-
year (< 15 cm) *Bairdiella chrysoura* during seasonal cruises of the National
Marine Fisheries Service groundfish survey. Details of sampling effort are
indicated in Fig. 3.4.

young-of-the-year < 40 mm SL feed mostly on copepods, but consume more mysids, amphipods, and other crustaceans as they grow; fishes become more important prey to individuals > 70 mm SL (Chao and Musick, 1977). At Beaufort, North Carolina, young-of-the-year (7 to 20 mm, n = 30) feed chiefly on copepods, supplemented by ostracods, a few amphipods, cladocerans, an occasional mysid, and chaetograths; young-of-the-year (25 to 50 mm, n = 79) feed more frequently on relatively larger crustaceans (mysids, decapod shrimp, crabs, chaetograths, and mollusks) and more sparingly on copepods, ostracods, and isopods, while young-of-the-year (50 to 80 mm, n = 15) largely eat mysids, shrimp, *Gammarus* spp., and chaetograths (Hildebrand and Cable, 1930).

MIGRATIONS

In Delaware Bay, most specimens leave the upper part of the bay by October and move into the deeper portions, where they have been collected in otter trawls (de Sylva et al., 1962). Movements out of estuaries and into the ocean may

be precipitated by declining temperatures because, at ambient winter estuarine temperatures, young-of-the-year experienced 100% mortality in a laboratory study (Able and Curran, 2008). However, some young-of-the-year are reported to overwinter in the deeper waters of Chesapeake Bay (Hildebrand and Schroeder, 1928).

The distribution of young-of-the-year (< 15 cm) on the continental shelf varies by season (Fig. 71.6). In the fall, young-of-the-year are distributed on the inner shelf in depths of 6 to 27 m from Massachusetts to North Carolina. There have been no individuals collected in the winter in the study area, but by spring they are common once again in similar depths (mean depth 16 m, range 12 to 25 m), at least in North Carolina and Virginia but extending, less abundantly, to New Jersey. The age 1+ (> 15 cm) juveniles and adults occur on the inner and mid-shelf in the fall (mean depth 13 m, range 8 to 33 m) from southern New Jersey to North Carolina. They are absent from the Middle Atlantic Bight in the winter. They reappear in the vicinity of Cape Hatteras during spring.

72

Cynoscion nebulosus (Cuvier)

SPOTTED SEATROUT

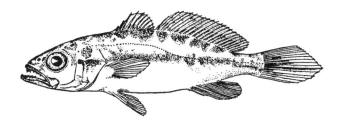

Estuary	Reproduction	Larvae/ Development	YOY Growth	YOY Overwinter
Ocean				
Both				
Jan				
Feb				
Mar				
Apr				
May				
Jun				
Jul				
Aug				
Sep				
Oct				
Nov				
Dec				

DISTRIBUTION

This species occurs in the western North Atlantic Ocean from Cape Cod to Florida and the Gulf of Mexico (Murdy et al., 1997). It is common in Chesapeake Bay, is uncommon in Delaware Bay, and rarely occurs north of Delaware Bay.

REPRODUCTION AND DEVELOPMENT

This species can mature as early as age 1 and at sizes as small as 292 mm for 50% of the females (Ihde, 2000). By age 2 and

417 mm, 100% are mature. Spawning occurs between May and August, with a single peak in late May–early June. Some females are capable of spawning multiple times in the same season (Ihde, 2000). The early development of this species has been well described and summarized (Fahay, 2007). Eggs are pelagic and spherical and measure 0.70 to 0.85 mm in diameter. The single oil globule is 0.18 to 0.27 mm in diameter. Larvae at hatching are 1.3 to 1.6 mm long and have pigmented eyes and well-formed mouth parts. Pigment is sparse in early larvae, but becomes pronounced along the lateral line in larger larvae. Juveniles are diagnosed by their large jaws, well-developed teeth, dark pigment along the dorsal margin and lateral surfaces, characteristic caudal fin shape (with central fin rays longest), and meristic characters.

LARVAL SUPPLY, SETTLEMENT, AND GROWTH

Most reproduction occurs in estuaries and bays (Mercer, 1984a), with larval development occurring in the same water bodies. Settlement occurs at sizes of 4 to 8 mm SL for a Florida population (Powell et al., 2007). Growth in Chesapeake Bay is fast, with young-of-the-year growing at the rate of 1.4 mm per day (Smith et al., 2008). Subsequent growth is fast as well, with the mean of age 1 individuals about 400 mm TL (Ihde, 2000). Ages 2 and 3 individuals continue this fast growth rate, approximately 100 mm per year, but growth slows beyond age 4. They reach a maximum length of 817 mm and an age of 10 years.

SEASONALITY AND HABITAT USE

The occurrence of this species in Chesapeake Bay is highly seasonal. All life history stages inhabit seagrass beds after settlement (Dorval et al., 2005). Their growth in these beds varies annually, perhaps as a function of freshwater flow to the estuary (Smith et al., 2008).

PREY AND PREDATORS

Adults are primarily piscivorous, but they also feed on decapod crustaceans (Mercer, 1984a). At sizes greater than 50 mm SL, fishes begin to dominate the diet, with *Lucania parva* leading the list (Hettler, 1989b). Young fish are preyed upon by other fishes. Predators on adults include *Morone saxatilis, Micropogonias undulatus, Megalops atlanticus,* and *Sphyraena barracuda* (Mercer, 1984a).

MIGRATIONS

The population in Chesapeake Bay is believed to be migratory, with adults migrating from the ocean to the bay in the spring and resident in the bay in the summer. The young-of-the-year migrate out of the bay and into the ocean during the fall (Murdy et al., 1997; Dorval et al., 2005).

73

Cynoscion regalis (Bloch and Schneider)

WEAKFISH

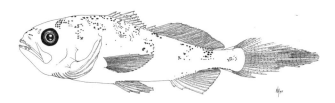

Estuary	Reproduction	Larvae/ Development	YOY Growth	YOY Overwinter
Ocean				
Both				
Jan				
Feb				
Mar				
Apr				
May				
Jun				
Jul				
Aug				
Sep				
Oct				
Nov				
Dec				

DISTRIBUTION

Cynoscion regalis occurs along the coast of North America from Nova Scotia to the Cape Canaveral region of Florida (Klein-MacPhee, 2002b). Its center of abundance lies between New York and North Carolina (Shepherd and Grimes, 1983). Eggs, larvae, and juvenile stages are found in most Middle Atlantic Bight estuaries (see Table 4.3).

REPRODUCTION AND DEVELOPMENT

This species, at least in Delaware Bay, matures at 168 mm; 97% of age 1 fish are mature (Nye et al., 2008). Spawning occurs from spring through summer, with most activity taking place close to the coast, near major inlets, or within bays or estuaries and involving homing to natal estuaries (Thorrold et al., 2001). Peak reproduction is during May through July (Welsh and Breder, 1923; Hildebrand and Schroeder, 1928; Pearson, 1941; Daiber, 1957; Harmic, 1958; Massmann et al., 1958; Merriner, 1976; Lowerre-Barbieri et al., 1996a, b; Berrien and Sibunka, 1999). Establishing the timing of spawning in discrete estuarine systems may be confounded by different groups (possibly composed of same-age individuals) migrating into and out of those systems and spawning at different times of the year (see above citations). For example, there are reports of two peaks of activity in the Delaware Bay region (Daiber, 1957; Goshorn and Epifanio, 1991). In Chesapeake Bay, all mature fish begin spawning together, but the cessation of spawning is asynchronous, varying among adults, and the duration (or resumption after an initial peak) of spawning varies between years, contributing to this variable pattern (Lowerre-Barbieri et al., 1996a, b). Variation in spawning time has also been found to be a function of the size of adults (Shepherd and Grimes, 1984). Judging from MARMAP egg and larval collections between 1977 and 1987, spawning reaches a peak in July along the central part of the Middle Atlantic Bight coast. Eleven-year average occurrences of eggs (Berrien and Sibunka, 1999) and occasional dense concentrations of larvae indicate that the inner continental shelf area adjacent to central New Jersey is an important spawning area (Able et al., 2006a).

The early development of *Cynoscion regalis*, including characters useful for distinguishing the eggs, larvae, and juveniles from those of other species in the family Sciaenidae, has been well described and summarized (Fahay, 2007). The pelagic eggs are spherical and measure 0.75 to 0.98 mm in diameter. Larvae hatch at sizes < 2.5 mm NL with pigmented eyes and fully formed mouth parts. Larvae are large headed with a series of pre-opercle spines. Pigment patterns are important characters for identification. All fin rays are completely formed by 9 to 10 mm SL. Optimum incubation temperatures are 18 to 24°C, and hatching occurs after 1000 degree hours. Thus, at 20°C, hatching occurs 50 hours after fertilization (Harmic, 1958). A report that newly hatched larvae sink to the bottom (Harmic, 1958) is contrary to recent laboratory observations, in which they were found to be buoyant (J. Duffy-Anderson, pers. comm.).

A total of 28 wild-caught individuals between 7.2 and 53.3 mm TL were examined for patterns of scale formation (see Fig. 71.1; Able et al., 2009a). Loci where scales originate include the lateral line on the caudal peduncle, on the head, and on the anterior portion of the operculum behind and below the eye. Individuals between 7.2 and 15.2 mm TL have no scales. Onset of scale formation occurs at about 14.9 mm

Passive Acoustics as a Key to Defining Spawning Habitat

Passive acoustics has been available for decades but it is only recently that it is being commonly applied to study the behavior of fishes (Luczovich et al., 2008; Rountree, 2008). One advantage of this approach is that it allows the detection of courtship sounds and, presumably, reproduction. One application has helped to verify that *Cynoscion regalis* spawns in the ocean and the estuary. During 2008, an acoustic recording system (Loggerhead Instruments) detected courtship sounds of this species, and others, on several occasions in approximately 15 m water depth at Beach Haven Ridge (C. Kennedy, T. Grothues, and K. W. Able, unpubl. data) just outside Little Egg Inlet. During the same summer courtship sounds were also detected at another regularly sampled location in Little Sheepshead Creek at depths of 3 to 4 m in the estuary behind Little Egg Inlet. Together these observations, along with the collection of larvae and juveniles (see species account), help to confirm that this species uses both the estuary and shallow ocean for reproduction and as larval and juvenile habitat.

TL (in at least one individual), where a few scales are observed along the lateral line of the caudal peduncle region (see also Szedlmayer et al., 1991). By approximately 18 mm TL, scale formation has spread anteriorly and laterally along the lateral line to just behind the pectoral fin. In addition, scales cover most of the caudal peduncle and extend onto the caudal fin. By 21 mm TL, scales occur on the operculum and cover the middle of the body posterior to the operculum, but do not extend onto the dorsal and ventral surfaces. Scale coverage extends onto the caudal fin in most specimens > 28 mm. By 28 mm TL, scales are present on a small area of the operculum and on the top of the head and they also cover the ventral surface of the body. By 34 mm TL, scales cover almost all of the body, including the dorsal and ventral surfaces, and they are further developed on the operculum. By 53 mm TL, scale formation has achieved the adult condition, where scales cover 76.2% of the body and 55.0% of the body and fins. However, the sample size was small, and completion may occur at smaller sizes. Specimens between 33 and 50.9 mm TL were not available. See Figures 71.1 and 75.1 for comparisons with other sciaenids. Estimated size and age when scales are first formed are 14.3 mm SL and 26 days, respectively (Szedlmayer et al., 1991).

LARVAL SUPPLY, GROWTH, AND MORTALITY

As with other sciaenid species, the population has declined precipitously in the past half century, and there are wide annual fluctuations in local abundances (Nye et al., 2008). The collection of larvae in bridge net sampling at our study site indicates that between 1989 and 2006, the only year with markedly increased abundance was 1999 (Fig. 73.1). Older stages collected in otter trawls between 1997 and 2006 indicate greater abundance in 1997.

Larvae, at least those off the coast of New Jersey, originate from spawning in the estuary and on the inner shelf (Witting et al., 1999; Able et al., 2006a). This is also evident from larval collections in the surfzone and immediately off ocean beaches on the coast of northern New Jersey, although their occurrence there varied annually (Able et al., 2009c, 2010). In those years when weakfish occurred, they were found both

in the surfzone and near shore. It is unclear if both the estuary and the inner shelf act as nurseries in the strict sense of the term (Beck et al., 2001).

Growth rates during the first year vary greatly between areas, and estimates are confounded by season, year, subpopulation, and investigative method (Wilk, 1976). In some years, bimodal distributions are evident in late summer and early fall due to the contributions of progeny from more than one spawning group, as has been described for Delaware Bay (Thomas, 1971). During some years, this phenomenon is not apparent, and instead a single cohort dominates (Thomas, 1971), as is evident from combined collections from the central part of the Middle Atlantic Bight. In early fall, young-of-the-year range from about 2 to 25 cm TL (Fig. 73.2). In early winter, the mode is 5 to 18 cm TL, possibly because the larger members of the cohort migrate out of the study area first. Several authors have reported potential lengths at age 1 of 18 to 20 cm (Welsh and Breder, 1923; Pearson, 1941; Thomas, 1971; Shepherd and Grimes, 1983). By the following

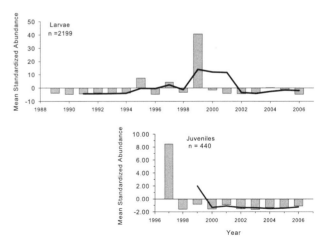

Fig. 73.1. Annual mean (bars) and three-year moving mean (line) of standardized abundance for *Cynoscion regalis* collected in larval sampling with plankton net (*upper*) and juvenile sampling with otter trawl (*lower*). Data were standardized by subtracting overall mean from annual abundance. Three-year moving mean was calculated by taking the mean of the standardized values from the previous two years and the year in which data are plotted.

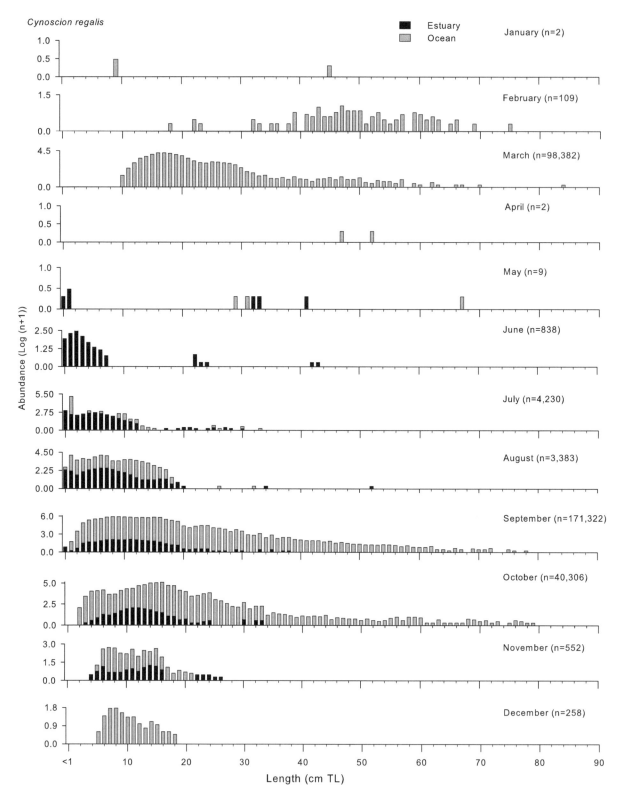

Fig. 73.2. Monthly length frequencies of *Cynoscion regalis* collected in estuarine and coastal ocean waters of the central Middle Atlantic Bight. Sources: National Marine Fisheries Service: otter trawl (n =305,581); Rutgers University Marine Field Station: 1 m beam trawl (n = 1); seine (n = 112); Methot trawl (n = 1); 1 m plankton net (n = 1378); 2 m beam trawl (n = 48); killitrap (n = 8); 4.9 m otter trawl (n = 1422); multi-panel gill net (n = 16); and Able and Fahay 1998 (n = 10,827).

spring (March) the modal size is from 10 to 30 cm. In Delaware Bay, growth rates increase with rising temperatures and vary from 0.29 mm per day at 20°C and salinities of 19 or 26 ppt to 1.49 mm per day at 28°C and 19 ppt (Lankford and Targett, 1994).

Recent studies have found that growth rates of young-of-the-year within Delaware and Chesapeake bays can vary between specific habitats located along a gradient of salinities and temperatures (Szedlmayer et al., 1990; Lankford and Targett, 1994). Young-of-the-year in mesohaline areas in Delaware Bay experienced increased growth rates and growth efficiencies compared to those in oligohaline areas. Young-of-the-year initially occur in low salinities, then move to higher salinities with growth (Chao and Musick, 1977). However, the opposite behavior is found in smaller larvae, with four of five cohorts moving upstream (into lower salinities) with growth (Szedlmayer et al., 1990) or no net movement with growth (Paperno, 1991). Given that the potential for growth is better in mesohaline habitats, the fact that Delaware Bay young-of-the-year migrate into oligohaline areas might imply an early life history trade-off, in which they sacrifice enhanced growth for more protection from predation (Lankford and Targett, 1994).

SEASONALITY AND HABITAT USE

Adults are usually found in shallow, coastal waters, including bays and estuaries, over sandy or muddy substrates. Because most spawning activity occurs in coastal waters, near inlets, or in estuaries themselves (Olney, 1983; Bourne and Govoni, 1988; Lankford and Targett, 1994; Berrien and Sibunka, 1999), eggs and larvae occur pelagically in both ocean and estuarine environments. Both of these early stages have also been found in Delaware Bay waters in temperatures of 17.0 to 26.5°C and salinities of 12.1 to 31.3 ppt (Harmic, 1958). Other studies in this bay have indicated that the young-of-the-year had an avoidance threshold at approximately 2 mg dissolved oxygen and demonstrated dynamic movements in response to this estuarine variability (Tyler and Targett, 2007). During a study in the Beach Haven Ridge area, larvae were commonly collected during July, with lingering, sparse occurrences into September (Able et al., 2006a). They were also commonly collected in the adjacent Great Bay–Little Egg Harbor estuary (Witting et al., 1999). In a more recent analysis, based on a longer time series, larvae occurred from June through August and reached a peak in abundance during July (Fig. 73.3). Recent studies suggest larvae utilize selective tidal stream transport in estuarine habitats such as Delaware Bay, whereby larvae migrate into the upper water column during flood tides and descend to near bottom during ebb, thus effecting retention (and net up-estuary transport) (Rowe and Epifanio, 1994a, b). Thus, some larvae originating in the ocean could move into the estuary.

Habitat use is variable because of extensive movements of young-of-the-year. We have found young-of-the-year at Beach Haven Ridge on the inner shelf off southern New Jersey during all months and suggest that the coastal ocean may also provide an important habitat (Fig. 73.4). However, young-of-the-year also occur in lower estuarine waters and can penetrate as far upstream as freshwater. A similar pattern is evident in Chesapeake Bay during summer. It has been assumed that young-of-the-year use a net up-estuary flow in deeper layers of channels to move upstream (Thomas, 1971). Then they return downstream to the ocean in the fall (Massmann et al., 1958). Other authors have remarked that young-of-the-year are more common in channel habitats than in adjacent grass beds, at least in lower bay (higher-salinity) areas (Olney and Boehlert, 1988), thus substantiating the possible use of channels for upstream or downstream migration. In habitat-specific sampling in Great Bay–Little Egg Harbor, young-of-the-year in the estuary were collected mostly from stations where the substrate was silty and depths were 2 to 3 m. Structural components at these locations varied from none to shell, rubble, or peat deposits. In Chesapeake Bay during summer, juveniles have been collected in temperatures of 21.8 to 28.4°C and salinities of 32.9 to 36.8 ppt (Cowan and Birdsong, 1985) and have also been reported to occur in freshwater habitats (Massmann, 1954). In Delaware Bay, they are found as far upstream as the intrusion of saltwater (Thomas, 1971).

During extensive sampling across 12 habitat types in the Delaware Bay estuary, abundance was low in river habitats (Able et al., 2007a). Larvae and small juveniles (10 to 20 mm) occurred in the bay water column in April–May. At larger sizes (10 to 120 m), young-of-the-year were abundant in all other bay habitats. By October–November, the young-of-the-year had migrated out of the river and bay. While in the bay, most young-of-the-year used the intertidal and subtidal marsh, bay-shallow, and bay-deep habitat, but became more apparent in bay-deep habitats as they became larger (> 100 mm) (Able et al., 2007a). Age 1+ individuals (> 130 mm) were evident in bay-deep habitat in the spring and bay-deep throughout the summer months. During several years of intensive sampling of shallow sub- and intertidal marsh creeks (where salinities ranged from 22 to 32 ppt) in the Great Bay–Little Egg Harbor system, Rountree et al. (1992a) collected no young-of-the-year, while a confamilial, *Leiostomus xanthurus*, was abundant. Thomas (1971), however, reported *Cynoscion regalis* from all four low-salinity tidal creeks he sampled in Delaware Bay.

PREY AND PREDATORS

Several studies (n = 11) have examined the food habits of young-of-the-year (3.5 to 200 mm, n = 2989) in Middle Atlantic Bight estuaries from Acushnet River, Massachusetts, to Morehead City, North Carolina (see Table 8.1). Important prey include a variety of invertebrates (mysids, decapod shrimp, amphipods, copepods, polychaetes, and eggs) and fishes (*Anchoa mitchilli*, *Anchoa* spp., *Fundulus heteroclitus*, *Menidia menidia*, *Alosa aestivalis*, and *Micropogonias*

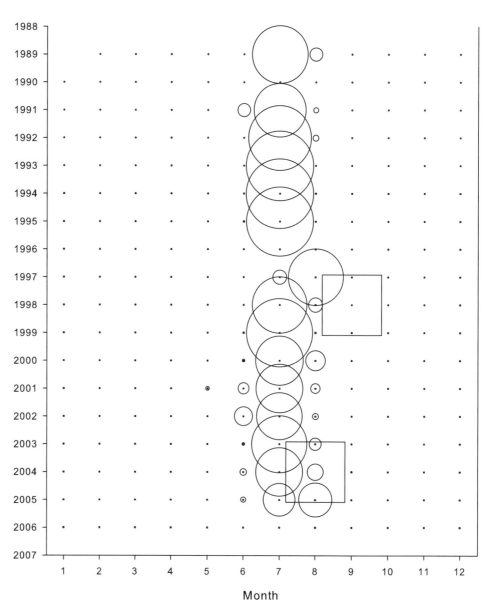

Fig. 73.3. Seasonal occurrence of larvae based on ichthyoplankton samples collected in Little Sheepshead Creek behind Little Egg Inlet from 1989 to 2006 (open circles represent the percentage of the mean number/1000 m³ of larval young-of-the-year captured by year) and juveniles of *Cynoscion regalis* based on trap collections in Rutgers University Marine Field Station boat basin in Great Bay (open squares represent the percentage of the mean catch per unit effort captured by year) from 1994 to 2006. Values range from 0 to 100% for ichthyoplankton and trap sampling. The smallest circles represent when samples were taken but no individuals were collected.

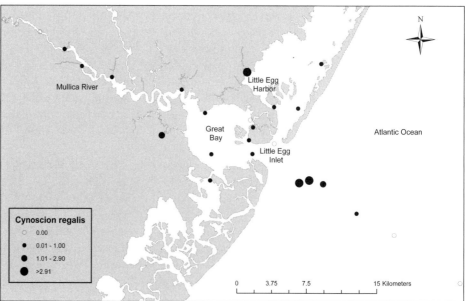

Fig. 73.4. Habitat use and average catch per tow (see key) for juvenile and adult *Cynoscion regalis* in the Ocean–Great Bay–Mullica River corridor in southern New Jersey based on small otter trawl collections during July and September from 1988 to 2006. See Table 3.2 for characteristics at each sampling collection.

Age 1+ *Cynoscion regalis*　　　　　　　**YOY *Cynoscion regalis***

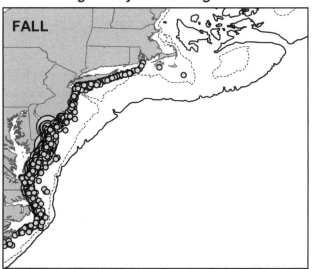

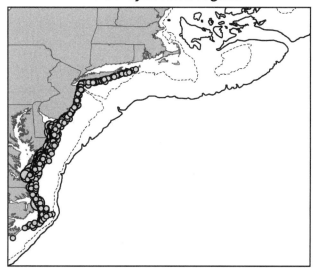

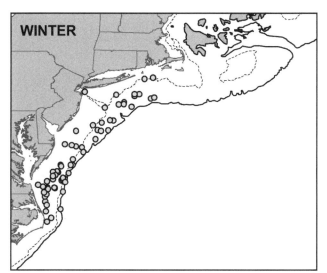

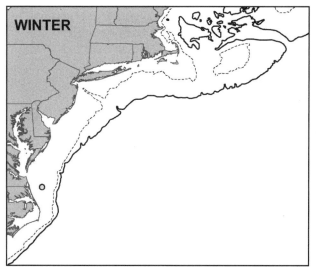

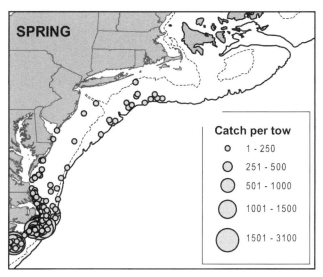

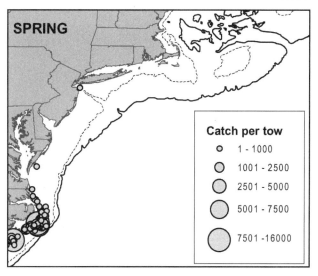

Fig. 73.5. Composite distribution of age 1+ (> 20 cm) and young-of-the-year (< 20 cm) *Cynoscion regalis* during seasonal cruises of the National Marine Fisheries Service groundfish survey. Details of sampling effort are indicated in Fig. 3.4.

undulatus). Also, young-of-the-year exhibit cannibalistic behavior (Thomas, 1971; Hartman and Brandt, 1995b; Nemerson, 2001). The diet of young-of-the-year shifts with ontogeny at approximately 60 mm, from a diet of small invertebrates (mysids and copepods) at smaller sizes to include more fishes and decapod shrimp at larger sizes (Thomas, 1971; Chao and Musick, 1977; Goshorn, 1990; Grecay, 1990; Nemerson, 2001). In the lower Delaware River, mysids and amphipods (68% frequency of occurrence [FO], 36% FO, respectively) are the most frequently occurring food items of young-of-the-year (< 180 mm, n = 494). Copepods (12% FO) are generally restricted to the smallest size group (0 to 60 mm), while fishes (51% FO) and *Crangon septemspinosa* (17% FO) are more important to larger young-of-the-year (121 to 180 mm) (Thomas, 1971). Geographical variation in diet has also been noted within the Delaware River–Delaware Bay estuary (Thomas, 1971; Grecay, 1990; Nemerson, 2001). In Delaware Bay marsh creeks, piscivory decreased with decreasing salinity, was highest at meso-polyhaline lower bay sites Dennis Township / Moores Beach (45.5% weight [W]), was intermediate at oligo-mesohaline Browns Run (20.4% W), and was almost absent at limnetic-oligohaline Mill Creek (1.9% W). Mysid consumption is similar across Dennis/Moores, Browns Run and oligo-mesohaline Mad Horse Creek (26.0% W, 31.1% W, 33.6% W, respectively), but negligible at Mill Creek (0.6% W). Crustacean consumption is significantly higher at Browns Run (66.2% W) than at any of the other sites (28.2 to 35.6% W) (Nemerson, 2001). Similarly, mysids dominate the diet in all areas of Delaware Bay, but are of lesser importance in the upper bay, where benthic invertebrates and detritus are included in the diet to a greater extent than in the lower and middle bay (Grecay, 1990). Relative importance of individual prey items also varies seasonally in the Delaware Bay, where fishes (35% FO) and decapod shrimp (15% FO) are more important in fall (September–October), especially *Cynoscion regalis* (23% FO) in September and decapod shrimp (*Palaeonetes* spp. and *Crangon* spp.) (19.7% FO) in October (Thomas, 1971). Diet also varies seasonally in Chesapeake Bay, such that *Anchoa mitchilli* (73.3 % W) is the main prey item, supplemented by mysids (13.0% W) and grass shrimp (11.1% W), in summer (July–August). In fall (September–October), young-of-the-year consume *Anchoa mitchilli* almost exclusively (97.9% W), and in winter (November–December) *Anchoa mitchilli* (58.2% W) and *Micropogonias undulatus* (37.4% W) are the main prey items (Hartman and Brandt, 1995a).

Adults are primarily piscivorous but also feed on a wide variety of invertebrates, including crabs, amphipods, mysids, shrimp, squid, mollusks, and annelid worms. Smaller adults (< 20 cm TL), feed mostly on crustaceans, but larger adults shift to a mostly fish diet. Chief among fish prey are engraulids, clupeids, atherinopsids, carangids, other sciaenids, *Peprilus*, *Ammodytes*, and various flatfishes. Diets vary with lo-

cality and generally include species most readily available, although small clupeids and engraulids are often the most important food items (Collette and Klein-MacPhee, 2002). In coastal areas, adults often feed in the surfzone, but they also do so in estuaries on bottom organisms or in the upper water column when selecting fish prey. In the Chesapeake Bay, they are top predators around the periphery of eelgrass beds, where they appear to forage in low-light crepuscular periods (Lascara, 1981).

Cynoscion regalis is preyed upon by two species (*Bairdiella chrysoura* and *Pomatomus saltatrix*) in temperate estuaries. In one study it was an important prey for other *Cynoscion regalis*, while it occurred in the diet of estuarine predators in four other studies (see Table 8.3; Mancini and Able, 2005).

MIGRATIONS

Young-of-the-year begin to leave estuarine and coastal ocean nursery areas and migrate offshore during October. In deeper parts of estuaries, however, young-of-the-year may remain until winter in our study area, although our data indicate these occurrences are irregular. At times they have been found to be among the most abundant components of trawl hauls in Delaware Bay as late as December (Abbe, 1967). After leaving the estuary, young-of-the-year undertake a longer winter migration than do older fish (Wilk, 1976; Mercer, 1983), eventually spending the colder months south of Cape Hatteras, while older fish may remain in the Cape Hatteras area for the winter.

In a further examination, the distribution of young-of-the-year (< 20 cm) changes with the season and reflects the extensive migrations of this species (Fig. 73.5). In the fall, the young-of-the-year are continuously abundant on the inner continental shelf (mean depth 16 m, range 6 to 40 m) from the eastern tip of Long Island to south of Cape Hatteras. During the winter, they are virtually absent from the Middle Atlantic Bight, with only a single collection (24 m) off northern North Carolina during the sampling period. In the spring, most individuals are collected along the inner shelf (mean depth 18 m, range 8 to 65 m) between North Carolina and Virginia.

The age 1+ (> 20 cm) juveniles and adults are found more frequently on the outer continental shelf, especially during the winter and spring (see Fig. 73.5). In the fall adults are concentrated between Rhode Island and North Carolina on the inner shelf (mean depth 17 m, range 5 to 49 m), with a few individuals collected on Nantucket Shoals. In the winter, they are distributed in deeper waters (mean depth 59 m, range 15 to 141 m) over the same general range. In the spring, they are found back in shallow depths (mean depth 29 m, range 8 to 159 m), especially off North Carolina and Virginia, but in the deeper portion of the depth range off southern New Jersey and north.

74

Leiostomus xanthurus Lacepède

SPOT

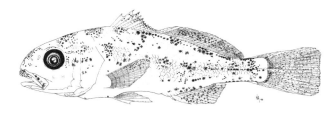

Estuary	Reproduction	Larvae/ Development	YOY Growth	YOY Overwinter
Ocean				
Both				
Jan				
Feb				
Mar				
Apr				
May				
Jun				
Jul				
Aug				
Sep				
Oct				
Nov				
Dec				

DISTRIBUTION

Leiostomus xanthurus is found along the east coast of North America from Massachusetts Bay to Mexico (Klein-MacPhee, 2002b). It is most abundant between North Carolina and Chesapeake Bay (Hildebrand and Schroeder, 1928; Thomas, 1971). Larvae and juveniles occur in Middle Atlantic Bight estuaries between the Hudson River and Cape Hatteras, whereas in most estuaries north and east of the Hudson River only juveniles have been found (see Table 4.3). We have

been unable to document the occurrences of eggs or larvae in Narragansett Bay as reported by Stone et al. (1994) because synoptic surveys of that bay have failed to collect these stages (Herman, 1963; Bourne and Govoni, 1988).

REPRODUCTION AND DEVELOPMENT

Spawning (by 2- or 3-year-old adults) occurs from winter to early spring in continental shelf waters, often in relatively warm water near the Gulf Stream Front off North Carolina (Norcross and Bodolus, 1991; Govoni, 1993, Govoni and Pietrafeso, 1994). In North Carolina, spawning is protracted, lasting from mid-October to mid-March, but 90% of larvae are produced in the two-month period of mid-November to mid-January (Flores-Coto and Warlen, 1993). During most years, there is no spawning within the Middle Atlantic Bight.

The early development of *Leiostomus xanthurus*, and characters useful for distinguishing the eggs, larvae, and juveniles from those of other sciaenids, have been described and summarized (Fahay, 2007). Eggs are pelagic and spherical and measure 0.72 to 0.87 mm in diameter. They have multiple oil globules that coalesce into a single one. Larvae hatch at sizes of 1.2 to 2.0 mm. They are relatively deep-bodied anteriorly and taper to a narrow caudal peduncle. Small spines form on the pre-opercle. Larvae are very similar to those of *Micropogonias undulatus*, but they are lightly pigmented and distinguishable by these patterns. Fin rays are complete at about 12 mm (except for pectoral fins). Juveniles at about 23 mm in length resemble adults, but have a characteristic pigment pattern unlike that in either larvae or adults.

Forty-seven juvenile *Leiostomus xanthurus* between 16.7 and 69.4 mm TL were examined for patterns of scale formation (see Fig. 71.1; Able et al., 2009a). Scales begin to form in juveniles after they enter estuaries but while remnants of larval and juvenile pigmentation remain. Scales were not observed on specimens from 16.7 to 19 mm TL. Scales originate at several loci, including the lateral line on the caudal peduncle, under the eye, on the dorsal surface of the head, and on the operculum. Scales originating on the lateral line first occur at 19.6 mm TL, with a few on the caudal peduncle region. By approximately 20 mm TL, scales spread anteriorly along the lateral line and expand dorsally and ventrally. By 25 mm TL, scales continue to expand along the lateral line and also cover the caudal peduncle. Also at this size, scales first appear under the eye. By approximately 28 mm TL, scales are observed on the dorsal surface of the head and on the operculum. At this size, scales also cover the body posterior to the operculum, but they are absent from the dorsal and ventral surfaces. By 33 mm TL, scales fully cover the body posterior to the operculum. Scales also appear elsewhere on the operculum, on much of the dorsal and lateral surfaces of the head, and on the caudal fin. By 51 mm TL, scales fully cover the body and head region as well as a larger proportion of the caudal fin, including that portion of the lateral line that extends onto that fin. By 69 mm TL, scales extend onto the base of the second dorsal fin. Larger specimens were not available to confirm the size at which scales reach the adult condition. At 69 mm TL, scales cover 96.5% of the body and 72.3% of the body and fins (see Figs. 71.1 and 75.1 for comparisons with other sciaenids). Previous studies have suggested scale formation is complete by 30 mm (Hildebrand and Cable, 1930), but examination of our series suggests this occurs at larger sizes.

LARVAL SUPPLY, SETTLEMENT, GROWTH, AND MORTALITY

Populations in the central part of the Middle Atlantic Bight demonstrate wide annual fluctuations in abundance. In the Delaware Bay, they ranked 12th in abundance in 1967, did not rank at all in 1968, and ranked 3rd in 1969 (Daiber and Smith, 1970). In a two-year (1988–1989) Great Bay study, nearly all *Leiostomus xanthurus* were taken during one year (1988) when they were abundant (Rountree et al., 1992a). Likewise, nearly all larvae collected in the Indian River Inlet, Delaware, during a three-year study were collected during a single year (Pacheco and Grant, 1973). Larvae entering our study site through Little Egg Inlet were collected during three of six years studied (Witting et al., 1999). Within Great Bay, variations in annual abundance of larvae occurred in 1994 and 1996 but seldom since then (Fig. 74.1).

Distributional data support the hypothesis that larvae spawned south of Cape Hatteras contribute to recruitment in Chesapeake Bay (Flores-Coto and Warlen, 1993), but larvae also occur in the southern part of the Middle Atlantic Bight (between the Chesapeake Bay mouth and Cape Hatteras). During ichthyoplankton surveys over the entire Middle Atlantic Bight between 1977 and 1987, larvae were infrequently collected; when larvae were present, they were restricted to the southern part of the bight and occurred primarily during January (Able and Fahay, 1998). These larvae might be the source that supplies the Great Bay, New Jersey, area in some years.

Growth of young-of-the-year has been well studied. Larvae collected off North Carolina increased from 1.2 mm at hatching to 16.1 mm in 80 days (Flores-Coto and Warlen, 1993). The smallest young-of-the-year collected in New Jersey estuaries are between 1 and 2 cm TL as they enter from the ocean in winter and spring (Fig. 74.2). In Delaware Bay, they range from about 9 to 20 mm (de Sylva et al., 1962; Thomas, 1971; Pacheco and Grant, 1973; Hettler and Barker, 1993). After growing through the first summer in the Mullica River–Great Bay estuary, young-of-the-year reach a mode of about 5 to 20 cm TL by August. Similar-size fish occur in September, but larger individuals are collected on the continental shelf in surveys there in September and October. They achieve comparable sizes in the upper Delaware Bay and lower Delaware River area (Able and Fahay, 1998).

In the James River (Chesapeake Bay) estuary, young-of-the-year grew 11.3 mm per month during spring and sum-

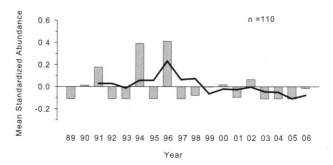

Fig. 74.1. Annual variation in abundance of *Leiostomus xanthurus* in Little Sheepshead Creek Bridge 1 m plankton net collections, 1989–2006. Data were standardized by subtracting the overall mean from annual abundance. Vertical bars represent standardized annual abundance; the solid line is a three-year running mean that was calculated by taking the mean of the standardized values from the previous two years and the year in which the data are plotted.

mer, but growth then leveled off during fall (McCambridge and Alden, 1984). Winter growth in North Carolina estuaries has been estimated at 0.14 to 0.16 mm per day (Weinstein and Walters, 1981). Previous conclusions concerning sizes at various ages are equivocal. Estimates of the sizes attained at age 1 range from 8 to 10 cm in New Jersey (Welsh and Breder, 1923) to an average of 15 to 17 cm in Chesapeake Bay (McCambridge and Alden, 1984; Pacheco, 1962; respectively) or up to approximately 20 cm based on back calculations from otoliths (Piner and Jones, 2004). Our own data from Great Bay–Little Egg Harbor indicate young-of-the-year reach between 19 and 25 cm TL as they begin their first winter (October), whereas in North Carolina, they range from 5 to 11 cm SL (6 to 14 cm TL) in October (Weinstein and Walters, 1981). There appears to be little growth during the winter, as the fish collected in March were similar in size to those collected in the previous fall (see Fig. 74.2). Most fish collected in several commercial fishing gears were dominated by age 1 individuals and ranged up to 4 years (Piner and Jones, 2004).

Habitat quality may affect growth rates. Weight-length relationships were compared between young-of-the-year from several locations between Chesapeake Bay and Texas; fish from the former location were found to be heavier at length than fish from other nurseries, suggesting that habitat quality was greater in the Chesapeake Bay system (McCambridge and Alden, 1984). Density-dependent growth and mortality occurred in young-of-the-year in caging experiments because of presumed competition for food (Craig et al., 2007).

SEASONALITY AND HABITAT USE

Leiostomus xanthurus is a euryhaline species that commonly occurs in freshwater or low-salinity water (Massmann, 1954), or in coastal ocean waters, usually over sandy or muddy substrates in depths to 132 m (Dawson, 1958; Musick, 1974). Larval development occurs in continental shelf waters during winter. The larvae enter Middle Atlantic Bight estuaries during the late winter and early spring. In Great Bay–Little Egg

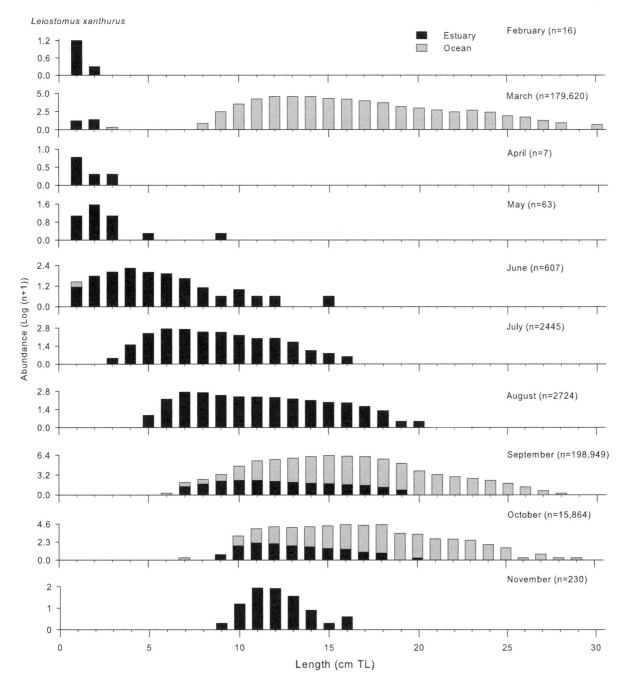

Fig. 74.2. Monthly length frequencies of *Leiostomus xanthurus* collected in estuarine and coastal ocean waters of the central Middle Atlantic Bight. Sources: National Marine Fisheries Service: otter trawl (n =392,492); Rutgers University Marine Field Station: 1 m beam trawl (n = 17); seine (n = 79); 1 m plankton net (n = 51); 4.9 m otter trawl (n = 110); multi-panel gill net (n = 273); and Able and Fahay 1998 (n = 7503).

Harbor larvae sporadically occurred from February through June (Fig. 74.3). These compare with a mean size of 17.2 mm and a mean age of 82 days in a North Carolina study (Flores-Coto and Warlen, 1993). In the Indian River Inlet–Delaware Bay–Little Egg Inlet regions, ingress reaches a peak in March (Pacheco and Grant, 1973; Witting et al., 1999), April, or May (de Sylva et al., 1962; Thomas, 1971). This ingress is followed

by movement into the upper estuary. Although larvae were not collected in the two years of collecting at Beach Haven Ridge and were only rarely found in central Middle Atlantic Bight continental shelf waters during the 11-year MARMAP ichthyoplankton study, young-of-the-year sometimes occur abundantly both in upper Delaware Bay habitats (Thomas, 1971) and in Great Bay (Rountree et al., 1992a; Witting et al.,

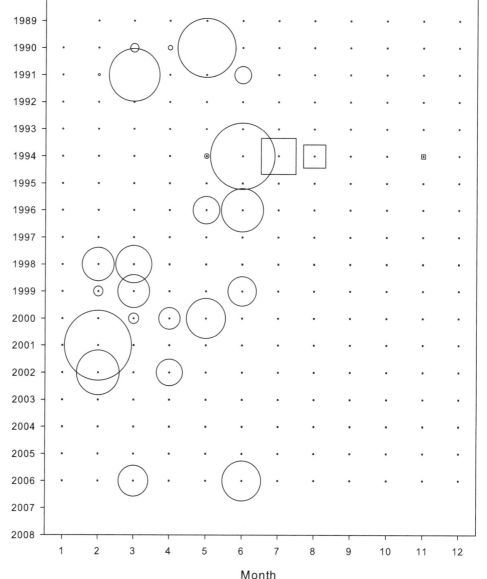

Fig. 74.3. Seasonal occurrence of larvae based on ichthyoplankton samples collected in Little Sheepshead Creek behind Little Egg Inlet from 1989 to 2006 (open circles represent the percentage of the mean number/1000 m³ of larval young-of-the-year captured by year) and juveniles of *Leiostomus xanthurus* based on trap collections in Rutgers University Marine Field Station boat basin in Great Bay (open squares represent the percentage of the mean catch per unit effort captured by year) from 1994 to 2006. Values range from 0 to 100% and 0 to 60% for ichthyoplankton and trap sampling, respectively. The smallest circles and squares represent when samples were taken but no individuals were collected.

1999). In Delaware Bay, it has been observed that young-of-the-year also commonly occur in the Chesapeake and Delaware (C&D) Canal. The possibility exists, therefore, that upper Delaware Bay young-of-the-year arrive there from the upper Chesapeake Bay via that canal.

Juveniles occupy a variety of estuarine habitats during their first year. During extensive sampling in Delaware Bay (Able et al., 2007a) the small young-of-the-year (< 230 mm) were abundant in bay-shallow and intertidal and subtidal creeks in the spring. Young-of-the-year also occurred in bay-deep habitats. Later in the summer, at larger sizes (approximately 140 to 180 mm), they were most abundant in intertidal and subtidal marsh habitats. In Great Bay–Little Egg Harbor, they are somewhat more common in subtidal creeks than in intertidal habitats (Rountree and Able, 1992a). In a trawl survey covering several different habitats other than tidal creeks, they were most abundant at a relatively

deep (5 m) station characterized by a shell, sponge, and peat substrate (Able and Fahay, 1998). This pattern has been repeated with additional data that indicated they occurred from the inner continental shelf off Little Egg Inlet to the upper limits of salt water intrusion in the Mullica River (Fig. 74.4). Young-of-the-year also were found in marsh creek and channel stations along a salinity gradient in the Mullica River–Great Bay estuary in New Jersey (Martino and Able, 2003). In a Virginia study, they were present in tidal creeks but moved between habitats, with individuals remaining within single creeks for an average of 81 days (Weinstein et al., 1984). They are also especially abundant over beds of thick, loose mud (Reid, 1955; Cowan and Birdsong, 1985.) Juveniles occur in a wide range of salinities and can survive prolonged periods in freshwater (Massmann, 1954). In 23 collections in Delaware Bay between August and October, young-of-the-year were collected in salinities ranging from 0 to 10 ppt, temperatures

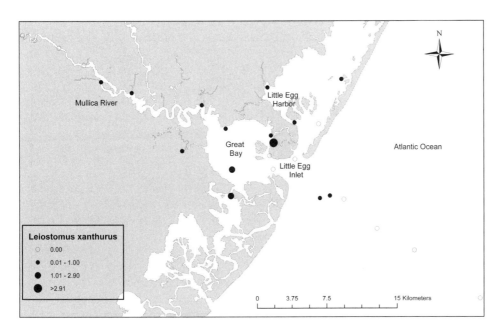

Fig. 74.4. Habitat use and average catch per tow (see key) for juvenile and adult *Leiostomus xanthurus* in the Ocean–Great Bay–Mullica River corridor in southern New Jersey based on small otter trawl collections during July and September from 1988 to 2006. See Table 3.2 for characteristics at each sampling collection.

from 14.0 to 27.2°C, and dissolved oxygen from 3.8 to 10.8 ppm (Thomas, 1971).

Several analyses of this species' response to hypoxia have found that it is capable of detecting and avoiding low (1 mg/l⁻¹) levels (Wannamaker and Rice, 2000) and experiences 100% mortality at 0.6 ppm O_2. This response varies with fish size and temperature (Shimps et al., 2005). Other studies illustrate that feeding is enhanced during hypoxia (Pihl et al., 1992). Thus the response of this species to hypoxia is varied (both negative and positive). Juveniles have also been collected at sites with salinities as high as 35.2 ppt (Cowan and Birdsong, 1985). Olney and Boehlert (1988) observed that larvae and juveniles became the most abundant taxa in their lower Chesapeake Bay grass-bed study site in the spring as the submerged vegetation was beginning to grow. Taken together, it is clear that the young-of-the-year can utilize a wide variety of habitats.

PREY AND PREDATORS

Several studies (n = 10) have examined the food habits of young-of-the-year (9 to 159 mm, n = 6151) *Leiostomus xanthurus* in estuaries from Delaware Bay to Cape Fear River, North Carolina (see Table 8.1). Important prey of young-of-the-year include a variety of invertebrates (copepods, polychaetes, oligochaetes, amphipods, nematodes, bivalves, mysids, and tanaids) and detritus. Sediment is also found frequently in stomachs. Fishes (*Atherinopsidae, Fundulus heteroclitus, Gobiidae,* and *Micropogonias undulatus*) are a minor dietary component. The diet of young-of-the-year shifts with ontogeny at approximately 40 mm, from small invertebrates (copepods and mysids) to more benthic prey (polychaetes, oligochaetes, cumaceans, and amphipods) (Hildebrand and Cable, 1930; Kimmel, 1973; Hodson et al., 1981; Smith et al., 1984; O'Neil and Weinstein, 1987; Nemerson, 2001). Dentary teeth

are lost as the juvenile diet shifts from planktivory to feeding on benthic organisms (Govoni, 1987). In Delaware Bay marsh creeks, mysid and meiobenthos consumption is highest by small individuals (20 to 40 mm); mysid consumption decreases to very low levels after 60 mm and meiobenthos consumption (mostly harpacticoid copepods) drops steadily over the entire 20 to 140 mm range, while annelid consumption increases from 20 to 80 mm before leveling off thereafter (Nemerson, 2001). Similarly, in the York River estuary in Virginia, small (16 to 20 mm) young-of-the-year eat primarily copepods, whereas slightly larger individuals (26 to 30 mm) consume more benthic prey (polychaetes and nematodes). Larger young-of-the-year (40 to 100 mm) feed on a wide variety of predominantly benthic organisms (polychaetes, amphipods, nematodes, and oligochaetes), while the largest young-of-the-year examined (> 100 mm) consume primarily cumaceans and amphipods (O'Neil and Weinstein, 1987).

Geographical variation in diet has also been noted within Delaware Bay (Nemerson, 2001) such that annelids strongly dominate diet at meso-polyhaline Dennis Township (77.9% W), while only 23.9% W and 31.5% W at oligo-mesohaline Browns Run and Mad Horse Creek, respectively. Meibenthos are common prey, but significantly higher at lower-salinity habitats (Browns Run) (29.2% by weight) than in higher-salinity, lower bay (Dennis and Mad Horse creeks) (12.4% by weight at each), while crustacean consumption is significantly higher at Browns Run (19.6% W) and Mad Horse (19.1% W) than at Dennis (3.3% W). Spatial diet variation was also noted at York River, Virginia, where dominant items consumed by young-of-the-year at meso-oligohaline Goalder's Creek included nereid polychaetes, clam siphons, gammarid amphipods, and harpacticoid copepods, while young-of-the-year from polyhaline Blevins Creek ate more nematodes, maldanid polychaetes and oligochaetes (O'Neil

Age 1+ *Leiostomus xanthurus*

YOY *Leiostomus xanthurus*

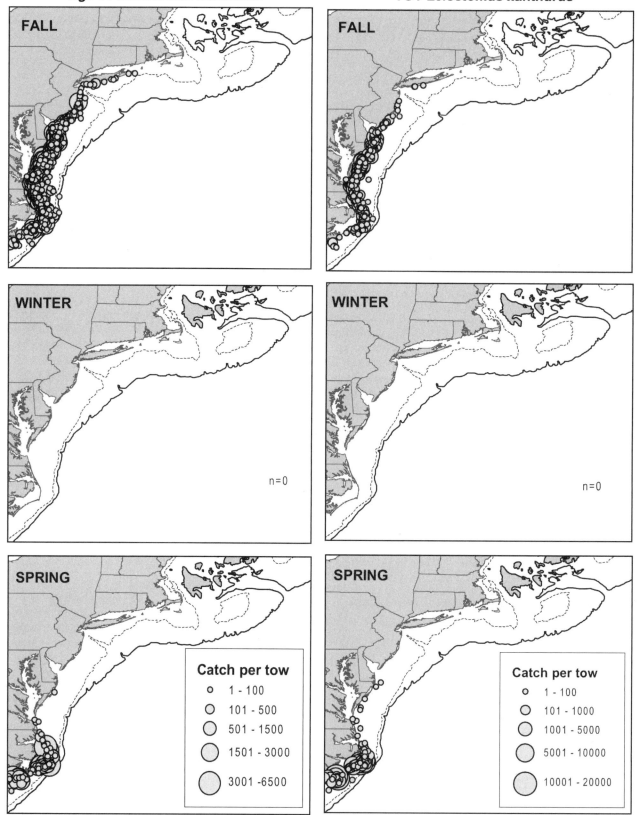

Fig. 74.5. Composite distribution of age 1+ (> 15 cm) and young-of-the-year (< 15 cm) *Leiostomus xanthurus* during seasonal cruises of the National Marine Fisheries Service groundfish survey. Details of sampling effort are indicated in Fig. 3.4.

and Weinstein, 1987). Relative importance of individual prey items also varies intra-annually (Hines et al., 1990; Nemerson, 2001). In the Rhode River sub-estuary of Chesapeake Bay, young-of-the-year (68 to 124 mm, n = 187) consume mainly amphipods and polychaetes early in the season, primarily clam (*Macoma balthica*), siphons, small crustaceans, and meiofauna in June, and a variety of polychaetes throughout the year (Hines et al., 1990).

In several studies in temperate estuaries this species is the important prey in one study and present in four others (see Table 8.3; Mancini and Able, 2005). It is also preyed upon by several species of sharks and bony fishes, including *Lophius*, *Pomatomus saltatrix*, *Cynoscion regalis*, and *Paralichthys dentatus* (Lascara, 1981; Bowman et al., 2000).

MIGRATIONS

As indicated in the length-frequency histogram (see Fig. 74.2) and elsewhere (Able and Fahay, 1998), young-of-the-year begin to vacate estuaries in the central part of the Middle Atlantic Bight in October or November, concurrent with fall cooling, and are common in ocean sampling in September and October. Larger individuals (> 15 cm) apparently leave estuarine habitats earliest, and all have migrated to oceanic habitats by December. There is no evidence that young-of-the-year overwinter in estuaries in the central part of the Middle Atlantic Bight, as they sometimes do in Chesapeake Bay. Tagging studies indicate a fall and winter migration (along with older year classes) to wintering grounds south of Cape Hatteras (Pearson, 1932; Pacheco, 1962).

After the young-of-the-year leave the estuary, their distribution varies as the result of extensive seasonal migrations (Fig. 74.5). In the fall, they are distributed on the inner continental shelf (mean depth 17 m, range 5 to 41 m), primarily between New Jersey and North Carolina. They are absent from the sampling area from Cape Hatteras north in the winter, but are found on the inner continental shelf (mean depth 16.8 m, range 8 to 45 m) in the spring from Maryland to off North Carolina, where they are the most abundant.

These seasonal patterns of the young-of-the-year are essentially identical to those of the age 1+ juveniles and adults except that the latter can be found in slightly deeper waters (see Fig. 74.5). In the fall, the adults can be distributed on the inner and mid-shelf (mean depth 18 m, 5 to 240 m) from Long Island south to North Carolina, but can also be found at the deeper portions of the shelf off North Carolina. The adults are absent from the portion of the Middle Atlantic Bight sampled in the winter. In the spring they are concentrated primarily in shallow waters of the inner shelf (mean depth 19 m, range 9 to 65 m) off North Carolina, but a few individuals have been collected off Virginia.

75

Menticirrhus saxatilis (Bloch and Schneider)

NORTHERN KINGFISH

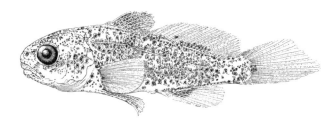

Estuary	Reproduction	Larvae/ Development	YOY Growth	YOY Overwinter
Ocean				
Both				
Jan				■
Feb				■
Mar				■
Apr	■			
May	■			
Jun	■			
Jul	■		▨	
Aug	■		▨	
Sep	■		▨	
Oct			▨	
Nov				■
Dec				■

DISTRIBUTION

Menticirrhus saxatilis occurs along the coast of North America from the Gulf of Maine to the Gulf of Mexico, but it is most abundant in the Middle Atlantic Bight, where it is common from late spring through fall (Klein-MacPhee 2002b). Adults and early life history stages are distributed in most es-

tuaries in our study area, but the eggs are only reported from estuaries from New Jersey and north (see Table 4.3).

REPRODUCTION AND DEVELOPMENT

Most spawners are age 1 individuals (Schaefer, 1965). The presence of larvae in May through September indicates that spawning occurs in the Middle Atlantic Bight from spring through late summer (Able and Fahay, 1998). Spawning takes place slightly earlier (April–May) off North Carolina (Hildebrand and Cable, 1934).

The early development of *Menticirrhus saxatilis*, and characters useful for distinguishing the larvae and juveniles from those of congeners, have been well described and summarized (Fahay, 2007). Eggs are pelagic and spherical and measure 0.80 to 0.85 mm in diameter. As many as 18 small oil globules coalesce to a single one, 0.19 to 0.26 mm in diameter. Larvae are deep-bodied anteriorly, tapering to a moderately deep caudal peduncle. Head ornamentation is limited to a series of very small spines along the edge of the pre-opercle. Pigment is dense over the head and body. The incubation period for eggs is 46 to 50 hours at a temperature of 20 to 21°C (Welsh and Breder, 1923).

The sequence of scale formation was examined in 37 specimens between 7.4 and 76.6 mm TL (Fig. 75.1; Able et al., 2009a). No scales were observed on specimens between 7.4 and 13.4 mm TL. Scales originate at several loci, including the lateral line at the caudal peduncle, on the operculum and on the head behind the eye, on the dorsal surface of the head, and on the ventral surface of the body anterior to the pelvic fin. Onset of scale formation was observed along the lateral line by approximately 15.0 mm TL. Scales spread anteriorly and laterally along the lateral line to the operculum by 18 mm TL. By 25 mm TL, scales cover much of the lateral surface of the body posterior to the operculum, and also appear on the operculum, anterior to the pelvic fin, and on the venter of the body. In some specimens at similar lengths, scales cover all of the body posterior to the operculum, and are forming on the dorsal surface of the head and behind the eyes. By 30 mm TL, scales extend onto the caudal fin and have achieved the adult condition, where scales cover 82.8% of the body and 50.3% of the body and fins. Scales are absent on the snout and on portions of the operculum. See Figures 71.1 and 75.1 for comparisons with other sciaenids.

LARVAL SUPPLY, SETTLEMENT, GROWTH, AND MORTALITY

We have collected larvae of this species during two of our sampling programs: bridge net sampling with a plankton net on flood tides inside Little Egg Inlet from 1989 to 2007; and surfzone ichthyoplankton sampling along the ocean beachfront from 1995 to 1999 (Able et al., 2009c, 2010). Larvae collected in both efforts apparently are products of spawning over the continental shelf of the Middle Atlantic Bight, as described in Able and Fahay (1998). Between 1977 and 1987, this spawn-

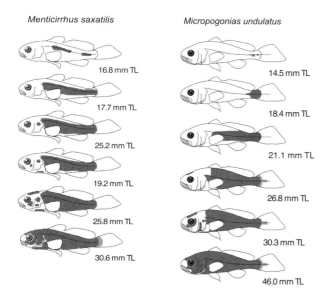

Menticirrhus saxatilis

16.8 mm TL
17.7 mm TL
25.2 mm TL
19.2 mm TL
25.8 mm TL
30.6 mm TL

Micropogonias undulatus

14.5 mm TL
18.4 mm TL
21.1 mm TL
26.8 mm TL
30.3 mm TL
46.0 mm TL

Pogonias cromis

18.7 mm TL
20.3 mm TL
24.3 mm TL
26.5 mm TL
30.4 mm TL
46.5 mm TL

Fig. 75.1. Description of scale formation relative to total length (TL) in *Menticirrhus saxatilis*, *Micropogonias undulatus*, and *Pogonias cromis*.

ing was most intense in that portion of the Middle Atlantic Bight located between the Delaware Bay mouth and Cape Hatteras, with only scattered and localized reproduction in the ocean off the New Jersey coast. There are no available data to suggest that spawning in more recent years has increased in more northern portions of the bight. However, between 1995 and 1999, the density of larvae identified as *Menticirrhus* sp. collected in the surfzone during 1999 increased more than 10-fold over the 4 earlier years of sampling (Able et al., 2010).

The monthly length-frequency distributions from New Jersey waters suggest that at least two year classes may be represented (Fig. 75.2), although these interpretations are somewhat equivocal. Collections in June suggest that young-of-the-year are represented by those smaller than 5 cm while age 1+ are represented by those between 17 and 30 cm. This interpretation assumes that young-of-the-year, from both the estuary and the ocean, are quite variable in size and may attain up to 24 cm by September. This interpretation of relatively fast growth is supported by similar estimates from Long Island, where extensive collections found young-of-the-year ranging from 2.5 to 31.5 cm TL, with a modal length of 25 cm in October (Schaefer 1965). These estimates were confirmed from back calculations of annuli from scales. If this interpretation is correct, then the large group of fish collected in the ocean in March may represent larger survivors of the previous year class, ranging from 11 to 30 cm TL. It might also imply that smaller individuals either do not survive the winter or migrate to other areas in the Middle Atlantic Bight.

SEASONALITY AND HABITAT USE

In samples taken from 1989 to 1993 in the Great Bay–Little Egg Harbor estuary, larvae (n = 50) were collected primarily in late June and July but occurred as late as September (Able and Fahay, 1998). Over an 18-year period of plankton collections in the same study area, larvae have usually reached peak abundances in July in years when they have occurred

at all (Fig. 75.3). In lower Delaware Bay, young-of-the-year occur from July through October (de Sylva et al., 1962).

Adults occur in shallow coastal or estuarine waters over sandy to sand-mud substrates. They are common in the surfzone as well as in sandy channels near inlets. The collection of eggs in every portion of Narragansett Bay indicates that spawning occurs throughout that embayment (Bourne and Govoni, 1988). The larvae are most consistently found south of Delaware Bay from May through September (Able and Fahay, 1998).

The habitats of young-of-the-year range from the estuary to the ocean. All sizes have been collected from sandy beaches in the vicinity of Little Egg Inlet (Haroski, 1998). Our recent observations of young-of-the-year in Corson's Inlet, New Jersey, duplicate those of Phillips (1914) from the same location some 90 years earlier, which suggests that their occurrence is regular. The preference for ocean beach and inlets in southern New Jersey is supported by the lack of young-of-the-year in extensive collections in salt marsh creeks (Rountree and Able, 1992a) and shallow muddy shorelines (Able et al., 1996b) throughout the estuary behind Little Egg Inlet. In polyhaline waters of lower Delaware Bay, young-of-the-year typically occur at salinities greater than 8.0 ppt, although some have been collected in as low as 5.2 ppt (Thomas, 1971). Abundance in the estuary declines through the fall, and by November relatively few individuals are present (Able and Fahay, 1998).

PREY AND PREDATORS

Menticirrhus saxatilis is a bottom feeder that locates prey by olfaction and tactile sense with its inferiorly located mouth and barbel (Chao and Musick, 1977). This species has been reported to feed on decapod crustaceans, fishes, and polychaetes in the ocean (Bowman et al., 2000). The most frequently eaten decapods include *Callianassa setimanus*, *Crangon septemspinosa*, *Ovalipes ocellatus*, and *Pinnixa chaetopterana*.

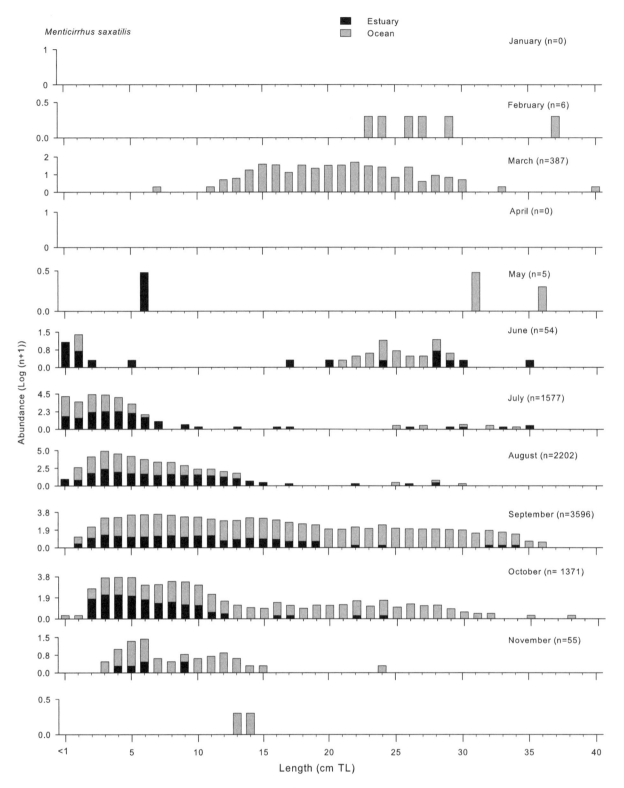

Fig. 75.2. Monthly length frequencies of *Menticirrhus saxatilis* collected in estuarine and coastal ocean waters of the central Middle Atlantic Bight. Sources: National Marine Fisheries Service: otter trawl (n = 2631); Rutgers University Marine Field Station: 1 m beam trawl (n = 2); seine (n = 4550); 1 m plankton net (n = 72); 2 m beam trawl (n = 4); killitrap (n = 2); 4.9 m otter trawl (n = 64); multi-panel gill net (n = 12); 1 m bongo net (n = 314); and Able and Fahay 1998 (n = 1614).

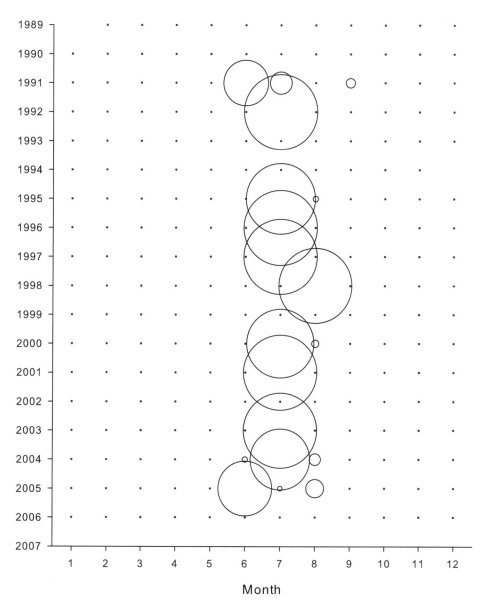

Fig. 75.3. Seasonal occurrence of the larvae of *Menticirrhus saxatilis* based on ichthyoplankton samples collected in Little Sheepshead Creek behind Little Egg Inlet from 1989 to 2006. Full circles represent the percentage of the mean number/1000 m³ of larval young-of-the year *Menticirrhus saxatilis* captured by given year. Values range from 0 to 100%. The smallest circles represent when samples were taken but no larvae were collected.

In the York River, Virginia, it preys mostly on crustaceans (*Neomysis, Crangon, Palaeomonetes*, amphipods, and copepods), polychaetes, and organic detritus (Chao and Musick, 1977). In a Cape May, New Jersey study, its prey was dominated by amphipods, polychaetes, detritus, decapod shrimp, and isopopds (Welsh and Breder, 1924). Two studies have examined the food habits of young-of-the-year (10 to 140 mm, n = 139) *Menticirrhus saxatilis* in the Delaware River estuary and York River, Virginia. Important prey of young-of-the-year include a variety of invertebrates (polychaetes, mysid shrimp, amphipods, and copepods), fishes (family Clupeidae), and detritus (de Sylva et al., 1962; Chao and Musick, 1977).

MIGRATIONS

This species has pronounced seasonal migrations in and out of the Middle Atlantic Bight. In the fall, adults and young-of-the-year are distributed on the inner continental shelf and in estuaries throughout our study area (Fig. 75.4). In the fall,

young-of-the-year (< 23 cm) are found on the inner continental shelf (mean depth 16 m, range 8 to 42 m) from Long Island, Rhode Island, and Massachusetts, south to North Carolina. While no individuals are collected in the sampling area during the winter, by spring they are evident in North Carolina inner shelf waters (mean depth 19 m, range 10 to 43 m), mostly south of Cape Hatteras.

Age 1+ (> 23 cm) juveniles and adults have a similar seasonal distribution except that a few individuals are collected at the edge of the continental shelf in the vicinity of Hudson Canyon (mean depth 115 m, range 76 to 128) during winter and spring (see Fig. 75.4). In the fall, they are found on the inner continental shelf (mean depth 16 m, range 7 to 43 m) from Massachusetts to North Carolina, with most individuals present from Long Island and south. In the spring, most older stages are off North Carolina on the inner shelf (mean depth 24 m, range 9 to 76 m), where they congregate with younger stages.

Age 1+ Menticirrhus saxatilis

YOY Menticirrhus saxatilis

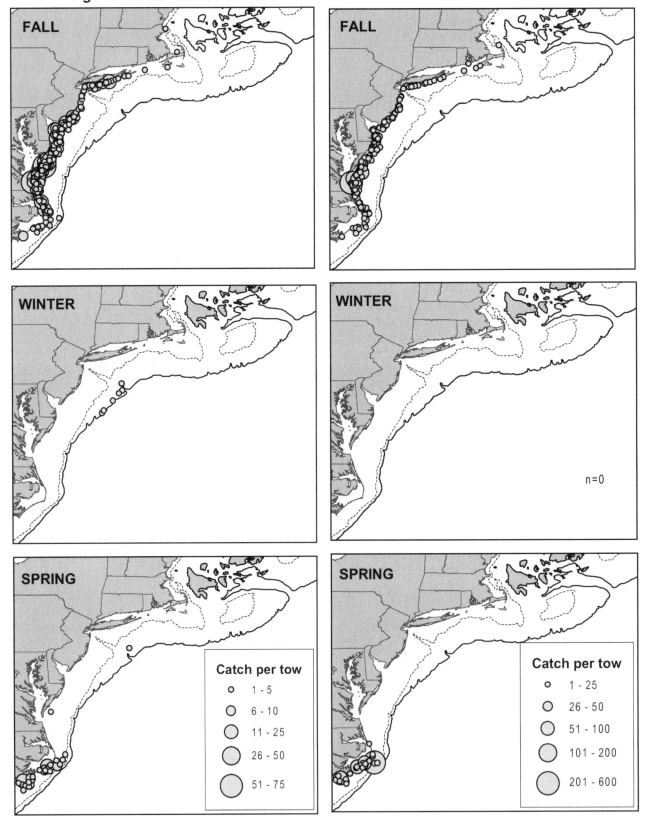

Fig. 75.4. Composite distribution of age 1+ (> 23 cm) and young-of-the-year (< 23 cm) *Menticirrhus saxatilis* during seasonal cruises of the National Marine Fisheries Service groundfish survey. Details of sampling effort are indicated in Fig. 3.4.

76

Micropogonias undulatus (Linnaeus)

ATLANTIC CROAKER

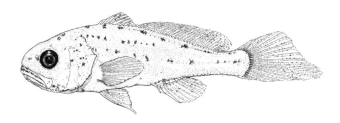

Estuary	Reproduction	Larvae/ Development	YOY Growth	YOY Overwinter
Ocean				
Both				
Jan				
Feb				
Mar				
Apr				
May				
Jun				
Jul				
Aug				
Sep				
Oct				
Nov				
Dec				

DISTRIBUTION

Micropogonias undulatus is found in the western North Atlantic Ocean between Cape Cod and the Gulf of Mexico (Welsh and Breder, 1923; Robins and Ray, 1986). Discrete populations are found along the Atlantic coast of the United States and in the Gulf of Mexico, but there is no evidence suggesting that Atlantic coast populations north and south of Cape Hatteras are discrete (Lankford et al., 1999). It was the most common species collected by A. E. Parr and associates from 1929 to 1933 in the ocean and bays of southern New Jersey and was once a frequent constituent of seine collections in Delaware Bay (Shuster, 1959); it failed to rank among the top 20 in a similar effort in 1972 (Thomas and Milstein, 1974). Subsequently, the population has rebounded in the Middle Atlantic Bight in recent years in response to climate change (Hare and Able, 2007). Larvae and juveniles occur in Middle Atlantic Bight estuaries from the Hudson River to Cape Hatteras (see Table 4.3), with only rare occurrences north and east of the Hudson River.

REPRODUCTION AND DEVELOPMENT

Spawning begins in early September, peaks in October, and ends in late December in the Middle Atlantic Bight (Morse, 1980) and lasts from July through October in the Chesapeake Bay region (Watkins, 2001). Spawning activity in the Middle Atlantic Bight is almost totally restricted to the continental shelf between Delaware Bay and Cape Hatteras (Morse et al., 1987; Barbieri et al., 1994; Able and Fahay, 1998) and apparently does not occur in Chesapeake Bay, at least to any significant degree (Watkins, 2001). The expansion of the distribution in response to milder winters may have allowed reproduction to occur farther north based on the occurrence of recently hatched larvae (2 to 3 mm TL) in Delaware Bay (Miller et al., 2003).

The eggs are undescribed, but the early development of *Micropogonias undulatus*, and characters useful for distinguishing the larvae and juveniles from those of other sciaenids have been well described and summarized (Fahay, 2007). Newly hatched larvae are about 2.5 mm long, and are deep-bodied anteriorly, tapering to the tip of the urostyle. The caudal peduncle soon deepens, but remains shallower than the anterior body. Pigment is light, but patterns of melanophores on the venter anterior to the vent are important for identification. Juveniles have large heads, rounded snouts, and small spines along the pre-opercle edge and post-temporal region. Early juveniles have the lower rays of the caudal fin elongate, a feature lost in adults, where the caudal fin is squared off. They also have relatively small eyes and a small, inferior mouth equipped with barbels. All of these transient features indicate that these juveniles are relatively fast swimmers that feed in the lower water column (Chao and Musick, 1977).

A total of 28 individuals between 12.7 and 89.0 mm TL were examined for patterns of scale formation (see Fig. 75.1). The smallest specimen available (12.7 mm TL) did not have scales. Scale formation occurred over the size range from 14.5 to 46.0 mm TL. Specimens between 36.0 and 45.9 mm TL were not available. There are several loci for scale formation, including the lateral surface of the caudal peduncle, the dorsal surface of the head, below and behind the eye, and at two locations on the operculum. Initial scale development occurs along the lateral line of the caudal peduncle region, and by 18 mm TL scales cover that region. Scale formation continues anteriorly and laterally along the lateral line. By 21 mm

TL, scales are observed anteriorly and laterally almost to the origin of the pectoral fin and onto the mid-caudal fin. By 25 mm TL, scales cover the dorsal surface of the head above the eye and most of the body posterior to the operculum, but they do not occur on the ventral surface of the chest. Scales also extend farther onto the base of the caudal fin at this size. By approximately 30 mm TL, scales originate at two locations on the operculum. By 46 mm TL, scale formation has achieved the adult condition, where scales fully cover the body and head area, tip of snout excluded, and extend onto the caudal fin. At this size, scales cover 92.9% of the body and 61.7% of the body and fins. See Figures 71.1 and 75.1 for comparisons with other sciaenids.

LARVAL SUPPLY, SETTLEMENT, GROWTH, AND MORTALITY

In the Middle Atlantic Bight between 1977 and 1987, larvae typically were found on the continental shelf between the mouth of Delaware Bay and Cape Hatteras, with only occasional occurrences off the New Jersey coast (Morse et al., 1987; Able and Fahay, 1998). Annual fluctuations in abundance were apparent between 1959 and 1961 at Indian River, Delaware, when 98% of the 3641 larvae collected were taken during one of three study years (Pacheco and Grant, 1973). The larvae, at ingress, appear to be influenced by both active and passive transport processes (Norcross, 1991). After a decade of relatively stable numbers in our study area, abundances of both larvae and older stages began to increase in 2001, then increased further in 2006 (Fig. 76.1) in both plankton net and otter trawl sampling. During a two-year study on the Beach Haven Ridge off Little Egg Inlet, New Jersey, pelagic larvae were rare in September 1991 and somewhat more

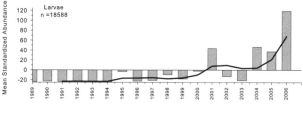

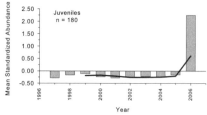

Fig. 76.1. Annual mean (bars) and three-year moving mean (line) of standardized abundance for *Micropogonius undulatus* collected in larval sampling with plankton net (*upper*) and juvenile sampling with otter trawls (*lower*). Data were standardized by subtracting overall mean from annual mean abundance. Three-year moving mean was calculated by taking the mean of the standardized values from the previous two years and the year in which data are plotted.

abundant in September and October 1992. None was larger than 9.5 mm SL (Able et al., 2006a). At about 8 to 20 mm, pelagic larvae leave ocean waters and enter estuaries. In collections inside Little Egg Inlet, this ingress occurs between September and January and usually reaches a peak in September, October, or November (Fig. 76.2). These larvae average 11.4 mm SL, and a large proportion are in the postflexion stage (Witting et al., 1999; Able et al., 2006a). This ingress occurs as early as August in Delaware Bay (Miller et al., 2003) and continues into November at inlets in Virginia (Thomas, 1971; Cowan and Birdsong, 1985). Ingress of larvae between 8 and 46 mm reaches a strong peak in December in the Indian River Inlet, Delaware (Pacheco and Grant, 1973).

A clear pattern of growth for young-of-the-year emerges from length-frequency distributions (Fig. 76.3). It is evident that very little growth occurs in fall or winter after ingress into the estuary. Other studies have reported a virtual lack of growth over the winter, with observations of young-of-the-year in the spring at sizes approximating those of the previous fall (Chao and Musick, 1977; Miller et al., 2003; Hare and Able, 2007). In Delaware Bay, the young-of-the-year can grow very quickly, at rates of 0.8 to 1.4 mm per day from spring through summer (Miller et al., 2003). As a result, they reach lengths of 135 to 140 mm at age 1 (Thomas, 1971). Those estimates are consistent with August observations from Great Bay–Little Egg Harbor, where the age 1+ class ranges from 80 to about 200 mm, with a mode at about 120 mm. Another analysis, based on samples from Chesapeake Bay, suggested they grew very quickly during the first year and reached a mean size of 247 mm, but only 387 mm by age 10 (Foster, 2001). The age and size composition has varied in recent years, with many more older fish (> 7 years) in Chesapeake Bay in 1998–2000, relative to 1988–1991 (Foster, 2001).

Minimum overwinter temperatures are critical to the survival of young-of-the-year in our study area (Norcross, 1983; Hare and Able, 2007). The density of this year class is often determined by minimum winter temperatures. At temperatures below 3°C survival is severely reduced, but this response varies with salinity (Lankford and Targett, 2001).

SEASONALITY AND HABITAT USE

Adults occur in coastal waters and estuaries over mud and sand-mud substrates, in maximum depths of 100 m. They are strongly euryhaline and can be found in salinities ranging from freshwater to 70 ppt (Simmons, 1957). During winter adults move offshore and south, perhaps as far as the South Atlantic Bight. In past decades, larvae have only rarely occurred in shelf waters in the central part of the Middle Atlantic Bight (Able and Fahay, 1998). In some years, however, young stages are relatively common in this area and contribute significantly to the Delaware Bay fauna (Thomas, 1971) and other estuaries (Hare and Able, 2007). In that estuary, ingress of larvae occurs in the fall; the young-of-the-year

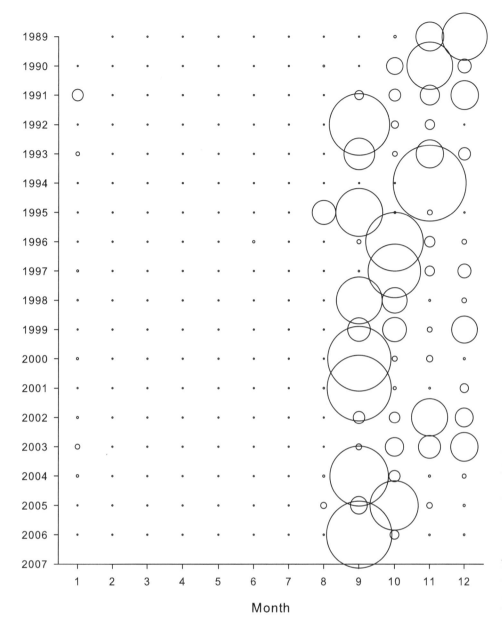

Fig. 76.2. Seasonal occurrence of the larvae of *Micropogonius undulatus* based on ichthyoplankton samples collected in Little Sheepshead Creek behind Little Egg Inlet from 1989 to 2006. Full circles represent the percentage of the mean number/1000 m³ of larval young-of-the year *Micropogonius undulatus* captured by given year. Values range from 0 to 100%. The smallest circles represent when samples were taken but no larvae were collected.

spend the winter and following spring and summer before egressing in late summer and fall (Miller et al., 2003).

Important habitats include coastal waters of the continental shelf for larvae and low-salinity habitats such as tributaries to major bay systems for the earliest settlement stages. These early settlement stages have been found to be more common in grass beds than in adjacent channel sites in the Chesapeake Bay (Olney and Boehlert, 1988). In extensive sampling across 12 habitat types in the Delaware Bay estuary, they occupied a variety of habitats as sampled with a variety of gears (Able et al., 2007a). Small young-of-the-year (< 20 mm) can be found in many habitats, including bay-water column, bay-shoreline, bay-shallow and bay-deep, and subtidal and intertidal marshes during the fall. They also occur in Delaware River water column, shoreline, and river-deep habitats. Larger young-of-the-year (> 20 mm) are col-

lected in all subtidal and intertidal marsh and bay habitats in the following spring. In many estuaries, the young-of-the-year are most abundant over muddy sediments in low-salinity waters (Miller et al., 2003). In other studies, young-of-the-year have been most commonly collected during the fall over a soft mud bottom at temperatures of 9.5 to 23.2°C and salinities of 24.0 to 33.7 ppt (Cowan and Birdsong, 1985). When juveniles occur in more southern estuaries in the summer they can be exposed to hypoxia, but this species is capable of detecting and avoiding dissolved oxygen levels of 1 mg/l (Wannamaker and Rice, 2000). Another study demonstrated that habitat quality, as reflected in larval and juvenile growth, was influenced by freshwater runoff (Searcy et al., 2007b).

As seasonal cooling proceeds, young-of-the-year gradually occupy the deeper parts of nursery creeks, where they are able to overwinter, at least in the Chesapeake Bay region.

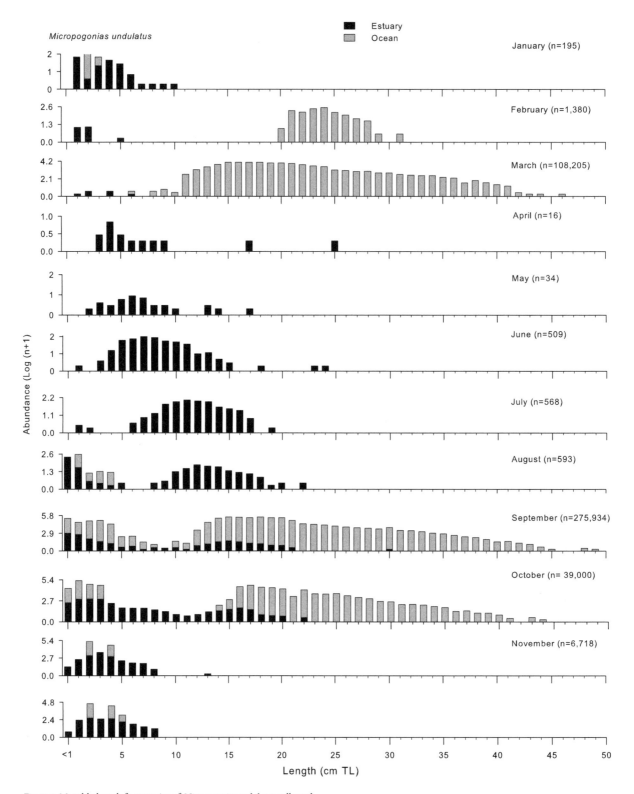

Fig. 76.3. Monthly length frequencies of *Micropogonias undulatus* collected
in estuarine and coastal ocean waters of the central Middle Atlantic Bight.
Sources: National Marine Fisheries Service: otter trawl (n = 417,853);
Rutgers University Marine Field Station: 1 m beam trawl (n = 2); seine
(n = 308); Methot trawl (n = 358); 1 m plankton net (n = 4226); 2 m beam
trawl (n = 124); killitrap (n = 5); 4.9 m otter trawl (n = 329); multi-panel gill
net (n = 5); and Able and Fahay 1998 (n = 11,728).

However, in unusually cold winters, mass mortalities of young-of-the-year have occurred in that bay (Chao and Musick, 1977) and in other estuaries (Hare and Able, 2007). The most intensive sampling effort in our study area is that reported by Thomas (1971) in upper Delaware Bay. Despite a three-year effort in that study, very few larvae or young-of-the-year were collected. Most of these were taken during December and none after that month. Those taken in December were collected concurrently with collections made in the Chesapeake and Delaware (C&D) Canal and upper Chesapeake Bay. The origin of these larvae is therefore in some question.

PREY AND PREDATORS

Important diet items change during ontogeny. Young stages concentrate on zooplankton, but fishes and macroinvertebrates become most important in adult stages (Darnell, 1961). Detritus is often included in the gut contents during all life history stages, perhaps attesting to their feeding method of skimming the substrate surface or foraging below the substrate surface (Roelofs, 1954). Barbels on the lower jaw tip facilitate finding food in adults. Younger stages lack these chin barbels, but they also feed near bottom (Hildebrand and Cable, 1930). The larvae eat tintinnids, pteropods, pelecypods, ostracods, and the egg, naupliar, copepodid, and adult stages of copepods (Govoni et al., 1983). Several studies (n = 5) have examined the food habits of young-of-the-year (17 to 152.4 mm, n = 2273) *Micropogonias undulatus* in Middle Atlantic Bight estuaries from Delaware Bay to North Carolina (see Table 8.1). Important prey items include a variety of invertebrates (copepods, polychaetes, mysids, amphipods, oligochaetes, gastropods, bivalves, decapod shrimp, and ostracods) and fishes (Hildebrand and Cable, 1930). Adults are known to eat crustaceans, annelids, mollusks, ascidians, ophiurans, and fishes (Hildebrand and Schroeder, 1928).

The diet of young-of-the-year shifts with ontogeny in Delaware Bay (Thomas, 1971) and Delaware Bay marsh creeks (Nemerson, 2001), such that they consume more fishes, crustaceans, and annelids and fewer mysids and zoo-plankton with growth. Spatial variation in diet has also been noted between marsh creeks within Delaware Bay, where annelids and crustaceans are important prey at all sites. Piscivory is highest at meso-polyhaline lower bay sites Dennis/Moores (9.2% W) (Nemerson, 2001). Mysids are more important at Dennis/Moores and oligo-mesohaline Mad Horse Creek (22.0% W and 23.8% W, respectively), while zooplankton and insects are important (19.8% W, 15.1% W) at limnetic-oligohaline Mill Creek. Relative importance of individual prey items also varies intra-annually in Delaware Bay marsh creeks, where amphipods become more important, while copepods and decapod shrimps decline in importance from September through November (Nemerson, 2001).

A study of stomach contents of striped bass during winter in Chesapeake Bay indicates heavy predation on overwintering young-of-the-year (Dovel, 1968).

MIGRATIONS

As juveniles reach their first birthday in the fall, most leave Middle Atlantic Bight embayments with older fish and migrate offshore and south for their second winter. After young-of-the-year (< 14 cm) leave the estuary, their distribution varies as a result of extensive seasonal migrations (Fig. 76.4). In the fall, they are distributed near bay mouths and on the inner continental shelf (mean depth 14 m, range 9 to 26 m) from Delaware Bay to the southern limit of sampling off North Carolina. They are not collected within the sampling limits during the winter, but can be found, primarily in shallow waters (mean depth 16 m, range 9 to 25 m) off North Carolina during the spring. The seasonal patterns of the age 1+ (> 14 cm) juveniles and adults are generally similar to the young-of-the-year, except the adults are more abundant in all seasons. In the fall, adults are concentrated on the inner and mid-shelf (mean depth 20 m, range 5 to 273 m) primarily from Long Island to North Carolina. Some adults have been captured on the continental shelf (mean depth 34 m, range 27 to 41 m) in the winter between Virginia and North Carolina. In the spring, they are found from Delaware Bay and south on the inner shelf (mean depth 19 m, range 9 to 65 m), with most individuals collected off North Carolina.

Age 1+ *Micropogonias undulatus*

YOY *Micropogonias undulatus*

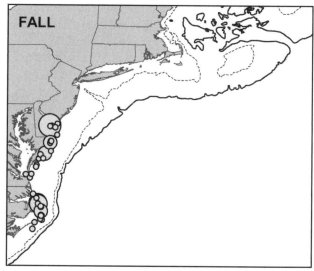

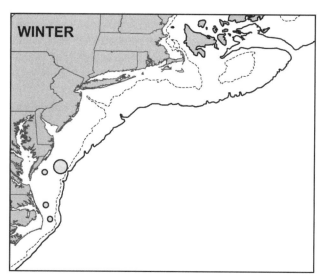

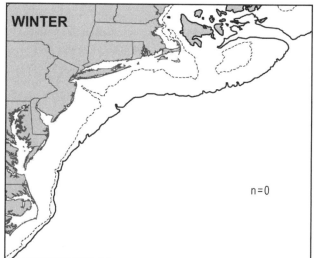

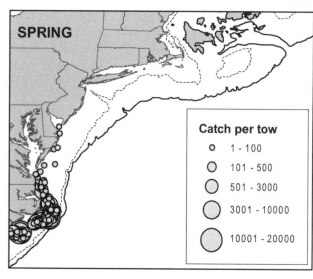

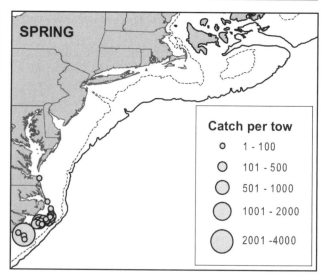

Fig. 76.4. Composite distribution of age 1+ (> 14 cm) and young-of-the-year (< 14 cm) *Micropogonias undulatus* during seasonal cruises of the National Marine Fisheries Service groundfish survey. Details of sampling effort are indicated in Fig. 3.4.

77

Pogonias cromis (Linnaeus)

BLACK DRUM

Estuary	Reproduction	Larvae/ Development	YOY Growth	YOY Overwinter
Ocean				
Both				
Jan				
Feb				
Mar				
Apr				
May				
Jun				
Jul				
Aug				
Sep				
Oct				
Nov				
Dec				

DISTRIBUTION

Pogonias cromis has been reported from the Gulf of Maine to Argentina (Hildebrand and Schroeder, 1928) and occasionally as far north as the Bay of Fundy (Bleakney, 1963). Despite intensive collections in our Great Bay–Little Egg Harbor study area during the late 1980s and early 1990s, very few young-of-the-year were collected (Able and Fahay, 1998). In Middle Atlantic Bight estuaries, eggs and larvae have been reported from Delaware Bay to Chesapeake Bay, whereas only juveniles have been recorded, infrequently, north of Delaware Bay (see Table 4.3). However, since around 1998, the adults, young-of-the-year, and larger juveniles have been observed and collected in Great Bay, New Jersey, with a variety of sampling gears as well as from authenticated accounts by recreational fishermen.

REPRODUCTION AND DEVELOPMENT

Spawning occurs during the spring, with a possible minor, secondary spawning in fall. Sexual maturity is reached at approximate sizes of 285 to 330 mm (Pearson, 1929; Simmons and Breuer, 1962) and females spawn multiple batches (Daniel, 1995; Macchi et al., 2002). Spawning is variously reported to occur in the coastal ocean near larger sounds and bays and in estuaries. Spawning peaks in the middle of May in lower Delaware Bay and adults are seldom captured there after the end of June (Wang and Kernehan, 1979; Able and Fahay, 1998). Egg occurrences show that spawning in Chesapeake Bay occurs in inshore areas (Joseph et al., 1964; Daniel and Graves, 1994) and lasts from April to June (Richards, 1973). Temperatures at the time of spawning are 17.5 to 19.0°C at the mouth of Chesapeake Bay (Joseph et al., 1964; Richards, 1973).

The early development of *Pogonias cromis*, and characters useful for distinguishing the eggs, larvae, and juveniles from those other species in the family Sciaenidae, have been well described and summarized (Fahay, 2007). Eggs are pelagic and spherical and measure 0.82 to 1.02 mm in diameter. Oil globules are multiple initially, then coalesce into a single one, 0.22 to 0.26 mm in diameter. Larvae hatch at about 2.5 mm, with a large head and mouth. Pigment on the body becomes concentrated into four or five bars as larvae transform into juveniles. Fin rays are complete at about 8 to 10 mm. Multiple mandibular barbels form on the lower jaw in early juveniles. Dense, black pigment forms on the membranes of the first dorsal fin.

A total of 18 individuals between 18.7 and 53.7 mm TL were examined for patterns of scale formation (see Fig. 75.1; Able et al., 2009a). Scale formation occurs over the size range from 18.7 to 41.8 mm TL, at several loci: on the lateral surface of the caudal peduncle, above the anal fin, on the anterior portion of the operculum, and behind the eye. Onset of scale formation occurs on the caudal peduncle, and in a small area over the anal fin base at about 18.7 mm TL. By 19 to 22 mm TL, scales spread anteriorly and laterally along the lateral line as far as the operculum. Over these same sizes, a few scales are also observed on the side of the head below the eye. By approximately 23 mm TL, scales cover most of the body posterior to the operculum except at the dorsal and ventral margins. By 26 mm TL, scales cover the entire body posterior to the operculum, and a few are also developed on the side of the head. By approximately 30 mm TL, scales cover the body,

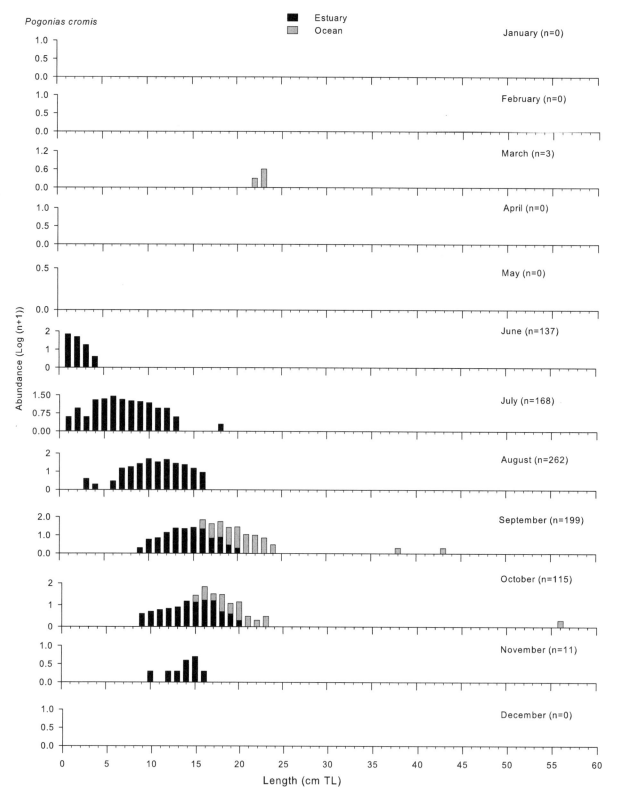

Fig. 77.1. Monthly length frequencies of *Pogonias cromis* collected in estuarine and coastal ocean waters of the central Middle Atlantic Bight and especially Delaware Bay. Sources: National Marine Fisheries Service: otter trawl
(n = 92); Rutgers University Marine Field Station: seine (n = 7); killitrap
(n = 4); 4.9 m otter trawl (n = 10); multi-panel gill net (n = 6); and Able and
Fahay 1998 (n = 776).

and are present on the dorsal surface of the head anterior to the dorsal fin, on the lateral surface of the head, and on the opercle. By 42 mm TL, scales cover the body and head and extend onto the caudal fin. No scales are observed on the other fins. Additional specimens need to be examined to confirm if this is the adult condition. At this size scales covered 88.8% of the body and 67.5% of the body and fins. See Figures 71.1 and 75.1 for comparisons with other sciaenids.

LARVAL SUPPLY, SETTLEMENT, AND GROWTH

In the Middle Atlantic Bight, spawning occurs in Delaware and Chesapeake bays, as well as coastal ocean waters near those bays. In recent years, the occurrence of large adults and young-of-the-year suggests that they may be spawning in Great Bay (K. W. Able, unpubl. data). Juveniles are 1 to 4 cm in June and increase rapidly in length until October, when they range from 9 to 23 cm (Fig. 77.1). The progression of modal lengths suggests an average growth rate of 1.1 mm per day. These estimates are similar to those reported for the coast of Virginia, with lengths of more than 200 mm by the end of the first year (Richards, 1973).

SEASONALITY AND HABITAT USE

Adults occur in coastal waters over a variety of habitats, including oyster reefs, clam shell deposits, sand, and softer substrates. They are typically found in the high-salinity portion of estuaries, while the early life history stages tolerate low salinities and even freshwaters. In Delaware Bay and Chesapeake Bay, larvae and small juveniles move into low-salinity nursery areas after spawning occurs in the lower bays (Joseph et al., 1964; Thomas and Smith, 1973). In Delaware Bay, some small individuals (5 to 10 mm) have been found as far up as the Chesapeake and Delaware (C&D) Canal.

Habitat use varies seasonally. The adults occur in Virginia coastal waters and Delaware Bay waters during spring and early summer (Richards, 1973). This is at the same time that large catches of adults are made in the commercial and recreational fisheries. Small juveniles are most abundant in the middle and upper parts of low-salinity (< 6.0 ppt) marsh creeks, with little current and mud substrate in Delaware Bay (Thomas and Smith, 1973) and Chesapeake Bay (Frisbie, 1961). In late June and early July, the largest juveniles (30 to 50 mm) begin to move out of the creeks and can be found both inshore and in deeper waters of the lower Delaware River and upper Delaware Bay (Thomas, 1971). By the end of August, there are few young in the Delaware River. Most are found in the lower bay by the end of October through De-

cember (Abbe, 1967; Daiber and Smith, 1970). A recent examination across 12 habitats in the Delaware Bay and River found most young-of-the-year in subtidal (54%) and intertidal (14.7%) marsh habitats (Able et al., 2007a). They also occur in other deep, shallow, and bay shoreline habitats in mesohaline salinities, but not in the river. After October, the average size in the estuary is smaller and larger individuals begin to be collected in coastal ocean waters, suggesting that the larger individuals are the first to leave the bay (see Fig. 77.1).

PREY AND PREDATORS

Adults prey mostly on mollusks and crustaceans. Two studies examined the food habits of young-of-the-year (30 to 211 mm, n = 268) in the Delaware River estuary (see Table 8.1). Important prey of young-of-the-year include a variety of invertebrates (polychaetes, oligochaetes, bivalves, insects, and amphipods). Fishes (*Menidia* spp., Family Clupeidae, and *Anchoa mitchilli*) are a minor dietary component. Habitat variation in diet has been noted within the lower Delaware Bay estuary, such that amphipods are of greatest numerical importance to young-of-the-year in both the river and creek locations, while annelids are of secondary importance in the river and copepods in creeks (Thomas, 1971).

MIGRATIONS

In general, members of the Atlantic Coast population of this species engage in long-range migrations, with a pattern of movement to the north and inshore during spring and to the south and offshore during fall (Richards, 1973; Jones and Wells, 2001). The seasonal occurrence of this species in the Middle Atlantic Bight is limited. After the young-of-the-year (< 20 cm) leave estuaries in the fall, their occurrence in samples on the continental shelf is restricted to inshore waters (mean depth 12 m, range 8 to 19 m) from Maryland to northern North Carolina, with most catches from Virginia and Maryland (Fig. 77.2). They have not been detected in these samples in the winter or spring, which suggests that they migrate out of the Middle Atlantic Bight during those seasons. Age 1+ (> 20 cm) juveniles and adults have a pattern similar to that of the young-of-the-year. In the fall, they have been collected on the inner shelf (mean depth 13 m, range 8 to 24 m) off Maryland, Virginia, and North Carolina. They are not present in the Middle Atlantic Bight during winter, and in the spring the few collections are limited to shallow waters (mean depth 14 m, range 13 to 14 m) off North Carolina.

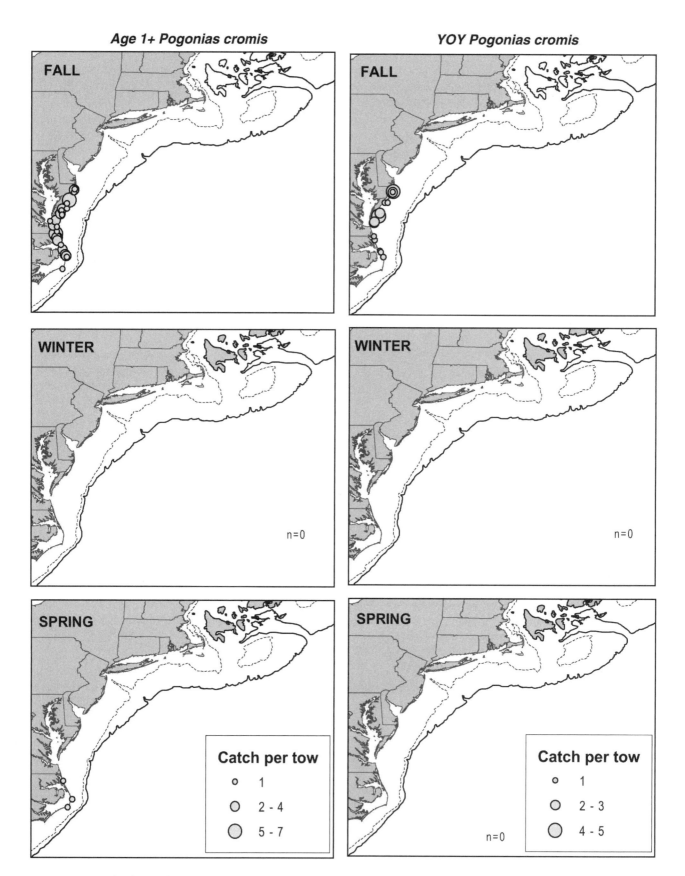

Fig. 77.2. Composite distribution of age 1+ (> 20 cm) and young-of-the-year (< 20 cm) *Pogonias cromis* during seasonal cruises of the National Marine Fisheries Service groundfish survey. Details of sampling effort are indicated in Fig. 3.4.

78

Sciaenops ocellatus (Linnaeus)

RED DRUM

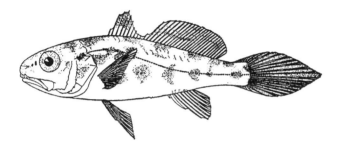

	Reproduction	Larvae/ Development	YOY Growth	YOY Overwinter
Estuary				
Ocean				
Both				
Jan				
Feb				
Mar				
Apr				
May				
Jun				
Jul				
Aug				
Sep				
Oct				
Nov				
Dec				

DISTRIBUTION

This species has been found along the coast of North America from Long Island to the western Gulf of Mexico, where it is most abundant (Fahay, 2007). Despite the historical accounts of its northern occurrence (Welsh and Breder, 1923), recent landings indicate that it rarely occurs north of Chesapeake Bay (Stewart and Scharf, 2008). Early stages are found in Atlantic coastal estuaries from Chesapeake Bay, North Carolina, and south. In this chapter we emphasize the results of studies from the Atlantic coast population.

REPRODUCTION AND DEVELOPMENT

Males become mature at an earlier age and at smaller sizes than females in North Carolina (Ross et al., 1995). In this study, more than 50% of males were mature at age 2 and 100% at age 3. More than 50% of females were mature at age 3 and 100% at age 4. Spawning occurs during the fall and winter, with peaks in August and September in North Carolina waters, but may take place from July through October (Ross et al., 1995; Stewart and Scharf, 2008). Spawning often occurs near the mouths of large embayments (Pearson, 1929; Yokel, 1966), but can also take place in the ocean, along beaches, or near the mouths of estuaries (Pearson, 1929; Yokel, 1966; Mercer, 1984b; Ross et al., 1995; Barrios, 2004). Thus, it is not surprising that larvae from the continental shelf were represented in the Middle Atlantic Bight by only 3 individuals in an 11-year MARMAP ichthyoplankton survey (1977–1987) (M. P. Fahay, pers. observ.). Larval fish holdings at Harvard University's Museum of Comparative Zoology include only two larvae, and both were collected south of Cape Hatteras (M. P. Fahay, pers. observ.). These observations suggest that if spawning occurs elsewhere in the Middle Atlantic Bight, it is within estuaries or bays and not over the continental shelf. Indeed, in North Carolina, spawning adults have been reported to be concentrated near the mouth of the Pamlico River, in smaller bays between the Pamlico and Neuse rivers, and in Pamlico Sound (Ross et al., 1995).

The early development of *Sciaenops ocellata*, including characters useful for distinguishing the larvae and juveniles from those of other sciaenids, has been well described and summarized in Fahay (2007). Eggs are pelagic and spherical and 0.86 to 0.98 mm in diameter. Larvae hatch at sizes of 1.7 to 1.8 mm SL (Holt et al., 1981). Pigment patterns are useful for distinguishing the early stages. Transformation to the juvenile stage occurs by 10.0 mm SL, when all fin rays are formed. Juveniles are diagnosed by meristic characters, shape of caudal fin, and series of pigment accumulations along the dorsum and midline of the body.

LARVAL SUPPLY, SETTLEMENT, GROWTH, AND MORTALITY

Young-of-the-year begin to occupy estuaries in September at sizes of 20 to 60 mm TL in Chesapeake Bay (Hildebrand and Schroeder, 1928; Mansueti, 1960); 27 to 62 mm TL in Neuse River, North Carolina (Tagatz and Dudley, 1961); and 12 to 40 mm TL in Cape Fear River, North Carolina (Weinstein, 1979; Stewart and Scharf, 2008). By November, these ranges have increased to 30 to 90 mm TL, 25 to 71 mm TL, and 23 to 47 mm TL, respectively. By December, fish in the Cape Fear and New rivers, North Carolina, attained sizes of 63 to 81 mm (Stewart and Scharf, 2008). Growth rates in larvae and juveniles are variable between estuaries and years. They have been estimated in studies in Texas and Louisiana, and range from 0.7 mm TL per day to 1.7 mm TL per day (Arnold et al., 1977; Crocker et al., 1981, respectively). Estimates for two North Carolina estuaries range from 0.45 to 0.75 mm per day (Stewart and Scharf, 2008). This species may reach 62 years (Ross et al., 1995).

One study concluded that mortality rates were high for age 0 and age 1 year classes (Ross et al., 1995). Survival rates for juveniles in North Carolina were lower than those reported from South Carolina or Georgia. In North Carolina, survival rates were 16 to 19%; in South Carolina they ranged from 6 to 29%; in Georgia they were from 15 to 43% (Vaughan and Helser, 1990). One important source of mortality for juveniles may be the low temperatures that occur during the winter (Bacheler et al., 2008; Stewart and Scharf, 2008).

SEASONALITY AND HABITAT USE

After young-of-the-year are transported into estuaries from nearshore or bay mouth spawning areas beginning in September, they have been reported to prefer grass bed or oyster shell habitats over muddy bottoms in shallow water (Pearson, 1929; Mansueti, 1960). Within Chesapeake Bay, a small-scale movement of young-of-the-year from these shoal waters during the fall to deeper channels during early winter has been reported (Mansueti, 1960). This same study noted that young-of-the-year are never collected from December to February, indicating an emigration out of the estuary during winter. A Texas study demonstrated the value of grass beds for protecting newly settled *Sciaenops ocellata* from predation by larger fishes (Rooker et al., 1998).

Adults are euryhaline, occurring in salinities ranging from freshwater to 50 ppt in the Gulf of Mexico (Mercer, 1984b; Bacheler et al., 2008). Juveniles are also tolerant of a wide range of salinities, but this tolerance varies with size and increases with age. In another Gulf of Mexico study, survival in fresh water ranged from 5% in larvae (6.2 mm), to 70% in larger larvae (16.2 to 19.7 mm), to 95% in juveniles (56.9 mm) (Crocker et al., 1981). They have also been reported from a wide range of temperatures (8.5 to 33.5°C) in North Carolina estuaries (Bacheler et al., 2008). Habitats favored by juveniles along the Atlantic coast are not well described. They have been noted over both sand and mud substrates in North Carolina, near oyster bars in South Carolina, and over muddy substrates in Florida (all unpubl. data, cited in Mercer, 1984b). Juveniles may have a high degree of site fidelity in estuaries at low tide based on tracking of ultrasonically tagged individuals (261 to 385 mm TL) in a Georgia estuary (Dresser and Kneib, 2007). A detailed study of acoustically tagged age 2 individuals in a North Carolina estuary found a negative

response to salinity, a positive response to dissolved oxygen and total prey, and a dome-shaped response to prey evenness (Bacheler et al., 2009a, b).

PREY AND PREDATORS

Diets apparently change with size of fish. In the size range of 30 to 1075 mm, smaller fish fed on *Gammarus* and *Mysis*, whereas larger fish fed mostly on shrimp (Hildebrand and Cable, 1930). In another study, fish < 15 mm fed on zooplankton, fish 15 to 75 mm consumed small benthic invertebrates, and those > 75 mm fed mostly on decapod crustaceans and other fishes (Bass and Avault, 1975). Other studies have confirmed that this species eats proportionately more crabs as they grow to larger sizes, and fishes become less important in the diets of the largest individuals (Yokel, 1966). Crustaceans (crabs and shrimp) and fishes are important diet items for populations on the Atlantic coast (Hildebrand and Schroeder, 1928; Overstreet and Heard, 1978). In North Carolina, the diet items included *Callinectes sapidus* and fishes such as *Mugil cephalus*, *Leiostomus xanthurus*, *Lagodon rhomboides*, *Paralichthys lethostigma*, *Anguilla rostrata*, *Bairdiella chrysoura*, and *Orthopristis chrysoptera* (Bacheler et al., in press). *Sciaenops ocellata* has a long lifespan and few predators after age 1 (Ross et al., 1995). One such predator includes the bottlenose dolphin (Gannon, 2003).

MIGRATIONS

This species migrates annually along the Atlantic coast. It arrives at the latitude of Cape Hatteras during March and April (Yokel, 1966); some of these enter Pamlico Sound, while others continue a migration to points farther north. They arrive off Virginia and the Chesapeake Bay about three to four weeks after passing Oregon Inlet, North Carolina. Historically, this migration has reached New Jersey waters during summer, perhaps in response to high population abundance and proper environmental conditions (Welsh and Breder, 1923). There are no recent reports of this, however. Conversely, some adults apparently leave Virginia estuaries in October and begin a southward fall migration along the coast (Bacheler et al., 2009b). Fishing for these migrating fish along the North Carolina coast extends from August to November (Yokel, 1966; Bacheler et al., 2009b).

In extensive studies of several age classes tagged with conventional and acoustic tags, movements were dependent on age, region, and season of tagging (Bacheler et al., 2009b). Ages 1 and 2 individuals tagged along the coast generally moved along the coast, while fish tagged in oligohaline waters were recovered along the coast in the fall. Adults (age 4+) moved from overwintering areas on the continental shelf into Pamlico Sound in spring and summer and departed again in the fall.

79

Chaetodon capistratus
Linnaeus

FOUREYE BUTTERFLYFISH

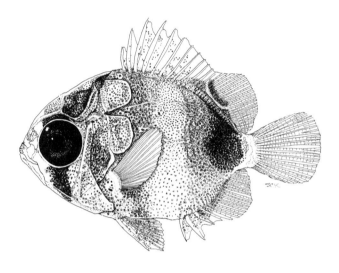

DISTRIBUTION

This species occurs in the western North Atlantic Ocean from North Carolina to Brazil, including Bermuda and the Gulf of Mexico (Burgess, 2002). Stragglers may reach Massachusetts in late summer. Most individuals in our study area are young-of-the-year.

REPRODUCTION AND DEVELOPMENT

Spawning takes place between February and April in the Bahamas and various Caribbean locations (Colin, 1989). Eggs are 0.76 to 0.77 mm in diameter, transparent, and spherical, and have a single oil globule 0.18 mm in diameter (Kelly, 2006). An early tholichthys stage, 5.4 mm SL, has been tentatively described (Kelly, 2006). The dermal plates on the head in this individual are larger than those shown in the present example. There is no other information available on early development.

LARVAL SUPPLY, SETTLEMENT, GROWTH, AND MORTALITY

There is no evidence that this species spawns in temperate portions of the East Coast. Larval specimens are presumably transported to the Middle Atlantic Bight from farther south via the Gulf Stream, as in *C. ocellatus* (McBride and Able, 1998). They do not occur in every year. When they do arrive in our study area, they are as small as 15 mm TL (Fig. 79.1). There is little evidence of growth, and length ranges remain at 20 to 50 mm TL during the summer and fall. We presume these young-of-the-year die with falling temperatures, as does *C. ocellatus* (McBride and Able, 1998).

SEASONALITY AND HABITAT USE

Tholichthys-stage individuals arrive in the Mullica River– Great Bay estuary as early as August and might occur as late as November (Fig. 79.2). Early stages are attracted to coral reefs, rocky substrates, pilings, and seawalls. In our study area, they also occur in marsh creeks. They remain in these habitats until temperatures cool in the fall and winter.

PREY AND PREDATORS

No information.

MIGRATIONS

No information except that early stages apparently are likely transported into the study area by the Gulf Stream.

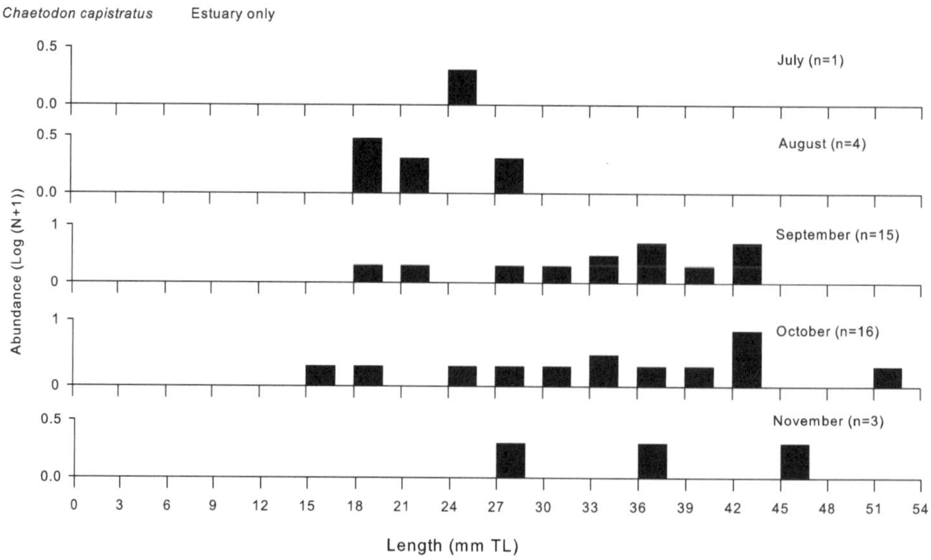

Chaetodon capistratus   Estuary only

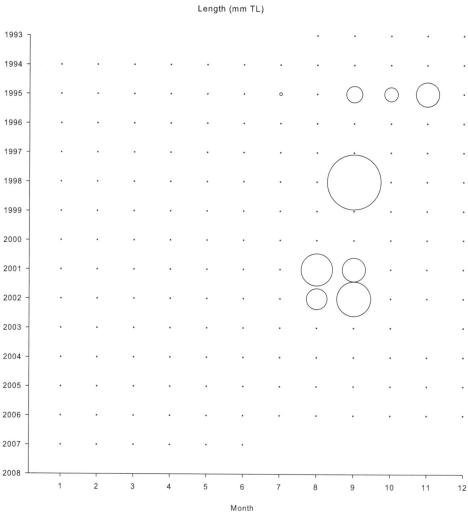

July (n=1)

August (n=4)

September (n=15)

October (n=16)

November (n=3)

Abundance (Log (N+1))

Length (mm TL)

Fig. 79.1. Monthly length frequencies of *Chaetodon capistratus* collected in estuarine waters of the Great Bay–Little Egg Harbor study area. Source: Rutgers University Marine Field Station: experimental trap (n = 22); killitrap (n = 17).

Fig. 79.2. Seasonal occurrence of the larvae of *Chaetodon capistratus* based on ichthyoplankton samples collected in Little Sheepshead Creek behind Little Egg Inlet from 1989 to 2006. Full circles represent the percentage of the mean number /1000 m³ of larval young-of-the year *Chaetodon capistratus* captured by given year. Values range from 0 to 100%. The smallest circles represent when samples were taken but no larvae were collected.

Month

80

Chaetodon ocellatus
Bloch

SPOTFIN BUTTERFLYFISH

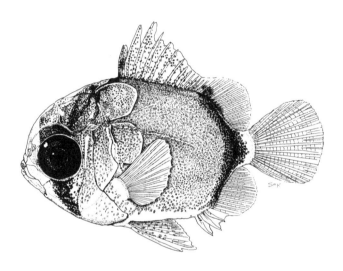

Estuary	Reproduction	Larvae/ Development	YOY Growth	YOY Overwinter
Ocean				
Both				
Jan				
Feb				
Mar				
Apr				
May				
Jun				
Jul				
Aug				
Sep				
Oct				
Nov				
Dec				

DISTRIBUTION

Chaetodon ocellatus occurs in the western North Atlantic Ocean from Nova Scotia to Brazil, including the Gulf of Mexico (Longley and Hildebrand, 1941; Scott and Scott, 1988). All of the individuals collected north of Cape Hatteras have been young-of-the-year (McBride and Able, 1998). These have been reported sporadically from Middle Atlantic Bight estuaries but primarily in those south of Long Island (see Table 4.3).

REPRODUCTION AND DEVELOPMENT

Few data are available on reproduction. Spawning probably occurs from winter into spring (January to May) (Colin, 1989), and presumably south of Cape Hatteras. Ripe females have been collected in May, one week before the full moon in North Carolina.

The early larval development of *Chaetodon ocellatus* is undescribed. Characters pertaining to eggs and transforming larvae/juveniles, including methods to distinguish this species from juveniles of *Chaetodon capistratus*, have been summarized (Fahay, 2007). The latter species occurs more rarely in our study area. Eggs of *Chaetodon ocellatus* are pelagic and spherical, and measure 0.60 to 0.75 mm in diameter. Other pertinent details of eggs and early larvae are unknown. Transforming larvae go through a tholichthys stage (Hubbs, 1963; Burgess, 1978), wherein the head is enclosed in bony shields and earlike flaps (Fig. 80.1). These bony structures are presumably absorbed shortly after settlement. Pigment patterns are important for distinguishing the juveniles of several species of *Chaetodon*.

Individuals from 19.2 to 48.3 mm TL (n = 19) were examined for scale formation (see Fig. 80.1). Scales are well developed over all of the body behind the operculum in the smallest specimen (19.6 mm TL) captured in the estuary. These individuals still had the large plates under the origin of the dorsal fin at the posterior margin of the head, typical of the tholichthys stage. By approximately 24 mm TL, these plates were still present but they were absent in all larger specimens examined. At larger sizes, scales spread onto the bases of the caudal, dorsal, and anal fins. Scales also extended onto the head, beginning with the dorsal surface and an area behind the eye. At the same size, some scales had formed on the anterior portion of the caudal fin and on the posterior portion of the dorsal fin adjacent to the body. Beginning at approximately 25 mm TL, some scales extended onto the dorsal surface of the head and the anal fin. At 26 mm TL, scales appeared on the operculum behind the eye and were further developed on the dorsal fin. By 35 mm TL, scale formation expanded to cover a larger proportion of the head and fin bases. By 48 mm TL, scales covered the entire body and most of the head, except on the snout, and extended on to the base of the pectoral fin. At this size, scales covered 96% of the body and 74.3% of the body and fins. This approximates the adult condition.

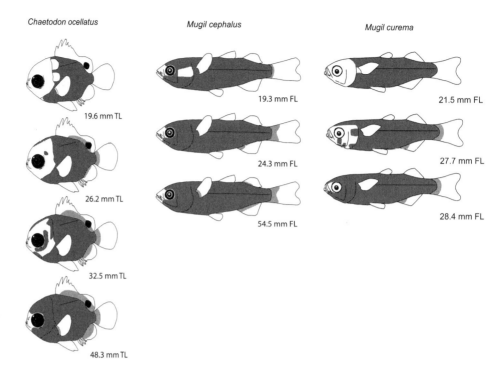

Fig. 80.1. Description of scale formation relative to length in *Chaetodon ocellatus*, *Mugil cephalus*, and *Mugil curema*.

LARVAL SUPPLY, SETTLEMENT, GROWTH, AND MORTALITY

The source of larvae to temperate estuaries, such as in New Jersey, is presumably from spawning south of Cape Hatteras. Individuals collected in the water column (under nightlights) are typically 17 to 22 mm TL in southern New Jersey waters. The narrow size range of these fish is indicative of the size at settlement, because fish of this size range were also collected in traps on the bottom (McBride and Able, 1998). Juveniles occur in trap samples in almost every year in our study area (Fig. 80.2). They appear to fluctuate from slightly more abundant to slightly less so at five-year intervals. In a single sampling site using traps, an average of 17 individuals is collected per year. This may occur over a long period of time because small individuals are evident from June through October (Fig. 80.3). Growth rates are difficult to interpret in the limited samples from our study area. Small fish ranging from 15 to 21 mm continue to be present from July through October, but fish in the study site can grow to as large as 69 mm by the fall. Based on collections from a variety of trap types and mark/recapture experiments, the growth rate ranged from 0.09 to 0.21 mm per day (McBride and Able, 1998).

Laboratory and field observations and tag/recapture experiments in southern New Jersey indicate that juveniles apparently remain in estuarine habitats until they succumb to seasonally lowering temperatures (McBride and Able, 1998). In the laboratory, young-of-the-year did not feed below 12°C, and no fish survived at temperatures below 10°C. In the field, individuals collected below 12°C were frequently disoriented and unable to remain upright. Thus, individuals of this species that settle and stay for the summer in Middle Atlantic

Bight estuaries cannot survive winter temperatures, do not migrate south for the winter, and are probably expatriates.

SEASONALITY AND HABITAT USE

Young-of-the-year *Chaetodon ocellatus* were common in the Great Bay estuary in midsummer and early fall (from early July through November) over a period of several years. Habitats of young-of-the-year are variable. During the summer, they occur in intertidal and subtidal areas in a variety of structured and unstructured habitats, including around dock pilings, marsh peat reefs, and tidal creeks (Szedlmayer and Able, 1996; McBride and Able, 1998). They show high fidelity to some sites, with small, marked individuals remaining in close proximity to the original release site. In one study, tagged individuals were recaptured 1 to 39 days after release (McBride and Able, 1998).

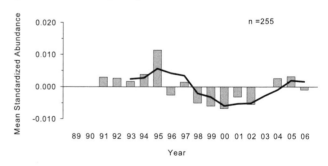

Fig. 80.2. Annual variation in abundance of *Chaetodon ocellatus* collected in trap sampling, 1991–2006. Data were standardized by subtracting the overall mean from annual abundance. Vertical bars represent standardized annual abundance; the solid line is a three-year running mean that was calculated by taking the mean of the standardized values from the previous two years and the year in which the data are plotted.

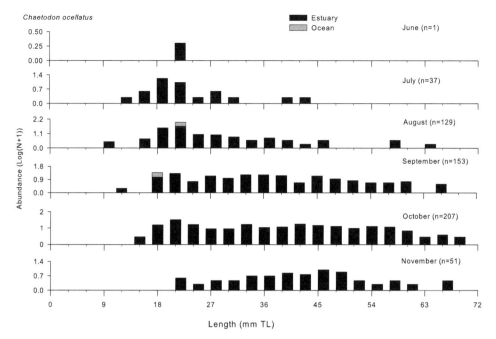

Fig. 80.3. Monthly length frequencies of *Chaetodon ocellatus* collected in estuarine and coastal ocean waters of the Great Bay–Little Egg Harbor study area. Sources: Rutgers University Marine Field Station: seine (n = 6); experimental trap (n = 141); killitrap (n = 141); and Able and Fahay 1998 (n = 291).

PREY AND PREDATORS

There have been no studies of the food habits of young-of-the-year *Chaetodon ocellatus* in Middle Atlantic Bight estuaries.

MIGRATIONS

There have been no reported collections of this species on the Middle Atlantic Bight continental shelf during fall or winter. Therefore, there is no evidence to support the notion that juveniles return to habitats south of Cape Hatteras. It must be assumed that habitats in southern New Jersey are sinks, and that all juveniles die as waters cool in the fall (McBride and Able, 1998).

81

Mugil cephalus Linnaeus

STRIPED MULLET

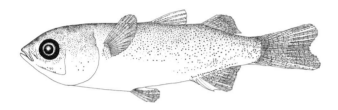

Estuary	Reproduction	Larvae/ Development	YOY Growth	YOY Overwinter
Ocean				
Both				
Jan				
Feb				
Mar				
Apr				
May				
Jun				
Jul				
Aug				
Sep				
Oct				
Nov				
Dec				

DISTRIBUTION

Mugil cephalus occurs worldwide in coastal waters and by one estimate is the most abundant and widespread inshore teleost (Odum, 1970). In the western Atlantic, it is found between Maine and Brazil, and a single juvenile specimen has been recorded from Nova Scotia (Scott and Scott, 1988). Other reports of juveniles from Canadian waters probably refer to *M. curema* (Gilhen, 1972). Pelagic juveniles are often collected in waters associated with the Gulf Stream or

Slope Sea. Juveniles have been recorded from most Middle Atlantic Bight estuaries (see Table 4.3), but neither eggs nor larvae have been reported from oceanic or estuarine waters in the Middle Atlantic Bight. Stages older than young-of-the-year are very rarely collected in the Middle Atlantic Bight or the Gulf of Maine (Hildebrand and Schroeder, 1928; K. Hartel, pers. comm., in Collette and Klein-MacPhee, 2002), although a few reports of fish exceeding 20 cm in the fall may refer to older year classes (Nichols and Breder, 1927; Schaefer, 1967).

REPRODUCTION AND DEVELOPMENT

Observations of spawning are few (Arnold and Thompson, 1958), and establishing locations where reproduction occurs is usually based on larval occurrences. After a migration from freshwater (often in large schools), spawning takes place in South Atlantic Bight continental shelf waters over a range of depths extending from about 36 m into the Gulf Stream (Breder, 1940; Günter, 1945; Broadhead, 1953; Anderson, 1958). The spawning season extends from October to February, with a peak in November and December (Anderson, 1958) or January through April (Collins and Stender, 1989), when continental shelf water temperatures are falling. Fish about 20 cm long (characterized as "ripe") have been collected during the fall near New York (Nichols and Breder, 1927), but the evidence for spawning is tenuous. Spawning in the Gulf of Mexico occurs from November to February and co-occurs with periods of northerly winds (Ibanez Aguirre and Gallardo-Cabello, 2004). In the Ibanez Aguirre and Gallardo-Cabello study, age at first maturation is 6 years for both sexes, although the authors include estimates of 1 to 3 years by other authors. Males reach sexual maturity at 373 mm TL (mean); females, at 377 mm TL (mean). Spawning in Cuba has been reported to take place in both summer and winter (Alvarez-Lajonchere, 1976).

Details of early development have been recently summarized (Fahay, 2007). Small larvae have only been collected in shelf waters of the South Atlantic Bight (Anderson, 1958), and these small stages (< 6 to 10 mm SL) were most common over the outer half of the shelf (Powles, 1981; Collins and Stender, 1989). In quarterly sampling in the South Atlantic Bight, pelagic juveniles (< 26.5 mm FL) were most common during the winter, with almost no occurrences in other seasons (Fahay, 1975). The querimana, or pelagic juvenile stage, ends at sizes between 35 and 45 mm. Information on scale formation in this species is limited because larvae and small juveniles are infrequently collected, and those few that are collected often lose their scales when contacted by nets. Scales begin to form at approximately 11 mm (Martin and Drewry, 1978). The distribution of scales on larger specimens (approximately 19 to 54 mm FL, n = 17) is quite advanced, with scales covering all surfaces of the body and head except for the anterior tip (see Fig. 80.1). At about 19 mm, some scales have formed at the base of the caudal fin, and this coverage extends posteriorly in larger sizes. At sizes of approximately 24 to 33 mm FL, a few scales begin to form on the anterior portion of the anal fin. By 54 to 59 mm FL, all of the body, except the snout, is covered with scales. The bases of the caudal, anal, and second dorsal fins are also scaled. In the adult condition 96.2% of the body and 73.7% of the body and fins are covered with scales.

LARVAL SUPPLY AND GROWTH

It is difficult to determine the source of pelagic juveniles to the Middle Atlantic Bight, but it is likely from south of this study area. In the South Atlantic Bight, a maximum size of 31 mm (mostly 20 to 25 mm) has been reported for pelagic juveniles occurring in offshore waters (Anderson, 1958; Fahay, 1975; Collins and Stender, 1989). Pelagic juveniles were most abundant near the Gulf Stream off the North Carolina coast during January in another study (Fahay, 1975) and were presumably then subject to advection north into Middle Atlantic Bight waters. A shoreward movement is then facilitated by favorable, wind-driven drift (Powles, 1981). Along the Georgia coast, young-of-the-year first arrive on the ocean side of barrier beaches in November and continue to occur there through April. They do not enter estuaries until January, and then become more abundant there through May and into June. They are never taken in abundance during summer and fall (Anderson, 1958). In North Carolina, pelagic juveniles enter estuaries and sounds beginning in February, at sizes between 18 and 25 mm TL (Jacot, 1920; Higgins, 1928). In sampling at Indian River Inlet, Delaware, young-of-the-year occurred December through May at sizes between 25 and 32 mm FL (Pacheco and Grant, 1973). In another Delaware estuary study, young-of-the-year were reportedly most common during fall and winter (Wang and Kernehan, 1979). When juveniles reach 40 mm SL, they are fully capable of osmoregulation and can then tolerate a range of salinities, from freshwater to full seawater (Nordlie et al., 1982).

Determining the growth rate of young-of-the-year in the central part of the Middle Atlantic Bight from available data is difficult, although all estimates indicate very rapid growth in the first year. The continued presence of fish 25 to 32 mm FL through the winter at Indian River Inlet (see above), combined with the progression of length modes later in the year in the Mullica River–Great Bay estuary (Fig. 81.1), indicates growth may initially be retarded in young-of-the-year entering estuaries during the winter, but rapid growth then ensues through the first summer; lengths of 19 to 25 cm may be reached by fall. Size at age 1 in several Gulf of Mexico areas varies from 13 to 18 cm FL (Broadhead, 1958) to about 20 cm TL (Ibanez-Aguirre and Gallardo-Cabello, 2004). Several growth curves derived from other areas in the South Atlantic Bight also suggest a size of about 15 cm as the year class enters its first fall (Anderson, 1958). Lengths of 20 cm in these other studies equate with age 2 fish (Broadhead, 1958). Assuming the modes shown in Figure 81.1 represent a single

year class, growth may be faster in the Middle Atlantic Bight, since these sizes are reached and exceeded by late summer. Alternately, perhaps the Middle Atlantic Bight is also visited by age 1 fish during the summer in some years, and the growth estimates represent more than a single year class, although there are no other datasets that suggest this might occur. We have also not found smaller juveniles (8 to 11 cm TL) occurring in October, as we reported previously (Able and Fahay, 1998), based on collections made in the shore zone of Delaware Bay by de Sylva et al. (1962). We have determined that these data are in error, or are based on misidentified juveniles of *Mugil curema*.

SEASONALITY AND HABITAT USE

Larvae have been collected in the northern Gulf of Mexico between October and March, with a peak in November–December (Ditty and Shaw, 1996). Larvae have also been collected off the southern United States between February and May (Collins and Stender, 1989), or October through May, most abundantly during January (Fahay, 1975). (See the *Mugil curema* chapter for comparative information on occurrences of larvae and pelagic juveniles in the South Atlantic Bight.) The seasonal occurrence in the Middle Atlantic Bight begins with collections in inlets and estuaries in the spring, followed by estuarine residency into the fall, when individuals leave estuaries (see Fig. 81.1).

Habitat varies with life history stage. Larvae and pelagic juveniles occur in the open ocean in near-surface layers (Fahay, 1975; Powles, 1981). Following the pelagic juvenile stage, they occur along barrier beaches and then enter estuaries throughout their range. Water temperatures probably regulate the time that young-of-the-year are able to remain in estuaries (Sylvester et al., 1974). Young-of-the-year smaller than 50 mm SL have been reported from extremely shallow waters, including the surfzone and tide pools, despite temperatures approaching the upper lethal limit. Juveniles larger than 50 mm SL occur in somewhat deeper waters but may move to shallower areas during flood tides (Major, 1978). Young are able to survive in marsh pools on Florida's Gulf Coast in temperatures as high as 34.5 C° (Kilby, 1949).

Juveniles of the two *Mugil* species appear to occur in different habitats during their first year. In the Great Bay–Little Egg Harbor study area, very few collections of this species were made in areas where *M. curema* were abundant, such as tidal creeks where salinities ranged from 23 to 33 ppt (Rountree and Able, 1992a), suggesting the two species do not co-occur during their first summer. In Delaware Bay and tributaries, young-of-the-year of *M. cephalus* may be present in freshwater (Smith, 1971; Wang and Kernehan, 1979) or in salinities as low as 4 ppt. These are lower than salinities where *M. curema* occurs (de Sylva et al., 1962). *M. cephalus* occurred in a beach seine study of the tidal Delaware River near Philadelphia where *M. curema* did not (Weisberg et al., 1996). Conversely, *M. curema* was commonly collected in nightlighting

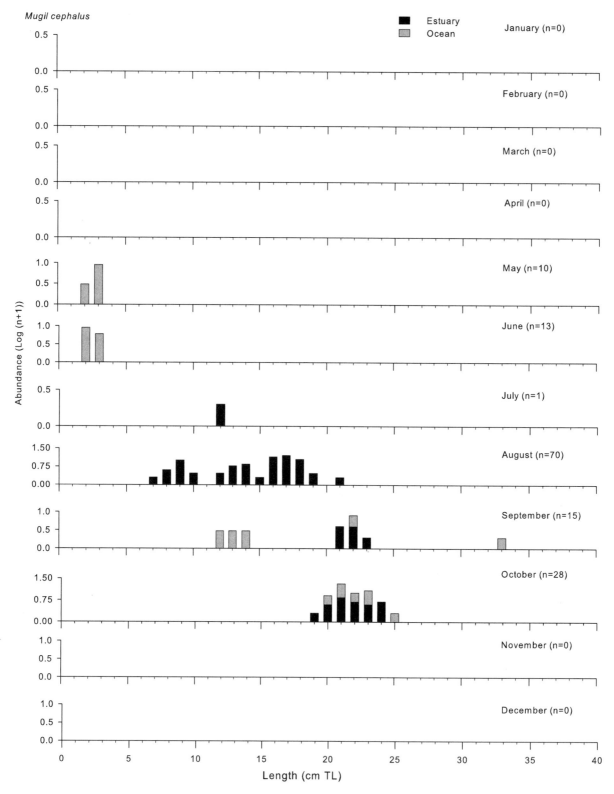

Fig. 81.1. Monthly length frequencies of *Mugil cephalus* collected in estuarine and coastal ocean waters of the central Middle Atlantic Bight. Sources: National Marine Fisheries Service: otter trawl (n = 4); Rutgers University Marine Field Station: seine (n = 128); 4.9 m otter trawl (n = 1); multi-panel gill net (n = 4).

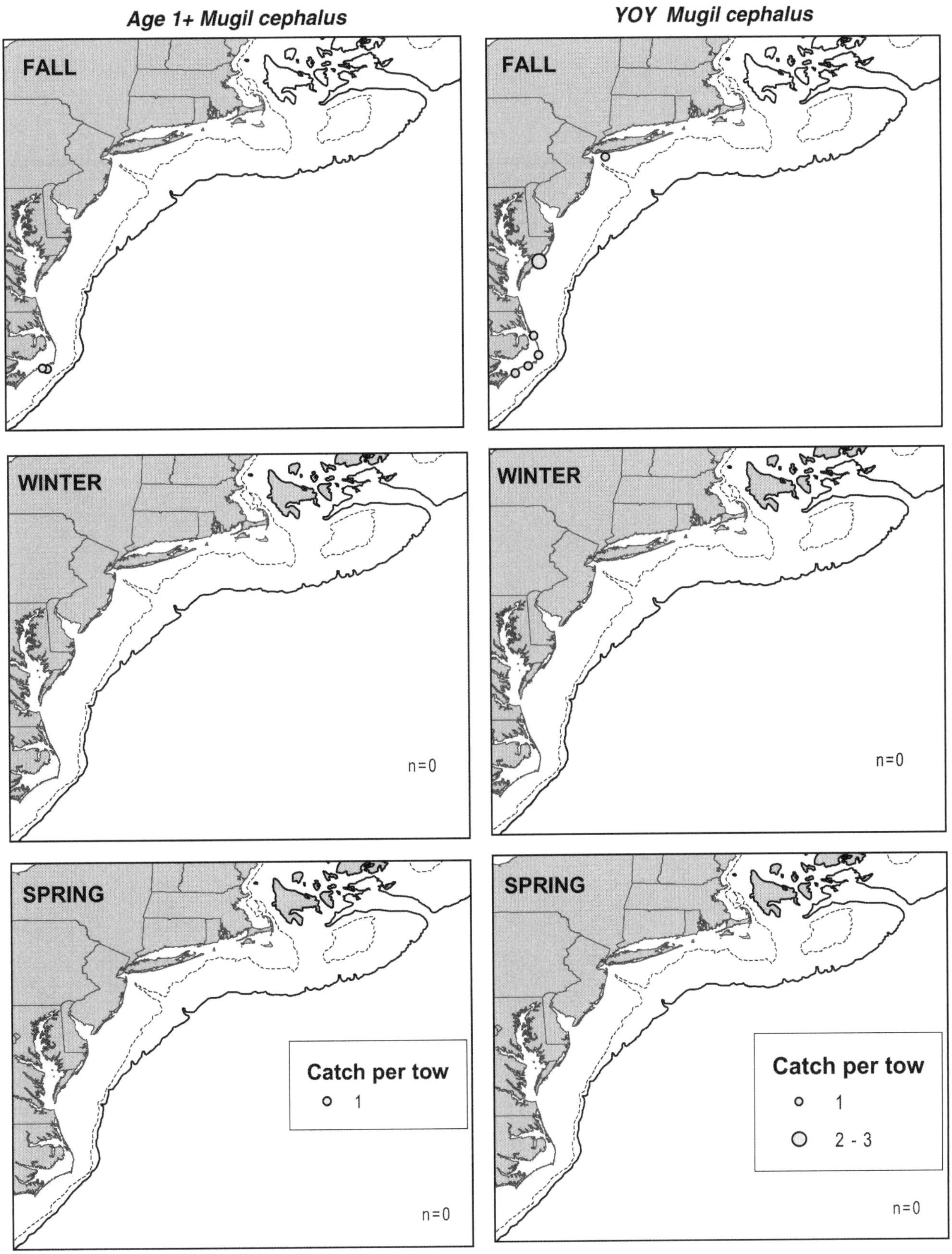

Age 1+ *Mugil cephalus*

YOY *Mugil cephalus*

Fig. 81.2. Composite distribution of age 1+ (> 19 cm) and young-of-the-year (< 19 cm) *Mugil cephalus* during seasonal cruises of the National Marine Fisheries Service groundfish survey. Details of sampling effort are indicated in Fig. 3.4.

in a high-salinity embayment in Great Bay, where *M. cephalus* was absent (Able et al., 1997).

Although large numbers of young-of-the-year overwinter in estuaries in the South Atlantic Bight (Nordlie et al., 1982), only a few have been recorded during winter months in Delaware Bay (de Sylva et al., 1962). The latter were small specimens (2 to 3 cm) and may have been early products of winter spawning in the South Atlantic Bight.

PREY AND PREDATORS

Juveniles are planktivorous; adults are herbivorous or feed on organic detritus (Harrison, 2002). They feed mostly on muddy bottoms in shallow water; gut contents of collected specimens contain diatoms and foraminifera as well as mud and plant detritus (Hildebrand and Schroeder, 1928). They also sometimes feed in surface layers on worms or other invertebrates or on bits of algae.

MIGRATIONS

There were very few young-of-the-year (< 19 cm) collected in National Marine Fisheries Service (NMFS) bottom trawl surveys in the Middle Atlantic Bight (Fig. 81.2), presumably because they occur most frequently in estuarine and coastal waters until such time as they migrate onto the continental shelf and south before winter. The few that have been collected were found in the fall in shallow waters (13 m) in western Long Island, Virginia, and in North Carolina. No young-of-the-year or older stages were collected in the winter or spring. The only age 1+ (> 19 cm) juveniles or adults collected during trawl surveys were found in shallow waters (17 m) in the fall off North Carolina. Older stages are rarely, if ever, reported from Middle Atlantic Bight waters.

82

Mugil curema Valenciennes

WHITE MULLET

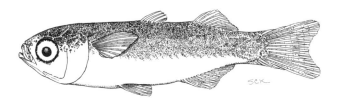

Estuary	Reproduction	Larvae/ Development	YOY Growth	YOY Overwinter
Ocean				
Both				
Jan				
Feb				
Mar				
Apr				
May				
Jun				
Jul				
Aug				
Sep				
Oct				
Nov				
Dec				

DISTRIBUTION

Mugil curema occurs in the western and eastern Atlantic Ocean. In the western Atlantic it is found from Nova Scotia to Brazil, although after the first year individuals of this species are rarely, if ever, collected north of Florida (Anderson, 1957). Juveniles have been reported from as far north as Nova Scotia (Scott and Scott, 1988) and from many estuarine locations within the Middle Atlantic Bight (see Table 4.3), but neither eggs nor larvae have been reported from oceanic or estuarine waters in our study area.

REPRODUCTION AND DEVELOPMENT

This species is a batch spawner, and the apparently long reproductive period is the result of non-synchrony within the population rather than prolonged production by individuals (Solomon and Ramnarine, 2007). Spawning off the southeastern United States occurs from February through October, with a strong peak between April and June (Anderson, 1957) or April and May (Collins and Stender, 1989). In the northern Gulf of Mexico, spawning peaks from February to May (Ibanez Aguirre and Gallardo-Cabello, 2004). Spawning also occurs in the southern Caribbean Sea, where the spawning season may extend from November to July (Solomon and Ramnarine, 2007). There is no evidence that it takes place in the Middle Atlantic Bight. The onset of spawning coincides with rising water temperatures in early spring in the South Atlantic Bight (Fahay, 1975). An immense school of spawning fish was once observed off the coast of southern Florida, close to the axis of the Gulf Stream (Anderson, 1957). Most observations of the smallest larvae have been across the South Atlantic Bight continental shelf, as far offshore as the Gulf Stream (Anderson, 1957; Fahay, 1975; Powles, 1981; Collins and Stender, 1989).

Details of development have been recently summarized (Fahay, 2007). Early stages are best distinguished from those of *Mugil cephalus* by counts of spines and rays in the anal fin. Specimens from 21.5 to 32.4 mm FL (n = 17) were examined for scale formation (see Fig. 80.1). The body is entirely covered with scales at 21.5 to 28.4 mm FL, and scales begin to cover the bases of caudal and anal fins at 21.5 mm. The head scales appear in discrete patches at sizes between 21 and 27 mm, until the entire head, except the anterior tip, is covered.

LARVAL SUPPLY AND GROWTH

In quarterly sampling over the continental shelf in the South Atlantic Bight, pelagic juveniles were most commonly collected during spring (May), although small numbers continued to be present through the remaining seasons (Fahay, 1975). In another study there, the smallest size class was distributed the farthest offshore and sizes were inversely related to distance from shore (Powles, 1981), suggesting offshore spawning and a migration toward shore with growth. Young-of-the-year occur in South Atlantic Bight waters in two size modes, possibly originating from reproduction in Caribbean and South Atlantic Bight waters. The results of these differences are that, during the spring, the largest pelagic juveniles in the South Atlantic Bight are developed well enough to be capable of actively swimming against the prevailing current and into estuaries there, but the smallest are subject to advection north via the Gulf Stream (Powles, 1981). Recent collections of larvae and pelagic juveniles (< 25 mm) have been made during May in the Slope Sea off the Middle Atlantic Bight, where they were strongly associated with the Gulf Stream and waters transitional between the Gulf Stream and the Slope Sea (Hare et al., 2001). Most of

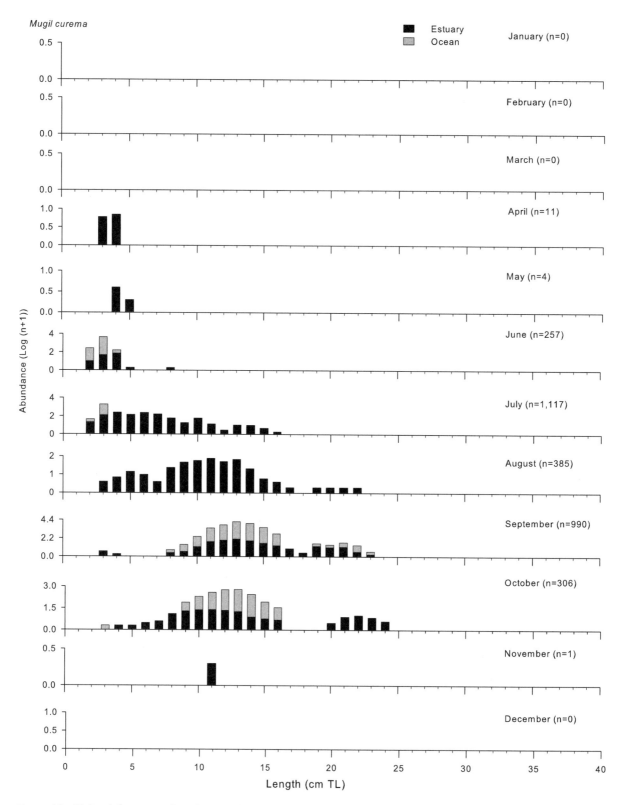

Fig. 82.1. Monthly length frequencies of *Mugil curema* collected in estuarine and coastal ocean waters of the central Middle Atlantic Bight. Sources: National Marine Fisheries Service: otter trawl (n = 11); Rutgers University Marine Field Station: seine (n = 1163); 4.9 m otter trawl (n = 1); multi-panel gill net (n = 29); and Able and Fahay 1998 (n = 1867).

these were collected in a neuston net, but smaller individuals were also taken in a plankton net, which sampled to a depth of 15 m. They occurred with an assemblage of other species, together categorized as "transients," members of which were spawned south of Cape Hatteras. Young-of-the-year of all members of this assemblage occupy Middle Atlantic Bight estuaries during the summer. In Great Bay, New Jersey, entering pelagic juveniles are apparently large enough to avoid capture by a passive plankton net, because only two individuals (< 21.2 mm SL) were collected in a six-year study (Witting et al., 1999). Large numbers have been collected by nightlighting in this system, however (Able et al., 1997; Able and Fahay, 1998).

Maximum size attained in offshore waters, and the size at which pelagic juveniles enter coastal habitats, is about 25 mm. Anderson (1957) took none larger than 25 mm in the ocean or when they began to enter a Georgia marsh. Maximum sizes in other continental shelf studies were 26 mm (Fahay, 1975) and 31 mm (Collins and Stender, 1989). Sizes at ingress in North Carolina are 20 to 21 mm, and this size continues to appear until September (Jacot, 1920). Our collections indicate growth resumes after ingress, and in the Great Bay–Little Egg Harbor estuary it is most rapid between June and September, when the observed size mode increases from 3 to 13 cm (Fig. 82.1). This rate equates with an increase of 1 mm per day, a value also estimated for Virginia young-of-the-year (Richards and Castagna, 1970). In a laboratory experiment, larvae were 36 mm long after 36 days (Houde et al., 1976). Two size modes are reflected in our length frequencies, and these are presumably the result of multiple size modes observed in shelf waters of the South Atlantic Bight (Powles, 1981). Young-of-the-year reach 9 to 16 cm by the fall in our study area. This is similar to an observation from the South Atlantic Bight, where young-of-the-year reached 12 cm by October and November (Anderson, 1957). This species is reported to reach 200 mm SL at age 1 (Alvarez-Lajonchere, 1976). We cannot verify this because the species does not occur in the Middle Atlantic Bight after its first fall. However, there is a singular report of young-of-the-year ranging from 10 to 20 cm in November in the discharge canal of the Indian River Power Plant (Wang and Kernehan, 1979).

SEASONALITY AND HABITAT USE

Larvae and pelagic juveniles in the South Atlantic Bight and Gulf of Mexico are most common during April and May (Fahay, 1975; Powles, 1981; Collins and Stender, 1989; Ditty and Shaw, 1996). Ingress into marshes in the South Atlantic Bight occurs primarily in April and May and continues through August (Anderson, 1957). Ingress at Beaufort, North Carolina, begins in late April, but peaks in May (Jacot, 1920). Individuals also first begin to occur in estuaries of the central part of the Middle Atlantic Bight in April, and small size classes continue to be found through June and into July. Results of nightlighting in Great Bay indicate they may enter

the system in pulses (Able and Fahay, 1998). A report of 41 specimens, 2 to 3 cm, collected during December from Delaware Bay (de Sylva et al., 1962) is inconsistent with other collections and is apparently in error.

Habitat use varies with life history stage. Larvae and pelagic juveniles occur in the open ocean and Gulf Stream, where they are primarily neustonic (Fahay, 1975; Powles, 1981; Hare et al., 2001). In the South Atlantic Bight, they are present over the entire shelf, from near shore to the proximity of the Gulf Stream, but the smallest size classes have been collected near the 180 m contour (Powles, 1981) or deeper (Collins and Stender, 1989). Observations from Great Bay–Little Egg Harbor indicate that young-of-the-year spend the summer in higher-salinity portions of the estuary, in contrast to *Mugil cephalus*, whose young stages are collected in lower salinities. In weir collections in high-salinity tidal creeks, *M. curema* (n = 1450) vastly outnumbered *M. cephalus* (n = 15) (Rountree and Able, 1992a). It has also been reported to be most abundant during summer and fall in higher-salinity portions of Delaware estuaries (Wang and Kernehan, 1979). Young-of-the-year occur in salinities greater than 13 ppt in Delaware Bay (de Sylva et al., 1962) and 10 to 15 ppt at Woodland Beach, Delaware (Wang and Kernehan, 1979), but none were found in low salinities or freshwater habitats in either of these studies.

PREY AND PREDATORS

Adults feed on organic detritus and small particulate matter, whereas juveniles feed mostly on planktonic organisms (Harrison, 2002). Our examination of food habits of this species (27 to 34 mm FL, n = 29) found that the stomach contents of these small young-of-the-year were dominated by sand (55.9% by weight) and detritus (12.8%), suggesting that they had already begun feeding on the bottom at these sizes. Stomachs also contained small amounts of unidentified crustaceans and zooplankton. These observations are consistent with those of others outside the study area (see Mancini and Able, 2005). Juveniles have been observed to form large schools when feeding, and depths over which they forage appear to be related to size (de Carvalho et al., 2007). The smallest size class (15 to 35 mm TL) forages in nearshore shallow water, whereas larger juveniles (40 to 100 mm TL) forage in deeper waters. These groups are known to feed on polychaetes near the surface at night, despite increased exposure to predation by fishing bats (*Noctilio leporinus*).

MIGRATIONS

Young-of-the-year leave estuaries in the central Middle Atlantic Bight in October. They also leave marshes in the Beaufort, North Carolina area in the fall (Jacot, 1920) and migrate offshore (Anderson, 1957). There were very few young-of-the-year (< 19 cm) collected in National Marine Fisheries Service (NMFS) bottom trawl surveys on the continental shelf (Fig. 82.2), presumably because they occur most fre-

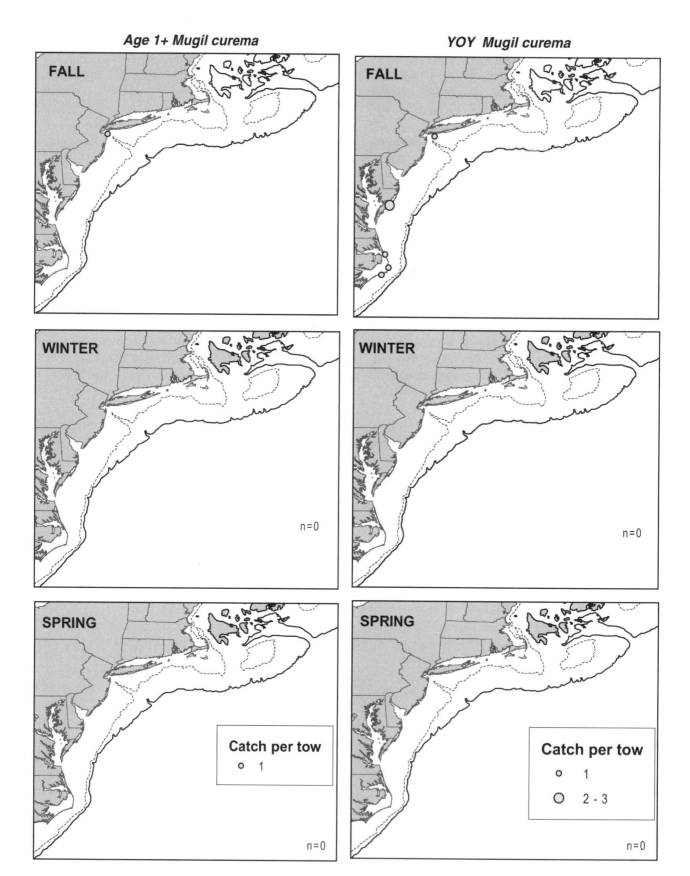

Fig. 82.2. Composite distribution of age 1+ (> 19 cm) and young-of-the-year (< 19 cm) *Mugil curema* during seasonal cruises of the National Marine Fisheries Service groundfish survey. Details of sampling effort are indicated in Fig. 3.4.

quently in estuarine and nearshore coastal waters. All young-of-the-year collected were found on the inner shelf between Long Island, New York, and Cape Hatteras (mean depth 15 m, range 9 to 25 m) in the fall. There was only one collection of an age 1+ (> 19 cm) individual from inner shelf waters of northern New Jersey in the fall. This individual was not aged by any method other than its size, and it may simply have been somewhat larger than others in its cohort.

83

Tautoga onitis (Linnaeus)

TAUTOG

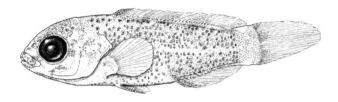

Estuary				
Ocean	Reproduction	Larvae/Development	YOY Growth	YOY Overwinter
Both				
Jan				
Feb				
Mar				
Apr				
May				
Jun				
Jul				
Aug				
Sep				
Oct				
Nov				
Dec				

DISTRIBUTION

Tautoga onitis occurs along the coast of North America from Nova Scotia to South Carolina but is most abundant from Cape Cod to Delaware Bay (Bigelow and Schroeder, 1953). Populations may be highly localized (Cooper, 1966), although seasonal inshore-offshore movements of large juveniles and adults are common (Olla et al., 1974). In the Middle Atlantic Bight, early life history stages are reported from most estuaries (see Table 4.3), but they are more abundant in the northern part of the bight.

REPRODUCTION AND DEVELOPMENT

Spawning occurs on the inner portion of the continental shelf and in estuaries in May through August. It appears to follow a northward progression through the summer, beginning as early as April in the southern part of the Middle Atlantic Bight and extending into the northern part by May. Peak spawning takes place in the central part of the bight in June and July, followed by a decline in August, based on egg occurrences (Berrien and Sibunka, 1999). This timing is consistent with estimates derived from gonadosomatic indices off Maryland and Virginia (Eklund and Targett, 1990) and the occurrence of larvae over the continental shelf (Able and Fahay, 1998). In the Mullica River–Great Bay estuary, eggs were collected from April through August, with peak abundances in June and July (Sogard et al., 1992). The initial occurrence and peak abundance were earlier in the river than in the bay and adjacent inlet, suggesting that spawning began earlier in the upper part of the estuary and continued later in the summer in the lower estuary. Back-calculation of spawning dates from sagittal otoliths of juveniles collected in Great Bay–Little Egg Harbor found a mean date of June 4 and a range of April 17–July 22 (Sogard et al., 1992). This study and those in other locations in the Middle Atlantic Bight suggest that there is more spawning in estuaries and bays than on the inner continental shelf. Fecundity and frequency of spawning events both increase with the size of females (LaPlante and Schultz, 2007). This study found that large females (500 mm) may spawn 24 to 86 times as many eggs as small individuals (250 mm). Average females in the 400 mm size range may spawn 10 to 16 million eggs during a single season. These authors predicted that production in the whole population would decrease as sizes of females decreased, but they also found that total annual production remained stable due to a shift toward a larger proportion of females.

The early development of *Tautoga onitis*, including characters useful for distinguishing the eggs, larvae, and juveniles from those of *Tautogolabrus adspersus*, has been well described and summarized (Fahay, 2007). Eggs are pelagic and spherical and measure 0.97 to 1.00 mm in diameter. Oil globules are lacking. Larvae at hatching are about 2.0 mm long and have unpigmented eyes and unformed mouth parts. Much of the anterior two-thirds of the larval body is heavily pigmented. Juveniles are diagnosed by their very small, terminal mouths and dense pigmentation covering the entire head and body, but not including the caudal peduncle.

Specimens examined for scale formation were between 6.1 and 36.4 mm TL (n = 23) (Fig. 83.1; Able et al., 2009a). The onset of scale formation occurs at approximately 13 to 17 mm TL, and continues until about 28 mm TL. Scales probably originate on the midline of the body, as for *Tautoga adspersus*. At approximately 17 mm TL, scales cover the midlateral portions of the body, and extend from the operculum to the origin of the caudal fin. At 23 to 25 mm TL, scales cover all of the body posterior to the operculum, except for

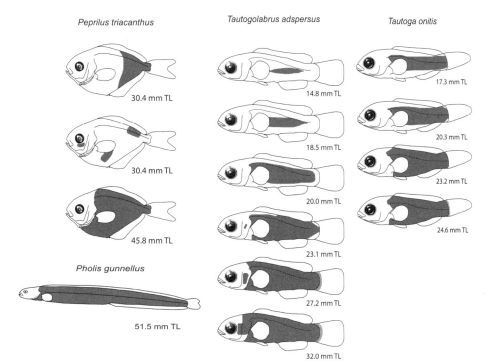

Peprilus triacanthus

30.4 mm TL

30.4 mm TL

45.8 mm TL

Pholis gunnellus

51.5 mm TL

Tautogolabrus adspersus

14.8 mm TL

18.5 mm TL

20.0 mm TL

23.1 mm TL

27.2 mm TL

32.0 mm TL

Tautoga onitis

17.3 mm TL

20.3 mm TL

23.2 mm TL

24.6 mm TL

Fig. 83.1. Description of scale formation relative to total length (TL) in *Tautoga onitis, Tautogolabrus adspersus, Pholis gunnellus,* and *Peprilus triacanthus.*

the base of the pectoral fin. At larger sizes, scales extended slightly onto the base of the caudal fin as well as onto the pectoral fin base. By 28.5 mm TL, a few scales form on the head behind the eye and coverage has reached the adult condition, where scales cover 85.1% of the body and 60.1% of the body and fins combined.

LARVAL SUPPLY, SETTLEMENT, GROWTH, AND MORTALITY

The larvae are likely from local sources, as spawning occurs on the inner shelf and in some estuaries. Some small larvae frequently are present in the surfzone and immediately near shore, but these are annually variable in abundance (Able et al., 2009c, 2010). The annual variation in abundance of early stages in our study area can be compared between larvae and early juveniles (Fig. 83.2). Ichthyoplankton sampling from 1989 to 2006 indicates that larvae in the estuary were rather consistently present between years and displayed a noticeable peak during 1992. Collections of small, post-settlement juveniles in wire mesh traps supported the presence of this abundant year class. Otter trawl samples based on larger individuals during the latter part of this period indicated peaks in abundance during 2000 and 2002.

Daily increment formation in the sagittal otoliths has been validated, and they have a well-defined settlement mark. Thus, several aspects of the early life history can be interpreted (Sogard et al., 1992). Individuals examined from Great Bay–Little Egg Harbor estuary (n = 37) spent approximately three weeks in the plankton before settlement. This is in agreement with laboratory studies in which settlement occurred 17 days after hatching (Schoedinger and Epifanio,

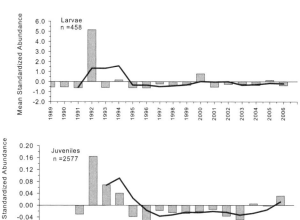

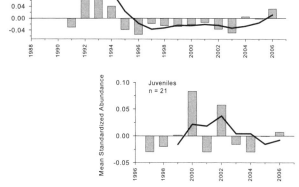

Fig. 83.2. Annual mean (bars) and three-year moving mean (line) of standardized abundance for *Tautoga onitis* collected in larval sampling with plankton net (*upper*), juvenile sampling with traps (*middle*), and juvenile sampling with otter trawls (*lower*). Data were standardized by subtracting overall mean from annual abundance. Three-year moving mean was calculated by taking the mean of the standardized values from the previous two years and the year in which data are plotted.

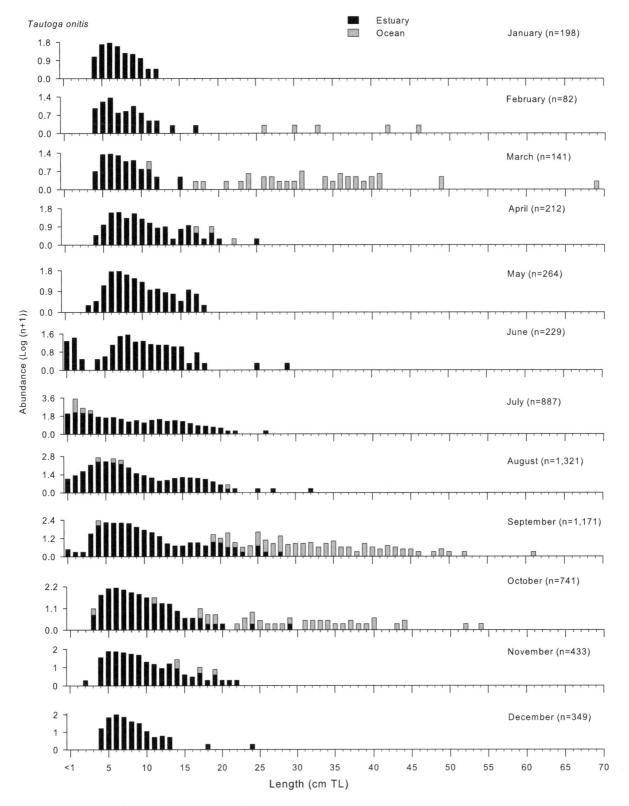

Fig. 83.3. Monthly length frequencies of *Tautoga onitis* collected in estuarine and coastal ocean waters of the central Middle Atlantic Bight. Sources: National Marine Fisheries Service: otter trawl (n = 214); Rutgers University Marine Field Station: 1 m beam trawl (n = 1); seine (n = 671); 1 m plankton net (n = 260); experimental trap (n = 170); killitrap (n = 850); 4.9 m otter trawl (n = 140); multi-panel gill net (n = 3); and Able and Fahay 1998 (n = 3719).

1997). Otoliths show that the smallest settled individuals (7.6 to 13.2 mm SL) collected in the field had been settled for 11 to 23 days. Date of settlement for these individuals ranged from May 6 to August 13, with a mean of June 25. This corresponds well with peak abundance of early demersal-stage individuals, which occurred in June through August.

Growth rates of planktonic larvae have been estimated in Buzzards Bay, Massachusetts, using otolith daily growth increment techniques (Gauthier et al., 2008). Growth rates in this study were about 0.23 mm per day and did not differ between three study sites. In Great Bay–Little Egg Harbor, the settled young-of-the-year were 1 to 3 cm TL in June, and age 1 fish were approximately 4 to 8 cm during the same month (Fig. 83.3). Most young-of-the-year attain lengths of 3 to 20 cm by October (Sogard et al., 1992). Comparison of these sizes with lengths of individuals older than age 1 the following June indicates very slow growth during the fall, winter, and spring. Age 1 fish reach a size of 11 to 17 cm TL by the end of their second summer, with a modal size in September of 15.5 cm SL. These size estimates of juveniles in New Jersey are larger than the mean lengths of individuals older than age 1 from Rhode Island (Cooper, 1967; Dorf and Powell, 1997) or from Virginia (Hostetter and Munroe, 1993). Analysis of modal length-frequency progressions and otolith ages indicates similar estimates of growth for settled tautog (0.52 mm per day and 0.47 mm per day, respectively) (Sogard et al., 1992). In caging experiments, growth rates varied from −0.47 to +0.84 mm per day, but growth in vegetated habitats averaged 0.45 mm per day. Growth was usually fastest for the smallest fish, although it was strongly influenced by location and habitat (Sogard, 1992; Phelan et al., 2000). Measurements of tagged fish were slightly lower, with rates of less than 0.3 mm per day for free-swimming fish (Able et al., 2005a). Similar caging studies in the Hudson River, based on smaller individuals, found comparable growth rates (Able et al., 1999).

Temperature has a profound effect on behavior and mortality of young-of-the-year based on laboratory observations (Hales and Able, 2001). As ambient seasonal temperatures decline, swimming frequency decreases sharply at 8°C, as does feeding at temperatures below 4°C. Burying occurs at low temperatures (2 to 7°C) and appears to be a short-term response to sudden decreases in temperature. There was relatively little mortality at the lowest temperatures (down to 2°C). However, differences in fall and spring lengths from field collections suggest size-selective winter mortality on the smallest fish (Hales and Able, 2001).

SEASONALITY AND HABITAT USE
This species is found primarily on the inner continental shelf and the polyhaline portions of estuaries, where it occupies a variety of structured habitats, from rocky reefs, pilings, jetties, boulders, rubble substrates, and mussel beds to eelgrass and the edges of deep channels, usually in depths < 30 m.

Several studies have indicated that small juveniles prefer vegetated habitats. In southern New Jersey, they have been found abundantly in beds of both sea lettuce (*Ulva lactuca*) and eelgrass (*Zostera marina*) (Nichols and Breder, 1927; Able et al., 1989; Sogard and Able, 1991) as well as in areas with shell and sponge (Szedlmayer and Able, 1996). In association with sea lettuce, the smaller juveniles (< 35 mm SL) are consistently a bright green, matching the color of the vegetation, while larger juveniles, which typically occur in unvegetated and deeper habitats, have a dark, mottled coloration similar to that of adults (Nichols and Breder, 1927; Sogard et al., 1992). Several other studies in the Middle Atlantic Bight have found higher abundances of juveniles in eelgrass habitats (Briggs and O'Connor, 1971; Orth and Heck, 1980; Heck et al., 1989; Szedlmayer and Able, 1996) and macroalgae (Dorf and Powell, 1997). More recent studies of heavily impacted and man-made habitats in the lower Hudson River estuary have found small juveniles associated with old pier pilings and inter-pier areas but not under large intact piers (Able et al., 1998). Larger juveniles are typically found in other types of structured habitats, such as rocks, jetties, and shipwrecks (Olla et al., 1974, 1979).

Embryonic and larval development occurs along the estuary-ocean gradient. In sampling along the Mullica River–Great Bay–Beach Haven Ridge corridor in southern New Jersey, the eggs are most abundant in Great Bay and the inner continental shelf and less so in the lower-salinity portions of the estuary (Sogard et al., 1992). In the ocean, larvae are most abundant off New Jersey, Long Island, New York, and Rhode Island during the summer (Able and Fahay, 1998), including in the surfzone and near shore (Able et al., 2009c, 2010). The larvae have been collected infrequently in Great Bay (n = 21; Sogard et al., 1992; Witting et al., 1999), and the same can be said of the nearby Beach Haven Ridge area on the inner continental shelf (15 m depth). Their monthly occurrences have been mostly during June and July (Fig. 83.4). Elsewhere in the Middle Atlantic Bight, larvae have been characterized as "abundant" in Buzzards Bay, Massachusetts (Gauthier et al., 2008), and have also been reported from other large bays (Herman, 1963; Croker, 1965; Bourne and Govoni, 1988) and barrier island estuaries (Allen et al., 1978; Monteleone, 1992). Larvae first occur at nearshore locations in late June in the Buzzards Bay study. As the summer progressed, larvae were found farther away from shore and their presence near shore decreased.

The infrequent occurrence of early demersal-stage individuals in collections showed little evidence of successful settlement at Beach Haven Ridge on the inner continental shelf, but the juveniles were often collected in the lower, polyhaline portion of the estuary (Fig. 83.5), with highest numbers in eelgrass (Able and Fahay, 1998). In Narragansett Bay the peak in young-of-the-year abundance occurred from July to August (Dorf and Powell, 1997). Small juveniles are present throughout the year in the Mullica River–Great Bay

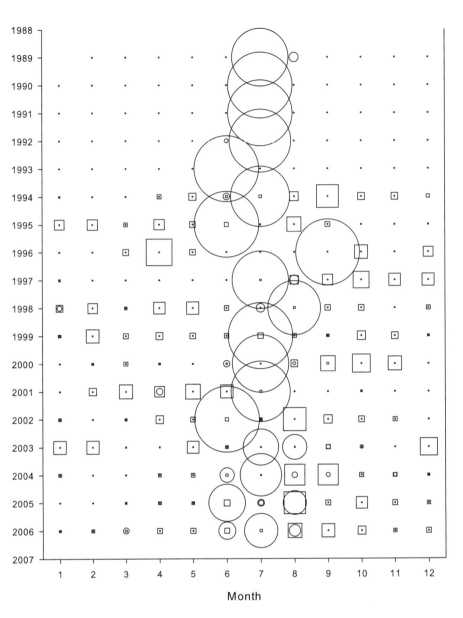

Fig. 83.4. Seasonal occurrence of larvae based on ichthyoplankton samples collected in Little Sheepshead Creek behind Little Egg Inlet from 1989 to 2006 (open circles represent the percentage of the mean number / 1000 m³ of larval young-of-the-year captured by year) and juveniles of *Tautoga onitis* based on trap collections in Rutgers University Marine Field Station boat basin in Great Bay (open squares represent the percentage of the mean catch per unit effort captured by year) from 1994 to 2006. Values range from 0 to 100% and 0 to 47% for ichthyoplankton and trap sampling, respectively. The smallest circles and squares represent when samples were taken but no individuals were collected.

Month

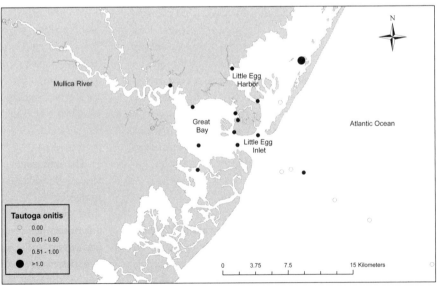

Fig. 83.5. Habitat use and average catch per tow (see key) for juvenile and adult *Tautoga onitis* in the Ocean–Great Bay–Mullica River corridor in southern New Jersey based on small otter trawl collections during July and September from 1988 to 2006. See Table 3.2 for characteristics at each sampling collection.

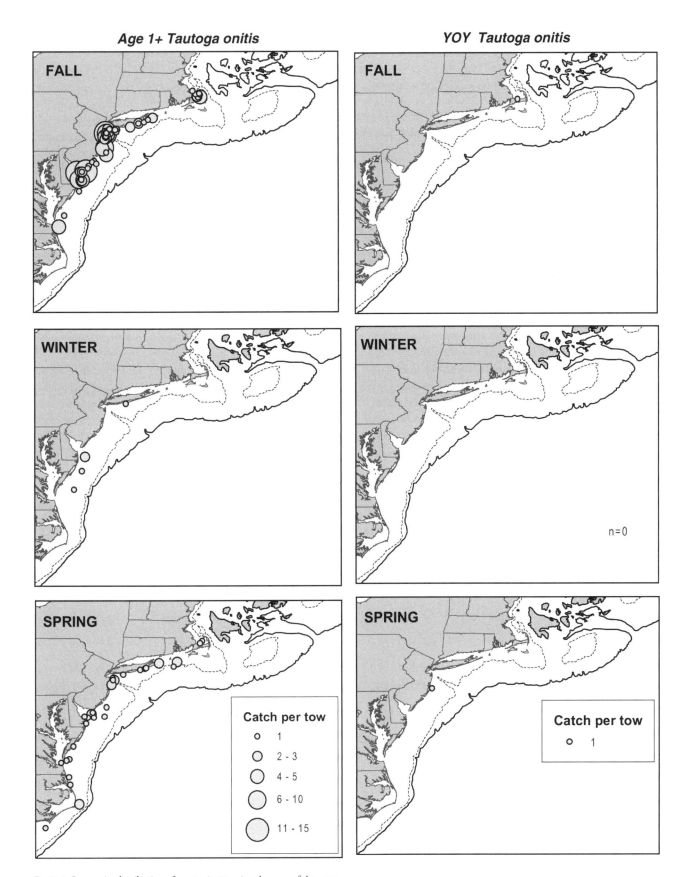

Age 1+ *Tautoga onitis*

FALL

WINTER

SPRING

YOY *Tautoga onitis*

FALL

WINTER

n = 0

SPRING

Catch per tow
- ○ 1
- ○ 2 - 3
- ○ 4 - 5
- ○ 6 - 10
- ○ 11 - 15

Catch per tow
- ○ 1

Fig. 83.6. Composite distribution of age 1+ (> 12 cm) and young-of-the-year (< 12 cm) *Tautoga onitis* during seasonal cruises of the National Marine Fisheries Service groundfish survey. Details of sampling effort are indicated in Fig. 3.4.

estuary, as measured by trap collections (see Fig. 83.4), although in most years they are most commonly collected in the fall.

Based on tag/recapture studies of juveniles (25 to 190 mm TL) in Great Bay, there is strong site fidelity after settlement and during the first summer (Able et al., 2005a). Of 1148 individuals tagged at a single location, there were 278 recaptures (14%), with some individuals caught more than once (up to 13 times) over a 9-month period. Average distance moved was 19 m during all seasons. Of 48 individuals captured 2 or more times, 96% were recaptured within 5 m of a previous capture location. Activity patterns indicate that this species is almost exclusively diurnal in habits, becoming quiescent at night (Arendt et al., 2001), although some increased night-time activity has been documented during the spawning season. This study also confirmed strong site fidelity, with 22 of 27 test animals remaining near monitored sites.

PREY AND PREDATORS

Adults feed opportunistically by sight during daylight (Olla et al., 1974). They consume a variety of invertebrates, including mollusks (bivalves and gastropods), barnacles, and crustaceans, such as amphipods, isopods, and decapods. They also consume echinoderms and occasionally small fishes (Lindquist et al., 1985). In some areas, mussels (*Mytilus edulis*) are the principal prey item (Olla et al., 1975). Two studies have examined the food habits of young-of-the-year (Grover, 1982; Dorf, 1994). Young-of-the-year consume primarily copepods (45.7% composition) and amphipods, but also include isopods and small decapod crustaceans in their diets (see Table 8.1). At sizes greater than about 120 mm, these fish gradually change their diets to include more small mussels.

Predators on eggs of this species include copepods (*Acartia tonsa*) (Perry, 1994). Juveniles and adults are consumed by a variety of larger fishes, including *Mustelus canis, Raja laevis, Urophycis chuss, Hemitripterus americanus,* and *Lophius americanus* (Bigelow and Schroeder, 1953). *Morone saxatilis* has also been implicated as an important predator of *Tautoga onitis* (Schaefer, 1970).

MIGRATIONS

Young-of-the-year (< 12 cm) are rarely collected in trawl surveys on the continental shelf. The age 1+ (> 12 cm) juveniles and adults are more frequently collected, with most collections from Massachusetts to North Carolina associated with the mouths of large embayments (Fig. 83.6). During all seasons these stages occur on the inner shelf in relatively shallow water (fall: mean depth 18 m, range 8 to 35 m; winter: mean depth 26 m, range 21 to 28 m; and spring: mean depth 21 m, range 6 to 60 m). The slightly deeper average occurrences in winter may reflect a minor inshore-offshore seasonal migration. Studies in the Chesapeake Bay have demonstrated that most (94%) individuals remain in estuarine habitats throughout the winter, residing in temperatures of 5 to 8°C (Arendt et al., 2001). This behavior may only apply to fish in the southern part of the Middle Atlantic Bight, because migrations offshore during winter have been documented in New York, Rhode Island, and Massachusetts studies.

84

Tautogolabrus adspersus (Walbaum)

CUNNER

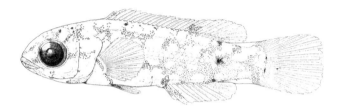

Estuary	Reproduction	Larvae/ Development	YOY Growth	YOY Overwinter
Ocean				
Both				
Jan				▓
Feb				▓
Mar				▓
Apr				▓
May	▓		▓	
Jun	▓		▓	
Jul	▓		▓	
Aug	▓		▓	
Sep	▓		▓	
Oct	▓		▓	
Nov				▓
Dec				▓

DISTRIBUTION

Tautogolabrus adspersus is found along the east coast of North America from Newfoundland and the Gulf of St. Lawrence to the Chesapeake Bay area (Robins and Ray, 1986). In our study area, it occurs more commonly northeast of the New York Bight apex. Thus, it is more common in Narragansett Bay than in the Hudson–Raritan estuary and more prevalent in the latter system than in Great Bay–Little Egg Harbor (Berg and Levinton, 1985). It is apparently rare to uncommon

in the Delaware Bay area, for only two specimens (adults) were captured in the surveys reported by de Sylva et al. (1962). Early life history stages have been reported from estuaries between Delaware Bay and Cape Cod (see Table 4.3), although all stages are rare in Delaware Bay (Stone et al., 1994). The lack of egg occurrences in Nauset Marsh is most likely due to inappropriate sampling (M. P. Fahay and K. W. Able, unpubl. observ.).

REPRODUCTION AND DEVELOPMENT

The reproductive season is relatively short for this species at any given location. In Newfoundland, they spawn in July and August, with most activity in a two- to four-week period in July (Pottle and Green, 1979). They also have short spawning seasons during early July in the Gulf of Maine (Levin et al., 1997) and in June in Connecticut (Dew, 1976). Spawning occurs from April to October in different inner continental shelf waters. It begins in spring in northern areas such as Massachusetts Bay (Collette and Hartel, 1988). Eggs were common constituents of MARMAP samples over the inner continental shelf from May to November (1977–1987), with a peak in June and July (Berrien and Sibunka, 1999). One center of abundance was between the New York Bight apex and Narragansett Bay, with a separate center over Georges Bank. There may also be some spawning in estuarine waters near Little Egg and Hereford inlets, New Jersey, however, as there have been reports of numerous eggs and larvae in both places during the early summer (Milstein and Thomas, 1977; Allen et al., 1978, respectively). This species becomes sexually mature at about 5.4 to 6.0 cm SL, a size they usually reach at age 1+ in the Middle Atlantic Bight (Dew, 1976). Fecundity is more closely related to length than it is to age or weight (Nitschke et al., 2001) but it has been estimated at 100,000 eggs for a 500 g female (Williams et al., 1973).

The early development of *Tautogolabrus adspersus*, including characters useful for distinguishing the eggs, larvae, and juveniles from those of *Tautoga onitis*, has been well described and summarized (Fahay, 2007). Eggs are pelagic and spherical and measure 0.84 to 0.92 mm in diameter. Oil globules are lacking. Larvae are about 2.2 mm at hatching and have unpigmented eyes and undeveloped mouth parts. Pigment is light and scattered, but soon aggregates into two corresponding blotches on the dorsal and ventral edges of the body in the mid-tail region. A prominent spot also forms on the nape. All fin rays (except those in the pelvic fin) are fully formed by 8.0 mm SL. Juveniles have a characteristic blotch of pigment on the anterior dorsal fin rays.

Specimens examined for scale formation were between 8.3 and 47.5 mm TL (n = 27) (see Fig. 83.1; Able et al., 2009a). Scales had not yet formed in individuals smaller than 13.8 mm TL. Scale formation occurs in individuals between 14.8 and 32 mm TL (see Fig. 83.1). Scales originate in two loci: on the midline of the body and on the operculum. Onset of scale formation occurs at approximately 15 mm TL along the

mid-lateral surface of the body, but is not associated with lateral line development, as observed in many other species. By 18.5 mm TL, rows of scales are observed along an oblique line between the operculum and caudal peduncle. Between 20 and 23 mm TL, scales spread laterally, anteriorly, and posteriorly, but do not occur on the dorsal or ventral edges of the body or on the posterior caudal peduncle. By 23 mm TL, scales begin to form on the operculum. By 27 mm TL, scales extend over the entire surface of the body and on to the base of the caudal fin. Completion of scale formation is achieved by 32.0 mm TL, with scales on the head behind the eye, but no scales on large portions of the head and most of the fins. At this size, the adult complement is well formed, whereby scales covered 84.8% of the body and 52.6% of the body and fins.

LARVAL SUPPLY, SETTLEMENT, GROWTH, AND MORTALITY

The abundance of larvae originating from spawning on the inner continental shelf was relatively stable in plankton net collections at Little Egg Inlet from 1989 to 2006 (Fig. 84.1), although an increase was noted from 2000 to 2002. During an earlier time period (1977–1987), the relative abundance of eggs off the New Jersey coast indicates that there is wide inter-annual variability in spawning activity in the Middle Atlantic Bight (Berrien and Sibunka, 1999) and this should be expected to affect the recruitment of this species in the Great Bay–Little Egg Inlet system.

As is the case with other labrids, *Tautogolabrus adspersus* settles at a relatively wide range of sizes (8 to 14 mm) (Tupper and Boutilier, 1995a). Growth during the first year is slow. The ingress of larvae and early benthic-stage juveniles is clearly depicted in collections near Little Egg Inlet during June and July (Fig. 84.2). The slow rate of growth (about 0.3 mm per day) is apparent, and the mode at the end of summer is only about 5 cm TL. Published growth rates from the Scotian Shelf and Gulf of Maine have also been estimated at about 0.5 mm per day (Bigelow and Schroeder,

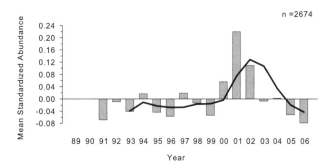

Fig. 84.1. Annual variation in abundance of *Tautogolabrus adspersus* collected in trap sampling, 1991–2006. Data were standardized by subtracting the overall mean from annual abundance. Vertical bars represent standardized annual abundance; the solid line is a three-year running mean that was calculated by taking the mean of the standardized values from the previous two years and the year in which the data are plotted.

1953; Tupper, 1994; Tupper and Boutilier, 1995a). In the region of Hereford Inlet, New Jersey, young (20 to 28 mm) first occur in seine hauls in August and only reach 40 to 60 mm before disappearing in early winter (Allen et al., 1978). There is little or no growth through the winter, and age 1 fish are apparently only 3 to 12 cm TL, with a mode between 6 and 8 cm TL. The observed length frequencies the following spring may reflect mortality during the winter, because it has been shown that this mortality affects smaller size classes more than larger ones (Hales and Able, 2001). During the summer, juveniles utilize a variety of habitats but suffer extreme post-settlement mortality in less structurally complex habitats (Levin, 1991, 1993; Tupper and Boutilier, 1995a).

SEASONALITY AND HABITAT USE

This is a marine species, only rarely penetrating low-salinity areas (Smith, 1985). It is apparently more abundant around structures such as ship wrecks, pilings, rocky reefs, oyster beds, and wharves. In all estuaries where individuals of this species occur, they are less common during winter, when they either move offshore as temperatures decline or simply become inactive, burrowing into the substrate (Smith, 1985). In Cape Cod and Gulf of Maine areas, juveniles enter metabolic torpor at temperatures below 5°C and then usually aestivate under rocks near their settlement sites, at least during their first winter (Dew, 1976; Curran, 1992; Tupper, 1994). Sampling at an inner continental shelf site in the New York Bight apex (Wilk et al., 1992) yielded large numbers consisting of fish age 1 and older at sizes between 10 and 20 cm. This size class was present nearly year-round, although they were rare or absent during June and July and were most abundant during the winter and early spring, suggesting that they occupy these shelf habitats during the coldest time of the year and near-coastal and estuarine habitats during summer.

Larvae are more abundant in July and August over the continental shelf northeast of the drowned Hudson River valley than they are in the southern part of the Middle Atlantic Bight (Morse et al., 1987; Malchoff, 1993). Larvae occur continuously over Georges Bank from July through September. In plankton collections made at the Beach Haven Ridge near Little Egg Inlet, larvae were first collected during July, which is consistent with collections made during the MARMAP surveys between 1977 and 1987. The larvae were also collected in surfzone and nearshore habitats off northern New Jersey, but their abundance varied from year to year (Able et al., 2009c, 2010). They were only rarely collected, however, during another study in this area (Milstein and Thomas, 1977), and then only in July, perhaps another indication of the inter-annual variability in spawning activity on the continental shelf off New Jersey.

In Great Bay–Little Egg Harbor, larvae usually appear in plankton collections during June or July at sizes between 5.2 and 15.6 mm SL (Witting, et al., 1999) (see Fig. 84.3). Collections of juveniles in traps can occur year round, but the small

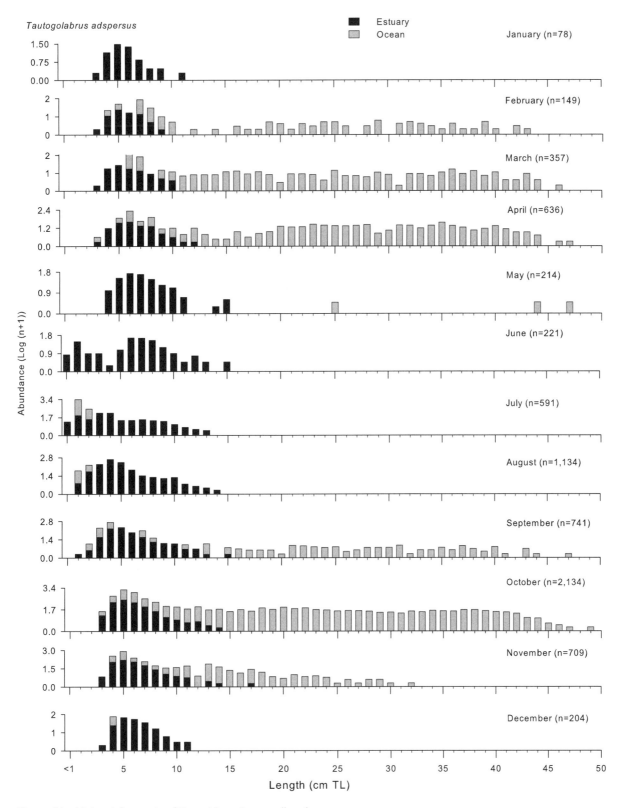

Fig. 84.2. Monthly length frequencies of *Tautogolabrus adspersus* collected
in estuarine and coastal ocean waters of the central Middle Atlantic Bight.
Sources: National Marine Fisheries Service: otter trawl (n = 2440); Rutgers
University Marine Field Station: seine (n = 34); Methot trawl (n = 4); 1 m
plankton net (n = 97); 2 m beam trawl (n = 1); experimental trap (n = 920);
killitrap (n = 1949); 4.9 m otter trawl (n = 27); multi-panel gill net (n = 1);
and Able and Fahay 1998 (n = 1695).

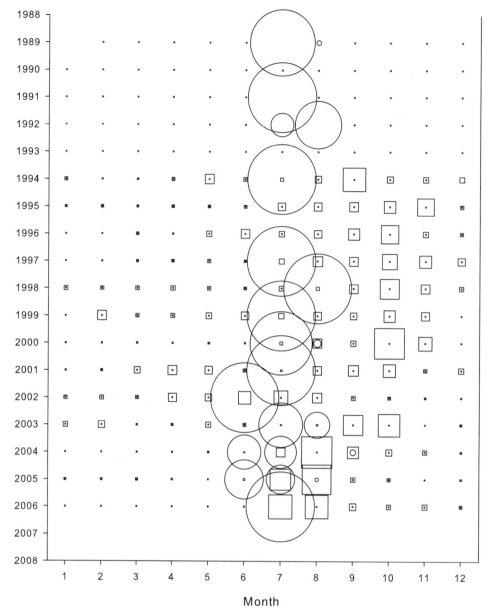

Fig. 84.3. Seasonal occurrence of larvae based on ichthyoplankton samples collected in Little Sheepshead Creek behind Little Egg Inlet from 1989 to 2006 (open circles represent the percentage of the mean number/1000 m³ of larval young-of-the-year captured by year) and juveniles of *Tautogolabrus adspersus* based on trap collections in Rutgers University Marine Field Station boat basin in Great Bay (open squares represent the percentage of the mean catch per unit effort captured by year) from 1994 to 2006. Values range from 0 to 100% and 0 to 53% for ichthyoplankton and trap sampling, respectively. The smallest circles and squares represent when samples were taken but no individuals were collected.

young-of-the-year begin to be abundant in July and August and continue through October (Fig. 84.3), although this pattern varies between years. In other study areas in the Middle Atlantic Bight, a large influx of larvae (all < 10 mm) appeared in Long Island waters (mostly along the ocean shore) during June (Perlmutter, 1939). Small larvae then decreased in abundance, whereas juveniles up to 34 mm continued to occur in July. In the few weeks following settlement, young-of-the-year are typically associated with rocky bottom, pilings, and debris, or seagrass or macroalgae beds. In localities where it is locally abundant, it may also settle in very high densities (> 6 individuals/m²) and is known to exhibit density-dependent growth and survival (Tupper and Boutilier, 1995a).

Mark/recapture studies indicate a high degree of habitat fidelity in this species during the summer (Tupper and Boutil-

ier, 1995a; Able et al., 2005c). In the Hudson River, all age classes (including young-of-the-year) are more abundant in pile field habitats than in inter-pier or under-pier habitats (Able and Duffy-Anderson, 2006). They have also been reported to take shelter in eelgrass beds during winter (Nichols and Breder, 1927). They are only rarely collected in Sandy Hook Bay and then only in October and November (Berg and Levinton, 1985). Traps are effective collecting devices, presumably because the young of this species occur in structured habitats. Some transforming juveniles may settle on the continental shelf and remain uncollected. Divers have observed large numbers of young-of-the-year around submerged wrecks on the continental shelf off New Jersey (D. Witting and K. W. Able, pers. observ.), and these sites may also provide critical habitat during the first year.

Age 1+ *Tautogolabrus adspersus*

YOY *Tautogolabrus adspersus*

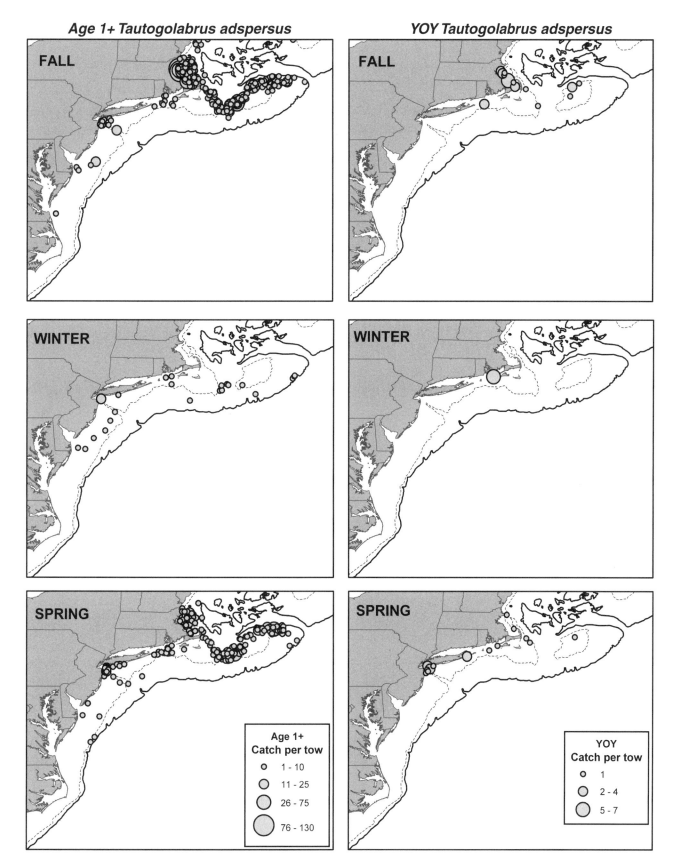

Fig. 84.4. Composite distribution of age 1+ (> 8 cm) and young-of-the-year (< 8 cm) *Tautogolabrus adspersus* during seasonal cruises of the National Marine Fisheries Service groundfish survey. Details of sampling effort are indicated in Fig. 3.4.

PREY AND PREDATORS

Adults are omnivorous, opportunistic predators that consume a wide variety of organisms, but mainly mollusks and crustaceans (Bowman et al., 2000). They use sight to find prey and feed continuously during daylight. Among the many studies of diet and feeding behavior in this species, two have described the effects of season on shifts in diet. Along the coast of Long Island, they feed mostly on *Mytilus edulis* from May to June, but then shift to an isopod, *Idotea baltica*, from July to September (Olla et al., 1975). In Narragansett Bay, these seasonal shifts in diet were found to be related to the relative stability of prey densities (Sand, 1982). At sites where relative prey densities remain stable through the season, diets also remain stable.

One study examined the food habits of young-of-the-year (11 to 64 mm, n = 54) in Great South Bay, New York (Grover, 1982). Young-of-the-year consumed predominantly copepods and amphipods (see Table 8.1). Fishes were a minor dietary component. Dependence on copepods decreases with growth, such that copepods comprise almost the entire diet (97.2%) at 15.5 mm, whereas at 45.9 mm copepods comprise only 32.1% of the diet.

MIGRATIONS

Judging from trap collections, *Tautogolabrus adspersus* occurs in the Great Bay–Little Egg Harbor study area throughout the year, but a decline during December and January (see Fig. 84.3) suggests that many either become inactive or migrate away from estuaries during the winter. Several marked fish were recaptured at intervals through the winter in a Great Bay study (Able et al., 2005c). Age 1+ fish (> 8 cm) are frequently collected on the inner continental shelf in depths from 9 to 150 m (mean depth 49.1 m) from the Gulf of Maine south to the mouth of Delaware Bay (Fig. 84.4). They are most abundant off Massachusetts to the Great South Channel and along the northern edge of Georges Bank. They are far less commonly found during the winter in any habitat sampled. Young-of-the-year < 8 cm are infrequently collected but when they occur they are found in shallow waters (mean depth 31.2 m, range 9 to 54 m) on the inner continental shelf from north of Cape Cod south to the New York–New Jersey line and on the central portion of Georges Bank.

85

Pholis gunnellus (Linnaeus)

ROCK GUNNEL

DISTRIBUTION

This species is found on both sides of the Atlantic Ocean. In the western Atlantic, it is distributed from Labrador to Delaware Bay but is rare south of southern New England (Collette, 2002). The distribution of early life history stages in Middle Atlantic Bight estuaries is usually not reported (see Table 4.3).

REPRODUCTION AND DEVELOPMENT

Spawning presumably takes place during winter, after a migration away from intertidal habitats. Within our study area, egg masses have been found in Long Island Sound and Peconic bays (Gudger, 1927). Nest sites are guarded by both parents and have been described from shallow waters where eggs have been found deposited in empty oyster shells (Nichols and Breder, 1927) to sites as deep as 73 m (Collette, 2002). Details of early development of eggs, larvae, and juveniles are summarized in Fahay (2007). All early stages are elongate and lightly pigmented until transformation and settlement, when an elaborate pigment pattern forms. Only a few specimens (26.9 to 51.5 mm TL, n = 4) were available to examine for scale formation (see Fig. 83.1). Individuals of 26.9 to 29.0 mm TL lacked scales, and a 51.5 mm TL specimen was the only individual with scales. In this specimen, the lateral, ventral, and dorsal surfaces were fully scaled from the base of the caudal fin to the operculum. At this size scales covered 90.3% of the body; this is the adult condition.

LARVAL SUPPLY, SETTLEMENT, AND GROWTH

Larvae are common constituents of the ichthyoplankton communities over the continental shelf and in certain large New England embayments, such as Beverly–Salem Harbor, Massachusetts (Elliott et al., 1979). Larvae collected during the MARMAP surveys (1977–1987) occurred throughout the Gulf of Maine, Georges Bank, and the northern part of the Middle Atlantic Bight (Able and Fahay, 1998). They have also been collected during January, March, and April in Newark Bay, New Jersey (S. J. Wilk, pers. comm.). Pelagic larvae at sea grow from a modal length of 12 mm NL in February to a modal length of about 26 to 36 mm NL in April (Fig. 85.1), at which size and time they settle to the bottom (Bigelow and Schroeder, 1953; Able and Fahay, 1998). Larvae collected concurrently in estuarine sampling during spring were consistently larger than their oceanic counterparts during February, March, and April. Settlement in this species occurs commonly on the continental shelf and more rarely in estuaries.

Juveniles collected in Long Island Sound in July and early August measured 46 to 103 mm and were presumed to be young-of-the-year (Perlmutter, 1939). Another study concluded that size at age 1 was 68 mm in the Bay of Fundy and 73 mm in New Hampshire (Sawyer, 1967).

SEASONALITY AND HABITAT USE

This species frequently occurs in tide pools and other intertidal habitats, where it is usually associated with structures such as stones, algae, and crevices. It has also been collected at considerable distances from shore, as at 183 m depth on Georges Bank (Schroeder, 1933; Bigelow and Schroeder, 1953). It inhabits the north shore of Long Island but avoids the generally warmer bays along the south shore (Perlmutter, 1939). It has also occurred in isolated inshore habitats such as Newark Bay (S. J. Wilk, pers. comm.). Larvae are common constituents of the ichthyoplankton communities over the continental shelf and in certain large New England embayments, as previously noted. Post-settlement young-of-the-year have also been collected in Long Island Sound in July and early August (Perlmutter, 1939). Because of their cryptic habits, the difficulty in sampling for this species prevents an appraisal of habitats used by young-of-the-year in bays and estuaries in the central part of the Middle Atlantic Bight.

PREY AND PREDATORS

Major items in this species' diet include amphipods, isopods, and polychaetes (Collette and Klein-MacPhee, 2002). Of minor importance are other crustaceans, gastropods, bivalves, echinoids, insects, and algae. There have been no studies of the food habits of young-of-the-year (< 73 mm) in Middle Atlantic Bight estuaries or elsewhere. Predators include a wide variety of fishes, seabirds, and seals (Collette, 2002). This species (presumably including young-of-the-year) are very important prey items in the diet of nesting double-crested cormorants in Maine (Blackwell et al., 1995).

MIGRATIONS

Adults apparently leave some intertidal areas and move offshore in the fall. This offshore migration is apparently

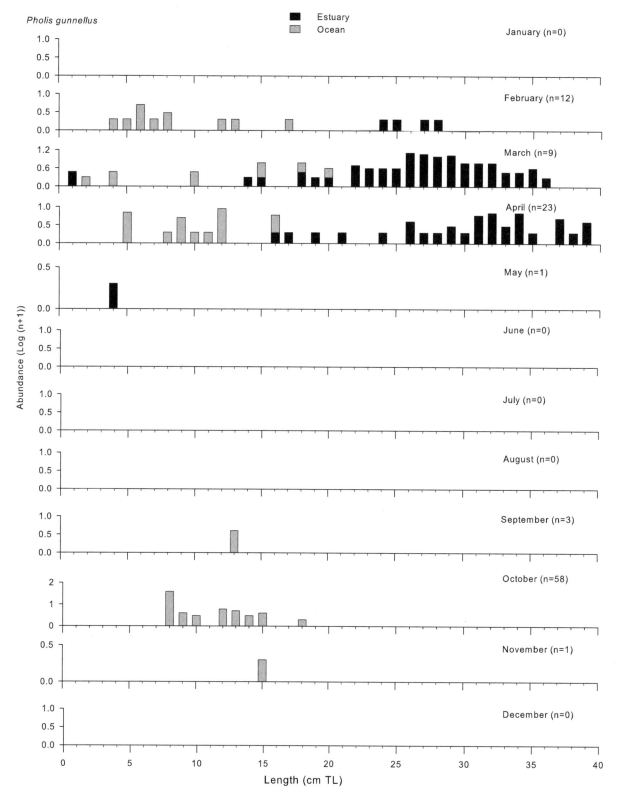

Fig. 85.1. Monthly length frequencies of *Pholis gunnellus* collected in estuarine and coastal ocean waters of the central Middle Atlantic Bight. Sources: National Marine Fisheries Service: otter trawl (n = 106); Rutgers University Marine Field Station: 1 m plankton net (n = 2); experimental trap (n = 1); and Able and Fahay 1998 (n = 130).

Age 1+ Pholis gunnellus   **YOY Pholis gunnellus**

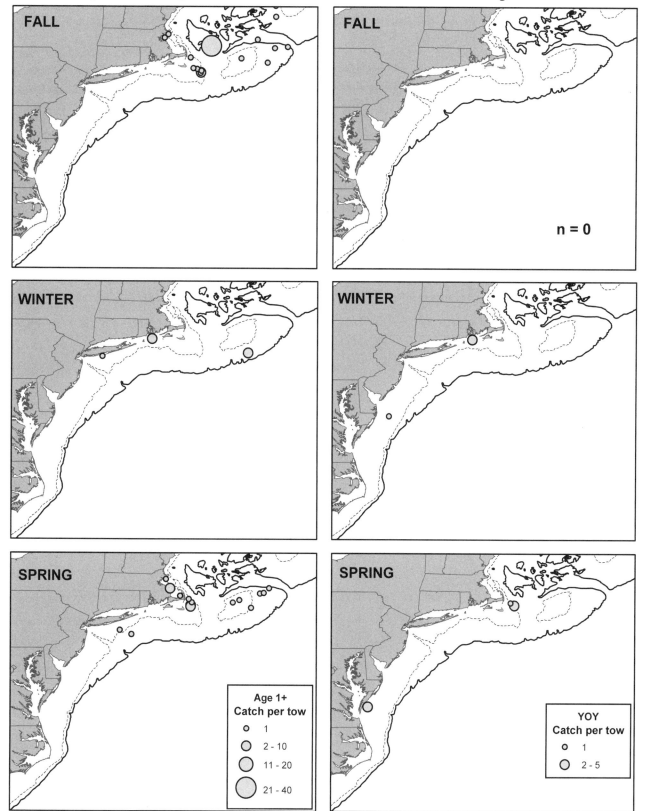

Fig. 85.2. Composite distribution of age 1+ (> 7 cm) and young-of-the-year
(< 7 cm) *Pholis gunnellus* during seasonal cruises of the National Marine
Fisheries Service groundfish survey. Details of sampling effort are indicated
in Fig. 3.4.

associated with reproduction. This conclusion is based on observations in New Hampshire where: (1) females in November contained eggs about 1.0 mm in diameter; (2) they disappeared during the winter; and (3) those returning to inshore waters in March were in a spent condition (Sawyer, 1967). In the National Marine Fisheries Service (NMFS) bottom trawl survey, young-of-the-year are only rarely collected during winter and spring, in isolated locations from southern New England to the southern Middle Atlantic Bight (Fig. 85.2). Since the range of this species is only as far south as Delaware Bay, some of these isolated occurrences may represent individuals that will not survive, possibly because of warmer temperatures. In the same trawl surveys, older fish occur from fall through spring, and occurrences are limited to Gulf of Maine, Georges Bank, and southern New England.

86

Ammodytes americanus DeKay

AMERICAN SAND LANCE

Estuary	Reproduction	Larvae/ Development	YOY Growth	YOY Overwinter
Ocean				
Both				
Jan				
Feb				
Mar				
Apr				
May				
Jun				
Jul				
Aug				
Sep				
Oct				
Nov				
Dec				

DISTRIBUTION

Ammodytes americanus occurs along the Atlantic coast of North America from Newfoundland and northern Labrador to Chesapeake Bay (Nizinski et al., 1990). Its congener, *A. dubius*, has a similarly wide distribution. Many aspects of the morphology and ecology of *Ammodytes* spp. along the East Coast of the United States are potentially confounded by the taxonomic problems in differentiating between *A. americanus* and *A. dubius* (Nizinski et al., 1990). We believe that while most estuarine collections of *Ammodytes* are *A. americanus*, special care should be taken with identification, especially

with collections from the inner continental shelf. Early life history stages have been reported from almost all estuaries in the Middle Atlantic Bight (see Table 4.3), except the southernmost estuaries such as Chesapeake Bay (larvae only) and Pamlico-Albemarle sounds (no reports).

REPRODUCTION AND DEVELOPMENT

Spawning occurs in the winter and spring based on timing of the maturation of field-collected individuals in the laboratory (Smigielski et al., 1984) and the presence of small larvae in the vicinity of Great Bay–Little Egg Harbor during March, April, and May. Hatching of yolk-sac larvae may occur one to two months after spawning at low winter temperatures (see below). Precise locations of spawning are unknown.

Many details of early development are summarized in Fahay (2007). The adhesive eggs are irregular in shape. In laboratory studies time to hatching varied in duration with temperature and ranged from 61 days at 2°C to 25 days at 10°C. Yolk sac absorption by the larvae is complete at 6.3 to 6.8 mm. Transformation to juvenile morphology occurred at 29 mm and 131 days after hatching at 4°C and 102 days at 7°C. These same laboratory-raised individuals begin schooling at 25 to 30 mm at 90 days after hatching. They begin burying in a sand substrate at 35 to 40 mm at 133 days after hatching (Smigielski et al., 1984).

LARVAL SUPPLY, SETTLEMENT, AND GROWTH

The abundance of this species in plankton collections at Little Egg Inlet varied relatively little across the period from 1989 to 2006 (Fig. 86.1). The exceptions were higher than average abundances in 1994 and 1995 and 2001. In the Great Bay–Little Egg Harbor estuary, larvae are a consistent, annual component of the ichthyoplankton in early spring, presumably from local spawning (Witting et al., 1999). They first appear in coastal ocean and estuarine waters in January (Fig. 86.2). Subsequently, small larvae have been collected in the estuary from February to April. In some years the peak in larval abundance may be delayed until May.

Most of the growth exhibited by young-of-the-year occurs between April and August based on field collections of a single cohort in the estuary (see Fig. 86.2). Modes increase in these individuals from 2 cm in April to 10 cm in August. This cohort appears to be represented by larger individuals (8 to 15 cm TL) from October through February, demonstrating very little, or no, growth during the winter. We also interpret the size at age 1 as 10 to 17 cm TL. Growth in the laboratory varies with temperature (2 to 10°C) and ranges from approximately 0.4 to 1.0 mm per day (Smigielski et al., 1984).

SEASONALITY AND HABITAT USE

The timing of peak larval abundance, as measured by plankton net sampling in our study site, is remarkably consistent from year to year, with most occurrences in April (see Fig.

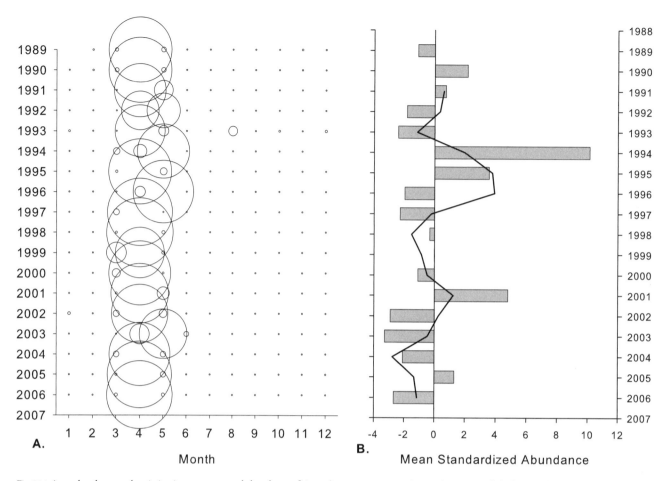

Fig. 86.1. Annual and seasonal variation in occurrence and abundance of *Ammodytes americanus*. (A) Seasonal variation of the larvae is based on ichthyoplankton samples collected in Little Sheepshead Creek behind Little Egg Inlet from 1989 to 2006. Open circles represent the percentage of the mean number/1000 m³ captured within each year. Values range from 0 to 96%. The smallest circles represent when samples were taken but no larvae were collected. (B) Annual variation in abundance for the same data was standardized by subtracting overall mean from annual abundance. Horizontal bars represent standardized annual abundance; the solid line is a three-year running mean that was calculated by taking the mean of the standardized values from the previous two years and the year in which the data are plotted.

86.1). After the larvae reach a peak in abundance in plankton samples in late spring, the young-of-the-year become available to seines towed in shallow water, especially in May and June (Able and Fahay, 1998), including on ocean beaches (K. W. Able, unpubl. data). The young-of-the-year become less available to our collecting gears in the estuary during summer but can become more abundant in the fall. The juveniles and adults are found in very shallow water of the inner continental shelf, bays, and estuaries (Nizinski et al., 1990). They typically occur over sandy substrates and are capable of burrowing into this substrate. This species is pelagic much of the time but can dive into sandy substrates, where it can reach densities of 8 individuals per m² based on routine collections (Heck et al., 1995).

PREY AND PREDATORS

Several studies (n = 4) have examined the food habits of young-of-the-year (8 to 159 mm, n = 740) in Long Island

Sound and Great South Bay, New York (see Table 8.1). Important prey of young-of-the-year include a variety of invertebrates (copepods, crabs, cirripedes, bivalves, gastropods, cnidarians, mysid shrimp, decapod shrimp, and cladocerans) and phytoplankton. Fishes are a minor dietary component, although some young-of-the-year have exhibited cannibalistic behavior (Richards, 1963). In Long Island Sound, the diet of larvae (< 8 to 24 mm, n = 175) shifts from phytoplankton to various stages of copepods with growth. For example, small larvae (< 8 mm) contain only phytoplankton, while larger larvae (8 to 11.9 mm) eat copepod nauplii. At larger sizes (12 to 18.9 mm), the larvae feed on both nauplii and copepodites, and at still larger sizes (19 to 23.9 mm) they consume greater amounts of copepodites and *Acartia hudsonica* adults than nauplii. The largest larvae examined feed on *A. hudsonica* adults almost exclusively. Relative importance of individual prey items also varies intra- and inter-annually in Long Island Sound (McKeown, 1984), such

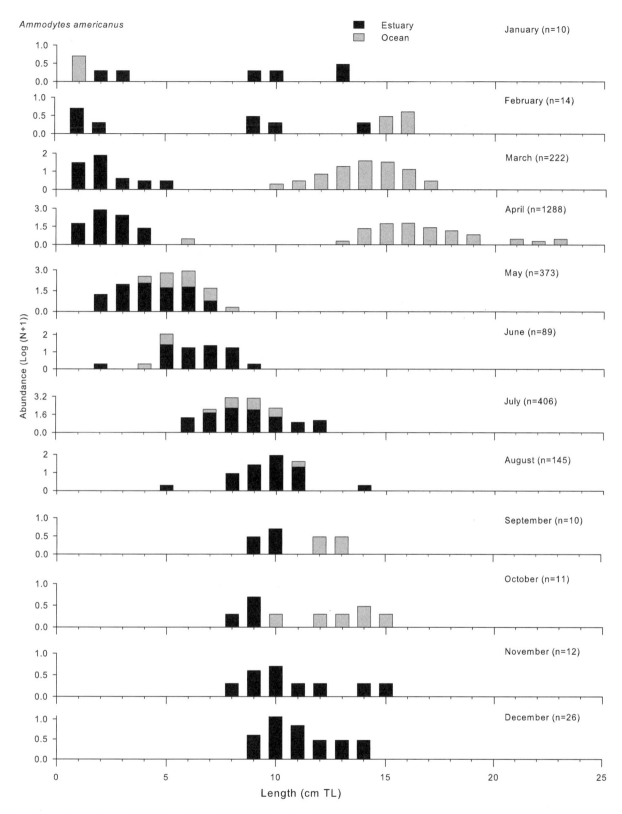

Fig. 86.2. Monthly length frequencies of *Ammodytes americanus* collected in
estuarine and coastal ocean waters of the central Middle Atlantic Bight.
Sources: National Marine Fisheries Service: otter trawl (n = 308); Rutgers
University Marine Field Station: seine (n = 62); 2 m beam trawl (n = 1); and
Able and Fahay 1998 (n = 2238).

that young-of-the-year examined from June 1982 contained large quantities of crab zoea, while those from June 1983 contained mostly cladocerans and snail veligers. *Acartia* spp. copepodites and adults comprised 60 and 10%, respectively, of the diet of young-of-the-year examined from July 1983, as well as small percentages of *Temora* adults and copepodites, harpacticoid copepods, and crab zoea. Adult *Labidocera* and adult *Acartia* spp. made up more than 30 and 20%, respectively, of the diet of individuals collected in December 1983.

MIGRATIONS

National Marine Fisheries Service (NMFS) trawl surveys on the continental shelf only rarely collect *Ammodytes americanus*. Isolated collections of young-of-the-year and older year classes have occurred in the area near Nantucket Shoals and at the mouths of Delaware and Chesapeake bays, mainly during spring when temperatures are at their lowest (K. W. Able, pers. observ.) This suggests that they may move out of estuaries and to the south in winter, but more data are necessary to confirm this.

87

Astroscopus guttatus
Abbott

NORTHERN STARGAZER

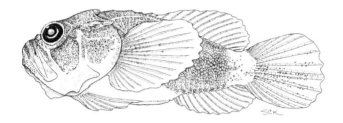

Estuary	Reproduction	Larvae/ Development	YOY Growth	YOY Overwinter
Ocean				
Both				
Jan				
Feb				
Mar				
Apr				
May	▓			
Jun	▓	▓		
Jul	▓	▓		
Aug	▓	▓		
Sep	▓	▓		
Oct		▓		
Nov				
Dec				

DISTRIBUTION

Astroscopus guttatus is one of a few identified endemic species in the Middle Atlantic Bight (Robins and Ray, 1986). It is infrequently observed throughout its range from New York to North Carolina perhaps because it buries in the substrate with only the eyes and top of head exposed (Smith, 1985). Juveniles have been reported from Hudson River–Raritan Bay south to Chesapeake Bay (see Table 4.3), but larvae identified as this species have only been reported from Great Bay and Chesapeake Bay.

REPRODUCTION AND DEVELOPMENT

Details of spawning are not well known and the eggs are undescribed. Based on collections of larvae (identified as "Uranoscopidae"), reproduction occurs in Middle Atlantic Bight continental shelf waters between June and October, with most activity in the mid-shelf region off the Delmarva Peninsula and Chesapeake Bay mouth (Able and Fahay, 1998). Spawning in lower Chesapeake Bay reportedly takes place during May and June (Murdy et al., 1997). Available aspects of the ontogeny have been summarized (Fahay, 2007). No scales were observed on specimens between 18.4 and 59.0 mm TL.

LARVAL SUPPLY, SETTLEMENT, GROWTH, AND MORTALITY

The smallest larvae occur over mid-depths of the continental shelf, whereas larger larvae are often collected closer to the coast, suggesting an inshore migration with development. Because of the scarcity of collections of this species, we can only make limited estimates of growth rate. On the basis of young-of-the-year specimens collected near Great Bay–Little Egg Inlet (Fig. 87.1), it appears to reach 14 cm by the first fall. Fish kills associated with low dissolved oxygen levels often result in mass mortalities of this species in Raritan Bay (M. P. Fahay, pers. observ.).

SEASONALITY AND HABITAT USE

This species occurs in bays, coastal areas, and the ocean to a depth of 200 m (Robins and Ray, 1986). Lengths of planktonic larvae collected by bongo net during the MARMAP surveys of the Middle Atlantic Bight continental shelf compared to lengths of larvae collected at Beach Haven Ridge adjacent to Little Egg Inlet suggest that larger larvae are present in coastal areas (Able and Fahay, 1998). Pelagic larvae (up to 11.5 mm SL) were collected from the Beach Haven Ridge study site between July and September of two years (Able et al., 2006a). Plankton net sampling near Little Egg Inlet collected larvae and early juveniles during the same time period as continental shelf collections (June–October) (Witting et al., 1999) and at the same sizes as those from Beach Haven Ridge (see Fig. 87.1).

Larvae (5 to 15 mm SL) collected from the water column under nightlights have been observed in laboratory conditions to settle to a sandy bottom. These individuals are able to bury themselves in the substrate in 1 to 2 seconds, using a combination of buccal pumping that liquefied the coarse sand below them, digging with the pectoral fins, and a swimming motion that redistributed the sand above them (D. Witting, pers. comm.). Further indications that larvae settle to the bottom at these lengths are provided by bottom trawl collections at the Beach Haven Ridge study site, where the smallest specimens were about 13 mm SL (see Fig. 87.1).

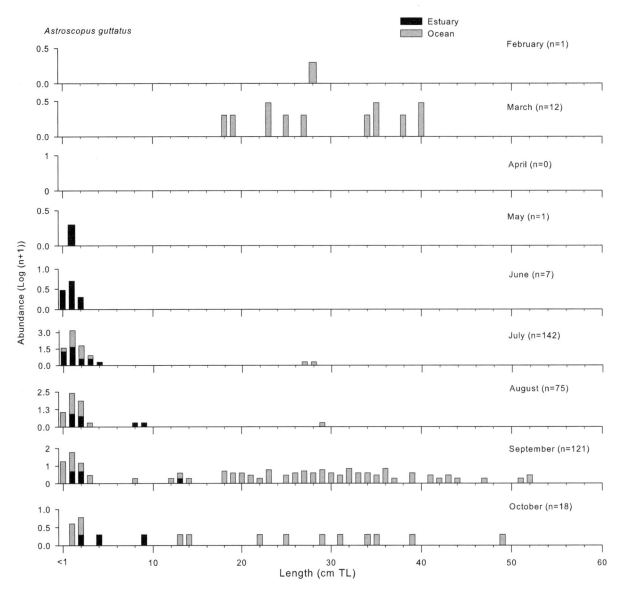

Fig. 87.1. Monthly length frequencies of *Astroscopus guttatus* collected in estuarine and coastal ocean waters of the central Middle Atlantic Bight. Sources: National Marine Fisheries Service: otter trawl (n = 101); Rutgers University Marine Field Station: seine (n = 59); Methot trawl (n = 62); 1 m plankton net (n = 73); 2 m beam trawl (n = 1); 4.9 m otter trawl (n = 8); and Able and Fahay 1998 (n = 73).

PREY AND PREDATORS

Our examination of food habits of this infrequently studied species (10.0 to 22.0 mm, n = 20) found that the diet was dominated by fishes (82.4% by weight) with much smaller amounts of mysids, zooplankton, and other crustaceans. Fishes and isopods have been reported for other individuals (104 to 130 mm, n = 6) (Hildebrand and Schroeder, 1928).

MIGRATIONS

There is no information on the seasonality of estuarine occupation or mechanisms associated with emigration from the estuary by this species. The young-of-the-year (< 16 cm) have

been frequently collected in National Marine Fisheries Service (NMFS) trawl surveys, with most individuals found on the inner continental shelf in the fall between western Long Island and Cape Hatteras but none during the winter (Fig. 87.2). In the spring, they are concentrated between the Delaware and Chesapeake bays. The age 1+ (> 16 cm) individuals are less frequently represented in these surveys in the fall from the shallow (mean depth 14.5 m, range 9 to 20 m) inner continental shelf from western Long Island to North Carolina. They are absent in the winter and spring, presumably because they are unavailable to the sampling gear.

Age 1+ *Astroscopus guttatus*   ## YOY *Astroscopus guttatus*

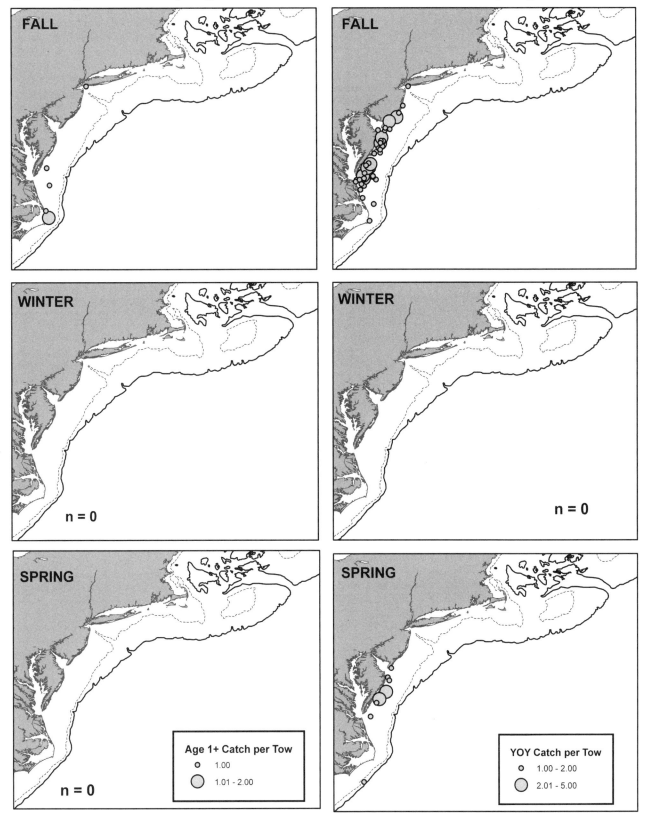

Fig. 87.2. Composite distribution of age 1+ (> 16 cm) and young-of-the-year (< 16 cm) *Astroscopus guttatus* during seasonal cruises of the National Marine Fisheries Service groundfish survey. Details of sampling effort are indicated in Fig. 3.4.

88

Chasmodes bosquianus (Lacepède)

STRIPED BLENNY

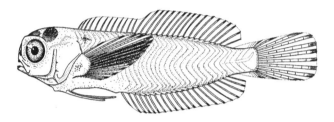

Estuary	Reproduction	Larvae/ Development	YOY Growth	YOY Overwinter
Ocean				
Both				
Jan				
Feb				
Mar				
Apr				
May				
Jun				
Jul				
Aug				
Sep				
Oct				
Nov				
Dec				

DISTRIBUTION

This species occurs along the Atlantic coast of the United States from New York to northeastern Florida (Williams, 2002). Early stages are rarely collected in our study area, but are more common in Chesapeake Bay (Musick, 1972).

REPRODUCTION AND DEVELOPMENT

Spawning occurs from April to August in the Chesapeake Bay and elsewhere (Hildebrand and Schroeder, 1928). Eggs are deposited in empty bivalve shells or other structured substrates. They are demersal and slightly oval and have an adhesive disk. The diameter is 0.92 to 1.1 mm along the major axis, and there are multiple oil globules. The juvenile pictured in this chapter was originally described as *Hypsoblennius hentz* but has since been identified as *Chasmodes bosquianus* (Ditty et al., 2006). Larvae feature early-developing, large pectoral fins, but lack the huge pre-opercle spines that characterize *Hypsobelnnius hentz* larvae. A summary of other ontogenetic characters may be found in Fahay (2007).

LARVAL SUPPLY, SETTLEMENT, AND GROWTH

Although we have only collected a few larvae from our Great Bay study area, we have no reason to conclude that these derived from spawning locations anywhere except the local source. We also have no information on whether larvae are pelagic, or whether there is a settlement stage. The monthly length frequencies make it difficult to interpret growth. One interpretation, based on limited data, is that the young-of-the-year are approximately 7 cm by fall of the first year (Fig. 88.1). This species is reported to reach approximately 10 cm (Hildebrand and Schroeder, 1928).

SEASONALITY AND HABITAT USE

This species is present throughout the year in bays and estuaries in the southern portion of our study area (Hildebrand and Schroeder, 1928). During the spring and summer, it is found on shallow sand and mud flats. It is particularly abundant on oyster reefs. Estimates from a Chesapeake Bay site ranged from 6 to 20 individuals per square meter (Harding and Mann, 2000). Their density was highest in areas with the greatest oyster shell cover. They move to deeper flats during the fall, and spend the winter in deeper channels with high salinities (Fritzsche, 1978).

PREY AND PREDATORS

Food items include small crustaceans, such as isopods and amphipods, small mollusks, and insect larvae (Hildebrand and Schroeder, 1928).

MIGRATIONS

No large-scale migrations have been reported for any life history stage of this species.

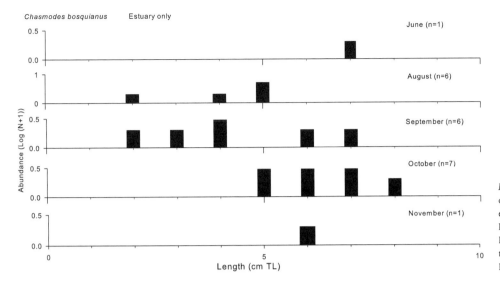

Fig. 88.1. Monthly length frequencies of *Chasmodes bosquianus* collected in estuarine waters of the Great Bay–Little Egg Harbor study area. Sources: Rutgers University Marine Field Station: seine (n = 12); and Able and Fahay 1998 (n = 9).

89

Hypsoblennius hentz (Lesueur)

FEATHER BLENNY

REPRODUCTION AND DEVELOPMENT

Spawning occurs from May through August, more rarely through September (Hildebrand and Schroeder, 1928; Hildebrand and Cable, 1938). The eggs are deposited on a hard substrate, such as oyster shell reefs or empty bivalve shells, and they are guarded by the male (Hildebrand and Cable, 1938). Incubation occupies 11 days at 26°C. The males guard the nests (often in bivalve shells) containing as many as 3750 eggs. Larvae are moderately abundant through October in Chesapeake Bay (Olney, 1983; Olney and Boehlert, 1988). In Great Bay–Little Egg Harbor, New Jersey, the larvae have been collected from June through November, suggesting a prolonged spawning period there as well.

The early development, including characters useful for distinguishing eggs and larvae from those of related species, has been well described and summarized (Fahay, 2007). Eggs are demersal, slightly elliptical, and flattened near an adhesive disk. There are multiple oil globules. Larvae hatch at lengths of 2.6 to 2.8 mm TL. They have large, bulbous heads and prominent spines on the pre-opercle. Early-forming pectoral fins are usually well pigmented. Small juveniles resemble adults, both in pigment pattern and in the possession of a feathery tentacle over each eye. Scales are lacking in all life history stages.

Estuary	Reproduction	Larvae/ Development	YOY Growth	YOY Overwinter
Ocean				
Both				
Jan				
Feb				
Mar				
Apr				
May				
Jun				
Jul				
Aug				
Sep				
Oct				
Nov				
Dec				

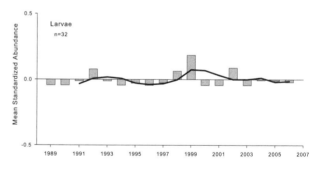

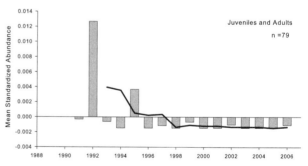

Fig. 89.1. Annual mean (bars) and three-year moving mean (line) of standardized abundance for *Hypsoblennius hentz* collected in larval sampling with plankton net (*upper*) and juvenile and adult sampling with traps (*lower*). Data were standardized by subtracting overall mean from annual abundance. Three-year moving mean was calculated by taking the mean of the standardized values from the previous two years and the year in which data are plotted.

DISTRIBUTION

Hypsoblennius hentz occurs in coastal waters of North America from Nova Scotia to the Yucatan Peninsula (Robins and Ray, 1986). It is abundant in Chesapeake Bay (Musick, 1972). The early life history stages occur in the southernmost estuaries of the Middle Atlantic Bight as far north as New Jersey (see Table 4.3).

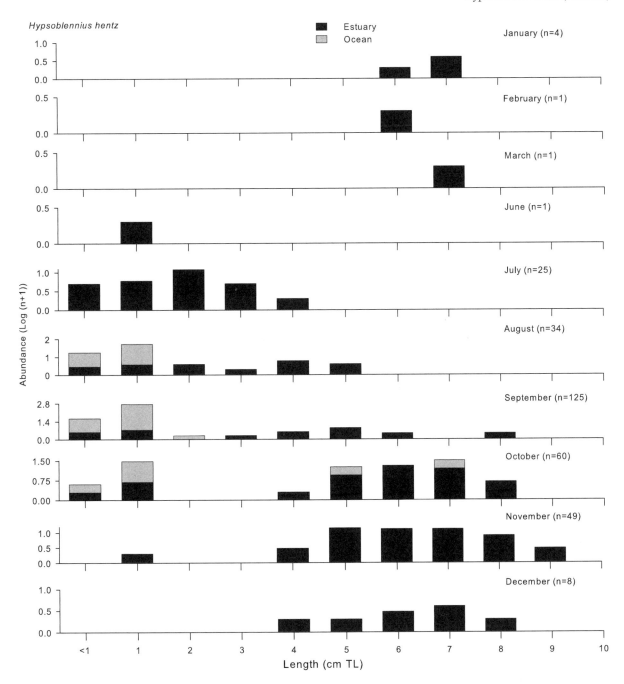

Fig. 89.2. Monthly length frequencies of *Hypsoblennius hentz* collected in the Great Bay–Little Egg Harbor and adjacent coastal ocean waters. Sources: Rutgers University Marine Field Station: seine (n = 20); Methot trawl (n = 125); 1 m plankton net (n = 30); killitrap (n = 7); and Able and Fahay 1998 (n = 126).

LARVAL SUPPLY, SETTLEMENT, AND GROWTH

The larvae are variable in occurrence, with none present in some years in estuarine sampling. The numbers of juveniles and adults are low during most years in southern New Jersey; however, there were peaks in abundance in 1992 and 1995 (Fig. 89.1). The larvae have also been collected in the ocean (Fig. 89.2), suggesting spawning there or that the larvae have been outwelled from the estuary. In Great Bay, larvae and juveniles grow at an approximate rate of 0.5 mm per day during the summer and fall. By November, they have reached 4 to 9 cm. These sparse data suggest that the population in New Jersey is made up primarily of a single year class. The reported maximum size is 104 mm for males and 84 mm for females (Hildebrand and Cable, 1938). Settlement occurs at

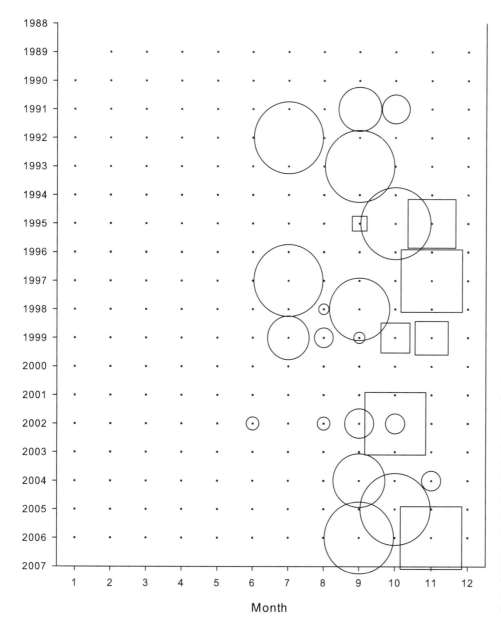

Fig. 89.3. Seasonal occurrence of larvae based on ichthyoplankton samples collected in Little Sheepshead Creek behind Little Egg Inlet from 1989 to 2006 (open circles represent the percentage of the mean number/1000 m³ of larval young-of-the-year captured by year) and juveniles of *Hypsoblennius hentz* based on trap collections in Rutgers University Marine Field Station boat basin in Great Bay (open squares represent the percentage of the mean catch per unit effort captured by year) from 1994 to 2006. Values range from 0 to 100% for ichthyoplankton and trap sampling. The smallest circles and squares represent when samples were taken but no individuals were collected.

sizes > 10 mm based on their reduced occurrence in ichthyoplankton collections (Hildebrand and Cable, 1938).

SEASONALITY AND HABITAT USE

The larvae are present in the estuary from June through November and in the ocean from August through October (Figs. 89.2, 89.3). Demersal young-of-the-year first appear in Great Bay estuarine trap collections in September through November (see Fig. 89.2). We lack data describing specific habitats where larvae and juveniles occur, but eggs are usually associated with oyster shell reefs or beds of empty bivalve shells (Hildebrand and Cable, 1938). The adults of this species have been reported to occur at higher salinities, where they are

found in oyster reefs and shallow flats during the summer and in deeper channels and holes during the winter (Hildebrand and Cable, 1938; Musick, 1972). Adults also occur on inner continental shelf habitats.

PREY AND PREDATORS

The food consists of small crustaceans, mollusks, and ascidians (Hildebrand and Schroeder, 1928).

MIGRATIONS

Unknown, but it has been found in deep water (23 to 45 m) during the winter (Hildebrand and Schroeder, 1928), suggesting a seasonal movement into deep water.

90

Gobiesox strumosus
Cope

SKILLETFISH

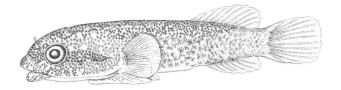

Estuary	Reproduction	Larvae/ Development	YOY Growth	YOY Overwinter
Ocean				
Both				
Jan				
Feb				
Mar				
Apr				
May				
Jun				
Jul				
Aug				
Sep				
Oct				
Nov				
Dec				

DISTRIBUTION

This species reaches the northern limit of its range in New Jersey, and occurs south to Brazil, including the Gulf of Mexico and West Indies (Runyan, 1961). Juveniles are only collected in bays and estuaries, but they are rarely, if ever, encountered in the extensively sampled Mullica River–Great Bay estuary. They are much more common in estuaries from Chesapeake Bay and farther south.

REPRODUCTION AND DEVELOPMENT

This species matures at 40 to 45 mm TL (Runyan, 1961). Spawning occurs from April to August, with a peak from late April through May (Martin and Drewry, 1978). Eggs are deposited in masses, often in high-energy habitats attached to the underside of rocks or shells. The early development has been well described (Dovel, 1963; Fahay, 2007). Eggs are demersal, slightly oval to round, with a diameter of 0.75 to 0.94 mm (if round). There are 70 to 80 oil globules, ranging from 0.02 to 0.10 mm in diameter. Larvae hatch at sizes of 2.4 to 3.4 mm TL, are heavily pigmented, and are sometimes in an advanced state of development. Their guts are typically thickened along the entire length. Juveniles resemble adults and possess a suction disk partially formed by the pelvic fins.

LARVAL SUPPLY, SETTLEMENT, GROWTH, AND MORTALITY

Larvae hatch from eggs deposited in estuaries; therefore they would not be expected to occur over the continental shelf or to be transported through inlets. They settle after the suction disk is fully developed at a size of about 9.0 mm TL (Runyan, 1961). One estimate reported a growth increase of 16 to 20 mm over a 4-month span from June to October (Runyan, 1961). Young-of-the-year may reach a modal size of 32 to 36 mm TL, with the largest individuals up to 64 mm TL (Runyan, 1961). We have no estimates of overwinter or other causes of mortality, but there are indications this species can survive temperatures as low as 4.3°C (Runyan, 1961).

SEASONALITY AND HABITAT USE

This species inhabits bays, estuaries, and coastal ocean habitats to a maximum depth of 33 m. It is often found near pilings, rocks, oyster reefs, sponges, and eelgrass beds. In oyster reefs individuals often attach to the undersurface of the shells with the adhesive disk (Runyan, 1961).

PREY AND PREDATORS

Unknown.

MIGRATIONS

There is no evidence that this species engages in migrations to or from its preferred estuarine habitats.

91

Ctenogobius boleosoma (Jordan and Gilbert)

DARTER GOBY

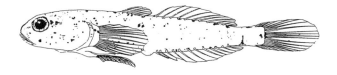

Estuary	Reproduction	Larvae/ Development	YOY Growth	YOY Overwinter
Ocean				
Both				
Jan				
Feb				
Mar				
Apr				
May				
Jun				
Jul				
Aug				
Sep				
Oct				
Nov				
Dec				

DISTRIBUTION

Ctenogobius boleosoma occurs from Massachusetts to Brazil, including Bermuda, the Gulf of Mexico, and the Caribbean Sea (Dawson, 1969). It is occasionally found in Chesapeake Bay (Massmann, 1954), at Indian River Inlet, Delaware (de Sylva et al., 1962), and in southern New Jersey (McDermott, 1971; Milstein and Thomas, 1976a). The early life history stages and adults have only rarely been collected in other Middle Atlantic Bight estuaries.

REPRODUCTION AND DEVELOPMENT

This species matures at approximately 25 to 30 mm (Hildebrand and Cable, 1938) and probably spawns at age 1. It has been reported to spawn in coastal ocean as well as estuarine waters, and its larvae have occurred in both environments. The presence of the smallest larvae in July and August in Great Bay–Little Egg Harbor estuary indicates that reproduction occurs in the summer in estuaries and may continue over a long period because other slightly larger larvae (< 12 mm) occur through December.

The early development, including characters for distinguishing eggs, larvae, and juveniles from those of closely related species, has been well described (Wyanski and Targett, 2000; Fahay, 2007). Eggs are irregularly shaped and tiny, and have a tuft of fibrous strands on the chorion. Larvae hatch at about 1.2 mm and have an elongate body. Pigment patterns on the developing air bladder are important for identification of the larvae. No scales were observed in specimens between 7.9 and 9.0 mm SL (n = 4), so development must occur later. Two ctenoid scales form at the base of the caudal fin rays in postflexion larvae of *Gobiosoma ginsburgi*, but those structures do not occur in the present species.

LARVAL SUPPLY, SETTLEMENT, GROWTH, AND MORTALITY

The occurrences of larvae of this species, as measured by plankton net sampling, are relatively uniform inter-annually, with some sporadic increases (Fig. 91.1). The smallest larvae have been collected in the estuary in July and August. Growth apparently is slow in young-of-the-year of this small species. Young-of-the-year reach a maximum size of about 42 mm by late fall (Fig. 91.2). There is no growth during the winter, although this is a conclusion based on very small sample sizes. Given that the maximum size is approximately 55 mm (Hildebrand and Cable, 1938), they probably spawn at the end of the first year.

SEASONALITY AND HABITAT USE

Although adults are rare, the larvae are among the common species collected in Great Bay–Little Egg Harbor (Witting et al., 1999). They first occur in July, but small larvae (< 10 mm) have been collected as late as December, with peak abundances usually in October (see Fig. 91.1). In North Carolina, larvae have been collected from May to November, with the peak in July and August (Hildebrand and Cable, 1938). Most of these were taken in the estuary, but a few were collected offshore, suggesting that the species may spawn in coastal waters as well or be outwelled onto the inner continental shelf. Larvae are reportedly more abundant in plankton collections near the bottom than at the surface (Hildebrand and Cable, 1938).

Little is known about adult habitat use in our study area, in part because *Ctenogobius boleosoma* is never abundant in collections. It occurs mostly in meso- to polyhaline estuarine

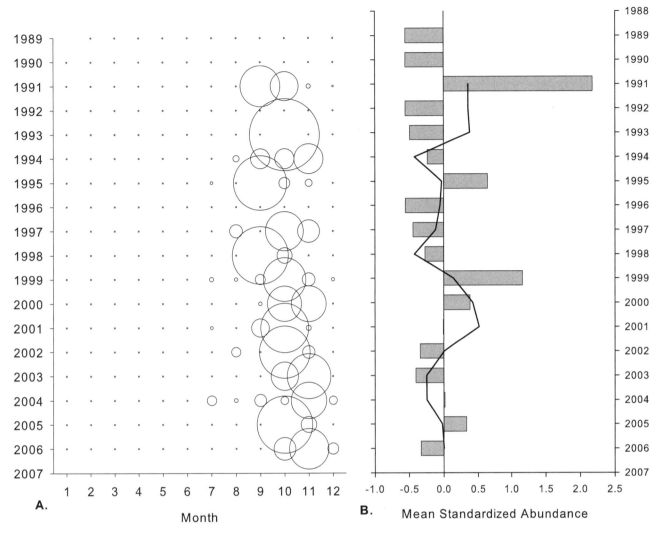

Fig. 91.1. Annual and seasonal variation in occurrence and abundance of *Ctenogobius boleosoma*. (A) Seasonal variation of the larvae is based on ichthyoplankton samples collected in Little Sheepshead Creek behind Little Egg Inlet from 1989 to 2006. Open circles represent the percentage of the mean number/1000 m³ captured within each year. Values range from 0 to 100%. The smallest circles represent when samples were taken but no larvae were collected. (B) Annual variation in abundance for the same data was standardized by subtracting overall mean from annual abundance. Horizontal bars represent standardized annual abundance; the solid line is a three-year running mean that was calculated by taking the mean of the standardized values from the previous two years and the year in which the data are plotted.

waters, often in grass beds. It occurs in a variety of shallow estuarine habitats over muddy bottoms in North Carolina (Hildebrand and Cable, 1938). The juveniles have been collected in the subtidal portions of a polyhaline marsh creek in Great Bay–Little Egg Harbor (Rountree and Able, 1992a).

PREY AND PREDATORS

There have been no studies of the food habits of young-of-the-year (< 40 mm) *Ctenogobius boleosoma* in the study area.

MIGRATIONS

Unknown.

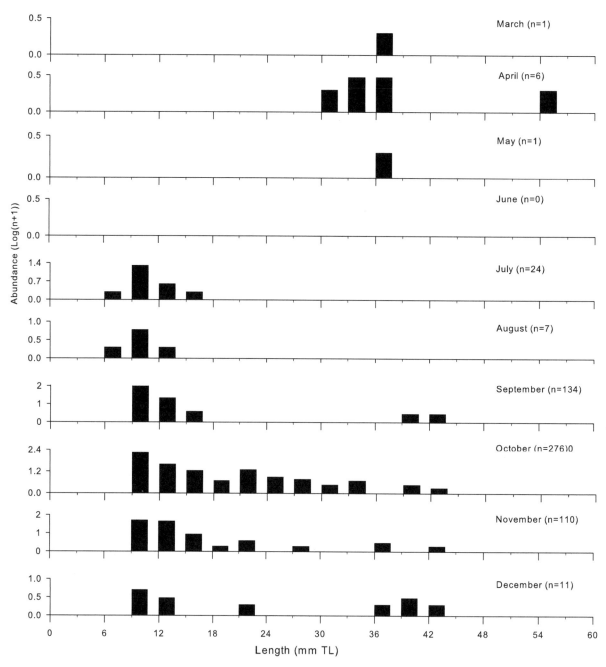

Fig. 91.2. Monthly length frequencies of *Ctenogobius boleosoma* collected
in estuarine and coastal ocean waters of the central Middle Atlantic
Bight. Sources: Rutgers University Marine Field Station: 1 m plankton net
(n = 338); experimental trap (n = 5); 4.9 m otter trawl (n = 3); and Able and
Fahay 1998 (n = 224).

92

Gobiosoma bosc (Lacepède)

NAKED GOBY

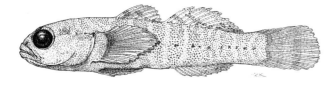

Estuary	Reproduction	Larvae/ Development	YOY Growth	YOY Overwinter
Ocean				
Both				
Jan				
Feb				
Mar				
Apr				
May				
Jun				
Jul				
Aug				
Sep				
Oct				
Nov				
Dec				

DISTRIBUTION

Gobiosoma bosc occurs in estuaries from Connecticut to Campeche, Mexico (Pearcy and Richards, 1962; Dawson, 1966). In the Middle Atlantic Bight, early life history stages are common in estuaries south of New Jersey and sporadically in estuaries to the north (see Table 4.3).

REPRODUCTION AND DEVELOPMENT

This species reaches sexual maturity at age 1 between 14.5 and 23 mm (Hildebrand and Cable, 1938; Dawson, 1966). Spawning occurs from April or May through August or September

in Middle Atlantic Bight estuaries (Lippson and Moran, 1974; Nero, 1976; Wang and Kernehan, 1979; Shenker et al., 1983). In New Jersey, the occurrence of small larvae and juveniles indicates that spawning peaks in the spring and early summer but may extend through the summer. Details of the development are relatively well known (Fahay, 2007). Eggs are deposited and guarded in nests, typically in oyster shells. The larvae hatch in about 4 to 5 days (Kuntz, 1916; Nelson, 1928).

LARVAL SUPPLY, SETTLEMENT, GROWTH, AND MORTALITY

The source of larvae to most estuaries is likely local but is variable between years. The larvae were relatively more abundant in the mid-1990s and again in 2001 and 2005 (Fig. 92.1). The juveniles were most plentiful in trap collections in 1993 during a period of high larval abundance. Prior to settlement, individuals begin schooling near the bottom at sizes of 6 to 10 mm SL. The overlap between the largest planktonic individuals and the smallest demersal individuals in southern New Jersey (Fig. 92.2) and Chesapeake Bay indicates that settlement occurs over a size range of 8 to 13 mm TL (Breitburg, 1991; Able and Fahay, 1998) and only in the estuary (Able et al., 2006a). Settled individuals have completed development of the pelvic fins (Breitburg, 1989, 1991). In Chesapeake Bay, settlement occurs during the day and night, and the transforming individuals aggregate in low-flow areas on the down-current sides of rocks. This active response influenced settlement sites (Breitburg, 1991; Breitburg et al., 1995). In Great Bay–Little Egg Harbor, settled individuals become increasingly abundant after the larval peak in July, and reach their greatest numbers in November.

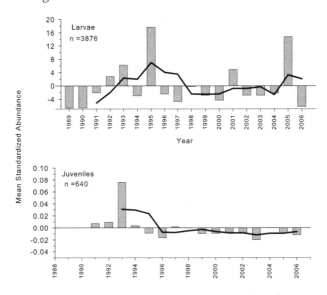

Fig. 92.1. Annual mean (bars) and three-year moving mean (line) of standardized abundance for *Gobiosoma bosc* collected in larval sampling with plankton net (*upper*) and juvenile sampling with otter trawls (*lower*). Data were standardized by subtracting overall mean from annual mean abundance. Three-year moving mean was calculated by taking the mean of the standardized values from the previous two years and the year in which data are plotted.

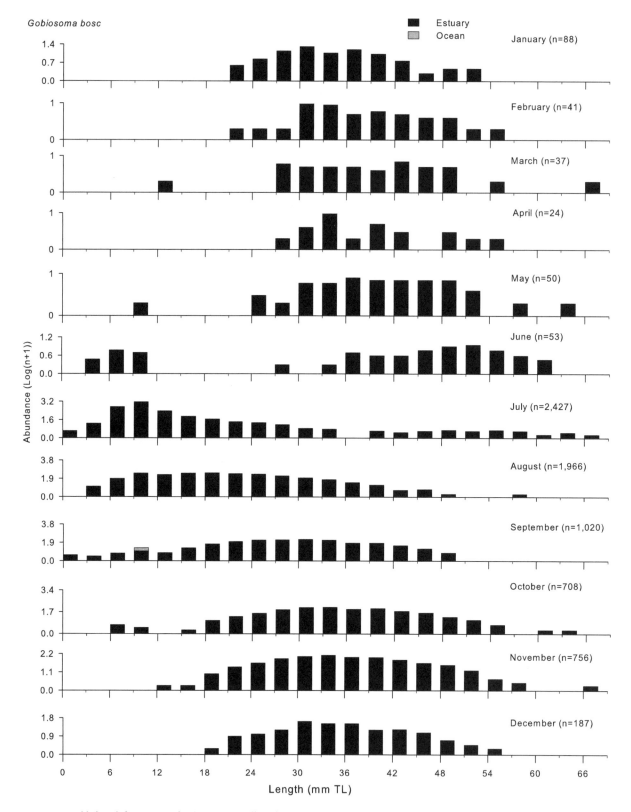

Fig. 92.2. Monthly length frequencies of *Gobiosoma bosc* collected in estuarine and coastal ocean waters of the central Middle Atlantic Bight. Sources: Rutgers University Marine Field Station: 1 m beam trawl (n = 347); seine (n = 492); Methot trawl (n = 1); 1 m plankton net (n = 1740); experimental trap (n = 579); killitrap (n = 270); 4.9 m otter trawl (n = 101); and Able and Fahay 1998 (n = 3827).

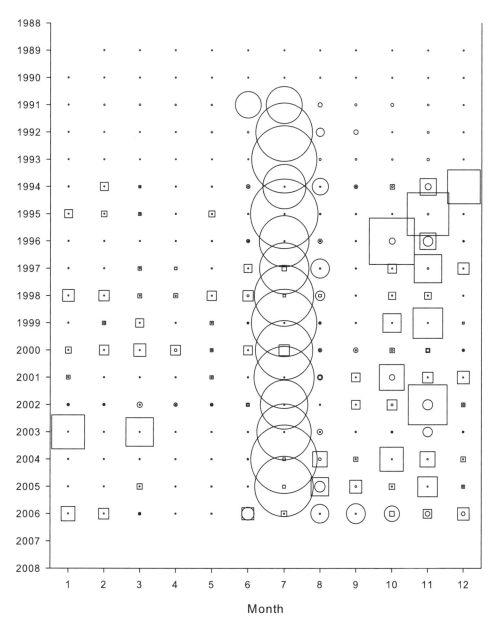

Month

Fig. 92.3. Seasonal occurrence of larvae based on ichthyoplankton samples collected in Little Sheepshead Creek behind Little Egg Inlet from 1989 to 2006 (open circles represent the percentage of the mean number/1000 m³ of larval young-of-the-year captured by year) and juveniles of *Gobiosoma bosc* based on trap collections in Rutgers University Marine Field Station boat basin in Great Bay (open squares represent the percentage of the mean catch per unit effort captured by year) from 1994 to 2006. Values range from 0 to 98% and 0 to 77% for ichthyoplankton and trap sampling, respectively. The smallest circles and squares represent when samples were taken but no individuals were collected.

Modal changes in length frequency show that growth of young-of-the-year averages approximately 0.25 mm per day. They reach approximately 21 to 48 mm TL by November (see Fig. 92.2). Little growth is evident during the winter. By the following May and June, approximately 12 months after hatching, the larger young-of-the-year are of the same size (27 to 60 mm TL) as reproductively mature adults (Nero, 1976). The recently hatched young-of-the-year are beginning to appear at this time as well; thus, this species is capable of spawning the first year, as observed for a Virginia population (Nero, 1976). There is little evidence that they survive their second summer because larger individuals are absent from the length-frequency data (see Fig. 92.2). Sources of mortality vary with life history stage. The larvae are susceptible to predation from jellyfishes, and the juveniles and adults are exposed to increased predation as the result of hypoxia (Breitburg et al., 1999).

SEASONALITY AND HABITAT USE

In Great Bay–Little Egg Harbor estuary, the planktonic larvae are usually more abundant in July but can be found through November (Fig. 92.3). In several Middle Atlantic Bight estuaries, the larvae move to low-salinity nursery areas soon after hatching (Massmann et al. 1963; Wang and Kernehan, 1979). This up-estuary movement may occur at the rate of 1 km per day and is presumably the result of tidal stream transport (Shenker et al., 1983).

The settled juveniles and adults can be found in every month, although in many years they are most abundant in traps in October through December (see Fig. 92.3). The habitat of this species is exclusively estuarine because intensive sampling on the adjacent inner continental shelf on Beach Haven Ridge has failed to collect this species, although *Gobiosoma ginsburgi* has been collected frequently (Duval and Able, 1998). It can also be found far up estuaries (Wang and

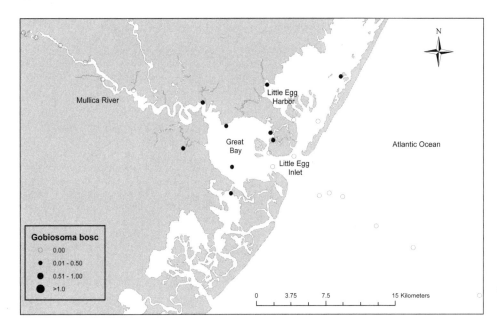

Fig. 92.4. Habitat use and average catch per tow (see key) for juvenile and adult *Gobiosoma bosc* in the Ocean–Great Bay–Mullica River corridor in southern New Jersey based on small otter trawl collections during July and September from 1988 to 2006. See Table 3.2 for characteristics at each sampling collection.

Kernehan, 1979) and has been collected at salinities as low as 8 ppt (de Sylva et al., 1962). A recent examination of the distribution relative to habitat type in the polyhaline portions of Great Bay–Little Egg Harbor estuary found that young-of-the-year collected with beam trawls were common across a variety of substrate types (Able and Fahay, 1998). Another examination in the same estuary found them at most sampling locations in the polyhaline Great Bay but not in the higher salinities in the ocean or in the lower salinities of the Mullica River (Fig. 92.4). They occur in sandy to very muddy substrate, where the dominant physical structure ranges from vegetation (eelgrass, sea lettuce; Sogard and Able, 1991) to other biogenic forms such as amphipod tubes and shell. The young-of-the-year also occur in inter- and subtidal marsh creeks (Rountree and Able, 1992a; Able et al., 1996b). Elsewhere in the Middle Atlantic Bight in the spring, summer, and fall, *Gobiosoma bosc* has been collected in areas with man-made structures such as piers and pile fields (Able et al., 1998), eelgrass beds (Orth and Heck, 1980), and marsh creeks (Musick, 1972; Sogard and Able, 1991; Able et al., 1996b), but it appears to be most abundant in oyster reefs (Nelson, 1928;

Dovel, 1971; Dahlberg and Conyers, 1973; Breitburg, 1991; Harding and Mann, 2000). A number of studies have identified dissolved oxygen as a habitat variable that negatively influences the behavior of larvae (Breitburg et al., 1994) and their survival when exposed to predators (Breitburg, 1992; Breitburg et al., 1994). For recently hatched larvae (2.4 to 2.6 mm), the critical dissolved oxygen level was 0.5 m/l (Saksena and Joseph, 1972).

PREY AND PREDATORS

Gobiosoma bosc is an important prey for many estuarine piscivores. In our review of the available literature for young-of-the-year predators, this species was preyed upon by six different species of temperate estuarine fishes (see Table 8.3; Mancini and Able, 2005). They were important in two studies of the diet of *Morone saxatilis* and were present in stomach contents in five other studies.

MIGRATIONS

During the winter, the species may move from shallows to channels and channel edges (Musick, 1972).

93

Gobiosoma ginsburgi
Hildebrand and
Schroeder

SEABOARD GOBY

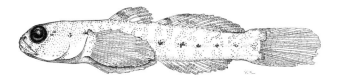

Estuary	Reproduction	Larvae/ Development	YOY Growth	YOY Overwinter
Ocean				
Both				
Jan				▓
Feb				▓
Mar				▓
Apr				▓
May				▓
Jun	░			
Jul	░	▓		▓
Aug	░	▓	▓	▓
Sep	░	▓	▓	▓
Oct			▓	
Nov				▓
Dec				▓

DISTRIBUTION

Gobiosoma ginsburgi occurs along the Atlantic coast of the United States from Massachusetts to Georgia (Ginsburg, 1933; Dawson, 1966; Lux and Nichy, 1971). In the Middle Atlantic Bight, early life history stages have been reported most consistently from New Jersey and south (see Table 4.3), although they occasionally occur as far north as Narragansett Bay and Massachusetts (Collette and Hartel, 1988).

REPRODUCTION AND DEVELOPMENT

This species presumably spawns at age 1 (see below). Spawning occurs throughout the summer, begins earlier in the southern part of its range, and continues progressively later to the north. In New Jersey, the occurrence of gravid females (19.0 to 38.7 mm SL) and larvae (Duval and Able, 1998) indicates that spawning takes place in estuaries from approximately June through September. This period overlaps with the presumed periods of reproduction in Rhode Island (Munroe and Lotspeich, 1979) and Long Island (Greeley, 1938) but it occurs slightly later than the May spawning in Chesapeake Bay (Hildebrand and Schroeder, 1928).

The early development of *Gobiosoma ginsburgi*, including comparisons with larvae and juveniles of related species, has been well described (Fahay, 2007). Eggs are undescribed. Length at hatching is unknown. Larvae are elongate and lightly pigmented. Two ctenoid scales form at the base of the caudal fin rays in postflexion larvae and continue to be present in adults. Pigment in juveniles consists of vague bars with a series of blotches along the lateral line. Scales (except for the two at the base of caudal fin) are lacking in this species.

LARVAL SUPPLY, SETTLEMENT, GROWTH, AND MORTALITY

The source of larvae varies among years and potentially between the estuary and ocean. Annual abundance in the estuary is variable with occasional strong year classes (e.g., 1995 and 1999) based on abundance in ichthyoplankton samples (Fig. 93.1). The larvae were collected from both the estuary and the ocean in most months. Settlement occurs at sizes of 9–12 mm and has taken place in both the estuary and the ocean (Able et al., 2006a).

Based on modal length-frequency progression and examination of sagittal otoliths, most individuals that occur in New Jersey are young-of-the-year (Duval and Able, 1998). In the summer, planktonic and demersal young-of-the-year in the estuary range from 3 to 33 mm SL but the largest individuals are in the ocean (Fig. 93.2). By October, the young-of-the-year are 6 to 42 mm SL. Little growth occurs during the winter so that by April they range from 24 to 48 mm SL. These individuals overlap in size with gravid females collected in the spring and summer; thus, this species apparently spawns at age 1. There is little evidence that adults survive through their second summer.

SEASONALITY AND HABITAT USE

The pelagic larvae (< 12 mm TL) are common through the summer, when they occur in both the coastal ocean and the estuary (see Figs. 93.1 and 93.2). Larvae have also been collected at similar times from estuaries throughout the range, including June through August in Rhode Island (Munroe and Lotspeich, 1979), July through October in Delaware (de Sylva et al., 1962), and June through December in Chesapeake Bay (Dovel, 1971; Olney, 1983). In the Great Bay–Little Egg Harbor

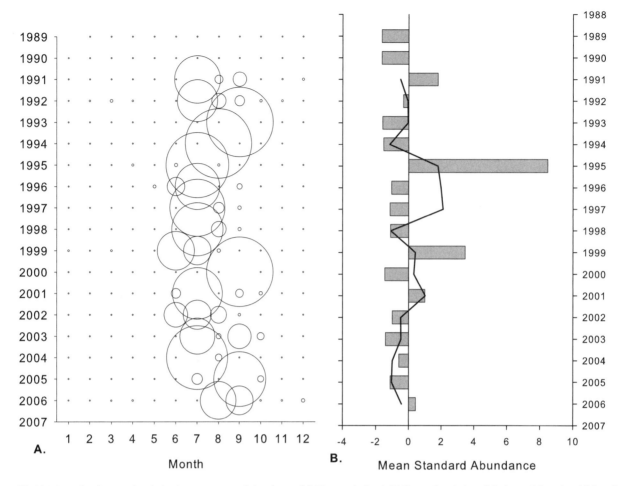

Fig. 93.1. Annual and seasonal variation in occurrence and abundance of *Gobiosoma ginsburgi*. (A) Seasonal variation of the larvae is based on ichthyoplankton samples collected in Little Sheepshead Creek behind Little Egg Inlet from 1989 to 2006. Open circles represent the percentage of the mean number/1000 m³ captured within each year. Values range from 0 to 100%. The smallest circles represent when samples were taken but no larvae were collected. (B) Annual variation in abundance for the same data was standardized by subtracting overall mean from annual abundance. Horizontal bars represent standardized annual abundance; the solid line is a three-year running mean that was calculated by taking the mean of the standardized values from the previous two years and the year in which the data are plotted.

area in southern New Jersey, planktonic larvae (3 to 9 mm SL) have been collected from the estuary from July through October (see Fig. 93.1). Recently settled juveniles (9 to 12 mm SL) are most abundant in July through October, both in the estuary and on an adjacent portion of the inner continental shelf (Duval and Able, 1998). During the winter and early spring the only collections available are from the estuary.

Eggs have only been collected from empty oyster valves in Georgia (Dahlberg and Conyers, 1973). These occur in subtidal areas and the males guard and aggressively defend them until hatching. In New Jersey, habitats of other life history stages range from the estuary to the inner continental shelf. Planktonic larvae, juveniles, and adults are reported from mid- to high-salinity (15 to 33 ppt) estuarine areas and the adjacent inner continental shelf in depths from 0.5 m (estuary) to 21 m (inner continental shelf) (Duval and Able, 1998). Habitats in Great Bay range from mud to sand substrate with shell, worm tubes, and hydroids over the substrate. Elsewhere this species has been reported from depths to 10 m off

Georgia (Dahlberg and Conyers, 1973) and to 45 m elsewhere (Dawson, 1966). In Chesapeake Bay, it has been collected at salinities between 15 and 31 ppt at oyster reefs and deeper flats during spring through summer and in higher-salinity channels in winter (Musick, 1972). The occurrence of adults in mollusk shells is reported throughout the species' range.

PREY AND PREDATORS
One study examined the food habits of young-of-the-year (18 to 42 mm SL, n = 49) in Sakonnet River, Rhode Island (Munroe and Lotspeich, 1979) (see Table 8.1). Crustaceans were the most frequent and abundant food items, predominantly harpacticoid copepods (60%) supplemented by gammaridean amphipods (18%), polychaetes (13%), and mollusks (11%).

MIGRATIONS
Evidence of seasonal migrations is lacking. Therefore it should be considered as a resident species that occurs in estuaries and coastal oceanic waters.

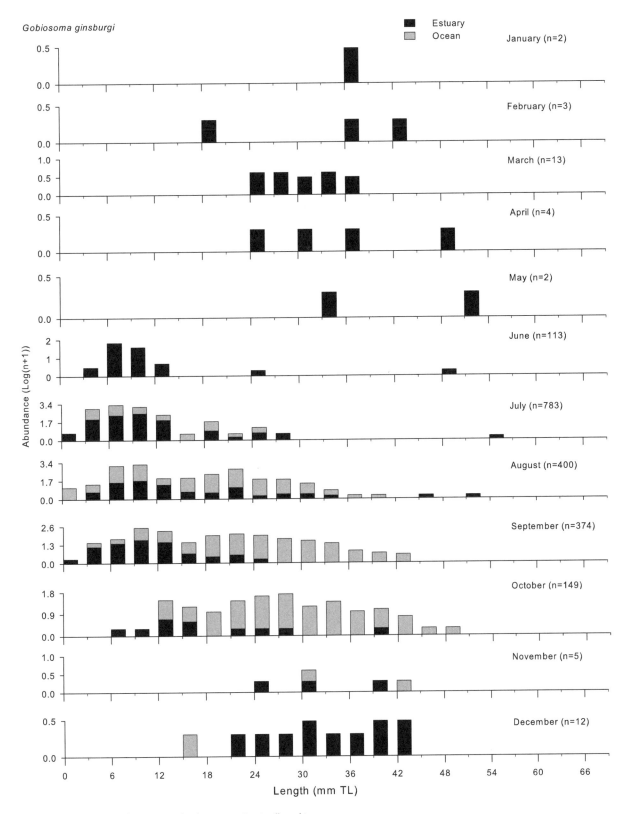

Fig. 93.2. Monthly length frequencies of *Gobiosoma ginsburgi* collected in estuarine and coastal ocean waters of the central Middle Atlantic Bight. Sources: Rutgers University Marine Field Station: 1 m plankton net (n = 892); 2 m beam trawl (n = 71); experimental trap (n = 2); killitrap (n = 2); 4.9 m otter trawl (n = 2); and Able and Fahay 1998 (n = 891).

94

Microgobius thalassinus (Jordan and Gilbert)

GREEN GOBY

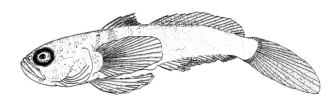

DISTRIBUTION

This species occurs along the coast of North America from Maryland to Texas (Murdy, 2002). Larvae have been collected in plankton nets in New Jersey, but neither juveniles nor adults have yet been reported from estuaries this far north.

REPRODUCTION AND DEVELOPMENT

Males and females are sexually mature during their second summer, and die after spawning (Schwartz, 1971). Spawning occurs from June to October in Chesapeake Bay (Fritzsche, 1978). Eggs are undescribed. The early development of *Microgobius thalassinus*, including characters useful for distinguishing the larvae and juveniles from those of other gobies, has been summarized (Fahay, 2007). Larvae hatch at < 2.0 mm SL. Air bladder pigment is restricted to the dorsal surface. Other pigment is light, but includes a line of melanophores along the anal fin base. Juveniles are diagnosed by relatively high pectoral fin ray counts and pigment evenly scattered over the entire body (Hildebrand and Cable, 1938). The body is robust and the mouth is terminal to slightly superior and oblique. Scales are first evident on the caudal peduncle at about 18 mm (Hildebrand and Cable, 1938). Scales then develop anteriorly and are complete by 23 mm.

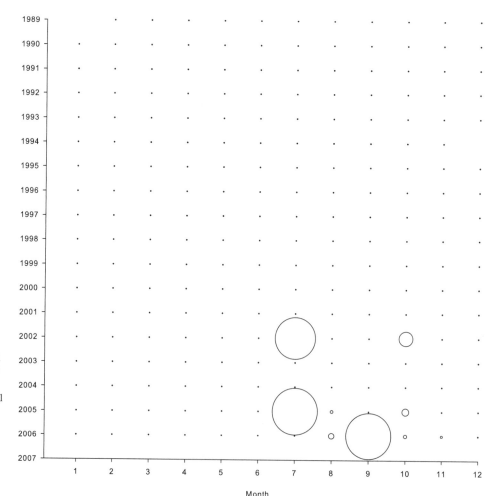

Fig. 94.1. Seasonal occurrence of the larvae of *Microgobius thalassinus* based on ichthyoplankton samples collected in Little Sheepshead Creek behind Little Egg Inlet from 1989 to 2006. Full circles represent the percentage of the mean number / 1000 m³ of larval young-of-the year *Microgobius thalassinus* captured by given year. Values range from 0 to 87%. The smallest circles represent when samples were taken but no larvae were collected.

LARVAL SUPPLY, SETTLEMENT, AND GROWTH

Spawning occurs within estuarine environments. There are no reports of spawning taking place in the ocean; however, larvae may be flushed out of estuaries and collected as far as 20 km offshore (Ginsburg, 1933). Larvae were not known to occur in our New Jersey study area until 2002. Since that year, they have also occurred in 2005 and 2006 (Fig. 94.1). This is another southern species that has become increasingly common, at least in its early stages, in recent years. There are no estimates of growth rates in early stages, but adults may reach lengths of 40 to 50 mm by age 1 (Hildebrand and Cable, 1938).

SEASONALITY AND HABITAT USE

Adults may live in shallow estuarine creeks over muddy substrates. They also inhabit the sponge *Microciona prolifera* in deep oyster reefs during most of the year, but occur mostly in deep channels during winter (Musick, 1972). They are found at depths of 3 to 6 m during the warmer months and up to 10 m during winter (Schwartz, 1971). Collections in our study area have been made in July through September, rarely into October (see Fig. 94.1). Larvae may occur at salinities > 16.5 ppt (Richardson and Joseph, 1975), and in depths up to 24 m (Hildebrand and Cable, 1938).

PREY AND PREDATORS

Unknown.

MIGRATIONS

There is no evidence that this species migrates away from estuarine environments during any season. Any seasonal movements are associated with occupation of deeper habitats during the winter.

95

Sphyraena borealis DeKay

NORTHERN SENNET

DISTRIBUTION

Sphyraena borealis occurs in the western Atlantic Ocean from Nova Scotia to 36 degrees south, including the northeastern Gulf of Mexico and the Caribbean Sea (Briggs, 1958; Murdy et al., 1997). *Sphyraena picudilla* occurs in the southern part of this range and has recently been established as a junior synonym of *S. borealis,* although some authors continue to treat it as a valid, separate species. In the Middle Atlantic Bight, early life history stages, represented only by juveniles, have been reported from most estuaries (see Table 4.3).

REPRODUCTION AND DEVELOPMENT

Spawning occurs during spring off the southern United States. The only known collections of eggs were off Florida (Houde, 1972); therefore spawning is assumed to take place in oceanic waters. Larvae have been collected in Slope Sea

waters north of Cape Hatteras during May (Hare et al., 2001). These were associated with a frontal assemblage of species, implying an origin in the Gulf Stream in the South Atlantic Bight and eventual transport into Middle Atlantic Bight continental shelf waters with increasing size.

The early development of *Sphyraena borealis*, including characters useful for distinguishing eggs, larvae, and juveniles from those of congeners, has been well described and summarized (Fahay, 2007). Eggs are pelagic and 1.22 to 1.24 mm in diameter, and have a single oil globule. Larvae are about 3.5 mm at hatching and have unpigmented eyes and ill-formed mouth parts. Larvae soon develop the characteristic, elongate body shape; the two dorsal fins are widely separated. The second dorsal and anal fins are positioned well back on the body. The mouth is large, the snout is long and pointed, and the lower jaw extends beyond the upper and has a fleshy tip. Transformation to the juvenile stage is gradual, with no sudden changes in morphology. In juveniles (> 14.5 mm SL), scales on the posterior part of the lateral line are ridged, forming a keel-like structure on the caudal peduncle. Juveniles also develop a characteristic, blotchy pigment pattern. Larvae start swimming two days after hatching in the laboratory and begin to feed by the third day (Houde, 1972).

LARVAL SUPPLY AND GROWTH

Although early stages of this species occur regularly in our study area, we lack sufficient data to describe their annual variation in abundance. Larvae average 5.5 mm SL at 7 days after hatching, 11 mm at 14 days, and about 13.5 mm at 21 days, although in this study growth may have been unnaturally slow because of a lack of appropriate food late in these observations (Houde, 1972). Larvae collected in Slope Sea waters north of Cape Hatteras during May 1993 ranged from 5.6 to 19.0 mm SL (mean 8.0 mm SL) (M. P. Fahay, pers. observ.). Juveniles first occur in Great Bay–Little Egg Harbor, New Jersey, in June at 4 to 9 cm TL (Fig. 95.1). By August, most range from 5 to 18 cm TL, and smaller larvae still may

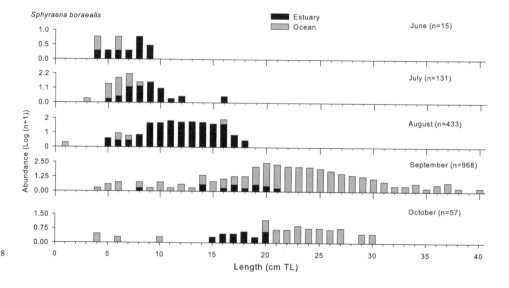

Fig. 95.1. Monthly length frequencies of *Sphyraena borealis* collected in estuarine and coastal ocean waters of the central Middle Atlantic Bight. Sources: National Marine Fisheries Service: otter trawl (n = 998); Rutgers University Marine Field Station: seine (n = 63); Methot trawl (n = 4); 1 m plankton net (n = 1); 4.9 m otter trawl (n = 5); and Able and Fahay 1998 (n = 533).

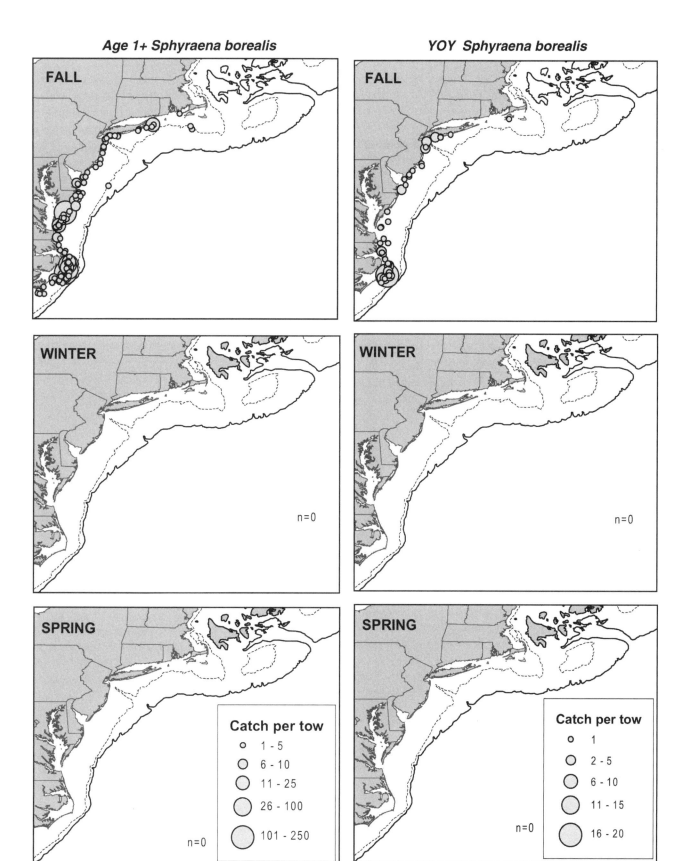

Fig. 95.2. Composite distribution of age 1+ (> 20 cm) and young-of-the-year
(< 20 cm) *Sphyraena borealis* during seasonal cruises of the National Marine
Fisheries Service groundfish survey. Details of sampling effort are indicated
in Fig. 3.4.

occur in adjacent ocean waters. Based on modal length-frequency progression, growth of young-of-the-year averages approximately 1.5 mm per day up to September. By October, they range from 15 to 30 cm TL.

SEASONALITY AND HABITAT USE

Adults often form large schools and occur in coastal waters in depths of 10 to 65 m, usually over muddy substrates. Young-of-the-year are pelagic and occur in estuarine marshes. They typically are found in southern New Jersey estuaries from June through October (see Fig. 95.1). Habitat use, like much of the life history, is poorly known. They are presumed to be pelagic because they are seldom collected in nets fished on the bottom but can be common in nets that enclose the entire water column (Rountree and Able, 1992a; Hagan and Able, 2003). In Great Bay–Little Egg Harbor, they have been collected during the day in subtidal marsh creeks (Rountree and Able, 1993) and have been observed to move onto the marsh surface during flood tides (K. Smith and K. W. Able, pers. observ.). They have been collected in an embayment off a marsh creek in Great Bay frequently at night (Hagan and Able, 2003, 2008). They are not collected in the estuary after October in our study area.

PREY AND PREDATORS

Little is known about the feeding habits of this species, but some observations suggest that they consume *Menidia* spp., other small fishes, mysids, and gastropods (Rountree, 2002).

One study examined the food habits of young-of-the-year (50 to 110 mm, n = 85) in the Delaware River estuary (de Sylva et al., 1962). Young-of-the-year primarily consumed fishes, most frequently *Menidia* spp., as well as *Alosa* spp., *Anchoa* spp., and *Gobiosoma* spp. with decreasing frequency. Young-of-the-year also exhibit cannibalistic behavior on a minor scale; mysids were also important prey. In a study in Great Bay, New Jersey, the young-of-the-year (40 to 170 mm FL, n = 387) ate primarily *Menidia menidia* and smaller numbers of *Anchoa mitchilli* (J. Soell and K. W. Able, unpubl. data). They may feed primarily at dusk and dawn based on field collections and stomach evacuation rates (9 to 12 hours) in the laboratory as part of the same study.

MIGRATIONS

Young-of-the-year are collected in bottom trawls during the fall as they migrate out of estuaries (Fig. 95.2). Most occur in shallow, coastal waters (mean depth 15 m, range 8 to 27 m) from Long Island to Cape Hatteras. Older fish have a similar distribution (mean depth 18 m, range 5 to 58 m) during fall trawl sampling, except that some individuals occur as far east as Cape Cod. Fish of all ages are absent in the Middle Atlantic Bight during winter and spring surveys.

96

Peprilus triacanthus (Peck)

BUTTERFISH

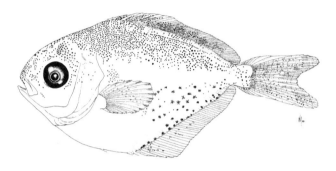

Cohort I, Spring					
Estuary	Reproduction	Larvae/Development	YOY Growth	YOY Overwinter	
Ocean					
Both					
Jan					
Feb					
Mar					
Apr					
May					
Jun					
Jul					
Aug					
Sep					
Oct					
Nov					
Dec					

Cohort II, Summer					
Estuary	Reproduction	Larvae/Development	YOY Growth	YOY Overwinter	
Ocean					
Both					
Jan					
Feb					
Mar					
Apr					
May					
Jun					
Jul					
Aug					
Sep					
Oct					
Nov					
Dec					

DISTRIBUTION

This species occurs along the Atlantic coast of North America from the Gulf of St. Lawrence to southern Florida (Carpenter, 2002). It is most abundant between Maine and Cape Hatteras. A small population is present off Newfoundland, but the species is absent from Bermuda and the Bahamas and only occurs as a stray in the Gulf of Mexico. Juveniles and adults are found in most estuaries in the Middle Atlantic Bight (see Table 4.3), but eggs and larvae have only been reported from the largest of these systems.

REPRODUCTION AND DEVELOPMENT

Spawning occurs during spring and summer in the Middle Atlantic Bight, and as early as February (or year-round, with a peak in spring) in the South Atlantic Bight (Fahay, 1975). There is evidence that spawning extends from February to late July along the coast of the United States, and, based on calculated hatch dates, this may be separated into a spring spawning (February to mid-April) in the South Atlantic Bight and a summer spawning (mid-May to late July) in the Middle Atlantic Bight (Rotunno, 1992; Rotunno and Cowen, 1997). In one New York Bight study, gonadosomatic indices were highest during May and June (Wilk et al., 1990), and a peak in June has also been reported for the Middle Atlantic Bight as a whole (Horn, 1970a). Eggs have been found during the spring along the edge of the continental shelf between Cape Hatteras and Georges Bank (Berrien and Sibunka, 1999) and have also been collected on Georges Bank from May to July (J. Sibunka et al., unpubl.). As seasonal water temperatures on the continental shelf rise, eggs are found increasingly close to the coast, in a south to north progression. Eggs reach a peak in July in the central part of the Middle Atlantic Bight (Berrien and Sibunka, 1999). Sampling of larvae during MARMAP surveys (1977–1987) also indicates a northward progression of spawning activity in the Middle Atlantic Bight (Able and Fahay, 1998). Other studies have reported the presence of larvae in the northern part of the Middle Atlantic Bight, Georges Bank, and the Gulf of Maine as late as October (Kawahara, 1978; Colton et al., 1979; Morse et al., 1987; Rotunno and Cowen, 1997). Although most summer spawning occurs in continental shelf waters of the Middle Atlantic Bight, reports of eggs and larvae (see Table 4.2) indicate that some reproductive activity also occurs in coastal and estuarine waters. Spawning in the spring in the South Atlantic Bight and summer in the Middle Atlantic Bight has the effect of extending the duration of reproduction, a strategy also employed by several other migratory fishes, including *Pomatomus saltatrix*.

The early development of *Peprilus triacanthus* has been well described and summarized (Fahay, 2007). The pelagic eggs are 0.68 to 0.83 mm in diameter and have a single oil globule. The larvae hatch at sizes smaller than 3.0 mm. They have deep bodies and large, rounded heads. They lack pelvic fins, but other fin spines and rays are well formed at sizes < 9.0 mm. Late larvae and juveniles strongly resemble adults while going through a gradual transformation. The formation of scales was examined in specimens from 15.1 to 70.5 mm TL (n = 11). In the smallest individuals that had scales (30.4 mm), scales either completely covered the posterior portion of the body or occurred in two patches on the body (see Fig. 83.1). A larger specimen (45.8 mm TL) had scales covering the entire body behind the operculum and the gular region below the branchiostegals. The size at which the adult condition of scale coverage occurs is unknown.

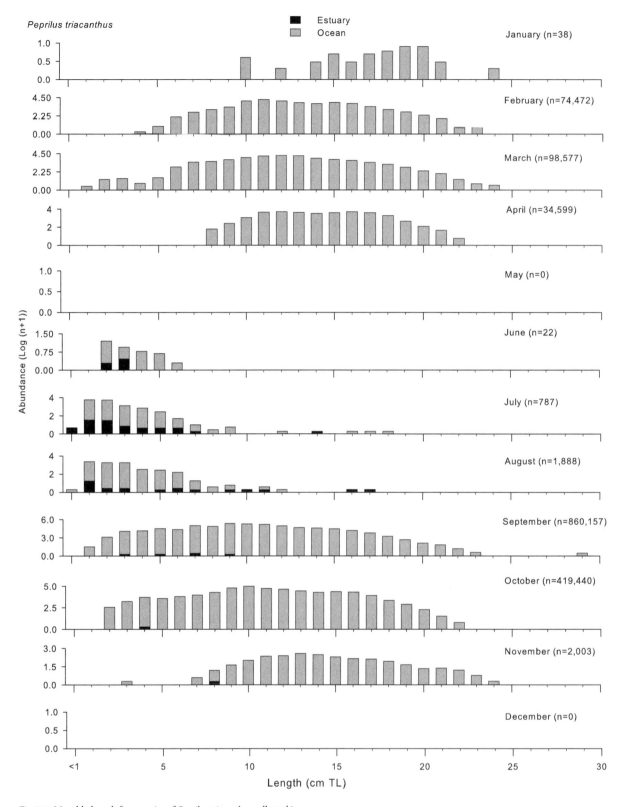

Peprilus triacanthus

Fig. 96.1. Monthly length frequencies of *Peprilus triacanthus* collected in estuarine and coastal ocean waters of the central Middle Atlantic Bight. Sources: National Marine Fisheries Service: otter trawl (n = 1,488,923); Rutgers University Marine Field Station: seine (n = 26); Methot trawl (n = 1788); 1 m plankton net (n = 59); 2 m beam trawl (n = 27); 4.9 m otter trawl (n = 907); and Able and Fahay 1998 (n = 253).

LARVAL SUPPLY, GROWTH, AND MORTALITY

An otolith study has discovered a bimodality in birthdates for young-of-the-year collected in the Middle Atlantic Bight (Rotunno, 1992), with an initial mode the result of spawning as early as February in the South Atlantic Bight. Larvae have been collected in May in the Slope Sea, where they were members of an assemblage that has its origins in oceanic and shelf regions south of Cape Hatteras (Hare et al., 2001). These larvae are transported north in the Gulf Stream and then to the Middle Atlantic Bight via filaments of discharged Gulf Stream water. This cohort is joined by a second one that is the result of summer spawning in the Middle Atlantic Bight. Therefore, juveniles collected in our study area are the result of spawning along the entire Atlantic coast of the United States.

Larvae and adults are both pelagic; therefore there is no identifiable settlement stage. Our estimates of growth during the first year are based on specimens collected from Great Bay–Little Egg Harbor and the nearby coastal ocean and continental shelf (Fig. 96.1). The smallest size classes (10 mm TL) occur in March and again in July and August. Larval growth rates of 0.23 mm per day for fish up to 30 mm have been reported, based on analyses of otolith increments (Rotunno, 1992). By November, the few remaining individuals of this cohort merge with larger fish of unknown age. Our length data do not display a bimodality resulting from two spawning periods as reported by Perlmutter (1939). Modal lengths of fish collected over the continental shelf during the spring are similar to those collected the previous fall, indicating very little growth during the winter. In a neighboring study, young-of-the-year collected in the New York Bight displayed a modal length of 2 cm FL during July and grew to a mode of 9 cm FL in November (Wilk et al., 1975). Growth then appeared to slow over the first winter. The mode in March for these fish was 11 cm, and by June (1 year after the peak in spawning) it had reached 14 cm. Other estimates of size at age 1 range from 9 to 13 cm FL (DuPaul and McEachran, 1973; Waring, 1975; Kawahara, 1978). The mean length at age 4 years has been reported at about 20 cm (Klein-MacPhee, 2002c), and the age composition of fish collected in abundance during spring and fall over the continental shelf probably spans four years. We have no estimates of the rates or causes of mortality in this species, although it reportedly does not survive temperatures below 10°C (Colton, 1972).

SEASONALITY AND HABITAT USE

Generally regarded as a pelagic species, *Peprilus triacanthus* also spends considerable time near sandy substrates on the continental shelf. When individuals occur in coastal waters, they are usually found near the surface over depths of 22 to 55 m (Klein-MacPhee, 2002c). Other evidence suggests they congregate near the bottom during the day and disperse upward at night (Waring, 1975). During early summer, young-of-the-year move from continental shelf waters, where they

are neustonic, to bays and other protected inshore areas (Horn, 1970b; Able and Fahay, 1998). Larger juveniles may dominate collections in some systems, but because the juveniles are pelagic and form schools (at times found near the bottom), they are often under-sampled, as they are in Narragansett Bay, for example (Oviatt and Nixon, 1973). Several authors have reported aggregations of young-of-the-year under the bells of coelenterates (Mansueti, 1963; Horn, 1970b; Scott and Scott, 1988). This relationship is presumably not an obligate one, because juveniles have also frequently been observed swimming freely near the surface (Bigelow and Schroeder, 1953) and juveniles are often collected in neuston nets without coelenterates (M. P. Fahay, pers. observ.).

A few larvae occur in the Cape Hatteras area during March and April, and they increase in abundance in the southern part (and near the continental shelf edge in the northern part) of the Middle Atlantic Bight during May and June. Larvae reach peak abundance in the central part of the bight during July and August (Able and Fahay, 1998), and reduced numbers linger into September and October. Pelagic juveniles (11.3 to 52.1 mm FL) occur year-round in the South Atlantic Bight, and spawning there may be continuous, with a peak in spring (Fahay, 1975). During May, larvae and pelagic juveniles have also been collected in the Slope Sea (Hare et al., 2001). These collections may be the result of advection from spawning in southern waters, and provide some evidence explaining the origin of the early age mode reported by Rotunno (1992).

PREY AND PREDATORS

Adults of this species feed mostly on urochordates and thecosome mollusks (Bowman et al., 2000). Juveniles consume a ctenophore (*Menmiopsis leidyi*) in Narragansett Bay, Rhode Island, and they have also been observed feeding on ctenophores in laboratory studies (Oviatt and Kremer, 1977). Another study examined the food habits of young-of-the-year (22.1 to 100 mm, n = 19) in Long Island Sound (see Table 8.1). Available crustaceans, most frequently copepods, were the majority of young-of-the-year prey.

MIGRATIONS

This species engages in movements that are reactions to temperature changes (Colton, 1972). In response to seasonal cooling, all age groups migrate toward the edge of the continental shelf in the fall and spend the winter near the edge of the continental shelf, where temperatures are warmer (Colton, 1972; Pentilla and Dery, 1988). They then disperse over the shelf during April and May. South of Delaware Bay, where winter cooling is not as severe, this offshore-inshore migration is not as pronounced (Waring and Murawski, 1982).

The young-of-the-year (< 14 cm) of this species are ubiquitous throughout much of the study area for most of the year (Fig. 96.2). In the fall, they are collected from the Gulf of Maine to south of Cape Hatteras across most depths (mean

Age 1+ *Peprilus triacanthus*   ## YOY *Peprilus triacanthus*

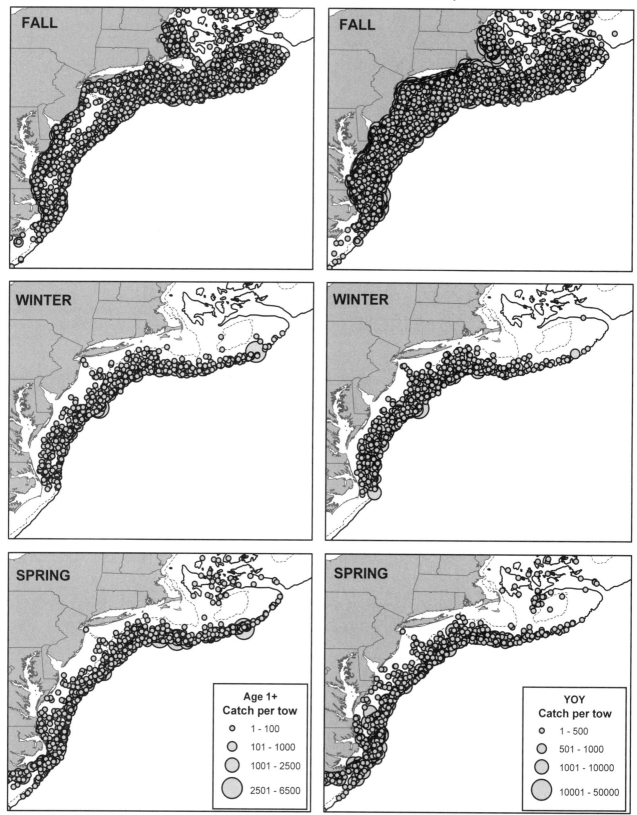

Age 1+
Catch per tow
○ 1 - 100
○ 101 - 1000
○ 1001 - 2500
○ 2501 - 6500

YOY
Catch per tow
○ 1 - 500
○ 501 - 1000
○ 1001 - 10000
○ 10001 - 50000

Fig. 96.2. Composite distribution of age 1+ (> 14 cm) and young-of-the-year (< 14 cm) *Peprilus triacanthus* during seasonal cruises of the National Marine Fisheries Service groundfish survey. Details of sampling effort are indicated in Fig. 3.4.

depth 48.0 m, range 5 to 380 m). In the winter, they are concentrated at the outer portion of the continental shelf (mean depth 86.6 m, range 20 to 388 m) from southern Georges Bank to Cape Hatteras. In the spring, they are still most frequently collected at the edge of the shelf, but some are found in shallower waters (mean depth 89.7 m, range 9 to 367 m) and some are also scattered on northern Georges Bank and into the Gulf of Maine. Age 1+ (> 14 cm) fish are distributed across much of the study area, from the Gulf of Maine to south of Cape Hatteras. In the fall they are collected over most depths (mean depth 64.7 m, range 6 to 382 m) across this range. In the winter, they are concentrated at the edge of

the continental shelf from southern Georges Bank to Cape Hatteras in deeper water (mean depth 95.0 m, range 20 to 413 m) in a pattern similar to that of the young-of-the-year. In the spring, they are found at similar depths, except that some individuals are collected in depths similar to those of winter (mean depth 102.7 m, 8 to 390 m), especially south of the mouth of Chesapeake Bay. Thus the population undergoes seasonal migrations triggered by temperature changes, whereby they are found in deep water in the southern part of their range during winter and summer movements are northward and toward shore (Fritz, 1965; Horn, 1970a; Waring, 1975).

97

Scophthalmus aquosus (Mitchill)

WINDOWPANE

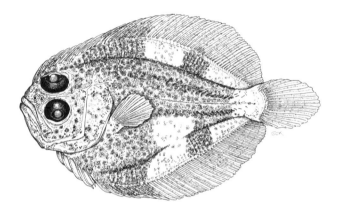

Cohort I, Spring					
Estuary	Reproduction	Larvae/ Development	YOY Growth	YOY Overwinter	
Ocean					
Both					
Jan					
Feb					
Mar					
Apr					
May					
Jun					
Jul					
Aug					
Sep					
Oct					
Nov					
Dec					

Cohort II, Fall					
Estuary	Reproduction	Larvae/ Development	YOY Growth	YOY Overwinter	
Ocean					
Both					
Jan					
Feb					
Mar					
Apr					
May					
Jun					
Jul					
Aug					
Sep					
Oct					
Nov					
Dec					

DISTRIBUTION

Scophthalmus aquosus is distributed from the Gulf of St. Lawrence to Florida (Scott and Scott, 1988) but is most abundant
from Georges Bank to Chesapeake Bay (Morse and Able,
1995). In the Middle Atlantic Bight most early life history
stages have been reported from all the estuaries for which we
have data (see Table 4.3).

REPRODUCTION AND DEVELOPMENT

A protracted spawning season begins in spring in the southern Middle Atlantic Bight, reaches a peak in summer on
Georges Bank, and then resumes as the second part of a split
season in the southern Middle Atlantic Bight in fall. Eggs
and larvae, therefore, occur in our immediate study area in
two cohorts. The entire reproductive period can stretch from
February to November, as shown by gonadosomatic indices
(Wilk et al., 1990; Morse and Able, 1995) and the distribution
of eggs (Berrien and Sibunka, 1999). In the vicinity of Little
Egg Inlet, the split spawning season is first evident when eggs
occur in the spring from the inner shelf into the estuary (Able
and Fahay, 1998). Spawning reoccurs in the fall, although very
few eggs are found in the estuary at this time. Spawning has
also been reported to take place in the high-salinity portions
of other estuaries in the Middle Atlantic Bight, including
Great South Bay (Monteleone, 1992), Sandy Hook Bay (Croker, 1965), Hereford Inlet (Allen et al., 1978), and other locations in the bight (see Morse and Able, 1995). This species is
reported to spawn at dusk or at night (Ferraro, 1980).

The early development of *Scophthalmus aquosus* and characters for distinguishing the eggs, larvae, and juveniles from
those of other flatfish species have been well described and
summarized (Fahay, 2007). Eggs are pelagic and spherical
and range from 0.88 to 1.22 mm in diameter. There may be a
single oil globule or multiple globules. The larvae hatch in 8
days at 10.6 to 13.3°C (Bigelow and Schroeder, 1953) and are
about 2.0 mm long. The body is deep and compressed, and
the gut bulges from the ventral outline. Dark pigment covers the anterior two-thirds of the head and body. All larval
stages are characterized by a very large mouth. Transformation to the juvenile stage begins at about 6.5 mm TL, as fin
rays complete formation and a characteristic pigment pattern forms on the body and extends onto the fins.

Scale formation occurs over a relatively large size range,
from 27 to 80 mm TL (Fig. 97.1; Able and Lamonaca, 2006).
It begins with a few scales on the midline of the caudal peduncle. By 33 mm TL, scales extend along the lateral line to
the posterior edge of the operculum as well as laterally. By
about 36 mm TL, scales have extended anteriorly beyond the
pectoral fins but are not present on the operculum or head.
By 41 mm TL, the lateral extent of scale formation has continued, with new loci for scale development at the dorsal and
ventral margins of the body near the middle of the dorsal
and anal fins. By approximately 46 to 54 mm TL, the body
is completely scaled and scales are beginning to form on the
operculum. Between sizes of 57 and 76 mm TL, scales are
starting to develop on the median fins. By 80 mm TL, scales
have formed on the dorsal, anal, and caudal fins and scale
formation has reached the adult condition. In an attempt
to determine if the extent or rate of development differed
between the ocular and blind sides of this species, scale formation on both sides was examined. Based on these observations, the blind side may lag slightly behind the eyed side at

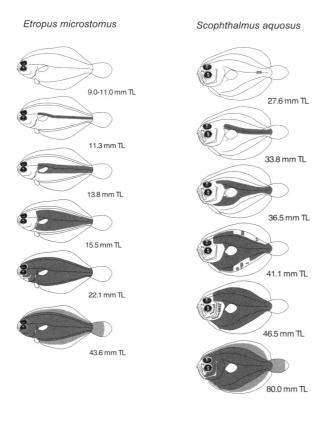

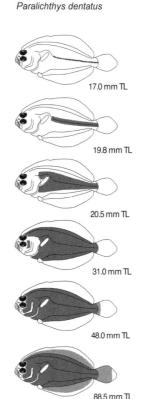

Fig. 97.1. Description of scale formation relative to total length (TL) in *Scophthalmus aquosus, Etropus microstomus*, and *Paralichthys dentatus*.

sizes between 35 and 55 mm TL, but at larger sizes they appear to form scales at a similar rate.

LARVAL SUPPLY, SETTLEMENT, GROWTH, AND MORTALITY

The relative abundance of larvae of the spring cohort was fairly uniform in our study area between 1989 and 2006, as measured by ichthyoplankton collections (Fig. 97.2). Increased abundance was observed in the period 1996–1998 and again between 2002 and 2004. Settlement occurs in the estuary and on the continental shelf for the spring cohort, primarily on the shelf for the fall cohort (Morse and Able, 1995). Eye migration during transformation occurs at sizes of 6.5 to 13.0 mm. Larvae begin to settle to the bottom at sizes of 10.0 mm, but some may remain pelagic to sizes of 20.0 mm, especially those on Georges Bank (Morse and Able, 1995). Otolith studies have demonstrated that in smaller, field-captured individuals belonging to both spring- and fall-spawned cohorts, the formation of accessory growth centers coincides with a transitional settlement period and the completion of eye migration, at approximately 8 to 20 mm SL (Neuman et al., 2001).

The growth patterns of young-of-the-year in southern New Jersey vary with the timing of spawning. Back-calculated growth rate estimates from otoliths for spring-spawned windowpane were significantly faster than those for autumn-spawned windowpane; these differences could produce differential rates of survival for the two cohorts

during the first year of life. The spring-spawned fish grow quickly (Fig. 97.3). Based on modal length-frequency progression, they achieve sizes of approximately 11 to 19 cm by September, approximately 4 months after spawning (Morse and Able, 1995). By the following spring, they are difficult to separate by length alone, but most of this cohort appear to be larger than 16 cm. The fall-spawned fish are most obvious in October, when they are 4 to 7 cm. Perhaps because they are exposed to winter temperatures soon after settlement, they grow very slowly during the winter and reach only 4 to 8 cm by March (Morse and Able, 1995). The fall-spawned

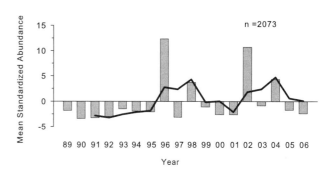

Fig. 97.2. Annual variation in abundance of the spring cohort *Scophthalamus aquosus* in Little Sheepshead Creek Bridge 1 m plankton net collections in the months of March–August, 1989–2006. Data were standardized by subtracting the overall mean from annual abundance. Vertical bars represent standardized annual abundance; the solid line is a three-year running mean that was calculated by taking the mean of the standardized values from the previous two years and the year in which the data are plotted.

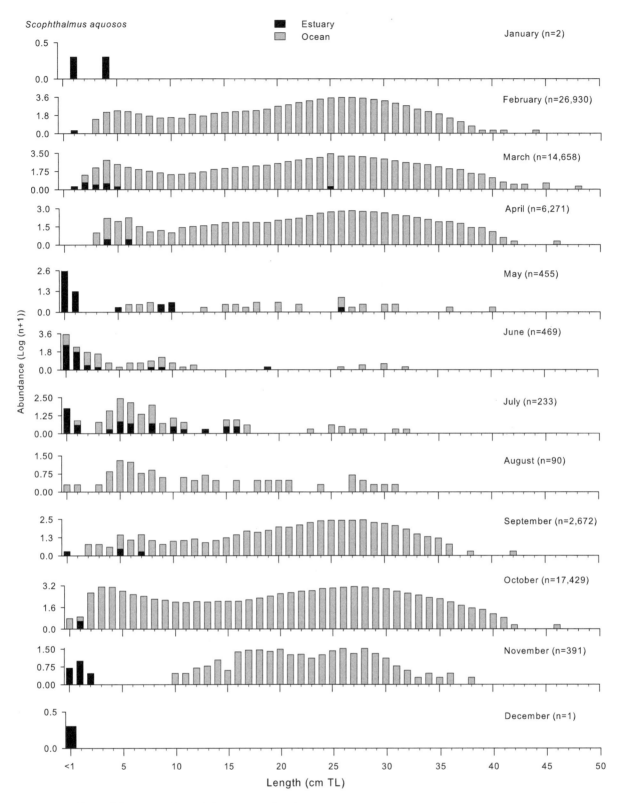

Fig. 97.3. Monthly length frequencies of *Scophthalmus aquosus* collected in
estuarine and coastal ocean waters of the central Middle Atlantic Bight.
Sources: National Marine Fisheries Service: otter trawl (n = 68,280);
Rutgers University Marine Field Station: 1 m beam trawl (n = 3); seine
(n = 323); Methot trawl (n = 20); 1 m plankton net (n = 887); 2 m beam
trawl (n = 5); killitrap (n = 1); 4.9 m otter trawl (n = 82).

fish increase in length through the following summer's growing season and reach approximately 18 to 26 cm by October, when they are approximately age 1. Other studies in the Middle Atlantic Bight indicate much slower rates (Moore, 1947) or are in close agreement (Grosslein and Azarovitz, 1982; Thorpe, 1991) to the rates we have estimated (Morse and Able, 1995).

The two cohorts differ in their initial response to overwinter conditions. An experiment was conducted to test whether differences in size-specific growth and mortality were apparent between or within cohorts in our study area (Neuman and Able, 2009). Sizes in the fall cohort were 15 to 37 mm TL, and sizes in the spring cohort were 83 to 140 mm TL. Members of both cohorts were subjected to ambient winter (November through April) temperatures of −2.0 to 14.0°C, and were provided food throughout the experiment. Growth continued through the winter period in both cohorts, but rates were lower than observed summer rates in this study. Both cohorts suffered some overwinter mortality, but it was higher in the fall cohort (75%) than in the summer cohort (31%). Furthermore, within the fall cohort, mortality was highest in the smallest individuals (< 24 mm TL). Therefore, winter mortality is size-selective and the fall cohort is more vulnerable than the summer cohort. Within the fall cohort, mortality is highest in the smallest sizes tested; only the largest individuals might be expected to survive their first winter.

SEASONALITY AND HABITAT USE

This species is found in shallow, sandy to sand/silt or mud substrates in inner continental shelf waters, usually in depths < 56 m (Wenner and Sedberry, 1989; Thorpe, 1991) but they can also be found as deep as 200 m. They also occur in bays and estuaries throughout their range.

In Great Bay–Little Egg Harbor, eggs may be found throughout the polyhaline portions of the estuary in the spring but are more concentrated on the continental shelf during the fall spawning (Morse and Able, 1995; Able and Fahay, 1998; Berrien and Sibunka, 1999). The larval distribution mirrors that of the eggs in both space and time, that is, they are found throughout the polyhaline portion of the estuary in the spring but primarily on the shelf in the fall (Morse and Able, 1995; Able and Fahay, 1998). On the continental shelf of the Middle Atlantic Bight the smallest larvae are distributed over the middle to inner shelf and onto the shallow portions of Georges Bank. With increasing size, larvae maintain a similar distribution pattern; by 11 to 20 mm they are infrequently collected in the Middle Atlantic Bight but remain abundant on Georges Bank.

Based on intensive collections in the Great Bay–Little Egg Inlet–Beach Haven Ridge corridor, it appears that both the inner shelf and adjacent estuaries serve as habitats for the spring-spawned young-of-the-year (Neuman and Able, 2003). The inner shelf appears to serve as the primary habitat for the fall-spawned young-of-the-year. Although the adults are commonly collected over sand bottoms (Bigelow and Schroeder, 1953; de Sylva et al., 1962), the habitats of the spring- or fall-spawned juveniles are not well defined. In extensive collections in estuarine shallows, juveniles were never collected in intertidal areas, but they occurred frequently along subtidal shores (Able et al., 1996b) and in a variety of deeper (< 1 to 8 m) habitats (Szedlmayer and Able, 1996). In a laboratory study, early demersal (8 to 18 mm SL) and larger (32 to 89 mm SL) juveniles preferred sand over mud substrate (Neuman and Able, 2003). During these observations, early demersal individuals buried less often and exhibited larval pigmentation more than juveniles in the larger size group.

PREY AND PREDATORS

Adult and juvenile *Scophthalmus aquosus* prey on a variety of small crustaceans, including mysids and decapod shrimp, as well as small fishes. The three major components of their diet include mysids, especially *Neomysis americana* (41.7%), small fishes (31.4 %), and decapod crustaceans, especially *Crangon septemspinosa* (14.3 %) (Bowman et al., 2000). They also occasionally consume chaetognaths, squid, mollusks, ascidians, polychaetes, cumaceans, isopods, amphipods, euphausiids, and salps (Hacunda, 1981). Adults (> 20 cm) may feed on juvenile fishes, including *Anchoa* spp., *Ophichthus cruentifer, Merluccius bilinearis, Microgadus tomcod, Brosme brosme,* various fundulids, *Menidia* spp., *Syngnathus fuscus, Helicolenus dactylopterus, Myoxocephalus octodecemspinosus, Morone saxatilis,* and *Ammodytes* spp., as well as various larval fishes (Langton and Bowman, 1980).

Larvae feed on copepods (*Temora* and *Centropages*), but mysids are the most important prey (Moore, 1947). Several studies (n = 3) have examined the food habits of young-of-the-year (2.8 to 170 mm, n = 520) in Middle Atlantic Bight estuaries from Long Island Sound to the Delaware River (see Table 8.1). Important prey of young-of-the-year include a variety of invertebrates (mysids, copepods, amphipods, decapod shrimp, and gastropods). Fishes are a minor dietary component. The diet of young-of-the-year in the Great Bay–Little Egg estuary, New Jersey, shifts with increasing size (Haberland, 2002), such that larvae (< 5 mm, n = 103) consume large amounts of gastropod veligers and calanoid copepods; small young-of-the-year (6 to 40 mm, n = 173) eat a mix of *Neomysis americana*, copepods, and amphipods; larger young-of-the-year (40 to 70 mm, n = 69) consume substantial amounts of *N. americana* along with increasing amounts of *Crangon*; and the largest young-of-the-year examined (70 to 90 mm, n = 38) increase their intake of *Crangon* and *Gammarus spp.* and reduce the amount of *N. americana*.

Major predators on both juveniles and adults include *Squalus acanthias, Lophius americanus, Amblyraja radiata, Gadus morhua, Cynoscion regalis,* and *Paralichthys dentatus* (Rountree, 1999).

Age 1+ *Scophthalmus aquosus*   ## YOY *Scophthalmus aquosus*

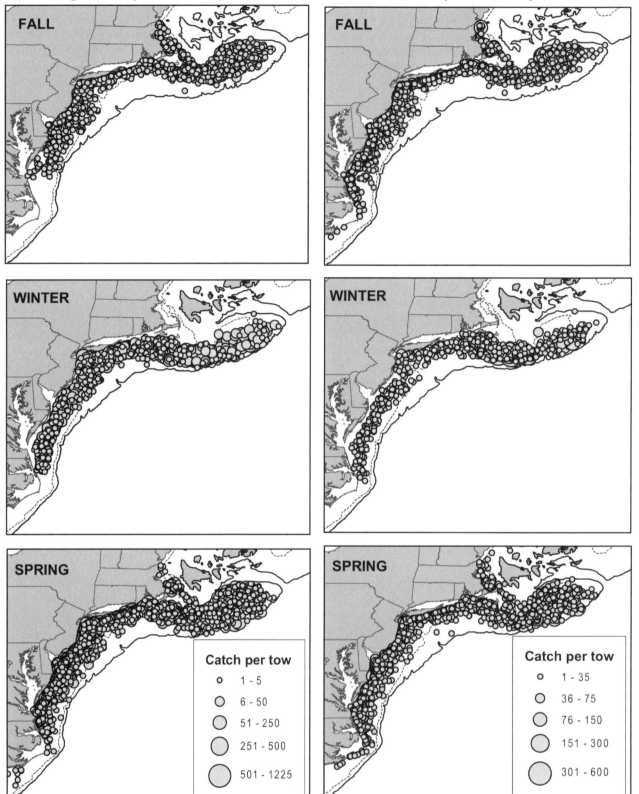

Fig. 97.4. Composite distribution of age 1+ (> 26 cm) and young-of-the-year (< 26 cm) *Scophthalmus aquosus* during seasonal cruises of the National Marine Fisheries Service groundfish survey. Details of sampling effort are indicated in Fig. 3.4.

MIGRATIONS

All benthic life history stages are well represented in all of the seasonal trawl surveys on the continental shelf, and there is no evidence that this species migrates away from the Middle Atlantic Bight during any stage (Fig. 97.4). The young-of-the-year (< 26 cm) have similar geographical distribution patterns, regardless of season, from the southern Gulf of Maine to North Carolina. It appears that both of the young-of-the-year cohorts move out of the estuary and offshore for the winter. Young-of-the-year are found primarily on the inner and mid-shelf in the fall (mean depth 30 m, range 6 to 111 m), winter (mean depth 48 m, range 12 to 151 m), and spring (mean depth 34 m, range 5 to 201 m). On Georges Bank, young-of-the-year are found primarily on the central portion during fall. In the winter, they are distributed farther to the south on the deeper portions, while in the spring they are found over most portions of the bank. During winter and spring they are seldom found on the eastern edge of Georges Bank.

The age 1+ (> 26 cm) juveniles and adults have a spatial and depth distribution very similar to that of the young-of-the-year in all seasons except that they are not present south of Chesapeake Bay in the fall (see Fig. 97.4). During all seasons they are found primarily on the inner to mid-shelf or slightly deeper (fall mean depth 36 m, range 8 to 180 m; winter mean depth 49 m, range 12 to 151 m; and spring mean depth 37 m, range 5 to 220 m).

98

Etropus microstomus (Gill)

SMALLMOUTH FLOUNDER

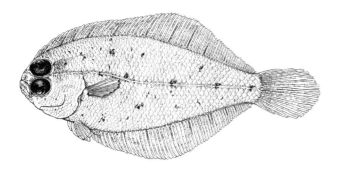

Estuary	Reproduction	Larvae/ Development	YOY Growth	YOY Overwinter
Ocean				
Both				
Jan				
Feb				
Mar				
Apr				
May				
Jun				
Jul				
Aug				
Sep				
Oct				
Nov				
Dec				

DISTRIBUTION

Etropus microstomus occurs along the east coast of North America from southern New England to South Carolina, but might stray as far south as Florida (Parr, 1931; Leslie and Stewart, 1986). The early life history stages have been col-lected in the polyhaline portions of adjacent estuaries from Long Island to North Carolina, but are uncommon in the northern portion of the Middle Atlantic Bight (see Table 4.3).

REPRODUCTION AND DEVELOPMENT

First reproduction occurs at sizes as small as 60 mm TL. Spawning occurs during the summer through early fall. Small larvae begin to appear in the spring off North Caro-lina, their occurrences increase northward during the sum-mer, and then late larvae remain in southern areas in the fall (Richardson and Joseph, 1973; Smith et al., 1975). Based on extensive collections over the continental shelf during 1977 to 1987, the smallest larvae (< 3.9 mm) occurred between May and October and were most abundant in July and Au-gust (Able and Fahay, 1998). In continental shelf collections, larvae are most dense in depths < 40 m. The presence of eggs in Great South Bay on Long Island (Monteleone, 1992) and Hereford Inlet in southern New Jersey (Keirans, 1977) in-dicates that spawning may also occur in estuaries.

The early development of this species, including charac-ters useful for distinguishing the larvae and juveniles from those of other flatfish species, has been well described and summarized (Fahay, 2007). Eggs have been reported to be pe-lagic and spherical and range from 0.56 to 0.74 mm in di-ameter. They have a single oil globule (Scherer and Bourne, 1980). However, characters are not available for distinguish-ing the eggs from those of *Citharichthys arctifrons*, a closely related species common on the outer continental shelf. Lar-vae hatch at a length of 1.4 mm NL. They are deep-bodied and have a characteristic pigment pattern, including a me-lanophore on the pectoral fin base, not present in larvae of *Citharichthys*. Larvae as small as 9.0 mm SL have complete complements of fin rays and vertebrae. Transformation to the juvenile stage occurs at 10 to 12 mm.

Scale formation occurs over a size range of approxi-mately 33 mm TL (see Fig. 97.1; Able and Lamonaca, 2006). The onset of scale formation occurs between 9.0 and 11.0 mm TL, based on our examination of 62 specimens between 10 and 60 mm. Scales begin forming on the lateral surface of the body and specimens 11.3 mm TL have a fully scaled lateral line. By about 13 mm TL, the area covered by the scales has expanded both above and below the lateral line and this expansion continues at 15 mm TL. By 17 to 18 mm TL, the body is completely covered and scales have begun to develop on the operculum. By approximately 22 mm TL, scales cover the head, except for the snout (as in the adult condition) and small areas of the operculum. At larger sizes, scales begin forming on the dorsal, anal, and caudal fins and they reach the adult condition by approximately 43 mm TL, when scales cover all except the snout and tips of the me-dian fins. See Figures 97.1 and 100.1 for comparisons with other flatfishes.

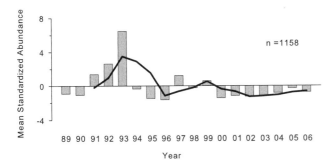

Fig. 98.1. Annual variation in abundance of *Etropus microstomus* in Little Sheepshead Creek Bridge 1 m plankton net collections, 1989–2006. Data were standardized by subtracting the overall mean from annual abundance. Vertical bars represent standardized annual abundance; the solid line is a three-year running mean that was calculated by taking the mean of the standardized values from the previous two years and the year in which the data are plotted.

LARVAL SUPPLY, SETTLEMENT, GROWTH, AND MORTALITY

Larvae are common, but not abundant, in estuarine sampling behind Little Egg Inlet. In an 18-year sampling period (1989–2006) larvae exceeded the average numbers during a single 4-year period (Fig. 98.1). It is unknown whether this increased abundance reflected more eggs and larvae on the adjacent continental shelf, because comparable sampling was not undertaken during those years.

Larvae of *Etropus microstomus* occur in the plankton at sizes of up to 20 mm (Able and Fahay, 1998). In the vicinity of Beach Haven Ridge, larval duration, based on examination of daily otolith increments, is up to 23 days at sizes less than 14.8 mm SL (L. S. Hales and K. W. Able, unpubl. data). Transformation, as indicated by eye migration, occurs over sizes of 10 to 12 mm (Richardson and Joseph, 1973) or at 8 to 15 mm SL at Beach Haven Ridge (L. S. Hales and K. W. Able, unpubl. data). Settlement in New Jersey waters occurs from June through November in the estuary and the ocean, based

on the presence of the smallest juveniles (< 2 cm TL; Fig. 98.2). More detailed studies at Beach Haven Ridge found the smallest early demersal individuals ranging from 11.8 to 22.0 mm (L. S. Hales and K. W. Able, unpubl. data).

Studies of the growth of demersal juveniles and adults (R. Bush and K. W. Able, unpubl. data) indicate that estuarine and inner continental shelf collections from southern New Jersey are dominated by young-of-the-year (12 to 90 mm TL; 73% of all collections). Age 1 fish ranged from 45 to 120 mm TL and make up 26% of the collections. These estimates are based on comparisons of length frequencies and sagittal otolith increments, which are in close agreement. The monthly progression of modal lengths indicates that after settlement, growth in the ocean and the estuary is relatively slow (0.2 mm per day) so that the young-of-the-year are approximately 2 to 11 cm TL by November (see Fig. 98.2). There is little evidence of growth through the winter. This lack of growth at low temperatures has been confirmed in laboratory experiments (Hales and Able, 2001). Growth resumes during spring, based on the occurrence of larger individuals after April. There is also evidence of size-selective mortality for young-of-the-year at low winter temperatures (Hales and Able, 2001).

SEASONALITY AND HABITAT USE

Adults of this species occur in estuaries, bays, and the coastal ocean, at depths generally shallower than 35 m (Klein-MacPhee, 2002d). Demersal young-of-the-year are present in Great Bay–Little Egg Harbor during the summer and fall but are most abundant in September and October; their numbers decline during the early winter. These individuals are found across an array of habitats in this estuary, but are conspicuously absent from some, including those in the mesohaline and oligohaline portions of the estuary (Fig. 98.3). They have been reported from salinities of 7 to 32.5 ppt in Chesapeake Bay (Musick, 1972). Also, they do not occur in shallow

Are Ctenophores Pre-settlement Habitat for Larval Fishes?

A single observation by Dave Witting during one summer suggested that metamorphosing *Etropus microstomus* may use ctenophores as habitat. While hanging on an anchor chain during a decompression stop at the end of a dive in the vicinity of Beach Haven Ridge in approximately 18 to 20 m of water off southern New Jersey, he observed numerous ctenophores drifting by. A single *Etropus* larva was also observed and as the diver moved, the larva seemingly reacted and swam quickly from the water column to a nearby ctenophore (*Mnemiopsis leidyi*) and attached to its side, where it remained stationary and almost invisible.

If this is a common phenomenon, does this behavior provide pre-settlement habitat? Does this habitat/behavior provide a predator refuge for flatfishes in the water column? Does it imply that this behavior is a precursor to settlement on bottom substrate for this benthic fish? If any of these are true, it should prompt us to think differently about fish habitat for larval fishes.

This behavior could occur in estuaries since both the larvae of this species (Witting et al. 1999, Chapter 6) and these ctenophores are common in many temperate estuaries. Certainly these types of symbiotic associations do occur between jellyfishes and *Peprilus triacanthus* (Mansueti 1963), although the nature of the association changes with ontogeny.

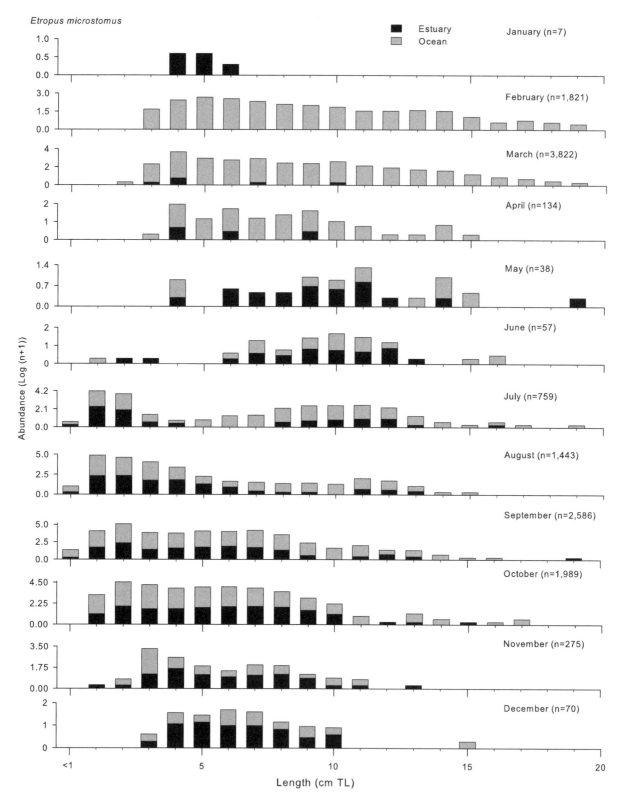

Fig. 98.2. Monthly length frequencies of *Etropus microstomus* collected in estuarine and coastal ocean waters of the central Middle Atlantic Bight. Sources: National Marine Fisheries Service: otter trawl (n = 6658); Rutgers University Marine Field Station: 1 m beam trawl (n = 9); seine (n = 496); Methot trawl (n = 499); 1 m plankton net (n = 559); 2 m beam trawl (n = 246); killitrap (n = 3); 4.9 m otter trawl (n = 211); and Able and Fahay 1998 (n = 4319).

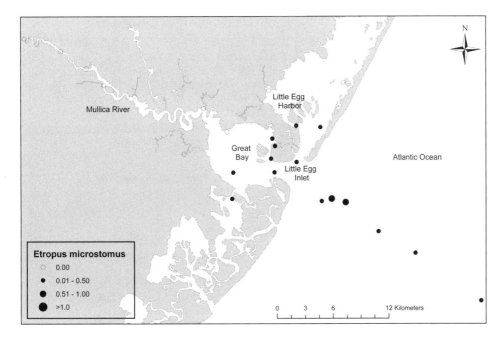

Fig. 98.3. Habitat use and average catch per tow (see key) for juvenile and adult *Etropus microstomus* in the Ocean–Great Bay–Mullica River corridor in southern New Jersey based on small otter trawl collections during July and September from 1988 to 2006. See Table 3.2 for characteristics at each sampling collection.

eelgrass or in areas with high silt content. Both of these latter habitat types are farthest from the inlet (Szedlmayer and Able, 1996). They are also not found in subtidal creeks (Rountree and Able, 1992a) but are collected along shallow subtidal shorelines (Able et al., 1996b). At a site on the inner continental shelf adjacent to Great Bay, juveniles and adults are one of the most abundant species collected by beam trawl (Able et al., 2006a).

PREY AND PREDATORS

The young-of-the-year and age 1 individuals (< 120 mm, n = 12) fed on 19 different types of prey, but the diet primarily consisted of polychaetes and crustaceans, including amphipods and hermit crabs (Richards, 1963).

On the continental shelf, *Etropus microstomus* are important prey items for at least seven species, especially for *Squalus acanthias* and *Pseudopleuronectes americanus* (Rountree, 1999).

MIGRATIONS

Etropus microstomus in any life history stage is infrequently collected during the winter in the estuary. This might be explained by movement into deeper water or avoidance of the collecting gear by burial in the substrate, which is known to occur at low temperatures in the laboratory (Hales and Able, 2001) and is inferred from seasonal collections near Cape

Hatteras (Leslie and Stewart, 1986). On the continental shelf, young-of-the-year (< 9 cm) are commonly collected from the vicinity of Cape Hatteras to Georges Bank although they are most consistently encountered south of Massachusetts in all seasons (Fig. 98.4). In the summer and fall, they frequently occur in estuaries. However, in the fall, they are distributed across the entire shelf to the edge in depths from 6 to 207 m (mean depth 38.5 m). In the winter, there appear to be more collections from the middle of the shelf in depths ranging from 16 to 296 m (mean depth 40.5 m). By the spring, the greatest frequency of individuals is in inner shelf waters but they still extend into deeper waters so that the depth range is 5 to 235 m (mean depth 23.4 m).

Juveniles and adults (age 1+, > 9 cm) are distributed between Cape Hatteras to off Massachusetts with a couple of individuals recorded from Georges Bank (see Fig. 98.4). There appear to be short seasonal migrations. In the fall, most individuals are collected on the inner continental shelf (mean depth 29.2 m, range 9 to 131 m). In the winter, when sampling is reduced, fewer individuals are collected on the inner shelf. Most collections occur on the middle of the shelf in depths from 16 to 78 m (mean depth 32.9 m). In the spring, the pattern of occurrence is similar to that in the fall, with most individuals on the inner shelf in shallow waters (mean depth 24.1 m, range 7 to 235 m).

Age 1+ *Etropus microstomus*   **YOY *Etropus microstomus***

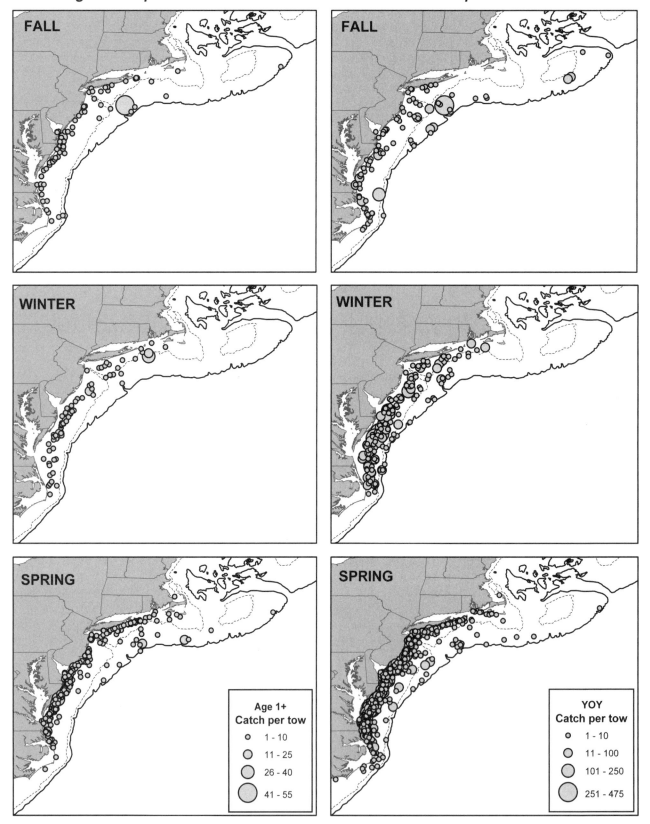

Fig. 98.4. Composite distribution of age 1+ (> 9 cm) and young-of-the-year (< 9 cm) *Etropus microstomus* during seasonal cruises of the National Marine Fisheries Service groundfish survey. Details of sampling effort are indicated in Fig. 3.4.

99

Paralichthys dentatus (Linnaeus)

SUMMER FLOUNDER

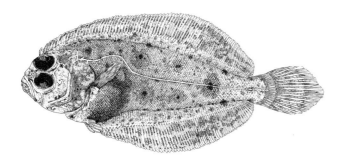

Estuary	Reproduction	Larvae/ Development	YOY Growth	YOY Overwinter
Ocean				
Both				
Jan				
Feb				
Mar				
Apr				
May				
Jun				
Jul				
Aug				
Sep				
Oct				
Nov				
Dec				

DISTRIBUTION

Paralichthys dentatus occurs in the western North Atlantic Ocean from Nova Scotia to the east coast of Florida (Gutherz, 1967; Gilbert, 1986; Scott and Scott, 1988). Despite considerable discussion concerning the occurrence of separate stocks or populations, the issue is largely unresolved (see Kraus and Musick, 2001). This species is most abundant on the conti-

nental shelf from Cape Cod, Massachusetts, to North Carolina (Able and Kaiser, 1994), but it also commonly occurs in estuaries within this range. In the Middle Atlantic Bight, larvae, juveniles, and adults are reported from most estuaries (see Table 4.3). However, some of these records may be questionable. Except for the report of nine larvae collected in the Hudson River estuary (Dovel, 1981) and a single report cited in Berg and Levinton (1985), which may pertain to a single larva, we have not found published evidence of larvae occurring in the Hudson/Raritan system, nor have we been able to find a reliable record of eggs ever being found there, despite the summary in Stone et al. (1994).

REPRODUCTION AND DEVELOPMENT

Length at maturity ranges from 240 to 270 mm TL for males and from 300 to 330 mm TL for females at age 2 (Morse, 1981). Males dominate the lengths between 210 to 350 mm TL, while females are more abundant at all lengths > 450 mm TL. Spawning occurs on the continental shelf in the Middle Atlantic Bight from September through January, with a peak in October and November, based on examination of gonads and the occurrence of eggs or larvae (Morse, 1981; Morse et al., 1987; Able et al., 1990; Wilk et al., 1990; Able and Kaiser, 1994; Berrien and Sibunka, 1999). During the fall, eggs initially occur near shore and then spread across the entire continental shelf, which suggests that spawning takes place as the adults are migrating offshore. By December and January, eggs are only found at the edge of the continental shelf. Egg distributions indicate that limited spawning continues in the southern part of the bight into the spring (Berrien and Sibunka, 1999).

The early development, and characters useful for distinguishing the eggs, larvae, and juveniles from those of other flatfishes, have been well described and summarized (Smith and Fahay, 1970; Fahay, 2007). Pigment patterns are critical for distinguishing the larvae from those of congeners. Eggs are pelagic, spherical, and 0.95 to 1.03 mm in diameter. The single oil globule is 0.17 to 0.23 mm in diameter. Eggs hatch in 2 to 9 days, depending upon temperature (Smith and Fahay, 1970; Smigielski, 1975; Johns et al., 1981). Larvae hatch at about 2.4 to 2.8 mm, with unpigmented eyes and unformed mouth parts. Larvae soon become deep-bodied, and typically have a bulge on top of the head. Transformation to the juvenile stage occurs between 9.5 and 13.0 mm SL (Able and Fahay, 1998). Low winter temperatures may influence stage duration, at least during metamorphosis (Keefe and Able, 1993). Individuals in the laboratory averaged 24.5 days (range 20 to 32 days) to complete metamorphosis at ambient spring temperatures of 16.6°C, but partial metamorphosis at colder average temperatures (6.6°C), required as much as 92.9 days (range 67 to 99 days).

Scale formation occurs over a larger size range than in other flatfish species, that is, 17 to 88 mm (see Fig. 97.1; Able and Lamonaca, 2006). By 17 to 19 mm TL, scales are present along the lateral line from the pectoral fin to the caudal fin.

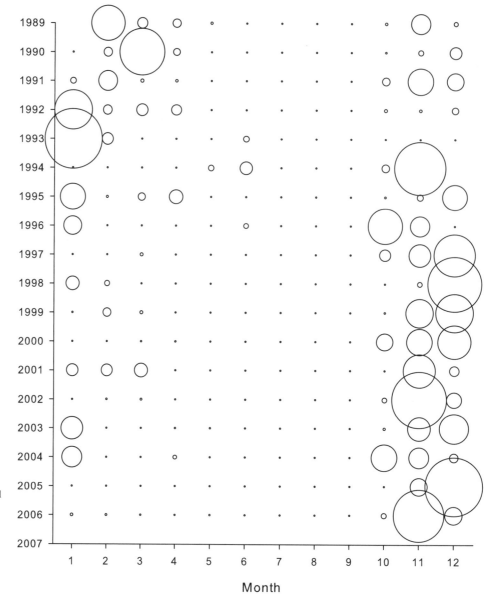

Fig. 99.1. Seasonal occurrence of the larvae of *Paralichthys dentatus* based on ichthyoplankton samples collected in Little Sheepshead Creek behind Little Egg Inlet from 1989 to 2006. Full circles represent the percentage of the mean number/1000 m³ of larval young-of-the year *Paralichthys dentatus* captured by given year. Values range from 0 to 79%. The smallest circles represent when samples were taken but no larvae were collected.

By approximately 20 mm TL, scales have expanded laterally but not anteriorly. By 22 mm TL, scales have expanded farther laterally but also anteriorly to the posterior margin of the operculum. By 31 mm TL, scale formation has extended to the dorsal and ventral limits of the body and also anteriorly to the dorsal portion of the head behind the eyes. By 40 mm TL, all scales are formed on the body and some are beginning to form on the median fins. By approximately 88 mm TL, all the scales are formed on the dorsal, anal, and caudal fins and scale formation has reached the adult condition. See Figures 97.1 and 100.1 for comparisons with other flatfishes.

LARVAL SUPPLY, SETTLEMENT, GROWTH, AND MORTALITY

Reproduction occurs over the continental shelf in the Middle Atlantic Bight (Able et al., 1990; Able and Fahay, 1998; Packer

et al., 1999) and, as a result, incubation, hatching, and early larval development occur at sea. This species is one of the few flatfishes for which we have some detailed information on transformation, immigration through inlets, and settlement. Larvae of 13 mm or larger have been collected from October through March (and rarely as late as June) on the continental shelf, in Little Egg Inlet and in adjacent estuaries (Olney, 1983; Olney and Boehlert, 1988; Able et al., 1990). Larvae at these larger sizes are initiating eye migration (Keefe and Able, 1993). Recent years have seen a shift in periodicity for this immigration (Fig. 99.1). In the earliest years of our sampling for larvae (1989–1993), they began to appear in October, but did not reach peak abundance until January through March. In subsequent years (1994–2007), most of this immigration occurred earlier, especially during November and December, which coincides with the time when larvae are most abundant on the shelf (Morse et al., 1987; Able

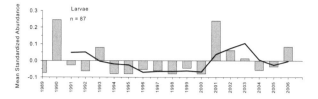

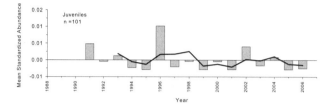

Fig. 99.2. Annual mean (bars) and three-year moving mean (line) of standardized abundance for *Paralichthys dentatus* collected in larval sampling with plankton net (*upper*) and juvenile sampling with otter trawls (*lower*). Data were standardized by subtracting overall mean from annual abundance. Three-year moving mean was calculated by taking the mean of the standardized values from the previous two years and the year in which data are plotted.

and Fahay, 1998). Broad-scale ichthyoplankton sampling over the continental shelf ended in 1987; thus we lack information on timing or distribution of eggs and larvae between 1989 and 2007. Perhaps this temporal shift in immigration is in response to the increased abundance of adults in recent years (Able et al., in review). Although larvae that enter Little Egg Inlet are relatively constant in abundance between years (1989–2006), there have been occasional years with noticeable increases, as in 2001–2003 (Fig. 99.2). Although it would be informative to compare these data with the same information from other inlets in the Middle Atlantic Bight, we presently lack these data.

Growth rates of young-of-the-year in the Mullica River–Great Bay estuary are rapid, after a fall and winter period of almost no growth. Little growth is apparent in larvae that arrive in the estuary as early as October (at sizes of 1 to 2 cm) until May, when a few may reach 5 cm (Fig. 99.3). Larger young-of-the-year (> 8 cm TL) are commonly collected during the summer and early fall. These grow very quickly, averaging 1.5 to 1.9 mm per day (Rountree and Able, 1992b; Szedlmayer et al., 1992). This year class therefore reaches modal lengths of about 27 cm by September and October. During the spring (May–June), growth rates of caged juveniles (12 to 41 mm) ranged from approximately 0.5 to more than 1.0 mm per day (Keefe and Able, 1992). At these rates, increments in the sagittal otoliths are deposited daily (Szedlmayer and Able, 1992) and thus may be useful in determining more detailed growth rates. Length frequencies derived from individuals collected in Delaware Bay are very similar to those from the Mullica River–Great Bay estuary (K. W. Able, unpubl. data).

Sources of mortality are difficult to determine. However, larval survival may be dependent on the length of delay after the mouth becomes functional but before first feeding com-

mences (Bisbal and Bengtson, 1995). The point of no return for non-feeding larvae, that is, the time at which feeding is no longer possible and mortality is certain, is six to seven days after hatching. In addition, low winter temperatures may have significant effects on early demersal individuals that enter the estuary in the winter. Small juveniles confined in natural habitats in New Jersey during the winter grew a negligible amount (−0.6 and 0.01 mm per day), and there was little change in development stage (M. Keefe and K. W. Able, unpubl. data). Transforming larvae and juveniles exposed to temperatures below 2 to 3 °C suffer significant mortality, at least in the laboratory (Malloy and Targett, 1991; Szedlmayer et al., 1992). The impact of low temperatures may vary between years, depending on the severity of the winter. In some years, the larvae enter much later, and thus may encounter rising temperatures during the spring warming and more favorable conditions for survival and growth.

SEASONALITY AND HABITAT USE

Adults are found on the continental shelf over sand or mud substrates, or in bays and estuaries, where they are associated with eelgrass beds or muddy and silty bottoms (Able and Kaiser, 1994; Able and Fahay, 1998). Eggs are shed and fertilized as adults undertake a cross-shelf fall migration toward deeper parts of the shelf for the winter. Under natural conditions, the pelagic larvae develop in continental shelf waters at sizes from 2 mm SL to approximately 13 mm SL (Able et al., 1990). Most larvae are present from October through January over the entire Middle Atlantic Bight shelf and Georges Bank (Morse et al., 1987; Able and Fahay, 1998). Peak larval abundance appears to occur in November and December. Movement of larvae into Great Bay–Little Egg Harbor occurs during the fall and winter. Movement into the estuary may involve intermittent settling to take advantage of tidal stream transport before permanent settlement (Hare et al., 2005a). When transforming larvae collected from Great Bay are held in aquaria in the absence of tidal currents, they swim up into the water column more often at night (Keefe and Able, 1994). These larvae are capable of periodically resting on the substrate on their right sides even before completing eye migration. Some transforming larvae are capable of partial burial at early stages but are not able to bury completely until late in eye migration (Keefe and Able, 1994).

Although planktonic larvae are common in New Jersey estuarine inlets, small, post-settlement juveniles (< 80 mm TL) have not been abundant there (Able et al., 1990; Szedlmayer et al., 1992; Keefe and Able, 1993; Able and Fahay, 1998). However, they have been collected frequently on Virginia's eastern shore (Wyanski, 1988) and more commonly in South Atlantic Bight estuaries (see Able and Kaiser, 1994). In New Jersey estuaries, transforming larvae and young-of-the-year can be found across a variety of high-salinity, subtidal habitats, including subtidal marsh creeks, coves, large bays, and inlets in both vegetated and unvegetated habitats (Fig.

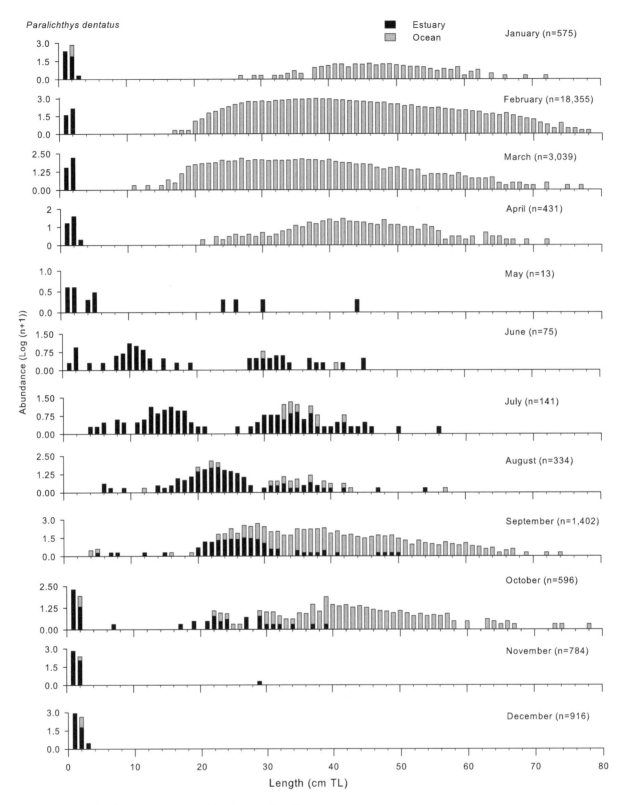

Fig. 99.3. Monthly length frequencies of *Paralichthys dentatus* collected in
estuarine and coastal ocean waters of the central Middle Atlantic Bight.
Sources: National Marine Fisheries Service: otter trawl (n = 23,130);
Rutgers University Marine Field Station: 1 m beam trawl (n = 3); seine
(n = 94); 1 m plankton net (n = 2022); 2 m beam trawl (n = 7); experimental trap (n = 2); killitrap (n = 5); 4.9 m otter trawl (n = 97); multi-panel gill
net (n = 6); and Able and Fahay 1998 (n = 1295).

99.4; Able et al., 1989; Rountree and Able, 1992b; Szedlmayer et al., 1992). A few have been collected on the inner continental shelf, but this occurred in the fall when they were presumably migrating out of the estuary. The temperature and salinity ranges where most life history stages are encountered are wide because of their extensive movement patterns (Able and Kaiser, 1994). In Great Bay–Little Egg Harbor, some of the recently settled juveniles (< 27 SL) have been found in many habitats, including salt marsh creeks, shallow coves, and shallow portions of bays. In North Carolina, these early settlement stages are most often found on shallow, high-energy sandy beaches, or shallow, low-energy muddy marshes (Kellison and Taylor, 2007). These early settlers demonstrate a high degree of site fidelity, and most remain in these same habitats for as long as 120 days, or until attaining lengths of 80 to 120 mm SL (Kellison, 2000). They can also be associated with tidal flats (Burke, 1995) or sand substrate or in transition zones from fine sand to silt and clay (Powell and Schwartz, 1977).

Although estuarine habitat quality might be expected to affect growth and survival, a North Carolina study measuring the effects of different levels of dissolved oxygen, salinity, temperature, and acidity (pH) on growth rates and survival found little variation in growth rates of caged juveniles between five mainland and barrier island habitats (Necaise et al., 2005). However, survival between the study sites did differ, and total mortality occurred in one of the barrier island sites. Hypoxia in certain estuarine habitats, especially coincident with higher temperatures (20 to 25°C), can result in lower feeding and depressed growth rates in juveniles (Stierhoff et al., 2006). This study observed a 25 to 60% decrease in growth rate under hypoxic conditions, depending on oxygen saturation levels. When salinity was reduced, the effect of hypoxia on growth rate was reduced from 25 to 15%.

During the summer, ultrasonically tagged individuals (210 to 254 mm TL) followed a regular pattern of movements in a 1 km-long subtidal creek in Great Bay–Little Egg Harbor (Szedlmayer and Able, 1993). These individuals spent most of the time at the mouth of the creek during the July–September study period. Movements up the creek typically occurred on night high tides, followed by a return down the creek on the following night ebb tide. Upstream movement appeared to be associated with periods of feeding on marsh creek fishes and crustaceans (Rountree and Able, 1992b). Movements down the creek may have been caused by the need to avoid low dissolved oxygen conditions in the upper portion of the creek on night low tides (Szedlmayer and Able, 1993). Larger juveniles (268 to 535 mm) in the same estuary resided there for most of the summer (mean 86 days) based on ultrasonic telemetry (Sackett et al., 2008). During this time they primarily used the lower bay, but moved into the upper portion of the estuary in a year when salinities averaged higher. Over diel and tidal periods these fish resided within small (mean 0.18 km²) areas for 3 to 6 hours but were moving 74% of the

time. Other studies demonstrated that they emigrated from the estuary between July (prompted by storms) and September, and this interpretation was supported by trawl surveys in the ocean off New Jersey (Sackett et al., 2007). Up to 34% of the ultrasonically tagged fish homed to the same estuary the next year (Sackett et al., 2007).

PREY AND PREDATORS

The major food items for adults are other bony fishes. Primary among these are *Ammodytes* spp., *Anchoa* spp., *Etrumeus teres, Merluccius bilinearis*, and other flatfishes (Klein-MacPhee, 2002d). Young-of-the-year *Pseudopleuronectes americanus* and sand shrimp (*Crangon septemspinosa*) are the dominant prey of age 1+ *Paralichthys dentatus* in the Navesink River, New Jersey, during June and July, but later in the summer this diet shifts to include blue crabs (*Callinectes sapidus*) and other fishes (*Menidia menidia* and *Brevoortia tyrannus*) (Manderson et al., 1999). Furthermore, a related laboratory study found that *P. dentatus* prefers the demersal *P. americanus* to pelagic prey (*Menidia menidia*) and the vulnerability of *P. americanus* increases with increasing size (Manderson et al., 2000). The vulnerability of this prey is not related to grain size of the sediments upon which it rested, thus negating the effect of burying as a survival technique, but it is reduced when vegetation (*Zostera marina*) is tested. Therefore, in this case, the prey's vulnerability is more related to the predator's visual and attack capabilities.

Pelagic, preflexion stage larvae collected over the continental shelf feed primarily during daylight hours on immature copepodites (Grover, 1998). As larvae grow through flexion stages and approach transformation, the diet gradually changes from copepod nauplii and tintinnids to larger prey, including calanoid copepods and appendicularians. Transforming larvae migrating through Little Egg Inlet at night feed at lower rates than earlier stages on the shelf. Once inside the estuary, these transforming juveniles consume mostly the calanoid copepod, *Temora longicornis* (Grover, 1998). The incidence of feeding decreases as transforming juveniles pass through eye migration (Keefe and Able, 1993). Another study documented the shift in early-stage diets from yolk dependency to first-feeding on zooplankton and from there to piscivory (Witting et al., 2004). Several other studies (n = 9) have examined the food habits of young-of-the-year (8.1 to 300 mm, n = 1211) in Middle Atlantic Bight estuaries from Great Bay–Little Egg Harbor, New Jersey, to Pamlico Sound, North Carolina (see Table 8.1). Important prey include a variety of invertebrates (mysid shrimp, decapod shrimp, polychaetes, copepods, and crabs) and fishes (*Menidia menidia, Fundulus heteroclitus*, and *Leiostomus xanthurus*). Detritus and sediment were also prominent constituents of the stomach contents. The diet of small young-of-the-year (9 to 20 mm, n = 173) shifted with growth during the immigration period (January–March 1988) to the Newport and North rivers of North Carolina (Burke, 1995), such that the

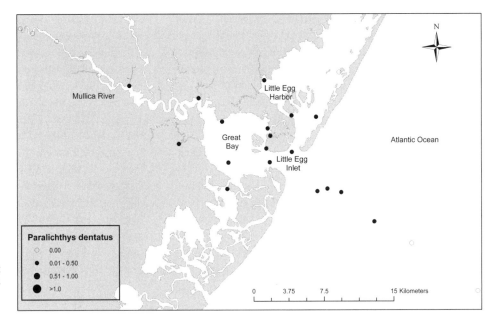

Fig. 99.4. Habitat use and average catch per tow (see key) for juvenile and adult *Paralichthys dentatus* in the Ocean–Great Bay–Mullica River corridor in southern New Jersey based on small otter trawl collections during July and September from 1988 to 2006. See Table 3.2 for characteristics at each sampling collection.

smallest individuals (9 to 12 mm, n = 110) consumed mainly a mixture of polychaetes, polychaete tentacles, mysids, and harpactacoid copepods. With growth (13 to 16 mm, n = 53), polychaete tentacles decreased in significance, harpacticoids were no longer consumed, and polychaetes and calanoid copepods became more important. With further growth (17 to 20 mm, n = 10), mysids became important prey in addition to polychaetes, supplemented by calanoid copepods and amphipods. In Pamlico Sound, North Carolina, diets also shifted such that smaller (100 to 200 mm, n = 470) individuals consumed predominantly mysids (42% by volume) and fishes (38%), while larger young-of-the-year (201 to 300 mm, n = 9) consumed more fishes (56%) and decapod shrimp (36%) (Powell and Schwartz, 1979). Some changes in feeding patterns related to temporal abundance of a food item in a discrete area were also noted in Pamlico Sound, such that during summer a relatively large amount of amphipods was consumed by fish at one collecting locality, while during winter and spring shrimp were relatively common in the diet near inlet stations. Laboratory experiments (Witting and Able, 1993, 1995; Barbeau, 2000) demonstrated that small individuals (11 to 16 mm SL) may be preyed upon by the sevenspine bay shrimp (*Crangon septemspinosa*), a common estuarine resident in our study area. *Callinectes sapidus* has also been identified as an important predator on juveniles (Kellison et al., 2000).

MIGRATIONS

This species is an abundant component of trawl surveys on the continental shelf, regardless of season (Fig. 99.5). How-

ever, a few individuals in the Great Bay estuary appear to lag behind in their fall emigration and continue to be present in the bay through December (Sackett et al., 2007). The collections of young-of-the-year (< 30 cm) vary along the coast, reflecting seasonal migrations. In the fall, they occur along the inner to mid-continental shelf (mean depth 17 m, range 5 to 172 m) from Rhode Island to North Carolina. By the winter they have migrated offshore and are found in deeper depths on the mid- and outer shelf (mean depth 54 m, range 15 to 238 m). In the spring, they are found over most of the continental shelf (mean depth 34 m, range 7 to 203 m), but most commonly south of Delaware Bay with fewer occurrences as far north as off Massachusetts. Age 1+ (> 30 cm) juveniles and adults exhibit a distribution similar to that of the young-of-the-year except that they are found farther to the north and east onto the central and southern portions of Georges Bank. In the fall, they are most abundant on the inner and mid-shelf (mean depth 24 m, range 6 to 172 m) from off Massachusetts and on Georges Bank to North Carolina. In the winter, they occur in deeper water (mean depth 80 m, range 15 to 429 m) over the same range but are not collected north of Cape Cod. In the spring, they are found at intermediate depths (mean depth 68 m, range 7 to 335 m) relative to the other two seasons, presumably because they are migrating back into the inner shelf and estuaries. At this time, a few can also be found north of Cape Cod.

Age 1+ *Paralichthys dentatus*

YOY *Paralichthys dentatus*

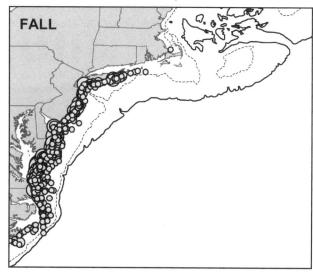

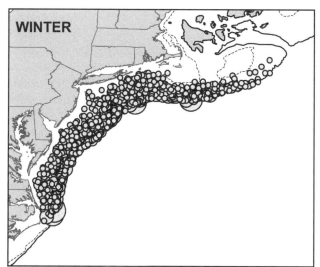

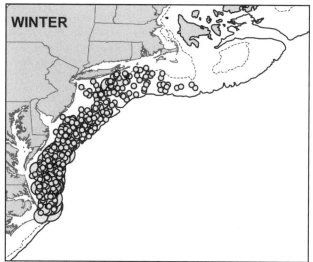

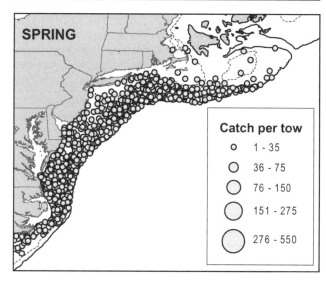

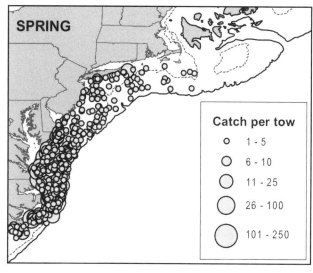

Catch per tow

○ 1 - 35
○ 36 - 75
○ 76 - 150
○ 151 - 275
○ 276 - 550

Catch per tow

○ 1 - 5
○ 6 - 10
○ 11 - 25
○ 26 - 100
○ 101 - 250

Fig. 99.5. Composite distribution of age 1+ (> 30 cm) and young-of-the-
year (< 30 cm) *Paralichthys dentatus* during seasonal cruises of the National
Marine Fisheries Service groundfish survey. Details of sampling effort are
indicated in Fig. 3.4.

100

Pseudopleuronectes americanus (Walbaum)

WINTER FLOUNDER

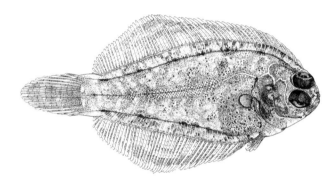

Estuary	Reproduction	Larvae/ Development	YOY Growth	YOY Overwinter
Ocean				
Both				
Jan				
Feb				
Mar				
Apr				
May				
Jun				
Jul				
Aug				
Sep				
Oct				
Nov				
Dec				

DISTRIBUTION

Pseudopleuronectes americanus occurs in the western North Atlantic Ocean from Labrador to Georgia (Klein-MacPhee, 2002e). It is most abundant between the Gulf of St. Lawrence and Chesapeake Bay. Within this range, there are numerous distinct populations, including one on Georges Bank, where it was once considered a separate species (Pierce and Howe, 1977). Early life history stages occur in estuaries throughout the Middle Atlantic Bight (see Table 4.3), but they are most abundant in the northern portion (Jeffries and Johnson, 1974; Howe et al., 1976). This trend is also evident along the coast of New Jersey, where they are quite abundant in Raritan Bay but become less so in the southern part of the state (Scarlett, 1991).

REPRODUCTION AND DEVELOPMENT

Spawning occurs in estuaries from late winter to early spring, but various populations may spawn during different periods. The exact timing is probably temperature dependent and thus varies with latitude. For example, in Massachusetts, spawning peaks in February and March (Bigelow and Schroeder, 1953; Howe et al., 1976), while it extends into April in Connecticut (Pearcy, 1962). In New Jersey, spawning takes place primarily in January to March (Scarlett and Allen, 1992). In a Rhode Island salt pond, spawning sites appear spatially distinct (Crawford and Carey, 1985). Spawning also occurs in the offshore waters of Georges Bank (Bigelow and Schroeder, 1953; Smith et al., 1975). There have been suggestions that spawning may take place in the ocean near Atlantic City (Tatham et al., 1974), but others propose that these larvae are spawned in estuaries and passively transported out on ebb tides (Pearcy, 1962; Smith et al., 1975). Recently, there are other indications that mature males and females can be found in the ocean during the winter, possibly indicating spawning may occur there (Wuenschel et al., 2009).

Spawning behavior has been observed in a large experimental aquarium during a 14-week period (Stoner et al., 1999). Spawning events occur throughout the night, but are concentrated between sunset and midnight and are always initiated by the males. These fish spawned over a 60-day period; females spawned an average of 40 times, whereas males spawned an average of 147 times. Spawning by an individual male elicited secondary spawning by up to six additional males. Concurrent observations in the field revealed that both sexes arrive in the estuary in ripe condition, and the highest gonadosomatic condition is observed in February and March. The ripe females in this study were > 20 cm in length and 2 years old or older. Most males were between 10 and 15 cm. Ripe males were encountered throughout the estuary, but ripe females were concentrated in the mid-reaches of the system (Stoner et al., 1999).

The early development of *Pseudopleuronectes americanus*, including characters useful for distinguishing the eggs, larvae, and juveniles from those of other flatfish species, has been summarized (Fahay, 2007). Eggs are demersal and adhesive, and measure 0.69 to 0.95 mm in diameter. Oil globules are lacking. Larvae are about 2.4 mm at hatching and have unpigmented eyes. Pigment is light and scattered, but

Pseudopleuronectes americanus   *Trinectes maculatus*   *Sphoeroides maculatus*

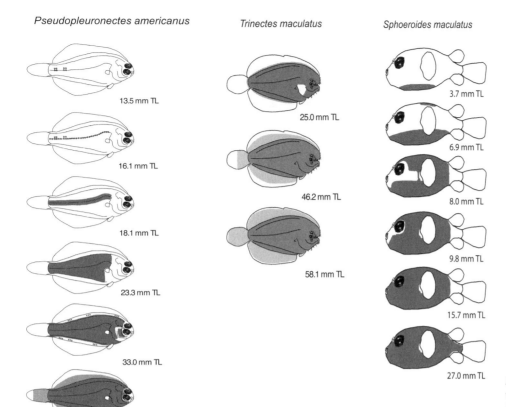

Fig. 100.1. Description of scale formation relative to total length (TL) in *Pseudopleuronectes americanus, Trinectes maculatus,* and *Sphoeroides maculatus.*

forms a vague band crossing the mid-tail region. All fin rays are fully formed by about 9.0 mm SL. After a brief pelagic stage immediately after hatching, all subsequent larval and juvenile stages are strongly bottom oriented. Dorsal and anal fin ray counts distinguish the juveniles from those of the very similar *Pleuronectes putnami*, a flatfish species that commonly occurs north of our study area. The duration of incubation varies with temperature (Williams, 1975). Hatching has been reported to occur at 11 to 63 days at 8.0 to −1.8°C and in 23 to 40 days at 3.5 to 0°C.

The sequence of scale formation was observed in specimens from 13.5 mm TL to 110.0 mm TL (Fig. 100.1; Able and Lamonaca, 2006). Scale formation begins with a few small scales in the middle of the caudal peduncle at about 13 mm TL. By 14 to 17 mm TL, scales are formed along the entire lateral line. By 16 to 20 mm TL, the scale rows have expanded dorsally and ventrally but have not yet reached the ventral margin of the body. At about 23 mm TL, the scales have expanded to cover a broader area extending to the base of the dorsal and anal pterygiophores. Scales begin to form on a small area of the operculum between 22 and 29 mm TL. Between 27 and 37 mm TL, new scales start developing at several locations over the bases of the dorsal and anal pterygiophores. By 44 mm TL, scales are forming on the median fins, and by 66 mm TL scale formation has reached the adult condition, where scales extend midway on to the dorsal and anal fins and slightly farther onto the caudal fin. In a comparison between the eyed and blind sides, scale formation on

the blind side seemed to lag slightly behind the eyed side at sizes between 20 to 40 mm TL and at sizes < 60 mm TL. See Figures 97.1 and 100.1 for comparisons with other flatfishes.

LARVAL SUPPLY, SETTLEMENT, GROWTH, AND MORTALITY

Various life history stages of this species fluctuate in abundance inter-annually, but their respective abundance levels are not always in synchrony. Larval abundance, on the other hand, exhibits a roughly 5- to 10-year pattern, where years showing low abundance alternate with those with high abundance. Juveniles occur in relatively constant numbers from year to year, with an occasional abundant year, as in 1997 (Fig. 100.2). Results from trapping of juveniles indicate little variation over a series of 15 years, although there are a few years where abundance is high.

Growth rates in our Great Bay study site, subsequent to settlement, ranged from 0.23 to 0.47 mm per day among 8 different habitats. The highest growth rates observed were in fish collected from one specific "settlement cove" (Witting, 1995), suggesting that habitat quality can influence growth and survival in early stages. Growth rates in experimental cages placed in a variety of habitats ranged from negative values to 1.3 mm per day (Sogard, 1992). Studies from other geographic areas within the Middle Atlantic Bight have observed similar growth rates (Pearcy, 1962; Mulkana, 1966). After settlement in March through May, modal lengths increase steadily through the summer and fall, ranging from

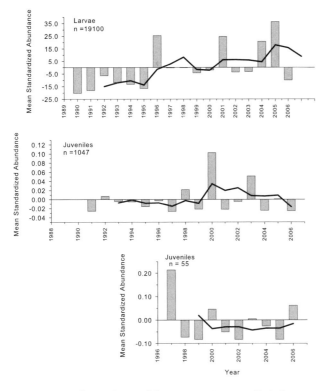

Fig. 100.2. Annual mean (bars) and three-year moving mean (line) of standardized abundance for *Pseudopleuronectes americanus* collected in larval sampling with plankton net (*upper*), juvenile sampling with traps (*middle*), and juvenile sampling with otter trawls (*lower*). Data were standardized by subtracting overall mean from annual abundance. Three-year moving mean was calculated by taking the mean of the standardized values from the previous two years and the year in which data are plotted.

about 6 cm in June to about 13 cm in November (Fig. 100.3). Growth slows during November and December, as young-of-the-year emigrate from estuarine to ocean habitats. There is little growth over the winter until it resumes the following April.

Dissolved oxygen levels may influence habitat quality as well (Bejda et al., 1992). This is reflected in caging studies of small juveniles in which growth was related to hypoxic conditions (< 4.0 mg/l^{-1}) (Meng et al., 2008). Laboratory studies of the effect of low dissolved oxygen on growth indicated that growth was significantly reduced under low dissolved oxygen (2.2 mg/l) conditions but was intermediate for individuals held at fluctuating levels (2.5 to 6.5 mg/l) (Stierhoff et al., 2006). Under fluctuating levels, 60% died when nighttime concentrations fell below 1.4 mg/l for several hours. In response to lowered dissolved oxygen levels in a range of temperatures, juvenile *P. americanus* growth rates were severely reduced at all levels tested. The lowered growth rates were always related to lowered consumption rates by the young-of-the-year. In other studies of mortality (Williams, 1975), survival was lower and more variable at temperatures above 10°C.

In a series of comprehensive studies of a population in the Mystic River estuary in Connecticut (Pearcy, 1962), it was determined that larvae were most common from March to June, where initially they were found in the upper estuary and subsequently, with development, became more abundant in the lower estuary. Larvae were typically more abundant near the bottom and, based on laboratory observations, were occasionally demersal. Even with this vertical distribution pattern, in a vertically stratified system, it was estimated that 3% per day were swept from the estuary and lost to the population. However, larvae in a Rhode Island salt pond were retained, apparently as a result of the hydrodynamics of the system (Crawford and Carey, 1985). See Chapter 8 for other discussions of mortality.

SEASONALITY AND HABITAT USE

In the Great Bay–Little Egg Harbor estuary planktonic larvae were a consistent component of the larval fish fauna and were abundant from mid-March to June (Fig. 100.4), but the timing of peak density varied from year to year (Witting, 1995; Sogard et al., 2001). The peaks occurred as early as April 10 and as late as May 14 and appeared to vary with water temperature during the period of spawning and egg development. In general, late peaks in larval density occurred when temperatures were primarily below 5°C from January to April. Early peaks occurred when temperatures were predominantly at or above 5°C during this period. Juveniles collected in traps are usually most abundant from June to August.

In Great Bay–Little Egg Harbor estuary juveniles (> 25 mm) occur across a variety of habitat types, regardless of sediment and structure. However, most are found in shallower depths (1 to 3 m) over sandy substrates (< 20% silt/clay) in the lower portion of the estuary (Fig. 100.5). In very shallow habitats (< 1 m), young-of-the-year are most common over unvegetated substrates (Sogard and Able, 1991), but this and other sampling also found them associated with macroalgae (Able et al., 1989). In the same system, young-of-the-year are also collected in the subtidal portions of polyhaline marsh creeks (Rountree and Able, 1992a). While there does not appear to be a distinct habitat preference for young-of-the-year (> 25 mm) during the first summer, the above observations suggest that loss of estuarine habitat may be as large a contributor to mortality as fishing (Boreman et al., 1993).

A detailed study of environmental parameters related to size classes of juveniles was conducted in the Sandy Hook Bay–Navesink River system (Stoner et al., 2001). Newly settled juveniles (< 25 mm) were strongly associated with low temperature and fine sediments, which placed them in deep, depositional habitats. Larger fish (25 to 55 mm) were more associated with high sediment organics, shallow depths (< 3 m), and salinities near 20 ppt. The largest juveniles tested (56 to 138 mm) were associated with shallow depths (< 2 m), temperatures near 22°C, and presence of macroalgae. This study revealed two centers for settlement within the system, and also suggested that habitat use shifts rapidly, with size of

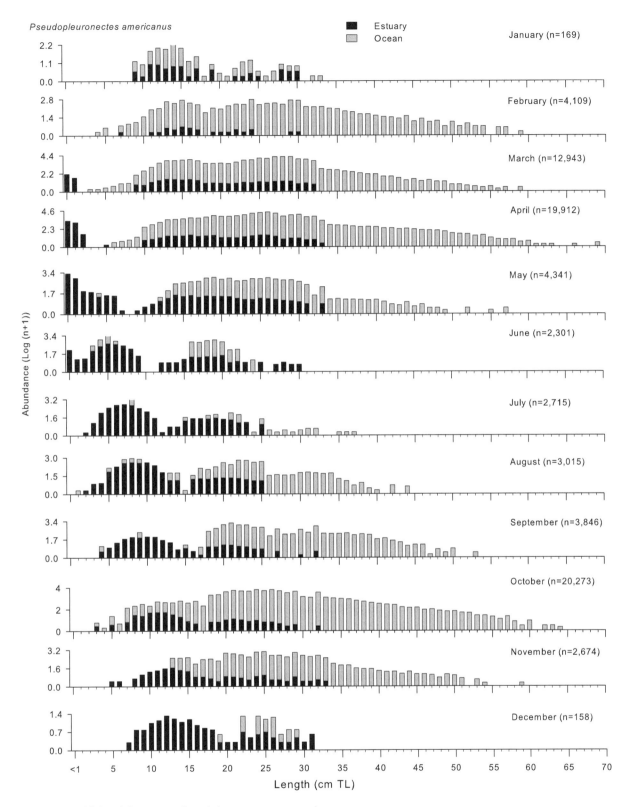

Fig. 100.3. Monthly length frequencies of *Pseudopleuronectes americanus* collected in estuarine and coastal ocean waters of the central Middle Atlantic Bight. Sources: National Marine Fisheries Service: otter trawl (n = 58,633); Rutgers University Marine Field Station: 1 m beam trawl (n = 176); seine (n = 916); 1 m plankton net (n = 4742); 2 m beam trawl (n = 141); killitrap (n = 696); 4.9 m otter trawl (n = 175); and Able and Fahay 1998 (n = 10,977).

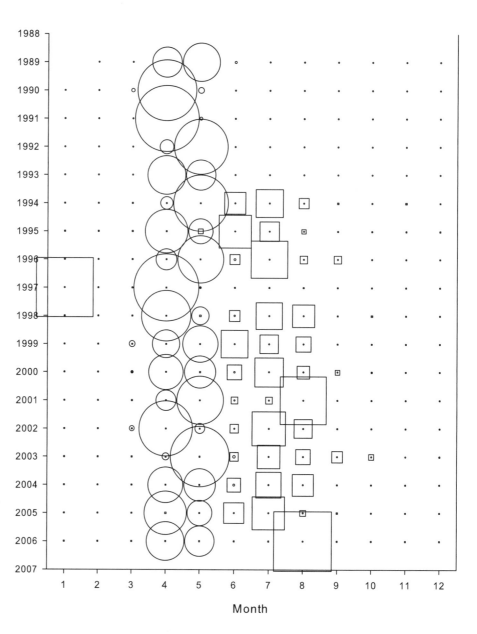

Fig. 100.4. Seasonal occurrence of larvae based on ichthyoplankton samples collected in Little Sheepshead Creek behind Little Egg Inlet (open circles represent the percentage of the mean number/1000 m³ of larval young-of-the-year captured by year) from 1989 to 2006 and juveniles of *Pseudopleuronectes americanus* based on trap collections in Rutgers University Marine Field Station boat basin in Great Bay (open squares represent the percentage of the mean catch per unit effort captured by year) from 1994 to 2006. Values range from 0 to 99% and 0 to 100% for ichthyoplankton and trap sampling, respectively. The smallest circles and squares represent when samples were taken but no individuals were collected.

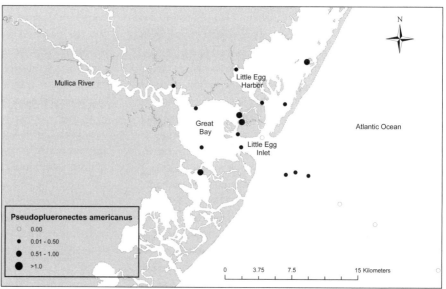

Fig. 100.5. Habitat use and average catch per tow (see key) for juvenile and adult *Pseudopleuronectes americanus* in the Ocean–Great Bay–Mullica River corridor in southern New Jersey based on small otter trawl collections during July and September from 1988 to 2006. See Table 3.2 for characteristics at each sampling collection.

fish, during the first year of life. It also demonstrated that distinct habitats are dynamic, and can change in size and/or position as environmental variables change (Stoner et al., 2001).

The habitat of young-of-the-year is influenced by their seasonal behavior. The young-of-the-year spend most of their first year in the estuary. They are most abundant in the spring as larvae and subsequently as settled juveniles during the summer. From spring through December, all young-of-the-year were collected in the estuary, but by January a large proportion of those captured had migrated to the ocean, as had larger and older fish. This wintertime occurrence in the ocean differs from the expected pattern, in which they are presumed to overwinter in estuaries (Bigelow and Schroeder, 1953). Large numbers have also been reported from the continental shelf in the New York Bight apex (Phelan, 1992) and outside the Great Bay Estuary (Able and Hagan, 1995). An alternate explanation for these observations is that larger individuals are reported to bury in the mud as deep as 12 to 15 cm during the winter (Fletcher, 1977), and this may account for the reduced numbers collected in the estuary. These contradictions in seasonal use of estuaries and the ocean may vary between estuarine systems and require further study (Wuenschel et al., 2009).

In Gulf of Maine estuaries, four habitats (kelp [*Laminaria longicruris*], eelgrass [*Zostera marina*], drift algae [*Phyllophora* sp.], and unvegetated sand/mud) were compared in three broad zones along the Maine coast for their importance as young-of-the-year essential fish habitat (Lazzari, 2008). In one year of this study, the abundance of young-of-the-year was similar in all four habitats in the mid-coast zone. However, in other years, higher numbers occurred in eelgrass beds. The overall results of this survey indicate that different types of habitat are important to young-of-the-year *P. americanus* in different years.

The relative depths of habitats are important to survival of young-of-the-year as well. Studies in Sandy Hook Bay, New Jersey, showed that as young-of-the-year grow, they occur in shallower habitats (Manderson et al., 2004). Ultimately, fish > 35 mm become concentrated in habitats about 1 m deep. Tethering experiments demonstrated that the risk of predation was low in these shallow depths and increased rapidly with greater depths tested. The important predator in these tests was *Paralichthys dentatus*, a species that occurs most commonly in deeper habitats. These shallow habitats therefore serve as predator refuges, and their loss (due to natural or anthropogenic factors) would negatively influence survival of young-of-the-year *P. americanus*.

PREY AND PREDATORS

Females begin feeding on ampeliscid amphipods and the siphons of *Mya arenaria* while in the estuary. Males begin feeding (in the laboratory) only after spawning events have ended (Stoner et al., 1999). Because of this species' small mouth size, diets are limited to small invertebrates, although occasionally small *Ammodytes* are consumed (Klein-MacPhee, 1978). Other important diet items include shrimp, small crabs, ascidians, holothurians, squid, and mollusks. They are sight feeders and feed primarily during the day (Bowman, 1981).

Several studies (n = 10) have examined the food habits of young-of-the-year (3 to 300 mm, n = 2681) in Middle Atlantic Bight estuaries from the lower Pettaquamscutt River, Rhode Island, to Rehoboth Bay, Delaware (see Table 8.1). Important prey of young-of-the-year include a variety of invertebrates (polychaetes, amphipods, copepods, isopods, bivalves, ostracods, gastropods, nemertines, decapod shrimp, oligochaetes, mysid shrimp, cumaceans, and crabs) and detritus. Fishes (*Menidia menidia*, *Anchoa mitchilli*, and *Gobiosoma bosc*) are a minor dietary component. The diet of young-of-the-year shifts with ontogeny at approximately 50 to 60 mm, from a diet of small polychaetes, copepods, and invertebrate eggs when smaller, to larger polychaetes, amphipods, and decapod shrimp when larger (Mulkana, 1966; Stehlik and Meise, 2000; Vivian et al., 2000). The diet of larvae (3 to 8 mm, n = 140) shifted in the Mystic River estuary, Connecticut (Pearcy, 1962), such that the smallest larvae (3 to 5 mm, n = 37) consumed copepod nauplii most commonly and invertebrate eggs and larval polychaetes with decreasing frequency, while phytoplankton (pinnate and filamentous diatoms) were noted occasionally in specimens 3 to 4 mm. Small polychaetes were the major identifiable food of slightly larger larvae (6 to 8 mm, n = 25), supplemented by nauplii, harpactacoid, and calanoid copepods. Young-of-the-year (15 to 299 mm, n = 1291) in the Hudson–Raritan estuary, New York/New Jersey, also showed an ontogenetic diet shift (Stehlik and Meise, 2000). Copepods and *Spionidae* polychaetes were prominent in the diet of small (15 to 49 mm) individuals and decreased in importance as size increased. At 50 mm, there was a sharp reduction in the importance of copepods in the diet. Amphipods were the most consistent prey in the much broader diet of larger (50 to 299 mm) individuals, supplemented by the bivalve *Mya arenaria* (whole and siphons), the gastropod *Nassarius spp.*, various polychaetes, isopods, amphipods, mysid shrimp, and the decapod shrimp *Crangon septemspinosa*. Vivian et al. (2000) found similar results in the Hudson River estuary. Young-of-the-year (10 to 80 mm, n = 123) also showed ontogenetic diet shifts in the lower Pettaquamscutt River estuary, Rhode Island, at approximately 60 mm, such that polychaetes, amphipods, and *Crangon* increased in frequency with increasing size, while copepods, ostracods, and invertebrate eggs decreased in frequency with size (Mulkana, 1966).

Pseudopleuronectes americanus are preyed upon by a variety of larger fishes, including *Gadus morhua*, *Squalus acanthias*, *Lophius americanus*, and *Raja* spp. (Rountree, 1999). They are also eaten by *Mustelus canis*, *Urophycis* spp., *Prionotus carolinus*, *Hemitripterus americanus*, *Morone saxatilis*, and *Pomatomus saltatrix* (Bowman and Michaels, 1984; Manderson et al., 2000). Non-fish predators include harp, harbor, and gray seals and the osprey (Scott and Scott, 1988; Bowen et al., 1993).

In a Mystic River estuary study, sources of natural mortality included predation on larvae by the medusae of *Sarsia tubulosa* (Pearcy, 1962). Together with hydrodynamic influences, this predation resulted in an estimated loss of 20% per day for small larvae and 4% per day for larger larvae. In a study of survival of juvenile (< 20 mm) *P. americanus* in estuarine habitats (Manderson et al., 2006), the demersal species *Prionotus evolans* and *Paralichthys dentatus* were found to be important predators. Co-occurring pelagic predators (*Pomatomus saltatrix, Cynoscion regalis,* and *Morone saxatilis*) consumed fewer newly settled *P. americanus*. In portions of the study estuary where salinities were < 20 ppt, predation by these demersal species was much reduced. The areal extent of this portion of the habitat was influenced by seasonal runoff and other climate conditions, including temperature, and these authors suggest that changes in climate might influence the amount of predator refuge available, and therefore might also affect year-class strength (Manderson et al., 2006).

The green crab, *Carcinus maenas*, has also been identified as an important predator on young-of-the-year *P. americanus* (Fairchild and Howell, 2000). Experiments demonstrated that all size classes of *P. americanus* are preyed upon by all sizes of the predator; however, the highest mortality involved the smallest prey and the largest predator. The number of *P. americanus* consumed per day was significantly higher in individuals < 20 mm than in all other size classes tested. Mortality decreases in larger size classes. A study of predation in New Jersey and Connecticut waters also implicated *Carcinus maenas*, as well as *Crangon septemspinosa*, as important predators on *P. americanus* eggs and juveniles (Taylor, 2003, 2004, 2005). The magnitude of egg predation was affected most by temporal and spatial overlap of predator and prey. The scale of juvenile predation was most impacted by size-dependent relationships between predator and prey. The temporal relationship was important in Narragansett Bay, Rhode Island, where large *Crangon* were found at the same time as newly settled *P. americanus*, and predation rates by the former were very high.

Studies in Sandy Hook Bay and the nearby Navesink River estuary show that *Prionotus evolans* is an important predator on young-of-the-year *P. americanus* (Manderson et al., 2000). These young-of-the-year were found in the diets of 69% of *P. evolans* stomachs examined during the month of June in Sandy Hook Bay. Observations in the laboratory revealed that *P. evolans* use their modified pectoral fin rays to detect, flush, and occasionally excavate buried *P. americanus*. When *P. evolans* 212 to 319 mm TL were provided with *P. americanus* 30 to 114 mm TL, they selected prey < 70 mm TL, and maximum prey size appeared to be limited by width of the predator's esophagus (Manderson et al., 2000). One conclusion from this study was that when these two species co-occur in estuarine habitats, *P. evolans* consume large quantities of *P. americanus* in the size range of 15 to 70 mm TL. Young-of-the-year *P. americanus* did not suffer this predation in the nearby Navesink River estuary, primarily because *P. evolans* did not extend into that habitat.

MIGRATIONS

Adults in southern New England are often reported to move offshore during the summer as estuarine temperatures warm and then back into estuaries in the fall to spend the winter (Bigelow and Schroeder, 1953). Some components of the populations in the New York Bight are known to spend the winter offshore (Phelan, 1992; Able and Hagan, 1995; Wuenschel et al., 2009). All life history stages of this species are common components of seasonal bottom trawl surveys (Fig. 100.6). The overall distribution of young-of-the-year (< 23 cm) on the continental shelf reflects their distribution in adjacent estuaries, with most individuals occurring between Massachusetts and Chesapeake Bay. There is a seasonal component to the distribution as well, with most young-of-the-year present on the inner continental shelf and on the shallow portions of Georges Bank (mean depth 34.7 m, range 8 to 75 m) in the fall, especially between Massachusetts and New Jersey. In the winter, the depth distribution is similar (mean depth 32.4 m, range 15 to 72 m). At this time the more northern parts of the distribution appear truncated but that is simply due to the lack of collections north of Cape Cod and on the northern portions of Georges Bank in the winter. By the spring, they have a distribution similar to that of the fall on the inner continental shelf (mean depth 26.8 m, range 7 to 80 m), except that they are found as far south as the mouth of Chesapeake Bay.

The age 1+ (> 23 cm) juvenile and adult distributions largely coincide with those of the young-of-the-year, except that they are found in slightly deeper waters and more consistently to the eastern tip of Georges Bank (see Fig. 100.6). In the fall (mean depth 44.3 m, range 10 to 209 m) and winter (mean depth 46.5 m, range 15 to 89 m) the average distribution is deepest, while it is somewhat shallower (mean depth 34.1 m, range 7 to 210 m) in the spring.

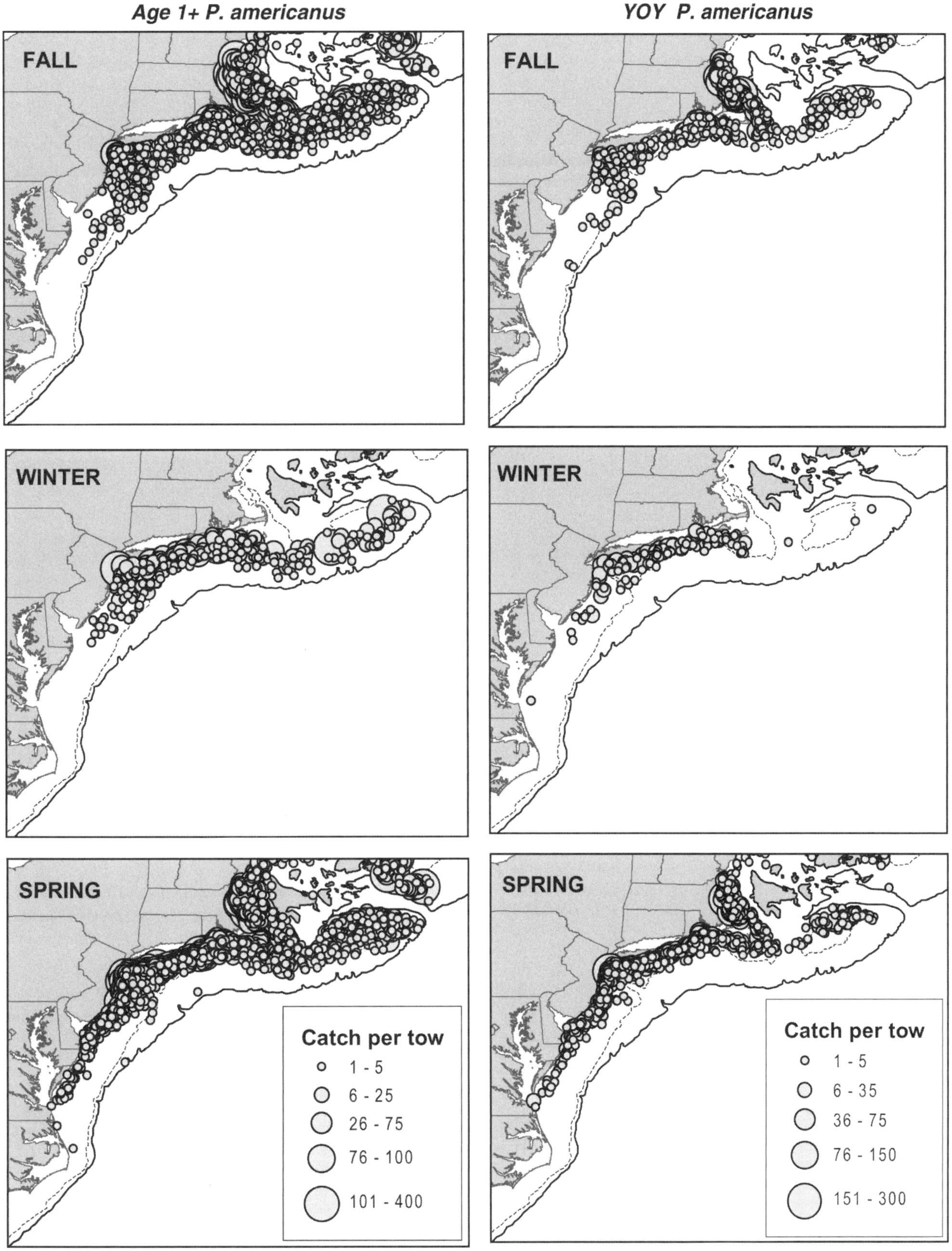

Age 1+ *P. americanus*

FALL

WINTER

SPRING

Catch per tow

○	1 - 5
○	6 - 25
○	26 - 75
○	76 - 100
○	101 - 400

YOY *P. americanus*

FALL

WINTER

SPRING

Catch per tow

○	1 - 5
○	6 - 35
○	36 - 75
○	76 - 150
○	151 - 300

Fig. 100.6. Composite distribution of age 1+ (> 23 cm) and young-of-the-year (< 23 cm) *Pseudopleuronectes americanus* during seasonal cruises of the National Marine Fisheries Service groundfish survey. Details of sampling effort are indicated in Fig. 3.4.

Trinectes maculatus (Bloch and Schneider)

HOGCHOKER

Estuary	Reproduction	Larvae/ Development	YOY Growth	YOY Overwinter
Ocean				
Both				
Jan				
Feb				
Mar				
Apr				
May				
Jun				
Jul				
Aug				
Sep				
Oct				
Nov				
Dec				

DISTRIBUTION

Trinectes maculatus occurs in estuarine and marine waters from Massachusetts to Florida and through the Gulf of Mexico to the coast of Venezuela (Bigelow and Schroeder, 1953). It is rarely found as far north as Maine (Peters and Boyd, 1972). Early life history stages occur in most estuaries in the Middle Atlantic Bight (see Table 4.3) but are most abundant from Chesapeake Bay (Musick, 1972) and south.

REPRODUCTION AND DEVELOPMENT

Maturation occurs at age 1 or 2 (Koski, 1978), but others report that sexual maturity may not occur until 4 years of age (Mansueti and Pauly, 1956; Dovel et al., 1969). Spawning can occur from May to October. Most spawning in the Hudson River takes place during June and July (Koski, 1978) and during the summer in Delaware Bay (Wockley, 1968). Sexually mature females (104 to 112 mm TL) with developed ovaries are most evident at that time, although there are no samples earlier than early June. Egg collection shows that spawning occurs throughout lower Chesapeake Bay (Olney, 1983). In the Patuxent River, a tributary of Chesapeake Bay, spawning peaks when water temperatures are greater than 25°C (Dovel et al., 1969). In intensive estuarine collections in the Patuxent River, eggs occurred in salinities from 0 to 24 ppt but were concentrated in 10 to 16 ppt (Dovel et al., 1969). Spawning may also occur in the ocean because eggs have been found as far as 9.7 km off the mouth of Chesapeake Bay (Hildebrand and Cable, 1938).

The early development was described more than 70 years ago (Hildebrand and Cable, 1938), but studies of its ontogeny have not been revisited. There are questions about the sizes of specimens in the early study, and there are other aspects of their development that need to be re-examined and described. What is known about the eggs and larvae has been summarized recently (Fahay, 2007). Eggs are spherical to slightly oval. Reportedly, they are buoyant in higher salinities, demersal in low salinities. Their diameters range from 0.67 to 1.22 mm, with the smallest diameters found in the highest salinities. Larvae hatch at 1.7 to 1.9 mm. They are deep-bodied, the eyes are unpigmented, and there is a prominent hump on top of the head. The body deepens further with development. The gut is large, initially bulging and prominent. The mouth is larger, proportionately, than it will be in adults. Scattered pigment in early larvae forms a barred pattern that extends onto the fins in larger larvae. During the early larval stage, the pectoral fins degenerate and are only visible as small flaps by approximately 6.0 mm. Transformation occurs at a small size, when the left eye migrates across the top of the head to the adult position on the right side at 6.0 to 10.0 mm.

Scale formation occurs before 18 mm. The series we examined was incomplete; therefore little is known about the sequence of scale formation, especially between sizes of 5 and 20 mm TL (see Fig. 100.1). Individuals < 5 mm TL have no scales. By 25 mm TL, scales cover almost all of the body

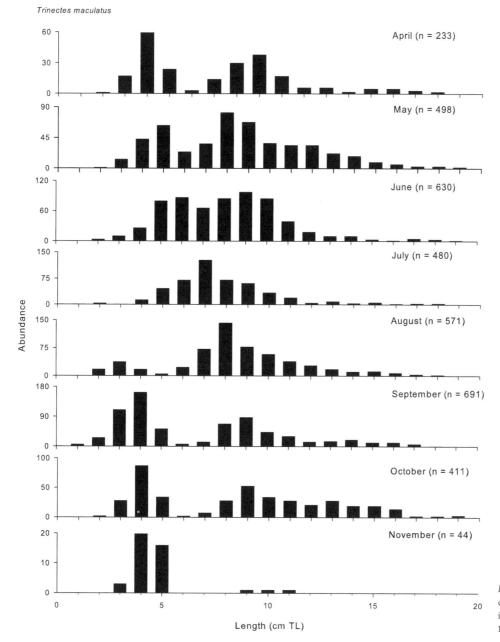

Trinectes maculatus

with the exception of a patch just anterior to the approximate location of the pectoral fin. At this same size some scales are beginning to form on the proximal portions of the dorsal, anal, and caudal fins. By 25 to 35 mm TL, scales on the body are complete and scales on the fins are more extensive. This pattern continues until 55 to 60 mm TL, at which size scales almost completely cover the fins, more so than in any of the other flatfishes examined (Able and Lamonaca, 2006; see Figs. 97.1 and 100.1), and achieve the adult condition.

LARVAL SUPPLY, SETTLEMENT, GROWTH, AND MORTALITY

The most intensive, recent studies in the Middle Atlantic Bight have been conducted in the Patuxent River, Maryland (Dovel et al., 1969), and these provide the basis for the following in-

terpretation of the early life history. After spawning occurs in the lower estuary, the eggs develop quickly (the eggs hatch in 26 to 36 hours at 23.3 to 24.5°C; Hildebrand and Cable, 1938), and the recently hatched larvae are most abundant in the same area. Larger larvae and recently transformed juveniles are reported to rapidly migrate upriver to low-salinity nursery areas, where the smallest juveniles are typically collected. The exact size at settlement is unknown, largely because collections of just-settled individuals are rarely made.

This species is among the slowest-growing flatfishes in estuaries in the Middle Atlantic Bight. Examination of length frequency modes in the Patuxent River (Dovel et al., 1969) suggests that young-of-the-year are approximately 10 to 50 mm by November, do not grow over the winter, and are approximately 20 to 60 mm by the following spring when they are

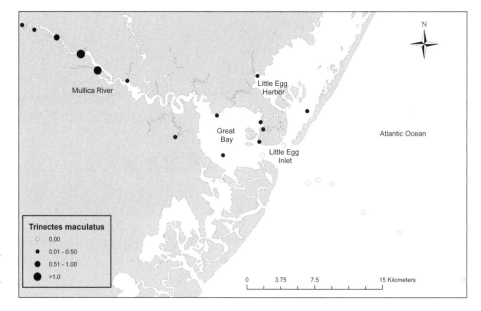

Fig. 101.2. Habitat use and average catch per tow (see key) for juvenile and adult *Trinectes maculatus* in the Ocean–Great Bay–Mullica River corridor in southern New Jersey based on small otter trawl collections during July and September from 1988 to 2006. See Table 3.2 for characteristics at each sampling collection.

age 1. Schwartz (1964) reported individuals of 27 to 34 mm from February through May in Isle of Wight Bay and Assawoman Bay, both in Maryland. Growth estimates based on scales from fish collected in the Patuxent River indicate an average back-calculated length of 40 mm at age 1 and 66 mm at age 2 (Mansueti and Pauly, 1956). Slightly larger sizes are reported from the Hudson River, based on analysis of scales, with average back-calculated sizes of 59 mm at age 1 and 81 mm at age 2, with a few individuals reaching age 5 and 6 (Koski, 1978). Collections in Delaware Bay more clearly demonstrate growth and the presence of discrete year classes (Fig. 101.1). The smaller individuals (1 to 2 cm) are evident in June through October. This year class appears to reach a modal size of approximately 4 cm by September, do not grow over the winter based on a summer size distribution in April, and by June at approximately 1 year of age are 5 to 7 cm. By October of the second year they have a modal size of 9 cm.

SEASONALITY AND HABITAT USE

Adults inhabit bays and estuaries, usually in brackish to freshwaters, on mud, sand, or silty substrates. They tolerate a wide range of temperatures and salinities as well as low oxygen concentrations. The eggs and larvae are dominant components of the Chesapeake Bay ichthyoplankton in the summer, especially during August, when they are distributed throughout the bay (Olney, 1983; Olney and Boehlert, 1988). Surprisingly little is known of the habitats of young-of-the-year of this abundant species, although the upstream limits of young-of-the-year nursery areas extend into freshwater (Hildebrand and Cable, 1938; Massmann, 1954). Sampling along a transect extending from riverine to oceanic conditions demonstrated that juveniles and older stages also occur in fresh or nearly freshwater conditions in our study area (Fig. 101.2). In Delaware Bay, juveniles are found over

mud substrates in creeks during the late summer and early fall (Smith, 1971). In Chesapeake Bay, young stages are present on shallow mud flats during the summer (Musick, 1972). However, a recent analysis across 12 different habitat types in subtidal marsh creeks, bay shoreline, and shallow and deep portions of the bay and extending into the Delaware River found the largest numbers of individuals were collected in the deeper portion of the bay (Able et al., 2007a).

PREY AND PREDATORS

Trinectes maculatus prey on amphipods, clam siphons, nereid worms, copepods, and small fishes (Hines et al., 1990). Barbels found around the mouth are chemosensory organs and indicate that this species probably depends on chemical stimuli to find prey (O'Connor, 1972). There is a dietary shift with increasing size in this species. Those smaller than 61 mm concentrate on amphipods, whereas larger fish focus on nereid worms and clam siphons. One study examined the food habits of young-of-the-year (21 to 60 mm, n = 83) in the York River, Virginia (Smith et al., 1984). Young-of-the-year consumed predominantly amphipods, supplemented by detritus, polychaetes, copepods, and clam siphons (Table 8.1). Little is known regarding predators on this species, although there is a single record of *Mustelus canis* containing it in their stomachs (Rountree, 1999).

MIGRATIONS

The conceptual interpretation of migrations in the Patuxent River estuary in upper Chesapeake Bay indicates that adults (ages 2 to 4) make down-estuary movements in the summer for spawning, followed by return movements in the fall (Dovel et al., 1969). The larvae move upstream into low-salinity areas, where they remain through the winter. A similar pattern has been reported in Florida (Castagna, 1955) and the Gulf of Mexico (Peterson, 1996).

102

Symphurus plagiusa (Linnaeus)

BLACKCHEEK TONGUEFISH

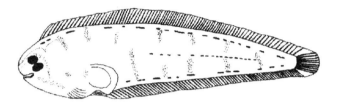

DISTRIBUTION

This species is found in the western North Atlantic Ocean from New York to the Campeche Peninsula, Mexico (Munroe, 1998). The distribution of early stages is not well described and reports may be confounded by possible misidentification of specimens as larvae of congeners, as suggested for the individuals caught off Beaufort Inlet by Hildebrand and Cable (1930) (Olney and Grant, 1976). Six other cynoglossid species may occur along the East Coast of the United States (Munroe, 1998), and some of these inhabit nearshore habitats along with *Symphurus plagiusa*. Methods for distinguishing the larvae of all of these congeners have not been resolved.

REPRODUCTION AND DEVELOPMENT

Males and females reach first maturity at age 0 or age 1 at sizes of 91 and 101 mm TL, respectively, in Chesapeake Bay (Terwilliger, 1996). In the same study, spawning likely occurred in the deeper, more saline portion of the bay based on gonad condition. Spawning takes place during the summer (Olney and Grant, 1976) but may extend from June through October, with multiple spawnings (up to 38 times every 3.4 days) (Terwilliger, 1996). Spawning over the inner continental shelf has also been reported from inshore, based on the occurrence of small larvae (2.1 to 4.0 mm NL), from near the mouth of Chesapeake Bay to Maryland (W. G. Smith, pers. comm., in Olney and Grant, 1976). When ichthyoplankton sampling occurred concurrently in both ocean and estuarine habitats, larvae were only collected in the estuary (Able et al., 2006a).

Eggs are undescribed, but those identified as *Symphurus* sp., possibly including those of *S. plagiusa*, have been reported from the continental shelf of the Middle Atlantic Bight as far north as New Jersey (Berrien and Sibunka, 1999). Larvae hatch at sizes < 1.3 mm. The early development has been described (Olney and Grant, 1976) and summarized in Fahay (2007). Transformation to the juvenile stage occurs at about 10 mm. Juveniles are diagnosed by a unique suite of meristic characters and more subtle pigment characters.

LARVAL SUPPLY AND GROWTH

Larvae that are collected in ichthyoplankton sampling at Little Egg Inlet may result from spawning over the continental shelf, or in the estuary, or both, although this species is also known to spawn in large estuaries such as Delaware and Chesapeake bays (de Sylva et al., 1962; Olney and Grant, 1976). The continental shelf off New Jersey is apparently the northernmost limit of reproduction, based on the distribution of *Symphurus* spp. eggs (Berrien and Sibunka, 1999). The occurrence of larvae varies annually. In the vicinity of Little Egg Inlet, New Jersey, larvae were most abundant during the early 1990s (Fig. 102.1). Most larvae collected in plankton nets ranged from 9 to 60 mm, with the largest individuals collected in November and December (Fig. 102.2).

Individuals in Chesapeake Bay reached approximately 80 mm TL in the first year based on back-calculated growth from otoliths (Terwilliger, 1996). Growth declines thereafter, with some individuals reaching an age of 5+ years and lengths of 190 and 202 mm for males and females, respectively.

SEASONALITY AND HABITAT USE

Most larvae occur in the deepest and most saline portions of Chesapeake Bay (Olney and Grant, 1976), and are most commonly collected in July and August. The timing of occurrence of larvae in the vicinity of Little Egg Inlet, New Jersey, is somewhat later, from August into December.

PREY AND PREDATORS

Unknown.

MIGRATIONS

Unknown.

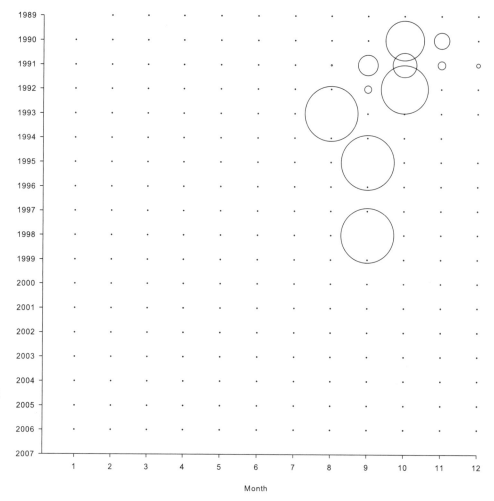

Fig. 102.1. Seasonal occurrence of the larvae of *Symphurus plagiusa* based on ichthyoplankton samples collected in Little Sheepshead Creek behind Little Egg Inlet from 1989 to 2006. Full circles represent the percentage of the mean number/1000 m³ of larval young-of-the year *Symphurus plagiusa* captured by given year. Values range from 0 to 90%. The smallest circles represent when samples were taken but no larvae were collected.

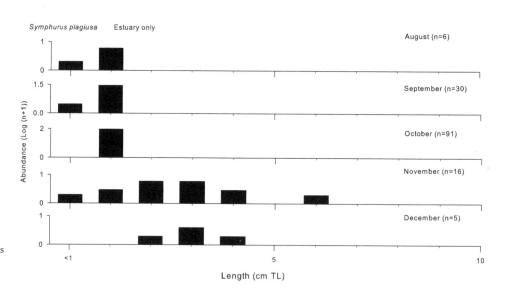

Fig. 102.2. Monthly length frequencies of *Symphurus plagiusa* collected in estuarine waters of the central Middle Atlantic Bight. Source: Rutgers University Marine Field Station: 1 m plankton net (n = 148).

103

Chilomycterus schoepfii (Walbaum)

STRIPED BURRFISH

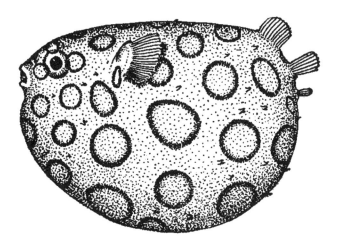

Estuary	Reproduction	Larvae/ Development	YOY Growth	YOY Overwinter
Ocean				
Both				
Jan				
Feb				
Mar				
Apr				
May				
Jun				
Jul				
Aug				
Sep				
Oct				
Nov				
Dec				

DISTRIBUTION

This species occurs from Nova Scotia to the Bahamas, Cuba, Belize, and southeastern Brazil, including the northern Gulf of Mexico (Klein-MacPhee, 2002f).

REPRODUCTION AND DEVELOPMENT

Spawning has not been well described. A 190 mm female has been reported with fully ripe eggs (Nichols and Breder, 1927). Estimates of spawning times are not well described. For the region, they range from early spring (Miller, 1965), to July (Nichols and Breder, 1927), to October (Hildebrand and Schroeder, 1928). The evidence from Florida Bay suggests that reproduction and recruitment may occur year-round (Powell et al., 2007). In the Middle Atlantic Bight, based on our own collections, reproduction likely occurs in mid-summer based on the presence of small individuals (< 30 mm) in July and August. Eggs are presumably pelagic (Leis, 1978). Unfertilized eggs are demersal, non-adhesive, and transparent, with a diameter of about 1.8 mm (Nichols and Breder, 1927). Early larval development is not described. However, this species goes through a specialized, pelagic, juvenile stage (the "lyosphaera") that differs from early larvae and older juveniles (Evermann and Kendall, 1898; Heck and Weinstein, 1978; Leis, 1984). During this stage (see above), elongated papillae develop over the body (instead of the spines that typify older stages), but these fail to develop into spines, or specialized scales, in their interiors. Instead, some of these papillae enlarge enormously. In these specialized pelagic juveniles, the body width exceeds the body depth. Spines develop over the body after pelagic juveniles settle to bottom habitats. See Fahay (2007) for a summary of other characters in early stages of this species.

LARVAL SUPPLY AND GROWTH

Spawning reportedly occurs well offshore (Springer and Woodburn, 1960), but precise locations are not described. Our estimates of growth rates are based on the few regularly occurring juveniles collected in our study area (Fig. 103.1). Juveniles increase in modal size from approximately 30 mm in July to about 110 mm, with some individuals as large as 130 mm, in September. The presumed age 1+ individuals are 16 to 24 cm TL in June of the following year. A maximum size of 33 cm TL has been reported for this species (Holmquist, 1997).

SEASONALITY AND HABITAT USE

Most of our collections of young-of-the-year of this species have been made in estuarine habitats during July through September, although a few have been made in the ocean in the fall (see Fig. 103.1). Some larger individuals (>15 cm), representing age 1 and perhaps older year classes, have also been collected in estuarine waters and inner continental shelf waters during the summer (Fig. 103.2). Adults are known to occupy eelgrass beds, deep flats, coral reefs, or sandy beaches in the Middle Atlantic Bight and elsewhere (Böhlke and Chaplin, 1968; Holmquist, 1997). They usually occur in depths < 18 m, but have been collected as deep as 91 m (Franks et al., 1972). In Chesapeake Bay, depth preference appears to change with season (Holmquist, 1997). During May–August they are most abundant in depths < 1.5 m. In September and October they are more frequently found in depths > 3.7 m.

PREY AND PREDATORS

We have no data on food habits of young-of-the-year while they occupy estuarine habitats. Adults are known to eat her-

mit crabs, *Pagurus longicarpus* (Kuhlmann, 1992), and snails, *Littorina irrorata* (Hamilton, 1976).

MIGRATIONS

Patterns of migrations for this species are undescribed. In Chesapeake Bay the movements from shallow to deep water appeared to precede their migration into the ocean as temperatures decline (Holmquist, 1997). The absence of all individuals from fall through spring in the Mullica River–Great Bay estuary in southern New Jersey suggests that they have migrated out of the area.

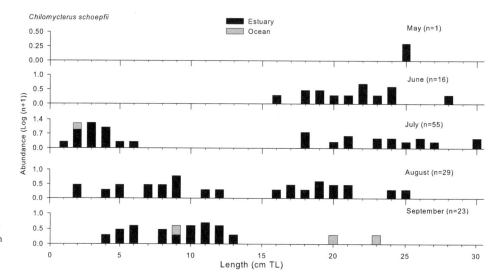

Fig. 103.1. Monthly length frequencies of *Chilomycterus schoepfii* collected in estuarine and coastal ocean waters of the Great Bay–Little Egg Harbor study area. Source: Rutgers University Marine Field Station: seine (n = 38); 1 m plankton net (n = 1); 2 m beam trawl (n = 1); 4.9 m otter trawl (n = 83).

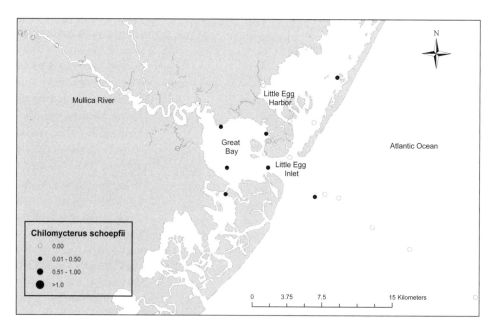

Fig. 103.2. Habitat use and average catch per tow (see key) for juvenile and adult *Chilomycterus schoepfii* in the Ocean–Great Bay–Mullica River corridor in southern New Jersey based on small otter trawl collections during July and September from 1988 to 2006. See Table 3.2 for characteristics at each sampling collection.

104

Sphoeroides maculatus (Bloch and Schneider)

NORTHERN PUFFER

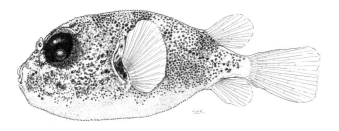

Estuary				
Ocean	Reproduction	Larvae/Development	YOY Growth	YOY Overwinter
Both				
Jan				■
Feb				■
Mar				■
Apr				
May				
Jun				
Jul				
Aug				
Sep				
Oct				
Nov				■
Dec				■

DISTRIBUTION

Sphoeroides maculatus occurs on the Atlantic coast of North America from Newfoundland to northern Florida (Shipp and Yerger, 1969). Larvae and juveniles have been reported from most estuaries in the Middle Atlantic Bight, but eggs have only been collected from Barnegat Bay (see Table 4.3).

REPRODUCTION AND DEVELOPMENT

Spawning occurs from May to August (rarely as late as October) in coastal ocean waters and in estuaries. Age 1 fish as small as 88 mm are capable of spawning (Laroche and Davis, 1973). Peak spawning in the New York area takes place in June (Perlmutter, 1939) and in Chesapeake Bay in June and July (Laroche and Davis, 1973). Observations of a captive pair indicate that eggs are partially buried in a circular depression in the substrate (Breder and Clark, 1947). Spawning continues through summer and into fall (Sibunka and Pacheco, 1981), and young-of-the-year as small as 15 mm FL have been collected as late as November in the Indian River Inlet area (Pacheco and Grant, 1973). In the Great Bay–Little Egg Inlet study area, the presence of small juveniles (< 2 cm) indicates that spawning is continuous from June to October.

The early development of *Sphoeroides maculatus* has been well described and summarized (Fahay, 2007). Eggs are demersal and adhesive with a diameter of 0.85 to 0.91 mm. The chorion is reticulated and there are numerous oil globules. Larvae are about 2.4 mm at hatching, with unpigmented eyes and unformed mouth parts. Hatching occurs after about 3.5 days' incubation at 19.4°C (Welsh and Breder, 1922). The body is stocky and the head profile is rounded. Larvae as small as 7.0 mm are capable of inflating their bodies. Early juveniles have small, terminal mouths and eyes that are situated near the dorsal outline. The pelagic juvenile stage is characterized by silvery sides and a deep blue dorsum.

Individuals from 3.7 to 109.7 mm TL (n = 25) were examined for scale formation (see Fig. 100.1; Able et al., 2009a). Modified scales (more accurately described as prickles) form over the size range of < 3.7 to 27 mm TL. Scales originate on the ventral surface anterior to the pectoral fin and on the dorsal surface above the pectoral fin. Scales are present on the ventral surface at 3.7 mm TL. By 7 mm TL, these have expanded in all directions to cover a large portion of the lower half of the body, and scales also begin to form on the dorsal surface. By 8 mm TL, scales expand to cover more of the surface, anteriorly to above the eye on the dorsal surface and forward of the eye on the ventral surface. They also occur just anterior to the pectoral fin. By 9.8 mm TL, the scales extend farther anteriorly to cover much more of the head. By 15.7 mm TL, the entire surface of the body is covered except the caudal peduncle behind the origin of the dorsal and anal fins. By 27 mm TL, scales cover 99% of the body and thus at this size the scales have reached the adult condition.

LARVAL SUPPLY, SETTLEMENT, GROWTH, AND MORTALITY

Sphoeroides maculatus was once much more common than it is today off the coast of New Jersey. During collections made in 1929 to 1933, it ranked fifth in frequency of occurrence and ninth by numerical abundance, but failed to rank among the top 20 in comparable collections in the 1970s (Thomas and Milstein, 1974). This species was also the most abundant

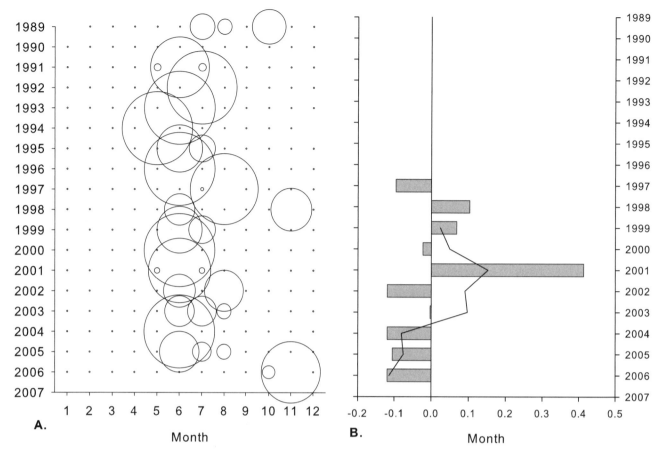

A.   Month

B.   Month

Fig. 104.1. Annual and seasonal variation in occurrence and abundance of *Sphoeroides maculatus*. (A) Seasonal variation of the larvae is based on ichthyoplankton samples collected in Little Sheepshead Creek behind Little Egg Inlet from 1989 to 2006. Open circles represent the percentage of the mean number/1000 m³ captured within each year. Values range from 0 to 100%. The smallest circles represent when samples were taken but no larvae were collected. (B) Annual variation in abundance is based on otter trawl sampling from 1997 to 2006. Data were standardized by subtracting overall mean from annual abundance. Horizontal bars represent standardized annual abundance; the solid line is a three-year running mean that was calculated by taking the mean of the standardized values from the previous two years and the year in which the data are plotted.

species collected during a two-year (1962–1963) survey of the surfzone of Long Island (Schaefer, 1967), but it was not present (in any life-history stage) in a six-year survey of the nearby Hudson River estuary (Kahnle and Hattala, 1988). It was reported to be one of the most abundant sport fishes in Great Bay in 1969 (Hamer, 1972) but declined drastically through the 1970s (Murawski and Festa, 1979). Otter trawling surveys in our study area indicate that this species displays an increase in abundance about 1 year out of 10 (Fig. 104.1).

Local spawning is apparently the source of most larvae. The prolonged spawning period in estuaries and coastal ocean habitats is reflected in the continued presence of the smallest size classes well into November (Fig. 104.2). Growth is rapid through the first summer and fall. In the Great Bay–Little Egg Harbor study area, young-of-the-year first appear in May and June when they are 3 cm or smaller. This cohort can be followed through October when the size mode is about 12 cm. Larger individuals apparently begin to leave the system first in October, and they can be detected returning to the estuary the following spring at sizes between 11 and

26 cm. Incursions of cold water, caused by upwelling, have been observed to slow down growth rates in Great Bay (Able and Fahay, 1998). In a study during 1988, juvenile growth was 0.47 mm per day between July and August (when temperatures were cold), but 1.5 mm per day during August and September when temperatures returned to normal. In a Barnegat Bay study, the growth rates of young-of-the-year have been calculated as 1.1 mm per day over 45 days and 0.93 mm per day over a 60-day period (Marcellus, 1972). In Chesapeake Bay, lengths of 158 to 187 mm TL have been attained by November, and sizes of 203 mm have been reported for fish reaching age 1 (Laroche and Davis, 1973). We have no information on winter mortality, but judging from the sizes of most returning young-of-the-year the following spring (March), the probability is high that the smallest larvae spawned late in the previous fall season do not return to our study area after their first winter. A massive kill involving this species was reported on May 10, 1969, from Sandy Hook to Manasquan Inlet, presumably due to an upwelling event that subjected the fish to a sudden drop in temperature (Wicklund, 1970).

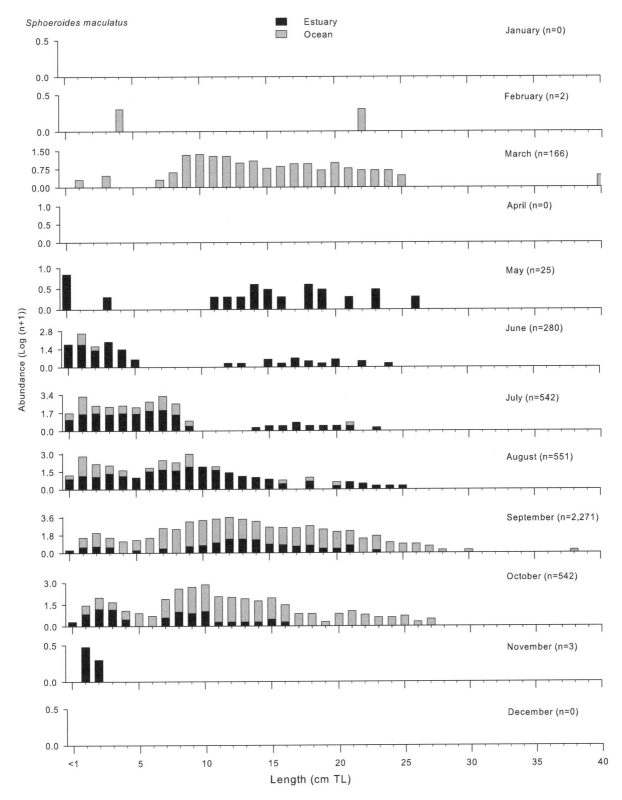

Fig. 104.2. Monthly length frequencies of *Sphoeroides maculatus* collected in estuarine and coastal ocean waters of the central Middle Atlantic Bight. Sources: National Marine Fisheries Service: otter trawl (n = 2675); Rutgers University Marine Field Station: 1 m beam trawl (n = 8); seine (n = 358); Methot trawl (n = 27); 1 m plankton net (n = 113); 2 m beam trawl (n = 11); killitrap (n = 3); 4.9 m otter trawl (n = 237); multi-panel gill net (n = 2); and Able and Fahay 1998 (n = 948).

SEASONALITY AND HABITAT USE

Adults inhabit bays, estuaries, and coastal ocean waters to a maximum depth of 60 m. They occur over a variety of substrates and are often associated with piers or other structured habitats. They are also commonly found in the surfzone. Adults may occur in dense aggregations, but only juveniles form schools. Larvae were rarely collected over the Middle Atlantic Bight continental shelf during 11 years of MARMAP sampling (1977–1987), although larger, pelagic-juveniles (identified only to the family level) occurred infrequently, particularly in the southern part of the bight (Able and Fahay, 1998). They have also recently been collected in the Slope Sea, where they were a member of a Gulf Stream assemblage. The members of this assemblage were likely spawned south of Cape Hatteras, transported into the Middle Atlantic Bight via the Gulf Stream, then advected with the aid of gyres or filaments onto the continental shelf from whence they ultimately migrated to estuarine nursery areas (Hare et al., 2001). Therefore, a portion of young-of-the-year that recruit to estuaries in the Middle Atlantic Bight originate from spawning in the South Atlantic Bight. Larvae are rarely collected in major bays and estuaries despite intensive sampling (Able and Fahay, 1998). However, bridge net sampling at Little Egg Inlet does detect immigrating larvae and small juveniles, and they occur mostly in May to August (see Fig. 104.1). In some years, spawning in coastal ocean or estuarine waters may extend into the fall, and the earliest stages resulting from this activity are collected into November.

Most habitat information is restricted to juveniles because, although larvae have been collected only rarely (see above), juveniles are common constituents of several studies undertaken in the Middle Atlantic Bight. In a study of habitat associations of estuarine fishes in Great Bay–Little Egg Harbor, S. maculatus was the sixth most abundant fish species collected (Szedlmayer and Able, 1996). Young-of-the-

year occurred most abundantly in a variety of depths (0.6 to 3.7 m), substrate types (3.8 to 54.4% silt), and structure (no structure to complex eelgrass beds) in the lower portion of the estuary and on the inner continental shelf (Fig. 104.3). Young-of-the-year have also been collected in subtidal marsh creeks in Great Bay (Rountree and Able, 1992a). One report describes juveniles as semidemersal over smooth bottoms (Merriman, 1947). Salinity may influence the distribution of young-of-the-year. Most collections of larvae and young in the Delaware Bay area have been made in higher salinities (Wang and Kernehan, 1979), although some have been made in salinities as low as 13 ppt (de Sylva et al., 1962). Juveniles have been found in salinities up to 32.2 ppt on the eastern shore of Virginia (Richards and Castagna, 1970). A Chesapeake Bay study found larvae and juveniles in salinities between 12 and 21 ppt and temperatures between 16 and 26°C (Dovel, 1971). Several studies have indicated that temperature exerts a strong influence on the distribution and abundance of this species, and this is particularly true for young-of-the-year. In one Virginia study, juveniles (13 to 90 mm) occurred only at temperatures between 20.4 and 29.4°C, whereas adults were present between 10 and 25.9°C (Richards and Castagna, 1970). In Barnegat Bay, young-of-the-year appeared as early as April when temperatures exceeded 15°C, but they did not appear until May in years when temperatures were colder (Vouglitois, 1983).

PREY AND PREDATORS

Sphoeroides maculatus preys on a variety of invertebrate taxa, including small crustaceans, mollusks, worms, barnacles, and sea urchins, as well as bryozoans, algae, and sponges (Collette and Klein-MacPhee, 2002). They feed during the day; groups have been known to attack and consume a blue crab (Nichols and Breder, 1927). One study examined the food habits of young-of-the-year (29.4 to 89.5 mm, n = 22) in Long

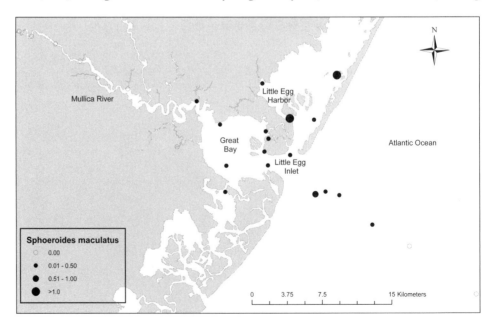

Fig. 104.3. Habitat use and average catch per tow (see key) for juvenile and adult Sphoeroides maculatus in the Ocean–Great Bay–Mullica River corridor in southern New Jersey based on small otter trawl collections during July and September from 1988 to 2006. See Table 3.2 for characteristics at each sampling collection.

Age 1+ *Sphoeroides maculatus*

YOY *Sphoeroides maculatus*

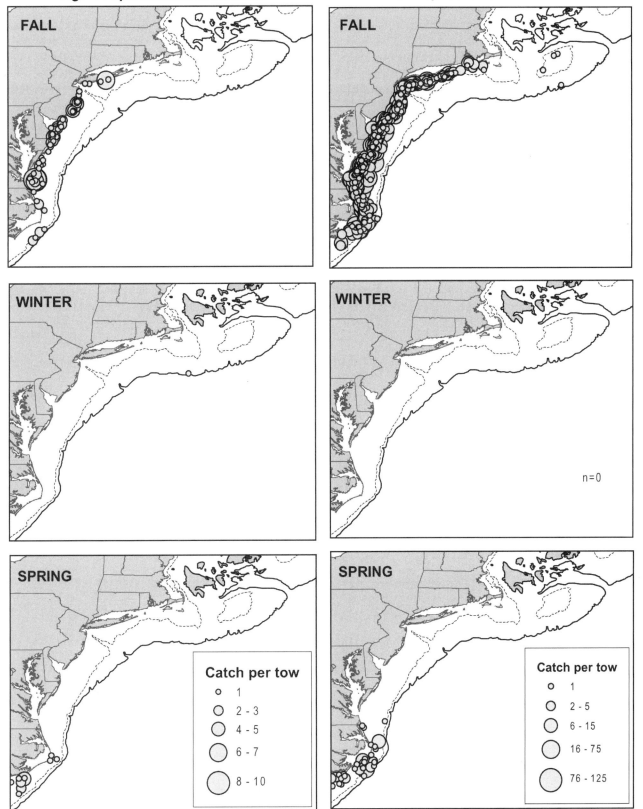

Fig. 104.4. Composite distribution of age 1+ (> 21 cm) and young-of-the-year (< 21 cm) *Sphoeroides maculatus* during seasonal cruises of the National Marine Fisheries Service groundfish survey. Details of sampling effort are indicated in Fig. 3.4.

Island Sound (see Table 8.1). Young-of-the-year consume predominantly polychaetes, crustaceans, most importantly amphipods, and pelecypod mollusks. Juveniles (and adults) are frequently included in the prey list of several larger fishes, including tunas, bluefish, and sharks. They have been observed avoiding predation by inflating their bodies.

MIGRATIONS

Sphoeroides maculatus makes annual coastal migrations, northward and inshore in spring, and southward and offshore in fall (Bigelow and Schroeder, 1953; Laroche and Davis, 1973; Shipp, 1974). All life history stages in trawl surveys are seasonal visitors to the Middle Atlantic Bight (Fig. 104.4). In the fall, the young-of-the-year (< 21 cm) are distributed on the inner continental shelf (mean depth 18 m, range 8 to 124 m) from southern Massachusetts to off North Carolina. A few stray individuals have been collected on Georges Bank at this time as well. In the winter, when sampling is restricted to waters north of Cape Hatteras they are not collected in the Middle Atlantic Bight. During the spring, they have been collected primarily on the inner continental shelf (mean depth 36 m, range 10 to 297 m) off North Carolina and up to the Virginia border.

In the surfzone of Long Island, adults are present throughout the summer and are joined by young-of-the-year (about 100 mm TL) in August and September (Schaefer, 1967). All year classes are rare in October and absent in November in these collections. These age 1+ (> 21 cm) individuals have a somewhat similar distribution to the young-of-the-year through the seasons (see Fig. 104.4). In the fall, they have been collected primarily on the inner continental shelf between Long Island and North Carolina (mean depth 18 m, range 8 to 88 m). The only individuals collected north of Cape Hatteras in the winter occurred in 124 m off Massachusetts. However, it should be noted that young-of-the-year and adults have been reported to migrate to deeper continental shelf waters, where they spend the winter in a quiescent state on the bottom (Bigelow and Schroeder, 1953). Those collected in the spring have been primarily from Cape Hatteras and Cape Lookout on the inner shelf (mean depth 22 m, 13 to 39 m) off North Carolina.

APPENDIX

Collection Data for Illustrations

Listings are by chapter number. Selected specimens have been deposited in the Academy of Natural Sciences of Philadelphia (ANSP).

CHAPTER 12 *Acipenser brevirostrum* larva, 24.4 mm SL. After Snyder 1988.

CHAPTER 13 *Acipenser oxyrinchus* larva, 28.9 mm SL. After Snyder 1988.

CHAPTER 17 *Anguilla rostrata* elver, 57.5 mm TL. Collected April 25, 1991, marsh pool, dip net, RUMFS, Great Bay, New Jersey. ANSP 175207. Illustrated by Susan Kaiser.

CHAPTER 18 *Myrophis punctatus* elver, 105 mm TL. Collected March 2003, beam trawl, White Oak River, North Carolina. Illustrated by Carolyn Hoss.

CHAPTER 19 *Conger oceanicus* elver, 62.0 mm TL. Collected May 14, 1990, dip net (nightlight), RUMFS boat basin, Great Bay, New Jersey. ANSP 175208. Illustrated by Susan Kaiser.

CHAPTER 20 *Alosa aestivalis* juvenile, 45 mm TL. After Mansueti and Hardy 1967, with permission of the University of Maryland, College Park.

CHAPTER 21 *Alosa mediocris* juvenile, 35.2 mm TL. After Mansueti 1962, with permission of Springer Science and Business Media.

CHAPTER 22 *Alosa pseudoharengus* juvenile, 42.0 mm TL. After Mansueti and Hardy 1967, with permission of the University of Maryland, College Park.

CHAPTER 23 *Alosa sapidissima* juvenile, 39.5 mm TL. After Mansueti and Hardy 1967, with permission of the University of Maryland, College Park.

CHAPTER 24 *Brevoortia tyrannus* juvenile, 41.0 mm TL. After Kuntz and Radcliffe 1917.

CHAPTER 25 *Clupea harengus* juvenile, 46.0 mm SL. Collected May 6, 1993, dip net (nightlight), RUMFS boat basin, Great Bay, New Jersey. ANSP 175209. Illustrated by Susan Kaiser.

CHAPTER 26 *Dorosoma cepedianum*, 24.2 mm TL pre-transformation larva. After Lippson and Moran 1974.

CHAPTER 27 *Opisthonema oglinum*, 30.8 mm TL juvenile. After Richards et al. 1974.

CHAPTER 28 *Anchoa hepsetus* juvenile, 34.0 mm TL. Collected August 2, 1995, pop net, RUMFS boat basin, Great Bay, New Jersey. ANSP 175210. Illustrated by Nancy Arthur-McGehee.

CHAPTER 29 *Anchoa mitchilli* juvenile, 33.0 mm SL. Collected September 27, 1990, 1 m plankton net, Little Sheepshead Creek Bridge, Great Bay, New Jersey. ANSP 175211. Illustrated by Susan Kaiser.

CHAPTER 30 *Engraulis eurystole*, 22.8 mm TL. After Lippson and Moran 1974.

CHAPTER 31 *Osmerus mordax* juvenile, 34 mm TL. Collected September 18, 2003, 2 m beam trawl, Penobscot Bay, Stockton Harbor, Maine. Illustrated by Nancy Arthur-McGehee.

CHAPTER 32 *Synodus foetens* early juvenile, 35.4 mm SL. Collected July 23, 1990, dip net (nightlight), RUMFS boat basin, Great Bay, New Jersey. ANSP 175212. Illustrated by Susan Kaiser.

CHAPTER 33 *Enchelyopus cimbrius*, 10.3 mm SL. Late larva prior to pelagic juvenile stage. After Fahay 1983.

CHAPTER 34 *Urophycis chuss* early demersal juvenile, 35.0 mm SL. Collected August 6, 1994, 2 m beam trawl, Beach Haven Ridge, New Jersey. ANSP 175215. Illustrated by Susan Kaiser.

CHAPTER 35 *Urophycis regia* late pelagic-early demersal juvenile, 29.0 mm SL. Collected November 3, 1993, 1 m plankton net, Little Sheepshead Creek Bridge, Great Bay, New Jersey. ANSP 175216. Illustrated by Susan Kaiser.

CHAPTER 36 *Urophycis tenuis* early demersal juvenile, 55.0 mm

SL. Collected June 7, 1988, 5.3 m otter trawl, Nauset Marsh, Cape Cod, Massachusetts. ANSP 175217. Illustrated by Susan Kaiser.

CHAPTER 37 *Microgadus tomcod* juvenile, 53.3 mm TL. Collected June 22, 1993, pyramid trap, Pier 40, Hudson River, New York. ANSP 175213. Illustrated by Nancy Arthur-McGehee.

CHAPTER 38 *Pollachius virens* juvenile, 39.5 mm SL. Collected April 15, 1988, seine, cove near RUMFS, Great Bay, New Jersey. ANSP 175214. Illustrated by Susan Kaiser.

CHAPTER 39 *Ophidion marginatum* early juvenile, 41.8 mm SL. Collected March 13, 1991, 1 m plankton net, Little Sheepshead Creek Bridge, Great Bay, New Jersey. ANSP 175218. Illustrated by Susan Kaiser.

CHAPTER 40 *Opsanus tau* early juvenile, 23.0 mm TL. Collected July 12, 1973, seine, Absecon Bay, New Jersey, ANSP 175219. Illustrated by Nancy Arthur-McGehee.

CHAPTER 41 *Strongylura marina* juvenile in "half beak" stage, 23.5 mm BL. After Collette 1966.

CHAPTER 42 *Cyprinodon variegatus* juvenile, 10.6 mm TL. Collected June 6, 1991, dip net, marsh pool, RUMFS, Great Bay, New Jersey. ANSP 175220. Illustrated by Susan Kaiser.

CHAPTER 43 *Fundulus confluentus*, 13.0 mm SL. Laboratory-reared from North River, North Carolina population; September 19, 2006. Illustrated by Mike Fahay.

CHAPTER 44 *Fundulus diaphanus* juvenile, 12.8 mm TL. After Jones and Tabery 1980, with permission of the U.S. Fish and Wildlife Service.

CHAPTER 45 *Fundulus heteroclitus* juvenile, 11.0 mm TL prior to formation of vertical bars on the body. Laboratory-raised, April 27, 1991. ANSP 175221. Illustrated by Susan Kaiser.

CHAPTER 46 *Fundulus luciae* juvenile, 6.9 mm SL. Collected June 23, 1995, dip net, marsh pool, RUMFS, Great Bay, New Jersey. ANSP 175222. Illustrated by Susan Kaiser.

CHAPTER 47 *Fundulus majalis* juvenile, 14.5 mm TL. Collected July 7, 1991, push net, cove near RUMFS, Great Bay, New Jersey. ANSP 175223. Illustrated by Susan Kaiser.

CHAPTER 48 *Lucania parva* juvenile, 11.1 mm TL. Collected June 12, 1991, throw trap, marsh pool near RUMFS, Great Bay, New Jersey. ANSP 175224. Illustrated by Susan Kaiser.

CHAPTER 49 *Gambusia holbrooki* juvenile, 6.6 mm SL, laboratory-raised from The Straits, North Carolina stock. May 24, 2001. Illustrated by Nancy Arthur-McGehee.

CHAPTER 50 *Membras martinica* juvenile, 16.0 mm SL. Collected October 12, 1991, dip net (nightlight), RUMFS boat basin, Great Bay, New Jersey, ANSP 175225. Illustrated by Susan Kaiser.

CHAPTER 51 *Menidia beryllina* juvenile, 10.3 mm SL. Collected June 25, 1993, dip net, Bass River, New Jersey. ANSP 175226. Illustrated by Susan Kaiser.

CHAPTER 52 *Menidia menidia* juvenile, 17.4 mm SL. Collected June 13, 1987, dip net (nightlight), RUMFS boat basin, Great Bay, New Jersey. ANSP 175227. Illustrated by Susan Kaiser.

CHAPTER 53 *Apeltes quadracus* juvenile, 17.9 mm SL. Collected May 30, 1991, 1 m beam trawl, small cove, Holgate, New Jersey. ANSP 175228. Illustrated by Susan Kaiser.

CHAPTER 54 *Gasterosteus aculeatus* juvenile, 18.6 mm TL.

Collected May 17, 1987, dip net (nightlight), RUMFS boat basin, Great Bay, New Jersey. ANSP 175229. Illustrated by Susan Kaiser.

CHAPTER 55 *Hippocampus erectus* juvenile, 3.3 mm TL. After Lippson and Moran 1974.

CHAPTER 56 *Syngnathus fuscus* juvenile, 13.2 mm TL. After Lippson and Moran 1974.

CHAPTER 57 *Prionotus carolinus* early juvenile, 12.0 mm SL. Collected September 23, 1991, 2 m beam trawl, Beach Haven Ridge, New Jersey. ANSP 175230. Illustrated by Susan Kaiser.

CHAPTER 58 *Prionotus evolans* early juvenile, 12.5 mm SL. Collected August 6, 1986, dip net (nightlight), boat basin, Shark River, New Jersey. ANSP 175231. Illustrated by Susan Kaiser.

CHAPTER 59 *Myoxocephalus aenaeus* juvenile, 19.4 mm SL. Collected May 12, 1989, 1 m plankton net, Little Sheepshead Creek Bridge, Great Bay, New Jersey. ANSP 175232. Illustrated by Susan Kaiser.

CHAPTER 60 *Morone americana* juvenile, 28 mm TL. Collected June 20, 2001, 4.9 m otter trawl, Brown's Run, Delaware Bay, New Jersey. Illustrated by Susan Kaiser.

CHAPTER 61 *Morone saxatilis* juvenile, 35 mm. Collected June 20, 2001, 4.9 m otter trawl, Brown's Run, Delaware Bay, New Jersey. Illustrated by Susan Kaiser.

CHAPTER 62 *Centropristis striata* juvenile, estimated TL of 16.5 mm. Collected July 20, 1994, experimental trap, Hudson River, New York. Illustrated by Nancy Arthur-McGehee.

CHAPTER 63 *Mycteroperca microlepis*, 22.6 mm SL, juvenile. After Kendall 1979.

CHAPTER 64 *Pomatomus saltatrix* juvenile, 24.3 mm SL. Collected August 4, 1993, seine, Great Bay, New Jersey. ANSP 175233. Illustrated by Susan Kaiser.

CHAPTER 65 *Caranx hippos* juvenile, 32.6 mm SL. After Berry 1959.

CHAPTER 66 *Trachinotus carolinus*, 14.8 mm SL juvenile. After Fields 1962.

CHAPTER 67 *Trachinotus falcatus*, 18.0 mm SL juvenile. After Fields 1962.

CHAPTER 68 *Lutjanus griseus* juvenile, 15.8 mm TL. Collected August 27, 1973, Absecon Inlet, New Jersey. ANSP 130329. Illustrated by Nancy Arthur-McGehee.

CHAPTER 69 *Lagodon rhomboides*, 27.0 mm SL juvenile. After Hildebrand and Cable 1938.

CHAPTER 70 *Stenotomus chrysops* juvenile, 26.0 mm TL. Collected August 14, 1929, off Broadkill Beach, Delaware. ANSP 128648. Illustrated by Nancy Arthur-McGehee.

CHAPTER 71 *Bairdiella chrysoura* juvenile, 19 mm TL. Collected July 6, 2006, otter trawl, Mott's Creek, Great Bay, New Jersey. Illustrated by Nancy Arthur-McGehee.

CHAPTER 72 *Cynoscion nebulosus*, 41.0 mm SL juvenile. After Welsh and Breder 1923.

CHAPTER 73 *Cynoscion regalis* juvenile, 31.0 mm TL. Collected July 16, 1996, mouth of Dennis Creek, Delaware Bay, otter trawl. ANSP 175234. Illustrated by Nancy Arthur-McGehee.

CHAPTER 74 *Leiostomus xanthurus* juvenile, 23.5 mm SL.

Collected June 13, 1995, seine, New Cove, Great Bay, New Jersey. ANSP 175235. Illustrated by Nancy Arthur-McGehee.

CHAPTER 75 *Menticirrhus saxatilis* juvenile, 16.4 mm TL. Collected July 7, 1991, seine, intertidal pool, Corson's Inlet, New Jersey. ANSP 175236. Illustrated by Susan Kaiser.

CHAPTER 76 *Micropogonias undulatus* juvenile, 19.0 mm SL. Collected October 22, 1992, experimental trap, RUMFS boat basin, Great Bay, New Jersey. ANSP 175237. Illustrated by Susan Kaiser.

CHAPTER 77 *Pogonias cromis* juvenile, 43.5 mm TL. Collected August 15, 1957, Absecon Island, Atlantic County, New Jersey. ANSP 122234. Illustrated by Nancy Arthur-McGehee.

CHAPTER 78 *Sciaenops ocellatus* juvenile, 42.0 mm TL. After Pearson 1929.

CHAPTER 79 *Chaetodon capistratus* "tholichthys" stage, 13.7 mm SL. Collected June 29, 1991, killitrap, RUMFS boat basin, Great Bay, New Jersey. ANSP 175239. Illustrated by Susan Kaiser.

CHAPTER 80 *Chaetodon ocellatus* "tholichthys" stage, 13.8 mm SL. Collected July 31, 1991, killitrap, RUMFS boat basin, Great Bay, New Jersey. ANSP 175238. Illustrated by Susan Kaiser.

CHAPTER 81 *Mugil cephalus* juvenile, 37.0 mm TL. Collected 1953, off Miami, Florida. ANSP 128648. Illustrated by Nancy Arthur-McGehee.

CHAPTER 82 *Mugil curema* "querimana" stage juvenile, 24.2 mm SL. Collected June 11, 1986, dip net (nightlight), RUMFS boat basin, Great Bay, New Jersey. ANSP 175240. Illustrated by Susan Kaiser.

CHAPTER 83 *Tautoga onitis* juvenile, 11.4 mm TL. Collected July 30, 1992, beam trawl, RUMFS boat basin, Great Bay, New Jersey. ANSP 175241. Illustrated by Susan Kaiser.

CHAPTER 84 *Tautogolabrus adspersus* juvenile, 13.8 mm SL. Collected July 16, 1992, Beach Haven Ridge, New Jersey. ANSP 175242. Illustrated by Susan Kaiser.

CHAPTER 85 *Pholis gunnellus* juvenile, 56 mm TL. Collected August 11, 1994, suction device, Submersible Delta, Dive no. 3309, 30 m depth, east of Little Egg Inlet, New Jersey. ANSP 175243. Illustrated by Nancy Arthur-McGehee.

CHAPTER 86 *Ammodytes americanus* juvenile, 45.5 mm TL. Collected June 6, 1995, seine, Tuckers Island, Little Egg Inlet, New Jersey. ANSP 175245. Illustrated by Nancy Arthur-McGehee.

CHAPTER 87 *Astroscopus guttatus* juvenile, 17.0 mm SL. Collected October 3, 1991, dip net (nightlight), RUMFS boat basin, Great Bay, New Jersey. ANSP 175244. Illustrated by Susan Kaiser.

CHAPTER 88 *Chasmodes bosquianus*, 9.8 mm TL early juvenile. After Hildebrand and Cable 1938. Originally described as *Hypsoblennius hentz*. See Fahay 2007 for correct developmental series.

CHAPTER 89 *Hypsoblennius hentz* early juvenile, 14.8 mm TL.

Collected June 22, 1942, Bird Shoal, Beaufort, North Carolina. ANSP 74120. Illustrated by Nancy Arthur-McGehee.

CHAPTER 90 *Gobiesox strumosus* juvenile, 12.0 mm TL. After Runyan 1961, with permission of Springer Science and Business Media.

CHAPTER 91 *Ctenogobius boleosoma*, 9.1 mm TL. After Wyanski and Targett 2000.

CHAPTER 92 *Gobiosoma bosc* juvenile, 15.0 mm SL. Collected August 3, 1990, dip net, RUMFS boat basin, Great Bay, New Jersey. ANSP 175246. Illustrated by Susan Kaiser.

CHAPTER 93 *Gobiosoma ginsburgi* juvenile, 14.8 mm SL. Collected August 1991, bottom grab sample, Delaware Bay. ANSP 175247. Illustrated by Susan Kaiser.

CHAPTER 94 *Microgobius thalassinus* juvenile, 15.1 mm SL. D. Ruple (unpubl.) used with permission.

CHAPTER 95 *Sphyraena borealis* juvenile, 56.0 mm TL. Collected June 27, 1973, Beasley's Point, New Jersey. ANSP 169550 (1 of 2). Illustrated by Nancy Arthur-McGehee.

CHAPTER 96 *Peprilus triacanthus* juvenile, 24.0 mm TL. Collected August 5, 1974, ca. 3.7 km off Holgate Peninsula, Long Beach Island, New Jersey. ANSP 175248 (CBM-74-151). Illustrated by Nancy Arthur-McGehee.

CHAPTER 97 *Scophthalmus aquosus* juvenile, 12.5 mm SL. Collected May 28, 1991, plankton net, Little Sheepshead Creek, Great Bay, New Jersey. ANSP 175249. Illustrated by Susan Kaiser.

CHAPTER 98 *Etropus microstomus* juvenile, 19.9 mm SL. Collected September 25, 1990, seine, Great Bay, New Jersey. ANSP 175250. Illustrated by Susan Kaiser.

CHAPTER 99 *Paralichthys dentatus* juvenile, 27.0 mm SL. Collected January 7 1990, 1 m plankton net, Little Sheepshead Creek, Great Bay, New Jersey. ANSP 175252. Illustrated by Susan Kaiser.

CHAPTER 100 *Pseudopleuronectes americanus* juvenile, 14.0 mm SL. Collected April 1991, beam trawl, small cove at Holgate, Great Bay, New Jersey. ANSP 175253. Illustrated by Susan Kaiser.

CHAPTER 101 *Trinectes maculatus* juvenile, 18.0 mm TL. After Hildebrand and Cable 1938.

CHAPTER 102 *Symphurus plagiusa* juvenile, 8.0 mm SL. After Farooqi et al. 2006.

CHAPTER 103 *Chilomycterus schoepfii*, measurement given as < 31.75 mm. "Lyosphaera globosa," or special juvenile stage. After Evermann and Kendall 1898.

CHAPTER 104 *Sphoeroides maculatus* juvenile, 9.7 mm SL. Pigmentation based on combination from two uncatalogued specimens. Collected June 4, 1991, nightlight, dip net, RUMFS boat basin, Great Bay, New Jersey. ANSP 175254. Illustrated by Susan Kaiser.

BIBLIOGRAPHY

Abbe, G. R. 1967. An evaluation of the distribution of fish populations of the Delaware River estuary. M.S. thesis. University of
Delaware, Newark. 64 pp.

Able, K. W. 1978. Ichthyoplankton of the St. Lawrence estuary:
composition, distribution, and abundance. J. Fish. Res. Board
Can. 35(12):1518–1531.

———. 1984. Variation in spawning site selection of the mummichog, *Fundulus heteroclitus*. Copeia 2:522–525.

———. 1990. Life history patterns of New Jersey salt marsh killifishes. Bull. N.J. Acad. Sci. 35(2):23–30.

———. 1992. Checklist of New Jersey saltwater fishes. Bull. N.J.
Acad. Sci. 37(1):1–11.

———. 1999. Measures of juvenile fish habitat quality: examples
from a National Estuarine Research Reserve. Pp. 134–147 *in*
L. R. Benaka, ed. Fish Habitat: Essential Fish Habitat and Rehabilitation. Am. Fish. Soc. Symp. 22, Bethesda, Md.

———. 2005. A reexamination of fish estuarine dependence: evidence for connectivity between estuarine and ocean habitats.
Estuar. Coast. Shelf Sci. 64:5–17.

Able, K. W., and R. Brown. 2005. Distribution and abundance of
young-of-the-year estuarine fishes: seasonal occurrence on the
Middle Atlantic Bight continental shelf. Rutgers Univ. Inst. Mar.
Coastal Sci. Tech. Rept. 2005-14.

Able, K. W., and M. Castagna. 1975. Aspects of an undescribed
reproductive behavior in *Fundulus heteroclitus* (Pisces: Cyprinodontidae) from Virginia. Chesapeake Sci. 16(4):282–284.

Able, K. W., and M. C. Curran. 2008. Winter mortality in some
temperate young-of-the-year fishes. Bull. N.J. Acad. Sci. 53(2):
1–5.

Able, K. W., and J. T. Duffy-Anderson. 2006. Impacts of piers on
juvenile fishes and selected invertebrates in the lower Hudson
River. Pp. 429–440 *in* J. Levinton and C. Hiles, eds. The Hudson
River. Cambridge University Press, New York.

Able, K. W., and M. P. Fahay. 1998. The First Year in the Life of Estuarine Fishes in the Middle Atlantic Bight. Rutgers University
Press, New Brunswick, N.J. 342 pp.

Able, K. W., and J. D. Felley. 1986. Geographical variation in *Fundulus heteroclitus*: tests for concordance between egg and adult
morphologies. Am. Zool. 26:1145–1157.

———. 1988. Bermuda *Fundulus* (Pisces: Fundulidae) revisited:
taxonomy of the endemic forms. Proc. Acad. Nat. Sci. Phila.
140(2):99–114.

Able, K. W., and T. M. Grothues. 2007a. An approach to understanding habitat dynamics of flatfishes: advantages of biotelem-
etry. J. Sea Res. 58:1–7.

———. 2007b. Diversity of estuarine movements of striped bass
(*Morone saxatilis*): a synoptic examination of an estuarine system in southern New Jersey. Fish. Bull. 105:426–435.

Able, K. W., and S. M. Hagan. 1995. Fishes in the vicinity of Beach
Haven Ridge: annual and seasonal patterns of abundance dur-
ing the early 1970s. Rutgers Univ. Inst. Mar. Coastal Sci. Tech.
Rept. 95-24.

———. 2000. Effects of common reed (*Phragmites australis*) invasion on marsh surface microfauna: response of fishes and decapod crustaceans. Estuaries 23(5):633–646.

———. 2003. The impact of common reed, *Phragmites australis*, on
essential fish habitat: influence on reproduction, embryological
development and larval abundance of mummichog (*Fundulus
heteroclitus*). Estuaries 26(1):40–50.

Able, K. W., and L. S. Hales Jr. 1997. Movements of juvenile black
sea bass, *Centropristis striata*, in a southern New Jersey estuary.
J. Exp. Mar. Biol. Ecol. 213:153–167.

Able, K. W., and D. Hata. 1984. Reproductive behavior in the *Fundulus heteroclitus–F. grandis* complex. Copeia 4:820–825.

Able, K. W., and S. C. Kaiser. 1994. Synthesis of summer flounder habitat parameters. NOAA Coastal Ocean Program Decision Analysis Series No. 1. NOAA Coastal Ocean Office, Silver
Spring, Md.

Able, K. W., and J. C. Lamonaca. 2006. Scale formation in selected western North Atlantic flatfishes. J. Fish Biol. 68:1679–1692.

Able, K. W., C. B. Grimes, R. A. Cooper, and J. R. Uzmann. 1982. Burrow construction and behavior of tilefish, *Lopholatilus chamaeleonticeps*, in Hudson Submarine Canyon. Env. Biol. Fish. 7(3):199–205.

Able, K. W., C. Talbot, and J. K. Shisler. 1983. The spotfin killifish, *Fundulus luciae*, is common in New Jersey salt marshes. Bull. N.J. Acad. Sci. 28(1):7–10.

Able, K. W., K. L. Heck Jr., M. P. Fahay, and C. T. Roman. 1988. Use of salt-marsh peat reefs by small juvenile lobsters on Cape Cod, Massachusetts. Estuaries 11(2):83–86.

Able, K. W., K. A. Wilson, and K. L. Heck Jr. 1989. Fishes of vegetated habitats in New Jersey estuaries: compositions, distributions, and abundance based on quantitative sampling. New Brunswick, N.J. Rutgers University. Center for Coastal and Environmental Studies, Tech Report No. 1041.

Able, K. W., R. E. Matheson, W. W. Morse, M. P. Fahay, and G. Shepherd. 1990. Patterns of summer flounder (*Paralichthys dentatus*) early life history in the Mid-Atlantic Bight and New Jersey estuaries. U.S. Fish. Bull. 88(1):1–12.

Able, K. W., R. Hoden, D. A. Witting, and J. B. Durand. 1992. Physical parameters of the Great Bay–Mullica River Estuary (with a list of research publications). Rutgers Univ. Inst. Mar. Coastal Sci. Tech. Rept. 92-06.

Able, K. W., M. P. Fahay, and G. R. Shepherd. 1995a. Early life history of black sea bass *Centropristis striata* in the Mid-Atlantic Bight and a New Jersey estuary. Fish. Bull. 93:429–445.

Able, K. W., A. L. Studholme, and J. P. Manderson. 1995b. Habitat quality in the New York / New Jersey Harbor estuary: an evaluation of pier effects in fishes. Final Report to the Hudson River Foundation. Hudson River Foundation, New York.

Able, K. W., R. Lathrop, and M. P. De Luca. 1996a. Background for research and monitoring in the Mullica River–Great Bay Estuary. Rutgers Univ. Inst. Mar. Coastal Sci. Tech. Rep. 96-07.

Able, K. W., D. A. Witting, R. S. McBride, R. A. Rountree, and K. J. Smith. 1996b. Fishes of polyhaline estuarine shores in Great Bay–Little Egg Harbor, New Jersey: a case study of seasonal and habitat influences. Pp. 335–353 *in* K. F. Nordstrom and C. T. Roman, eds. Estuarine Shores: Evolution, Environments and Human Alterations. John Wiley & Sons, Chichester, England.

Able, K. W., A. Kustka, D. Witting, K. Smith, R. Rountree, and R. McBride. 1997. Fishes of Great Bay, New Jersey: larvae and juveniles collected by nightlighting. Rutgers Univ. Inst. Mar. Coastal Sci. Tech. Rept. 97-05.

Able, K. W., J. P. Manderson, and A. L. Studholme. 1998. The distribution of shallow water juvenile fishes in an urban estuary: the effects of man-made structures in the lower Hudson River. Estuaries 21(4b):731–744.

Able, K. W., J. P. Manderson, and A. L. Studholme. 1999. Habitat quality for shallow water fishes in an urban estuary: the effects of manmade structures on growth. Mar. Ecol. Prog. Ser. 187:227–235.

Able, K. W., D. Nemerson, R. Bush, and P. Light. 2001. Spatial variation in Delaware Bay (U.S.A.) marsh creek fish assemblages. Estuaries 24(3):441–452.

Able, K. W., M. P. Fahay, K. L. Heck Jr., C. T. Roman, M. A. Lazzari, and S. C. Kaiser. 2002. Seasonal distribution and abundance of fishes and decapod crustaceans in a Cape Cod estuary. Northeast. Nat. 9(3):285–302.

Able, K. W., S. M. Hagan, and S. A. Brown. 2003a. Mechanisms of marsh habitat alteration due to *Phragmites*: response of young-of-the-year mummichog (*Fundulus heteroclitus*) to treatment for *Phragmites* removal. Estuaries 26(2b):484–494.

Able, K. W., P. Rowe, M. Burlas, and D. Byrne. 2003b. Use of ocean and estuarine habitats by young-of-year bluefish (*Pomatomus saltatrix*) in the New York Bight. Fish. Bull 101:201–214.

Able, K. W., M. J. Neuman, and H. Wennhage. 2005a. Ecology of the adult and juvenile stages: distribution and dynamics of habitat associations. Pp. 164–184 *in* R. N. Gibson, ed. Flatfishes: Biology and Exploitation. Blackwell Publishing, Oxford, U.K.

Able, K.W., K. J. Smith, and S. M. Hagan. 2005b. Fish composition and abundance in New Jersey salt marsh pools: sampling technique effects. Northeastern Naturalist 12(4):485–502.

Able, K. W., L. S. Hales Jr., and S. M. Hagan. 2005c. Movement and growth of juvenile (age 0 and 1+) tautog (*Tautoga onitis*) and cunner (*Tautogolabrus adspersus*) in a southern New Jersey estuary. J. Exp. Mar. Biol. Ecol. 327(1):22–35.

Able, K. W., M. P. Fahay, D. A. Witting, R. S. McBride, and S. M. Hagan. 2006a. Fish settlement in the ocean vs. estuary: comparison of pelagic larval and settled juvenile composition and abundance from southern New Jersey, U.S.A. Estuar. Coast. Shelf Sci. 66:280–290.

Able, K. W., S. M. Hagan, and S. A. Brown. 2006b. Habitat use, movement and growth of young-of-the-year *Fundulus* spp. in southern New Jersey salt marshes: comparisons based on tag/recapture. J. Exp. Mar. Biol. Ecol. 335:177–187.

Able, K. W., J. H. Balletto, S. M. Hagan, P. R. Jivoff, and K. Strait. 2007a. Linkages between salt marshes and other nekton habitats in Delaware Bay, USA. Rev. Fish. Sci. 15:1–61.

Able, K. W., S. M. Hagan, K. Kovitvongsa, S. A. Brown, and J. C. Lamonaca. 2007b. Piscivory by the mummichog *Fundulus heteroclitus*: evidence from the laboratory and salt marshes. J. Exp. Mar. Biol. Ecol. 345:26–37.

Able, K. W., T. M. Grothues, S. M. Hagan, M. E. Kimball, D. M. Nemerson, and G. L. Taghon. 2008. Long-term response of fishes and other fauna to restoration of former salt hay farms: multiple measures of restoration success. Rev. Fish Biol. Fish. 18:65–97.

Able, K. W., G. Sakowicz, and J. Lamonaca. 2009a. Scale formation in selected fundulid and cyprinodontid fishes. Ichthyol. Res. 56:1–9.

Able, K. W., K. M. M. Jones, and D. A. Fox. 2009b. Large nektonic fishes in nearshore and marsh creek habitats in the Delaware Bay estuary: distribution and abundance. Northeast. Nat. 16(1):27–44.

Able, K. W., D. Clarke, J. Lamonaca, and A. Muzeni-Corino. 2009c. Fishes of temperate estuaries: aspects of larval supply and development. IMCS Technical Report 2009–11.

Able, K. W., D. H. Wilber, A. Corino, and D. G. Clarke. 2010. Spring and summer larval fish assemblages in the surf zone and nearshore off northern New Jersey, USA. Estuaries and Coasts. 33:211–222.

Able, K. W., J. Dobarro, and A. Muzeni-Corina. In review. An evaluation of boat basin dredging effects response of fishes and crabs in a New Jersey estuary. N. Amer. J. Fish. Mgt.

Adamowicz, S. C., and C. T. Roman. 2005. New England salt marsh pools: a quantitative analysis of geomorphic and geographic features. Wetlands 25(2):279–288.

Adams, S. M. 1976. Feeding ecology of eelgrass communities. Trans. Am. Fish. Soc. 105(4):514–519.

Adams, S. M., J. B. Lindmeier, and D. D. Duvernell. 2006. Microsatellite analysis of the phylogeography, Pleistocene history and secondary contact hypotheses for the killifish, *Fundulus heteroclitus.* Mol. Ecol. 15:1109–1123.

Adey, W. H., and R. S. Steneck. 2001. Thermogeography over time creates biogeographic regions: a temperature / space / time-integrated model and an abundance-weighted test for benthic marine algae. J. Phycol. 37:677–698.

Ahlstrom, E. H., J. L. Butler, and B. Y. Sumida. 1976. Pelagic stromateoid fishes (Pisces, perciformes) of the eastern Pacific, kinds, distributions, and early life histories and observations on five of these from the northwest Atlantic. Bull. Mar. Sci. 26(3): 285–402.

Ahrenholz, D. W., J. F. Guthrie, and C. W. Krouse. 1989. Results of abundance surveys of juvenile Atlantic and Gulf menhaden, *Brevoortia tyrannus* and *B. patronus.* NOAA Tech. Rep. NMFS 84. Beaufort, N.C.: Southeast Fisheries Center, Beaufort Laboratory, National Marine Fisheries Service, NOAA, 14 pp.

Ahrenholz, D. W., D. D. Squires, J. A. Rice, S. W. Nixon, and G. R. Fitzhugh. 2000. Periodicity of increment formation in otoliths of overwintering postlarval and prejuvenile Atlantic menhaden, *Brevoortia tyrannus.* Fish. Bull. 98:421–426.

Alemadi, S. D., and D. G. Jenkins. 2008. Behavioral constraints for the spread of the eastern mosquitofish, *Gambusia holbrooki* (Poeciliidae). Biological Invasions 10:59–66.

Allen, D. M., J. P. Clymer, and S. S. Herman. 1978. Fishes of the Hereford Inlet Estuary, Southern New Jersey. The Wetlands Institute, Lehigh University, Bethlehem, Pa.

Allen, D. M., S. K. Service, and M. V. Ogburn-Matthews. 1992. Factors influencing the collection efficiency of estuarine fishes. Trans. Am. Fish. Soc. 121:234–244.

Allman, R. J., and C. B. Grimes. 2002. Temporal and spatial dynamics of spawning, settlement, and growth of gray snapper (*Lutjanus griseus*) from the West Florida shelf as determined from otolith microstructures. Fish. Bull. 100(3):391–403.

Alvarez-Lajonchere, L. 1976. Contribucion al estudio del ciclo de vida de *Mugil curema* Valenciennes in Cuvier and Valenciennes, 1836 (Pisces, Mugilidae). Cienc. Ser. 8 Invest. Mar. (Havana). No. 28.

Anderson, E. D. 1982. Red hake, *Urophycis chuss.* MESA (Mar. Ecosyst. Anal.) N.Y. Bight Atlas Monogr. 15:74–76.

Anderson, W. W. 1957. Early development, spawning, growth and occurrence of the silver mullet (*Mugil curema*) along the south Atlantic coast of the United States. U.S. Fish. Bull. 57:397–414.

———. 1958. Larval development, growth and spawning of striped mullet (*Mugil cephalus*) along the south Atlantic coast of the United States. U.S. Fish. Bull. 58:501–519.

———. 1968. Fishes taken during shrimp trawling along the south Atlantic coast of the United States, 1931–1935. U.S. Fish Wildl. Serv. Spec. Sci. Rep. Fish 570.

Anderson, K. A., R. A. Rountree, and F. Juanes. 2008. Soniferous fishes in the Hudson River. Trans. Am. Fish. Soc. 137:616–626.

Andrews, A. K. 1970. Squamation chronology of the fathead minnow, *Pimephales promelas.* Trans. Am. Fish. Soc. 99:429–432.

Anthony, E. J. 1994. Natural and artificial shores of the French Riviera: an analysis of their interrelationship. J. Coast. Res. 10(1): 48–58.

Anthony, V. C. 1982. Atlantic herring, *Clupea harengus harengus.* MESA (Mar. Ecosyst. Anal.) N.Y. Bight Atlas Monogr. 15:63–65.

April, J., and J. Turgeon. 2006. Phylogeography of the banded killifish (*Fundulus diaphanus*): glacial races and secondary contact. J. Fish Biol. 69(b):212–228.

Arcos, D. F., L. A. Cubillos, and S. P. Nunez. 2001. The jack mackerel fishery and El Niño 1997–98 effects off Chile. Progr. Oceanogr. 49:597–617.

Arendt, M. D., J. A. Lucy, and D. A. Evans. 2001. Diel and seasonal activity patterns of adult tautog, *Tautoga onitis,* in lower Chesapeake Bay, inferred from ultrasonic telemetry. Environ. Biol. Fishes 62:379–391.

Arias, A. M., and P. Drake. 1990. Estados juveniles de la ictiofauna en los Canos de las Salinas de la Bahia de Cadiz. Instituto de Ciencias Marinas de Andalucia. Cadiz, Spain.

Armstrong, J. C. 1973. Squamation chronology of the zebrafish (Cyprinidae), *Brachydanio rerio.* Copeia 4:823–824.

Arndt, R. G. 2004. Annotated checklist and distribution of New Jersey freshwater fishes, with comments on abundance. Bull. N.J. Acad. Sci. 49(1):1–33.

Arnold, C. R., W. H. Bailey, T. D. Williams, A. Johnson, and J. L. Lasswell. 1977. Laboratory spawning and larval rearing of red drum and southern flounder. Proc. 31st Annual Conf. SE Assoc. Fish Wildl. Ag. P. 437–440.

Arnold, E. L., and J. R. Thompson. 1958. Offshore spawning of the striped mullet *Mugil cephalus* in the Gulf of Mexico. Copeia: 705–707.

Arve, J. 1960. Preliminary report on attracting fish by oyster-shell plantings in Chincoteague Bay, Maryland. Chesapeake Sci. 1: 58–65.

Ashizawa, D., and J. J. Cole. 1994. Long-term temperature trends of the Hudson River: a study of the historical data. Estuaries 17:166–171.

Ashton, D. E., M. T. Barbour, J. R. Fletcher, V. R. Kranz, B. D. Lorenz, and G. C. Slawson. 1975. Overview environmental study of electric power development in the Delaware River Basin, 1975–1989: aquatic ecology baseline. Prepared for Delaware River Basin Electric Utility Group. NUS Corp., Pittsburgh, Pa.

Atkinson, D. 1994. Temperature and organism size—a biological law for ectotherms? Adv. Ecol. Res. 25:1–58.

Atkinson, R. J. A., and A. C. Taylor. 1991. Burrows and burrowing behaviour of fish. *In* P. S. and A. Meadows, eds. The Environmental Impact of Burrowing Animals and Animal Burrows. Symp. Zool. Soc. Lond. 63:133–155.

Attrill, M. J., and M. Power. 2002. Climatic influence on a marine fish assemblage. Nature 417:275–278.

Attrill, M. J., and S. D. Rundle. 2002. Ecotone or ecocline: ecological boundaries in estuaries. Estuar. Coast. Shelf Sci. 55:929–936.

August, S. M., and B. J. Hicks. 2008. Water temperature and upstream migration of glass eels in New Zealand: implications of climate change. Environ. Biol. Fishes 81:195–205.

Austin, H. M. 2002. Decadel oscillations and regime shifts, a characterization of the Chesapeake Bay marine climate. Pp. 155–170 *in* N. A. McGinn, ed. Fisheries in a Changing Climate. Am. Fish. Soc. Symp. 32.

Auth, T. D. 2003. Interannual and regional patterns of abundance, growth, and feeding ecology of larval bay anchovy (*Anchoa mitchilli*) in Chesapeake Bay. M.S. thesis. University of Maryland, College Park, Md.

Ayvazian, S. G., L. A. Deegan, and J. T. Finn. 1992. Comparison of habitat use by estuarine fish assemblages in the Acadian and Virginian zoogeographic provinces. Estuaries 15(3):368–383.

Azarovitz, T. R. 1981. A brief historical review of the Woods Hole Laboratory trawl survey time series. Bottom trawl surveys. Can. Spec. Publ. Fish. Aquat. Sci. 58:1–273.

Bacheler, N. M., L. M. Paramore, J. A. Buckel, and F. S. Scharf.

2008. Recruitment of juvenile red drum in North Carolina: spatiotemporal patterns of year-class strength and validation of a seine survey. N. Am. J. Fish. Manag. 28:1086–1098.

Bacheler, N. M., L. M. Paramore, J. A. Buckel, and J. E. Hightower. 2009a. Abiotic and biotic factors influence the habitat use of an estuarine fish. Mar. Ecol. Prog. Ser. 377:263–277.

Bacheler, N. M., L. M. Paramore, S. M. Burdick, J. A. Buckel, and J. E. Hightower. 2009. Seasonal variation in age-specific movement patterns of red drum *Sciaenops ocellatus* inferred from conventional tagging and telemetry. Fish. Bull. 107:405–419.

Bailey, K. M., and E. D. Houde. 1989. Predation on eggs and larvae of marine fishes and the recruitment problem. Adv. Mar. Biol. 25:1–83.

Baird, D., and R. E. Ulanowicz. 1989. The seasonal dynamics of the Chesapeake Bay ecosystem. Ecol. Monogr. 59(4):329–364.

Baird, D., R. R. Christian, C. H. Peterson, and G. A. Johnson. 2004. Consequences of hypoxia on estuarine ecosystem function: energy diversion from consumers to microbes. Ecol. Appl. 14(3):805–822.

Baird, F. T., Jr. 1967. The smelt (*Osmerus mordax*). Fish. Educ. Ser. Unit No. 5, Bull. Maine Dept. Sea and Shore Fisheries. 7 pp.

Baird, R. C. 1999. Part one: essential fish habitat perspectives. Pp. 1–2 *in* L. R. Benaka, ed. Fish Habitat: Essential Fish Habitat and Rehabilitation. Am. Fisheries Soc. Symp. 22, American Fisheries Society, Bethesda, Md.

Baker, R. R. 1978. The Evolutionary Ecology of Animal Migration. Holmes and Meier, New York.

Baker, R., and M. Sheaves. 2005. Redefining the piscivore assemblage of shallow estuarine nursery habitats. Mar. Ecol. Prog. Ser. 291:197–213.

———. 2006. Visual surveys reveal high densities of large piscivores in shallow estuarine nurseries. Mar. Ecol. Prog. Ser. 323: 75–82.

———. 2007. Shallow-water refuge paradigm: conflicting evidence from tethering experiments in a tropical estuary. Mar. Ecol. Prog. Ser. 349:13–22.

Balon, E. K. 1981. Saltatory processes and altricial to precocial forms in the ontogeny of fishes. Am. Zool. 21:573–596.

———. 1984. Reflections on some decisive events in the early life of fishes. Trans. Am. Fish. Soc. 133:178–185.

———. 1985. The theory of salutatory ontogeny and life history models revisited. Pp. 13–28 *in* E. K. Balon, ed. Early Life Histories of Fishes. New Developmental, Ecological and Evolutionary Perspectives. W. Junk Publishers, Dordrecht, The Netherlands.

———. 1990. Epigenesis of an epigeneticist: the development of some alternative concepts on the early ontogeny and evolution of fishes. Guelph Ichthyol. Rev. 1:1–48.

———. 1999. Alternative ways to become a juvenile or definitive phenotype (and on some persistent linguistic offenses). Environ. Biol. Fishes 56:17–36.

Baranes, A., and J. Wendling. 1981. The early stages of development in *Carcharhinus plumbeus*. J. Fish Biol. 18:159–175.

Barans, C. A. 1969. Distribution, growth and behavior of the spotted hake in the Chesapeake Bight. M.S. thesis. College of William and Mary, Williamsburg, Va. 53 pp.

———. 1972. Spotted hake, *Urophycis regia* of the York River and Lower Chesapeake Bay. Chesapeake Sci. 13(1):59–62.

Barbeau, S. 2000. The influence of temperature-dependent growth of summer flounder, *Paralichthys dentatus*, on their period of vulnerability to two invertebrate predators. M.S. thesis. Rutgers University, New Brunswick, N.J. 70 pp.

Barber, I., D. Hoare, and J. Krause. 2000. Effects of parasites on fish behaviour: a review and evolutionary perspective. Rev. Fish Biol. Fish. 10:131–165.

Barbieri, L. R., M. E. Chittenden Jr., and S. K. Lowerre-Barbieri. 1994. Maturity, spawning, and ovarian cycle of Atlantic croaker, *Micropogonias undulatus*, in the Chesapeake Bay and adjacent coastal waters. Fish. Bull. 92:671–685.

Bardach, J. E., and J. Case. 1965. Sensory capabilities of the modified fins of squirrel hake (*Urophycis chuss*) and searobins (*Prionotus carolinus* and *P. evolans*). Copeia:194–206.

Barkman, R. C., D. A. Bengtson, and A. D. Beck. 1981. Daily growth of juvenile fish (*Menidia menidia*) in the natural habitat compared with juveniles reared in the laboratory. Rapp. P.-V. Reun. Cons. Int. Explor. Mer. 178:324–326.

Barletta, M., U. Saint-Paul, A. Barletta-Bergan, W. Ekau, and D. Schories. 2000. Spatial and temporal distribution of *Myrophis punctatus* (Ophichthidae) and associated fish fauna in a northern Brazilian intertidal mangrove forest. Hydrobiologia 426: 65–74.

Barrios, A. T. 2004. Use of passive acoustic monitoring to resolve spatial and temporal patterns of spawning activity for red drum, *Sciaenops ocellatus*, in the Neuse River Estuary, North Carolina. M.S. thesis. North Carolina State University, Raleigh.

Barros, N. B., and R. S. Wells. 1998. Prey and feeding patterns of resident bottlenose dolphins (*Tursiops truncatus*) in Sarasota Bay, Florida. J. Mammal. 79:1045–1059.

Barse, A. M., and D. H. Secor. 1999. An exotic nematode parasite of the American eel. Fisheries 24(2):6–10.

Barse, A. M., S. A. McGuire, M. A. Vinores, L. E. Eierman, and J. A. Weeder. 2001. The swimbladder parasite nematode *Anguillicola crassus* in American eels (*Anguilla rostrata*) from middle and upper regions of Chesapeake Bay. J. Parasitol. 87:1366–1370.

Barthem, R., and M. Goulding. 1997. The Catfish Connection: Ecology, Migration, and Conservation of Amazon Predators. Columbia University Press, New York.

Bass, C. S., S. Khan, and J. S. Weis. 2007. Morphological changes to the gills of killifish associated with severe parasite infection. J. Fish Biol. 71:920–925.

Bass, R. J., and J. W. Avault Jr. 1975. Food habits, length-weight relationship, condition factor, and growth of juvenile red drum, *Sciaenops ocellata*, in Louisiana. Trans. Am. Fish. Soc. 104:35–45.

Bassista, T. P., and K. J. Hartman. 2005. Reproductive biology and egg mortality of bay anchovy, *Anchoa mitchilli*, in the Hudson River estuary. Environ. Biol. Fishes 73:49–59.

Bate, G. C., A. K. Whitfield, J. B. Adams, P. Huizinga, and T. H. Wooldridge. 2002. The importance of the river-estuary interface (REI) zone in estuaries. Water S. A. 28(3):271–279.

Bath, D. W., and J. M. O'Connor. 1982. The biology of white perch, *Morone americana*, in the Hudson River estuary. U.S. Fish. Bull. 80:599–610.

———. 1985. Food preferences of white perch in the Hudson river estuary. N.Y. Fish Game J. 32(1):63–70.

Bath, D. W., J. M. O'Connor, J. B. Alber, and L. G. Arvidson. 1981. Development and identification of larval Atlantic sturgeon (*Acipenser oxyrhynchus*) and shortnose sturgeon (*A. brevirostrum*) from the Hudson River Estuary. Copeia:711–717.

Beardsley, R. C., and C. N. Flagg. 1976. The water structure, mean currents and shelf water / slope water front on the New England continental shelf. Mem. Soc. R. Sci. Liège 6(10):209–225.

Beasley, C. A., and J. E. Hightower. 2000. Effects of a low-head dam on the distribution and characteristics of spawning habitat used by striped bass and American shad. Trans. Am. Fish. Soc. 129:1316–1330.

Beck, A. J., J. K. Cochran, and S. A. Sañudo-Wilhelmy. 2009. Temporal trends of dissolved trace metals in Jamaica Bay, NY: importance of wastewater input and submarine groundwater discharge in an urban estuary. Estuaries and Coasts 32(3):535–550.

Beck, M. W., K. L. Heck Jr., K. W. Able, D. L. Childers, D. B. Eggleston, B. M. Gillanders, B. Halpern, C. G. Hays, K. Hoshino, T. J. Minello, R. J. Orth, P. F. Sheridan, and M. P. Weinstein. 2001. The identification, conservation, and management of estuarine and marine nurseries for fish and invertebrates. BioScience 51(8):633–641.

———. 2003. The role of nearshore ecosystems as fish and shellfish nurseries. Issues Ecol. 11:1–12.

Beck, S. 1995. White perch. Pp. 235–243 *in* L. E. Dove and R. M. Nyman, eds. Living Resources of the Delaware Estuary. Delaware Estuary Program.

Beck, W. R., and W. H. Massmann. 1951. Migratory behavior of the rainwater fish, *Lucania parva*, in the York River, Virginia. Copeia 2:176.

Beckley, L. E. 1984. The ichthyofauna of the Sundays Estuary, South Africa, with particular reference to the juvenile marine component. Estuaries 7(3):248–258.

———. 1985. Coastal zone utilization by juvenile fish in the eastern cape, South Africa. Ph.D. diss. University of Cape Town, Cape Town, South Africa.

———. 1986. The ichthyoplankton assemblage of Algoa Bay nearshore region in relation to coastal zone utilization by juvenile fish. S. Afr. J. Zool. 21(3):245–252.

Behrens, J. W., H. J. Stahl, J. F. Steffensen, and R. N. Glud. 2007. Oxygen dynamics around buried lesser sandeels *Ammodytes tobianus* (Linnaeus 1785): mode of ventilation and oxygen requirements. J. Exp. Biol. 210:1006–1014.

Bejda, A. J., A. L. Studholme, and B. L. Olla. 1987. Behavioral responses of red hake, *Urophycis chuss*, to decreasing concentrations of dissolved oxygen. Environ. Biol. Fishes 19(4):261–268.

Bejda, A. J., B. A. Phelan, and A. L. Studholme. 1992. The effect of dissolved oxygen on the growth of young-of-the-year winter flounder, *Pseudopleuronectes americanus*. Environ. Biol. Fishes 34: 321–327.

Bell, G. W., J. A. Buckel, and A. W. Stoner. 1999. Effects of alternative prey on cannibalism in age-1 bluefish. J. Fish. Biol. 55:990–1000.

Bell, G. W., D. A. Witting, and K. W. Able. 2003. Aspects of metamorphosis and habitat use in the conger eel, *Conger oceanicus*. Copeia 3:544–552.

Bell, J. D., A. S. Steefe, and M. Westoby. 1985. Artificial seagrass: how useful is it for field experiments on fish and macroinvertebrates? J. Exp. Mar. Biol. Ecol. 90:171–177.

Bell, J. D., M. Westoby, and A. S. Steffe. 1987. Fish larvae settling in seagrass: do they discriminate between beds of different leaf density. J. Exp. Mar. Biol. Ecol. 111:133–144.

Benaka, L. R. 1999. Summary of panel discussions and steps toward an agenda for habitat policy and science. Pp. 455–459 *in* L. R. Benaka, ed. Fish Habitat: Essential Fish Habitat and Rehabilitation. Am. Fish. Soc. Symp. 22, American Fisheries Society, Bethesda, Md.

Bengston, D. A. 1984. Resource partitioning by *Menidia menidia* and *Menidia beryllina* (Osteichthyes: Atherinidae). Mar. Ecol. Prog. Ser. 18:21–30.

Bengtson, D. A., and R. C. Barkman. 1981. Growth of postlarval Atlantic silversides in four temperature regimes. Prog. Fish-Cult. 43:146–148.

Bennett, B. A. 1989. The fish community of a moderately exposed beach on the southwestern cape coast of South Africa and an assessment of this habitat as a nursery for juvenile fish. Estuar. Coast. Shelf Sci. 28:293–305.

Bennett, W. A., and T. L. Beitinger. 1997. Temperature tolerance of the sheepshead minnow, *Cyprinodon variegatus*. Copeia 1:77–87.

Bennett, W. A., W. J. Kimmerer, and J. R. Burau. 2002. Plasticity of vertical migration by native and exotic estuarine fishes in a dynamic low salinity zone. Limnol. Oceanogr. 47:1496–1507.

Berg, D. L., and J. S. Levinton. 1985. The biology of the Hudson-Raritan estuary, with emphasis on fishes. U.S. Dept. Commer. Natl. Ocean. Atmos. Adm. / Natl. Ocean Serv. Tech. Memo NOS OMA 16:1–170.

Berghahn, R. 2000. Response to extreme conditions in coastal areas: biological tags in flatfish otoliths. Mar. Ecol. Prog. Ser. 192:277–285.

Berrien, P. L., and J. D. Sibunka. 1999. Distribution patterns of fish eggs in the U.S. Northeast continental shelf ecosystem, 1977–1987. NOAA Tech. Rep. No. 145 NMFS. Seattle, Wash., p. 310.

———. 2006. A Laboratory Guide to the Identification of Marine Fish Eggs Collected on the Northeast Coast of the United States, 1977–1994. Northeast Fisheries Science Center Reference Document 06-21.

Berry, F. H. 1959. Young jack crevalles (*Caranx* species) off the southeastern Atlantic coast of the United States. U.S. Fish. Bull. 59:417–535.

———. 1960. Scale and scute development of the carangid fish *Caranx crysos* (Mitchill). Quart. J. Florida Acad. Sci. 23(1):59–66.

———. 1964. Review and emendation of family Clupeidae. Copeia:720–730.

Berryman, A. A., and B. A. Hawkins. 2006. The refuge as an integrating concept in ecology and evolution. Oikos 115:192–196.

Bertin, L. 1956. Eels: A Biological Study. Cleaver-Hume Press, London.

Bertness, M. D. 1999. The Ecology of Atlantic Shorelines. Sinauer Associates, Inc., Sunderland, Mass.

Bertness, M. D., C. Holdredge, and A. H. Altieri. 2009. Substrate mediates consumer control of salt marsh cordgrass on Cape Cod, New England. Ecology 90(8):2108–2117.

Beverton, R. J. H., and T. C. Iles. 1992. Mortality rates of 0-group plaice (*Pleuronectes platessa* L.), dab (*Limanda limanda* L.) and turbot (*Scophthalmus maximus* L.) in European waters. III. Density-dependence of mortality rates of 0-group plaice and some demographic implications. Neth. J. Sea Res. 29(1–3): 61–79.

Biernacki, E. 1979. Fish Kills Caused by Pollution. Washington, D.C.: Environmental Protection Agency, 0-277-208.

Bigelow, H. B., and W. C. Schroeder. 1953. Fishes of the Gulf of Maine. U.S. Fish Wildl. Serv. Fish. Bull. 74:1–576.

Bigelow, H. B., and M. Sears. 1935. Studies of the waters on the continental shelf, Cape Cod to Chesapeake Bay. II. Salinity. Pap. Phys. Oceanogr. Meterol. 4. 94 pp.

Bigelow, H. B., and W. W. Welsh. 1925. Fishes of the Gulf of Maine. Bull. U.S. Bur. of Fish. 40.

Bilkovic, D. M., and M. M. Roggero. 2008. Effects of coastal development on nearshore estuarine nekton communities. Mar. Ecol. Prog. Ser. 358:27–39.

Bilkovic, D. M., J. E. Olney, and C. H. Hershner. 2002. Spawning of American shad (*Alosa sapidissima*) and striped bass (*Morone saxatilis*) in the Mattaponi and Pamunkey Rivers, Virginia. Fish. Bull. 100:632–640.

Bisbal, G. A., and D. A. Bengtson. 1995. Effects of delayed feeding on survival and growth of summer flounder *Paralichthys dentatus* larvae. Mar. Ecol. Prog. Ser. 121:301–306.

Blaber, S. J. M. 1980. Fish of the Trinity Inlet system of north Queensland with notes on the ecology of fish faunas of tropical Indo-Pacific estuaries. Aust. J. Mar. Freshw. Res. 31:137–146.

———. 1981. The zoogeographical affinities of estuarine fishes in south-east Africa. S. Afr. J. Sci. 77:305–307.

———. 2000. Mangroves and estuarine dependence. Pp. 185–201 *in* S. J. M. Blaber, ed. Tropical Estuarine Fishes, Ecology, Exploitation and Conservation. Blackwell Science, Oxford, England.

Blaber, S. J. M., and T. G. Blaber. 1980. Factors affecting the distribution of juvenile estuarine and inshore fish. J. Fish Biol. 17: 143–162.

Blaber, S. J. M., D. T. Brewer, and J. P. Salini. 1989. Species composition and biomasses of fishes in different habitats of a tropical northern Australian estuary: their occurrence in the adjoining sea and estuarine dependence. Estuar. Coast. Shelf Sci. 29: 509–531.

———. 1995. Fish communities and the nursery role of the shallow inshore waters of a tropical bay in the Gulf of Carpentaria, Australia. Estuar. Coast. Shelf Sci. 40:177–193.

Blackwell, B. F., W. B. Krohn, and R. B. Allen. 1995. Foods of nestling double-crested cormorants in Penobscot Bay, Maine, USA: temporal and spatial comparisons. Colonial Waterbirds 18(2): 199–208.

Blake, R. W., P. Y. L. Kwok, and K. H. S. Chan. 2006. Effects of two parasites, *Schistocephalus solidus* (Cestoda) and *Bunodera* spp. (Trematoda), on the escape fast-start performance of threespined sticklebacks. J. Fish Biol. 69:1345–1355.

Blaxter, J. H. S. 1988. Pattern and variety in development. Pp. 1–58 *in* W. S. Hoard and D. J. Randall, eds. Fish Physiology, Vol. 11, Part A. Egg and Larvae. Academic Press, New York.

Blaxter, J. H. S., and J. R. Hunter. 1982. The biology of the clupeoid fishes. Adv. Mar. Biol. 20:1–223.

Blaylock, R. A. 1989. A massive school of cownose rays, *Rhinoptera bonasus* (Rhinopteridae), in Lower Chesapeake Bay, Virginia. Copeia 3:744–748.

Bleakney, J. S. 1963. First record of the fish *Pogonias cromis* from Canadian waters. Copeia 1:173.

Boehlert, G. W. 1997. Application of acoustic and archival tags to assess estuarine, nearshore, and offshore habitat utilization and movements by salmonids. NOAA Tech. Memo. NOAA-TM-NMFS-SWFSC-236.

Boehlert, G. W., and B. C. Mundy. 1988. Roles of behavioral and physical factors in larval and juvenile fish recruitment to Estuarine Nursery Areas. Am. Fish. Soc. Symp. 3:51–67.

Boesch, D., and R. E. Turner. 1984. Dependence of fishery species on salt marshes: the role of food and refuge. Estuaries 7: 460–468.

Böhlke, J. E., and C. C. G. Chaplin. 1968. Fishes of the Bahamas and Adjacent Tropical Waters. University of Texas Press, Austin.

———. 1993. Fishes of the Bahamas and Adjacent Tropical Waters. 2nd ed. University of Texas Press, Austin. 771 pp.

Boicourt, W. C. 1982. Estuarine larval retention mechanisms on two scales. Pp. 445–457 *in* V. S. Kennedy, ed. Estuarine Comparisons. Academic Press, New York.

Boisneau, C., F. Moatar, M. Bodin, and Ph. Boisneau. 2008. Does

global warming impact on migration patterns and recruitment of Allis shad (*Alosa alosa* L.) young of the year in the Loire River, France? Hydrobiologia 602:179–186.

Bonhommeau, S., E. Chassot, B. Planque, E. Rivot, A. H. Knap, and O. Le Pape. 2008. Impact of climate on eel populations of the Northern Hemisphere. Mar. Ecol. Prog. Ser. 373:71–80.

Booth, D. J. 1995. Juvenile groups in a coral-reef damselfish: density-dependent effects on individual fitness and population demography. Ecology 76(1):91–106.

Booth, D. J., W. F. Figueira, M. A. Gregson, L. Brown, and G. Beretta. 2007. Occurrence of tropical fishes in temperate southeastern Australia: role of the east Australian current. Estuar. Coast. Shelf Sci. 72:102–114.

Booth, R. A. 1967. A description of the larval stages of the tomcod, *Microgadus tomcod*, with comments on its spawning ecology. Ph.D. diss. University of Connecticut, Storrs. 53 pp.

Boreman, J., and R. R. Lewis. 1987. Atlantic coastal migration of striped bass. *In* M. J. Dadswell et al., eds. Common Strategies of Anadromous and Catadromous Fishes. Am. Fish. Soc. Symp. 1:331–339.

Boreman, J., S. J. Correia, and D. B. Witherell. 1993. Effects of changes in age-0 survival and fishing mortality on egg production of winter flounder in Cape Cod Bay. Am. Fish. Soc. Symp. 14:39–45.

Boschung, H. T., Jr. 1957. The fishes of Mobile Bay and the Gulf Coast of Alabama. Ph.D. diss. University of Alabama, Tuscaloosa. 626 pp.

Boumans, R. M. J., D. M. Burdick, and M. Dionne. 2002. Modeling habitat change in salt marshes after tidal restoration. Rest. Ecol. 10(3):543–555.

Bourne, D. W., and J. J. Govoni. 1988. Distribution of fish eggs and larvae and patterns of water circulation in Narragansett Bay, 1972–1973. Am. Fish Soc. Symp. 3:132–148.

Bowen, B. W., and J. C. Avise. 1990. Genetic structure of Atlantic and Gulf of Mexico populations of sea bass, menhaden, and sturgeon: influence of zoogeographic factors and life-history patterns. Mar. Biol. 107:371–381.

Bowen, W. D., J. W. Lawson, and B. Beck. 1993. Seasonal and geographic variation in the species composition and size of prey consumed by grey seals (*Halichoerus grypus*) on the Scotian Shelf. Can. J. Fish. Aquat. Sci. 50:1768–1778.

Bowers-Altman, J. 1993. Wildlife in New Jersey: the striped cusk eel. New Jersey Outdoors Fall:64–65.

Bowman, R. E. 1981. Food of 10 species of northwest Atlantic juvenile groundfish. Fish. Bull. 79(1):200–206.

Bowman, R. E., and W. L. Michaels. 1984. Food of seventeen species of northwest Atlantic fish. NOAA Tech. Memo. NMFS-F/NEC-28. 183 pp.

Bowman, R. E., T. R. Azarovitz, E. S. Howard, and B. P. Hayden. 1987. Food and distribution of juveniles of seventeen northwest Atlantic fish species, 1973–1976. NOAA Tech. Memo. NMFS-F/NEC-45. 57 pp.

Bowman, R. E., C. E. Stillwell, W. L. Michaels, and M. D. Grosslein. 2000. Food of northwest Atlantic fishes and two common species of squid. NOAA Tech. Memo. NMFS 155.

Boynton, W. R., T. T. Polgar, and H. H. Zion. 1981. Importance of juvenile striped bass food habits in the Potomac estuary. Trans. Am. Fish. Soc. 110:56–63.

Boynton, W. R., E. M. Setzler, K. V. Wood, H. H. Zion, and M. Homer. 1987. Final report of Potomac River fisheries study: ichthyoplankton and juvenile investigations. Univ. Md. Center of Environmental and Estuarine Studies, Chesapeake Biol. Lab. Ref. No. 77-169.

Bradford, M. J., and G. Cabana. 1997. Interannual variability in stage-specific survival rates and the causes of recruitment variation. Pp. 469–493 *in* R. C. Chambers and E. A. Trippel, eds. Early Life History and Recruitment in Fish Populations. Fish and Fisheries Series 21, Chapman and Hall, London.

Branstetter, S. 1990. Early life-history implications of selected carcharhinoid and lamnoid sharks of the northwest Atlantic. *In* H. L. Pratt Jr., S. H. Gruber, and T. Taniuchi, eds. Elasmobranchsas Living Resources: Advances in the Biology, Ecology, Systematics, and the Status of the Fisheries, pp. 17–28. U.S. Dep. Commer., NOAA Tech. Rep. NMFS 90.

———. 2002. Smooth dogfishes. Family Traikidae. Pp. 37–38 *in* B. B. Collette and G. Klein-MacPhee, eds. Bigelow and Schroeder's Fishes of the Gulf of Maine. 3rd ed. Smithsonian Institution Press, Washington and London.

Breder, C. M., Jr. 1936. "All Modern Conveniences," a note on the nest architecture of the four-spined stickleback. Bull. N.Y. Zool. Soc. 39:72–76.

———. 1940. The spawning of *Mugil cephalus* on the Florida west coast. Copeia:138–139.

———. 1948. Field Book of Marine Fishes of the Atlantic Coast from Labrador to Texas. Putnam, New York.

———. 1962. Effects of a hurricane on the small fishes of a shallow bay. Copeia 2:459.

Breder, C. M., Jr., and E. Clark. 1947. A contribution to the visceral anatomy, development, and relationship of the Plectognathi. Bull. Am. Mus. Nat. Hist. 88(5):291–319.

Breitburg, D. L. 1989. Demersal schooling prior to settlement by larvae of the naked goby. Environ. Biol. Fishes 26:97–103.

———. 1991. Settlement patterns and presettlement behavior of the naked goby, *Gobiosoma bosci*, a temperate oyster reef fish. Mar. Biol. 109:213–221.

———. 1992. Episodic hypoxia in Chesapeake Bay: interacting effects of recruitment, behavior, and physical disturbance. Ecol. Monogr. 62(4):525–546.

———. 1999. Are three-dimensional structure and healthy oyster populations the keys to an ecologically interesting and important fish community? Pp. 239–250 *in* M. W. Luckenback., R. Mann, and J. A. Wesson, eds. Oyster Reef Habitat Restoration. A Synopsis and Synthesis of Approaches. Virginia Institute of Marine Science Press, Gloucester Point, Va.

Breitburg, D. L., N. Steinberg, S. DuBeau, C. Cooksey, and E. D. Houde. 1994. Effects of low dissolved oxygen on predation on estuarine fish larvae. Mar. Ecol. Prog. Ser. 104:235–246.

Breitburg, D. L., M. A. Palmer, and T. Loher. 1995. Larval distributions and the spatial patterns of settlement of an oyster reef fish: responses to flow and structure. Mar. Ecol. Prog. Ser. 125:45–60.

Breitburg, D. L., K. A. Rose, and J. H. Cowan Jr. 1999. Linking water quality to larval survival: predation mortality of fish larvae in an oxygen-stratified water column. Mar. Ecol. Prog. Ser. 178:39–54.

Breitburg, D. L., L. D. Coen, M. W. Luckenbach, R. Mann, M. Posey, and J. A. Wesson. 2000. Oyster reef restoration: convergence of harvest and conservation strategies. J. Shellfish Res. 19(1):371–377.

Briggs, J. C. 1958. A list of Florida fishes and their distribution. Bull. Fla. State. Mus. Biol. Sci. 2:223–318.

———. 1960. Fishes of worldwide (circumtropical) distribution. Copeia:171–180.

———. 1974. Marine Zoogeography. McGraw-Hill Book Company, New York.

———. 1995. Global Biogeography. Developments in Palaeontology and Stratigraphy Series (14), Elsevier Publishing, Amsterdam.

Briggs, P. T., and J. S. O'Connor. 1971. Comparison of shore zone fishes over naturally vegetated and sand-filled bottoms in Great South Bay. N.Y. Fish Game J. 18:15–41.

Briggs, P. T., and J. R. Waldman. 2002. Annotated list of fishes reported from the marine waters of New York. Northeast. Nat. 9(1):47–80.

Broadhead, G. C. 1953. Investigations of the black mullet, *Mugil cephalus*, L., in northwest Florida. Fla. Bd. Conserv. Tech. Ser. No. 7.

———. 1958. Growth of the black mullet, (*Mugil cephalus* L.) in west and northwest Florida. Fla. Bd. Conserv. Tech. Ser. No. 25.

Brodeur, R. D. 1979. Guide to Otoliths of Some Northwest Atlantic Fishes. National Marine Fisheries Service, Northeast Fisheries Center, Woods Hole Laboratory, Laboratory Reference No. 79-36. Woods Hole, Mass.

Brodziak, J., and L. O'Brien. 2005. Do environmental factors affect recruits per spawner anomalies of New England groundfish? ICES J. Mar. Sci. 62:1394–1407.

Brosnan, T. M., and M. L. O'Shea. 1996. Long-term improvements in water quality due to sewage abatement in the lower Hudson River. Estuaries 19:890–900.

Brousseau, D. J., A. E. Murphy, N. P. Enriquez, and K. Gibbons. 2008. Foraging by two estuarine fishes, *Fundulus heteroclitus* and *Fundulus majalis*, on juvenile Asian shore crabs (*Hemigrapsus sanguineus*) in Western Long Island Sound. Estuaries and Coasts: J. CERF 31:144–151.

Browman, H. I. 1989. Embryology, ethology and ecology of ontogenetic critical periods in fish. Brain Behav. Evol. 34(1):5–12.

Brown, C. A., G. A. Jackson, S. A. Holt, and G. J. Holt. 2005. Spatial and temporal patterns in modeled particle transport to estuarine habitat with comparisons to larval fish settlement patterns. Estuar. Coast. Shelf Sci. 64:33–46.

Brundage, H. M., III, and R. E. Meadows. 1982a. Occurrence of the endangered shortnose sturgeon, *Acipenser brevirostrum*, in the Delaware River Estuary. Estuaries 5(3):203–208.

———. 1982b. The Atlantic sturgeon, *Acipenser oxyrinchus*, in the Delaware River Estuary. Fish. Bull., U.S. 80(2):337–343.

Brunel, T., and J. Boucher. 2007. Long-term trends in fish recruitment in the north-east Atlantic related to climate change. Fish. Oceanogr. 16(4):336–349.

Brush, G. S. 2009. Historical land use, nitrogen, and coastal eutrophication: a paleoecological perspective. Estuaries and Coasts 32(1):18–28.

Brzorad, J. N. 1994. Resource predictability and the information center hypothesis. Ph.D. diss. Rutgers University, New Brunswick, N.J. 229 pp.

Buckel, J. A., and D. O. Conover. 1997. Movements, feeding periods, and daily ration of piscivorous young-of-the-year bluefish, *Pomatomus saltatrix*, in the Hudson River estuary. Fish. Bull. 95: 665–679.

Buckel, J. A., and K. A. McKown. 2002. Competition between juvenile striped bass and bluefish: resource partitioning and growth rate. Mar. Ecol. Prog. Ser. 234:191–204.

Buckel, J. A., and A. W. Stoner. 2000. Functional response and switching behavior of young-of-the-year piscivorous bluefish. J. Exp. Mar. Biol. Ecol. 245:25–41.

Buckel, J. A., M. J. Fogarty, and D. O. Conover. 1999a. Foraging habits of bluefish, *Pomatomus saltatrix*, on the U.S. east coast continental shelf. Fish. Bull. 97:758–775.

Buckel, J. A., D. O. Conover, N. D. Steinberg, and K. A. McKown. 1999b. Impact of age-0 bluefish (*Pomatomus saltatrix*) predation on age-0 fishes in the Hudson River estuary: evidence for density-dependent loss of juvenile striped bass (*Morone saxatilis*). Can. J. Fish. Aquat. Sci. 56:275–287.

Buckel, J. A., B. L. Sharack, and V. S. Zdanowicz. 2004. Effect of diet on otolith composition in *Pomatomus saltatrix*, an estuarine piscivore. J. Fish. Biol. 64:1469–1484.

Buckley, J., and B. Kynard. 1981. Spawning and rearing of shortnose sturgeon from the Connecticut River. Prog. Fish-Culturist 43(2):74–76.

———. 1985. Yearly movements of shortnose sturgeons in the Connecticut River. Trans. Am. Fish. Soc. 114(6):813–820.

Budikova, D. 2008. Effect of the Arctic Oscillation on precipitation in the eastern USA during ENSO winters. Clim. Res. 37:3–16.

Bulger, A. J., M. E. Monaco, D. M. Nelson, and M. G. McCormick-Ray. 1993. Biologically-based estuarine salinity zones derived from a multivariate analysis. Estuaries 16(2):311–322.

Bullock, A. M., and R. J. Roberts. 1974. The dermatology of marine teleost fish. I. The normal integument. Oceanogr. Mar. Biol. Ann. Rev. 13:383–411.

Bumpus, D. F. 1973. A description of the circulation on the continental shelf of the East Coast of the U.S. Prog. Oceanogr. 6: 111–157.

Burbridge, R. G. 1974. Distribution, growth, selective feeding, and energy transformation of young-of-the-year blueback herring, *Alosa aestivalis* (Mitchill), in the James River, Virginia. Trans. Am. Fish. Soc. 103:297–311.

Burger, J., K. W. Able, S. M. Hagan, and C. Jeitner. In prep. The effect of fish prey base on foraging in Forster's (*Sterna forsteri*) and Common Terns (*S. hirundo*) and Black Skimmers (*Rynchops niger*) in an east coast estuary, USA.

Burgess, W. E. 1978. Butterflyfishes of the world. T. F. H. Publications, Jersey City, N.J.

———. 2002. Chaetodontidae: butterflyfishes, Pomacanthidae: Angelfishes, and Ephippidae: spadefishes. Pp. 1663–1683 and 1799–1800 *in* K. E. Carpenter, ed. The Living Marine Resources of the Western Central Atlantic. Vol. 3, Bony Fishes, Part 2 (Opisthognathidae to Molidae). Food and Agriculture Organization of the United Nations, Rome.

Burke, J. S. 1995. Role of feeding and prey distribution of summer and southern flounder in selection of estuarine nursery habitats. J. Fish Biol. 47:355–366.

Burke, J. S., J. M. Miller, and D. E. Hoss. 1991. Immigration and settlement pattern of *Paralichthys dentatus* and *P. lethostigma* in an estuarine nursery ground, North Carolina, USA. Neth. J. Sea. Res. 27:393–405.

Burkholder, J. M., H. B. Glasgow Jr., and C. W. Hobbs. 1995. Fish kills linked to a toxic ambush-predator dinoflagellate: distribution and environmental conditions. Mar. Ecol. Prog. Ser. 124: 43–61.

Burnett, J., S. H. Clark, and L. O'Brien. 1984. A preliminary assessment of white hake in the Gulf of Maine–Georges Bank area. Natl. Mar. Fish Serv., Northeast Fish. Sci. Cent. Woods Hole Lab. Ref. Doc. No. 84-31:1–33.

Burr, B. M., and F. J. Schwartz. 1986. Occurrence, growth, and food habits of the spotted hake, *Urophycis regia*, in the Cape Fear estuary and adjacent Atlantic Ocean, North Carolina. Northeast Gulf Sci. 8(2):115–127.

Burrell, V. G., Jr., and W. A. Van Engel. 1976. Predation by and distribution of a ctenophore, *Mnemiopsis leidyi* A. Agassiz, in the York River estuary. Estuar. Coast. Mar. Sci. 4:235–242.

Burreson, E. M., and D. E. Zwerner. 1984. Juvenile summer flounder, *Paralichthys dentatus*, mortalities in the western Atlantic Ocean caused by the hemoflagellate *Trypanoplasma bullocki*: evidence from field and experimental studies. Helgoländer Meeresunters 37:343–352.

Burton, M. L. 2001. Age, growth, and mortality of gray snapper, *Lutjanus griseus*, from the east coast of Florida. Fish. Bull. 99(2): 254–265.

Buzzelli, C. P., R. A. Luettich Jr., S. P. Powers, C. H. Peterson, J. E. McNich, J. L. Pinckney, and H. W. Paerl. 2002. Estimating the spatial extent of bottom-water hypoxia and habitat degradation in a shallow estuary. Mar. Ecol. Prog. Ser. 230:103–112.

Byrne, D. M. 1978. Life history of the spotfin killifish, *Fundulus luciae* (Pisces: Cyprinodontidae), in Fox Creek Marsh, Virginia. Estuaries 4:211–227.

Caddy, J. F. 2008. The importance of "cover" in the life histories of demersal and benthic marine resources: a neglected issue in fisheries assessment and management. Bull. Mar. Sci. 83(1):7–52.

Cadigan, K. M., and P. E. Fell. 1985. Reproduction, growth and feeding habits of *Menidia menidia* (Atherinidae) in a tidal marsh-estuarine system in southern New England. Copeia 1:21–26.

Calder, W. A. 1984. Size, function, and life history. Harvard University Press, Cambridge, Mass.

Caldwell, G. S. 1981. Attraction to tropical mixed-species heron flocks: proximate mechanism and consequences. Behav. Ecol. Sociobiol. 8:99–103.

Callaway, J. C. 2005. The challenge of restoring functioning salt marsh ecosystems. J. Coast. Res. 40:24–36.

Cameron, W. M., and D. W. Pritchard. 1963. Estuaries. Pp. 306–324 *in* M. N. Hill, ed. The Sea: Ideas and Observations on Progress in the Study of Seas, Vol. 2, The Composition of Sea-Water Comparative and Descriptive Oceanography. Interscience Publishers, New York.

Campana, S. E. 1996. Year-class strength and growth rate in young Atlantic cod *Gadus morhua*. Mar. Ecol. Prog. Ser. 135:21–26.

Campana, S. E., K. T. Frank, P. C. F. Hurley, P. A. Koeller, F. H. Page, and P. C. Smith. 1989. Survival and abundance of young Atlantic cod (*Gadus morhua*) and haddock (*Melanogrammus aeglefinus*) as indicatory of year-class strength. Can. J. Fish. Aquat. Sci. 46 (suppl. 1):171–182.

Campbell, B. C., and K. W. Able. 1998. Life history characteristics of the northern pipefish, *Syngnathus fuscus* in southern New Jersey. Estuaries 21(3):470–475.

Cantelmo, F. R., and C. H. Wahtola. 1992. Aquatic habitat impacts of pile-supported and other structures in the lower Hudson River. Pp. 59–75 *in* W. Wise, D. J. Suszkowski, and J. R. Waldman, eds. Proceedings: Conference of Impacts of New York Harbor Development on Aquatic Resources. Hudson River Foundation, New York.

Carlson, D. M., and K. W. Simpson. 1987. Gut contents of juvenile shortnose sturgeon in the upper Hudson estuary. Copeia 3: 796–802.

Carpenter, K. E., ed. 2002. The Living Marine Resources of the Western Central Atlantic. Vols. 1–3. Food and Agriculture Organization of the United Nations, Rome.

Carr, M. H. 1991. Habitat selection and recruitment of an assemblage of temperate zone reef fishes. J. Exp. Mar. Biol. Ecol. 146: 113–137.

Carr, M. H., and M. A. Hixon. 1995. Predation effects on early post-settlement survivorship of coral reef fishes. Mar. Ecol. Prog. Ser. 124:31–42.

Carr, W. E. S., and C. A. Adams. 1972. Food habits of juvenile marine fishes: evidence of the cleaning habit in the leatherjacket, *Oligoplites saurus*, and the spottail pinfish, *Diplodus holbrooki*. Fish. Bull. 70(4):1111–1120.

———. 1973. Food habitats of juvenile marine fishes occupying seagrass beds in the estuarine zone near Crystal River, Florida. Trans. Am. Fish. Soc. 102(3):511–540.

Carscadden, J. E., and W. C. Leggett. 1975. Life history variations in populations of American shad, *Alosa sapidissima* (Wilson), spawning tributaries of the St. John River, New Brunswick. J. Fish Biol. 7(5):595–609.

Casey J. G., and N. E. Kohler. 1990. Long distance movements of Atlantic sharks from NMFS cooperative shark tagging program. Pp. 87–91 *in* H. L. Pratt Jr., S. H. Gruber, and T. Taniuchi, eds. *Elasmobranchs as Living Resources: Advances in the Biology, Ecology, Systematics, and the Status of the Fisheries*. U.S. Department of Commerce, NOAA Technical Report NMFS 90.

Casey, J. G., and L. J. Natanson. 1992. Revised estimates of age and growth of the sandbar shark (*Carcharhinus plumbeus*) from the Western North Atlantic. Can. J. Fish. Aquat. Sci. 49:1474–1477.

Castagna, M. 1955. A study of the hogchoker, *Trinectes maculatus* (Bloch and Schneider), in the Wakulla River, Florida. M.S. thesis. Florida State University, Tallahassee. 39 pp.

Casterlin, M. E., and W. W. Reynolds. 1979. Diel activity patterns of the smooth dogfish shark, *Mustelus canis*. Bull. Mar. Sci. 29(3):440–442.

Castillo-Rivera, M., G. Moreno, and R. Iniestra. 1994. Spatial, seasonal, and diel variation in abundance of the bay anchovy, *Anchoa mitchilli* (Teleostei: Engraulidae), in a tropical coastal lagoon of Mexico. SW Nat. 39(2):263–268.

Castonguay, M., and J. D. McCleave. 1987. Vertical distributions, diel and ontogenetic vertical migrations and net avoidance of leptocephali of *Anguilla* and other common species in the Sargasso Sea. J. Plankton Res. 9(1):195–214.

Castonguay, M., P. V. Hodson, C. Moriarity, K. F. Drinkwater, and B. M. Jessop. 1994. Is there a role of ocean environment in American and European eel decline? Fish. Oceanogr. 3:197–203.

Castro, J. I. 1983. The Sharks of North American Waters. Texas A&M University Press, College Station.

———. 1993. The shark nursery of Bulls Bay, South Carolina, with a review of the shark nurseries of the southeastern coast of the United States. Environ. Biol. Fishes 38:37–48.

Castro, L. R., and R. K. Cowen. 1991. Environmental factors affecting the early life history of bay anchovy *Anchoa mitchilli* in Great South Bay, New York. Mar. Ecol. Prog. Ser. 76:235–247.

Cau, A., and P. Manconi. 1984. Relationship of feeding reproductive cycle and bathymetric distribution in *Conger conger*. Mar. Biol. 81:147–151.

Chacko, P. I. 1949. Bionomics of the Indian paradise fish, *Macropodus cupanus*, in the waters of Madras. Proc. 36th Indian Sci. Congr. Pt. 3:164.

Chambers, R. C., W. C. Leggett, and J. A. Brown. 1988. Variation in and among life history traits of laboratory-reared winter flounder *Pseudopleuronectes americanus*. Mar. Ecol. Prog. Ser. 47:1–15.

Chambers, R. C., and E. A. Trippel, eds. 1997. Early Life History and Recruitment of Fish Populations. Chapman and Hall, London.

Chambers, R. M., L. A. Meyerson, and K. Saltonstall. 1999. Expansion of *Phragmites australis* into tidal wetlands of North America. Aquat. Bot. 64:261–273.

Chant, R. J., M. C. Curran, K. W. Able, and S. M. Glenn. 2000. Delivery of winter flounder (*Pseudopleuronectes americanus*) larvae to settlement habitats in coves near tidal inlets. Estuar. Coast. Shelf Sci. 51:529–541.

Chao, L. N., and J. A. Musick. 1977. Life history, feeding habits, and functional morphology of juvenile sciaenid fishes in the York River Estuary, Virginia. U.S. Fish. Bull. 75:657–702.

Chao, L. N., L. E. Pereira, J. P. Vieira, M. A. Bemvenuti, and L. P. R. Cunha. 1982. Relacao preliminar dos peixes estuarinos e marinhos da lagoa dos patos e regiao costeira adjacente. Atlantica, Rio Grande 5:67–75.

Chapleau, F. 1993. Pleuronectiform relationships: a cladistic reassessment. Bull. Mar. Sci. 52(1):516–540.

Charnell, R. L., and D. V. Hansen. 1974. Summary and analysis of physical oceanography data collected in the New York Bight. Apex. Mar. Ecosyst. Anal. Rep. No. 74-3.

Chen, Y. L., H. Y. Chen, and W. N. Tzeng. 1994. Reappraisal of the importance of rainfall in affecting catches of *Anguilla japonica* elvers in Taiwan. Aust. J. Mar. Freshw. Res. 45:185–190.

Cheney, R. E. 1978. Oceanographic observations in the western North Atlantic during FOX 1, May 1978. Tech. Note 3700-79-78, V. 5. U.S. Naval Oceanographic Office, Washington, D.C.

Chernoff, B. 2002. New world silversides. Family Atherinopsidae. Pp. 280–284 *in* B. B. Collette and G. Klein-MacPhee, eds. Bigelow and Schroeder's Fishes of the Gulf of Maine. 3rd ed. Smithsonian Institution Press, Washington and London.

Cheung, W. W. L., C. Close, V. Lam, R. Watson, and D. Pauly. 2008. Application of macroecological theory to predict effects of climate change on global fisheries potential. Mar. Ecol. Prog. Ser. 365:187–197.

Chiarella, L. A., and D. O. Conover. 1990. Spawning season and first-year growth of adult bluefish from the New York Bight. Trans. Am. Fish. Soc. 119(3):455–462.

Chidester, F. E. 1916. A Biological Study of the More Important of the Fish Enemies of the Salt Marsh Mosquitoes. Bull. 300, N.J. Agr. Exp. Sta., pp. 1–16.

Childs, A.-R., P. D. Cowley, T. F. Næsje, A. J. Booth, W. M. Potts, E. B. Thorstad, and F. Økland. 2008. Do environmental factors influence the movement of estuarine fish? A case study using acoustic telemetry. Estuar. Coast. Shelf Sci. 78:227–236.

Chittenden, M. E., Jr. 1969. Life history and ecology of the American shad, *Alosa sapidissima*, in the Delaware River. Ph.D. diss. Rutgers University, New Brunswick, N.J. 458 pp.

———. 1971. Status of the striped bass, *Morone saxatilis*, in the Delaware River. Chesapeake Sci. 12:131–136.

———. 1972a. Salinity tolerance of young blueback herring, *Alosa aestivalis*. Trans. Am. Fish. Soc. 101(1):123–125.

———. 1972b. Response of young American shad, *Alosa sapidissima*, to low temperatures. Trans. Am. Fish. Soc. 101(4):680–685.

———. 1973a. Salinity tolerance of young American shad, *Alosa sapidissima*. Chesapeake Sci. 14(3):207–210.

———. 1973b. Effects of handling on oxygen requirements of American shad (*Alosa sapidissima*). J. Fish. Res. Board Can. 30:105–110.

———. 1975. Dynamics of American shad, *Alosa sapidissima*, run in the Delaware River. U.S. Fish. Bull. 73(3):487–494.

———. 1976. Weight loss, mortality, feeding and duration of residence of adult American shad, *Alosa sapidissima*, in fresh water. U.S. Fish. Bull. 74(1):151–157.

Chittenden, M. E., L. R. Barbieri, and C. M. Jones. 1993. Fluctuations in abundance of Spanish mackerel in Chesapeake Bay and the mid-Atlantic region. N. Am. J. Fish. Manag. 13:450–458.

Chitty, J. 2000. The response of a resident marsh killifish, *Cyprinodon variegatus*, to marsh restoration in southern New Jersey. M.S. thesis. Rutgers University, New Brunswick, N.J.

Chitty, J. D., and K. W. Able. 2004. Habitat use, movements and growth of the sheepshead minnow, *Cyprinodon variegatus*, in a restored salt marsh. Bull. N.J. Acad. Sci. 49(2):1–8.

Chizmadia, P. A., M. J. Kennish, and V. L. Ohori. 1984. Physical description of Barnegat Bay. Pp. 1–28 *in* M. J. Kennish and R. A. Lutz, eds. Ecology of Barnegat Bay, New Jersey. Lecture Notes on Coastal and Estuarine Studies. Springer, New York.

Chrobot, R. J. 1951. The feeding habitats of the toadfish (*Opanus tau*) based on an analysis of the contents of the stomach and intestine. M.S. thesis. University of Maryland. 139 pp.

Churchill, J. H., R. B. Forward, R. A. Luettich, J. L. Hench, W. F. Hettler, L. B. Crowder, and J. O. Blanton. 1999a. Circulation and larval fish transport within a tidally dominated estuary. Fish. Oceanogr. 8(2):173–189.

Churchill, J. H., J. L. Hench, R. J. Luettich, J. O. Blanton, and F. E. Werner. 1999b. Flood tide circulation near Beaufort Inlet North Carolina: implications for larval recruitment. Estuaries 22: 1057–1070.

Cianci, J. M. 1969. Larval development of the alewife and the glut herring. M.S. thesis. University of Connecticut, Storrs. 62 pp.

Ciannelli, L., G. E. Dingsor, B. Bogstad, G. Ottersen, K. Chan, H. Gjpseter, J. E. Stiansen, and N. C. Stenseth. 2007. Spatial anatomy of species survival: effects of predation and climate-driven environmental variability. Ecology 83:635–646.

Clark, J. 1967. Fish and Man: Conflict in the Atlantic Estuaries. Am. Littoral Soc. Spec. Pub. 5.

Clark, J. C. 1994. Impoundment estuarine interactions. Federal aid in Fisheries Restoration Project F-44-R Annual Report. Delaware Division of Fish and Wildlife, Dover, Del.

Clark, W. G., and S. R. Hare. 2002. Effects of climate and stock size on recruitment and growth of Pacific halibut. N. Am. J. Fish. Mgmt. 22:852–862.

Clay, D., W. T. Stobo, B. Beck, and P. C. F. Hurley. 1989. Growth of juvenile pollock (*Pollachius virens* L.) along the Atlantic coast of Canada with inferences of inshore-offshore movements. J. Northw. Atl. Fish. Sci. 9:37–43.

Clayton, G. R. 1976. Reproduction, first year growth, and distribution of anadromous rainbow smelt, *Osmerus mordax,* in the Parker River and Plum Island Sound estuary, Massachusetts. M.S. thesis, University of Massachusetts, Amherst. 105 pp.

Clayton, G. R., C. F. Cole, S. Murawski, and J. Parrish. 1976. Common marine fishes of coastal Massachusetts. Mass. Coop. Fish. Res. Unit Contrib. No. 54. 231 pp.

Clemmer, G. H., and F. J. Schwartz. 1964. Age, growth and weight relationships of the striped killifish, *Fundulus majalis*, near Solomons, Maryland. Trans. Am. Fish. Soc. 93(2):197–198.

Clymer, J. P. 1978. The distribution, trophic dynamics and competitive interactions of three salt marsh killifishes (Pisces: Cyprinodontidae). Ph.D. diss. Lehigh University, Bethlehem, Pa.

Coen, L. D., and M. W. Luckenbach. 2000. Developing success criteria and goals for evaluating oyster reef restoration: ecological function or resource exploitation? Ecol. Eng. 15:323–343.

Coen, L. D., M. W. Luckenbach, and D. L. Breitburg. 1999. The role of oyster reefs as essential fish habitat: a review of current knowledge and some new perspectives. Pp. 438–454 *in* L. R.

Benaka, ed. Fish Habitat: Essential Fish Habitat and Rehabilitation. Am. Fish. Soc. Symp. 22, Bethesda, Md.

Cohen, D. M., and J. L. Russo. 1979. Variations in the fourbeard rockling, *Enchelyopus cimbrius*, a North Atlantic gadid fish, with comments on the genera of rocklings. Fish. Bull. U.S. 2(6): 144–148.

Cole, C. F., R. J. Essig, and O. R. Sainelle. 1980. Biological investigation of the alewife population, Parker River, Massachusetts. U.S. Natl. Mar. Fish. Serv. Final Rep. No. 11, 884-32-50-80-CR.

Coleman, F. C., and J. Travis. 1998. Phenology of recruitment and infection patterns of *Ascocotyle pachycystis*, a digenean parasite in the sheepshead minnow, *Cyprinodon variegatus*. Environ. Biol. Fishes 51:87–96.

Colgan, P., and N. Costeloe. 1980. Plasticity of burying behavior by the banded killifish, *Fundulus diaphanus*. Copeia 2:349–351.

Colin, P. L. 1989. Aspects of the spawning of western Atlantic butterflyfishes (Pisces: Chaetodontidae). Env. Biol. Fishes 25:131–141.

Collette. B. B. 1966. *Belonion*, a new genus of fresh-water needlefishes from South America. Amer. Mus. Nat. Hist. Novitates, No. 2274. 22 pp.

———. 1968. *Strongylura timucu* (Walbaum): a valid species of western Atlantic needlefish. Copeia 1:189–192.

———. 2002. Gunnels. Family Pholidae. Pp. 481–483 *in* B. B. Collette and G. Klein-MacPhee, eds. Bigelow and Schroeder's Fishes of the Gulf of Maine. 3rd ed. Smithsonian Institution Press, Washington, D.C.

Collette, B. B., and K. E. Hartel. 1988. An annotated list of the fishes of Massachusetts Bay. Natl. Ocean. Atmos. Adm. Tech. Memo. Natl. Mar. Fish. Serv., Northeast Fish. Sci. Cent. 51.

Collette, B. B., and G. Klein-MacPhee, eds. 2002. Bigelow and Schroeder's Fishes of the Gulf of Maine. 3rd ed. Smithsonian Institution Press, Washington, D.C.

Collie, J. S., A. D. Wood, and H. P. Jeffries. 2008. Long-term shifts in the species composition of a coastal fish community. Can. J. Fish. Aquat. Sci. 65:1352–1365.

Collins, M. R., and B. W. Stender. 1987. Larval king mackerel (*Scomberomorus cavalla*), Spanish mackerel (*S. maculatus*) and bluefish (*Pomatomus saltatrix*) off the southeast coast of the United States, 1973–1980. Bull. Mar. Sci. 41:822–834.

———. 1989. Larval striped mullet (*Mugil cephalus*) and white mullet (*Mugil curema*) off the southeastern United States. Bull. Mar. Sci. 45(3):580–589.

Collins, M. R., C. W. Waltz, W. A. Roumillat, and D. L. Stubbs. 1987. Contributions to the life history and reproductive biology of gag, *Mycteroperca microlepis* (Serranidae) in the south Atlantic bight. Fish. Bull. 85:648–653.

Collins, M. R., T. I. J. Smith, W. C. Post, and O. Pashuk. 2000. Habitat utilization and biological characteristics of Atlantic sturgeon in two South Carolina rivers. Trans. Am. Fish. Soc. 129(4):982–988.

Colton, J. B., Jr. 1972. Temperature trends and the distribution in continental shelf waters, Nova Scotia to Long Island. U.S. Fish. Bull. 70(3):637–658.

Colton, J. B., Jr., W. G. Smith, A. W. Kendall Jr., P. L. Berrien, and M. P. Fahay. 1979. Principal spawning areas and times of marine fishes, Cape Sable to Cape Hatteras. U.S. Fish. Bull. 76:911–915.

Colvocoresses, J. A., and J. A. Musick. 1984. Species associations and community composition of Middle Atlantic Bight continental shelf demersal fishes. Fish. Bull. 82(2):295–313.

Compagno, L. J. V. 2002. Sharks. *In* K. E. Carpenter, ed. The Living Marine Resources of the Western Central Atlantic. Vol. 2, Bony

Fishes, Part 1 (Acipenseridae to Grammatidae). FAO Species Identification Guide for Fishery Purposes and American Society of Ichthyologists and Herpetologists Special Publications, No. 5. Roma, FAO.

Comyns, B. H., and G. C. Grant. 1993. Identification and distribution of *Urophycis* and *Phycis* (Pisces: Gadidae) larvae and pelagic juveniles in the U.S. Middle Atlantic Bight. U.S. Fish. Bull. 91: 210–223.

Conover, D. O. 1990. The relation between capacity for growth and length of growing season: evidence for and implications of countergradient variation. Trans. Am. Fish. Soc. 119:416–430.

———. 1992. Seasonality and the scheduling of life history at different latitudes. J. Fish Biol. 41(b):161–178.

Conover, D. O., and B. E. Kynard. 1984. Field and laboratory observations of spawning periodicity and behavior of a northern population of the Atlantic silverside, *Menidia menidia* (Pisces: Atherinidae). Environ. Biol. Fishes 11:161–171.

Conover, D. O., and S. A. Murawski. 1982. Offshore winter migration of the Atlantic silverside, *Menidia menidia*. U.S. Fish. Bull. 80:145–150.

Conover, D. O., and T. M. C. Present. 1990. Countergradient variation in growth rate: compensation for length of the growing season among Atlantic silversides from different latitudes. Oecologia 83:316–324.

Conover, D. O., and M. R. Ross. 1982. Patterns in seasonal abundance, growth, and biomass of the Atlantic silverside, *Menidia menidia*, in a New England estuary. Estuaries 5:275–286.

Conover, D. O., T. Gilmore, and S. B. Munch. 2003. Estimating relative contribution of spring- and summer-spawned cohorts to the Atlantic coast bluefish stock. Trans. Am. Fish. Soc. 132: 1117–1124.

Conrath, C. L., and J. A. Musick. 2002. Reproductive biology of the smooth dogfish, *Mustelus canis*, in the northwest Atlantic Ocean. Environ. Biol. Fishes 64:367–377.

———. 2007. The sandbar shark summer nursery within bays and lagoons of the eastern shore of Virginia. Trans. Am. Fish. Soc. 136:999–1007.

———. 2008. Investigations into depth and temperature habitat utilization and overwintering grounds of juvenile sandbar sharks, *Carcharhinus plumbeus*: the importance of near shore North Carolina waters. Environ. Biol. Fishes 82:123–131.

Conrath, C. L., J. Gelsleichter, and J. A. Musick. 2002. Age and growth of the smooth dogfish (*Mustelus canis*) in the northwest Atlantic Ocean. Fish. Bull. 100:674–682.

Cook, S. K. 1988. Physical oceanography of the Middle Atlantic Bight. Pp. 1–49 *in* A. L. Pacheco, ed. Characterization of the Middle Atlantic Water Management Unit of the Northeast Regional Action Plan. NOAA Tech. Memo. NMFS-F/NEC-56. Woods Hole, Mass. 322 pp.

Cook, T., M. Folli, J. Klinck, S. Ford, and J. Miller. 1998. The relationship between increasing sea-surface temperature and northward spread of *Perkinsus marinus* (Dermo) disease epizootics in oysters. Estuar. Coast. Shelf Sci. 46:587–597.

Cooper, J. A., and F. Chapleau. 1998. Monophyly and relationships of the family Pleuronectidae (Pleuronectiformes), with a revised classification. U.S. Fish. Bull. 96(4):686–726.

Cooper, J. E. 1966. Migration and population estimation of the tautog, *Tautoga onitis* (Linnaeus), from Rhode Island. Trans. Am. Fish. Soc. 95:239–247.

———. 1967. Age and growth of tautog, *Tautoga onitis* (Linnaeus), from Rhode Island. Trans. Am. Fish. Soc. 96:134–142.

———. 1978. Identification of eggs, larvae, and juveniles of the rainbow smelt, *Osmerus mordax*, with comparisons to larval alewife, *Alosa pseudoharengus*, and gizzard shad, *Dorosoma cepedianum*. Trans. Am. Fish. Soc. 107(1):56–62.

Cooper, J. E., R. A. Rulifson, J. J. Isley, and S. E. Winslow. 1998. Food habits and growth of juvenile striped bass, *Morone saxatilis*, in Albemarle Sound, North Carolina. Estuaries 21(2):307–317.

Cooper, M. J. P., M. D. Beevers, and M. Oppenheimer. 2008. The potential impacts of sea level rise on the coastal region of New Jersey, U.S.A. Clim. Change 90:475–492.

Cooper, R. A. 1961. Early life history and spawning migration of the alewife. M.S. thesis. University of Rhode Island, Kingston. 58 pp.

Cooper, R. A., P. Valentine, J. R. Uzmann, and R. A. Slater. 1987. Submarine canyons. Pp. 52–63 *in* R. H. Backus and D. W. Bourne, eds. Georges Bank. The MIT Press, Cambridge, Mass.

Cooper, S. R., and G. S. Brush. 1991. Long-term history of Chesapeake Bay anoxia. Science 254:992–996.

Coorey, D. N. 1981. Aspects of the life history of the atherinid fish *Menidia beryllina* (Cope), in New Jersey salt marshes. M.S. thesis. Rutgers University, New Brunswick, N.J. 80 pp.

Coorey, D. N., K. W. Able, and J. K. Shisler. 1985. Life history and food habits of *Menidia beryllina* in a New Jersey salt marsh. Bull. N.J. Acad. Sci. 30(1):29–37.

Copp, G. H., and V. Kovac. 1996. When do fish with indirect development become juveniles? Can. J. Fish. Aquat. Sci. 53:746–752.

Coppack, T., I. Tindemans, M. Czisch, A. Van der Linden, P. Berthold, and F. Pulido. 2008. Can long-distance migratory birds adjust to the advancement of spring by shortening migration distance? The response of the pied flycatcher to latitudinal photoperiodic variation. Global Change Biol. 14:2516–2522.

Correia, A. T., K. W. Able, C. Antunes, and J. Coimbra. 2004. Early life history of the American conger eel (*Conger oceanicus*) as revealed by otolith microstructure and microchemistry of metamorphosing leptocephali. Mar. Biol. 145:477–488.

Costantini, M., S. A. Ludsin, D. M. Mason, X. Zhang, W. C. Boicourt, and S. B. Brandt. 2008. Effect of hypoxia on habitat quality of striped bass (*Morone saxatilis*) in Chesapeake Bay. Can. J. Fish. Aquat. Sci. 65:989–1002.

Coull, B. C. 1990. Are members of the meiofauna food for higher trophic levels? Trans. Am. Microscopical Soc. 109(3):233–246.

Coull, B. C., J. G. Greenwood, D. R. Fielder, and B. A. Coull. 1995. Subtropical Australian juvenile fish eat meiofauna: experiments with winter whiting *Sillago maculate* and observations on other species. Mar. Ecol. Prog. Ser. 125:13–19.

Courrat, A., J. Lobry, D. Nicolas, P. Laffargue, R. Amara, M. Lepage, M. Girardin, and O. Le Pape. 2009. Anthropogenic disturbance on nursery function of estuarine areas for marine species. Est. Coast. Shelf Sci. 81:179–190.

Cowan, J. H., Jr., and R. S. Birdsong. 1985. Seasonal occurrence of larval and juvenile fishes in a Virginia Atlantic coast estuary with emphasis on drums (Family Sciaenidae). Estuaries 8(1): 48–59.

Cowan, J. H., Jr., C. B. Grimes, and R. F. Shaw. 2008. Life history, history, hysteresis, and habitat changes in Louisiana's coastal ecosystem. Bull. Mar. Sci. 83(1):197–215.

Cowardin, L. M., V. Carter, F. C. Golet, and E. T. LaRoe. 1997. Classification of Wetlands and Deepwater Habitats of the United States, FWS/OBS-79/31. U.S. Fish and Wildlife Service, Office of Biological Services, Washington, D.C.

Cowen, R. K. 1985. Large scale pattern of recruitment by the labrid, *Semicossyphus pulcher*: causes and implications. J. Mar. Res. 43:719–742.

Cowen, R. K., and S. Sponaugle. 1997. Relationships between early life history traits and recruitment among coral reef fishes. Pp. 423–450 *in* R. C. Chambers and E. A. Trippel, eds. Early Life History and Recruitment in Fish Populations. Chapman and Hall, London.

Cowen, R. K., L. Chiarella, C. Gomez, and M. Bell. 1991. Distribution, age and lateral plate variation of larval sticklebacks (*Gasterosteus*) off the Atlantic coast of New Jersey, New York and southern New England. Can. J. Fish. Aquat. Sci. 48:1679–1984.

Cowen, R. K., J. A. Hare, and M. P. Fahay. 1993. Beyond hydrography: Can physical processes explain larval fish assemblages within the Middle Atlantic Bight? Bull. Mar. Sci. 53(2):567–587.

Cowen, R. K., J. A. Hare, and T. M. Grothues. 1998. Transport of larval fish in the vicinity of Cape Hatteras. Ocean Sciences Meeting 1998.

Cox, J., K. Macdonald, and T. Rigert. 1994. Engineering and geotechnical techniques for shoreline erosion management in Puget Sound. Coastal Erosion Management Studies, Vol. 4, Shorelands and Coastal Zone Management Program, Washington Department of Ecology, Olympia.

Cox, P. 1916. Are migrating eels deterred by a range of lights? Report on experimental tests. Contrib. Can. Biol. Sessional Pap. 38a:115–118.

Crabtree, R. E., P. B. Hood, and D. Snodgrass. 2002. Age, growth and reproduction of permit (*Trachinotus falcatus*) in Florida waters. Fish. Bull. 100:26–34.

Craig, J. K., B. J. Burke, L. B. Crowder, and J. A. Rice. 2006. Prey growth and size-dependent predation in juvenile estuarine fishes: experimental and model analyses. Ecology 87(9):2366–2377.

Craig, J. K., J. A. Rice, L. B. Crowder, and D. A. Nadeau. 2007. Density-dependent growth and mortality in an estuary-dependent fish: an experimental approach with juvenile spot *Leiostomus xanthurus*. Mar. Ecol. Prog. Ser. 343:251–262.

Crain, C. M., K. B. Gedan, and M. Dionne. 2009. Tidal restrictions and mosquito ditching in New England marshes: case studies of the biotic evidence, physical extent, and potential for restoration of altered tidal hydrology. Pp. 149–169 *in* B. R. Silliman, E. D. Grosholz, and M. D. Bertness, eds. Human Impacts on Salt Marshes, A Global Perspective. University of California Press, Berkeley.

Crawford, R. E., and C. G. Carey. 1985. Retention of winter flounder larvae within a Rhode Island salt pond. Estuaries 8(2b): 217–227.

Creaser, E. P., and H. C. Perkins. 1994. The distribution, food and age of juvenile bluefish, *Pomatomus saltatrix*, in Maine. Fish. Bull. 92(3):494–508.

Creel, S., and D. Christianson. 2008. Relationships between direct predation and risk effects. Trends Ecol. Evol. 23:194–201.

Crocker, P. A., C. R. Arnold, J. A. DeBoer, and J. D. Holt. 1981. Preliminary evaluation of survival and growth of juvenile red drum (*Sciaenops ocellata*) in fresh and salt water. J. World Maricult. Soc. 12(1):122–134.

Croker, R. A. 1965. Planktonic fish eggs and larvae of Sandy Hook estuary. Chesapeake Sci. 6(2):92–95.

Crowder, L. B., and W. F. Figueira. 2006. Metapopulation ecology and marine conservation. Pp. 491–515 *in* J. P. Kritzer and P. F. Sale, eds. Marine Metapopulations. Elsevier Academic Press Publications, San Diego, Calif.

Crowder, L. B., D. D. Squires, and J. A. Rice. 1997. Nonadditive

effects of terrestrial and aquatic predators on juvenile estuarine fish. Ecology 78(6):1796–1804.

Crowley, M. F., and S. M. Glenn. 1994. Remote sensing and in situ observations of coastal upwelling/downwelling offshore New Jersey. Trans. Am. Geophys. Union 75(3):156.

Csanady, G. T. 1978. Wind effects on surface to bottom fronts. J. Geophy. Res. 83(c9):4633–4640.

Csanady, G. T., and P. Hamilton. 1988. Circulation of slopewater. Continent. Shelf Res. 8(5–7):565–624.

Curran, M. C. 1992. The behavioral physiology of labroid fishes. Ph.D. diss. Massachusetts Institute of Technology / Woods Hole Oceanographic Institution, WHOI-92-41.

Curran, M. C., and K. W. Able. 2002. Annual stability in the use of coves near inlets as settlement areas for winter flounder (*Pseudopleuronectes americanus*). Estuaries 25(2): 227–234.

Currin, C. A., S. C. Wainright, K. W. Able, M. P. Weinstein, and C. M. Fuller. 2003. Determination of food web support and trophic position of the mummichog (*Fundulus heteroclitus*) in New Jersey smooth cordgrass (*Spartina alterniflora*), common reed (*Phragmites australis*) and restored salt marshes. Estuaries 26(2b):495–510.

Cushing, D. H. 1974. The possible density-dependence of larval mortality and adult mortality in fishes. Pp. 103–111 *in* J. H. S. Blaxter, ed. The Early Life History of Fish. Springer, New York.

———. 1996. Towards a Science of Recruitment in Fish Populations. Inter-Research, Oldendorf/Luhe, Germany. 175 pp.

Cyrus, D. P., and S. J. M. Blaber. 1992. Turbidity and salinity in a tropical northern Australian estuary and their influence on fish distribution. Estuar. Coast. Shelf Sci. 35:545–563.

Dadswell, M. J. 1976. Biology of the shortnose sturgeon (*Acipenser brevirostrum*) in the Saint John River estuary, New Brunswick, Canada. Trans. Atl. Chap. Can. Soc. Environ. Biol. Annu. Meet. 1975:20–72.

———. 1979. Biology and population characteristics of the shortnose sturgeon, *Acipenser brevirostrum* LeSueur 1818 (Osteichthys: Acipenseridae), in the Saint John River Estuary, New Brunswick, Canada. Can. J. Zool. 57:2186–2210.

Dadswell, M. J., B. D. Taubert, T. S. Squiers, D. Marchette, and J. Buckley. 1984. Synopsis of biological data on shortnose sturgeon, *Acipenser brevirostrum* LeSeur 1818. FAO Fisheries Synopsis No. 140. 51 pp.

Dadswell, M. J., G. D. Melvin, P. J. Williams, and D. E. Themelis. 1987. Influences of origin, life history, and change on the Atlantic coast migration of American shad. *In* M. J. Dadswell et al., eds. Common Strategies of Anadromous and Catadromous Fishes. Am. Fish. Soc. Symp. 1:313–330.

Dahlberg, M. D. 1975. Guide to Coastal Fishes of Georgia and Nearby States. University of Georgia Press, Athens.

Dahlberg, M. D., and J. C. Conyers. 1973. An ecological study of *Gobiosoma bosci* and *G. ginsburgi* (Pisces, Gobiidae) on the Georgia coast. U.S. Fish. Bull. 71(1):279–287.

Dahlgren, C. P., G. T. Kellison, A. J. Adams, B. M. Gillanders, M. S. Kendall, C. A. Layman, J. A. Ley, I. Nagelkerken, and J. E. Serafy. 2006. Marine nurseries and effective juvenile habitats: concepts and applications. Mar. Ecol. Prog. Ser. 312:291–295.

Daiber, F. C. 1957. Sea trout research. Estuarine Bull. 2(5):1–6.

Daiber, F. C., and R. W. Smith. 1970. An analysis of fish populations in the Delaware Bay area. Annu. Dingell-Johnson Rep., 1969–1970, Proj. F-13-R-12 (unpubl. ms.).

Daly, R. J. 1970. Systematics of southern Florida anchovies (Pisces: Engraulidae). Bull. Mar. Sci. 20(1):70–104.

Dame, R. F., M. Alber, D. Allen, M. Mallin, C. Montague, A. Lewitus, A. Chalmers, R. Gardner, C. Gilman, B. Kjerfve, J. Pinckney, and N. Smith. 2000. Estuaries of the south Atlantic coast of North America: their geographical signatures. Estuaries 23(6):793–819.

Dando, P. R. 1984. Reproduction in estuarine fish. Pp. 155–170 *in* W. Potts and R. J. Wootton, eds. Fish Reproduction: Strategies and Tactics. Academic Press, New York.

Daniel, L. B., III. 1995. Spawning and ecology of early life stages of black drum, *Pogonias cromis*, in lower Chesapeake Bay. Ph.D. diss. College of William and Mary, Williamsburg, Va. 166 pp.

Daniel, L. B., III, and J. E. Graves. 1994. Morphometric and genetic identification of eggs and spring-spawning sciaenids in lower Chesapeake Bay. U.S. Fish. Bull. 92:254–261.

Daniels, R. A. 1996. Guide to the Identification of Inland Fishes of Northeastern North America. New York State Museum Bulletin, 488. 97 pp.

Daniels, R. A., K. E. Limburg, R. E. Schmidt, D. L. Strayer, and C. Chambers. 2005. Changes in fish assemblages in the tidal Hudson River, New York. Pp. 471–503 *in* J. N. Rinne, R. M. Hughes, and B. Calamusso, eds. Historical Changes in Large River Fish Assemblages of America. Am. Fish. Soc. Symp. 45.

Darcy, G. H. 1985. Synopsis of biological data on the pinfish, *Lagodon rhomboides* (Pisces: Sparidae). Seattle, Wash.: National Marine Fisheries Service, National Oceanic and Atmospheric Administration; FAO Fisheries Synopsis No. 141. (NOAA Tech. Rpt. NMFS 23).

Darnell, R. M. 1958. Food habits of fishes and larger invertebrates of Lake Pontchartrain, Louisiana, an estuarine community. Inst. Mar. Sci. Univ. Tex. Publ. No. 5:353–416.

———. 1961. Trophic spectrum of an estuarine community, based on studies of Lake Pontchartrain, Louisiana. Ecology 42(3): 553–568.

Darnell, R. M., and T. M. Soniat. 1979. The estuary/continental shelf as an interactive system. Pp. 487–525 *in* R. J. Livingston, ed. Ecological Processes in Coastal and Marine Systems. Marine Sciences 10. Plenum Press, New York.

Daverat, F., K. E. Limburg, I. Thibault, J. Shiao, J. J. Dodson, F. Caron, W. Tzeng, Y. Iizuka, and H. Wickström. 2006. Phenotypic plasticity of habitat use by three temperate eel species, *Anguilla anguilla*, *A. japonica* and *A. rostrata*. Mar. Ecol. Prog. Ser. 308:231–241.

Dawson, C. E. 1958. A study of the biology and life history of the spot, *Leiostomus xanthurus* Lacepede, with special reference to South Carolina. Contrib. Bears Bluff Lab. No. 28, 48 pp.

———. 1966. Studies of the gobies (Pisces: Gobiidae) of Mississippi Sound and adjacent waters. I. *Gobiosoma*. Am. Midl. Nat. 76(2): 379–409.

———. 1982. Family Syngnathidae. Pp. 1–172 *in* Fishes of the Western North Atlantic. Sears Found. Mar. Res. Mem. 1 (Part 8). Yale University, New Haven, Conn.

Day, J. H. 1981. Estuarine Ecology with Particular Reference to Southern Africa. A. A. Balkema, Rotterdam.

Day, J. W., Jr., C. A. S. Hall, W. M. Kemp, and A. Yáñez-Arancibia. 1989. Estuarine Ecology. Wiley, New York.

Day, J. W., R. R. Christian, D. M. Boesch, A. Yáñez-Arancibia, J. Morris, R. R. Twilley, L. Naylor, L. Schaffner, and C. Stevenson. 2008. Consequences of climate change on the ecogeomorphology of coastal wetlands. Estuaries and Coasts 31: 477–491.

de Carvalho, C. D., C. M. Corneta, and V. S. Uieda. 2007. School-ing behavior of *Mugil curema* (Perciformes: Mugilidae) in an estuary in southeastern Brazil. Neotrop. Ichthyol. 5(1):81–83.

Deckert, R. F. 1914. Further notes on the Salientia of Jacksonville, FL. Copeia 18:3–5.

Deegan, L., and J. W. Day. 1986. Coastal fishery habitat requirements. Pp. 44–52 *in* A. Yáñez-Arancibia and D. Pauly, eds. Recruitment Processes in Tropical Coastal Demersal Communities. Ocean Science in Relationship to Living Resources (OSLR), International Recruitment Project (IREP). IOC-FAO-UNESCO Workshop OSLR/IREP Project. Vol. 44. UNESCO, Paris.

Deegan, L. A., J. E. Hughes, and R. A. Rountree. 2000. Salt marsh ecosystem support of marine transient species. Pp. 333–365 *in* M. P. Weinstein, and D. A. Kreeger, eds. Concepts and Controversies in Tidal Marsh Ecology. Kluwer Academic Publishers, Dordrecht, The Netherlands.

de Lafontaine, Y., T. Lambert, G. R. Lilly, W. D. McKone, and R. J. Miller, eds. 1992. Juvenile stages: the missing link in fisheries research. Report of a Workshop. Can. Tech. Rep. Fish. Aquatic Sci. 1890.

Delbeek, J. C., and D. D. Williams. 1987. Food resource partitioning between sympatric populations of brackishwater sticklebacks. J. Anim. Ecol. 56:949–967.

De Morais, L. T., and J. Y. Bodiou. 1984. Predation on meiofauna by juvenile fish in a Western Mediterranean flatfish nursery ground. Mar. Biol. 82:209–215.

Denno, R. F., C. Gratton, H. Döbel, and D. L. Finke. 2003. Predation risk affects relative strength of top-down and bottom-up impacts on insect herbivores. Ecology 84(4):1032–1044.

Denoel, M., and B. Demars. 2008. The benefits of heterospecific oophagy in a top predator. Acta Oecologica 34:74–79.

Denoncourt, R. F., J. C. Fisher, and K. M. Rapp. 1978. A freshwater population of the mummichog, *Fundulus heteroclitus*, from the Susquehanna River drainage in Pennsylvania. Estuaries 1:269–272.

Deree, H. L. 1999. Age and growth, dietary habits, and parasitism of the fourbeard rockling, *Enchelyopus cimbrius*, from the Gulf of Maine. Fish. Bull. 97:39–52.

Derickson, W. K., and K. S. Price. 1973. The fishes of the shore zone of Rehoboth and Indian River bays, Delaware. Trans. Am. Fish. Soc. 102:552–562.

de Sylva, D. P., F. A. Kalber Jr., and C. N. Shuster. 1962. Fishes and ecological conditions in the shore zone of the Delaware River Estuary, with notes on other species collected in deeper water. Univ. Del. Mar. Lab. Inf. Ser. Publ. 5:1–164.

Deuel, D. G., J. R. Clark, and A. J. Mansueti. 1966. Description of embryonic and early larval stages of the bluefish, *Pomatomus saltatrix*. Trans. Am. Fish. Soc. 95:264–271.

Dew, C. B. 1976. A contribution to the life history of the cunner, *Tautogolabrus adspersus*, in Fishers Island Sound, Connecticut. Chesapeake Sci. 17:101–113.

———. 1991. Early life history and population dynamics of Atlantic tomcod (*Microgadus tomcod*) in the Hudson River estuary, New York. Ph.D. diss. City University of New York, New York.

———. 1995. The nonrandom size distribution and size-selective transport of age-0 Atlantic tomcod (*Microgadus tomcod*) in the lower Hudson River estuary. Can. J. Fish. Aquat. Sci. 52:2353–2366.

Dew, C. B., and J. H. Hecht. 1994a. Hatching, estuarine transport, and distribution of larval and early juvenile Atlantic tomcod, *Microgadus tomcod*, in the Hudson River. Estuaries 17(2): 472–488.

Dew, C. B., and J. H. Hecht. 1994b. Recruitment, growth, mortality, and biomass production of larval and early juvenile Atlantic tomcod in the Hudson River Estuary. Trans. Am. Fish. Soc. 123: 681–702.

Diaz, R. J., R. J. Neubauer, L. C. Schaffner, L. Pihl, and S. P. Baden. 1992. Continuous monitoring of dissolved oxygen in an estuary experiencing periodic hypoxia and the effects of hypoxia on macrobenthos and fish. Pp. 1055–1068 in R. A. Vollenveider, R. Marchetti, and R. Viviani, eds. Marine Coastal Eutrophication: The Response of Marine Transitional Systems to Human Impacts—Problems and Perspectives for Restoration. Elsevier, Amsterdam.

Diaz, R. J., G. R. Cutter, and K. W. Able. 2003. The importance of physical and biogenic structure to juvenile fishes on the shallow inner continental shelf. Estuaries 26(1):12–20.

DiMichele, L., and D. A. Powers. 1982. LDH-B genotype-specific hatching times of Fundulus heteroclitus embryos. Nature 296: 563–564.

DiMichele, L., and M. H. Taylor. 1980. The environmental control of hatching in Fundulus heteroclitus. J. Exp. Zool. 214:181–187.

Dingle, H., and V. A. Drake. 2007. What is migration? BioScience 57:113–121.

Ditty, J. G., and R. F. Shaw. 1996. Spatial and temporal distribution of larval striped mullet (Mugil cephalus) and white mullet (M. curema, family Mugilidae) in the northern Gulf of Mexico, with notes on mountain mullet, Agonostomus monticola. Bull. Mar. Sci. 59:271–288.

Ditty, J. G., M. R. Cavalluzzi, and J. E. Olney. 2006. Blenniidae: Combtooth blennies. Pp. 1969–1993 in W. J. Richards, ed. Early Stages of Atlantic Fishes: An Identification Guide for the Western Central North Atlantic. CRC Press, Taylor and Francis Group, Boca Raton, Fla.

Dixon, P. A., M. J. Milicich, and G. Sugihara. 1999. Episodic fluctuations in larval supply. Science 283:1528–1529.

Dodson, J. J., J. C. Dauvin, R. G. Ingram, and B. d'Anglejan. 1989. Abundance of larval rainbow smelt (Osmerus mordax) in relation to the maximum turbidity zone and associated macroplanktonic fauna of the middle St. Lawrence estuary. Estuaries 12:66–81.

Doherty, P. J. 1991. Spatial and temporal patterns in recruitment. Pp. 271–287 in P. F. Sale, ed. The Ecology of Fishes on Coral Reefs. Academic Press, San Diego, Calif.

Doherty, P. J., V. Dufour, R. Galzin, M. A. Hixon, M. G. Meekan, and S. Planes. 2004. High mortality during settlement is a population bottleneck for a tropical surgeonfish. Ecology 85: 2422–2428.

Dolbeth, M., F. Martinho, R. Leitao, H. Cabral, and M. A. Pardal. 2008. Feeding patterns of the dominant benthic and demersal fish community in a temperate estuary. J. Fish. Biol. 72:2500–2517.

Domeir, M. C., and P. L. Colin. 1997. Tropical reef fish spawning aggregations: defined and reviewed. Bull. Mar. Sci. 60:698–726.

Domermuth, R. B., and R. J. Reed. 1980. Food of juvenile American shad, Alosa sapidissima, juvenile blueback herring, Alosa aestivalis, and pumpkinseed, Lepomis gibbosus, in the Connecticut River below Holyoke Dam, Massachusetts. Estuaries 3:65–68.

Dorf, B. A. 1994. Ecology of juvenile tautog (Tautoga onitis, Family Labridae) in Narragansett Bay, Rhode Island, 1988–1992. Estuaries 20:589–600.

Dorf, B. A., and J. C. Powell. 1997. Distribution, abundance, and habitat characteristics of juvenile tautog (Tautoga onitis, Family Labridae) in Narragansett Bay, Rhode Island, 1988–1992. Estuaries 20(3):589–600.

Dorsey, S. E., E. D. Houde, and J. C. Gamble. 1996. Cohort abundances and daily variability in mortality of eggs and yolk-sac larvae of bay anchovy, Anchoa mitchilli, in Chesapeake Bay. U.S. Fish. Bull. 94:257–267.

Dorval, E., C. M. Jones, R. Hannigan, and J. van Montfrans. 2005. Can otolith chemistry be used for identifying essential seagrass habitats for juvenile spotted seatrout, Cynoscion nebulosus, in Chesapeake Bay? Mar. Freshwat. Res. 56:645–653.

Dovel, W. L. 1960. Larval development of the oyster toadfish, Opsanus tau. Chesapeake Sci. 1(3–4):187–195.

———. 1963. Larval development of Clingfish, Gobiesox strumosus, 4.0 to 12.0 millimeters total length. Chesapeake Sci. 4(3): 161–166.

———. 1968. Fish eggs and larvae. In Biological and geological research on the effects of dredging spoil disposal in the upper Chesapeake Bay. Eighth Prog. Rep. No. 68-2-B. To: U.S. Army Corps Eng., Contr. No. 14-16-005-2096. Univ. Md. Nat. Resour. Inst. Contrib. Ref. No. 68-2-B.

———. 1971. Fish eggs and larvae of the upper Chesapeake Bay. Univ. Md. Nat. Resour. Inst. Contrib. No.460, Spec. Sci. Rep. 4.

———. 1978. Biology and management of Atlantic sturgeons, Acipenser oxyrhynchyus (Mitchill), and shortnose sturgeons, Acipenser brevirostrum (LeSueur), of the Hudson Estuary. Wapora Inc. 181 pp.

———. 1981. Ichthyoplankton of the lower Hudson Estuary, New York. N.Y. Fish Game J. 28(1):21–39.

———. 1992. Movements of Immature Striped Bass in the Hudson Estuary. Pp. 276–300 in Estuarine Research in the 1980's. State University of New York Press, Albany.

Dovel, W. L., T. A. Milhursky, and A. J. McErlean. 1969. Life history and aspects of the hogchoker, Trinectes maculatus, in the Patuxent River estuary, Maryland. Chesapeake Sci. 10(2):104–119.

Dresser, B. K., and R. T. Kneib. 2007. Site fidelity and movement patterns of wild sub-adult red drum, Sciaenops ocellatus (Linnaeus), within a salt marsh-dominated estuarine landscape. Fisheries Manag. Ecol. 14:183–190.

Drinkwater, K. F., A. Belgrano, A. Borja, A. Conversi, M. Edwards, C. H. Greene, G. Ottersen, A. J. Pershing, and H. Walker. 2003. The response of marine ecosystems to climate variability associated with the North Atlantic oscillation. Pp. 211–234 in The North Atlantic Oscillation: Climatic Significance and Environmental Impact. Am. Geophys. Union, Geophys. Monogr. 134.

Duffy-Anderson, J. T., and K. W. Able. 1999. Effects of municipal piers on the growth of juvenile fish in the Hudson River estuary: a study across the pier edge. Mar. Biol. 133:409–418.

———. 2001. An assessment of the feeding success of young-of-the-year winter flounder (Pseudopleuronectes americanus) near a municipal pier in the Hudson River Estuary, U.S.A. Estuaries 24(3):430–440.

Duffy-Anderson, J. T., J. P. Manderson, and K. W. Able. 2003. A characterization of juvenile fish assemblages around man-made structures in the New York–New Jersey harbor estuary, U.S.A. Bull. Mar. Sci. 72(3):877–889.

Duffy-Anderson, J. T., K. Bailey, L. Ciannelli, P. Cury, A. Belgrano, and N. C. Stenseth. 2005. Phase transitions in marine fish recruitment processes. Ecol. Comp. 2:205–218.

Dugan, J. E., D. M. Hubbard, I. F. Rodil, D. L. Revell, and S. Schroeter. 2008. Ecological effects of coastal armouring on sandy beaches. Mar. Ecol. 29(1):160–170.

Duggins, C. F., Jr., A. A. Karlin, and K. G. Relyea. 1983. Electrophoretic comparison of *Cyprinodon variegatus* Lacépede and *Cyprinodon hubbsi* Carr, with comments on the genus *Cyprinodon* (Atheriniformes: Cyprinodontidae). Northeast Gulf Sci. 6(2): 99–107.

Dulvy, N. K., S. I. Rogers, S. Jennings, V. Stelzenmuller, S. R. Dye, and H. R. Skjoldal. 2008. Climate change and deepening of the North Sea fish assemblage: a biotic indicator of warming seas. J. Appl. Ecol. 45(4):1029–1039.

Dunning, D. J., J. R. Waldman, Q. E. Ross, and M. T. Mattson. 1997. Use of Atlantic tomcod and other prey by striped bass in the lower Hudson River estuary during winter. Trans. Am. Fish. Soc. 126(5):857–861.

Dunstan, P. K., and N.J. Bax. 2007. How far can marine species go? Influence of population biology and larval movement on future range limits. Mar. Ecol. Prog. Ser. 344:15–28.

DuPaul, W. D., and J. D. McEachran. 1973. Age and growth of the butterfish, *Peprilus triacanthus*, in the lower York River. Chesapeake Sci. 14(3):205–207.

Durand, J. B. 1984. Nitrogen distribution in New Jersey coastal bays. Pp. 29–51 *in* M. J. Kennish and R. A. Lutz, eds. Ecology of Barnegat Bay, New Jersey. Lecture Notes on Coastal and Estuarine Studies. Springer, New York.

Durant, J. M., D. Ø. Hjermann, G. Ottersen, and N. C. Stenseth. 2007. Climate and the match or mismatch between predator requirements and resource availability. Clim. Res. 33:271–283.

Durbin, A. G., and E. G. Durbin. 1975. Grazing rates of the Atlantic menhaden *Brevoortia tyrannus* as a function of particle size and concentration. Mar. Biol. 33(3):265–277.

———. 1998. Effects of menhaden predation on plankton populations in Narragansett Bay, Rhode Island. Estuaries 21:449–465.

Durski, S. 1996. Detection of the Hudson-Raritan Estuary plume by satellite remote sensing. Bull. N.J. Acad. Sci. 41(1):15–20.

Duryea, R., J. Donnelly, D. Guthrie, O. O. C. Malley, M. Romanowski, and R. Schmidt. 1996. *Gambusia affinis* effectiveness in New Jersey mosquito control. Proceedings of the Eighty-Third Annual Meeting of the New Jersey Mosquito Control Association, 95–102.

Duval, E. J., and K.W. Able. 1998. Aspects of the life history of the seaboard goby, *Gobiosoma ginsburgi*, in estuarine and inner continental shelf waters. Bull. N.J. Acad. Sci. 43(1):5–10.

Eales, J. G. 1968. The eel fisheries of eastern Canada. Fish. Res. Board Can. Bull. 166.

Edeline, E., L. Beaulaton, R. Le Barh, and P. Elie. 2007. Dispersal in metamorphosing juvenile eel *Anguilla rostrata*. Mar. Ecol. Prog. Ser. 344:213–218.

Edgar, G. J., N. S. Barrett, D. J. Graddon, and P. R. Last. 2000. The conservation significance of estuaries: a classification of Tasmanian estuaries using ecological, physical and demographic attributes as a case study. Biol. Conservat. 92:383–397.

Edwards, R. L., and K. O. Emory. 1968. The view from a storied sub. The *Alvin* off Norfolk, VA. Commerc. Fish. Rev. 30(8–9):48–55.

Edwards, R. L., R. Livingstone Jr., and P. E. Hamer. 1962. Winter water temperatures and an annotated list of fishes—Nantucket Shoals to Cape Hatteras. *Albatross III* Cruise No. 126. Spec. Sci. Rep. Fisheries. 397:1–31.

Ehrlich, P. R., D. S. Dobkin, and D. Wheye. 1988. The Birder's Handbook. A Field Guide to the Natural History of North American Birds. Simon and Schuster, Inc. New York.

Eigenmann, C. H. 1902. The egg and development of the conger eel. Bull. U.S. Fish Comm. 21(1901):37–44.

Eklund, A. M., and T. E. Targett. 1990. Reproductive seasonality of fishes inhabiting hard bottom areas in the Middle Atlantic Bight. Copeia:1180–1194.

———. 1991. Seasonality of fish catch rates and species composition from the hard bottom trap fishery in the Middle Atlantic Bight (U.S. east coast). Fish. Res. 12(1991):1–22.

Ekman, S. 1953. Zoogeography of the Sea. Sidgwick and Jackson, London. 417 pp.

Eldred, B. 1966. The early development of the spotted worm eel, *Myrophis punctatus* Lutken (Ophichthidae). Fla. Board Conserv. Mar. Res. Lab. Leafl., Ser. 4 (Pt. 1 No. 1). 13 pp.

Elliott, E. M., D. Jimenez, C. O. Anderson Jr., and D. J. Brown. 1979. Ichthyoplankton abundance and distribution in Beverly–Salem Harbor, March 1975 through February 1977. Massachusetts Division of Marine Fisheries, Boston.

Elliott, J. M. 1989. The critical-period concept for juvenile survival and its relevance for population regulation in young sea trout, *Salmo trutta*. J. Fish. Biol. 35a:91–98.

Elliott, M., and K. L. Hemingway. 2002. Fishes in Estuaries. Blackwell Science Ltd., London.

Elliott, M., and D. S. McLusky. 2002. The need for definitions in understanding estuaries. Estuar. Coast. Shelf Sci. 55:815–827.

Ellis, J. K. 2003. Diet of the sandbar shark, *Carcharhinus plumbeus*, in Chesapeake Bay and adjacent waters. M.S. thesis. College of William and Mary, Williamsburg, Va. 90 pp.

Ellis, T., B. R. Howell, and R. N. Hughes. 1997. The cryptic responses of hatchery-reared sole to a natural sand substratum. J. Fish Biol. 51:389–401.

Elson, P. F. 1939. Order of appearance of scales in speckled trout. J. Fish. Res. Board of Can. 4(4):302–308.

Enghoff, I. B., B. R. MacKenzie, and E. E. Nielsen. 2007. The Danish fish fauna during the warm Atlantic period (ca. 7000–3900 BC): forerunner of future changes? Fish. Res. 87:167–180.

Engle, V. D., and J. K. Summers. 1999. Latitudinal gradients in benthic community composition in Western Atlantic estuaries. J. Biogeogr. 26:1007–1023.

Ennis, G. P. 1969. Occurrences of the little sculpin, *Myoxocephalus aenaeus*, in Newfoundland waters. J. Fish. Res. Board Can. 27: 1689–1694.

EPA (U.S. Environmental Protection Agency). 1998. Condition of the Mid-Atlantic Estuaries. EPA 600-R-98-147.

Epifanio, C. E. 1995. Transport of blue crab (*Callinectes sapidus*) larvae in the waters off Mid-Atlantic states. Bull. Mar. Sci. 57(3): 713–725.

Epifanio, C. E., and R. W. Garvine. 2001. Larval transport on the Atlantic continental shelf of North America: a review. Estuar. Coast. Shelf Sci. 52:51–77.

Epifanio, C. E., A. K. Masse, and R. W. Garvine. 1989. Transport of blue crab larvae by surface currents off Delaware Bay, USA. Mar. Ecol. Prog. Ser. 54:35–41.

Everett, R. A., and G. M. Ruiz. 1993. Coarse woody debris as a refuge from predation in aquatic communities. Oecologia 93: 457–486.

Evermann, B. W., and W. C. Kendall. 1898. Descriptions of new or little-known genera and species of fishes from the United States. Bull. U.S. Fish. Comm. Vol. 17 for 1897:125–133.

Ewart, K. V., and G. L. Fletcher. 1990. Isolation and characterization of antifreeze proteins from smelt (*Osmerus mordax*) and Atlantic herring (*Clupea harengus harengus*). Can. J. Zool. 68: 1652–1658.

Fabrizio, M. C. 1987. Contribution of Chesapeake Bay and Hudson

River stocks of striped bass to Rhode Island coastal waters as estimated by isoelectric focusing of eye lens proteins. Trans. Am. Fish. Soc. 116(4):588–593.

Fahay, M. P. 1975. An annotated list of larval and juvenile fishes captured with surface-towed meter net in the South Atlantic Bight during four RV *Dolphin* cruises between May 1967 and Feb 1968. NOAA Technical Report NMFS SSRF-685.

———. 1983. Guide to the early stages of marine fishes occurring in the western North Atlantic ocean, Cape Hatteras to the southern Scotian Shelf. J. Northwest Atl. Fish. Sci. 4:1–423.

———. 1987. Larval and juvenile hakes (*Phycis-Urophycis* sp.) examined during a study of the white hake, *Urophycis tenuis* (Mitchill), in the Georges Bank–Gulf of Maine area. Dep. Comm. Natl. Ocean. Atmos. Adm. NMFS Northeast Fish. Cent., Sandy Hook Lab. Rep. No. 87-03.

———. 1992. Development and distribution of cusk eel eggs and larvae in the Middle Atlantic Bight with a description of *Ophidion robinsi* n. sp. (Teleostei: Ophidiidae). Copeia 3:799–819.

———. 1993. The early life history stages of New Jersey's saltwater fishes: sources of information. Bull. N.J. Acad. Sci. 38(2):1–16.

———. 2007. Early Stages of Fishes in the Western North Atlantic Ocean. (Davis Strait, Southern Greenland and Flemish Cap to Cape Hatteras). Vols. 1 and 2. Northwestern Atlantic Fisheries Organization, Dartmouth, Nova Scotia.

Fahay, M. P., and K. W. Able. 1989. The white hake, *Urophycis tenuis* in the Gulf of Maine: spawning seasonality, habitat use and growth in young-of-the-year, and relationships to the Scotian Shelf population. Can. J. Zool. 67:1715–1724.

Fahay, M. P., and D. F. Markle. 1984. Gadiformes: development and relationships. Pp. 265–283 in H. G. Moser, W. J. Richards, D. M. Cohen, M. P. Fahay, A. W. Kendall Jr., and S. L. Richardson, eds. Ontogeny and Systematics of Fishes. Am. Soc. Ichthyol. Herpetol. Spec. Publ. No. 1. Allen Press, Lawrence, Kans.

Fahay, M. P., and C. L. Obenchain. 1978. Leptocephali of the Ophichthid Genera *Ahlia, Myrophis, Ophichthus, Pisodonphis, Callechelys, Letharchus, and Apterichtus* on the Atlantic continental shelf of the United States. Bull. Mar. Sci. 28(3):442–486.

Fahy, W. E. 1978. The influence of crowding upon the total number of vertebrae developing in *Fundulus majalis* (Walbaum). J. Cons. Int. Explor. Mer. 38(2):252–256.

———. 1979. The influence of temperature change on number of anal fin rays developing in *Fundulus majalis*. J. Cons. Int. Explor. Mer. 38(3):280–285.

———. 1980. The influence of temperature change on number of dorsal fin rays developing in *Fundulus majalis* (Walbaum). J. Cons. Int. Explor. Mer. 39(1):104–109.

———. 1982. The influence of temperature change on number of pectoral fin rays developing in *Fundulus majalis* (Walbaum). J. Cons. Int. Explor. Mer. 40(1):21–26.

Fairchild, E. A., and W. H. Howell. 2000. Predator-prey size relationships between *Pseudopleuronectes americanus* and *Carcinus maenas*. J. Sea Res. 44:81–90.

Farooqi, T. W., R. F. Shaw, J. G. Ditty, and J. Lyczkowski-Shultz. 2006. Cynoglossidae: tongue fishes. Pp. 2367–2379 in W. J. Richards, ed. Early Stages of Atlantic Fish: An Identification Guide for the Western Central North Atlantic. CRC Press, Taylor and Francis Group, Boca Raton.

Fay, C. W., R. J. Neves, and G. B. Pardue. 1983. Striped bass. Species profiles: life histories and environmental requirements of coastal fishes and invertebrates (Mid-Atlantic). National Coastal Ecosystem Team, U.S. Fish and Wildlife Service, Washington, D.C.

Feller, R. J., B. C. Coull, and B. T. Hentschel. 1990. Meiobenthic copepods: tracers of where juvenile *Leiostomus xanthurus* (Pisces) feed? Can. J. Fish. Aquat. Sci. 47:1913–1919.

Ferraro, S. P. 1980. Daily time of spawning of 12 fishes in the Peconic Bays, New York. U.S. Fish. Bull. 78(2):455–464.

Festa, P. J. 1979. Analysis of the fish forage base in the Little Egg Harbor estuary. NJDEP Div. Fish, Game, and Shellfisheries. Bur. Fisheries. Nacote Creek Res. Station. Tech. Rep. No. 24M. 271 pp.

Ficke, A. D., C. A. Myrick, and L. J. Hansen. 2007. Potential impacts of global climate change on freshwater fisheries. Rev. Fish Biol. Fish. 17:581–613.

Fields, H. M. 1962. Pompanos (*Trachinotus* spp.) of the South Atlantic coast of the United States. U.S. Fish. Bull. 62:189–222.

Fields, P. A., J. B. Graham, R. H. Rosenblatt, and G. N. Somero. 1993. Effects of expected global climate change on marine faunas. Trends Ecol. Evol. 8(10):361–367.

Fine, M. L. 1978. Seasonal and geographical variation of the mating call of the oyster toadfish *Opsanus tau* L. Oecologia 36: 45–57.

Finkelstein, S. L. 1969a. Age and growth of scup in the waters of eastern Long Island Sound. N.Y. Fish Game J. 16(1):84–110.

———. 1969b. Age at maturity of scup from New York waters. N.Y. Fish Game J. 16(2):224–237.

Finne, K. L. 2001. Phylogeographic structure of the Atlantic pupfish, *Cyprinodon variegatus* (Cyprinodontidae), along the eastern coast of North America: evidence from mitochondrial nucleotide sequences. M.S. thesis. Virginia Polytechnic Institute and State University, Blacksburg. 61 pp.

Finucane, J. H. 1969. Ecology of the pompano (*Trachinotus carolinus*) and the permit (*T. falcatus*) in Florida. Trans. Am. Fish. Soc. 3:478–486.

Fisher, R., D. R. Bellwood, and S. D. Job. 2000. Development of swimming abilities in reef fish larvae. Mar. Ecol. Prog. Ser. 202: 163–173.

Fitzgerald, J., and J. L. Chamberlin. 1981. Anti-cyclonic warm core Gulf Stream eddies off the northeastern United States during 1979. Ann. Biol. Copenh. 36:44–51.

Fitzhugh, G. R., and J. W. Fleeger. 1985. Goby (Pisces: Gobiidae) interactions with meiofauna and small macrofauna. Bull. Mar. Sci. 36(3):436–444.

Fitzhugh, G. R., S. W. Nixon, D. W. Ahrenholz, and J. A. Rice. 1997. Temperature effects on otolith microstructure and birth month estimation from otolith increment patterns in Atlantic menhaden. Trans. Am. Fish. Soc. 126:579–593.

Fletcher, G. L. 1977. Circannual cycles of blood plasma freezing point and Na+ and Cl− concentrations in Newfoundland winter flounder (*Pseudopleuronectes americanus*): correlation with water temperature and photoperiod. Can. J. Zool. 55: 789–795.

Flores-Coto, C., and S. M. Warlen. 1993. Spawning time, growth, and recruitment of larval spot *Leiostomus xanthurus* into a North Carolina estuary. U.S. Fish. Bull. 91(1):8–22.

Foerster, J. W., and S. L. Goodbred. 1978. Evidence for a resident alewife population in the northern Chesapeake Bay. Estuar. Coast. Mar. Sci. 7:437–444.

Fogarty, M. J., M. P. Sissenwine, and E. B. Cohen. 1991. Recruitment variability and the dynamics of exploited marine populations. Trends Ecol. Evol. 6(8):241–245.

Forrester, G. E. 1990. Factors influencing the juvenile demography of a coral reef fish. Ecology 71(5):1666–1681.

Fortier, L., and W. C. Leggett. 1983. Vertical migrations and transport of larval fish in a partially mixed estuary. Can. J. Fish. Aquat. Sci. 40:1543–1555.

Forward, R. J., and R. A. Tankersley. 2001. Selective tidal-stream transport of marine animals. Oceanogr. Mar. Biol. Ann. Rev. 39:305–353.

Foster, A. M., and J. P. Clugston. 1997. Seasonal migration of Gulf sturgeon in the Suwannee River, Florida. Trans. Am. Fish. Soc. 126(2):302–308.

Foster, J. R. 2001. Age, growth, and mortality of Atlantic croaker, *Micropogonias undulatus,* in the Chesapeake Bay region. M.S. thesis. College of William and Mary. Williamsburg, Va. 85 pp.

Foster, N. R. 1967. Comparative studies on the biology of killifishes (Pisces: Cyprinodontidae). Ph.D. diss. Cornell University, Ithaca, N.Y.

———. 1974. Belonidae—needlefishes. Pp. 125–126 *in* A. J. Lippson and R. L. Moran, eds. Manual for identification of early developmental stages of fishes of the Potomac River Estuary. Md. Power Plant Siting Prog. Misc. Rep. No. 13.

Fowler, H. W. 1906. The fishes of New Jersey. Annu. Rep. N.J. State Mus. 1905 (2):35–477.

Fox, M. G., and A. Keast. 1990. Effects of winterkill on population structure, body size, and prey consumption patterns of pumpkinseed in isolated beaver ponds. Can. J. Zool. 68:2489–2498.

———. 1991. Effects of overwinter mortality on reproductive life history characteristics of pumpkinseed (*Lepomis gibbosus*) populations. Can. J. Fish. Aquat. Sci. 48(9):1792–1799.

Franco A., M. Elliot, P. Franzoi, and P. Torricelli. 2008. Life strategies of fishes in European estuaries: the functional guild approach. Mar. Ecol. Prog. Ser. 354:219–228.

Frank, K. T., R. I. Perry, and K. F. Drinkwater. 1990. Predicted response of northwest Atlantic invertebrate and fish stocks to CO_2-induced climate change. Trans. Am. Fish. Soc. 119: 353–365.

Frank, K. T., B. Petrie, J. S. Choi, and W. C. Leggett. 2005. Trophic cascades in a formerly cod-dominated ecosystem. Science 308: 1621–1623.

Franks, J. S., J. Y. Christmas, W. L. Siler, R. Combs, R. Waller, and C. Burns. 1972. A study of nektonic and benthic faunas of the shallow Gulf of Mexico off the state of Mississippi as related to some physical, chemical and geographical factors. Gulf Res. Rept. No. 4:1–148.

Fraser, T. H. 1997. Abundance, seasonality, community indices, trends and relationships with physicochemical factors of trawled fish in upper Charlotte Harbor, Florida. Bull. Mar. Sci. 60(3):739–763.

Freeman, S., S. Mackinson, and R. Flatt. 2004. Diel patterns in the habitat utilisation of sandeels revealed using integrated acoustic surveys. J. Exp. Mar. Biol. Ecol. 305:141–154.

Freeman, B. L., and S. C. Turner. 1977. The effects of anoxic water on the summer flounder (*Paralichthys dentatus*), a bottom dwelling fish. NOAA Tech. Ser. No. 3:451–462.

Friedland, K. D., D. W. Ahrenholz, and J. F. Guthrie. 1996. Formation and seasonal evolution of Atlantic menhaden juvenile nurseries in coastal estuaries. Estuaries 19(1):105–114.

Friedland, K. D., and L. W. Haas. 1988. Emigration of juvenile Atlantic menhaden, *Brevoortia tyrannus* (Pisces: Clupeidae), from the York River Estuary. Estuaries 11:45–50.

Friedland, K. D., and J. A. Hare. 2007. Long-term trends and regime shifts in sea surface temperature on the continental shelf of the northeast United States. Cont. Shelf Res. 27:2313–2328.

Friedland, K. D., D. W. Ahrenholz, and J. F. Guthrie. 1989. Influence of plankton on distribution patterns of the filterfeeder *Brevoortia tyrannus* (Pisces: Clupeidae). Mar. Ecol. Prog. Ser. 54:111.

Friedland, K. D., M. J. Miller, and B. Knights. 2007. Oceanic changes in the Sargasso Sea and declines in recruitment of the European eel. ICES J. Mar. Sci. 64:519–530.

Fries, L. T., D. J. Williams, and S. K. Johnson. 1996. Occurrence of *Anguillicola crassus*, an exotic parasitic swim bladder nematode of eels, in the southeastern United States. Trans. Am. Fish. Soc. 125:794–797.

Frisbie, C. M. 1961. Young black drum, *Pogonias cromis*, in tidal fresh and brackish waters, especially in the Chesapeake and Delaware Bay areas. Chesapeake Sci. 2(1–2):94–100.

Fritz, E. S., W. H. Meredith, and V. A. Lotrich. 1975. Fall and winter movements and activity level of the mummichogs, *Fundulus heteroclitus*, in a tidal creek. Chesapeake Sci. 2(1–2):94–100.

Fritz, R. L. 1965. Autumn distribution of groundfish species in the Gulf of Maine and adjacent waters, 1955–1961. Am. Geol. Soc. Ser. Atlas Mar. Environ. Folio 10:1–48.

Fritzsche, R. A. 1978. Development of Fishes of the Mid-Atlantic Bight: An Atlas of Egg, Larval and Juvenile Stages. Vol. 5, Chaetodonitdae through Ophidiidae. Biological Services Program, U.S. Fish and Wildlife Service FWS/OBS-78/12. U.S. Government Printing Office, Washington, D.C.

Fuda, K. M., B. M. Smith, M. P. Lesser, B. J. Legare, H. R. Breig, R. B. Stack, and D. L. Berlinsky. 2007. The effects of environmental factors on rainbow smelt *Osmerus mordax* embryos and larvae. J. Fish Biol. 71:539–549.

Fuiman, L. A. 1994. The interplay of ontogeny and scaling in the interactions of fish larvae and their predators. J. Fish Biol. 45: 55–79.

———. 1997. What can flatfish ontogenies tell us about pelagic and benthic lifestyles? J. Sea Res. 37:257–267.

Fuiman, L. A., and J. H. Cowan. 2003. Behavior and recruitment success in fish larvae: responsibility and covariation of survival skills. Ecology 84:53–67.

Fuiman, L. A., K. R. Poling, and D. M. Higgs. 1998. Quantifying developmental progress for comparative studies of larval fishes. Copeia 3:602–611.

Funderburk, S. L., S. J. Jordan, J. A. Mihursky, and D. Riley. 1991. Habitat requirements for Chesapeake Bay living resources. Living Research Subcommittee, Chesapeake Bay Program, Solomons, Md.

Gagliano, M., M. I. McCormick, and M. G. Meekan. 2007a. Temperature-induced shifts in selective pressure at a critical developmental transition. Oecologia 152:219–225.

———. 2007b. Survival against the odds: ontogenetic changes in selective pressure mediate growth-mortality trade-offs in a marine fish. Proc. R. Soc. B. 274:1575–1582.

Ganger, M. T. 1999. The spatial and temporal distribution of young-of-the-year *Osmerus mordax* in the Great Bay estuary. Environ. Biol. Fishes 54:253–261.

Gannon, D. P. 2003. Behavioral ecology of an acoustically mediated predator-prey system: bottlenose dolphins and sciaenid fishes. Ph.D. diss. Duke University, Durham, N.C.

Gannon, D. P., N. B. Barros, D. P. Nowacek, A. J. Read, D. M. Waples, and R. S. Wells. 2005. Prey detection by bottlenose dolphins, *Tursiops truncatus*: an experimental test of the passive listening hypothesis. Anim. Behav. 69:709–720.

Gardinier, M. N., and T. B. Hoff. 1982. Diet of striped bass in the Hudson River estuary. N.Y. Fish Game J. 29:152–165.

Garlo, E. V., C. B. Milstein, and A. E. Jahn. 1979. Impact of hypoxic conditions in the vicinity of Little Egg Inlet, New Jersey in summer, 1976. Estuar. Coast. Mar. Sci. 8:421–432.

Gartland, J. 2002. Diet composition of young-of-the-year bluefish, *Pomatomus saltatrix*, in the lower Chesapeake Bay and Virginia's coastal ocean. M.S. thesis. College of William and Mary, Williamsburg, Va. 129 pp.

Gauthier, D. A., J. A. Lincoln, and J. T. Turner. 2008. Distribution, abundance, and growth of larval tautog, *Tautoga onitis*, in Buzzards Bay, Massachusetts, USA. Mar. Ecol. 29:76–88.

Gedan, K. B., and B. R. Silliman. 2009. Using facilitation theory to enhance mangrove restoration. J. Human Env. 38(2):109.

Gee, J. M. 1989. An ecological economic review of meiofauna as food for fish. Zool. J. Linn. Soc. 96:243–261.

Gelsleichter, J., J. A. Musick, and S. Nichols. 1999. Food habits of the smooth dogfish, *Mustelus canis*, dusky shark, *Carcharhinus obscurus*, Atlantic sharpnose shark, *Rhizoprionodon terraenovae*, and the sand tiger, *Carcharias taurus*, from the northwest Atlantic ocean. Environ. Biol. Fishes 54:205–217.

Gibbs, R. H., Jr. 1959. A synopsis of the post larvae of western Atlantic lizard-fishes (Synodontidae). Copeia 3:232–236.

Gibson, R. N., and L. Robb. 1992. The relationship between body size, sediment grain size and the burying ability of juvenile plaice, *Pleuronectes platessa* L. J. Fish Biol. 40:771–778.

———. 2000. Sediment selection in juvenile plaice and its behavioural basis. J. Fish Biol. 56:1258–1275.

Gibson, R. N., L. Robb, M. T. Burrows, and A. D. Ansell. 1996. Tidal, diel and longer term changes in the distribution of fishes on a Scottish sandy beach. Mar. Ecol. Prog. Ser. 130:1–17.

Gibson, R. N., L. Pihl, M. T. Burrows, J. Modin, H. Wennhage, and L. A. Nickell. 1998. Diel movements of juvenile plaice (*Pleuronectes platessa*) in relation to predators, competitors, food availability and abiotic factors on a microtidal nursery ground. Mar. Ecol. Prog. Ser. 165:145–159.

Gilbert, C. R. 1986. Species profiles: life histories and environmental requirements of coastal fishes and invertebrates (South Florida) southern, Gulf, and summer flounders. U.S. Fish Wildl. Serv. Biol. Rep. 82(11,54).

Gilhen, J. 1972. The white mullet, *Mugil curema*, added to and the striped mullet, *M. cephalus*, deleted from the Canadian Atlantic fish fauna. Can. Field-Nat. 86:74–77.

Gillanders, B. M. 2002. Connectivity between juvenile and adult fish populations: do adults remain near their recruitment estuaries? Mar. Ecol. Prog. Ser. 240:215–223.

———. 2005. Using elemental chemistry of fish otoliths to determine connectivity between estuarine and coastal habitats. Estuar. Coast. Shelf Sci. 64:47–57.

Gillanders, B. M., K. W. Able, J. A. Brown, D. E. Eggleston, and P. F. Sheridan. 2003. Evidence of connectivity between juvenile and adult habitats for mobile marine fauna: an important component of nurseries. Mar. Ecol. Prog. Ser. 247:281–295.

Gilliers, C., O. Le Pape, Y. Desaunay, J. Morin, D. Guerault, and R. Amara. 2006. Are growth and density quantitative indicators of essential fish habitat quality? An application to the common sole *Solea solea* nursery grounds. Estuar. Coast. Shelf Sci. 69:96–106.

Gilmore, R. G. 1977. Fishes of the Indian River Lagoon and adjacent waters, Florida. Bull. Fla. St. Mus. Biol. Sci. 22(3):101–147.

Gilmore, R. G., L. H. Bullock, and F. H. Berry. 1978. Hypothermal mortality in marine fishes, south-central Florida, January, 1977. Northeast Gulf Sci. 2(2):77–97.

Ginsburg, I. 1933. A revision of the genus *Gobiosoma* (Family Gobiidae). Bull. Bingham Oceanogr. Coll. Yale Univ. 4(5):1–59.

———. 1952. Fishes of the family Carrangidae of the northern Gulf of Mexico and three related species. Publ. Inst. Mar. Sci., Univ. Texas 2(2):43–117.

Gleason, T. R., and D. A. Bengston. 1996. Growth, survival and size-selective predation mortality of larval and juvenile inland silversides, *Menidia beryllina* (Pisces: Atherinidae). J. Exp. Mar. Biol. Ecol. 199:165–177.

Glenn, S. M., M. F. Crowley, D. B. Haidvogel, and Y. T. Song. 1996. Underwater observatory captures coastal upwelling events off New Jersey. EOS Trans. Am. Geophys. Union 77(25):233–236.

Goldberg, R., B. Phelan, J. Pereira, S. Hagan, P. Clark, A. Bejda, A. Calabrese, A. Studholme, and K. W. Able. 2002. Variability in the distribution and abundance of young-of-the-year winter flounder, *Pseudopleuronectes americanus*, across habitat types in three northeastern U.S. estuaries. Estuaries 25:215–226.

Gonzalez-Villasenor, L. I., and D. A. Powers. 1990. Mitochondrial DNA restriction site polymorphisms in the teleost *Fundulus heteroclitus* supports secondary intergradation. Evolution 44(1):27–37.

Good, R. E., and N. F. Good. 1984. The Pinelands National Reserve: an ecosystem approach to management. Bioscience 34(3):169–173.

Goode, G. B., and T. H. Bean. 1896. Oceanic ichthyology: deep-sea and pelagic fishes of the world. Spec. Bull. U.S. Natl. Mus. 2:1–553.

Gooley, B. R., and F. H. Lesser. 1977. The history of the use of *Gambusia affinis* (Baird and Girard) in New Jersey. Proc. Annu. Meet. N.J. Mosq. Control Assoc. 64:154–159.

Gordo, O. 2007. Why are bird migration dates shifting? A review of weather and climate effects on avian migratory phenology. Clim. Res. 35:37–58.

Goshorn, D. M. 1990. Distribution of larval weakfish (*Cynoscion regalis*) in Delaware Bay and the relationship of field prey concentration to laboratory determined growth and mortality rates. Ph.D. diss. University of Delaware, Newark. 173 pp.

Goshorn, D. M., and C. E. Epifanio. 1991. Diet of larval weakfish and prey abundance in Delaware Bay. Trans. Am. Fish. Soc. 120:684–692.

Gosline, W. A. 1948. Speciation in the fishes of the genus *Menidia*. Evolution 2(4):306–313.

Govoni, J. J. 1987. The ontogeny of dentition in *Leiostomus xanthurus*. Copeia 4:1041–1046.

———. 1993. Flux of larval fishes across frontal boundaries: examples from the Mississippi River plume front and the western Gulf Stream front in winter. Bull. Mar. Sci. 53(2):538–566.

Govoni, J. J., and D. E. Hoss. 2001. Comparison of the development and function of the swimbladder of *Brevoortia tyrannus* (Clupeidae) and *Leiostomus xanthurus* (Sciaenidae). Copeia 2:430–442.

Govoni, J. J., and J. E. Olney. 1991. Potential predation on fish eggs by the lobate ctenophore *Mnemiopsis leidyi* within and outside the Chesapeake Bay Plume. Fish. Bull. 89:181–186.

Govoni, J. J., and L. J. Pietrafesa. 1994. Eulerian views of layered water currents, vertical distribution of some larval fishes, and inferred advective transport over the continental shelf off North Carolina, USA, in winter. Fish. Oceanogr. 3(2):120–132.

Govoni, J. J., D. E. Hoss, and A. J. Chester. 1983. Comparative feeding of three species of larval fishes in the northern Gulf of Mexico: *Brevoortia patronus*, *Leiostomus xanthurus*, and *Micropogonias undulatus*. Mar. Ecol. Prog. Ser. 13:189–199.

Gozlan, R. E., G. H. Copp, and J. N. Tourenq. 1999. Comparison of growth plasticity in the laboratory and field, and implications for the onset of juvenile development in sofie, *Chondrostoma toxostoma*. Environ. Biol. Fishes 56:153–165.

Grabe, S. A. 1978. Food and feeding habits of juvenile Atlantic tomcod, *Microgadus tomcod*, from Haverstraw Bay, Hudson River. U.S. Fish. Bull. 76(1):89–94.

———. 1996. Feeding chronology and habits of *Alosa* spp. (Clupeidae) juveniles from the lower Hudson River estuary, New York. Environ. Biol. Fishes 47:321–326.

Graham, J. J. 1956. Observations on the alewife, *Pomolobus pseudoharengus* (Wilson), in fresh water. Univ. Toronto Stud., Biol. Ser., Ont. Fish. Res. Lab. Publ. No. 74:1–43.

Graham, J. J., B. J. Joule, and C. L. Crosby. 1984. Characteristics of the Atlantic herring (*Clupea harengus* L.) spawning population along the Maine coast, inferred from larval studies. J. Northw. Atl. Fish. Sci. 5(2):131–142.

Granados-Dieseldorff, P., and D. Baltz. 2008. Habitat use by nekton along a stream-order gradient in a Louisiana estuary. Estuaries and Coasts 31:572–583.

Grange, N., A. K. Whitfield, C. J. de Villiers, and B. R. Allanson. 2000. The response of two South African east coast estuaries to altered flow regimes. Aquat. Conserv.: Mar. Freshwat. Ecosyst. 10:155–177.

Gratton, C., and R. F. Denno. 2003. Seasonal shift from bottom-up to top-down impact in phytophagous insect populations. Oecologia 134:487–495.

Gray, C. A., R. C. Chick, and D. J. McElligott. 1998. Diel changes in assemblages of fishes associated with shallow seagrass and bare sand. Estuar. Coast. Shelf Sci. 46:849–859.

Gray, G. A., and H. E. Winn. 1961. Reproductive ecology and sound production of the toadfish, *Opsanus tau*. Ecology 42(2):274–282.

Gray, J. S, R. S. Wu, and Y. Or. 2002. Effects of hypoxia and organic enrichment on the coastal marine environment. Mar. Ecol. Prog. Ser. 238:249–279.

Grecay, P. A. 1990. Factors affecting spatial patterns of feeding success and condition of juvenile weakfish (*Cynoscion regalis*) in Delaware Bay: field and laboratory assessment. Ph.D. diss. University of Delaware, Newark. 179 pp.

Grecay, P. A., and T. E. Targett. 1996. Effects of turbidity, light level and prey concentration on feeding of juvenile weakfish *Cynoscion regalis*. Mar. Ecol. Prog. Ser. 131:11–16.

Greeley, J. R. 1938. Section II. Fishes and habitat conditions of the shore zone based upon July and August seining investigations. Pp. 72–91 *in* A biological survey of the salt waters of Long Island, Part II. N.Y. State Conserv. Dep. Suppl. 28th Annu. Rep.

Greene, C. H., A. J. Pershing, T. M. Cronin, and N. Ceci. 2008. Arctic climate change and its impacts on the ecology of the North Atlantic. Ecology 89(11):S24–S38.

Greene, C. H., B. A. Block, D. Welch, G. Jackson, G. L. Lawson, and E. L. Rechisky. 2009. Advances in conservation oceanography: new tagging and tracking technologies and their potential for transforming the science underlying fisheries management. Oceanography 22(1):210–223.

Greenfield, D. W. 1968. Observations on the behavior of the basketweave cusk-eel *Otophidium scrippsi* Hubbs. Cal. Fish Game 54:108–114.

Greenwood, M. F. D. 2007. Nekton community change along estuarine salinity gradients: can salinity zones be defined? Estuaries and Coasts 30(3):537–542.

Gregory, W. K. 1933. Fish Skulls: A Study of the Evolution of Natural Mechanisms. Krieger Publishing Company. Malabar, Fla. 481 pp.

Grimes, C. B., and J. A. Mountain. 1971. Effects of thermal effluent upon marine fishes near the Crystal River stream electric station. Florida DNR 17:1–64.

Grimes, C. B., K. W. Able, and R. S. Jones. 1986. Tilefish, *Lopholatilus chamaeleonticeps*, habitat, behavior and community structure in Mid-Atlantic and southern New England waters. Environ. Biol. Fishes 15(4):273–292.

Griswold, C. A., and T. W. McKenney. 1984. Larval development of the scup, *Stenotomus chrysops* (Pisces: Sparidae). U.S. Fish. Bull. 82(1):77–84.

Grosslein, M. D., and T. R. Azarovitz, eds. 1982. Fish distribution. (Mar. Ecosyst. Anal.) N.Y. Bight Atlas Monogr. 15.

Grothues, T. M., and K. W. Able. 2003a. Response of juvenile fish assemblages in tidal salt marsh creeks treated for *Phragmites* removal. Estuaries 26(2b):563–573.

———. 2003b. Discerning vegetation and environmental correlates with subtidal marsh fish assemblage dynamics during *Phragmites* eradication efforts: interannual trend measures. Estuaries 26(2b):574–586.

———. 2007. Scaling acoustic telemetry of bluefish in an estuarine observatory: detection and habitat use patterns. Trans. Am. Fish. Soc. 136:1511–1519.

Grothues, T. M., R. K. Cowen, L. J. Pietrafesa, F. Bignami, G. L. Weatherly, and C. N. Flagg. 2002. Flux of larval fish around Cape Hatteras. Limnol. Oceanogr. 47(1):165–175.

Grothues, T. M., K. W. Able, J. McDonnell, and M. M. Sisak. 2005. An estuarine observatory for real-time telemetry of migrant macrofauna: design, performance and constraints. Limnol. Oceanogr. Meth. 3:275–289.

Grothues, T. M., K. W. Able, J. Carter, and T. W. Arienti. 2009. Migration patterns of striped bass through nonnatal estuaries of the U.S. Atlantic Coast. Pp. 135–160 *in* A. J. Haro, K. L. Smith, R. A. Rulifson, C. M. Moffitt, R. J. Klauda, M. J. Dadswell, R. A. Cunjak, J. E. Cooper, K. L. Beal, and T. S. Avery, eds. 2009. Challenges for Diadromous Fishes in a Dynamic Global Environment. Am. Fish. Soc. Symp. 69, Bethesda, Md.

Grothues, T. M., J. Dobarro, A. Higgs, J. Ladd, G. Niezgoda, D. Miller. In press. Use of a multi-mensored AUV to telemeter tagged Atlantic sturgeon and map their spawning habitat in the Hudson River, USA. Proceedings of the 2008 Autonomous Underwater Vehicle (AUV) workshop. Woods Hole Oceanographic Institute, Woods Hole, Mass.

Grover, J. J. 1982. The comparative feeding ecology of five inshore marine fishes off Long Island, New York. Ph.D. diss. Rutgers University, New Brunswick, N.J. 197 pp.

———. 1998. Feeding habits of pelagic summer flounder, *Paralichthys dentatus*, larvae in oceanic and estuarine habitats. Fish. Bull. 96(2):248–257.

Grubbs, R. D., and J. A. Musick. 2007. Spatial delineation of summer nursery areas for juvenile sandbar sharks in Chesapeake Bay, Virginia. Pp. 63–86 *in* C. T. McCandless, N. E. Kohler, and H. L. Pratt Jr., eds. Shark Nursery Grounds of the Gulf of Mexico and the East Coast Waters of the United States. Am. Fish. Soc. Symp. 50, Bethesda, Md.

Grubbs, R. D., J. A. Musick, C. L. Conrath, and J. G. Romine. 2007. Long-term movements, migration, and temporal delineation of a summer nursery for juvenile sandbar sharks in the Chesapeake Bay region. Pp. 87–108 in C. T. McCandless, N. E. Kohler, and H. L. Pratt Jr., eds. Shark Nursery Grounds of the Gulf of Mexico and the East Coast Waters of the United States. Am. Fish. Soc. Symp. 50, Bethesda, Md.

Gruchy, C. G., and B. Parker. 1980. *Acipenser oxyrhynchus* Mitchill, Atlantic sturgeon. P. 41 in D. S. Lee et al., eds. Atlas of North American Freshwater Fishes. North Carolina State Mus. Nat. Hist.

Grusha, D. S. 2005. Investigation of the life history of the cownose ray, *Rhinoptera bonasus* (Mitchill 1815). M.S. thesis. College of William and Mary, Williamsburg, Va. 116 pp.

Gubbins, C. 2002. Use of home ranges by resident bottlenose dolphins (*Tursiops truncates*) in a South Carolina estuary. J. Mammalogy 83:178–187.

Gudger, E. W. 1910. Habits and life history of the toadfish (*Opsanus tau*). Bull. U.S. Bur. Fish. 28:1071–1109.

———. 1927. The nest and nesting habits of the butterfish or gunnel, *Pholis gunnellus*. Am. Mus. Nat. Hist. 27:65–72.

Gulland, J. A. 1965. Survival of the youngest stages of fish and its relation to year-class strength. Spec. Publ. Int. Comm. Northwest Atl. Fish. 6:365–371.

Günter, G. 1941. Relative numbers of shallow water fishes of the northern Gulf of Mexico, with some records of rare fishes from the Texas coast. Am. Midl. Nat. 26(1):194–200.

———. 1945. Studies on marine fishes of Texas. Publ. Inst. Mar. Sci. Univ. Tex. 1:1–190.

———. 1947. Differential rate of death for large and small fishes caused by hard cold waves. Science U.S. 106:472.

———. 1950. Distribution and abundance of fishes on the Aransas National Wildlife Refuge, with life history notes. Publ. Inst. Mar. Sci. Univ. Tex. 1(2):89–101.

———. 1956. Some relations of faunal distributions to salinity in estuarine waters. Ecology 37(3):616–619.

———. 1961. Some relations of estuarine organisms to salinity. Limnol. Oceanogr. 6:182–190.

———. 1967. Some relationships of estuaries to the fisheries of the Gulf of Mexico. Am. Assoc. Adv. Sci., Publ. 83:621–638.

Gutherz, E. J. 1967. Field guide to the flatfishes of the family Bothidae in the Western North Atlantic. U.S. Fish Wildl. Serv. Bur. Commer. Fish. Circ. 263.

Haas, H. L., C. J. Freeman, J. M. Logan, L. Deegan, and E. F. Gaines. 2009. Examining mummichog growth and movement: are some individuals making intra-season migrations to optimize growth? J. Exp. Mar. Biol. Ecol. 369:8–16.

Haberland M. S. 2002. Feeding habits of windowpane, *Scophthalmus aquosus*. M.S. thesis. Rutgers University, New Brunswick, N.J.

Hackney, C. T., W. D. Burbanck, and O. P. Hackney. 1976. Biological and physical dynamics of a Georgia tidal creek. Chesapeake Sci. 17(4):271–280.

Hacunda, J. S. 1981. Trophic relationships among demersal fishes in a coastal area of the Gulf of Maine. Fish. Bull. 79(4):775–788.

Haedrich, R. L. 1983. Estuarine fishes. Pp. 183–207 in B. H. Ketchum, ed. Ecosystems of the World. Vol. 26, Estuaries and Enclosed Seas. Elsevier, New York.

Haedrich, R. L., and S. O. Haedrich. 1974. A seasonal survey of the fishes in the Mystic River, a polluted estuary in downtown Boston, Massachusetts. Estuar. Coast. Mar. Sci. 2:59–73.

Haedrich, R. L., and C. A. S. Hall. 1976. Fishes and estuaries. Oceanus 19(5):55–63.

Haedrich, R. L., G. T. Rowe, and P. T. Polloni. 1980. The megabenthic fauna in the deep sea south of New England, USA. Mar. Biol. 57:165–179.

Haeseker, S. L., J. T. Carmichael, and J. E. Hightower. 1996. Summer distribution and condition of striped bass within Albemarle Sound, North Carolina. Trans. Am. Fish Soc. 125:690–704.

Hagan, S. M., and K. W. Able. 2003. Seasonal changes of the pelagic fish assemblage in a temperate estuary. Estuar. Coast. Shelf Sci. 56:15–29.

———. 2008. Diel variation in the pelagic fish assemblage in a temperate estuary. Estuaries and Coasts 31:33–42.

Hagan, S. M., S. A. Brown, and K. W. Able. 2007. Production of mummichog *Fundulus heteroclitus*: response in marshes treated for common reed *Phragmites australis* removal. Wetlands 27(1):54–67.

Hales, L. S., Jr., and K. W. Able. 1995. Effects of oxygen concentration on somatic and otolith growth rates of juvenile black sea bass, *Centropristis striata*. Pp. 135–153 in D. H. Secor et al., eds. Recent Developments in Fish Otolith Research. University of South Carolina Press, Columbia.

———. 2001. Winter mortality, growth, and behavior of young-of-the-year of four coastal fishes in New Jersey (USA) waters. Mar. Biol. 139:45–54.

Hall, D. J., and T. J. Ehlinger. 1989. Perturbation, planktivory, and pelagic community structure: the consequence of winterkill in a small lake. Can. J. Fish. Aquat. Sci. 46:2203–2209.

Hall, J. W., T. I. J. Smith, and S. D. Lamprecht. 1991. Movements and habitats of shortnose sturgeon, *Acipenser brevirostrum* in the Savannah River. Copeia 1991(3):695–702.

Hall, L. S., P. R. Krausman, and M. L. Morrison. 1997. The habitat concept and a plea for standard terminology. Wildl. Soc. Bull. 25(1):173–182.

Halliday, I. A., J. B. Robins, D. G. Mayer, J. Staunton-Smith, and M. J. Sellin. 2008. Effects of freshwater flow on the year-class strength of a non-diadromous estuarine finfish, king threadfin (*Polydactylus macrochir*), in a dry-tropical estuary. Mar. Freshwat. Res. 59:157–164.

Halpin, P. M. 2000. Habitat use by an intertidal salt-marsh fish: trade-offs between predation and growth. Mar. Ecol. Prog. Ser. 198:203–214.

Hamer, P. E. 1959. Age and growth studies of the bluefish (*Pomatomus Saltatrix* [Linnaeus]) of the New York Bight. M.S. thesis. Rutgers University, New Brunswick, N.J. 27 pp.

———. 1970. Studies of the scup, *Stenotomus chrysops*, in the Middle Atlantic Bight. N.J. Div. Fish Game Shellfish Misc. Rep. No. 5M.

———. 1972. Phase III. Use studies. Pp. 115–141 in Studies of the Mullica River–Great Bay Estuary. N.J. Dept. Environ. Protect. Div. Fish Game Shellfish Bur. Fish. Misc. Rep. No. 6M-1.

Hamer, P. E., and F. E. Lux. 1962. Marking experiments on fluke (*Paralichthys dentatus*) in 1961. Presented at the 28th Meeting of the North Atlantic Section, Atlantic States Marine Fisheries Commission, Atlanta, Ga.

Hamilton, P. V. 1976. Predation on *Littorina irrorata* (Mollusca: Gastropoda) by *Callinectes sapidus* (Crustacea: Portunidae). Bull. Mar. Sci. 26:403–409.

Haney, R. A., B. R. Silliman, A. J. Fry, C. A. Layman, and D. M. Rand. 2007. The Pleistocene history of the sheepshead minnow (*Cyprinodon variegatus*): non-equilibrium evolutionary dynamics within a diversifying species complex. Mol. Phylogen. Evol. 43:743–754.

Harding, J. M., and R. Mann. 1999. Fish species richness in relation to restored oyster reefs, Piankatank River, Virginia. Bull. Mar. Sci. 65:289–300.

———. 2000. Estimates of naked goby (*Gobiosoma bosc*), striped blenny (*Chasmodes bosquianus*) and eastern oyster (*Cassostrea virginica*) larval production around a restored Chesapeake Bay oyster reef. Bull. Mar. Sci. 66(1):29–45.

———. 2001. Diet and habitat use by bluefish, *Pomatomus saltatrix*, in a Chesapeake Bay estuary. Environ. Biol. Fish. 60:401–409.

———. 2003. Influence of habitat on diet and distribution of striped bass (*Morone saxatilis*) in a temperate estuary. Bull. Mar. Sci. 72(3):841–851.

Hardy, C. C., E. R. Baylor, and P. Moskowitz. 1976. Sea surface circulation in the northwest apex of the New York Bight—with appendix: bottom drift over the continental shelf. NOAA Tech. Memo. No. 13.

Hardy, J. D., Jr. 1978a. Development of Fishes of the Mid-Atlantic Bight: An Atlas of Egg, Larval and Juvenile Stages. Vol. 2, Anquillidae through Syngnathidae. Biological Services Program, U.S. Fish and Wildlife Service FWS/OBS-789/12.455. U.S. Government Printing Office, Washington, D.C.

———. 1978b. Development of Fishes of the Mid-Atlantic Bight: An Atlas of Egg, Larval and Juvenile Stages. Vol. 3, Aphredoderidae through Rachycentridae. Biological Services Program, U.S. Fish and Wildlife Service FWS/OBS-78/12. U.S. Government Printing Office, Washington, D.C.

Hardy, J. D., Jr., and L. L. Hudson. 1975. Descriptions of the eggs and juveniles of the Atlantic tomcod, *Microgadus tomcod*. Univ. MD Nat. Resour. Inst. Chesapeake Biol. Lab. Ref. 75–11.

Hardy, R. S., and M. K. Livak. 2004. Effects of temperature on the early development, growth, and survival of shortnose sturgeon, *Acipenser brevirostrum*, and Atlantic sturgeon, *Acipenser oxyrinchus*, yolk-sac larvae. Environ. Biol. Fishes 70:145–154.

Hare, J. A., and K. W. Able. 2007. Mechanistic links between climate and fisheries along the east coast of the United States: explaining population outbursts of Atlantic croaker (*Micropogonias undulatus*). Fish. Oceanogr. 16(1):31–45.

Hare, J. A., and R. K. Cowen. 1993. Ecological and evolutionary implications of the larval transport and reproductive strategy of bluefish (*Pomatomus saltatrix*). Mar. Ecol. Prog. Ser. 98:1–16.

———. 1994. Ontogeny and otolith microstructure of bluefish (*Pomatomus saltatrix*) (Pisces: Pomatomidae). Mar. Biol. 118:541–550.

———. 1996. Transport mechanisms of larval and pelagic juvenile bluefish (*Pomatomus saltatrix*) from South Atlantic Bight spawning grounds to Middle Atlantic Bight nursery habitats. Limnol. Oceanogr. 41:1264–1280.

Hare, J. A., J. A. Quinlan, F. E. Werner, B. O. Blanton, J. J. Govoni, R. B. Forward, L. R. Settle, and D. E. Hoss. 1999. Larval transport during winter in the SABRE study area: results of a coupled vertical larval behavior—three-dimensional circulation model. Fish. Oceanogr. 8(suppl. 2):57–76.

Hare, J. A., M. P. Fahay, and R. K. Cowen. 2001. Springtime ichthyoplankton of the slope region off the north-eastern United States of America: larval assemblages, relation to hydrography and implications for larval transport. Fish. Oceanogr. 10(2):164–192.

Hare, J. A., J. H. Churchill, R. K. Cowen, T. J. Berger, P. C. Cornillon, P. Dragos, S. M. Glenn, J. J. Govoni, and T. N. Lee. 2002. Routes and rates of larval fish transport from the southeast to the northeast United States continental shelf. Limnol. Oceanogr. 47:1774–1789.

Hare, J. A., S. Thorrold, H. Walsh, C. Reiss, A. Valle-Levinson, and C. Jones. 2005a. Bio-physical mechanisms of larval fish ingress into Chesapeake Bay. Mar. Ecol. Prog. Ser. 303:295–310.

Hare, J. A., H. J. Walsh, and M. J. Wuenschel. 2005b. Sinking rates of late-stage fish larvae: implications for larval ingress into estuarine nursery habitats. J. Exp. Mar. Biol. Ecol. 330(2):493–504.

Harmic, J. L. 1958. Some aspects of the development and ecology of the pelagic phase of the gray squeteague, *Cynoscion regalis* (Bloch and Schneider), in the Delaware estuary. Ph.D. diss. University of Delaware, Newark. 84 pp.

Harnden, C. W., R. E. Crabtree, and J. M. Shenker. 1999. Onshore transport of elopomorph leptocephali and glass eels (Pisces: Osteichthyes) in the Florida Keys. Gulf Mex. Sci. 17:17–26.

Haro, A. J., and W. H. Krueger. 1988. Pigmentation, size, and migration of elvers (*Anguilla rostrata* [Lesueur]) in a coastal Rhode Island stream. Can. J. Zool. 66:2528–2533.

Haro, A., W. Richkus, K. Whalen, A. Hoar, W. Dieter Busch, S. Lary, T. Brush, and D. Dixon. 2000. Population decline of the American eel: implications for research and management. Fisheries 25(9):7–16.

Haroski, D. M. 1998. Fish assemblages of inlet beaches in a southern New Jersey estuary. M.S. thesis. Rutgers University, New Brunswick, N.J. 97 pp.

Harrington, R. W., Jr. 1959a. Effects of four combinations of temperature and daylength on the ovogenetic cycle of a low latitude fish, *Fundulus confluentus* (Goode and Bean). Zoologica 44(4):149–168.

———. 1959b. Delayed hatching in stranded eggs of marsh killifish, *Fundulus confluentus*. Ecology 40(3):430–437.

Harrington, R. W., Jr., and J. S. Haeger. 1958. Prolonged natural deferment of hatching in killifish. Science 128(3337):1511.

Harrington, R. W., Jr., and E. S. Harrington. 1972. Food of female marsh killifish, *Fundulus confluentus* (Goode and Bean), in Florida. Am. Midl. Nat. 87(2):492–502.

Harris, J. E., R. S. McBride, and R. O. Williams. 2007. Life history of hickory shad in the St. Johns River, Florida. Trans. Am. Fish. Soc. 136:1463–1471.

Harris, S. A., D. P. Cyrus, and L. E. Beckley. 2001. Horizontal trends in larval fish diversity and abundance along an ocean-estuarine gradient on the northern KwaZulu–Natal coast, South Africa. Estuar. Coast. Shelf Sci. 53:221–235.

Harrison, I. J. 2002. Order Mugiliformes: Mugilidae: mullets. Pp. 1071–1085 *in* K. E. Carpenter, ed. The Living Marine Resources of the Western Central Atlantic. Vol. 2, Bony Fishes, Part 1 (Acipenseridae to Grammatidae). Food and Agriculture Organization of the United Nations, Rome, Italy.

Hartel, K. E., D. B. Halliwell, and A. E. Launer. 2002. Inland fishes of Massachusetts. Natural History of New England Series, Massachusetts Audubon Society Press, Lincoln, Mass.

Hartman, K. J., and S. B. Brandt. 1995a. Predatory demand and impact of striped bass, bluefish, and weakfish in the Chesapeake Bay: applications of bioenergetics models. Can. J. Fish. Aq. Sci. 52(8):1667–1687.

———. 1995b. Comparative energetics and the development of bioenergetics models for sympatric estuarine piscivores. Can. J. Fish. Aq. Sci. 52(8):1647–1666.

———. 1995c. Trophic resource partitioning, diets and growth of sympatric estuarine predators. Trans. Am. Fish. Soc. 124: 520–537.

Harvell, C. D., K. Kim, J. M. Burkholder, R. R. Colwell, P. R. Epstein, D. J. Grimes, E. E. Hofmann, E. K. Lipp, A. D. M. E. Osterhaus, R. M. Overstreet, J. W. Porter, G. W. Smith, and G. R. Vasta. 1999. Emerging marine diseases—climate links and anthropogenic factors. Science 285:1505–1510.

Harvey, C. J. 1998. Use of sandy beach habitat by *Fundulus majalis*, a surf-zone fish. Mar. Ecol. Prog. Ser. 164:307–310.

Hastings, R. W., and R. E. Good. 1977. Population analysis of the fishes of a freshwater tidal tributary of the lower Delaware River. Bull. N.J. Acad. Sci. 22(2):13–20.

Hastings, R. W., J. C. O'Herron II, K. Schick, and M. A. Lazzari. 1987. Occurrence and distribution of shortnose sturgeon, *Acipenser brevirostrum*, in the upper tidal Delaware River. Estuaries 10(4):337–341.

Hauser, W. J. 1975. Occurrence of two leptocephali in an estuary. U.S. Fish. Bull. 77:444–445.

Havey, K. A. 1973. Production of juvenile alewives at Love Lake, Washington County, Maine. Trans. Am. Fish. Soc. 102(2): 434–347.

Hazel, J. E. 1970. Atlantic continental shelf and slope of the United States—ostracode zoogeography in the southern Nova Scotian and northern Virginian Faunal Provinces. Geological Survey Professional Paper 529-E, U.S. Government Printing Office, Washington, D.C.

Heck, K. L., Jr., and J. F. Valentine. 2007. The primacy of top-down effects in shallow benthic ecosystems. Estuaries and Coasts 30(3):371–381.

Heck, K. L., Jr., and M. P. Weinstein. 1978. Mimetic relationships between tropical burrfishes and opisthobranchs. Biotropika 10: 78–79.

Heck, K. L., Jr., K. W. Able, M. P. Fahay, and C. T. Roman. 1989. Fishes and decapod crustaceans of Cape Cod eelgrass meadows: species composition and seasonal abundance patterns. Estuaries 12:59–65.

Heck, K. L., Jr., K. W. Able, C. T. Roman, and M. P. Fahay. 1995. Composition, abundance, biomass, and production of macrofauna in a New England estuary: comparisons among eelgrass meadows and other nursery habitats. Estuaries 18(2):379–389.

Heck, K. L., Jr., G. Hays, and R. J. Orth. 2003. Critical evaluation of the nursery role hypothesis for seagrass meadows. Mar. Ecol. Prog. Ser. 253:123–136.

Heck, K. L., Jr., T. J. B. Carruthers, C. M. Duarte, A. R. Hughes, G. Kendrick, R. J. Orth, and S. W. Williams. 2008. Trophic transfers from seagrass meadows subsidize diverse marine and terrestrial consumers. Ecosystems 11:1198–1210.

Hedenström, A., Z. Barta, B. Helm, A. I. Jouston, J. M. McNamara, and N. Jonzén. 2007. Migration speed and scheduling of annual events by migrating birds in relation to climate changes. Clim. Res. 35:79–91.

Hedgepeth, J. W. 1957. Marine biogeography. Pp. 359–382 *in* Treatise on Marine Ecology and Paleoecology, 1. Mem. Geol. Soc. Am. 67.

———. 1967. Treatise on Marine Ecology and Paleoecology. National Research Council, National Academy of Science. Washington, D.C.

Hedgepeth, M. Y. 1983. Age, growth and reproduction of American eels, *Anguilla rostrata*, (Lesueur), from the Chesapeake Bay area. M.S. thesis. College of William and Mary, Williamsburg, Va. 61 pp.

Heemstra, P. C., W. D. Anderson Jr., and P. S. Lobel. 2002. Serranidae: Groupers (seabasses, creolefish, coney, hinds, hamlets, anthiines, and soapfishes). Pp. 1308–1369 *in* K. E. Carpenter, ed. The Living Marine Resources of the Western Central Atlantic.

Vol. 2, Bony Fishes, Part 1 (Acipenseridae to Grammatidae). Food and Agriculture Organization of the United Nations, Rome. 2:601–1374.

Heins, D. C., and J. A. Baker. 2008. The stickleback-*Schistocephalus* host-parasite system as a model for understanding the effect of a macroparasite on host reproduction. Behaviour 145:625–645.

Heintzelman, D. S., ed. 1971. Rare or Endangered Fish and Wildlife of New Jersey. N.J. State Mus. Sci. Notes 4.

Heist, E. J., J. E. Graves, and J. A. Musick. 1995. Population genetics of the sandbar shark (*Carcharhinus plumbeus*) in the Gulf of Mexico and Mid-Atlantic Bight. Am. Soc. Ich. Herp. 3:555–562.

Heithaus, M. R., A. Frid, A. J. Wirsing, and B. Worm. 2007. Predicting ecological consequences of marine top predator declines. Trends Ecol. Evol. 23(4):202–210.

Helfman, G. S. 1978. Patterns of community structure in fishes: summary and overview. Environ. Biol. Fishes 3(1):129–148.

Helfman, G. S., B. B. Collette, and D. E. Facey. 1997. The Diversity of Fishes. Blackwell Science, Inc., Malden, Mass.

Hempel, G. 1965. On the importance of larval survival for the population dynamics of marine fish food. Calif. Coop. Ocean. Fish. Invest. Rep. 10:13–23.

Henderson, P. A., R. H. A. Holmes, and R. N. Bamber. 1988. Size-selective wintering mortality in the sand smelt, *Atherina boyeri*, and its role in population regulation. J. Fish Biol. 33:221–233.

Herke, W. H. 1977. Life History Concepts of Motile Estuarine-dependent Species Should Be Re-evaluated. Self-published, Baton Rouge, La. Library of Congress No. 77-90015.

Herman, S. S. 1963. Planktonic fish eggs and larvae of Narragansett Bay. Limnol. Oceanogr. 8:103–109.

Herzka, S. Z. 2005. Assessing connectivity of estuarine fishes based on stable isotope ratio analysis. Estuar. Coast. Shelf Sci. 64: 58–69.

Hettler, W. F., Jr. 1989a. Nekton use of regularly-flooded saltmarsh cordgrass habitat in North Carolina, USA. Mar. Ecol. Prog. Ser. 56:111–118.

———. 1989b. Food habits of juveniles of spotted seatrout and gray snapper in western Florida bay. Bull. Mar. Sci. 44(1): 155–162.

Hettler, W. F., Jr., and D. L. Barker. 1993. Distribution and abundance of larval fishes at two North Carolina inlets. Estuar. Coast. Shelf Sci. 37:161–179.

Hettler, W. F., Jr., D. S. Peters, D. R. Colby, and E. H. Laban. 1997. Daily variability in abundance of larval fishes inside Beaufort Inlet. Fish. Bull. 95:477–493.

Heupel, M. R, J. K. Carlson, and C. A. Simpfendorfer. 2007. Shark nursery areas: concepts, definition, characterization and assumptions. Mar. Ecol. Prog. Ser. 337:287–297.

Hickey, C. R., Jr., A. D. Sosnow, and J. W. Lester. 1975. Pound net catches of warm-water fishes at Montauk, New York. N.Y. Fish Game J. 22(1):38–50.

Hicks, D. C., and J. R. Miller. 1980. Meteorological forcing and bottom water movement off the northern New Jersey coast. Estuar. Coast. Mar. Sci. 2:563–571.

Higgins, E. 1928. Progress report in biological inquiries, 1926. U.S. Comm. Fish. Rep. 1927 8:515–559.

Higgs, D. M., and L. A. Fuiman. 1996. Ontogeny of visual and mechnanosensory structure and function in Atlantic Menhaden, *Brevoortia tyrannus*. J. Exp. Biol. 199:2619–2629.

Higham, J. R., and W. R. Nicholson. 1964. Sexual maturation and spawning of Atlantic menhaden. Fish. Bull. 63:255–271.

Hildebrand, S. F. 1916. The United States fisheries biological station at Beaufort, North Carolina, during 1914 and 1915. Science (n.s.) 43:303–307.

———. 1941. An annotated list of salt and brackish water fishes, with a new menhaden, found in North Carolina since the publication of "The Fishes of North Carolina" by Hugh M. Smith in 1907. Copeia 1:220–232.

———. 1943. A review of the American anchovies (Family Engraulidae). Bull. Bingham Oceanogr. Collect. Yale Univ. 8(2):1–165.

———. 1963. Part 3. Family Clupeidae. Pp. 257–454 *in* Y. H. Olsen, ed. Fishes of the Western North Atlantic. Sears Foundation for Marine Research, Yale University, New Haven, Conn.

Hildebrand, S. F., and L. E. Cable. 1930. Development and life history of fourteen teleostean fishes at Beaufort, N.C. U.S. Bur. Fish. Bull. 46(1093):383–488.

———. 1934. Reproduction and development of whitings or kingfishes, drums, spot, croaker, and weakfishes or sea trouts (Family Sciaenidae), of the Atlantic Coast of the United States. U.S. Bur. Fish. Bull. 48(16):41–117.

———. 1938. Further notes on the development and life history of some teleosts at Beaufort, N.C. U.S. Bur. Fish. Bull. 48:505–642.

Hildebrand, S. F., and W. C. Schroeder. 1928. Fishes of Chesapeake Bay. Bull. U.S. Bur. Fish. 43(1):1–366.

Himchak, P. J. 1981. Monitoring of the striped bass population in New Jersey: spawning and recruitment of the striped bass, *Morone saxatilis*, in the Delaware River. N.J. Dept. Environ. Protect. Div. Fish Game Wildl. Bur. Mar. Fish. Final Rep. April 1, 1980 to March 31, 1981.

———. 1982a. Monitoring of the spring ichthyoplankton in the Delaware River. N.J. Dep. Environ. Protect. Fiv. Fish Game Wildl. Bur. Mar. Fish. Nacote Creek Res. Stn. Misc. Rep. No. 55M-1-13.

———. 1982b. Distribution and abundance of larval and young finfishes in the Maurice River and in waterways near Atlantic City, New Jersey. M.S. thesis. Rutgers University, New Brunswick, N.J. 78 pp.

———. 1983. Spring ichthyoplankton in the Raritan/South and the Great Egg Harbor Rivers, New Jersey. N.J. Dep. Environ. Protect. Div. Fish Game Wildl. Bur. Mar. Fish. Tech. Serv. 83-2:1–17.

Himchak, P. J., and R. Allen. 1985. Spring ichthyoplankton in the Navesink/Swimming River, New Jersey. N.J. Dep. Environ. Protect. Div. Fish Game Wildl. Bur. Mar. Fish. Tech. Ser. 85-1:1–11.

Hines, A. H., K. E. Osgood, and J. J. Miklas. 1985. Semilunar reproductive cycles in *Fundulus heteroclitus* (Pisces: Cyprinodontidae) in an area without lunar tidal cycles. U.S. Fish. Bull. 83(3): 467–472.

Hines, A. H., A. M. Haddon, and L. A. Wiechert. 1990. Guild structure and foraging impact of blue crabs and epibenthic fish in a subestuary of Chesapeake Bay. Mar. Ecol. Prog. Ser. 67:105–126.

Hjort, J. 1914. Fluctuations in the great fisheries of northern Europe. Cons. Perm. Int. Explor. Mer., Rapp. P.-V. Vol. 20.

———. 1926. Fluctuations in the year classes of important food fishes. J. Cons. Int. Explor. Mer. 1:5–38.

Hobbs, R. J., and J. A. Harris. 2001. Restoration ecology: repairing the earth's ecosystems in the new millennium. Rest. Ecol. 9: 239–246.

Hodson, R. G., J. O. Hackman, and C. R. Bennett. 1981. Food habits of young spots in nursery areas of the Cape Fear River estuary, North Carolina. Trans. Am. Fish. Soc. 110:495–501.

Hoese, H. D. 1965. Spawning of marine fishes in the Port Aransas, Texas area as determined by the distribution of young and larvae. Ph.D. diss. University of Texas, Austin. 144 pp.

Hoese, H. D., and R. H. Moore. 1977. Fishes of the Gulf of Mexico—Texas, Louisiana, and Adjacent Waters. Texas A&M University Press, College Station.

Hoese, H. D., C. E. Richards, and M. Castagna. 1961. Appearance of the gag, *Mycteroperca microlepis*, in coastal waters of Virginia. Chesapeake Sci. 2(1–2):104–105.

Hoff, H. K. 1976. The life history of the striped bass, *Morone saxatilis* (Walbaum), in the Great Bay–Mullica River estuary and the vicinity of Little Egg Inlet. Pp. 43–53 *in* D. L. Thomas et al., eds. Ecological studies in the bays and waterways near Little Egg Inlet and in the ocean in the vicinity of the proposed site for the Atlantic Generating Station, New Jersey. Progress Report for the period January–December 1975. Ichthyological Associates, Ithaca, N.Y.

Hoff, J. G. 1965. Two shortnose sturgeon, *Acipenser brevirostrium*, from the Delaware River, Scudder's Falls, New Jersey. Bull. N.J. Acad. Sci. 10:23.

———. 1971. Mass mortality of the crevalle jack, *Caranx hippos* (Linnaeus) on the Atlantic coast of Massachusetts. Chesapeake Sci. 12(1):49.

———. 1980. Review of the present status of the stocks of the Atlantic sturgeon *Acipenser oxyrhynchus,* Mitchill. Prepared for the National Marine Fisheries Service, Northeast Region, Gloucester, Mass.

Hoffman, E. E., and T. M. Powell. 1998. Environmental variability effects on marine fisheries: four case histories. Ecol. Appl. 8(1): Supplement 1998, pp. S23–S32.

Hoffman, J. C., D. A. Bronk, and J. E. Olney. 2007a. Tracking nursery habitat use in the York River estuary, Virginia, by young American shad using stable isotopes. Trans. Am. Fish. Soc. 136: 1285–1297.

———. 2007b. Contribution of allochthonous carbon to American shad production in the Mattaponi River, Virginia, using stable isotopes. Estuaries and Coasts 30:1034–1048.

Hoffman, J. C., K. E. Limburg, D. A. Bronk, and J. E. Olney. 2008. Overwintering habitats of migratory juvenile American shad in Chesapeake Bay. Environ. Biol. Fishes 81:329–345.

Holbrook, S. J., and R. J. Schmitt. 1989. Resource overlap, prey dynamics and the strength of competition. Ecology 70:1943–1953.

Holliday, S. E., and C. J. Klein III. 1993. Surficial sediments in Mid-Atlantic Estuaries. NOAA National Estuarine Inventory. 170 pp.

Hollis, E. H. 1967. An investigation of striped bass in Maryland. Final Rep. Fed. Aid. Fish Rest. Proj. F-3-R.

Hollister, J. W., P. V. August, and J. F. Paul. 2008. Effects of spatial extent on landscape structures and sediment metal concentration relationships in small estuarine systems of the United States' Mid-Atlantic Coast. Landscape Ecol. 23:91–106.

Hollowed, A. B., S. R. Hare, and W. S. Wooster. 2001. Pacific Basin climate variability and patterns of Northeast Pacific marine fish production. Prog. Oceanogr. 49:257–282.

Holmes, N. T. H. 2006. The importance of long-term data sets in science and river management Aquat. Conserv.: Mar. Freshwat. Ecosyst. 16:329–333.

Holmquist, R. K. 1997. Maximum size revision and Chesapeake Bay distribution for striped burrfish, *Chilomycterus schoepfi*. VA J. Sci. 48(3):225–230.

Holt, G. J., R. Goodbout, and C. R. Arnold. 1981. Effects of temperature and salinity on egg hatching and larval survival of red drum, *Sciaenops ocellatus*. Fish. Bull. 79(3):569–573.

Holt, R. D., and T. H. Keitt. 2005. Species borders: a unifying theme in ecology. Oikos 108:3–6.

Hood, P. B., K. W. Able, and C. B. Grimes. 1988. Biology of the conger eel *Conger oceanicus* in the Mid-Atlantic Bight. I. Distribution, age, growth, and reproduction. Mar. Biol. 98:587–596.

Horn, M. H. 1970a. Systematics and biology of the stromateoid fishes of the genus *Peprilus*. Bull. Mus. Comp. Zool. Harv. Univ. 149(5):165–261.

———. 1970b. The swimbladder as a juvenile organ in stromateoid fishes. Breviora-Museum of Comparative Zoology 359:1–9.

Horn, M. H., and L. G. Allen. 1976. Numbers of species and faunal resemblance of marine fishes in California bays and estuaries. Bull. So. Calif. Acad. Sci. 75(2):159–170.

———. 1985. Fish community ecology in southern California bays and estuaries. Pp. 169–189 in A. Yáñez-Arancibia, ed. Fish Community Ecology in Estuaries and Coastal Lagoons: Towards an Ecosystem Integration. UNAM Press, Distrito Federal Mexico.

Hoss, D. E., and G. W. Thayer. 1993. The importance of habitat to the early life history of estuarine dependent fishes. Am. Fish. Soc. Symp. 14:147–158.

Hostetter, E. B., and T. A. Munroe. 1993. Age, growth and reproduction of tautog *Tautoga onitis* (Labridae: Perciformes) from coastal waters of Virginia. U.S. Fish. Bull. 91:45–64.

Houde, E. D. 1972. Development and early life history of the northern sennet, *Sphyraena borealis* Dekay (Pisces: Sphyraenidae) reared in the laboratory. U.S. Fish. Bull. 70:185–196.

———. 1987. Fish early life dynamics and recruitment variability. Amer. Fish. Soc. Symp. 2:17–29.

———. 2008. Emerging from Hjort's Shadow. J. Northw. Atl. Fish. Sci. 41:53–70.

Houde, E. D., and C. E. Zastrow. 1991. Bay anchovy, *Anchoa mitchilli*. Pp. 8.1–8.14 in S. L. Funderburk, S. L. Mihursky, J. A. Jordan, and S. D. Riley, eds. Habitat Requirements for Chesapeake Bay Living Resources. 2nd ed. Living Resources Subcommittee, Chesapeake Bay Program, Annapolis, Md.

Houde, E. D., S. A. Berkeley, J. J. Klinovsky, and R. C. Schekter. 1976. Culture of larvae of the white mullet (*Mugil curema*). Aquaculture 8:365–370.

Houde, E. D., J. C. Gamble, S. E. Dorsey, and J. H. Cowan Jr. 1994. Drifting mesocosms: the influence of gelatinous zooplankton on mortality of bay anchovy, *Anchoa mitchilli*, eggs and yolk-sac larvae. Int. Counc. Explor. Sea J. Mar. Sci. 51:383–394.

Houghton, R. W., F. Aikman III, and H. W. Ou. 1988. Shelf-slope frontal structure and cross-shelf exchange at the New England shelf-break. Cont. Shelf Res. 8(5–7):687–710.

Howe, A. B. 1971. Biological investigation of the Atlantic tomcod, *Microgadus tomcod* (Walbaum), in the Weweantic River estuary, Massachusetts, 1967. M.S. thesis. University of Massachusetts, Amherst. 82 pp.

Howe, A. B., P. G. Coates, and D. E. Pierce. 1976. Winter flounder estuarine year-class abundance, mortality and recruitment. Trans. Am. Fish. Soc. 105(6):647–657.

Hsieh, H. J., Y. L. Hsien, W. S. Tsai, C. A. Chen, W. C. Su, and M. S. Jeng. 2008. Tropical fishes killed by the cold. Coral Reefs 27:599.

Hubbs, C. L. 1931. Studies of the fishes of the order Cyprinodontes. X. Four nominal species of *Fundulus* placed in synonym. Occ. Papers. Mus. Zool. University of Michigan No. 3231:1–8.

———. 1936. Fishes of the Yucatan Peninsula. Carn. Inst. Wash., Publ. 457:157–287.

———. 1963. *Chaetodon aya* and related deep-living butterflyfishes: their variation, distribution and synonymy. Bull. Mar. Sci. 13(1):133–192.

Hubbs, C. L., and R. R. Miller. 1965. Studies on cyprinodont fishes. XXII. Variation in *Lucania parva*, its establishment in western United States, and description of a new species from an interior basin in Coabuila, Mexico. Misc. Publ. Mus. Zool. Univ. Mich. 127:1–104.

Huber, M. E. 1978. Adult spawning success and emigration of juvenile alewives (*Alosa pseudoharengus*) from the Parker River, Massachusetts. M.S. thesis. University of Massachusetts, Amherst. 75 pp.

Hughes, L. 2000. Biological consequences of global warming: is the signal already apparent? Tree 15(2):56–61.

Humphries, C. J., and L. R. Parenti. 1999. Cladistic Biogeography. 2nd ed.: Interpreting Patterns of Plant and Animal Distributions. Oxford University Press, Oxford, England.

Hunter, K. L., D. A. Fox, L. M. Brown, and K. W. Able. 2006. Responses of resident marsh fishes to stages of *Phragmites australis* invasion in three mid-Atlantic estuaries, U.S.A. Estuaries and Coasts 29(3):487–498.

Hunter, K. L., M. Fox, and K. W. Able. 2007. Habitat influences on reproductive allocation and growth of the mummichog (*Fundulus heteroclitus*) in a coastal salt marsh. Mar. Biol. 151:617–627.

Hunter, K. L., M. G. Fox, and K. W. Able. 2009. Influence of flood frequency, temperature and population density on migration of *Fundulus heteroclitus* in semi-isolated marsh pond habitats. Mar. Ecol. Prog. Ser. 391:85–96.

Hunter-Thomson, K. 2002. Estuarine-open-water comparison of fish community structure in eelgrass (*Zostera marina* L.) habitats of Cape Cod. Biol. Bull. 203:247–248.

Huntsman, A. G. 1922. The fishes of the Bay of Fundy. Contrib. Can. Biol. 1921(3):49–72.

Hunt von Herbing, I., and W. Hunte. 1991. Spawning and recruitment of the bluehead wrasse *Thalassoma bifasciatum* in Barbados. Mar. Ecol. Prog. Ser. 72:49–58.

Hurrell, J. W., Y. Kushnir, G. Ottersen, and M. Visbeck. 2003. An overview of the North Atlantic oscillation. Pp. 1–35 in J. W. Hurrell, Y. Kushnir, G. Ottersen, and M. Visbeck, eds. The North Atlantic Oscillation: Climatic Significance and Environmental Impact. American Geophysical Union, Geophysical Monograph 134, Washington, D.C.

Hurst, T. P. 2007. Causes and consequences of winter mortality in fishes. J. Fish Biol. 71:315–345.

Hurst, T. P., and D. O. Conover. 1998. Winter mortality of young-of-the-year Hudson River striped bass (*Morone saxatilis*): size-dependent patterns and effects on recruitment. Can. J. Fish. Aquat. Sci. 55:1122–1130.

———. 2001. Activity-related constraints on overwintering young-of-the-year striped bass (*Morone saxatilis*). Can. J. Zool. 79(1):129–136.

Hurst, T. P., E. T. Schultz, and D. O. Conover. 2000. Seasonal energy dynamics of young-of-the-year Hudson River striped bass. Trans. Am. Fish. Soc. 129:145–157.

Hutchins, J. B. 1991. Dispersal of tropical fishes to temperate seas in the Southern Hemisphere. J. R. Soc. West. Aust. 74:79–84.

Hutchins, J. B., and A. F. Pearce. 1994. Influence of the Leeuwin current on recruitment of tropical reef fishes at Rottnest Island, western Australia. Bull. Mar. Sci. 54:245–255.

Iafrate, J., and K. Oliveira. 2008. Factors affecting migration patterns of juvenile river herring in a coastal Massachusetts stream. Environ. Biol. Fishes 81:101–110.

Ibanez Aguirre, A. L., and M. Gallardo-Cabello. 2004. Reproduction of *Mugil cephalus* and *M. curema* (Pisces: Mugilidae) from a coastal lagoon in the Gulf of Mexico. Bull. Mar. Sci. 75(1): 37–49.

Ihde, T. F. 2000. Biology of the spotted seatrout, *Cynoscion nebulosus*, in the Chesapeake Bay region. M.S. thesis. College of William and Mary, Williamsburg, Va. 121 pp.

Iles, T. D., and M. Sinclair. 1982. Atlantic herring: stock discreteness and abundance. Science 215:627–633.

Ingham, M., ed. 1982. Summary of the physical oceanographic processes and features pertinent to pollution distribution in the coastal and offshore waters of the northeastern United States, Virginia to Maine. NOAA Tech. Mem. NMFS NE 17:1–165.

Inman, D. L., and C. E. Nordstrom. 1971. On the tectonic and morphologic classification of coasts. J. Geol. 79(1):1–21.

Islam, Md. S., M. Hibino, and M. Tanaka. 2006. Distribution and diets of larval and juvenile fishes: influence of salinity gradient and turbidity maximum in a temperate estuary in upper Ariake Bay, Japan. Est. Coast. Shelf Sci. 68:62–74.

Jackson, C. F. 1953. Northward occurrence of southern fishes (*Fundulus, Mugil, Pomatomus*) in coastal waters of New Hampshire. Copeia 3:192.

Jackson, J. B. C., M. X. Kirby, W. H. Berger, K. A. Bjorndal, L. W. Botsford, B. J. Bourque, R. H. Bardbury, R. Cooke, J. Erlandson, J. A. Estes, T. P. Hughes, S. Kidwell, C. B Lange, H. S. Lenihan, J. M. Pandolfi, C. H. Peterson, R. S. Steneck, M. J. Tegner, and R. R Warner. 2001. Historical overfishing and the recent collapse of coastal ecosystems. Science 293:629–637.

Jackson, L. F., and C. V. Sullivan. 1995. Reproduction of the white perch: the annual gametogenic cycle. Trans. Am. Fish. Soc. 124: 563–577.

Jacot, A. P. 1920. Age, growth and scale characters of the mullets, *Mugil cephalus* and *Mugil curema*. Trans. Am. Microscop. Soc. 34(3):199–229.

James, N. C., P. D. Cowley, A. K. Whitfield, and S. J. Lamberth. 2007. Fish communities in temporarily open/closed estuaries from the warm- and cool-temperate regions of South Africa: a review. Rev. Fish. Biol. Fisheries 17:565–580.

James-Pirri, M. J., K. B. Raposa, and J. G. Catena. 2001. Diet composition of mummichogs, *Fundulus heteroclitus*, from restoring and unrestricted regions of a New England (U.S.A.) salt marsh. Estuar. Coast. Shelf. Sci. 53:205–213.

Jeffries, H. P. 1960. Winter occurrences of *Anguilla rostrata* elvers in New England and Middle Atlantic estuaries. Limnol. Oceanogr. 5(3):338–340.

———. 1972. Fatty acid ecology of a tidal marsh. Limnol. Oceanogr. 17:433–440.

———. 1975. Diets of juvenile Atlantic menhaden (*Brevoortia tyrannus*) in three estuarine habitats as determined from fatty acid composition of gut contents. J. Fish. Res. Bd. Can. 32:587–592.

Jeffries, H. P., and W. C. Johnson. 1974. Seasonal distributions of bottom fishes in the Narragansett Bay area: seven-year variations in the abundance of winter flounder (*Pseudopleuronectes americanus*). J. Fish. Res. Board Can. 31:1057–1066.

Jellyman, D. J., and P. W. Lambert. 2003. Factors affecting recruitment of glass eels into the Grey River, New Zealand. J. Fish Biol. 63:1–13.

Jenkins, G. P., K. P. Black, M. J. Wheatley, and D. N. Hatton. 1997. Temporal and spatial variability in recruitment of a temperate, seagrass-associated fish is largely determined by physical processes in the pre- and post-settlement phases. Mar. Ecol. Prog. Ser. 148:23–35.

Jesien, R. V., C. H. Hocutt, and S. K. Gaichas. 1992. Tagging studies

and stock characterization of summer flounder (*Paralichthys dentatus*) in Maryland's coastal waters near Ocean City, MD. Progress report submitted to Tidewater Administration, Maryland Department of Natural Resources, Cambridge, Md. 11+ pp.

Jessop, B. M., J. C. Shiao, Y. Iizuka, and W. N. Tzeng. 2004. Variation in the annual growth, by sex and migration history, of silver American eels *Anguilla rostrata*. Mar. Ecol. Prog. Ser. 272: 231–244.

Jessop, B. M., D. K. Cairns, I. Thibault, and W. N. Tzeng. 2008. Life history of American eel *Anguilla rostrata*: new insights from otolith microchemistry. Aquat. Biol. 1:205–216.

Johns, D. M., W. H. Howell, and G. Klein-MacPhee. 1981. Yolk utilization and growth to yolk-sac absorption in summer flounder (*Paralichthys dentatus*) larvae at constant and cyclic temperatures. Mar. Biol. 63:301–308.

Johnson, D. D., D. Rotherham, and C. A. Gray. 2008. Sampling estuarine fish and invertebrates using demersal otter trawls: effects of net height, tow duration and diel period. Fish. Res. 93:315–323.

Johnson, G. D. 1978. Development of Fishes of the Mid-Atlantic Bight: An Atlas of Egg, Larval and Juvenile Stages. Vol. 4, Carangidae through Ephippidae. Chesapeake Biological Laboratory, Center for Environmental and Estuarine Studies, University of Maryland. Prepared for U.S. Fish and Wildlife Service. U.S. Government Printing Office, Washington, D.C.

Johnson, M. S. 1974. Comparative geographic variation in *Menidia*. Evolution 28:607–618.

Johnson, T. B., and D. O. Evans. 1996. Temperature constraints on overwinter survival of age-0 white perch. Trans. Am. Fish. Soc. 125:466–471.

Johnson, W. S., D. M. Allen, M. V. Ogburn, and S. E. Stancyk. 1990. Short-term predation responses of adult bay anchovies *Anchoa mitchilli* to estuarine zooplankton availability. Mar. Ecol. Prog. Ser. 64:55–68.

Johnston, R., M. Sheaves, and B. Molony. 2007. Are distributions of fishes in tropical estuaries influenced by turbidity over small spatial scales? J. Fish Biol. 71:657–671.

Jones, C. M. 2006. Estuarine and diadromous fish metapopulations. Pp. 119–156 *in* J. P. Kritzer and P. F. Sale., eds. Marine Metapopulations. Elsevier Academic Press Publications, San Diego, Calif.

Jones, C. M., and B. K. Wells. 2001. Yield-per-recruit analysis for black drum, *Pogonias cromis*, along the east coast of the United States and management strategies for Chesapeake Bay. Fish. Bull. 99:328–337.

Jones, G. G., and M. A. Tabery. 1980. Larval development of the banded killifish (*Fundulus diaphanus*) with notes on the distribution in the Hudson River estuary. Larval Fish Conference, 4th, University of Mississippi, 1980. Proceedings of the Fourth Annual Larval Fish Conference. U.S. Fish and Wildlife Service, Biological Services Program, National Power Plant Team, Ann Arbor, Mich., FWS/OBS-80/43. 179 pp.

Jones, G. P. 1987a. Competitive interactions among adults and juveniles in a coral reef fish. Ecology 68:1534–1547.

———. 1987b. Some interactions between residents and recruits in two coral reef fishes. J. Exp. Mar. Biol. Ecol. 114:169–182.

———. 1990. The importance of recruitment to the dynamics of a coral reef fish population. Ecology 71:1691–1698.

———. 1991. Postrecruitment processes in the ecology of coral reef fish populations: a multifactorial perspective. Pp. 294–328 *in* P. F. Sale, ed. The Ecology of Fishes on Coral Reefs. Academic Press, San Diego, Calif.

Jones, K. M., and K. W. Able. In review. Effects of restoration in marshes dominated by the common reed *Phragmites australis* in Delaware Bay, New Jersey: distribution, abundance and diets of large piscivorous fishes. Estuaries and Coasts.

Jones, P. W., F. D. Martin, and J. D. Hardy Jr. 1978. Development of Fishes of the Mid-Atlantic Bight: An Atlas of Egg, Larval and Juvenile Stages. Vols. 1–6. Chesapeake Biological Laboratory, Center for Environmental and Estuarine Studies, University of Maryland. Prepared for U.S. Fish and Wildlife Service. U.S. Government Printing Office, Washington, D.C.

Joseph, E. B. 1972. Status of sciaenid stocks of the middle Atlantic coast. Chesapeake Sci. 13(2):87–100.

Joseph, E. B., and J. Davis. 1965. A preliminary assessment of the river herring stocks of lower Chesapeake Bay. Va. Inst. Mar. Sci. Spec. Sci. Rep. No. 51.

Joseph, E. B., W. H. Massmann, and J. J. Norcross. 1964. The pelagic eggs and larval stages of the black drum from Chesapeake Bay. Copeia:425–434.

Joyce, T. M. 2002. One hundred plus years of wintertime climate variability in the eastern United States. J. Clim. 15:1076–1086.

Joyeux, J.-C. 1998. Spatial and temporal entry patterns of fish larvae into North Carolina estuaries: comparisons among one pelagic and two demersal species. Estuar. Coast. Shelf Sci. 47: 731–752.

Juanes, F. 2007. Role of habitat in mediating mortality during the post-settlement transition phase of temperate marine fishes. J. Fish Biol. 70:661–677.

Juanes, F., and D. O. Conover. 1994a. Piscivory and prey size selection in young-of-the-year bluefish: predator preference or size-dependent capture success? Mar. Ecol. Prog. Ser. 114:59–69.

———. 1994b. Rapid growth, high feeding rates, and early piscivory in the young-of-the-year bluefish (*Pomatomus saltatrix*). Can. J. Fish. Aquat. Sci. 51(8):1752–1761.

———. 1995. Size-structured piscivory: advection and the linkage between predator and prey recruitment in young-of-the-year bluefish. Mar. Ecol. Prog. Ser. 128:287–304.

Juanes, F., R. E. Marks, K. A. McKown, and D. O. Conover. 1993. Predation by age-0 bluefish on age-0 anadromous fishes in the Hudson River estuary. Trans. Am. Fish. Soc. 122:348–356.

Juanes, F., J. A. Buckel, and D. O. Conover. 1994. Accelerating the onset of piscivory: intersection of predator and prey phenologies. J. Fish Biol. 45:41–54.

Juanes, F., J. A. Hare, and A. G. Miskiewicz. 1996. Comparing early life history strategies of *Pomatomus saltatrix*: a global approach. Mar. Freshwat. Res. 47:365–379.

Juanes, F., J. A. Buckel, and F. S. Scharf. 2001. Predatory behavior and selectivity of a primary piscivore: comparison of fish and non-fish prey. Mar. Ecol. Prog. Ser. 217:157–165.

———. 2002. Feeding ecology of piscivorous fishes. Pp. 267–283 *in* P. J. B. Hart and J. D. Reynolds, eds. Handbook of Fish Biology and Fisheries. Vol. 1, Fish Biology. Blackwell Publishing, Malden, Mass.

June, F. C., and F. T. Carlson. 1971. Food of young Atlantic menhaden *Brevoortia tyrannus* in relation to metamorphosis. U.S. Fish. Wildl. Serv. Fish. Bull. 68:493–512.

June, F. C., and J. L. Chamberlain. 1959. The role of the estuary in the life history and biology of the Atlantic menhaden. Proc. Gulf Caribb. Fish Inst. 11(1958):41–45, 55–57.

Jung, S., and E. D. Houde. 2004a. Recruitment and spawning-stock biomass distribution of bay anchovy (*Anchoa mitchilli*) in Chesapeake Bay. Fish. Bull. 102:63–77.

———. 2004b. Production of bay anchovy *Anchoa mitchilli* in Chesapeake Bay: application of size-based theory. Mar. Ecol. Prog. Ser. 281:217–232.

Kahnle, A., and K. Hattala. 1988. Bottom trawl survey of juvenile fishes in the Hudson River estuary. Summary Report for 1981–1986. Hudson River Fisheries Research Unit, Bureau of Fisheries, Region 3, N.Y. State Department of Environmental Conservation, New Paltz, N.Y.

Karpouzi, V. S., and K. I. Stergiou. 2003. The relationships between mouth size and shape and body length for 18 species of marine fishes and their trophic implications. J. Fish Biol. 62:1353–1365.

Katselis, G., K. Koukou, E. Dimitriou, and C. Koutsikopoulos. 2007. Short-term seaward fish migration in the Messolonghi-Etoliko lagoons (Western Greek coast) in relation to climatic variables and the lunar cycle. Est. Coast. Shelf Sci. 73:571–582.

Kaufman, L., J. Ebersole, J. Beets, and C. C. McIvor. 1992. A key phase in the recruitment dynamics of coral reef fishes: post-settlement transition. Environ. Biol. Fishes 34:109–118.

Kawahara, S. 1978. Age and growth of butterfish, *Peprilus triacanthus* (Peck), in ICNAF Subarea 5 and Statistical Area 6. Int. Comm. Northwest Atl. Fish. Select. Pap. 3:73–78.

Keane, J. P., and F. J. Neira. 2008. Larval fish assemblages along the south-eastern Australian shelf: linking mesoscale non-depth-discriminate structure and water masses. Fish. Oceanogr. 17(4): 263–280.

Keefe, M., and K. W. Able. 1992. Habitat quality in New Jersey estuaries: habitat-specific growth rates of juvenile summer flounder in vegetated habitats. Final Rept. New Jersey Dept. Environ. Protection & Energy.

———. 1993. Patterns of metamorphosis in summer flounder, *Paralichthys dentatus*. J. Fish. Biol. 42:713–728.

———. 1994. Contributions of abiotic and biotic factors on settlement in summer flounder, *Paralichthys dentatus*. Copeia 2: 458–465.

Keefer, M. L., C. A. Peery, N. Wright, W. R. Daigle, C. C. Caudill, T. S. Clabough, D. W. Griffith, and M. A. Zacharias. 2008. Evaluating the NOAA coastal and marine ecological classification standard in estuarine systems: a Columbia River estuary case study. Estuar. Coast. Shelf Sci. 78:89–106.

Keeling, C. D. 1998. Rewards and penalties of monitoring the earth. Annu. Rev. Energy Environ. 23:25–82.

Keener, P. G., D. J. Bruce, W. Stender, E. B. Brothers, and H. R. Beatty. 1988. Ingress of postlarval gag, *Mycteroperca microlepis* (Pisces: Serranidae), through a South Carolina Barrier Island Inlet. Bull. Mar. Sci. 42(3):376–396.

Keenlyside, N. S., M. Laatif, J. Jungclaus, L. Kornblueh, and E. Roeckner. 2008. Advancing decadal-scale climate prediction in the North Atlantic sector. Nature 453:84–88.

Keirans, W. J., Jr. 1977. An immunochemically assisted ichthyoplankton survey with elaboration of species specific antigens of fish egg vitellins: southern New Jersey barrier island-lagoon complex. Ph.D. diss. Lehigh University, Bethlehem, Pa. 159 pp.

Keirans, W. J., Jr., S. S. Herman, and R. G. Marlsberger. 1986. Differentiation of *Prionotus carolinus* and *Prionotus evolans* eggs in Hereford Inlet Estuary, southern New Jersey, using immunodiffusion. U.S. Fish. Bull. 84(1):63–68.

Keller, G. H., D. Lamberg, G. Rowe, and N. Staresinic. 1973. Bottom currents in the Hudson Canyon. Science 180:191–193.

Kelley, D. 2002. Abundance, growth and first-winter survival of young bass in nurseries of south-west England. J. Mar. Biol. Assoc. U.K. 82:307–319.

Kellison, G. T. 2000. Evaluation of stock enhancement potential for summer flounder (*Paralichthys dentatus*): an integrated laboratory, field and modeling study. Ph.D. diss., North Carolina State University, Raleigh. 98 pp.

Kellison, G. T., and J. C. Taylor. 2007. Demonstration and implications of habitat-specific chemical signatures in otoliths of juvenile summer flounder (*Paralichthys dentatus* Linnaeus) in North Carolina. J. Fish Biol. 71:350–359.

Kellison, G. T., D. B. Eggleston, and J. B. Burke. 2000. Comparative behavior and survival of hatchery-reared versus wild summer flounder (*Paralichthys dentatus*). Can. J. Fish. Aquat. Sci. 57: 1870–1877.

Kellogg, R. L. 1982. Temperature requirements for the survival and early development of the anadromous alewife. Prog. Fish-Cult. 44:63–73.

Kelly, S. 2006. Chaetodontidae: Butterflyfishes. Pp. 1763–1766 *in* W. J. Richards, ed. Early Stages of Atlantic Fishes: An Identification Guide for the Western Central North Atlantic. CRC Press, Boca Raton, Fla.

Kendall, A. W., Jr. 1972. Description of black sea bass, *Centropristis striata*, (Linnaeus), larvae and their occurrence north of Cape Lookout, North Carolina. U.S. Fish. Bull. 70:1243–1259.

———. 1979. Morphological comparisons of North American sea bass larvae (Pisces: Serranidae). (Natl. Ocean. Atmos. Adm.) Tech. Rep. NMFS Circ. 428.

Kendall, A. W., Jr., and L. P. Mercer. 1982. Black sea bass, *Centropristis striata*. MESA (Mar. Ecosyst. Anal.) New York Bight Atlas Monogr. 15:82–83.

Kendall, A. W., Jr., and N. A. Naplin. 1981. Diel-depth distribution of summer ichthyoplankton in the middle Atlantic bight. U.S. Fish. Bull. 79(4):705–726.

Kendall, A. W., Jr., and J. W. Reintjes. 1975. Geographic and hydrographic distribution of Atlantic menhaden eggs and larvae along the Middle Atlantic coast from R/V *Dolphin* cruises, 1965–66. U.S. Fish. Bull. 73:317–335.

Kendall, A. W., Jr., and L. A. Walford. 1979. Sources and distribution of bluefish, *Pomatomus saltatrix*, larvae and juveniles off the east coast of the United States. U.S. Fish. Bull. 77:213–227.

Kendall, A. W., Jr., E. H. Ahlstrom, and H. G. Moser. 1984. Early life history stages of fishes and their characters. Pp. 11–22 *in* H. G. Moser, W. J. Richards, D. M. Cohen, M. P. Fahay, A. W. Kendall Jr., and S. L. Richardson, eds. Ontogeny and Systematics of Fishes. Am. Soc. of Ichthyolo. and Herpetolo., Spec. Publ. No. 1. Allen Press, Lawrence, Kans.

Kendall, W. C. 1914. The fishes of Maine. Pp. 1–198 *in* Proceedings of the Portland Society of Natural History. Vol. 3. Portland, Me.

Kennish, M. J. 1992. Ecology of Estuaries: Anthropogenic Effects. CRC Press, Boca Raton, Fla. 494 pp.

———. 1998. Pollution Impacts on Marine Biotic Communities. CRC Press, Boca Raton, Fla. 310 pp.

———. 2002. Environmental threats and environmental future of estuaries. Environ. Conservat. 29(1):78–107.

———. 2004. Estuarine Research, Monitoring, and Resource Protection. CRC Press, Boca Raton, Fla. 273 pp.

Kernehan, R. J., R. E. Smith, S. L. Tyler, and M. L. Brewster. 1977. Ichthyoplankton. Vol. 2 in Ecological studies in the vicinity of the proposed Summit Power Station. January 1975 through December 1975. Ichthyological Associates, Summit, Del.

Kerr, L. A., D. H. Secor, and P. M. Piccoli. 2009. Partial migration of fishes as exemplified by the estuarine-dependent white perch. Fisheries 34(3):114–123.

Ketchum, B. H., and N. Corwin. 1964. The persistence of "winter" water on the continental shelf south of Long Island, New York. Limnol. Oceanogr. 9(4):467–475.

Kieffer, M. C., and B. Kynard. 1993. Annual movements of shortnose and Atlantic sturgeons in the Merrimack River, Massachusetts. Trans. Am. Fish. Soc. 122:1088–1103.

———. 1996. Spawning of the shortnose sturgeon in the Merrimack River, Massachusetts. Trans. Am. Fish. Soc. 125:179–186.

Kilby, J. D. 1949. A preliminary report on the young striped mullet *Mugil cephalus* L. in two gulf coastal areas of Florida. Q. J. Fla. Acad. Sci. 11:7–23.

Kimball, M. E., and K. W. Able. 2007a. Nekton utilization of intertidal salt marsh creeks: tidal influences in natural *Spartina*, invasive *Phragmites*, and marshes treated for *Phragmites* removal. J. Exp. Mar. Biol. Ecol. 346:87–101.

———. 2007b. Tidal utilization of nekton in Delaware Bay restored and reference intertidal salt marsh creeks. Estuaries and Coasts 30(6):1075–1087.

———. In review. Intertidal creek migrations of nekton in invasive *Phragmites australis* (Common Reed) salt marshes and marshes treated for *Phragmites* removal: small scale video observations. Estuaries and Coasts.

Kimball, M. E., K. W. Able, and T. M. Grothues. In press. Evaluation of long-term response of intertidal creek nekton to *Phragmites autralis* (Common Reed) removal in oligohaline Delaware Bay salt marshes. Restoration Ecology.

Kimmel, J. J. 1973. Food and feeding of fishes from Magothy Bay, Virginia. M.S. thesis. Old Dominion University, Norfolk, Va. 190 pp.

Kimura, R., D. H. Secor, E. D. Houde, and P. M. Piccoli. 2000. Up-estuary dispersal of young-of-the-year bay anchovy *Anchoa mitchilli* in the Chesapeake Bay: inferences from microprobe analysis of strontium in otoliths. Mar. Ecol. Prog. Ser. 208:217–227.

Kinch, J. C. 1979. Trophic habits of the juvenile fishes within artificial waterways—Marco Island, Florida. Contrib. Mar. Sci. 22: 77–90.

King, J. R., and G. A. McFarlane. 2006. A framework for incorporating climate regime shifts into the management of marine resources. Fish. Manage. Ecol. 13:93–102.

King, N. J., D. M. Bailey, and I. G. Priede. 2007. Role of scavengers in marine ecosystems—introduction. Mar. Ecol. Prog. Ser. 350: 175–178.

Kissil, G. W. 1969. Contribution to the life history of the alewife, (*Alosa pseudoharengus*) (Wilson), in Connecticut. Ph.D. diss. University of Connecticut, Storrs.

———. 1974. Spawning of the anadromous alewife in Bride Lake, Connecticut. Trans. Am. Fish. Soc. 103:312–317.

Klauda, R. J., J. B. McLaren, R. E. Schmidt, and W. P. Dey. 1988. Life history of white perch in the Hudson River Estuary. Am. Fish. Soc. Monogr. 4:69–88.

Klein-MacPhee, G. 1978. Synopsis of biological data for the winter flounder, *Pseudopleuronectes americanus* (Walbaum). NOAA Tech. Rep. NMFS Circ., 414. 43 pp.

———. 1979. Growth, activity and metabolism studies of summer flounder, *Paralichthys dentatus* (L.) under laboratory conditions. Ph.D. diss. University of Rhode Island, Kingston. 99 pp.

———. 2002a. Smelts. Family Osmeridae. Pp. 162–170 *in* B. B.

Collette and G. Klein-MacPhee, eds. Bigelow and Schroeder's Fishes of the Gulf of Maine. 3rd ed. Smithsonian Institution Press, Washington, D.C.

———. 2002b. Croakers, drums, and weakfishes. Family Sciaenidae. Pp. 435–446 in B. B. Collette and G. Klein-MacPhee, eds. Bigelow and Schroeder's Fishes of the Gulf of Maine. 3rd ed. Smithsonian Institution Press, Washington, D.C.

———. 2002c. Butterfishes. Family Stromateidae. Pp. 540–545 in B. B. Collette and G. Klein-MacPhee, eds. Bigelow and Schroeder's Fishes of the Gulf of Maine. 3rd ed. Smithsonian Institution Press, Washington, D.C.

———. 2002d. Sand Flounders. Family Paralichthyidae. Pp. 551–560 in B. B. Collette and G. Klein-MacPhee, eds. Bigelow and Schroeder's Fishes of the Gulf of Maine. 3rd ed. Smithsonian Institution Press, Washington, D.C.

———. 2002e. Righteye flounders. Family Pleuronectidae. Pp. 560–587 in B. B. Collette and G. Klein-MacPhee, eds. Bigelow and Schroeder's Fishes of the Gulf of Maine. 3rd ed. Smithsonian Institution Press, Washington, D.C.

———. 2002f. Porcupinefishes. Family Didontidae. Pp. 601–607 in B. B. Collette and G. Klein-MacPhee, eds. Bigelow and Schroeder's Fishes of the Gulf of Maine. 3rd ed. Smithsonian Institution Press, Washington, D.C.

Kleppel, G. S., R. M. DeVoe, and M. V. Rawson. 2006. Changing Land Use Patterns in the coastal zone: Managing Environmental Quality in Rapidly Developing Regions. Springer: New York. 305 pp.

Klima, E. F. 1971. Distribution of some coastal pelagic fishes in the western Atlantic. Commerc. Fish. Rev. 33(6):21–34.

Kneib, R. T. 1978. Habitat, diet, reproduction and growth of the spotfin killifish, *Fundulus luciae*, from a North Carolina salt marsh. Copeia:164–168.

———. 1982. The effects of predation by wading birds (Ardeidae) and blue crabs (*Callinectes sapidus*) on the population size structure of the common mummichog, *Fundulus heteroclitus*. Estuar. Coast. Shelf Sci. 14:159–165.

———. 1984. Patterns in the utilization of the intertidal salt marsh by larvae and juveniles of *Fundulus heteroclitus* (Linnaeus) and *Fundulus luciae* (Baird). J. Exp. Mar. Biol. Ecol. 83:41–51.

———. 1986. Size-specific patterns in the reproductive cycle of the killifish, *Fundulus heteroclitus* (Pisces: Fundulidae) from Sapelo Island, Georgia. Copeia:342–351.

Kneib, R. T., and A. E. Stiven. 1978. Growth, reproduction and feeding of *Fundulus heteroclitus* (L.) in a North Carolina salt marsh. J. Exp. Mar. Biol. Ecol. 31:121–140.

Kohut, J. T., S. M. Glenn, and R. J. Chant. 2004. Seasonal current variability on the New Jersey inner shelf. J. Geophys. Res. 109: C07S07. 16 pp.

Kosa, J. T., and M. E. Mather. 2001. Processes contributing to variability in regional patterns of juvenile river herring abundance across small coastal systems. Trans. Am. Fish. Soc. 130:600–619.

Koski, T. K. 1978. Age, growth, and maturity of the hogchoker, *Trinectus maculatus*, in the Hudson River, New York. Trans. Am. Fish. Soc. 107(3):449–453.

Kraus, R. T., and J. A. Musick. 2001. A brief interpretation of summer flounder, *Paralichthys dentatus*, movements and stock structure with new tagging data on juveniles. Mar. Fish. Rev. 63(3):1–6.

Kraus, R. T., and D. H. Secor. 2004a. Application of the nursery-role hypothesis to an estuarine fish. Mar. Ecol. Prog. Ser. 291: 301–305.

———. 2004b. Dynamics of white perch *Morone americana* population contingents in the Patuxent River estuary, Maryland, USA. Mar. Ecol. Prog. Ser. 279:247–259.

———. 2005. Connectivity in estuarine white perch populations of Chesapeake Bay: evidence from historical fisheries data. Estuar. Coast. Shelf Sci. 64:108–118.

Krause, J., E. M. A. Hensor, and G. D. Ruxton. 2002. Fish as prey. Pp. 284–298 in P. J. B. Hart and J. D. Reynolds, eds. Handbook of Fish Biology and Fisheries. Vol. 1, Fish Biology. Blackwell Publishing, Malden, Mass.

Kremer, P. 1994. Patterns of abundance for *Mnemiopsis* in U.S. coastal waters: a comparative overview. ICES J. Mar. Sci. 51: 347–354.

Kritzer, J. P., and P. F. Sale. 2006. Marine Metapopulations. Elsevier Academic Press Publications, San Diego, Calif. 544 pp.

Krumholz, L. A. 1948. Reproduction in the western mosquito fish *Gambusia affinis affinis* (Baird and Girard) and its use in mosquito control. Ecol. Monogr. 18:1–43.

Kuhlmann, M. L. 1992. Behavioral avoidance of predation in an intertidal hermit crab. J. Exper. Mar. Biol. Ecol. 157:143–158.

Kuntz, A. 1914. Notes on the habits, morphology of the reproductive organ and embryology of the viviparous fish *Gambusia affinis*. Bull. U.S. Bur. Fish. 33(1913):181–190.

———. 1916. Notes on the embryology and larval development of five species of teleostean fishes. Bull. U.S. Bur. Fish. 34(1914): 407–429.

Kuntz, A., and L. Radcliffe. 1917. Notes on the embryology and larval development of twelve teleostean fishes. Bull. U.S. Bur. Fish. 35:89–134.

Kuroki, M., M. Kawai, B. Jónsson, J. Aoyama, M. J. Miller, D. L. G. Noakes, and K. Tsukamoto. 2008. Inshore migration and otolith microstructure / microchemistry of anguillid glass eels recruited to Iceland. Environ. Biol. Fish. 83:309–325.

Kynard, B., and M. Horgan. 2002. Ontogenetic behavior and migration of Atlantic sturgeon, *Acipenser oxyrinchus oxyrinchus*, and shortnose sturgeon, *A. brevirostrum*, with notes on social behavior. Environ. Biol. Fishes 63:137–150.

Laegdsgaard, P. 2006. Ecology, disturbance and restoration of coastal saltmarsh in Australia: a review. Wetlands Ecol. Manag. 14:379–399.

Lafferty, K. D., S. Allesina, M. Arim, C. J. Briggs, G. DeLeo, A. P. Dobson, J. A. Dunne, P. T. J. Johnson, A. M. Kuris, D. J. Marcogliese, N. D. Martinez, J. Memmott, P. A. Marquet, J. P. McLaughlin, E. A. Mordecai, M. Pascual, R. Poulin, and D. W. Thieltges. 2008. Parasites in food webs: the ultimate missing links. Ecol. Lett. 11:533–546.

Lake, T. 1983. Poor man's marlin—Atlantic needlefish. Conservationist 38(1):32–36.

Lambert, T. C. 1985. Gastric emptying time and assimilation efficiency in Atlantic mackerel (*Scomber scombrus*). Can. J. Zool. 63: 817–820.

Lang, K. L., F. P. Almeida, G. R. Bolz, and M. P. Fahay. 1996. The use of otolith microstructure in resolving issues of first year growth and spawning seasonality of white hake, *Urophycis tenuis*, in the Gulf of Maine–Georges Bank region. U.S. Fish. Bull. 94(1):170–175.

Langton, R. W., and R. E. Bowman. 1980. Food of fifteen northwest Atlantic gadiform fishes. NOAA Tech. Rept. NMFS SSRF No. 740, 23 pp.

Lankford, T. E., Jr. 1997. Estuarine recruitment processes and stock structure in Atlantic croaker, *Micropogonia undulatus* (Linnaeus).

Ph.D. diss. Graduate College of Marine Studies, University of Delaware, Newark.

Lankford, T. E., Jr., and T. E. Targett. 1994. Suitability of estuarine nursery zones for juvenile weakfish (*Cynoscion regalis*): effects of temperature and salinity on feeding, growth, and survival. Mar. Biol. 119:611–620.

———. 2001. Low-temperature tolerance of age-0 Atlantic croakers: recruitment implications for U.S. Mid-Atlantic estuaries. Trans. Am. Fish. Soc. 130:236–249.

Lankford, T. E., Jr., T. E. Targett, and P. M. Gaffney. 1999. Mitochondrial DNA analysis of population structure in the Atlantic croaker, *Micropogonias undulatus* (Perciformes: Sciaenidae). Fish. Bull. 97:884–890.

LaPlante, L. H., and E. T. Schultz. 2007. Annual fecundity of tautog in Long Island Sound: size effects and long-term changes in a harvested population. Trans. Am. Fish. Soc. 136:1520–1533.

Lapolla, A. E. 2001a. Bay anchovy *Anchoa mitchilli* in Narragansett Bay, Rhode Island. I. Population structure, growth and mortality. Mar. Ecol. Prog. Ser. 217:93–102.

———. 2001b. Bay anchovy *Anchoa mitchilli* in Narragansett Bay, Rhode Island. II. Spawning season, hatch-date distribution and young-of-the-year growth. Mar. Ecol. Prog. Ser. 217:103–109.

Laprise, R., and J. J. Dodson. 1989. Ontogenetic changes in the longitudinal distribution of two species of larval fish in a turbid well-mixed estuary. J. Fish. Biol. 35(suppl. A):39–47.

Laroche, J. L., and J. Davis. 1973. Age, growth, and reproduction of the northern puffer, *Sphoeroides maculatus*. U.S. Fish. Bull. 71(4): 955–963.

Lascara, J. 1981. Fish predator-prey interactions in areas of eelgrass (*Zostera marina*). M.S. thesis, College of William and Mary, Williamsburg, Va. 81 pp.

Lasiak, T. A. 1986. Juveniles, food and the surf zone habitat: implications for teleost nursery areas. S. Afr. J. Zool. 21:51–56.

Lassiter, R. R. 1962. Life history aspects of the bluefish, *Pomatomus saltatrix* (Linnaeus), from the coast of North Carolina. M.S. thesis. North Carolina State College, Raleigh. 103 pp.

Lathrop, R. G., L. Windham, and P. Montesano. 2003. Does *Phragmites* alter the structure and function of marsh landscapes? Patterns and processes revisited. Estuaries 26:423–435.

Lauff, G. H. 1967. Estuaries. American Association for the Advancement of Science, Washington, D.C.

Laughlin, R. A. 1982. Feeding habits of the blue crab, *Callinectes sapidus* Rathbun, in the Apalachicola estuary, Florida. Bull. Mar. Sci. 323(4):807–822.

Lavenda, N. 1949. Sexual differences and normal protogynous hermaphroditism in the Atlantic sea bass, *Centropristis striata*. Copeia 3:185–194.

Lawler, Matusky and Skelly Engineers. 1975. 1974 Hudson River aquatic ecology studies. Bowline Point and Lovett Generating Stations. Prepared for Orange and Rockland Utilities, Inc.

Layman, C. A., D. E. Smith, and J. D. Herod. 2000. Seasonally varying importance of abiotic and biotic factors in marsh-pond fish communities. Mar. Ecol. Progr. Ser. 207:155–169.

Layman, C. A., C. P. Dahlgren, G. T. Kellison, A. J. Adams, B. M. Gillanders, M. S. Kendall, J. A. Ley, I. Nagelkerken, and J. E. Serafy. 2006. Marine nurseries and effective juvenile habitats. Mar. Ecol. Prog. Ser. 318:307–308.

Lazzari, M. A. 2008. Habitat variability in young-of-the-year winter flounder, *Pseudopleuronectes americanus*, in Maine estuaries. Fish. Res. 90:296–304.

Lazzari, M. A., and K. W. Able. 1990. Northern pipefish, *Syngna-*

thus fuscus, occurrences over the Mid-Atlantic Bight continental shelf: evidence of seasonal migration. Environ. Biol. Fishes 27: 177–185.

Lazzari, M. A., J. C. O'Herron II, and R. W. Hastings. 1986. Occurrence of juvenile Atlantic sturgeon, *Acipenser oxyrinchus*, in the upper tidal Delaware River. Estuaries 9(48):356–361.

Lazzari, M. A., K. W. Able, and M. P. Fahay. 1989. Life history and food habits of the grubby, *Myoxocephalus aenaeus* (Cottidae), in a Cape Cod estuary. Copeia 1:7–12.

Lazzari, M. A., S. Sherman, and J. K. Kanwitt. 2003. Nursery use of shallow habitats by epibenthic fishes in Maine nearshore waters. Estuar. Coast. Shelf Sci. 56:73–84.

Leck, C. F. 1984. The Status and Distribution of New Jersey's Birds. Rutgers University Press, New Brunswick, N.J.

Lecomte, F., and J. J. Dodson. 2004. Role of early life-history constraints and resource polymorphism in the segregation of sympatric populations of an estuarine fish. Evol. Ecol. Res. 6: 631–658.

Lee, D. S., C. R. Gilbert, C. H. Hocutt, R. E. Jenkins, D. E. McAllister, and J. R. Stauffer Jr. 1980. Atlas of North American Freshwater Fishes. North Carolina State Museum of Natural History, Raleigh, N.C.

Leggett, W. C. 1976. The American shad (*Alosa sapidissima*), with special reference to its migration and population dynamics in the Connecticut River. Am. Fish. Soc. Monogr. 1:169–225.

———. 1977. The ecology of fish migrations. Ann. Rev. Ecol. System. 8:285–308.

———. 1986. The dependence of fish larval survival on food and predator densities. Pp. 117–137 *in* S. Skreslet, ed. The Role of Freshwater Outflow in Coastal Marine Ecosystems. NATO ASI Series. Vol. G7. Springer, Berlin.

Leggett, W. C., and J. E. Carscadden. 1978. Latitudinal variation in reproductive characteristics of American shad (*Alosa sapidissima*): evidence for population specific life history strategies in fish. J. Fish. Res. Board Can. 35:1469–1478.

Leiby, M. M. 1989. Family Ophichthidae: Leptocephali. Pp. 764–897 *in* Fishes of the Western North Atlantic. Mem. Sears Found. Mar. Res. 1(9):657–1055.

Leim, A. H. 1924. The life history of the shad (*Alosa sapidissima*) (Wilson) with special reference to the factors limiting its abundance. Contrib. Can. Biol. 2:163–284.

Leis, J. M. 1978. Systematics and zoogeography of the porcupinefishes (Diodon, Diodontidae, Tetraodontiformes) with comments on egg and larval development. U.S. Fish. Bull. 76(3): 535–567.

———. 1984. Tetraodontoidei: development. Pp. 447–450 *in* H. G. Moser, W. J. Richards, D. M. Cohen, M. P. Fahay, A. W. Kendall Jr., and S. L. Richardson, eds. Ontogeny and Systematics of Fishes. Am. Soc. Ich. Herp. Spec. Publ. No. 1.

———. 2006. Are larvae of demersal fishes plankton or nekton? Adv. Mar. Biol. 51:59–141.

Leis, J. M., and B. M. Carson-Ewart. 2002. *In situ* settlement behaviour of damselfish (*Pomacentridae*) larvae. J. Fish Biol. 61:325–346.

Leis, J. M., H. P. A. Sweatman, and S. E. Reader. 1996. What the pelagic stages of coral reef fishes are doing out in blue water daytime field observation of larval behavioural capabilities. Mar. Freshwat. Res. 47:401–411.

Lenanton, R. C. J. 1982. Alternative non-estuarine nursery habitats for some commercially and recreationally important fish species of south-western Australia. Aust. J. Mar. Freshw. Res. 33: 881–900.

Lenanton, R. C. J., and E. P. Hodgkin. 1985. Life history strategies of fish in some temperate Australian estuaries. Pp. 267–284 *in* A. Yáñez-Arancibia, ed. Fish Community Ecology in Estuaries and Coastal Lagoons: Towards an Ecosystem Integration. UNAM Press, Mexico City, Mexico.

Lenanton, R. C. J., and I. C. Potter. 1987. Contributions of estuaries to commercial fisheries in temperate western Australia and the concept of estuarine dependence. Estuaries 10(1):28–35.

Lenarz, W. H., F. B. Schwing, D. A. Ventresca, F. Chavez, and W. M. Graham. 1995. Explorations of El Niño events and associated biological population dynamics off central California. Calif. Coop. Oceanic Fish. Invest. Rep. 36:106–119.

Lenihan, H. S., and C. H. Peterson. 1998. How habitat degradation through fishery disturbance enhances impacts of hypoxia on oyster reefs. Ecol. Appl. 8:128–140.

Lenihan, H. S., C. H. Peterson, J. E. Byers, J. H. Grabowski, G. W. Thayer, and D. R. Colby. 2001. Cascading of habitat degradation: Oyster reefs invaded by refugee fishes escaping stress. Ecol. Appl. 11(3):764–782.

Leonard, J. B. K., and S. D. McCormick. 1999. Effects of migration distance on whole-body and tissue-specific energy use in American shad (*Alosa sapidissima*). Can. J. Fish. Aquat. Sci. 56: 1159–1171.

Le Pape, O., J. Holley, D. Guerault, and Y. Desausay. 2003. Quality of coastal and estuarine essential fish habitats: estimations based on the size of juvenile common sole (*Solea solea* L.). Estuar. Coast. Shelf Sci. 58:793–803.

Lercari, D., and E. A. Chávez. 2007. Possible causes related to historic stock depletion of the totoaba, *Totoaba macdonaldi* (Perciformes: Sciaenidae), endemic to the Gulf of California. Fish. Res. 86(2007):136–142.

Leslie, A. J., Jr., and D. J. Stewart. 1986. Systematics and distribution ecology of *Etropus* (Pisces, Bothidae) on the Atlantic coast of the United States with a description of a new species. Copeia 1:140–156.

Levin, P. S. 1991. Effects of microhabitat on recruitment in a Gulf of Maine reef fish. Mar. Ecol. Prog. Ser. 75:183–189.

———. 1993. Habitat structure, conspecific presence and spatial variation in the recruitment of a temperate reef fish. Oecologia 94:176–185.

Levin, P. S., and G. W. Stunz. 2005. Habitat triage for exploited fishes: can we identify essential "Essential Fish Habitat"? Estuar. Coast. Shelf Sci. 64:70–78.

Levin, P. S., W. Chiasson, and J. M. Green. 1997. Geographic differences in recruitment and population structure of a temperate reef fish. Mar. Ecol. Prog. Ser. 161:23–25.

Levinton, J. S., and J. R. Waldman. 2006. The Hudson River Estuary. Cambridge University Press, New York.

Levitus, S., J. I. Antonov, T. P. Boyer, and C. Stephens. 2000. Warming of the world ocean. Science 287(5461):2225–2229.

Levy, A., K. W. Able, C. B. Grimes, and P. Hood. 1988. Biology of the conger eel *Conger oceanicus* in the Mid-Atlantic Bight. II. Foods and feeding ecology. Mar. Biol. 98:597–699.

Lewis, R. M. 1965. The effect of minimum temperature on the survival of larval Atlantic menhaden, *Brevoortia tyrannus*. Trans. Am. Fish. Soc. 94(4):409–412.

Lewis, R. M., E. P. H. Wilkens, and H. R. Gordy. 1972. A description of young Atlantic menhaden, *Brevoortia tyrannus*, in the White Oak Estuary, North Carolina. Fish. Bull. 70:115–118.

Li, X., M. K. Litvak, and J. E. H. Clarke. 2007. Overwintering habitat use of shortnose sturgeon (*Acipenser brevirostrum*): de-

fining critical habitat using a novel underwater video survey and modeling approach. Can. J. Fish. Aquat. Sci. 64:1248–1257.

Limburg, K. E. 1995. Otolith strontium traces environment history of subyearling American shad *Alosa sapidissima*. Mar. Ecol. Prog. Ser. 119:25–35.

———. 2001. Through the gauntlet again: demographic restructuring of American shad by migration. Ecology 82(6):1584–1596.

Limburg, K. E., I. Blackburn, R. Schmidt, T. Lake, J. Hasse, M. J. Elfman, and P. Kristiansson. 2001. Otolith microchemistry indicated unexpected patterns of residency and anadromy in blueback herring, *Alosa aestivalis,* in the Hudson and Mohawk Rivers. Bull. Fr. Peche Piscic. 362/363:931–938.

Limburg, K. E., and R. E. Schmidt. 1990. Patterns of fish spawning in Hudson River tributaries: response to an urban gradient? Ecology 71:1238–1245.

Lindsey, C. C. 1988. Factors controlling meristic variation. Pp. 197–274 *in* W. S. Hoar and D. J. Randall, eds. Fish Physiology Vol. 11B, Chapter 3. Academic Press, San Diego, Calif.

Lindquist, D. G., M. V. Ogburn, W. B. Stanley, H. L. Troutman, and S. M. Pereira. 1985. Fish utilization patterns on temperate rubble-mound jetties in North Carolina. Bull. Mar. Sci. 37(1): 244–251.

Link, J. 2002. What does ecosystem-based fisheries management mean? Fisheries 27(4):18–21.

Linton, J. R., and B. L. Soloff. 1964. The physiology of the brood pouch of the male seahorse, *Hippocampus erectus*. Bull. Mar. Sci. Gulf Carib. 14:45–61.

Lippson, A. J., and R. L. Moran. 1974. Manual for identification of early developmental stages of fishes of the Potomac River Estuary. Md. Dep. Nat. Resour. Power Plant Siting Program PPSP-MP-13.

Litvin, S. Y., and M. P. Weinstein. 2003. Life history strategies of estuarine nekton: the role of marsh macrophytes, benthic microalgae, and phytoplankton in the trophic spectrum. Estuaries 26(2b):552–562.

Litzow, M. A., K. M. Bailey, F. G. Prahl, and R. Heintz. 2006. Climate regime shifts and reorganization of fish communities: the essential fatty acid limitation hypothesis. Mar. Ecol. Prog. Ser. 315:1–11.

Livingston, R. J. 1975. Diurnal and seasonal fluctuations of organisms in a North Florida estuary. Est. Coast. Mar. Sci. 4:373–400.

———. 1988. Inadequacy of species-level designation for ecological studies of coastal migratory fishes. Environ. Biol. Fishes 22(3): 225–234.

———. 2003. Trophic Organization in Coastal Systems. CRC Press. Boca Raton, Fla.

Livingston, R. J., X. Nui, G. Lewis III, and G. C. Woodsum. 1997. Freshwater input to a gulf estuary: long-term control of trophic organization. Ecol. Appl. 7:277–299.

Ljunggren, L., and A. Sandstrom. 2007. Influence of visual conditions on foraging and growth of juvenile fishes with dissimilar sensory physiology. J. Fish Bio. 70:1319–1334.

Loesch, J. G. 1969. A study of the blueback herring in Connecticut waters. Ph.D. diss. University of Connecticut, Storrs. 31 pp.

Loesch, J. G., and W. A. Lund, Jr. 1977. A contribution to the life history of the blueback herring, *Alosa aestivalis*. Trans. Am. Fish. Soc. 106:583–589.

Long, W. C., and R. D. Seitz. 2008. Trophic interactions under stress: hypoxia enhances foraging in an estuarine food web. Mar. Ecol. Prog. Ser. 362:59–68.

———. 2009. Hypoxia in Chesapeake Bay tributaries: worsening

effects of macrobenthic community structure in the York River. Estuaries and Coasts 32:287–297.

Longhurst, A. R., and D. Pauly. 1987. Tropical estuarine fish assemblages. Pp. 181–183 *in* A. R. Longhurst and D. Pauly, eds. Ecology of Tropical Oceans. Academic Press, Inc., San Diego, Calif.

Longley, W. H., and S. F. Hildebrand. 1941. Systematic catalogue of the fishes of Tortugas, Florida. Papers from Tortugas Laboratory. Carnegie Inst. Wash. Publ. 34:1–317.

Lookabaugh, P. S., and P. L. Angermeier. 1992. Diet patterns of American eel, *Anguilla rostrata,* in the James River drainage, Virginia. J. Fresh. Eco. 7(4):425–431.

Loos, J. J., and E. S. Perry. 1991. Larval migration and mortality rates of bay anchovy in the Patuxent River. Pp. 65–76 *in* R. D. Hoyt, ed. Larval fish recruitment and research in the Americas. Natl. Ocean. Atmos. Adm. Tech. Rep. NMFS 95.

Lotrich, V. A. 1975. Summer home range and movements of *Fundulus heteroclitus* in a tidal creek. Ecology 56:191–198.

Lotze, H. K., H. S. Lenihan, B. J. Bourque, R. H. Bradbury, R. G. Cooke, M. C. Kay, S. M. Kidwell, M. X. Kirby, C. H. Peterson, and J. B. C. Jackson. 2006. Depletion, degradation, and recovery potential of estuaries and coastal seas. Science 312:1806–1809.

Love, J. W., D. F. Luers, and B. D. Williams. 2009. Spatio-temporal patterns of larval fish ingress to Chincoteague Bay, Maryland, USA during winter and spring 2004 to 2007. Mar. Ecol. Prog. Ser. 377:203–212.

Lowe, J. A., D. R. G. Farrow, A. S. Pait, S. J. Arenstam, and E. F. Lavan. 1991. Fish kills in coastal waters: 1980–1989. National Oceanic and Atmospheric Administration.

Lowerre-Barbieri, S. K., M. E. Chittenden Jr., and L. R. Barbieri. 1996a. The multiple spawning pattern of weakfish in the Chesapeake Bay and Middle Atlantic Bight. J. Fish Biol. 48:1139–1163.

———. 1996b. Variable spawning activity and annual fecundity of weakfish in Chesapeake Bay. Trans. Am. Fish. Soc. 125:532–545.

Lowery, T. A., M. E. Monaco, and A. J. Bulger. 1995. Analysis of relative abundance estimates of estuarine-catadromous larvae and their utilization of coastal inlets. Pp. 45–110 *in* A. J. Bulger, T. A. Lowery, and M. E. Monaco, eds. Estuarine-Catadromy: A Life History Strategy Coupling Marine and Estuarine Environments Via Coastal Inlets. Chapter 2, ELMR Rep. No. 14. NOAA/NOS Strategic Environmental Assessments Division, Silver Spring, Md.

Luczkovich, J. J., and B. L. Olla. 1983. Feeding behavior, prey consumption, and growth of juvenile red hake. Trans. Am. Fish. Soc. 112:629–637.

Luczkovich, J. J., M. W. Sprague, S. E. Johnson, and R. C. Pullinger. 1999. Delimiting spawning areas of weakfish, *Cynoscion regalis* (family Sciaenidae), in Pamlico South, North Carolina using passive hydroacoustic surveys. Bioacoustics 10:143–160.

Luczkovich, J. J., D. A. Mann, and R. Rountree. 2008. Passive acoustics as a tool in fisheries science. Trans. Am. Fish. Soc. 137:533–541.

Lund, W. A., Jr., and B. C. Marcy Jr. 1975. Early development of the grubby, *Myoxocephalus aenaeus* (Mitchill). Biol. Bull. (Woods Hole) 149:373–383.

Luo, J., and J. A. Musick. 1991. Reproductive biology of the bay anchovy *Anchoa mitchilli,* in Chesapeake Bay. Trans. Am. Fish. Soc. 120:701–710.

Lux, F. E., and F. E. Nichy. 1971. Number and lengths, by season, of fishes caught with an otter trawl near Woods Hole, Massachusetts, September 1961 to December 1962. U.S. Dep. Comm. Spec. Sci. Rep. 622:1–14.

Maccarone, A. D., and K. C. Parsons. 1994. Factors affecting the use of a freshwater and an estuarine foraging site by egrets and ibises during the breeding season in New York City. Colonial Waterbirds 17(1):60–68.

Macchi, G. J., E. M. Acha, and C. A. Lasta. 2002. Reproduction of black drum (*Pogonias cromis*) in the Rio de la Plata estuary, Argentina. Fish. Res. 59:83–92.

MacKenzie, C. L., Jr. 1992. The Fisheries of Raritan Bay. Rutgers University Press, New Brunswick, N.J. 304 pp.

MacKenzie, R. A., and M. Dionne. 2008. Habitat heterogeneity: importance of salt marsh pools and high marsh surfaces to fish production in two Gulf of Maine salt marshes. Mar. Ecol. Prog. Ser. 368:217–230.

Magnuson, J. J. 2002. A future of adapting to climate change and variability. Pp. 273–282 *in* N. A. McGinn, ed. Fisheries in a Changing Climate. Am. Fish. Soc. Symp. 32.

Mahoney, J. B., and J. K. McNulty. 1992. Disease-associated blood changes and normal seasonal haematological variation in winter flounder in the Hudson-Raritan estuary. Trans. Am. Fish. Soc. 121:261–268.

Major, P. F. 1978. Aspects of estuarine intertidal ecology of juvenile striped mullet, *Mugil cephalus,* in Hawaii. U.S. Fish. Bull. 76: 299–313.

Maki, K. L., J. M. Hoenig, J. E. Olney, and D. M. Heisey. 2006. Comparing historical catch rates of American shad in multifilament and monofilament nets: a step toward setting restoration targets for Virginia stocks. N. Am. J. Fish. Manage. 26:282–288.

Malchoff, M. H. 1993. Age, growth and distribution of cunner (*Tautogolabrus adspersus*) and tautog (*Tautoga onitis*) larvae in the New York Bight: a single season analysis. M.S. thesis. Bard College, Annandale-on-Hudson, New York. 75 pp.

Malloy, K. D., and T. E. Targett. 1991. Feeding, growth and survival of juvenile summer flounder *Paralichthys dentatus*: experimental analysis of the effects of temperature and salinity. Mar. Ecol. Prog. Ser. 72(3):213–223.

Mancini, F., and K. W. Able. 2005. Food habits of young-of-the-year estuarine fishes in Middle Atlantic Bight estuaries: a synthesis. Rutgers University Institute of Marine and Coastal Sciences Technical Report 2005-15.

Manderson, J. P. 2008. The spatial scale of phase synchrony in winter flounder (*Pseudopleuronectes americanus*) production increased among southern New England nurseries in the 1990s. Can. J. Fish. Aquat. Sci. 65:340–351.

Manderson, J. P., A. W. Stoner, and L. L. Stehlik. 1997. Species responses to environmental gradients and the description of estuarine landscape pattern. ICES Annual Science Conference, Baltimore, ICES CM.S15.

Manderson, J. P., B. A. Phelan, A. J. Bejda, L. L. Stehlik, and A. W. Stoner. 1999. Predation by striped searobin (*Prionotus evolans,* Triglidae) on young-of-the-year winter flounder (*Pseudopleuronectes americanus,* Walbaum): examining prey size selection and prey choice using field observations and laboratory experiments. J. Exp. Mar. Biol. Ecol. 242: 211–231.

Manderson, J. P., B. A. Phelan, A. W. Stoner, and J. Hilbert. 2000. Predator-prey relations between age-1+ summer flounder (*Paralichthys dentatus,* Linnaeus) and age-0 winter flounder (*Pseudopleuronectes americanus,* Walbaum): predator diets, prey selection, and effects of sediments and macrophytes. J. Exp. Mar. Biol. Ecol. 251:17–39.

Manderson, J. P., J. Pessutti, J. G. Hilbert, and F. Juanes. 2004. Shallow water predation risk for a juvenile flatfish (winter flounder; *Pseudopleuronectes americanus*, Walbaum) in a northwest Atlantic estuary. J. Exp. Mar. Biol. Ecol. 304:137–157.

Manderson, J. P., J. Pessutti, P. Shaheen, and F. Juanes. 2006. Dynamics of early juvenile winter flounder predation risk on a North West Atlantic estuarine nursery ground. Mar. Ecol. Prog. Ser. 328:249–265.

Mangel, M., Levin, P., and A. Patil. 2006. Using life history and persistence criteria to prioritize habitats for management and conservation. Ecol. Appl. 16:797–806.

Mann, D. A., J. Bowers-Altman, and R. A. Rountree. 1997. Sounds produced by the striped cusk-eel *Ophidion marginatum* (Ophidiidae) during courtship and spawning. Copeia 3:610–612.

Manooch, C. S., III. 1984. Fishes of the Southeastern United States. A Fisherman's Guide. North Carolina State Museum of Natural History. Raleigh, N.C.

Mansueti, A. J., and J. D. Hardy. 1967. Development of Fishes of the Chesapeake Bay Region: An Atlas of Egg, Larval, and Juvenile Stages. Part I. Natural Resources Institute, University of Maryland, Port City Press, Baltimore.

Mansueti, R. J. 1960. Restriction of very young red drum, *Sciaenops ocellata*, to shallow estuarine waters of Chesapeake Bay during late autumn. Chesapeake Sci. 1:207–210.

———. 1962. Eggs, larvae, and young of the hickory shad, *Alosa mediocris*, with comments on its ecology in the estuary. Chesapeake Sci. 3(30):173–205.

———. 1963. Symbiotic behavior between small fishes and jellyfishes, with new data on that between the stromateid, *Peprilus alepidotus*, and the scyphomedusa, *Chrysaora quinquecirrha*. Copeia 1:40–80.

———. 1964. Eggs, larvae, and young of the white perch, *Roccus americanus*, with comments on its ecology in the estuary. Chesapeake Sci. 5:3–45.

Mansueti, R. J., and R. Pauly. 1956. Age and growth of the northern hogchoker, *Trinectus maculatus*, in the Patuxent River, Maryland. Copeia 1:60–62.

Mantua, N.J., and P. W. Mote. 2002. Uncertainty in scenarios of human-caused climate change. Pp. 263–272 in N. A. McGinn, ed. Fisheries in a Changing Climate. Amer. Fish. Soc. Symp. 32.

Marak, R. R. 1960. Food habits of larval cod, haddock, and coalfish in the Gulf of Maine and Georges Bank area. Journal Du Conseil 25:147–157.

Marcellus, K. L. 1972. Fishes of Barnegat Bay, New Jersey, with particular reference to seasonal influence and the possible effect of thermal discharge. Ph.D. diss. Rutgers University, New Brunswick, N.J. 172 pp.

Marcy, B. C., Jr. 1969. Age determination from scales of *Alosa pseudoharengus* and *Alosa aestivalis* in Connecticut waters. Trans. Am. Fish. Soc. 98:621–630.

———. 1972. Spawning of the American shad, *Alosa sapidissima*, in the lower Connecticut River. Chesapeake Sci. 13:116–119.

———. 1973. Vulnerability and survival of young Connecticut River fish entrained at a nuclear power plant. J. Fish. Res. Board Can. 30(8):1195–1203.

———. 1976. Fishes of the lower Connecticut River and the effects of the Connecticut Yankee plant. Pp. 61–113 in D. Merriman and L. M. Thorpe, eds. The Connecticut River Ecological Study. American Fisheries Society, Monograph No. 1. Washington, D.C.

Marcy, B. C., Jr., and P. M. Jacobson. 1976. Early life history studies of American shad in the lower Connecticut River and the effects of the Connecticut Yankee Plant. Pp. 141–168 in D. Merriman

and L. M. Thorpe, eds. The Connecticut River Ecological Study. American Fisheries Society, Monogr. 1. Washington, D.C.

Marcy, B. C., Jr., and F. P. Richards. 1974. Age and growth of the white perch, *Morone americana*, in the lower Connecticut River. Trans. Am. Fish. Soc. 103:117–120.

Marguiles, D. 1988. Effects of food concentrations and temperature on development, growth and survival of white perch, *Morone americana*, eggs and larvae. U.S. Fish. Bull. 87(1):63–72.

Markle, D. F., and G. C. Grant. 1970. The summer food habits of young-of-the year striped bass in three Virginia rivers. Chesapeake Sci. 11(1):50–54.

Markle, D. F., and J. A. Musick. 1974. Benthic-slope fishes found at 900m depth along a transect in the western N. Atlantic Ocean. Mar. Biol. 26:225–233.

Markle, D. F., W. B. Scott, and A. C. Kohler. 1980. New and rare records of Canadian fishes and the influence of hydrography on resident and non-resident Scotian Shelf ichthyofauna. Can. J. Fish. Aqua. Sci. 37:49–65.

Markle, D. F., D. A. Methven, and L. J. Coates-Markle. 1982. Aspects of spatial and temporal cooccurrence in the life history stages of the sibling hakes, *Urophycis chuss* (Walbaum 1792) and *Urophycis tenuis* (Mitchill 1815) (Pisces: Gadidae). Can. J. Zool. 60(9):2057–2078.

Markov, K. P. 1978. Adhesiveness of egg membranes in sturgeons (Family Acipenseridae). J. Ichthyol. 18:437–446.

Marks, R. E., F. Juanes, J. A. Hare, and D. O. Conover. 1996. Occurrence and effect of the parasitic isopod *Lironeca ovalis* (Isopoda: Cymothoidae), on young-of-the-year bluefish, *Pomatomus saltatrix* (Pisces: Pomatomidae). Can. J. Fish. Aquat. Sci. 53:2052–2057.

Marshall, N. 1946. Observations on the comparative ecology and life history of two sea robins, *Prionotus carolinus* and *Prionotus evolans strigatus*. Copeia 3:118–144.

Marteinsdottir, G. 1991. Early life history of the mummichog (*Fundulus heteroclitus*): egg size variation and its significance in reproduction and survival of eggs and larvae. Ph.D. diss., Rutgers University, New Brunswick, N.J.

Marteinsdottir, G., and K. W. Able. 1988. Geographic variation in egg size among populations of the mummichog, *Fundulus heteroclitus* (Pisces: Fundulidae). Copeia 2:471–478.

Martin, F. D., and G. E. Drewry. 1978. Development of Fishes of the Mid-Atlantic Bight: An Atlas of Egg, Larval and Juvenile Stages. Vol. 6, Stromateidae through Ogcocephalidae. Chesapeake Biological Laboratory, Center for Environmental and Estuarine Studies, University of Maryland. Prepared for U.S. Fish and Wildlife Service. FWS/OBS-78/12. U.S. Government Printing Office, Washington.

Martin, M. J. 1995. The effects of temperature, river flow, and tidal cycles on the onset of glass eel and elver migration into fresh water in the American eel. J. Fish Biol. 46:891–902.

Martino, E. 2001. Variation in fish assemblages across the marine to low-salinity transition zone of a temperate estuary. M.S. thesis. Rutgers University, New Brunswick, N.J. 134 pp.

———. 2008. Environmental controls and biological constraints on recruitment of striped bass *Morone saxatilis* in Chesapeake Bay. Ph.D. diss. University of Maryland, College Park. 273 pp.

Martino, E., and K. W. Able. 2003. Fish assemblages across the marine to low salinity transition zone of a temperate estuary. Estuar. Coast. Shelf Sci. 56(5–6):967–985.

Massmann, W. H. 1954. Marine fishes in fresh and brackish waters of Virginia rivers. Ecology 35(1):75–78.

Massmann, W. H., J. P. Whitcomb, and A. L. Pacheco. 1958. Distribution and abundance of gray weakfish in the York River system, Virginia. Trans. N. Am. Wildl. Natl. Resour. Conf. 23: 361–369.

Massmann, W. H., J. J. Norcross, and E. B. Joseph. 1961. Fishes and fish larvae collected from Atlantic plankton cruises of R/V *Pathfinder*, Dec. 1959–Dec. 1960. Va. Fish. Lab. Spec. Rep. 26.

Massmann, W. H., E. B. Joseph, and J. J. Norcross. 1962. Fishes and fish larvae collected from Atlantic plankton cruises of R/V *Pathfinder*, March 1961–March 1962. Va. Inst. Mar. Sci. Spec. Sci. Rep. No. 33.

Massmann, W. H., J. J. Norcross, and E. B. Joseph. 1963. Distribution of larvae of the naked goby, *Gobiosoma bosc*, in the York River. Chesapeake Sci. 4(3):120–125.

Matallanas, J., and G. Riba. 1980. Aspects biologicos de *Ophidion barbatum* Linnaeus, 1758 y *O. rochei* Muller, 1845 (Pisces, Ophidiidae) de la costa catalana. Invest. Pesq. 44:399–406.

Maurer, R. O., and R. E. Bowman. 1975. Food chain investigations: food habits of marine fishes of the northwest Atlantic: data report. NMFS Woods Hole Lab. Ref. Doc. No. 75-3. 90 pp.

May, R. C. 1974. Larval mortality in marine fishes and the critical period concept. Pp. 3–19 *in* J. S. Blaxter, ed. The Early Life History of Fish. Springer, New York.

Mayo, R. K. 1974. Population structure, movement, and fecundity of the anadromous alewife, *Alosa pseudoharengus* (Wilson), in the Parker River, Massachusetts, 1971–1972. M.S. thesis. University of Massachusetts, Amherst.

———. 1982. Blueback herring, *Alosa aestivalis*. (Mar. Ecosyst. Anal.) New York Bight Atlas Monogr. 15:54–57.

Mayo, R. K., J. M. McGlade, and S. H. Clark. 1989. Patterns of exploitation and biological status of pollock (*Pollachius virens* L.) in the Scotian Shelf, Georges Bank, and Gulf of Maine area. J. Northw. Atl. Fish. Sci. 9:13–36.

McAllister, D. E. 1960. Sand-hiding behavior in young white hake. Canad. Field Natur. 74(4):177.

McBride, R. A., and T. F. Moslow. 1991. Origin, evolution and distribution of shoreface sand ridges, Atlantic inner shelf, U.S.A. Mar. Geol. 97:57–85.

McBride, R. S. 1995. Perennial occurrence and fast growth rates by crevalle jacks (Carangidae: *Caranx hippos*) in the Hudson River estuary. Section VI: pp. 34 *in* E. A. Blair and J. R. Waldman, eds. Final reports of the Tibor T. Polgar Fellowship Program, 1994. Hudson River Foundation, New York.

McBride, R. S., and K. W. Able. 1994. Reproductive seasonality, distribution, and abundance of *Prionotus carolinus* and *P. evolans* (Pisces: Triglidae) in the New York Bight. Estuar. Coast. Shelf Sci. 38:173–188.

———. 1998. Ecology and fate of butterflyfishes, *Chaetodon* spp., in the temperate, western north Atlantic. Bull. Mar. Sci. 63(2): 401–416.

McBride, R. S., and D. O. Conover. 1991. Recruitment of young-of-the-year bluefish *Pomatomus saltatrix* to the New York Bight: variation in abundance and growth of spring- and summer-spawned cohorts. Mar. Ecol. Prog. Ser. 78:205–216.

McBride, R. S., and K. A. McKown. 2000. Consequences of dispersal of subtropically spawned crevalle jacks, *Caranx hippos*, to temperate estuaries. Fish. Bull. 98:528–538.

McBride, R. S., J. L. Ross, and D. O. Conover. 1993. Recruitment of bluefish *Pomatomus saltatrix* to estuaries of the U.S. South Atlantic Bight. Fish. Bull. U.S. 91:389–395.

McBride, R. S., M. D. Scherer, and J. D. Powell. 1995. Correlated variations in abundance, size, growth and loss rates of age-0 bluefish in a southern New England estuary. Trans. Am. Fish. Soc. 124:898–910.

McBride, R. S., M. P. Fahay, and K. W. Able. 2002. Larval and settlement periods of the northern searobin (*Prionotus carolinus*) and the striped searobin (*P. evolans*). Fish. Bull. 100:63–73.

McCambridge, J. T., Jr., and R. W. Alden III. 1984. Growth of juvenile spot, *Leiostomus xanthurus* Lacepede, in the nursery region of the James River, Virginia. Estuaries 7(4b):478–486.

McCandless, C. T., H. L. Pratt Jr., N. E. Kohler, R. R. Merson, and C. W. Recksiek. 2007. Distribution, localized abundance, movements, and migrations of juvenile sandbar sharks tagged in Delaware Bay. Pp. 45–62 *in* C. T. McCandless, N. E. Kohler, and H. L. Pratt Jr., eds. Shark Nursery Grounds of the Gulf of Mexico and the East Coast Waters of the United States. Am. Fish. Soc. Symp. 50, Bethesda, Md.

McCleave, J. D. 2008. Contrasts between spawning times of *Anguilla* species estimated from larval sampling at sea and from otolith analysis of recruiting glass eels. Mar. Biol. 155:249–262.

McCleave, J. D., and D. J. Jellyman. 2002. Discrimination of New Zealand stream waters by glass eels of *Anguilla australis* and *Anguilla dieffenbachii*. J. Fish Biol. 61:785–800.

McCleave, J. D., and R. C. Kleckner. 1982. Selective tidal stream transport in the estuarine migration of glass eels of the American eel (*Anguilla rostrata*). Journal du Conseil International pour l'Exploration de la Mer 40:262–271.

McCleave, J. D., and M. J. Miller. 1994. Spawning of *Conger oceanicus* and *Conger triporiceps* (Congridae) in the Sargasso Sea and subsequent distribution of leptocephali. Env. Biol. Fishes 39: 339–355.

McCleave, J. D., S. M. Fried, and A. K. Towt. 1977. Daily movements of shortnose sturgeon, *Acipenser brevirostrum*, in a Maine estuary. Copeia 1:149–157.

McCleave, J. D., G. P. Arnold, J. J. Dodson, and W. H. Neill, eds. 1984. Mechanisms of Migration in Fishes. Plenum Press, New York.

McDermott, V. 1971. Study of the ichthyoplankton associated with two of New Jersey's coastal inlets. N.J. Dep. Environ. Protect. Div. Fish, Game, Shellfish Bur. Misc. Rep. No. 7M.

McDowall, R. M. 2008. Diadromy, history and ecology: a question of scale. Hydrobiologia 602:5–14.

McEachran, J. D. 2002. Cownose rays. Family Rhinopteridae. Pp. 78–79 *in* B. B. Collette and G. Klein-MacPhee, eds. Bigelow and Schroeder's Fishes of the Gulf of Maine. 3rd ed. Smithsonian Institution Press, Washington, D.C.

McEachran, J. D., and J. Davis. 1970. Age and growth of the striped searobin. Trans. Am. Fish. Soc. 99(2):343–352.

McEachran, J. D., and J. D. Fechhelm. 1998. Fishes of the Gulf of Mexico: Myxiniformes to Gasterosteiformes. Vol. 1. University of Texas Press, Austin. 1112 pp.

McErlean, A. J. 1963. A study of the age and growth of the gag, *Mycteroperca microlepis* Goode and Bean (Pisces: Serranidae) on the west coast of Florida. Florida State Bd. Conserv. Mar. Lab. 41:1–29.

McFarlane, G. A., J. R. King, and R. J. Beamish. 2000. Have there been recent changes in climate? Ask the fish. Prog. Oceanogr. 47:147–169.

McGovern, J. C., and J. E. Olney. 1996. Factors affecting survival of early life stages and subsequent recruitment of striped bass on the Pamunkey River, Virginia. Can. J. Fish. Aquat. Sci. 53: 1713–1726.

McGrath, P., and H. A. Austin. 2009. Site fidelity, home range and tidal movements of white perch during the summer in two small tributaries of the York River, Virginia. Trans. Am. Fish. Soc. 138:966–974.

McGurk, M. S. 1984. Effects of delayed feeding and temperature on the age of irreversible starvation and on the rates of growth and mortality of Pacific herring larvae. Mar. Biol. 84:13–26.

McHugh, J. L. 1967. Estuarine nekton. Am. Assoc. Adv. Sci. Publ. 83:581–620.

McHugh, J. L., R. T. Oglesby, and A. L. Pacheco. 1959. Length, weight and age composition of the menhaden catch in Virginia waters. Limnology and Oceanography 4:145–162.

McIvor, C. C., and W. E. Odum. 1988. Food, predation risk and microhabitat selection in a marsh fish assemblage. Ecology 69:1341–1351.

McKeown, B. A. 1984. Fish Migration. Croom Helm, London.

McKinnon, L. J., and G. J. Gooley. 1998. Key environmental criteria associated with the invasion of *Anguilla australis* glass eels into estuaries of South-Eastern Australia. Bulletin Francais de la Peche et de la Pisciculture 349:117–128.

McKown, K. A. 1991. An investigation of the movements and growth of the 1989 Hudson River year class. A study of the striped bass in the marine district of New York VI. U.S. Dept. of Commer. Natl. Mar. Fish. Serv. Natl. Ocean. Atmos. Adm. Anadromous Fish Conservation Act P.L. 89-304. Completion Report. New York, Project AFC-14.

McLaren, J. B., T. H. Peck, W. P. Dey, and M. Gardinier. 1988. Biology of Atlantic tomcod in the Hudson River estuary. Trans. Am. Fish. Soc. Monogr. 4:102–112.

McLean, A. J. 2006. The shoreward transport of warm-water transient fish into the southern New England estuaries of Narragansett Bay and Long Island Sound, U.S.A. M.S. thesis. University of Rhode Island, Kingston.

McNatt, R. A., and J. A. Rice. 2004. Hypoxia-induced growth rate reduction in two juvenile estuary-dependent fishes. J. Exp. Mar. Biol. Ecol. 311:147–156.

Medved, R. J., and J. A. Marshall. 1981. Feeding behavior and biology of young sandbarsharks, *Carcharhinus plumbeus* (Pisces, Carcharhinidae), in Chincoteague Bay Virginia. Fish. Bull. 79(3):441–447.

———. 1983. Short-term movements of young sandbar sharks, *Carcharhinus plumbeus* (Pisces, Carcharhinidae). Bull. Mar. Sci. 33(1):87–93.

Meldrim, J. W., and J. J. Gift. 1971. Temperature preference, avoidance and shock experiments with estuarine fishes. Ichthyol. Assoc. Bull. No. 7. Ichthyological Associates, Ithaca, N.Y.

Meng, L., D. L. Taylor, J. Serbst, and J. C. Powell. 2008. Assessing habitat quality of Mount Hope Bay and Narragansett Bay using growth, RNA:DNA, and feeding habits of caged juvenile winter flounder (*Pseudopleuronectes americanus* Walbaum). Northeastern Naturalist 15(1):35–56.

Menhinick, E. F. 1991. The freshwater fishes of North Carolina. Pp. 128–137. North Carolina Wildlife Resources Commission, Raleigh.

Mercer, L. P. 1973. The comparative ecology of two species of pipefish (Syngnathidae) in the York River, Virginia. M.S. thesis. College of William and Mary, Gloucester Point, Va. 37 pp.

———. 1978. The reproductive biology and population dynamics of black sea bass, (*Centropristis striata*). Ph.D. diss. College of William and Mary, Gloucester Point, Va.

———. 1983. A biological and fisheries profile of weakfish, *Cy-*

noscion regalis. Div. Mar. Fish. Spec. Sci. Rep. No. 39. North Carolina Department of Natural Resources and Community Development, Morehead City.

———. 1984a. A biological and fisheries profile of spotted seatrout, *Cynoscion nebulosus*. Div. Mar. Fish. Spec. Sci. Rep. No. 40. North Carolina Department of Natural Resources and Community Development, Morehead City.

———. 1984b. A biological and fisheries profile of red drum, *Sciaenops ocellatus*. Div. Mar. Fish. Spec. Sci. Rep. No. 41, North Carolina Department of Natural Resources and Community Development, Raleigh.

Meredith, W. H., and V. A. Lotrich. 1979. Production dynamics of a tidal creek population of *Fundulus heteroclitus* (Linnaeus). Estuarine Coastal Mar. Sci. 8:99–118.

Merriman, D. 1947. Notes on the mid-summer ichthyofauna of a Connecticut beach at different tide levels. Copeia 4:281–286.

Merriman, D., and R. C. Sclar. 1952. The pelagic fish eggs and larvae of Block Island Sound. Bull. Bingham Oceanogr. Collect. Yale Univ. 13:165–219.

Merriman, D., and H. E. Warfel. 1948. Studies on the marine resources of southern New England. VII. Analysis of a fish population. Bull. Bingham Oceanogr. Collect. Yale Univ. 11(4):131–164.

Merriner, J. V. 1975. Food habits of the weakfish, *Cynoscion regalis*, in North Carolina waters. Chesapeake Sci. 16:74–76.

———. 1976. Aspects of the reproductive biology of the weakfish, *Cynoscion regalis* (Sciaenidae), in North Carolina. U.S. Fish. Bull. 74:18–26.

Merson, R. R., and H. L. Pratt Jr. 2001. Distribution, movements and growth of young sandbar sharks, *Carcharhinus plumbeus*, in the nursery grounds of Delaware Bay. Environ. Bio. Fish. 61:13–24.

———. 2007. Sandbar shark nurseries in New Jersey and New York: evidence of northern pupping grounds along the United States East Coast. Pp. 35–44 in C. T. McCandless, N. E. Kohler, and H. L. Pratt Jr., eds. Shark Nursery Grounds of the Gulf of Mexico and the East Coast Waters of the United States. Am. Fish. Soc. Symp. 50, Bethesda, Md.

Metcalfe, J. D., and G. P. Arnold. 1997. Tracking fish with electronic tags. Nature 387:665–666.

Metcalfe, J. D., G. Arnold, and R. McDowall. 2002. Migration. Pp. 175–199 in P. J. B. Hart and J. D. Reynolds, eds. Handbook of Fish Biology and Fisheries. Vol. 1, Fish Biology. Blackwell Publishing, Malden, Mass.

Metzger, C. V., J. T. Duffy-Anderson, and K. W. Able. 2001. Effects of a municipal pier on growth of young-of-the-year Atlantic tomcod (*Microgadus tomcod*): a study in the Hudson River Estuary. Bull. N.J. Acad. Sci. 46(1):5–10.

Meyer, T. L., R. A. Cooper, and R. W. Langton. 1979. Relative abundance, behavior and food habits of the American sand lance, *Ammodytes americanus*, from the Gulf of Maine. Fish. Bull. 77(1):243–253.

Meyerson, L. A., and H. A. Mooney. 2007. Invasive alien species in an era of globalization. Front. Ecol. Environ. 5:199–208.

Meynecke, J. O., S. Y. Lee, and N. C. Duke. 2008. Linking spatial metrics and fish catch reveals the importance of coastal wetland connectivity to inshore fisheries in Queensland, Australia. Biol. Conserv. 141:981–996.

Middaugh, D. P., and T. Takita. 1983. Tidal and diurnal spawning cues in the Atlantic silverside, *Menidia menidia*. Environ. Biol. Fishes 8:97–104.

Miller, J. M. 1965. A trawl survey of the shallow Gulf fishes near Port Aransas, Texas. Publ. Inst. Mar. Sci., Univ. Tex. 10:80–107.

———. 1994. An overview of the second flatfish symposium: recruitment in flatfish. Neth. J. Sea Res. 32(2):103–106.

Miller, L. W. 1963. Growth, reproduction and food habits of the white perch, *Roccus americanus* (Gmelin) in the Delaware River estuary. M.S. thesis. University of Delaware, Newark. 62 pp.

Miller, M. J. 1995. Species assemblages of leptocephali in the Sargasso Sea and Florida Current. Mar. Ecol. Prog. Ser. 121:11–26.

Miller, M. J., and J. D. McCleave. 1994. Species assemblages of leptocephali in the Subtropical Convergence Zone of the Sargasso Sea. J. Mar. Res. 52:743–772.

———. 2007. Species assemblages of leptocephali in the southwestern Sargasso Sea. Mar. Ecol. Prog. Ser. 344:197–212.

Miller, M. J., D. N. Nemerson, and K. W. Able. 2003. Seasonal distribution, abundance and growth of young-of-the-year Atlantic croaker, *Micropogonias undulatus*, in Delaware Bay and adjacent marshes. Fish. Bull. 101(1):100–115.

Miller, R. J. 1959. A review of the seabasses of the genus *Centropristis* (Serranidae). Tulane Studies Zool. 7(2):35–68.

Miller, R. R. 1972. Threatened freshwater fishes of the United States. Trans. Am. Fish. Soc. 101:239–252.

Miller, R. V. 1969. Continental migrations of fishes. Underw. Natur. 6(1):15–24.

Milstein, C. B. 1981. Abundance and distribution of juvenile *Alosa* species off southern New Jersey. Trans. Amer. Fish. Soc. 110:306–309.

Milstein, C. B., and D. L. Thomas. 1976a. Fishes new or uncommon to the New Jersey coast. Chesapeake Sci. 17(3):198–204.

———. 1976b. First record of the cubera snapper (*Lutjanus cyanopterus*) for New Jersey. Bull. N.J. Acad. Sci. 21(1):13.

———. 1977. Summary of ecological studies for 1972–1975 in the bays and other waterways near Little Egg Inlet and in the ocean in the vicinity of the proposed site for the Atlantic Generation Station, New Jersey. Ich. Assoc. Inc., Bull. No. 18.

Minello, T. J., R. J. Zimmerman, and R. J. Martinez. 1987. Fish predation on juvenile brown shrimp, *Penaeus aztecus* Ives: effects of turbidity and substratum on predation rates. U.S. Fish. Bull. 85:59–70.

Minello, T. J., K. W. Able, M. P. Weinstein, and C. G. Hays. 2003. Salt marshes as nurseries for nekton: testing hypotheses on density, growth, and survival through meta-analysis. Mar. Ecol. Prog. Ser. 246:39–59.

Mitchell, S. C. 2005. How useful is the concept of habitat?—a critique. Oikos 110(3):634–638.

Mitsch, W. J., and R. F. Wilson. 1996. Improving the success of wetland creation and restoration with know-how, time, and self-design. Ecol. Appl. 6:77–83.

Moles, A., and B. L. Norcross. 1995. Sediment preference in juvenile pacific flatfishes. Neth. J. Sea Res. 34(1–3):177–182.

Moller, A. P. 2008. Distribution of arrival dates in a migratory bird in relation to environmental conditions, natural selection and sexual selection. Ethol. Ecol. Evol. 20:193–210.

Monaco, M. E., T. A. Lowery, and R. L. Emmett. 1992. Assemblages of U.S. west coast estuaries based on the distribution of fishes. J. Biogeogr. 19:251–267.

Monteleone, D. M. 1992. Seasonality and abundance of ichthyoplankton in Great South Bay, New York. Estuaries 15(2):230–238.

Montgomery, J. C., N. Tolimieri, and O. S. Haine. 2001. Active habitat selection by pre-settlement reef fishes. Fish Fish. 2:261–277.

Montgomery, J. C., A. Jeffs, S. D. Simpson, M. Meekan, and C. Tindle. 2006. Sound as an orientation cue for the pelagic larvae of reef fishes and decapod crustaceans. Adv. Mar. Biol. 51:143–196.

Montolio, M. 1978. Algunos aspectos sobre el desove y las concentraciones larvarias de las especies de la familia Carangidae en al Mar Caribe. Rev. Cubana Invest. Pesqueras 3:29–49.

Moody, W. D. 1950. A study of the natural history of the spotted sea trout, *Cynoscion nebulosus*, in the Cedar Key, Florida, area. Q. J. Fla. Acad. Sci. 12:147–171.

Moore, E. 1947. Studies on the marine resources of southern New England. VI: The sand flounder, *Lophosetta aquosa* (Mitchill); a general study of the species with special emphasis on age determination by mean of scales and otoliths. Bull. Bingham Oceanogr. Collect. Yale Univ. 11(3):1–79.

Mora, C., and A. F. Ospina. 2002. Experimental effect of cold, La Nina temperatures on the survival of reef fishes from Gorgona Island (east Pacific Ocean). Mar. Biol. 141:789–793.

Morgan, R. P., II, and V. J. Rasin Jr. 1982. Influence of temperature and salinity on development of white perch eggs. Trans. Am. Fish Soc. 111(3):396–398.

Morgan, S. G. 1990. Impact of planktivorous fishes on dispersal, hatching, and morphology of estuarine crab larvae. Ecology 71(5):1639–1652.

Morin, R. P., and K. W. Able. 1983. Patterns of geographic variation in the egg morphology of the fundulid fish, *Fundulus heteroclitus*. Copeia 3:726–740.

Moring, J. R., and L. H. Mink. 2002. Anadromous alewives, *Alosa pseudoharengus*, as prey for white perch, *Morone americana*. Hydrobiologia 479:125–130.

Moring, J. R., and M. E. Moring. 1986. A late leptocephalus stage of a conger eel. *Conger oceanicus*, found in a tidepool. Copeia 1:222–223.

Morioka, T. 2005. Onset of burying behavior concurrent with growth and morphological changes in hatchery-reared Japanese sandfish *Arctoscopus japonicus*. Fish. Sci. 71:242–244.

Morley, J. W. 2004. Ecology of juvenile bluefish (*Pomatomus saltatrix*) overwintering off North Carolina. M.S. thesis. North Carolina State University, Raleigh.

Morley, J. W., J. A. Buckel, and T. E. Lankford Jr. 2007. Winter energy storage dynamics and cohort structure of young-of-the-year bluefish *Pomatomus saltatrix* off North Carolina. Mar. Ecol. Progr. Ser. 334:273–286.

Morrill, A. D. 1895. The pectoral appendages of *Prionotus* and their innervation. J. Morphol. 11:17–192.

Morrison, W. E., and D. H. Secor. 2003. Demographic attributes of yellow-phase American eels (*Anguilla rostrata*) in the Hudson River estuary. Can. J. Fish. Aquat. Sci. 60:1487–1501.

Morrow, J. E., Jr. 1951. Studies on the marine resources of southern New England. VIII. The biology of the longhorn sculpin, *Myoxocephalus octodecimspinosus* Mitchill, with a discussion of the southern New England "trash" fishery. Bull. Bingham Oceanogr. Coll. Yale Univ. 13(2):1–89.

Morse, W. W. 1978. Biological and fisheries data on scup, *Stenotomus chrysops* (Linnaeus). Sandy Hook Laboratory, N.E. Fish. Cent. Nat. Mar. Fish. Serv. Natl. Ocean. Atmos. Adm. Tech. Ser. Rep. No. 12:1–41.

———. 1980. Maturity, spawning, and fecundity of Atlantic croaker, *Micropogonias undulatus*, occurring north of Cape Hatteras, North Carolina. U.S. Fish. Bull. 78(1):190–195.

———. 1981. Reproduction of the summer flounder, *Paralichthys dentatus* (L.). J. Fish. Biol. 19:189–203.

————. 1982. Scup, *Stenotomus chrysops*. (Mar. Ecosyst. Anal.) New York Bight Atlas Monogr. 15:89–91.

Morse, W. W., and K. W. Able. 1995. Distribution and life history of windowpane, *Scophthalmus aquosus*, off the northeastern United States. U.S. Fish. Bull 93:675–693.

Morse, W. W., M. P. Fahay, and W. G. Smith. 1987. MARMAP surveys of the continental shelf from Cape Hatteras, North Carolina to Cape Sable, Nova Scotia (1977–1984). Atlas No. 2. Annual distribution patterns of fish larvae. NOAA Tech. Memo. NMFS NE 47:1–215.

Morsell, J. W., and C. R. Norden. 1968. Food habits of the alewife, *Alosa pseudoharengus* (Wilson), in Lake Michigan. Pp. 96–102. Proc. 11th Conf. Great Lakes Res.

Morton, T. 1989. Species profiles: life histories and environmental requirements of coastal fishes and invertebrates (Mid-Atlantic)-bay anchovy. U.S. Fish Wildl. Serv. Biol. Rep. 82(11.97):1–13.

Moser, H. G. 1981. Morphological and functional aspects of marine fish larvae. Pp. 89–131 *in* R. Lasker, ed. Marine Fish Larvae, Morphology, Ecology and Relation to Fisheries. University of Washington Press, Seattle.

Moser, M. L., and S. W. Ross. 1994. Effects of changing current regime and river discharge on the estuarine phase of anadromous fish migration. Pp. 343–347 *in* K. R. Dyer and R. J. Orth, eds. Changes in Fluxes in Estuaries. Fredensborg, Denmark, Olsen and Olsen, Fredensborg.

————. 1995. Habitat use and movements of shortnose and Atlantic sturgeons in the lower Cape Fear River, North Carolina. Trans. Amer. Fish. Soc. 124:225–234.

Moss, S. A. 1972. Tooth replacement and body growth rates in the smooth dogfish, *Mustelus canis* (Mitchill). Copeia 4:808–811.

————. 1973. The responses of planehead filefish, *Monacanthus hispidus* (Linnaeus), to low temperature. Chesapeake Sci. 14: 300–303.

Mountain, D. G. 2002. Potential consequences of climate change for the fish resources in the Mid-Atlantic Region. Pp. 185–194 *in* N. A. McGinn, ed. Fisheries in a Changing Climate. Am. Fish. Soc. Symp. 32.

————. 2003. Variability in the properties of shelf water in the Middle Atlantic Bight, 1977–1999. J. Geophys. Res. 108(3014): 14-1–14-11.

Moyle, P. B., and J. J. Cech Jr. 1996. Fishes: An Introduction to Ichthyology. 3rd ed. Prentice-Hall, Upper Saddle River, N.J.

Mulkana, M. S. 1966. The growth and feeding habits of juvenile fishes in two Rhode Island estuaries. Gulf Res. Rep. 2:97–167.

Mullen, D. M., C. W. Fay, and J. R. Moring. 1986. Species profiles: life histories and environmental requirements of coastal fishes and invertebrates (North Atlantic). Alewife/blueback herring. U.S. Fish Wildl. Serv. Coastal Ecology Group, Waterways Exp. Stn. Biol. Rep. 82(11.56).

Mulligan, T. J., and R. W. Chapman. 1989. Mitochondrial DNA analysis of Chesapeake white perch, *Morone americana*. Copeia: 679–688.

Munch, S. B., and D. O. Conover. 2000. Recruitment dynamics of bluefish (*Pomatomus saltatrix*) from Cape Hatteras to Cape Cod, 1973–1995. ICES J. Mar. Sci. 57:393–402.

Münchow, A., and R. J. Chant. 2000. Kinematics of inner shelf motions during the summer stratified season off New Jersey. J. Phys. Oceanogr. 30:247–268.

Munroe, T. A. 1998. Systematics and ecology of tonguefishes of the genus *Symphurus* (Cynoglossidae: Pleuronectiformes) from western North Atlantic Ocean. Fish. Bull. U.S. 96(1):1–182.

————. 2002. Herrings. Family Clupeidae. Pp. 111–160 *in* B. B. Collette and G. Klein-MacPhee, eds. Bigelow and Schroeder's Fishes of the Gulf of Maine. 3rd ed. Smithsonian Institution Press, Washington and London.

Munroe, T. A., and R. A. Lotspeich. 1979. Some life history aspects of the seaboard goby (*Gobiosoma ginsburgi*) in Rhode Island. Estuaries 2(1):22–27.

Munroe, T. A., and M. S. Nizinski. 2002. Clupeidae. Pp. 804–830 *in* K. E. Carpenter, ed. The Living Marine Resources of the Western Central Atlantic. Vol. 2, Bony Fishes, Part 1 (Acipenseridae to Grammatidae). Food and Agriculture Organization of the United Nations, Rome.

Murawski, S. A. 1993. Climate change and marine fish distributions: forecasting from historical analogy. Trans. Am. Fish. Soc. 122:647–658.

Murawski, W. S. 1969. The distribution of striped bass, *Roccus saxatilis*, eggs and larvae in the lower Delaware River. N.J. Dept. Conserv. Econ. Develop. Bur. Fish. Misc. Rep. No. 1M, Nacote Creek, N.J.

Murawski, W. S., and P. J. Festa. 1979. Creel census of the summer flounder, *Paralichthys dentatus* sportfishery in Great Bay, New Jersey. N.J. Dep. Environ. Protect. Div. Fish Game Shellfish. Nacote Creek Res. Stn. N.J. Tech. Rep. No. 19M. Nacote Creek, N.J.

Murchelano, R. A., and J. Ziskowski. 1982. Finrot disease in the New York Bight (1973–1977). Pp. 347–358 *in* Mayer, G. F., ed. Ecological Stress and the New York Bight: Science and Management. Estuarine Research Federation, Columbia, S.C.

Murdy, E. O. 2002. Gobiidae. Pp. 1781–1796 *in* K. E. Carpenter, ed. The Living Marine Resources of the Western Central Atlantic. Vol. 2, Bony Fishes, Part 2 (Opistognathidae to Molidae). FAO species identification guide for fishery purposes and American Society of Ichthyologists and Herpetologists Special Publ. No. 5. Rome, FAO.

Murdy, E. O., R. S. Birdsong, and J. A. Musick. 1997. Fishes of Chesapeake Bay. Smithsonian Institute Press, Washington, D.C. 324 pp.

Musick, J. A. 1969. The comparative biology of two American Atlantic hakes, *Urophycis chuss* and *U. tenuis* (Pisces: Gadidae). Ph.D. diss. Harvard University, Cambridge, Mass. 150 pp.

————. 1972. Fishes of Chesapeake Bay and the adjacent coastal plain. Pp. 175–212 *in* M. L. Wass et al. A check list of the biota of lower Chesapeake Bay, with inclusions from the upper bay and the Virginian Sea. Va. Inst. Mar. Sci. Spec. Sci. Rep. No. 65. Gloucester Point, Va.

————. 1974. Seasonal distribution of sibling hakes, *Urophycis chuss* and *U. tenuis* (Pisces: Gadidae) in New England. U.S. Fish. Bull. 72(2):481–495.

Musick, J. A., J. A. Colvocoresses, and E. J. Foell. 1989. Seasonality and the distribution, availability and composition of fish assemblages in Chesapeake Bight. International Symposium on Utilization of Coastal Ecosystems: Planning Pollution and Productivity, Contribution No. 1168 from the Virginia Institute of Marine Science, Gloucester Point, Va.

Myers, R. A., and B. Worm. 2003. Rapid worldwide depletion of predatory fish communities. Nature 423:280–283.

Myers, R. A., J. K. Baum, T. D. Shepherd, S. P. Powers, and C. H. Peterson. 2007. Cascading effects of the loss of apex predatory sharks from a coastal ocean. Science 315:1846–1850.

Naplin, A., and C. L. Obenchain. 1980. A description of eggs and larvae of the snake eel, *Pisodonophis cruentifer* (Ophichthidae). Bull. Mar. Sci. 30:413–423.

National Research Council. 1992. Restoration of Aquatic Ecosystems. National Academy Press, Washington, D.C. 552 pp.

———. 2001. Compensating for Wetland Losses Under the Clean Water Act. National Academy Press, Washington, D.C. 322 pp.

Necaise, A. S., S. W. Ross, and J. M. Miller. 2005. Estuarine habitat evaluation measured by growth of juvenile summer flounder *Paralichthys dentatus* in a North Carolina estuary. Mar. Ecol. Prog. Ser. 285:157–168.

Neckles, H. A., M. Dionne, D. M. Burdick, C. T. Roman, R. Buchsbaum, and E. Hutchins. 2002. A monitoring protocol to assess tidal restoration of salt marshes on local and regional scales. Rest. Ecol. 10:556–563.

Nelson, G. A., C. B. Chase, and J. Stockwell. 2003. Food habits of striped bass (*Morone saxatilis*) in coastal waters of Massachusetts. J. Northw. Atl. Fish. Sci. 23:1–25.

Nelson, J. S., E. J. Crossman, H. Espinosa-Perex, L. T. Findley, C. R. Gilbert, R. N. Lea, and J. D. Williams. 2004. Common and scientific names of fishes from the United States, Canada, and Mexico. Am. Fish. Soc. Special Pub. 29, Bethesda, Md.

Nelson, T. A., P. E. Gadd, and T. L. Clarke. 1978. Wind-induced current flow in the upper Hudson Valley. J. Geophys. Res. 83(c12):6073–6082.

Nelson, T. C. 1928. On the association of the common goby (*Gobiosoma bosci*) with the oyster, including a case of parasitism. Nat. Hist. 28(1):78–84.

Nemerson, D. M. 2001. Trophic dynamics and habitat ecology of the dominant fish of Delaware Bay (USA) marsh creeks. Ph.D. diss. Rutgers University, New Brunswick, N.J.

Nemerson, D. M., and K. W. Able. 2003. Spatial and temporal patterns in the distribution and feeding habits of *Morone saxatilis*, in marsh creeks of Delaware Bay, USA. Fish. Manag. Ecol. 20: 337–348.

———. 2004. Spatial patterns in diet and distribution of juveniles of four fish species in Delaware Bay marsh creeks: factors influencing fish abundance. Mar. Ecol. Prog. Ser. 276:249–262.

———. 2005. Juvenile sciaenid fishes respond favorably to marsh restoration in Delaware Bay. Ecol. Eng. 25:260–274.

Nemtzov, S. C. 1994. Intraspecific variation in sand-diving and predator avoidance behavior of green razorfish, *Xyrichtys splendens* (Pisces, Labridae): effect on courtship and mating success. Environ. Biol. Fishes 41:403–414.

Nero, L. 1976. The natural history of the naked goby (*Gobiosoma bosci*) (Perciformes: Gobiidae). M.S. thesis. Old Dominion University, Norfolk, Va.

Netzel, J., and E. Stanek. 1966. Some biological characteristics of blueback and alewife from Georges Bank, July and October, 1964. Int. Comm. Northwest Atl. Fish. Res. Bull. No. 3.

Neuman, M. 1996. Evidence of upwelling along the New Jersey coastline and the south shore of Long Island, New York. Bull. N.J. Acad. Sci. 41(1):7–13.

Neuman, M. J., and K. W. Able. 1998. Experimental evidence of sediment preference by early life history stages of windowpane, (*Scophthalmus aquosus*). J. Sea Res. 40:33–41.

———. 2003. Inter-cohort differences in spatial and temporal settlement patterns of young-of-the-year windowpane, *Scophthalmus aquosus*, in southern New Jersey. Estuar. Coast. Shelf Sci. 56:527–538.

———. 2009. Overwinter growth and mortality of young-of-the-year windowpanes: cohort-specific responses. Marine and Coastal Fisheries: Dynamics, Management, and Ecosystem Science 1:133–142.

Neuman, M. J., D. A. Witting, and K. W. Able. 2001. Relationships between otolith microstructure, otolith growth, sematic growth, otogentic transitions in two cohorts of windowpane. Journ. Fish. Biol. 58:967–984.

Neuman, M. J., K. W. Able, and S. M. Glenn. 2002. The effects of upwelling on larval fish occurrence and abundance in the Jacques Cousteau National Estuarine Research Reserve at Mullica River–Great Bay (JCNERR). Rutgers Univ. Inst. Mar. Coast. Sci. Tech. Rep. #100–17.

Neuman, M. J., G. Ruess, and K. W. Able. 2004. Species composition and food habits of dominant predators in salt marshes in an urbanized estuary, the Hackensack Meadowlands, New Jersey. Urban Hab. 2(1):3–22.

Neves, R. J. 1981. Offshore distribution of alewife, *Alosa pseudoharengus*, and blueback herring, *Alosa aestivalis*, along the Atlantic Coast. U.S. Fish. Bull. 79:473–485.

Neville, W. C., and G. B. Talbot. 1964. The fishery for scup with special reference to fluctuations in yield and their causes. U.S. Fish Wildl. Serv. Spec. Sci. Rep. Fish. 459.

Newberger, T. A., and E. D. Houde. 1995. Population biology of bay anchovy *Anchoa mitchilli* in the mid-Chesapeake. Mar. Ecol. Prog. Ser. 116:25–37.

Newman, H. H. 1908. The process of heredity as exhibited by the development of *Fundulus* hybrids. J. Exp. Zool. 5:503–561.

———. 1909. The question of viviparity in *Fundulus majalis*. Science 30:769–771.

———. 1914. Modes of inheritance in teleost hybrids. J. Exp. Zool. 16:447–499.

Ng, C., K. W. Able, and T. M. Grothues. 2007. Habitat use, site fidelity and movement of adult striped bass in a southern New Jersey estuary based on mobile acoustic telemetry. Trans. Am. Fish. Soc. 136:1344–1355.

Nichols, J. T., and C. M. Breder. 1927. The marine fishes of New York and southern New England. Zoologica, NY 9(1):1–192.

Nicholson, W. R. 1971. Coastal movements of Atlantic menhaden as inferred from changes in age and length distributions. Trans. Am. Fish. Soc. 100:708–716.

Niklitschek, E. J., and D. H. Secor. 2005. Modeling spatial and temporal variation of suitable nursery habitats for Atlantic sturgeon in the Chesapeake Bay. Estuar. Coast. Shelf Sci. 64:135–148.

Nitschke, P., M. Mather, and F. Juanes. 2001. A comparison of length-, weight-, and age-specific fecundity relationships for cunner in Cape Cod Bay. N. Am. J. Fish. Manage. 21(1):86–95.

Nixon, S. W. 1980. Between coastal marshes and coastal waters—a review of twenty years of speculation and research on the role of salt marshes in estuarine productivity and water chemistry. Pp. 437–525 *in* P. Hamilton and K. B. MacDonald, eds. Estuarine and Wetland Processes. Plenum, New York.

Nixon, S. W., S. Granger, B. A. Buckley, M. Lamont, and B. Rowell. 2004. A one hundred and seventeen year coastal water temperature record from Woods Hole, Massachusetts. Estuaries 27(3):397–404.

Nizinski, M. S., and T. A. Munroe. 2002. Order Clupeiformes: Engraulidae: Anchovies. Pp. 764–794 *in* K. E. Carpenter, ed. The Living Marine Resources of the Western Central Atlantic. Vol. 2, Bony Fishes, Part 1 (Acipenseridae to Grammatidae). Food and Agriculture Organization of the United Nations, Rome.

Nizinski, M. S., B. B. Collette, and B. B. Washington. 1990. Separation of two species of sand lances, *Ammodytes americanus* and *A. dubius*, in the western North Atlantic. U.S. Fish. Bull. 88: 241–255.

NOAA (National Oceanic and Atmospheric Administration). 1990. National coastal wetlands inventory: the distribution and areal extent of coastal wetlands in estuaries of the mid-Atlantic region.

———. 1996a. Magnuson-Stevens Fishery Conservation and Management Act amended through 11 October 1996. NMFS, NOAA Tech. Memo. NMFS-F/SPO-23. U.S. Dept. Commerce, Washington, D.C.

———. 1996b. NOAA's Estuarine Eutrophication Survey. Vol. 1, South Atlantic Region. Silver Spring, Md. Office of Ocean Resources Conservation Assessment. 50 pp.

———. 1997a. NOAA's Estuarine Eutrophication Survey. Vol. 2, Mid-Atlantic Region. Silver Spring, Md. Office of Ocean Resources Conservation Assessment. 51 pp.

———. 1997b. NOAA's Estuarine Eutrophication Survey. Vol. 3, North Atlantic Region. Silver Spring, Md. Office of Ocean Resources Conservation Assessment. 45 pp.

Norcross, B. L. 1983. Climate scale environmental factors affecting year-class fluctuations of Atlantic croaker (*Micropogonias undulatus*) in the Chesapeake Bay. Ph.D. diss. College of William and Mary, Williamsburg, Va. 388 pp.

———. 1991. Estuarine recruitment mechanisms of larval Atlantic croakers. Trans. Am. Fish. Soc. 120:673–683.

Norcross, B. L., and D. A. Bodolus. 1991. Hypothetical northern spawning limit and larval transport of spot. Natl. Ocean. Atmos. Adm. Tech. Rep. Natl. Mar. Fish. Serv. 95:77–88.

Norcross, J. J., S. L. Richardson, W. H. Massmann, and E. B. Joseph. 1974. Development of young bluefish (*Pomatomus saltatrix*) and the distribution of eggs and young in Virginia coastal waters. Trans. Am. Fish. Soc. 103:477–497.

Nordlie, F. G. 2003. Fish communities of estuarine salt marshes of eastern North America, and comparisons with temperate estuaries of other continents. Rev. Fish Biol. Fish. 13:281–325.

Nordlie, F. G., W. A. Szelistowski, and W. C. Nordlie. 1982. Ontogenesis of osmotic regulation in the striped mullet, *Mugil cephalus* L. J. Fish. Biol. 20:79–86.

North, E. W., and E. D. Houde. 2001. Retention of white perch and striped bass larvae: biological-physical interactions in Chesapeake Bay Estuarine Turbidity Maximum. Estuaries 24:756–769.

———. 2003. Linking ETM physics, zooplankton prey, and fish early-life histories to white perch *Morone Americana* and striped bass *M. saxatilis* recruitment success. Mar. Ecol. Prog. Ser. 260:219–236.

———. 2004. Distribution and transport of bay anchovy (*Anchoa mitchilli*) eggs and larvae in Chesapeake Bay. Est. Coast. Shelf Sci. 60:409–429.

———. 2006. Retention mechanisms of white perch (*Morone americana*) and striped bass (*Morone saxatilis*) early-life stages in an estuarine turbidity maximum: an integrative fixed-location and mapping approach. Fish. Oceanog. 15(6):429–450.

Northcote, T. G. 1978. Migratory strategies and production in freshwater fishes. Pp. 326–359 *in* E. D. Gerking, ed. Ecology of Freshwater Fish Populations. John Wiley and Sons, New York.

Nye, J. A., T. E. Targett, and T. E. Helser. 2008. Reproductive characteristics of weakfish in Delaware Bay: implications for management. N. Am. J. Fish. Manage. 28:1–11.

Nyman, R. M., and D. O. Conover. 1988. The relation between spawning season and the recruitment of young-of-the-year bluefish, *Pomatomus saltatrix*, to New York. U.S. Fish. Bull. 86:237–250.

O'Brien, L., J. Burnett, and R. K. Mayo. 1993. Maturation of nineteen species of finfish off the northeast coast of the United States, 1985–1990. Natl. Ocean. Atmos. Adm. Tech. Rep. Natl. Mar. Fish. Serv. 113:1–66.

O'Connell, M. T., R. C. Cashner, and C. S. Schieble. 2004. Fish assemblage stability over fifty years in the Lake Pontchartrain estuary; comparisons among habitats using canonical correspondence analysis. Estuaries 27(5):807–817.

O'Connor, J. M. 1972. Tidal activity rhythm in the hogchoker *Trinectes maculatus* (Bloch and Schneider). J. Exp. Mar. Biol. Ecol. 9:173–177.

O'Connor, N., and J. Bruno. 2007. Predatory fish loss affects the structure and functioning of a model marine food web. Oikos 116:2027–2038.

Odum, E. P. 1959. Fundamentals of Ecology. 2nd ed. Saunders, Philadelphia.

Odum, W. E. 1970. Utilization of the direct grazing and plant detritus food chains by the striped mullet *Mugil cephalus*. Pp. 222–240 *in* J. J. Steele, ed. Marine Food Chains. Oliver and Boyd, Edinburgh.

Officer, C. B., R. B. Biggs, J. L. Taft, L. E. Cronin, M. A. Tyler, and W. R. Boynton. 1984. Chesapeake Bay anoxia: origin, development and significance. Science 223:22–27.

Ogburn-Matthews, M. V., and D. M. Allen. 1993. Interactions among some dominant estuarine nekton species. Estuaries 16(4):840–850.

Ogden, J. C. 1970. Relative abundances, food habits, and age of the American eel, *Anguilla rostrata* (LeSueur), in certain New Jersey streams. Trans. Am. Fish. Soc. 99:54–59.

O'Herron, J. C., K. W. Able, and R. W. Hastings. 1993a. Seasonal distribution and movements of shortnose sturgeon (*Acipenser brevirostrum*) in the Delaware River. Estuaries 16(2):235–240.

O'Herron, J. C., II, T. Lloyd, and K. Laidig. 1994. A survey of fish in the Delaware Estuary from the area of the Chesapeake and Delaware Canal to Trenton. Prepared for Delaware Estuary Program, EPA-Region III, Philadelphia.

Oliveira, K. 1999. Life history characteristics and strategies of the American eel, *Anguilla rostrata*. Can. J. Fish. Aquat. Sci. 56(5):795–802.

Olla, B. L., C. E. Samet, and A. L. Studholme. 1972. Activity and feeding behavior of the summer flounder (*Paralichthys dentatus*) under controlled laboratory conditions. Fish. Bull. 70:1127–1136.

Olla, B. L., A. J. Bejda, and A. D. Martin. 1974. Daily activity, movements, feeding, and seasonal occurrence in the tautog, *Tautoga onitis*. U.S. Fish. Bull. 72:27–35.

———. 1975. Activity, movements, and feeding behavior of the cunner, *Tautogolabrus adspersus*, and comparison of food habits with young tautog, *Tautoga onitis*, off Long Island, New York. Fish. Bull. 73(4):895–900.

———. 1979. Seasonal dispersal and habitat selection of cunner, *Tautogolabrus adspersus*, and young tautog, *Tautoga onitis*, in Fire Island inlet, Long Island, New York. U.S. Fish. Bull. 77:255–261.

Olney, J. E. 1983. Eggs and early larvae of the bay anchovy, *Anchoa mitchilli*, and the weakfish, *Cynoscion regalis*, in lower Chesapeake Bay with notes on associated ichthyoplankton. Estuaries 6:20–35.

Olney, J. E., and G. W. Boehlert. 1988. Nearshore ichthyoplankton associated with seagrass beds in the lower Chesapeake Bay. Mar. Ecol. Progr. Ser. 45:33–43.

Olney, J. E., and G. C. Grant. 1976. Early planktonic larvae of the Blackcheek Tonguefish, *Symphurus plagiusa* (Pisces: Cynoglossidae), in the Lower Chesapeake Bay. Chesapeake Sci. 17(4): 229–237.

Olney, J. E., G. C. Grant, F. E. Schultz, C. L. Cooper, and J. Hageman. 1983. Pterygiophore-interdigitation patterns in larvae of four *Morone* species. Trans. Am. Fish. Soc. 112:525–531.

O'Neil, S. P., and M. P. Weinstein. 1987. Feeding habits of spot, *Leiostomus xanthurus*, in polyhaline versus meso-oligohaline tidal creeks and shoals. Fish. Bull. 85(4):785–796.

Orth, R. J., and K. L. Heck Jr. 1980. Structural components of eelgrass (*Zostera marina*) meadows in the lower Chesapeake Bay-fishes. Estuaries 3(4):278–288.

Orth, R. J., and K. A. Moore. 1993. Chesapeake Bay: an unprecedented decline in submerged aquatic vegetation. Science 222: 51–53.

Osgood, D. T., D. J. Yosso, R. M. Chambers, D. Jacobson, T. Hoffman, and J. Wnek. 2003. Tidal hydrology and habitat utilization by resident nekton in *Phragmites* and non-*Phragmites* marshes. Estuaries 26:523–534.

Ottersen, G. 2008. Pronounced long-term juvenation in the spawning stock of Arcto-Norwegian cod (*Gadus morhua*) and possible consequences for recruitment. Can. J. Fish. Aquat. Sci. 65: 523–534.

Overholtz, W. J. 2002. The Gulf of Maine–Georges Bank Atlantic herring (*Clupea harengus*): spatial pattern analysis of the collapse and recovery of a large marine fish complex. Fish. Res. 57:237–254.

Overholtz, W. J., and K. D. Friedland. 2002. Recovery of the Gulf of Maine–Georges Bank Atlantic herring (*Clupea harengus*) complex: perspectives based on bottom trawl survey data. Fish. Bull. 100:593–608.

Overholtz, W. J., and J. S. Link. 2006. Consumption impacts by marine mammals, fish and seabirds on the Gulf of Maine–Georges Bank Atlantic herring (*Clupea harengus*) complex during the year 1977–2002. ICES J. Mar. Sci. 64:83–96.

Overstreet, R. M., and R. W. Heard. 1978. Food of the red drum, *Sciaenops ocellata*, from Mississippi Sound. Gulf Res. Rep. 6(2): 131–135.

Oviatt, C. A. 2004. The changing ecology of temperate coastal waters during a warming trend. Estuaries 27(6):895–904.

Oviatt, C. A., and P. M. Kremer. 1977. Predation on the ctenophore, *Mnemiopsis leidyi*, by butterfish, *Peprilus triacanthus*, in Narragansett Bay, Rhode Island. Chesapeake Sci. 18(2):236–240.

Oviatt, C. A., and S. W. Nixon. 1973. The demersal fish of Narragansett Bay: an analysis of community structure, distribution and abundance. Est. Coast. Mar. Sci. 1:361–378.

Pacheco, A. L. 1962. Age and growth of spot in lower Chesapeake Bay, with notes on distribution and abundance of juveniles in the York River system. Chesapeake Sci. 3(1):18–28.

———. 1973. Proceedings of a workshop on egg, larval, and juvenile stages of fish in Atlantic coast estuaries. Held at Bears Bluff Laboratories, Wadmalaw Island, South Carolina, June 1968. Natl. Mar. Fish. Serv. Mid. Atl. Coast. Fish. Cent. Tech. Publ. No. 1. Highlands, N.J.

———. 1983. Seasonal occurrence of finfish and larger invertebrates at three sites in lower New York Harbor, 1981–1982. Natl. Ocean. Atmos. Adm. Natl. Mar. Fish. Serv. Northeast Fish. Cent., Sandy Hook Laboratory. Final Report to N.Y. District Corps of Engineers. Support Agreement NYD82-88(C).

———. 1984. Seasonal occurrence of finfish and larger invertebrates at eight locations in Lower and Sandy Hook Bays, 1982–1983. Natl. Ocean. Atmos. Adm. Natl. Mar. Fish. Serv.

Northeast Fish. Cent., Sandy Hook Laboratory. Final Report to N.Y. District Corps of Engineers. Support Agreement NYD83-46(C).

Pacheco, A. L., and G. C. Grant. 1965. Studies of the early life history of Atlantic menhaden in estuarine nurseries. U.S. Fish. Wildl. Serv. Spec. Sci. Rep. Fish. 504.

———. 1973. Immature fishes associated with larval Atlantic menhaden at Indian River inlet, Delaware, 1958–61. Pp. 78–87 *in* A. L. Pacheco, ed. Proceedings of a workshop on egg, larval and juvenile stages of fish in Atlantic coast estuaries. Natl. Mar. Fish. Serv., Mid. Atl. Coast. Cent. Tech. Publ. No. 1. Highlands, N.J.

Packer, D. B., and T. Hoff. 1999. Life history, habitat parameters, and essential habitat of mid-Atlantic summer flounder. Pp. 76–92 *in* Benaka, L. R., ed. Proc. Sea Grant Symp. Fish Habitat: essential fish habitat and rehabilitation. American Fisheries Society, Bethesda, Md.

Packer, D. B., S. J. Griesbach, P. L. Berrien, C. A. Zetlin, D. L. Johnson, and W. W. Morse. 1999. Essential fish habitat source document: summer flounder, *Paralichthys dentatus*, life history and habitat characteristics. U.S. National Oceanic and Atmospheric Administration, NOAA Technical Memorandum NMFS-NE-151.

Palmer, M. A. 2009. Reforming watershed restoration: science in need of application and applications in need of science. Estuaries and Coasts 32:1–17.

Palstra, A. P., D. F. M. Heppener, V. J. T. van Ginneken, C. Székely, and G. E. E. J. M. van den Thillart. 2007. Swimming performance of silver eels is severely impaired by the swim-bladder parasite *Anguillicola crassus*. J. Exp. Mar. Biol. Ecol. 352:244–256.

Pape, E., III. 1981. A drifter study of the Lagrainian mean circulation of Delaware Bay and adjacent shelf waters. M.S. thesis. University of Delaware, Newark.

Paperno, R. 1991. Spatial and temporal patterns of growth and mortality of juvenile weakfish (*Cynoscion regalis*) in Delaware Bay: assessment using otolith microincrement analysis. Ph.D. diss. University of Delaware, Newark.

Pardue, G. B. 1983. Habitat suitability index models: alewife and blueback herring. U.S. Dep. Int. Fish Wildl. Serv. FWS/OBS-82/10.58.

Parker, R. O., and R. L. Dixon. 1998. Changes in a North Carolina reef fish community after 15 yr of intense fishing-global warming implications. Trans. Am. Fish. Soc. 127:908–920.

Parmesan, C., and G. Yohe. 2003. A globally coherent fingerprint of climate change impacts across natural systems. Nature 421: 37–42.

Parr, A. E. 1931. A practical revision of the western Atlantic species of the genus *Citharichthys* (including *Etropus*). With observations on the Pacific *Citharichthys crossotus* and *C. spilopterus*. Bull. Bingham Oceanogr. Coll. Yale Univ. 41:1–24.

———. 1933. A geographical ecological analysis of the seasonal changes in water along the Atlantic coast of the U.S. Bull. Bingham Oceanogr. Collect. Yale Univ. 4:1–90.

Paterson, A. W., and A. K. Whitfield. 2000. Do shallow-water habitats function as refugia for juvenile fishes? Estuar. Coast. Shelf Sci. 51:359–364.

Patrick, R. 1994. Rivers of the United States. Vol. 1, Estuaries. Wiley, New York.

Paul, R. W. 2001. Geographical signatures of Middle Atlantic estuaries: historical layers. Estuaries 24(2):151–166.

Pauly, D. 1995. Anecdotes and the shifting baseline syndrome in fisheries. Trends Ecol. Evol. 10:420.

Pearce, A., and M. Feng. 2007. Observations of warming on the Western Australian continental shelf. Mar. Freshwat. Res. 58: 914–920.

Pearce, J. B. 2000. The New York Bight. Mar. Pollut. Bull. 41(1–6): 44–55.

Pearcy, W. G. 1962. Ecology of an estuarine population of winter flounder, *Pseudopleuronectes americanus* (Walbaum). II. Distribution and dynamics of larvae. Bull. Bingham Oceanogr. Coll. Yale Univ. 18(1):16–37.

Pearcy, W. G., and S. W. Richards. 1962. Distribution and ecology of fishes of the Mystic River estuary, Connecticut. Ecology 43: 248–259.

Pearse, A. S., and G. Günter. 1957. Salinity. Pp. 129–157 *in* J. W. Hedgepeth, ed. Ecology. Treatise on Marine Ecology and Paleoecology. Vol. 1. Memoir Geological Society of America, New York.

Pearson, J. C. 1929. Natural history and conservation of the redfish and other commercial sciaenids on the Texas coast. U.S. Bur. Fish. Bull. 44:129–214.

———. 1932. Winter trawl fishing off the Virginia and North Carolina coast. U.S. Bur. Fish. Invest. Rep. 10.

———. 1941. The young of some marine fishes taken in lower Chesapeake Bay, Virginia, with special reference to the gray sea trout, *Cynoscion regalis* (Bloch). U.S. Fish. Bull. 36:77–102.

Pekovitch, A. W. 1979. Distribution and Some Life History Aspects of the Shortnose Sturgeon (*Acipenser brevirostrium*) in the Upper Hudson River Estuary. Hazelton Environmental Science Corp., Northbrook, Ill. 67 pp.

Pelster, B., C. R. Bridges, and M. K. Grieshaber. 1988. Respiratory adaptations of the burrowing marine teleost *Lumpenus lampretaeformis* (Walbaum). II. Metabolic adaptations. J. Exp. Mar. Biol. Ecol. 124:43–55.

Penaz, M. 1974. Early development of the nase carp, *Chondrostoma nasus* (Linnaeus, 1758). Zool. Listy. 23:275–288.

———. 2001. A general framework of fish ontogeny: a review of the ongoing debate. Folia Zool. 50(4):241–256.

Penfold, R., A. A. Grutter, A. M. Kuris, M. I. McCormick, and C. M. Jones. 2008. Interactions between juvenile marine fish and gnathiid isopods: predation versus micropredation. Mar. Ecol. Prog. Ser. 357:111–119.

Pentilla, J., and L. M. Dery. 1988. Age determination methods for Northwest Atlantic species. Natl. Ocean. Atmos. Adm. Tech. Rep. Natl. Mar. Fish. Serv. 72.

Pepin, P. 1991. Effect of temperature and size on development, mortality, and survival rates of the pelagic early life history stages of marine fish. Can. J. Fish. Aquat. Sci. 48:503–518.

Perlmutter, A. 1939. A biological survey of the salt waters of Long Island, 1938. Part II, Sect. I. An ecological survey of young fishes and eggs identified from tow-net collections. N.Y. Conserv. Dept. Saltwater Survey, 1938(15):11–71.

———. 1963. Observations on fishes of the genus *Gasterosteus* in the waters of Long Island, New York. Copeia:168–174.

Perlmutter, A., E. E. Schmidt, and E. Leff. 1967. Distribution and abundance of fish along the shores of the lower Hudson River during the summer of 1965. N.Y. Fish and Game J. 14(1):47–75.

Perry, A. L., P. J. Low, J. R. Ellis, and J. D. Reynolds. 2005. Climate change and distribution shifts in marine fishes. Science 308:1912.

Perry, D. M. 1994. Artificial spawning of tautog under laboratory conditions. Prog. Fish-Cult. 56:33–36.

Persson, L. 1986a. Temperature-induced shift in foraging ability in two fish species, roach (*Rutilus rutilus*) and perch (*Perca fluviatilis*): implications for coexistence between poikilotherms. J. Anim. Ecol. 55:829–839.

———. 1986b. Patterns of food evacuation in fishes: a critical review. Environ. Biol. Fishes 16(1–3):51–58.

Peters, D. S., and M. T. Boyd. 1972. The effect of temperature, salinity, and availability of food on the feeding and growth of the hogchoker *Trinectes maculatus* (Bloch and Schneider). J. Exp. Mar. Biol. Ecol. 9:201–207.

Petersen, C. W., S. Salinas, R. L. Preston, and G. W. Kiddler III. In review. Spawning periodicity and reproductive behavior of *Fundulus heteroclitus* in a New England salt marsh. Copeia.

Peterson, C. H., and N. M. Peterson. 1979. The ecology of intertidal flats of North Carolina: a community profile. Rep. prepared for Natl. Coast. Ecosyst. Team, U.S. Fish Wildl. Serv. FWS/OBS-79/39.

Peterson, M. S., and S. T. Ross. 1991. Dynamics of littoral fishes and decapods along a coastal river-estuarine gradient. Estuar. Coast. Shelf Sci. 33:467–483.

Peterson, R. H., P. H. Johansen, and J. L. Metcalfe. 1980. Observations on early life stages of Atlantic tomcod, *Microgadus tomcod*. U.S. Fish. Bull. 78(1):147–158.

Peterson, T. L. 1996. Seasonal migration of the southern hogchoker, *Trinectes maculatus fasciatus* (Achiridae). Gulf Res. Rep. 9(3):169–176.

Phelan, B. A. 1992. Winter flounder movements in the inner New York Bight. Trans. Am. Fish. Soc. 121:777–784.

Phelan, B. A., R. Goldberg, A. J. Bejda, J. Pereira, S. Hagan, P. Clark, A. L. Studholme, A. Calabrese, and K. W. Able. 2000. Estuarine and habitat-related differences in growth rates of young-of-the-year winter flounder (*Pseudopleuronectes americanus*) and tautog (*Tautoga onitis*) in three northeastern U.S. estuaries. J. Exper. Mar. Biol. Ecol. 247:1–28.

Philipp, K. R. 2005. History of Delaware and New Jersey salt marsh restoration sites. Ecol. Eng. 25:214–230.

Picard, P., J. J. Dodson, and G. L. FitzGerald. 1990. The comparative ecology of the threespine (*Gasterosteus aculeatus*) and blackspotted sticklebacks (*G. wheatlandi*) (Pisces: Gasterosteidae) in three sub-habitats of the middle Saint Lawrence estuary, Canada. Can. J. Zool. 68:1202–1208.

Pierce, D. E., and A. B. Howe. 1977. A further study on winter flounder group identification off Massachusetts. Trans. Amer. Fish. Soc. 106:131–139.

Pihl, L., S. P. Baden, R. J. Diaz, and L. C. Schaffner. 1992. Hypoxia-induced structural changes in the diet of bottom-feeding fish and crustacea. Mar. Biol. 112:349–361.

Piner, K. R., and C. M. Jones. 2004. Age, growth and the potential for growth overfishing of spot (*Leiostomus xanthurus*) from the Chesapeake Bay, eastern USA. Mar. Fresh. Res. 55:553–560.

Pinnegar, J. K., and G. H. Engelhard. 2008. The 'shifting baseline' phenomenon: a global perspective. Rev. Fish Biol. Fish. 18: 1–16.

Pittman, S. J., and C. A. McAlpine. 2001. Movements of marine fish and decapod crustaceans: process, theory and application. Adv. Mar. Biol. 44:205–294.

Polunin, N. V. C., and J. K. Pinnegar. 2002. Trophic ecology and the structure of marine food webs. Pp. 301–320 *in* P. J. B. Hart and J. D. Reynolds, eds. Handbook of Fish Biology and Fisheries. Vol. 1, Fish Biology. Blackwell Publishing, Malden, Mass.

Poole, J. C. 1962. The fluke population in Great South Bay in relation to the sport fishery. N.Y. Fish Game J. 9:93–116.

Porter, S. M., and K. M. Bailey. 2007. The effect of early and late hatching on the escape response of walleye Pollock (*Theragra chalcogramma*) larvae. J. Plankton Res. 29(3):291–300.

Pörtner, H., C. Bock, R. Knust, G. Lannig, M. Lucassen, F. C. Mark, and F. J. Sartoris. 2008. Cod and climate in a latitudinal cline: physiological analyses of climate effects in marine fishes. Clim. Res. 37:253–270.

Pörtner, H. O., and R. Knust. 2007. Climate change affects marine fishes through the oxygen limitation of thermal tolerance. Science 315:95–97.

Portnoy, D. S. 2008. Understanding the reproductive behavior and population condition of the sandbar shark (*Carcharhinus plumbeus*) in the Western North Atlantic: a molecular approach to conservation and management. Ph.D. diss. College of William and Mary, Williamsburg, Va.

Post, J. R., and D. O. Evans. 1989. Size-dependent overwinter mortality of young-of-the-year yellow perch (*Perca flavescens*): laboratory, in situ enclosure, and field experiments. Can. J. Fish. Aquat. Sci. 46:1958–1968.

Potter, I. C., L. E. Beckley, A. K. Whitfield, and R. C. J. Lenanton. 1990. Comparisons between the roles played by estuaries in the life cycles of fishes in temperate Western Australia and Southern Africa. Environ. Biol. Fishes 28:143–178.

Potter, I. C., D. Tiivel, F. J. Valesini, and G. A. Hyndes. 1997. Comparisons between the ichthyofaunas of a temperate lagoonal-like estuary and the embayment into which that estuary discharges. Int. J. Salt Lake Res. 5:337–358.

Potthoff, M. T., and D. M. Allen. 2003. Site fidelity, home range, and tidal migrations of juvenile pinfish, *Lagodon rhomboides*, in salt marsh creeks. Environ. Biol. Fishes 67:231–240.

Potthoff, T. 1984. Clearing and staining techniques. Pp. 35–37 in H. G. Moser, W. J. Richards, D. M. Cohen, M. P. Fahay, A. W. Kendall, Jr., and S. L. Richardson, eds. Ontogeny and systematics of fishes. La Jolla, Spec. Publ. Amer. Soc. Ichthyol. Herpetol. 760 pp.

Pottle, R., and M. J. Dadswell. 1979. Studies on larval and juvenile shortnose sturgeon. Report to Northeast Utilities Service Co., Hartford, Conn. 87 pp.

Pottle, R. A., and J. M. Green. 1979. Territorial behavior of the north temperate labrid, *Tautogolabrus adspersus*. Can. J. Zool. 57:2337–2347.

Poulard, J., and F. Blanchard. 2005. The impact of climate change on the fish community structure of the eastern continental shelf of the Bay of Biscay. ICES J. Mar. Sci. 62:1436–1443.

Poulin, R. 1995. Phylogeny, ecology, and the richness of parasite communities in vertebrates. Ecol. Monog. 65(3):283–302.

Poulin, R., and G. J. FitzGerald. 1989. Early life histories of three sympatric sticklebacks in a salt-marsh. J. Fish. Biol. 34:207–221.

Powell, A. B., and M. D. Greene. 2002. Preliminary guide to the early life history stages of sparid fishes of the western central North Atlantic. NOAA Tech. Memo. NMFS-SEFSC. 480:1–20.

Powell, A. B., and R. E. Robbins. 1998. Ichthyoplankton adjacent to live-bottom habitats in Onslow Bay, North Carolina. U.S. Dep. Commer., NOAA Tech. Rep. NMFS 133. 32 pp.

Powell, A. B., and F. J. Schwartz. 1977. Distribution of paralichthid flounders (Bothidae: *Paralichthys*) in North Carolina estuaries. Chesapeake Sci. 18:334–339.

———. 1979. Food of *Paralichthys dentatus* and *P. lethostigma* (Pisces: Bothidae) in North Carolina estuaries. Estuaries and Coasts 2(4):276–279.

Powell, A. B., G. Thayer, M. Lacroix, and R. Cheshire. 2007. Juvenile and small resident fishes of Florida Bay, a critical habitat

in the Everglades National Park, Florida. NOAA Professional Paper NMFS 6.210.

Power, J. H., and J. D. McCleave. 1983. Simulation of the North Atlantic Ocean drift of *Anguilla leptocephali*. U.S. Fish. Bull. 81: 483–500.

Powers, D. A., M. Smith, I. Gonzalez-Villasenor, L. DiMichele, D. Crawford, G. Bernardi, and T. Lauerman. 1993. A multidisciplinary approach to the selection/neutralist controversy using the model teleost, *Fundulus heteroclitus*. Pp. 43–107 in D. Futuyma and J. Antonovics, eds. Oxford Surveys in Evolutionary Biology. Vol. 9. Oxford University Press, New York.

Powles, H. 1981. Distribution and movements of neustonic young estuarine dependant (*Mugil* spp.) and estuarine independent (*Coryphaena* spp.) fishes off the southeastern United States. Rapp. P.-V. Reun. Cons. Int. Explor. Mer 178:207–210.

Powles, P. M., and S. M. Warlen. 2002. Recruitment season, size, and age of young American eels (*Anguilla rostrata*) entering an estuary near Beaufort, North Carolina. Fish. Bull. 100:299–306.

Powles, P. M., J. A. Hare, E. H. Laban, and S. M. Warlen. 2006. Does eel metamorphosis cause a breakdown in the tenets of otolith applications? A case study using the speckled worm eel (*Myrophis punctatus*, Ophichthidae). Can. J. Fish. Aquat. Sci. 63: 1460–1468.

Pringle, C. M. 2000. Threats to U.S. public lands from cumulative hydrologic alterations outside of their boundaries. Ecol. Appl. 10:971–989.

Pritchard, D. W. 1967. What is an estuary: physical viewpoint. Pp. 3–8 in G. H. Lauff, ed. Estuaries. American Association for the Advancement of Science, Washington, D.C.

Psuty, N. P., M. P. DeLuca, R. Lathrop, K. W. Able, S. Whitney, and J. F. Grassle. 1993. The Mullica River–Great Bay National Estuarine Research Reserve: a unique opportunity for research, preservation and management. Proceedings of Coastal Zone '93, July 1993. New Orleans, La.

Public Service Electric and Gas Co. 1984. White perch (*Morone americana*): a synthesis of information on natural history, with reference to occurrence in the Delaware River and Estuary and involvement with the Salem Generating Station. Salem Nuclear Generating Station 316(b) Demonstration Appendix X. Newark, N.J.

Purcell, J. E., D. A. Neinazie, S. E. Dorsey, C. D. Houde, and J. C. Gamble. 1994. Predation mortality of bay anchovy *Anchoa mitchilli* eggs and larvae due to scyphomedusae and ctenophores in Chesapeake Bay. Mar. Ecol. Prog. Ser. 114:47–58.

Pyle, A. B. 1964. Some aspects of the life history of *Cyprinodon variegatus* Lacepède 1803, in New Jersey and its reaction to environmental changes. M.S. thesis. Rutgers University, New Brunswick, N.J.

Quattrini, A. M., S. W. Ross, K. J. Sulak, A. M. Necaise, T. L. Casazza, and G. D. Dennis. 2004. Marine fishes new to continental United States waters, North Carolina, and the Gulf of Mexico. S. E. Nat. 3(1):155–172.

Rachlin, J. W., B. E. Warkentine, and A. Pappantoniou. 2007. An evaluation of the ichthyofauna of the Bronx River, a resilient urban waterway. Northeast. Nat. 14(4):531–544.

Radtke, R. L., and J. M. Dean. 1982. Increment formation in the otoliths of embryos, larvae and juveniles of the mummichog, *Fundulus heteroclitus*. U.S. Fish. Bull. 80(2):201–215.

Radtke, R. L., M. L. Fine, and J. Bell. 1985. Somatic and otolith growth in the oyster toadfish (*Opsanus tau* L.). J. Exp. Mar. Biol. Ecol. 90:259–275.

Raichel, D. L., K. W. Able, and J. M. Hartman. 2003. The influence of *Phragmites* (common reed) on the distribution, abundance, and potential prey of a resident marsh fish in the Hackensack Meadowlands, New Jersey. Estuaries 26(2b):511–521.

Rakocinski, C., D. M. Baltz, and J. W. Fleeger. 1992. Correspondence between environmental gradients and the community structure of marsh-edge fishes in a Louisiana estuary. Mar. Ecol. Prog. Ser. 80:135–148.

Ramenofsky, M., and J. C. Wingfield. 2007. Regulation of migration. BioScience 57:135–143.

Rangely, R. W., and D. L. Kramer. 1995a. Use of rocky intertidal habitats by juvenile pollock *Pollachius virens*. Mar. Ecol. Prog. Ser. 126:9–17.

———. 1995b. Tidal effects on habitat selection and aggregation by juvenile pollock *Pollachius virens* in the rocky intertidal zone. Mar. Ecol. Prog. Ser. 126:19–29.

Raposa, K. B., and C. A. Oviatt. 2000. The influence of contiguous shoreline type, distance from shore, and vegetation biomass on nekton community structure in eelgrass beds. Estuaries 23(1): 46–55.

Raposa, K. B., and C. T. Roman. 2001. Seasonal habitat-use patterns of nekton in a tide-restricted and unrestricted New England salt marsh. Wetlands 21(4):451–461.

Rasband, W. S. 2003. ImageJ. US Nat. Inst. Health, Bethesda, MD. http://rsb.info.nih.gov/ij/.

Rathjen, W. F., and L. C. Miller. 1957. Aspects of the early life history of the striped bass (*Roccus saxatilis*) in the Hudson River. N.Y. Fish Game J. 4(1):43–60.

Ray, G. C. 1991. Coastal-zone biodiversity patterns: principles of landscape ecology may help explain the processes underlying coastal diversity. BioScience 41(7):490–498.

———. 1997. Do the metapopulation dynamics of estuarine fishes influence the stability of shelf ecosystems? Bull. Mar. Sci. 60: 1040–1049.

———. 2005. Connectivities of estuarine fishes to the coastal realm. Estuar. Coast. Shelf Sci. 64:18–32.

Ray, G. C., and B. P. Hayden. 1992. Coastal zone ecotones. Pp. 403–420 *in* A. J. Hansen and F. di Castri, eds. Landscape Boundaries. Vol. 92. Springer-Verlag, New York.

Ray, G. C., B. P. Hayden, M. G. McCormick-Ray, and T. M. Smith. 1997. Land-seascape diversity of the USA east coast coastal zone with particular reference to estuaries. Pp. 337–371 *in* R. F. G. Ormond, J. D. Gage, and M. V. Angel, eds. Marine Biodiversity: Patterns and Processes. Cambridge University Press, Cambridge, U.K., and New York.

Raymond, J. A. 1992. Glycerol is a colligative antifreeze in some northern fishes. J. Exp. Zool. 262:347–352.

Razinkovas, A., Z. Gasiūnaitė, P. Viaroli, and J. M. Zladívar. 2008. Preface: European lagoons—need for further comparison across spatial and temporal scales. Hydrobiologia 611:1–4.

Rebstock, G. A. 2003. Long-term change and stability in the California Current System: lessons from CalCOFI and other long-term data sets. Deep-Sea Res. II 50:2583–2594.

Reid, G. K., Jr. 1954. An ecological study of the Gulf of Mexico fishes, in the vicinity of Cedar Key, Florida. Bull. Mar. Sci. Gulf Carib. 4:1–94.

———. 1955. A summer study of the biology and ecology of East Bay, Texas. II. The fish fauna of East Bay, the Gulf Beach, and summary. Texas J. Sci. 7(4):430–453.

Reintjes, J. W. 1969. Synopsis of biological data on the Atlantic menhaden *Brevoortia tyrannus*. FAO Species Synopsis, 42. U.S. Dept. Inter. Fish. Wildl. Serv. Circul. 320. 30 pp.

Reintjes, J. W., and A. B. Pacheco. 1966. The relation of menhaden to estuaries. Am. Fish. Soc. Spec. Publ. 3:50–58.

Reis, R. R., and J. M. Dean. 1981. Temporal variation in the utilization of an intertidal creek by the bay anchovy (*Anchoa mitchilli*). Estuaries 4:16–23.

Reisman, H. 1963. Reproductive behavior of *Apeltes quadracus*, including some comparisons with other gasterosteid fishes. Copeia:191–192.

Reiss, C. S., and J. R. McConaugha. 1999. Cross-frontal transport and distribution of ichthyoplankton associated with Chesapeake Bay plume dynamics. Cont. Shelf Res. 19:151–170.

Relyea, K. G. 1965. Taxonomic studies of the Cyprinodont fishes, *Fundulus confluentus* Goode and Bean, and *Fundulus pulvereus* (Evermann). M.S. thesis. Florida State University. 73 pp.

———. 1983. A systematic study of two species complexes of the genus *Fundulus* (Pisces: Cyprinodontidae). Bull. Fla. State Mus. Biol. Sci. 29(1):1–64.

Rhode, M. 2008. Dynamics of the larval fish assemblage at two coastal Delaware inlets. M.S. thesis, University of Delaware, Newark.

Richards, C. E. 1973. Age, growth and distribution of the black drum (*Pogonias cromis*) in Virginia. Trans. Am. Fish. Soc. 102(3): 584–590.

Richards, C. E., and M. Castagna. 1970. Marine fishes of Virginia's eastern shore (inlet and marsh, seaside waters). Chesapeake Sci. 11(4):235–248.

Richards, S. W. 1959. Pelagic fish eggs and larvae. VI. Oceanography of Long Island Sound. Bull. Bingham Oceanogr. Collect. Yale Univ. 17:95–124.

———. 1963. The demersal fish population of Long Island Sound. III. Food of juveniles from a mud locality (Station 3A). Bull. Bingham Oceanogr. Coll. Yale Univ. 18:73–93.

Richards, S. W., J. M. Mann, and J. A. Walker. 1979. Comparison of spawning seasons, age, growth rates and food of two sympatric species of searobins, *Prionotus carolinus* and *Prionotus evolans*, from Long Island Sound. Estuaries 2(4):255–268.

Richards, W. J. 1999. Problems with unofficial and inaccurate geographical names in the fisheries literature. Mar. Fish. Rev. 61(3): 56–57.

Richards, W. J., R. V. Miller, and E. D. Houde. 1974. Egg and larval development of the Atlantic thread herring, *Opisthonema oglinum*. Fish Bull., U.S. 72:1123–1126.

Richards, W. J., and V. P. Saksena. 1980. Description of larvae and early juveniles of laboratory-reared gray snapper, *Lutjanus griseus* (Linnaeus) (Pisces, Lutjanidae). Bull. Mar. Sci. 30:516–521.

Richardson, L. R. 1939. The spawning behaviour of *Fundulus diaphanus* (Le Sueur). Copeia 3:165–167.

Richardson, S. L., and E. B. Joseph. 1973. Larvae and young of western North Atlantic bothid flatfishes *Etropus microstomus* and *Citharichthys arctifrons* in the Chesapeake Bight. U.S. Fish. Bull. 71:735–767.

———. 1975. Occurrence of larvae of the green goby, *Microgobius thallasinus*, in the York River, Virginia. Chesapeake Sci. 16: 215–218.

Richkus, W. A. 1974. Factors influencing the seasonal and daily patterns of alewife (*Alosa pseudoharengus*) migration in a Rhode Island river. J. Fish. Res. Board Can. 31:1485–1497.

———. 1975. Migratory behavior and growth of juvenile anadromous alewives, *Alosa pseudoharengus,* in a Rhode Island drainage. Trans. Am. Fish. Soc. 104:483–493.

Ritter, A. F., K. Wasson, S. I. Lonhart, R. K. Preisler, A. Woolfolk, K. A. Griffith, S. Connors, and K. W. Heiman. 2008. Ecological signatures of anthropogenically altered tidal exchange in estuarine ecosystems. Estuaries and Coasts 31:554–571.

Rivas, R. 1963. Subgenera and species groups in the poeciliid fish genus *Gambusia Poey.* Copeia:331–347.

Robbins, T. W. 1969. A systematic study of the silversides, *Membras* (Bonaparte) and *Menidia* (Linnaeus), (Atherinidae, Teleostei). Ph.D. diss., Cornell University, Ithaca, N.Y.

Roberts, C. M. 1997. Connectivity and management of Caribbean coral reefs. Science 278:1454–1456.

Roberts, S. C. 1978. Biological and fisheries data on northern searobin, *Prionotus carolinus* (Linnaeus). U.S. Dep. Commer. Natl. Ocean. Atmos. Adm. Natl. Mar. Fish. Serv. Northeast Fish. Cent. Sandy Hook Lab. Tech. Ser. Rep. No. 13.

Roberts-Goodwin, S. C. 1981. Biological and fisheries data on striped searobin, *Prionotus evolans* (Linnaeus). U.S. Dep. Commer. Natl. Ocean. Atmos. Adm. Natl. Mar. Fish. Serv. Northeast Fish. Cent. Sandy Hook Lab. Tech. Ser. Rep. No. 25.

Robertson, D. R. 1988a. Abundances of surgeonfishes on patch-reef in Caribbean Panama: due to settlement, or post-settlement events? Mar. Biol. 97:495–501.

———. 1988b. Extreme variation in settlement of the Caribbean triggerfish *Balistes vetula* in Panama. Copeia:698–703.

Robillard, E., C. S. Reiss, and C. M. Jones. 2008. Reproductive biology of bluefish (*Pomatomus saltatrix*) along the East Coast of the United States. Fish. Res. 90:198–208.

———. 2009. Age-validation and growth of bluefish (*Pomatomus saltatrix*) along the East Coast of the United States. Fish. Res. 95:65–75.

Robins, C. R., and G. C. Ray. 1986. A Field Guide to Atlantic Coast Fishes of North America. Houghton Mifflin, Boston.

Robins, C. R., D. M. Cohen, and C. H. Robins. 1979. The eels, *Anguilla* and *Histiobranchus*, photographed on the floor of the deep Atlantic in the Bahamas. Bull. Mar. Sci. 29(3):401–405.

Rodney, W. S., and K. T. Paynter. 2006. Comparisons of macro-faunal assemblages on restored and non-restored oyster reefs in mesohaline regions of Chesapeake Bay in Maryland. J. Exp. Mar. Biol. Ecol. 335:39–51.

Roelke, D. L., and S. M. Sogard. 1993. Gender-based differences in habitat selection and activity level in the northern pipefish (*Syngnathus fuscus*). Copeia 2:528–532.

Roelofs, E. L. 1954. Food studies of young sciaenid fishes, *Micropogon* and *Leiostomus*, from North Carolina. Copeia 2:151–153.

Roessig, J. M., C. M. Woodley, J. J. Cech Jr., and L. J. Hansen. 2004. Effects of global climate change on marine and estuarine fishes and fisheries. Rev. Fish Biol. Fish. 14:251–275.

Roff, D. A. 1988. The evolution of migration and some life history parameters in marine fishes. Environ. Biol. Fishes 22(2):133–146.

———. 1992. The Evolution of Life Histories: Theory and Analysis. Chapman and Hall, New York.

Rogers, S. G., T. E. Targett, and S. B. VanSant. 1984. Fish-nursery use in Georgia salt-marsh estuaries: the influence of springtime freshwater conditions. Trans. Am. Fish. Soc. 113:595–606.

Rohde, F. C., and V. J. Schuler. 1974. Abundance and distribution of fishes in Appoquinimink and Alloway creeks. Pp. 395–453 *in* V. J. Schuler, ed. An ecological study of the Delaware River in the vicinity of Artificial Island. Progress report for the period January through December 1972. Ichthyological Associates, Ithaca, N.Y.

Roman, C. T., N. Jaworski, F. T. Short, S. Findlay, and R. S. Warren. 2000. Estuaries of the Northeastern United States: habitat and land use signatures. Estuaries 23(6):743–764.

Roman, C. T., K. B. Roposa, S. C. Adamowicz, and M. James-Pirri. 2002. Quantifying vegetation and nekton response to tidal restoration of a New England salt marsh. Rest. Ecol. 10(3):450–460.

Rooker, J. R., G. J. Holt, and S. A. Holt. 1998. Vulnerability of newly settled red drum (*Sciaenops ocellatus*) to predatory fish: is early-life survival enhanced by seagrass meadows. Mar. Biol. 131:145–151.

Rose, K. A. 2000. Why are quantitative relationships between environmental quality and fish populations so elusive? Ecol. Appl. 10(2):367–385.

Rosenzweig, C., D. Karoly, M. Vicarelli, P. Neofotis, Q. Wu, G. Casassa, A. Menzel, T. L. Root, N. Estrella, B. Seguin, P. Tryjanowski, C. Liu, S. Rawlins, and A. Imeson. 2008. Attributing physical and biological impacts to anthropogenic climate change. Nature 453:353–358.

Ross, J. L., T. M. Stevens, and D. S. Vaughan. 1995. Age, growth, mortality, and reproductive biology of red drums in North Carolina waters. Trans. Am. Fish. Soc. 124:37–54.

Ross, S. W. 2003. The relative value of different estuarine nursery areas in North Carolina for transient juvenile marine fishes. Fish. Bull. 101:384–404.

Ross, S. W., and S. P. Epperly. 1985. Utilization of shallow estuarine nursery areas by fishes in Pamlico Sound and adjacent tributaries, North Carolina. Pp. 207–232 *in* A. Yáñez-Arancibia, ed. Fish Community Ecology in Estuaries and Coastal Lagoons: Towards an Ecosystem Integration. UNAM Press, Mexico City.

Ross, S. W., and M. L. Moser. 1995. Life history of juvenile gag, *Mycteroperca microlepis*, in North Carolina estuaries. Bull. Mar. Sci. 56(1):222–237.

Rotherham, D., C. A. Gray, M. K. Broadhurst, D. D. Johnson, L. M. Barnes, and M. V. Jones. 2006. Sampling estuarine fish using multi-mesh gill nets: effects of panel length and setting times. J. Exp. Mar. Biol. Ecol. 331:226–239.

Rotherham, D., C. A. Gray, D. D. Johnson, and P. Lokys. 2008. Effects of diel period and tow duration on estuarine fauna sampled with a beam trawl over bare sediment: consequences for designing more reliable and efficient surveys. Est. Coast. Shelf Sci. 78:179–189.

Rothschild, B. J. 1986. Dynamics of Marine Fish Populations. Harvard University Press, Cambridge, Mass.

Rotunno, T. K. 1992. Species identification and temporal spawning patterns of butterfish, *Peprilus* spp., in the south and mid-Atlantic bights. M.S. thesis. Marine Science Research Center, State University of New York, Stony Brook.

Rotunno, T., and R. K. Cowen. 1997. Temporal and spatial spawning patterns of the Atlantic butterfish, *Peprilus triacanthus,* in the South and Middle Atlantic Bights. U.S. Fish. Bull. 95:785–799.

Rountree, R. A. 1992. Fish and macroinvertebrate community structure and habitat use patterns in salt marsh creeks of southern New Jersey, with a discussion of marsh carbon export. Ph.D. diss. Rutgers University, New Brunswick, N.J.

———. 1999. Nov. Diets of NW Atlantic fishes and squid. www.fishecology.org.

———. 2008. Passive acoustics. Marine Tech. Reporter 51(9):40–46.

Rountree, R. A., and K. W. Able. 1992a. Fauna of polyhaline subtidal marsh creeks in southern New Jersey: composition, abundance and biomass. Estuaries 15(2):171–185.

———. 1992b. Foraging habits, growth, and temporal patterns of salt-marsh creek habitat use by young-of-year summer flounder in New Jersey. Trans. Am. Fish. Soc. 121:765–776.

———. 1993. Diel variation in decapod crustacean and fish assemblages in New Jersey polyhaline marsh creeks. Estuar. Coast. Shelf Sci. 37:181–201.

———. 1996. Seasonal abundance, growth and foraging habits of juvenile smooth dogfish, *Mustelus canis*, in a New Jersey estuary. U.S. Fish. Bull. 94(3):522–534.

———. 1997. Nocturnal fish use of New Jersey marsh creek and adjacent bay shoal habitats. Estuar. Coast. Shelf Sci. 44:703–711.

———. 2007. Spatial and temporal habitat use patterns for salt marsh nekton: implications for ecological functions. Aquat. Ecol. 41:25–45.

Rountree, R. A., K. J. Smith, and K. W. Able. 1992. Length frequency data for fishes and turtles from polyhaline subtidal and intertidal marsh creeks in southern New Jersey. IMCS Technical Report #92–34. Rutgers University, New Brunswick, N.J.

Rowe, C. L., and W. A. Dunson. 1995. Individual and interactive effects of salinity and initial fish density on a salt marsh assemblage. Mar. Ecol. Prog. Ser. 128:271–278.

Rowe, P. M., and C. E. Epifanio. 1994a. Tidal stream transport of weakfish larvae in Delaware Bay, USA. Mar. Ecol. Progr. Ser. 110:105–114.

———. 1994b. Flux and transport of larval weakfish in Delaware Bay, USA. Mar. Ecol. Progr. Ser. 110:115–120.

Rowell, K, K. W. Flessa, D. L. Dettman, M. J. Roman, L. R. Gerber, and L. T. Findley. 2008. Diverting the Colorado River leads to dramatic life history shift in an endangered marine fish. Biol. Conserv. 141:1138–1148.

Roy, P. S, R. J. Williams, A. R. Jones, I. Yassini, P. J. Gibbs, B. Coates, R. J. West, P. R. Scanes, J. P. Hudson, and S. Nichol. 2001. Structure and function of South East-Australian estuaries. Estuar. Coast. Shelf Sci. 53:351–384.

Rozas, L., T. Caldwell, and T. J. Minello. 2005. The fishery value of salt marsh restoration projects. J. Coast. Res. 40:37–50.

Rozas, L. P., and W. E. Odum. 1987. Use of tidal freshwater marshes by fishes and macrofaunal crustaceans along a marsh stream-order gradient. Estuaries 10(1):36–43.

Rubec, P. J., J. M. Lewis, M. A. Shirley, P. O'Donnell, and S. D. Locker. 2006. Relating changes in freshwater inflow to species distribution in Rookery Bay, Florida, via habitat suitability modeling and mapping. 57th Gulf Car. Fish. Ins. Pp. 61–76.

Ruehl, C. B., and T. J. DeWitt. 2007. Trophic plasticity and foraging performance in red drum, *Sciaenops ocellatus* (Linnaeus). J. Exp. Mar. Biol. Ecol. 349:284–294.

Ruiz, G. M., A. H. Hines, and M. H. Posey. 1993. Shallow water as a refuge habitat for fish and crustaceans in non-vegetated estuaries: an example from Chesapeake Bay. Mar. Ecol. Prog. Ser. 99:1–16.

Runyan, S. 1961. Early development of the clingfish, *Gobiesox strumosus* cope. Chesapeake Sci. 2:113–141.

Russell, B. C. 2002. Synodontidae; Lizardfishes. Pp. 923–930 *in* K. E. Carpenter, ed. The Living Marine Resources of the Western Central Atlantic. Vol. 2, Bony Fishes, Part 2 (Acipenseridae to Grammatidae). Food and Agriculture Organization of the United Nations, Rome.

Rutherford, E. S., T. W. Schmidt, and J. T. Tilmant. 1989. Early life history of spotted seatrout (*Cynoscion nebulosus*) and gray snapper (*Lutjanus griseus*) in Florida Bay, Everglades National Park, Fla.

Rutten, O. C. 1998. Size and age of juvenile gag (*Mycteroperca microlepis*) at egress from estuary to offshore hardbottom in North Carolina. M.S. thesis. University of North Carolina, Wilmington. 32 pp.

Ryder, J. A. 1887. On the development of osseous fishes, including marine freshwater forms. U.S. Comm. Fish. Rep. 13(1885): 488–604.

Ryer, C. H., and R. J. Orth. 1987. Feeding ecology of the northern pipefish, *Syngnathus fuscus*, in a seagrass community of the lower Chesapeake Bay. Estuaries 10(4):330–336.

Rypel, A. L., C. A. Layman, and D. A. Arrington. 2007. Water depth modifies relative predation risk for a motile fish taxon in Bahamian tidal creeks. Estuaries and Coasts 30(3):518–525.

Sackett, D. K., K. W. Able, and T. M. Grothues. 2007. Dynamics of summer flounder, *Paralichthys dentatus*, seasonal migrations based on ultrasonic telemetry. Est. Coast. Shelf Sci. 74:119–130.

———. 2008. Habitat dynamics of summer flounder, *Paralichthys dentatus*, within a shallow USA estuary, based on multiple approaches using acoustic telemetry. Mar. Ecol. Prog. Ser. 364: 199–212.

Saenger, C., T. M. Cronin, D. Willard, J. Halka, and R. Kerhin. 2008. Increased terrestrial to ocean sediment and carbon fluxes in the northern Chesapeake Bay associated with twentieth century land alteration. Estuaries and Coasts 31:492–500.

Safina, C., R. H. Wagner, D. A. Witting, and K. J. Smith. 1990. Prey delivered to roseate and common tern chicks: composition and temporal variability. J. Field Ornithol. 61(3):331–338.

Saila, S. B., and R. G. Lough. 1981. Mortality and growth estimation from size data—an application to some Atlantic herring larvae. Rapp. P.-V. Reun. Cons. Int. Explor. Mer 178:7–14.

Sakowicz, G. P. 2003. Comparative morphology and behavior of larval salt marsh fishes: *Fundulus heteroclitus* and *Cyprinodon variegatus*. M.S. thesis. Rutgers University, New Brunswick, N.J. 57 pp.

Saksena, V. P., and E. B. Joseph. 1972. Dissolved oxygen requirements of newly-hatched larvae of the striped blenny (*Chasmodes bosquianus*), the naked goby (*Gobiosoma bosci*) and the skilletfish (*Gobiesox strumosus*). Chesapeake Sci. 13(1):23–28.

Sale, P. F., K. Hanski, and J. P. Kritzer. 2006. The merging of metapopulation theory and marine ecology: establishing the historical context. Pp. 3–30 *in* J. P. Kritzer and P. F. Sale, ed. Marine Metapopulations. Elsevier Academic Press Publications. Burlington, Mass.

Salerno, D. J., J. Burnett, and R. M. Ibara. 2001. Age, growth, maturity, and spatial distribution of bluefish, *Pomatomus saltatrix* (Linnaeus), off the northeast coast of the United States, 1985–96. J. Northw. Fish. Sci. 29:31–39.

Saltonstall, K. 2002. Cryptic invasion of a non-native genotype of the common reed, *Phragmites australis*, into North America. Proc. of the Nat. Acad. Sci. USA. 99:2445–2449.

Samaritan, J. M., and R. E. Schmidt. 1982. Aspects of the life history of a freshwater population of the mummichog, *Fundulus heteroclitus* (Pisces: Cyprinodontidae), in the Bronx River, New York, U.S.A. Hydrobiologia 94:149–154.

Sand, R. L. 1982. Aspects of the feeding ecology of the cunner, *Tautogolabrus adspersus*, in Narragansett Bay. M.S. thesis, University of Rhode Island, Kingston. 94 pp.

Saucerman, S. E., and L. A. Deegan. 1991. Lateral and cross-channel movement of young-of-the-year winter flounder (*Pseduopleuronectes americanus*) in Waquoit Bay, Massachusetts. Estuaries 14:440–446.

Sawyer, P. J. 1967. Intertidal life-history of the rock gunnel, *Pholis gunnellus*, in the western Atlantic. Copeia 1:55–61.

Scarlett, P. G. 1991. Relative abundance of winter flounder (*Pseudopleuronectes americanus*) in nine inshore areas of New Jersey. Bull. N.J. Acad. Sci. 36(2):1–5.

Scarlett, P. G., and R. L. Allen. 1992. Temporal and spatial distribution of winter flounder (*Pleuronectes americanus*) spawning in Manasquan River, New Jersey. Bull. N.J. Acad. Sci. 37(1):13–17.

Scavia, D., J. C. Field, D. F. Boesch, R. W. Buddemeier, V. Burkett, D. R. Cayan, M. Fogarty, M. A. Harwell, R. W. Howarth, C. Mason, D. J. Reed, T. C. Royer, A. H. Sallenger, and J. G. Titus. 2002. Climate change impacts on U.S. coastal and marine ecosystems. Estuaries 25(2):149–164.

Schaefer, R. H. 1965. Age and growth of the northern kingfish in New York waters. N.Y. Fish Game J. 12(2):191–216.

———. 1967. Species composition, size and seasonal abundance of fish in the surf waters of Long Island. N.Y. Fish Game J. 14(1):1–46.

———. 1970. Feeding habits of striped bass from the surf waters of Long Island. N.Y. Fish Game J. 17:1–17.

Scharf, F. S., J. A. Buckel, and F. Juanes. 2002. Size-dependent vulnerability of juvenile bay anchovy *Anchoa mitchilli* to bluefish predation: does large body size always provide a refuge? Mar. Ecol. Prog. Ser. 233:241–252.

Scherer, M. D. 1972. The biology of the blueback herring, *Alosa aestivalis* (Mitchill) in the Connecticut River above the Holyoke Dam, Holyoke, Massachusetts. M.S. thesis. University of Massachusetts, Amherst.

Scherer, M. D., and D. W. Bourne. 1980. Eggs and early larvae of the smallmouth flounder, *Etropus microstomus*. U.S. Fish. Bull. 77:708–712.

Schmelz, G. W. 1964. A natural history study of the mummichog, *Fundulus heteroclitus* (Linnaeus), in Canary Creek marsh. M.S. thesis. University of Delaware. 65 pp.

———. 1970. Some effects of temperature and salinity on the life processes of the striped killifish, *Fundulus majalis* (Walbaum). Ph.D. diss. University of Delaware, Newark.

Schmidt, J. 1931. Eels and conger eels of the North Atlantic. Nature 128:602–604.

Schmidt, N., and M. E. Luther. 2002. ENSO impacts on salinity in Tampa Bay, Florida. Estuaries and Coasts 25(5):976–984.

Schmidt, R. E. 2007. Young striped searobins (Triglidae: *Prionotus evolans*) in the Hudson River. J. Northw. Atl. Fish. Sci. 38:67–71.

Schmitten, R. A. 1999. Essential fish habitat: opportunities and challenges for the next millennium. Pp. 3–10 *in* L. R. Benaka, ed. Fish Habitat: Essential Fish Habitat and Rehabilitation. American Fisheries Society, Bethesda, Md.

Schoedinger, S. E., and C. E. Epifanio. 1997. Growth, development and survival of larval *Tautoga onitis* (Linnaeus) in large laboratory containers. J. Exp. Mar. Biol. Ecol. 210:143–155.

Schroeder, E. H. 1966. Average surface temperatures of the western North Atlantic. Bull. Mar. Sci. 16:302–323.

Schroeder, W. C. 1933. Unique records of the brier skate and rock eel from New England. Bull. Boston Soc. Nat. Hist. 66:5–6.

Schubel, J. R. 1968. Turbidity maximum of the northern Chesapeake Bay. Science 161:1013–1015.

Schubel, J. R., C. F. Smith, and T. S. Y. Koo. 1977. Thermal effects of powerplant entrainment on survival of larval fishes: a laboratory assessment. Chesapeake Sci. 18(3):290–298.

Schuler, V. J. 1974. An ecological study of the Delaware River in the vicinity of Artificial Island. Progress Report for the period January–December 1972. Ichthyological Associates, Ithaca, N.Y.

Schultz, E. T., K. E. Reynolds, and D. O. Conover. 1996. Countergradient variation in growth among newly hatched *Fundulus heteroclitus*: geographic differences revealed by common-environment experiments. Funct. Ecology 10:366–374.

Schultz, E. T., D. O. Conover, and A. Ehtisham. 1998. The dead of winter: size dependent variation and genetic differences in seasonal mortality among Atlantic silverside (Atherinidae: *Menidia menidia*) from different latitudes. Can. J. Fish. Aquat. Sci. 55:1149–1157.

Schultz, E. T., R. K. Cowen, K. M. M. Lwiza, and A. M. Gospodarek. 2000. Explaining advection: do larval bay anchovy (*Anchoa mitchilli*) show selective tidal-stream transport? Jour. Mar. Sci. 57:360–371.

Schwartz, F. J. 1960. Additional comments on adult bull sharks, *Carcharhinus leucas* (Muller and Henle), from Chesapeake Bay, Maryland. Chesapeake Sci. 1:68–71.

———. 1964. Effects of winter water conditions on fifteen species of captive marine fishes. Am. Midl. Nat. 71:434–444.

———. 1965a. Inter-American migrations and systematics of the western Atlantic cownose ray, *Rhinoptera bonasus. In* Association of Island Marine Laboratories of the Caribbean, Sixth Meeting, Isla Margarita, Venezuela.

———. 1965b. Movements of the oyster toadfish (Pisces: Batrachoididae) about Solomons, Maryland. Chesapeake Sci. 5(1):155–159.

———. 1965c. Age, growth, and egg complement of the stickleback, *Apeltes quadracus*, at Solomons, Maryland. Chesapeake Sci. 6(2):116–118.

———. 1971. Biology of *Microgobius thalassinus* (Pisces: Gobiidae) a sponge-inhabiting goby of the Chesapeake Bay, with range extensions of two goby associates. Chesapeake Sci. 12(3):156–166.

———. 1992. Fishes and ecology of freshwater ponds on North Carolina's Outer Banks. Pp. 93–118 *in* C. Cole and K. Turner, eds. Barrier Island Ecology of the Mid-Atlantic Coast: A Symposium. Tech. Rept. NPS/SERCAHA/NRTR-93/04.

———. 1997. Biology of the striped cusk-eel, *Ophidion marginatum*, from North Carolina. Bull. Mar. Sci. 61:327–342.

Schwartz, F. J., and B. W. Dutcher. 1963. Age, growth and food of the oyster toadfish near Solomons, Maryland. Trans. Am. Fish. Soc. 92:170–173.

Scott, J. S. 1982. Depth, temperature and salinity preferences of common fishes of the Scotian Shelf. J. Northw. Atl. Fish. Sci. 3:29–39.

Scott, J., and G. Csanady. 1976. Nearshore currents off Long Island. J. Geophys. Res. 18(30):5401–5409.

Scott, W. B., and E. J. Crossman. 1973. Freshwater fishes of Canada. Fish. Res. Board Can. Bull. 184.

Scott, W. B., and M. C. Scott. 1988. Atlantic Fishes of Canada. Can. Bull. Fish. Aquat. Sci. 219.

Searcy, S. P. 2005. Is growth a reliable indicator of essential fish habitat? Ph.D. diss. Marine, Earth, and Atmospheric Sciences, North Carolina State University, Raleigh, N.C.

Searcy, S. P., D. B. Eggleston, and J. A. Hare. 2007a. Environmental influences on the relationship between juvenile and larval growth of Atlantic croaker, *Micropogonias undulatus*. Mar. Ecol. Prog. Ser. 349:81–88.

———. 2007b. Is growth a reliable indicator of habitat quality and essential fish habitat for a juvenile estuarine fish? Can. J. Fish. Aquat. Sci. 64:681–691.

Secor, D. H. 1999. Specifying divergent migrations in the concept of stock: the contingent hypothesis. Fish. Res. 43:13–34.

Secor, D. H., and T. E. Gunderson. 1998. Effects of hypoxia and temperature on survival, growth, and respiration of juvenile Atlantic sturgeon, *Acipenser oxyrinchus*. Fish. Bull. 96:603–613.

Secor, D. H., and E. D. Houde. 1995. Temperature effects on the timing of striped bass egg production, larval viability, and recruitment potential in the Patuxent River (Chesapeake Bay). Estuaries 18(3):527–544.

Secor, D. H., and L. A. Kerr. 2009. Lexicon of life cycle diversity in diadromous and other fishes. Am. Fish. Soc. Symp. 69:1–20.

Secor, D. H., and P. M. Piccoli. 2007. Oceanic migration rates of upper Chesapeake Bay striped bass (*Morone saxatilis*), determined by otolith microchemical analysis. Fish. Bull. 105: 62–73.

Secor, D. H., and J. R. Rooker. 2005. Connectivity in the life histories of fishes that use estuaries. Estuar. Coast. Shelf Sci. 64:1–3.

Secor, D. H., et al. 1997. Restoration of Atlantic sturgeon, *Acipenser oxyrhynchus* in the Chesapeake Bay: habitat considerations. Chesapeake Biological Lab, University of Maryland Center of Environmental Science, Solomons, Md.

Secor, D. H., T. E. Gunderson, and K. Karlsson. 2000. Effect of temperature and salinity on growth performance in anadromous (Chesapeake Bay) and nonanadromous (Santee-Cooper) strains of striped bass *Morone saxatilis*. Copeia 1:291–296.

Secor, D. H., J. R. Rooker, E. Zlokovitz, and V. S. Zdanowicz. 2001. Identification of riverine, estuarine, and coastal contingents of Hudson River striped bass based upon otolith elemental fingerprints. Mar. Ecol. Prog. Ser. 211:245–253.

Sedberry, G. R. 1983. Food habits and trophic relationships of a community of fishes on the outer continental shelf. NOAA Technical Report NMFS SSRF-773.

Seligman, E. G., Jr. 1951. *Cyprinodon variegatus riverendi* (Poey) and other aquatic notes. Aquarium 20:234–236.

Serafy, J. E., and R. M. Harrell. 1993. Behavioural response of fishes to increasing pH and dissolved oxygen: field and laboratory observations. Freshwat. Biol. 30:53–61.

Setzler-Hamilton, E. A. 1991. White perch, *Morone americana*. In S. L. Funderburk, S. J. Jordan, J. A. Mihursky, and D. Riley, eds. Habitat Requirements for Chesapeake Bay Living Resources. Chesapeake Research Consortium, Solomons, Md.

Sheaves, M. 2001. Are there really few piscivorous fishes in shallow estuarine habitats? Mar. Ecol. Prog. Ser. 222:279–290.

Sheaves, M., and R. Johnston. 2008. Influence of marine and freshwater connectivity on the dynamics of subtropical estuarine wetland fish metapopulations. Mar. Ecol. Prog. Ser. 357:225–243.

Sheaves, M., R. Baker, and R. Johnston. 2006. Marine nurseries and effective juvenile habitats: an alternative view. Mar. Ecol. Prog. Ser. 318:303–306.

Shenker, J. M., and J. M. Dean. 1979. The utilization of an intertidal salt marsh creek by larval and juvenile fishes: abundance, diversity and temporal variation. Estuaries and Coasts 2(3):154–163.

Shenker, J. M., D. J. Hepner, P. E. Frere, L. E. Currence, and W. W. Wakefield. 1983. Upriver migration and abundance of naked goby (*Gobiosoma bosci*) larvae in the Patuxent River Estuary, Maryland. Estuaries 6(1):35–42.

Shepherd, G. R. 1991. Meristic and morphometric variation in black sea bass north of Cape Hatteras, North Carolina. N. Am. J. Fish. Manage. 11:139–148.

Shepherd, G. R., and C. B. Grimes. 1983. Geographic and historic variations in growth of weakfish, *Cynoscion regalis*, in the Middle Atlantic Bight. U.S. Fish. Bull. 81:803–813.

———. 1984. Reproduction of weakfish, *Cynoscion regalis*, in the New York Bight and evidence for geographically specific life history characteristics. U.S. Fish. Bull. 82:501–511.

Shepherd, G. R., and M. Terceiro. 1994. The summer flounder, scup, and black sea bass fishery of the Middle Atlantic Bight and southern New England waters. Natl. Ocean. Atmos. Adm. Tech. Rep. Natl. Mar. Fish. Serv. 122:1–13.

Shepherd, G. R., J. Moser, D. Deuel, and P. Carlson. 2006. The migration patterns of bluefish (*Pomatomus saltatrix*) along the Atlantic coast determined from tag recoveries. Fish. Bull. 104: 559–570.

Sheridan, P., and C. Hays. 2003. Are mangroves nursery habitat for transient fishes and decapods? Wetlands 23(2):449–458.

Sherman, K., W. Smith, W. Morse, M. Berman, J. Green, and L. Esjymont. 1984. Spawning strategies of fishes in relation to circulation, phytoplankton production, and pulses in zooplankton off the northeastern United States. Mar. Ecol. Prog. Ser. 18: 1–19.

Shields, M. 2002. Brown pelican (*Pelecanus occidentalis*). In A. Poole and F. Gill, eds. The Birds of North America, No. 609. The Birds of North America, Inc. Philadelphia.

Shima, M. 1989. Oceanic transport of the early life history stages of bluefish (*Pomatomus saltatrix*) from Cape Hatteras to the Mid-Atlantic Bight. M.S. thesis, State University of New York, Stony Brook.

Shimps, E. L., J. A. Rice, and J. A. Osborne. 2005. Hypoxia tolerance in two juvenile estuary dependent fishes. Mar. Biol. Ecol. 325:146–162.

Shipp, R. L. 1974. The puffer fishes (Tetraodontidae) of the Atlantic Ocean. Publ. Gulf Coast Res. Lab. Mus. No. 4, Gulf Coast Research Laboratory, Ocean Springs, Miss.

Shipp, R. L., and R. W. Yerger. 1969. Status, characters, and distribution of the northern and southern puffers of the genus *Sphoeroides*. Copeia:425–433.

Shoji, J., E. W. North, and E. D. Houde. 2005. The feeding ecology of *Morone americana* larvae in the Chesapeake Bay estuarine turbidity maximum: the influence of physical conditions and prey concentrations. J. Fish Biol. 66:1328–1341.

Short, F. T., D. M. Burdick, J. S. Wolf, and G. E. Jones. 1993. Eelgrass in estuarine reserves along the East Coast, USA. Part I: Declines from pollution and disease and Part II: Management of eelgrass meadows. NOAA Coastal Ocean Program Publication, University of New Hampshire, Durham.

Shulman, M. J., and J. C. Ogden. 1987. What controls tropical reef fish populations: recruitment or benthic mortality: an example in the Caribbean reef fish *Haemulon flavolineatum*. Mar. Ecol. Prog. Ser. 39:233–242.

Shuster, C. N., Jr. 1959. A biological evaluation of the Delaware River estuary. Univ. Del. Mar. Lab. Info. Ser. Publ. (3).

Shuter, B. J., and J. R. Post. 1990. Climate, population variability, and the zoogeography of temperate fishes. Trans. Am. Fish. Soc. 119:314–336.

Sibley, D. A. 2001. The Sibley Guide to Bird Life and Behavior. Natural Audubon Society. Alfred A. Knopf, New York. 588 pp.

Sibunka, J. D., and A. L. Pacheco. 1981. Biological and fisheries data on northern puffer, *Sphoeroides maculatus* (Bloch and

Schneider). U.S. Dep. Commer. Natl. Ocean. Atmos. Adm. Natl. Mar. Fish. Serv. Northeast Fish. Cent. Sandy Hook Lab. Tech. Ser. Rep. No. 26.

Sibunka, J. D., and M. J. Silverman. 1984. MARMAP surveys of the continental shelf from Cape Hatteras, North Carolina to Cape Sable, Nova Scotia (1977–1983). Atlas No. 1. Summary of operations. Natl. Ocean. Atmos. Adm. Tech. Memo. Natl. Mar. Fish. Serv.—F/Northeast Fish Cent.-33.

———. 1989. MARMAP surveys of the continental shelf from Cape Hatteras, North Carolina to Cape Sable, Nova Scotia (1984–1987). Atlas No. 3. Summary of operations. Natl. Ocean. Atmos. Adm. Tech. Memo. Natl. Mar. Fish Serv.—F/NEC-68.

Silverman, M. J. 1975. Scale development in the bluefish, *Pomatomus saltatrix*. Trans. Am. Fish. Soc. 104(4):773–774.

Simmons, E. G. 1957. An ecological survey of the upper Laguna Madre of Texas. Publ. Inst. Mar. Sci. Univ. Tex. 4(2):156–202.

Simmons, E. G., and J. P. Breuer. 1962. A study of redfish, *Sciaenops ocellatus* Linnaeus, and black drum, *Pogonias cromis* Linnaeus. Publ. Inst. Mar. Sci. Univ. Tex. 8:184–211.

Simonin, P. W., K. E. Limburg, and L. S. Machut. 2007. Bridging the energy gap: anadromous blueback herring feeding in the Hudson and Mohawk rivers, New York. Trans. Am. Fish. Soc. 136:1614–1621.

Sindermann, C. J. 1966. Diseases of Marine Fishes. Academic Press, London. 89 pp.

Sirabella, P., A. Guiliani, A. Colosimo, and J. W Dippner. 2001. Breaking down the climate effects on cod recruitment by principal component analysis and canonical correlation. Mar. Ecol. Prog. Ser. 216:213–222.

Sire, J.-Y. 1981. La scalation (apparition et mise en place des ecailles) chez *Hemichromis bimaculatus* (Gill, 1862) (Teleosteans, Perciformes, Cichlides). Cybium 5(3):51–66.

Sire, J.-Y., and I. Arnulf. 1990. The development of squamation in four teleostean fishes with a survey of the literature. Jpn. J. Ichthyol. 37(2):133–143.

Sirois, P., and J. J. Dodson. 2000. Influence of turbidity, food density, and parasites on the ingestion and growth of larval rainbow smelt, *Osmerus mordax*, in an estuarine turbidity maximum. Mar. Ecol. Prog. Ser. 193:167–179.

Sissenwine, M. P. 1984. Why do fish populations vary? Pp. 59–94 *in* R. M. May, ed. Exploitation of Marine Communities. Dahlem Konferenzen. Springer, New York.

Sissenwine, M. P., E. B. Cohen, and M. D. Grosslein. 1984. Structure of the Georges Bank ecosystem. Rapp. R.-V. Reun. Cons. Int. Explor. Mer. 183:243–254.

Sisson, R. T. 1974. The growth and movement of scup (*Stenotomus chrysops*) in Narragansett Bay, Rhode Island and along the Atlantic coast. Completion Rep. PL88-309, Project 3-K38-R-3. Rhode Island Department of Natural Resources, Providence.

Skomal, G. B. 2007. Shark nursery areas in the coastal waters of Massachusetts. Am. Fish. Soc. 50:17–33.

Slater, J. J., T. E. Lankford Jr., and J. A. Buckel. 2007. Overwintering ability of young-of-the-year bluefish (*Pomatomus saltatrix*): effect of ration and cohort of origin on survival. Mar. Ecol. Prog. Ser. 339:259–269.

Slocum, C. J., A. Ferland, N. Furina, and S. Evert. 2005. What do harbor seals eat in New Jersey? A first report from the Mid-Atlantic region (USA). Proceedings of the 16th Biennial Conference on the Biology of Marine Mammals, San Diego, Calif. December 12–16.

Smigielski, A. S. 1975. Hormone induced spawnings of the summer flounder and rearing of the larvae in the laboratory. Prog. Fish-Cult. 37:3–8.

Smigielski, A. S., T. A. Halavik, L. J. Buckley, S. M. Drew, and G. C. Laurence. 1984. Spawning, embryo development and growth of the American sand lance *Ammodytes americanus* in the laboratory. Mar. Ecol. Prog. Ser. 14:287–292.

Smith, B. A. 1971. The fishes of four low-salinity tidal tributaries of the Delaware River estuary. M.S. thesis, Cornell University. 304 pp.

Smith, C. L. 1985. The inland fishes of New York State. New York State Dep. Environment and Conservation, Albany.

Smith, D. G. 1968. The occurrence of larvae of the American eel, *Anguilla rostrata*, in the Straits of Florida and nearby areas. Bull. Mar. Sci. 18:280–293.

———. 1989a. Introduction to Leptocephali. Pp. 657–668 *in* E. B. Böhlke, ed. Fishes of the Western North Atlantic. Vol. 2, Part 9. Sears Foundation for Marine Research, New Haven, Conn.

———. 1989b. Family Anguillidae: Leptocephali. Pp. 898–899 *in* E. B. Böhlke, ed. Fishes of the Western North Atlantic. Vol. 2, Part 9. Sears Foundation for Marine Research. Yale University, New Haven, Conn.

———. 1989c. Family Congridae: Leptocephali. Pp. 723–763 *in* E. B. Böhlke, ed. Fishes of the Western North Atlantic. Vol. 2, Part 9. Sears Foundation for Marine Research, New Haven, Conn.

Smith, D. G., and K. A. Tighe. 2002. Conger eels. Family Congridae. Pp. 101–102 *in* B. B. Collette and G. Klein-MacPhee, eds. Bigelow and Schroeder's Fishes of the Gulf of Maine. 3rd ed. Smithsonian Institution Press, Washington and London.

Smith, H. M. 1907. The Fishes of North Carolina. North Carolina Geological and Economic Survey. Vol. 2. Raleigh. 435 pp.

Smith, J. W. 1994. Biology and fishery for Atlantic thread herring, *Opisthonema oglinum*, along the North Carolina coast. Mar. Fish Rev. 56(4):1–7.

———. 1999. A large fish kill of Atlantic Menhaden, *Brevoortia tyrannus*, on the North Carolina coast. The Journ. Elisha. Mitchell Sci. Soc. 115(3):157–163.

Smith, J. W., and J. V. Merriner. 1985. Food habits and feeding behavior of the cownose ray, *Rhinoptera bonasus*, in Lower Chesapeake Bay. Estuaries 8(3):305–310.

———. 1986. Observations on the reproductive biology of the cownose ray, *Rhinoptera bonasus*, in Chesapeake Bay. Fish. Bull. 84(4):871–877.

———. 1987. Age and growth, movements and distribution of the cownose ray, *Rhinoptera bonasus*, in Chesapeake Bay. Estuaries 10(2):153–164.

Smith, K. A., and I. M. Suthers. 1999. Displacement of diverse ichthyoplankton assemblages by a coastal upwelling event on the Sydney Shelf. Mar. Ecol. Prog. Ser. 176:49–62.

Smith, K. J. 1995. Processes regulating habitat use by salt marsh nekton in a southern New Jersey estuary. Ph.D. diss. Rutgers University, New Brunswick, N.J.

Smith, K. J., and K. W. Able. 1994. Salt-marsh tide pools as winter refuges for the mummichogs, *Fundulus heteroclitus*, in New Jersey. Estuaries 17(1b): 226–234.

———. 2003. Dissolved oxygen dynamics in salt marsh pools and its potential impacts on fish assemblages. Mar. Ecol. Prog. Ser. 258:223–232.

Smith, K. J., G. L. Taghon, and K. W. Able. 2000. Trophic linkages in marshes: ontogenetic changes in diet for young-of-the-year mummichog, *Fundulus heteroclitus*. Pp. 221–237 *in* M. P. Weinstein and D. A. Kreeger, eds. Concepts and Controversies in Tidal Marsh Ecology. Kluwer Academic Publishers, The Netherlands.

Smith, M. W., and J. W. Saunders. 1955. The American eel in certain freshwaters of the maritime provinces of Canada. J. Fish. Res. Board. Can. 12:238–269.

Smith, M. W., M. C. Glimcher, and D. A. Powers. 1992. Differential introgression of nuclear alleles between subspecies of the teleost *Fundulus heteroclitus*. Mol. Mar. Biol. Biotechnol. 1:226–238.

Smith, M. W., R. W. Chapman, and D. A. Powers. 1998. Mitochondrial DNA analysis of Atlantic Coast, Chesapeake Bay, and Delaware Bay populations of the teleost *Fundulus heteroclitus* indicated temporally unstable distributions over geologic time. Mol. Mar. Biol. Biotechnol. 7(2):79–87.

Smith, N. G., C. M. Jones, and J. Van Montfrans. 2008. Spatial and temporal variability of juvenile spotted seatrout *Cynoscion nebulosus* growth in Chesapeake Bay. J. Fish. Biol. 73:597–607.

Smith, P. E., and H. G. Moser. 2003. Long-term trends and variability in the larvae of Pacific sardine and associated fish species of the California Current region. Deep. Sea. Res. II 50:2519–2536.

Smith, R. E. 1966. Foreword. Am. Fish. Soc. Spec. Publ. No. 3:vii–viii.

Smith, S. M., J. G. Hoff, S. O'Neil, and M. P. Weinstein. 1984. Community and trophic organization of nekton utilizing shallow marsh habitats, York River, Virginia. Fish. Bull. 82(4):455–467.

Smith, T. I., J. D. E. Marchette, and R. A. Smiley. 1982. Life history, ecology, culture and management of Atlantic sturgeon, *Acipenser oxyrhynchus oxyrhynchus*, Mitchill, in South Carolina. South Carolina Wildlife Marine Resources. Resources Department, Final Report to U.S. Fish and Wildlife Service Project AFS-9. 75 pp.

Smith, W. G., and M. P. Fahay. 1970. Description of eggs and larvae of the summer flounder, *Paralichthys dentatus*. U.S. Fish. Wildl. Serv. Res. Rep. 75:1–21.

Smith, W. G., and W. W. Morse. 1988. Seasonal distribution, abundance and diversity patterns of fish eggs and larvae in the Middle Atlantic Bight. Pp. 177–189 *in* A. Pacheco, ed. Characterization of the Middle Atlantic Water Management Unit of the Northeast Regional Action Plan. NOAA. Tech. Memo. NMFS NE 56.

———. 1990. Larval distribution patterns: evidence for the collapse/recolonization of Atlantic herring on Georges Bank. Int. Counc. Explor. Sea C.M. 1990/H:17:1–16.

———. 1993. Larval distribution patterns: early signals for the collapse/recovery of Atlantic herring *Clupea harengus* in the Georges Bank area. Fish. Bull. 91:338–347.

Smith, W. G., J. D. Sibunka, and A. Wells. 1975. Seasonal distribution of larval flatfishes (Pleuronectiformes) on the continental shelf between Cape Cod, Massachusetts, and Cape Lookout, North Carolina, 1965–1966. Natl. Ocean. Atmos. Adm. Tech. Rep. Natl. Mar. Fish. Serv. Spec. Sci. Rep. Fish. 691

Smith, W. G., P. Berrien, and T. Potthoff. 1994. Spawning patterns of bluefish, *Pomatomus saltatrix*, in the northeast continental shelf ecosystem. Bull. Mar. Sci. 54(1):8–16.

Snyder, D. E. 1988. Description and identification of shortnose and Atlantic sturgeon larvae. Amer. Fish. Soc. Symp. 5:7–30.

Snyder, R. J., and H. Dingle. 1990. Effects of freshwater and marine wintering environments on life histories of threespine sticklebacks: evidence for adaptive variation between anadromous and resident freshwater populations. Oecologia 84:386–390.

Sogard, S. M. 1989. Colonization of artificial seagrass by fishes and decapod crustaceans: importance of proximity to natural eelgrass. J. Exp. Mar. Biol. Ecol. 133:15–37.

———. 1991. Interpretation of otolith microstructure in juvenile winter flounder (*Pseudopleuronectes americanus*): ontogentic development, daily increment validation, and somatic growth relationships. Can. J. Fish. Aq. Sci. 48(10):1862–1871.

———. 1992. Variability in growth rates of juvenile fishes in different estuarine habitats. Mar. Ecol. Prog. Ser. 85:35–53.

———. 1997. Size-selective mortality in the juvenile stage of teleost fishes: a review. Bull. Mar. Sci. 60(3):1129–1157.

Sogard, S. M., and K. W. Able. 1991. A comparison of eelgrass, sea lettuce macroalgae, and marsh creeks as habitat for epibenthic fishes and decapod crustaceans. Estuar. Coast. Shelf Sci. 33:501–519.

———. 1994. Diel variation in immigration of fishes and decapod crustaceans to artificial seagrass habitat. Estuaries 17(3):622–630.

Sogard, S. M., K. W. Able, and M. P. Fahay. 1992. Early life history of the tautog *Tautoga onitis* in the Mid-Atlantic Bight. U.S. Fish. Bull. 90:529–539.

Sogard, S. M., K. W. Able, and S. M. Hagan. 2001. Long-term assessment of settlement and growth of juvenile winter flounder (*Pseudopleuronectes americanus*) in New Jersey estuaries. J. Sea Res. 45:189–204.

Sola, C. 1995. Chemoattraction of upstream migrating glass eels *Anguilla anguilla* to earthy and green odorants. Environ. Biol. Fishes 43:179–185.

Sola, C., and P. Tongiorgi. 1996. The effect of salinity on the chemotaxis of glass eels, *Anguilla anguilla*, to organic earthy and green odorants. Environ. Biol. Fishes 47:213–218.

Solomon, F. N., and I. W. Ramnarine. 2007. Reproductive biology of white mullet, *Mugil curema* (Valenciennes) in the southern Caribbean. Fish. Res. 88:133–138.

Spraker, H., and H. M. Austin. 1997. Diel feeding periodicity of Atlantic silverside, *Menidia menidia*, in the York River, Chesapeake Bay, Virginia. J. Elisha Mitchell Sci. Soc. 113:171–182.

Springer, S. 1960. Natural history of the sandbar shark *Eulamia milberti*. Fish. Bull. 61:1–36.

Springer, S., and H. R. Bullis Jr. 1956. Collections by the *Oregon* in the Gulf of Mexico. List of crustaceans, mollusks, and fishes identified from collections made by the exploratory fishing vessel *Oregon* in the Gulf of Mexico and adjacent seas 1950 through 1955. U.S. Fish Wildl. Serv. Spec. Sci. Rep. Fish. 196.

Springer, V. G., and K. D. Woodburn. 1960. An ecological study of the fishes of the Tampa Bay area. Profess. Pap. Ser. No. 1. Florida State Board of Conservation Marine Lab.

Squires, I. S. 1983. Evaluation of the spawning run of shortnose sturgeon (*Acipenser brevirostrum*) in the Androscoggin River, ME. Maine Dept. Marine Resources, Augusta. 22 pp.

St. Pierre, R. A., and J. Davis. 1972. Age, growth and mortality of the white perch, *Morone americana*, in the James and York rivers, Virginia. Chesapeake Sci. 13(4):272–281.

Stahl, L., J. Koczan, and D. Swift. 1974. Anatomy of a shoreface-connected sand ridge on the New Jersey shelf: implications for the genesis of the shelf surficial sand sheet. Geology 2:117–120.

Stanley, J. C., and D. S. Danie. 1983. Species profile: Life histories and environmental requirements of coastal fishes and invertebrates (North America): white perch. U.S. Fish Wildl. Serv. Div. Biol. Serv. FWS/OBS-82/11.7.

Starck, W. A. 1971. Biology of the gray snapper, *Lutjanus griseus*

(Linnaeus), in the Florida Keys. Pp. 11–150 *in* W. A. Starck and R. E. Schroeder, eds. Investigations on the Gray Snapper, *Lutjanus griseus*. Stud. Trop. Oceangr. (Miami) 10:1–224.

Staunton-Smith, J., J. B. Robins, D. G. Mayer, M. J. Sellin, and I. A. Halliday. 2004. Does the quantity and timing of fresh water flowing into a dry tropical estuary affect year-class strength of barramundi (*Lates calcarifer*). Mar. Freshwat. Res. 55:787–797.

Stehlik, L. L. 2009. Effects of seasonal change on activity rhythms and swimming behavior of age-0 bluefish (*Pomatomus saltatrix*) and a description of gliding behavior. Fish. Bull. 107(1):1–12.

Stehlik, L. L., and C. J. Meise. 2000. Diet of winter flounder in a New Jersey estuary: ontogenic and spatial variation. Estuaries 23(3):381–391.

Stehlik, L. L., R. A. Pikanoswki, and D. G. McMillan. 2004. The Hudson-Raritan estuary as a crossroads for distribution of blue (*Callinectes sapidus*), lady (*Ovalipes ocellatus*), and Atlantic rock (*Cancer irroratus*) crabs. Fish. Bull. 102:693–710.

Steimle, F. W., and C. J. Sindermann. 1978. Review of oxygen depletion and associated mass mortalities of shellfish in the Middle Atlantic Bight in 1976. Mar. Fish. Rev. 40(12):17–26.

Steiner, W. W., and B. L. Olla. 1985. Behavioral responses of prejuvenile red hake, *Urophycis chuss*, to experimental thermoclines. Environ. Biol. Fishes 14(2–3):167–173.

Steiner, W. W., J. J. Luczkovich, and B. L. Olla. 1982. Activity, shelter usage, growth and recruitment of juvenile red hake, *Urophycis chuss*. Mar. Ecol. Progr. Ser. 7:125–135.

Steneck, R. S. 2005. An ecological context for the role of large carnivorous animals in conserving biodiversity. Pp. 9–33 *in* J. K. Ray, J., K. Redford, R. Steneck, and J. Berger, eds. Large Carnivores and the Conservation of Biodiversity. Island Press. Washington, D.C.

Stephenson, T. A., and A. Stephenson. 1954. Life between tidemarks in North America. IIIA. Nova Scotia and Prince Edward Island: the geographical features of the region. J. Ecol. 42:46–70.

Stevenson, D. K., and M. L. Scott. 2005. Essential fish habitat source document: Atlantic herring, *Clupea harengus,* life history and habitat characteristics. NOAA Tech. Mem. NMFS-NE-192. 2nd ed. 90 pp.

Stevenson, J. C., and M. S. Kearney. 2009. Impacts of global climate change and sea-level rise on tidal wetlands. Pp. 171–206 *in* B. R. Silliman, E. D. Grosholz, and M. D. Bertness, eds. Human Impacts on Salt Marshes, A Global Perspective. University of California Press, Berkeley.

Stevenson, J. C., J. E. Rooth, M. S. Kearney, and K. L. Sundberg. 2000. The health and long term stability of natural and restored marshes in Chesapeake Bay. Pp. 709–735 *in* M. P. Weinstein and D. A. Kreeger, eds. Concepts and Controversies in Tidal Marsh Ecology. Kluwer Academic Publishers, Dordrecht, The Netherlands.

Stevenson, J. C., M. S. Kearney, and E. W. Koch. 2002. Impacts of sea-level rise on tidal wetlands and shallow water habitats: a case study from Chesapeake Bay. *In* N. McGinn, ed. Fisheries in a Changing Environment. American Fisheries Society, Bethesda, Md.

Stevenson, J. T., and D. H. Secor. 2000. Age determination and growth of Hudson River Atlantic sturgeon, *Acipenser oxyrinchus*. Fish. Bull. 97:153–166.

Stevenson, R. A., Jr. 1958. The biology of the anchovies *Anchoa mitchilli mitchilli* Cuvier and Valenciennes 1848 and *Anchoa hepsetus hepsetus* Linnaeus 1758 in Delaware Bay. M.A. thesis. University of Delaware. 56 pp.

Stewart, C. B., and F. S. Scharf. 2008. Estuarine recruitment, growth, and first-year survival of juvenile red drum in North Carolina. Trans. Am. Fish. Soc. 137:1089–1103.

Stierhoff, K. L., T. E. Targett, and P. A. Grecay. 2003. Hypoxia tolerance of the mummichog: the role of access to the water surface. J. Fish. Biol. 63:580–592.

Stierhoff, K. L., T. E. Targett, and K. Miller. 2006. Ecophysiological responses of juvenile summer and winter flounder to hypoxia: experimental and modeling analyses of effects on estuarine nursery quality. Mar. Ecol. Prog. Ser. 325:255–266.

Stillwell, C. E., and N. E. Kohler. 1993. Food habits of the sandbar shark, *Carcharhinus plumbeus*, off the U.S. northeast coast with estimates of daily ration. Fish. Bull. 91(1):138–150.

Stobutzki, I. C., and D. R. Bellwood. 1997. Sustained swimming abilities of the late pelagic stages of coral reef fishes. Mar. Ecol. Prog. Ser. 149:35–41.

Stoecker, R. R., J. Collura, and P. J. Fallon Jr. 1992. Aquatic studies at the Hudson River Center site. Pp. 407–427 *in* C. L. Smith, ed. Estuarine Research in the 1980s. The Hudson River Environmental Society, 7th Symposium on Hudson River Ecology. State University of New York Press, Albany.

Stone, S. L., T. A. Lowery, J. D. Field, C. D. Williams, D. M. Nelson, S. H. Jury, M. E. Monaco, and L. Andreasen. 1994. Distribution and abundance of fishes and invertebrates in Mid-Atlantic estuaries. Estuarine Living Marine Resources Rep. No. 12. Natl. Ocean. Atmos. Adm. / NOS Strategic Environmental Assessments Division, Silver Spring, Md.

Stoner, A.W. 1980. Feeding ecology of *Lagodon rhomboides* (Pisces: Sparidae): variation and functional responses. Fish. Bull. 78(2): 337–352.

———. 1991. Diel variation in the catch of fishes and penaeid shrimps in a tropical estuary. Estuar. Coast. Shelf Sci. 33:57–69.

Stoner, A. W., and M. L. Ottmar. 2003. Relationships between size-specific sediment preferences and burial capabilities in juveniles of two Alaska flatfishes. J. Exp. Mar. Biol. Ecol. 282:85–101.

Stoner, A. W., A. J. Bejda, J. P. Manderson, B. A. Phelan, L. L. Stehlik, and J. P. Pessutti. 1999. Behavior of winter flounder, *Pseudopleuronectes americanus*, during the reproductive season: laboratory and field observations on spawning, feeding, and locomotion. Fish. Bull. 97:999–1016.

Stoner, A. W., J. P. Manderson, and J. P. Pessutti. 2001. Spatially explicit analysis of estuarine habitat for juvenile winter flounder: combining generalized additive models and geographic information systems. Mar. Ecol. Prog. Ser. 213:253–271.

Straile, D., and N. C. Stenseth. 2007. The North Atlantic Oscillation and ecology: links between historical time-series, and lessons regarding future climate warming. Clim. Res. 34:259–262.

Stroud, R. H. 1971. Introduction to symposium. Pp. 3–8 *in* P. A. Douglas and R. H. Stroud, eds. A Symposium on the Biological Significance of Estuaries. Sport Fishing Institute, Washington, D.C.

Strydom, N. A. 2003. Occurrence of larval and early juvenile fishes in the surf zone adjacent to two intermittently open estuaries, South Africa. Environ. Biol. Fishes 66:349–359.

Strydom, N. A., A. K. Whitfield, and A. W. Paterson. 2002. Influence of altered freshwater flow regimes on abundance of larval and juvenile *Gilchristella aestuaria* (Pisces: Clupeidae) in the upper reaches of two South African estuaries. Mar. Freshwat. Res. 53:431–438.

Subrahmanyam, C. B. 1964. Eggs and early development of a carnagid from Madras. J. Mar. Biol. Assoc. India 6(10):142–146.

Sugeha, H. Y., T. Arai, M. J. Miller, D. Limbong, and K. Tsuka-moto. 2001. Inshore migration of the tropical eels *Anguilla* spp. recruiting to the Poigar River estuary on north Sulawesi Island. Mar. Ecol. Prog. Ser. 221:233–243.

Sullivan, M. C., R. K. Cowen, and B. P. Steves. 2005. Evidence for atmosphere-ocean forcing of yellowtail flounder (*Limanda fer-ruginea*) recruitment in the Middle Atlantic Bight. Fish. Ocean-ogr. 14(5):386–399.

Sullivan, M. C., K. W. Able, J. A. Hare, and H. J. Walsh. 2006a. *An-guilla rostrata* glass eel ingress into two U.S. east coast estuar-ies: patterns, processes and implications for adult abundance. J. Fish. Biol. 69:1081–1101.

Sullivan, M. C., R. K. Cowen, K. W. Able, and M. P. Fahay. 2006b. Assessing habitat suitability of young-of-the-year benthic fishes on the NY Bight continental shelf using a research submersible. Continent. Shelf Res. 26:1551–1570.

Sullivan, M. C., M. J. Wuenschel, and K. W. Able. 2009. Inter- and intra-estuary variability in ingress, condition, and settlement of the American eel *Anguilla rostrata*: implications for estimating and understanding recruitment. J. Fish Biol. 74:1949–1969.

Sumner, F. B., R. C. Osburn, and L. J. Cole. 1913. A catalogue of the marine fauna. *In* A biological survey of the waters of Woods Hole and vicinity. Bull. Bur. Fish. 31, 1911, Part II, Sec-tion III:549–794.

Sures, B., and K. Knopf. 2004. Parasites as a threat to freshwater eels? Science 304:209–211.

Swanson, R., and C. Parker. 1988. Physical environmental fac-tors contributing to recurring hypoxia in the New York Bight. Trans. Am. Fish. Soc. 117:37–47.

Swanson, R. L., and C. J. Sindermann. 1979. Oxygen depletion and associated benthic mortalities in the New York Bight, 1976. Natl. Ocean. Atmos. Adm. Prof. Pap. 11.

Swiecicki, D. P., and T. R. Tatham. 1977. Ichthyoplankton. Pp. 115–147 *in* C. B. Milstein, D. L. Thomas, et al., eds. Summary of Ecological Studies for 1972–1975 in the Bays and other Water-ways near Little Egg Inlet and in the Ocean in the Vicinity of the Proposed Site for the Atlantic Generation Station, New Jersey. New Jersey Bull. No. 18. Ichthyological Associates, Ithaca, N.Y.

Sylvester, J. R., C. E. Nash, and C. E. Emberson. 1974. Preliminary study of temperature tolerance in juvenile Hawaiian mullet (*Mugil cephalus*). Prog. Fish-Cult. 36:99–100.

Szedlmayer, S. T., and K. W. Able. 1992. Validation studies of daily increment formation for larval and juvenile summer flounder, *Paralichthys dentatus*. Can. J. Fish. Aquat. Sci. 49(9):1856–1862.

———. 1993. Ultrasonic telemetry of age-0 summer flounder, *Paralichthys dentatus*, movements in a southern New Jersey estu-ary. Copeia 3:728–736.

———. 1996. Patterns of seasonal availability and habitat use by fishes and decapod crustaceans in a southern New Jersey estu-ary. Estuaries 19(3):697–707.

Szedlmayer, S. T., M. P. Weinstein, and J. A. Musick. 1990. Differen-tial growth among cohorts of age-0 weakfish, *Cynoscion regalis*, in Chesapeake Bay. U.S. Fish. Bull. 88:745–752.

Szedlmayer, S. T., K. W. Able, J. A. Musick, and M. P. Weinstein. 1991. Are scale circuli deposited daily in juvenile weakfish, *Cy-noscion regalis*? Environ. Biol. Fishes 31:87–94.

Szedlmayer, S. T., K. W. Able, and R. A. Rountree. 1992. Growth and temperature-induced mortality of young of the year sum-mer flounder (*Paralichthys dentatus*) in southern New Jersey. Copeia 1:120–128.

Tagatz, M. E., and D. L. Dudley. 1961. Seasonal occurrence of marine fishes in four shore habitats near Beaufort, N.C., 1957–1960. U.S. Fish. Wildl. Serv., Spec. Sci. Rep. Fish. No. 390.

Talbot, C. W., and K. W. Able. 1984. Composition and distribution of larval fishes in New Jersey high marshes. Estuaries 7(4a): 434–443.

Talbot, G. B., and J. E. Sykes. 1958. Atlantic coastal migrations of American shad. U.S. Fish. Bull. 58:473–490.

Talbot, C. W., K. W. Able, and J. K. Shisler. 1986. Fish species com-position in New Jersey salt marshes: effects of marsh altera-tions for mosquito control. Trans. Am. Fish. Soc. 115:269–278.

Talbot, C. W., K. W. Able, J. K. Shisler, and D. Coorey. 1980. Sea-sonal variation in the composition of fresh and brackish water fishes of New Jersey mosquito control impoundments. Proc. Annu. Meet. N.J. Mosq. Control. Assoc. 67:50–63.

Tanda, M. 1990. Studies on burying ability in sand and selection to the grain size for hatchery reared marbled sole and Japanese flounder. Nippon Suisan Gakkaishi (Jpn. Soc. Grassl. Sci.) 56: 1543–1548.

Tatham, T. R., D. P. Swiecicki, and F. C. Stoop. 1974. Ichthyo-plankton. Pp. 88–106 *in* D. L. Thomas et al., eds. Ecological Studies in the Bays and Other Waterways near Little Egg Inlet and in the Ocean in the Vicinity of the Proposed Site for the Atlantic Generating Station, N.J. Vol. 1, Fishes. Ichthyological Associates, Ithaca, N.Y.

Tatham, T. R., D. L. Thomas, and D. J. Danila. 1984. Fishes of Bar-negat Bay, New Jersey. Pp. 241–280 *in* M. J. Kennish and R. A. Lutz, eds. Ecology of Barnegat Bay, New Jersey. Springer, New York.

Taubert, B. D. 1980a. Reproduction of the shortnose sturgeon (*Acipenser brevirostrium*) in Holyoke Pool, Connecticut River, Massachusetts. Copeia 1980:114–117.

———. 1980b. Biology of shortnose sturgeon (*Acipenser brevirostrum*) in the Hollyoke Pool, Connecticut River, Massachusetts. Ph.D. diss., University of Massachusetts, Amherst. 136 pp.

Taylor, D. L. 2003. Predation on the Early Life History Stages of Winter Flounder (*Pseudopleuronectes americanus*) by the Sand Shrimp (*Crangon septemspinosa*). University of Rhode Island. 354 pp.

———. 2004. Immunological detection of winter flounder (*Pseudo-pleuronectes americanus*) eggs and juveniles in the stomach con-tents of crustacean predators. J. Exp. Mar. Biol. Ecol. 301:55–73.

———. 2005. Predation on post-settlement winter flounder *Pseudo-pleuronectes americanus* by sand shrimp *Crangon septemspinoa* in NW Atlantic Estuaries. Mar. Ecol. Prog. Ser. 289:245–262.

Taylor, D. L., and K. W. Able. 2006. Cohort dynamics of summer-spawned bluefish as determined by length-frequency and oto-lith microstructure analyses. Trans. Am. Fish. Soc. 135:955–969.

Taylor, D. L., P. M. Rowe, and K. W. Able. 2006. Habitat use of the inner continental shelf off southern New Jersey by summer-spawned bluefish (*Pomatomus saltatrix*). Fish. Bull. 104:593–604.

Taylor, D. L., R. S. Nichols, and K. W. Able. 2007. Habitat selection and quality for multiple cohorts of young-of-the-year blue-fish (*Pomatomus saltatrix*): comparisons between estuarine and ocean beaches in southern New Jersey. Estuar. Coast. Shelf. Sci. 73:667–679.

Taylor, J. C., and P. S. Rand. 2003. Spatial overlap and distribution of anchovies (*Anchoa* spp.) and copepods in a shallow stratified estuary. Aquat. Living Resour. 16:191–196.

Taylor, M. H. 1999. A suite of adaptations for intertidal spawning. American Zoologist 39(2):313–320.

Taylor, M. H., and L. DiMichele. 1983. Spawning site utilization in a Delaware population of *Fundulus heteroclitus* (Pisces: Cyprinidontidae). Copeia:719–725.

Taylor, M. H., L. DiMichele, and G. J. Leach. 1977. Egg stranding in the life cycle of the mummichog, *Fundulus heteroclitus*. Copeia:397–399.

Taylor, M. H., G. J. Leach, L. DiMichele, W. M. Levitan, and W. F. Jacob. 1979. Lunar spawning cycle in the mummichog, *Fundulus heteroclitus* (Pisces: Cyprinodontidae). Copeia:291–297.

Taylor, M. H., L. DiMichele, R. T. Kneib, and S. Bradford. 1982. Comparison of reproductive strategies in Georgia and Massachusetts populations of *Fundulus heteroclitus*. Am. Zool. 191:821.

Taylor, W. R. 1967. An enzyme method of clearing and staining small vertebrates. Proc. U.S. Nat. Museum 122(3596):1–17.

Teal, J. M. 1986. The ecology of regularly flooded salt marshes of New England: a community profile. U.S. Fish. Wild. Serv. Biol. Rept. 85(7.4).

Teal, L. R., J. J. de Leeuw, H. W. van der Veer, and A. D. Rijnsdorp. 2008. Effects of climate change on growth of 0-group sole and plaice. Mar. Ecol. Prog. Ser. 358:219–230.

Teixeira, R. 1995. Reproductive and feeding biology of selected syngnathids (Pisces: Teleostei) of the Western Atlantic. Ph.D. diss. College of William and Mary, Williamsburg, Va.

Teo, S. L. H., and K. W. Able. 2003. Growth and production of the mummichog (*Fundulus heteroclitus*) in a restored salt marsh. Estuaries 26(1):51–63.

Ter Morshuizen, L. D., A. K. Whitfield, and A. W. Paterson. 1996. Influence of freshwater flow regime of fish assemblages in the Great Fish River and estuary. S. Afr. J. Aquat. Sci. 22:52–61.

Terwilliger, M. R. 1996. Age, growth, and reproductive biology of blackcheek tonguefish, *Symphurus plagiusa* (Cynoglossidae: Pleuronectiformes), in Chesapeake Bay, Virginia. M.S. thesis, College of William and Mary, Williamsburg, Va. 114 pp.

Thayer, G. W., T. A. McTigue, R. J. Salz, D. H. Merkey, F. M. Burrows, and P. F. Gayaldo, eds. 2005. Science-Based Restoration Monitoring of Coastal Habitats. Vol. 2, Tools for Monitoring Coastal Habitats. NOAA Coastal Open Program Decision Analysis Series No. 23. NOAA National Centers for Coastal Ocean Science, Silver Spring, Md. 628 pp plus appendixes.

Thibault, I., J. J. Dodson, F. Caron, W.-N. Tzeng, Y. Iizuka, and J.-C. Shiao. 2007. Facultative catadromy in American eels: testing the conditional strategy hypothesis. Mar. Ecol. Prog. Ser. 344: 219–229.

Thomas, D. L. 1971. The early life history and ecology of six species of drum (Sciaenidae) in the lower Delaware River, a brackish tidal estuary. An ecological study of the Delaware River in the vicinity of Artificial Island, pt. III. Bull. No. 3. Ichthyological Associates, Ithaca, N.Y.

Thomas, D. L., and C. B. Milstein. 1973. Ecological studies in the bays and other waterways near Little Egg Inlet and in the ocean in the vicinity of the proposed site for the Atlantic Generating Station, New Jersey. Progress Report for Period January–December 1972, Part I. Ichthyological Associates, Ithaca, N.Y.

———. 1974. A comparison of fishes collected by trawl along the coast of southern NJ from 1929 to 1933 with those taken in the vicinity of Little Egg Inlet in 1972 and 1973. Pp. 70–75 *in* D. L. Thomas et al., eds. Ecological Studies in the Bays and Other Waterways near Little Egg Inlet and in the Ocean in the Vicinity of the Proposed site for the Atlantic Generating Station, NJ. Vol. 1, Fishes. Ichthyological Associates, Ithaca, N.Y.

Thomas, D. L., and B. A. Smith. 1973. Studies of young of the

black drum, *Pogonias cromis*, in low salinity waters of the Delaware Estuary. Chesapeake Sci. 14(2):124–130.

Thomas, D. L., C. B. Milstein, T. R. Tatham, E. Van Eps, D. K. Stauble, and H. K. Hoff. 1972. Ecological Considerations for Ocean Sites off New Jersey for Proposed Nuclear Generating Stations. Vol. 1, Part 1, Ecological Studies in the Vicinity of Ocean Sites 7 and 8. Ichthyological Associates, Ithaca, N.Y.

Thomas, D. L., C. B. Milstein, T. R. Tatham, R. C. Bieder, F. J. Margraf, D. J. Danila, H. K. Hoff, E. A. Illjes, M. M. McCullough, and F. A. Swiecicki. 1974. Fishes. Pp. 115–131 *in* Ecological Studies in the Bays and Other Waterways near Little Egg Inlet and in the Ocean in the Vicinity of the Proposed Site for the Atlantic Generating Station, NJ. Vol. 1. Ichthyological Associates, Ithaca, N.Y.

Thompson, J. M., E. P. Bergerson, C. A. Carlson, and L. R. Kaeding. 1991. Role of size, condition, and lipid content in wintering survival of age-0 Colorado squawfish. Trans. Am. Fish. Soc. 120:346–353.

Thomson, K. S., W. H. Weed III, and A. G. Taruski. 1971. Saltwater Fishes of Connecticut. Bulletin 105. State Library, Hartford, State Librarian.

Thorman, S. 1986. Physical factors affecting the abundance and species richness of fishes in the shallow waters of the southern Bothnian Sea (Sweden). Est. Coast. Shelf Sci. 22:357–369.

Thorpe, L. A. 1991. Aspects of the biology of windowpane flounder, *Scophthalmus aquosus*, in the northwest Atlantic Ocean. M.S. thesis. University of Massachusetts, Amherst.

Thorrold, S. R., C. Latkoczy, P. K. Swart, and C. M. Jones. 2001. Natal homing in a marine fish metapopulation. Science 291: 297–299.

Thunberg, B. E. 1971. Olfaction in parent stream selection by the alewife. Anim. Behav. 19:217–225.

Todd, C. D., S. L. Hughes, C. T. Marshall, J. C. MacLean, M. E. Longergan, and E. M. Biuw. 2008. Detrimental effects of recent ocean surface warming on growth condition of Atlantic salmon. Global Change Biol. 14:958–970.

Tolan, J. M. 2008. Larval fish assemblage response to freshwater inflows: a synthesis of five years of ichthyoplankton monitoring within Nueces Bay, Texas. Bull. Mar. Sci. 82(3):275–296.

Tolan, J. M., and M. Fisher. 2009. Biological response to changes in climate patterns: population increases of gray snapper (*Lutjanus griseus*) in Texas bays and estuaries. Fish. Bull. 107(1):36–44.

Toth-Brown, J. 2007. Aspects of the ecology of bottlenose dolphins, *Tursiops truncatus*, in New Jersey. M.S. thesis. Rutgers University, New Brunswick, N.J. 92 pp.

Townsend, D. W., A. C. Thomas, L. M. Mayer, and M. A. Thomas. 2004. Oceanography of the northwest Atlantic continental shelf (1,W) Ch. 5 *in* A. R. Robinson and K. H. Brink, eds. The Sea: The Global Coastal Ocean: Interdisciplinary Regional Studies and Syntheses. Harvard University Press.

Tracy, H. C. 1959. Stages in the development of the anatomy of motility of the toadfish (*Opsanus tau*). J. Comp. Neurol. 11:27–82.

Tsou, T., and R. E. Matheson, Jr. 2002. Seasonal changes in the nekton community of Swannee River estuary and the potential impacts of freshwater withdrawal. Estuaries 25:1372–1381.

Tucker, J. W., Jr. 1989. Energy utilization in bay anchovy, *Anchoa mitchilli*, and black sea bass, *Centropristis striata*, eggs and larvae. Fish. Bull. 87(2):279–293.

Tulp, I., L. J. Bolle, and A. D. Rijnsdorp. 2008. Signals from the shallows: in search of common patterns in long-term trends in Dutch estuarine and coastal fish. J. Sea Res. 60:54–73.

Tupper, M. 1994. Settlement and post-settlement processes in the population regulation of a temperate reef fish: the role of energy. Ph.D. diss. Dalhousie University, Halifax, Canada.

Tupper, M., and K. W. Able. 2000. Movements and food habits of striped bass (*Morone saxatilis*) in Delaware Bay (USA) salt marshes: comparison of a restored and a reference marsh. Mar. Biol. 137:1049–1058.

Tupper, M., and R. G. Boutilier. 1995a. Effects of conspecific density on settlement, growth and post-settlement survival of a temperate reef fish. J. Exp. Mar. Biol. Ecol. 191:209–222.

———. 1995b. Size and priority at settlement determine growth and competitive success of newly settled Atlantic cod. Mar. Ecol. Prog. Ser. 118:295–300.

Tupper, M., and W. Hunte. 1994. Recruitment dynamics of coral reef fishes in Barbados. Mar. Ecol. Prog. Ser. 108:225–235.

Tyler, A. V. 1971. Period and resident components in communities of Atlantic fishes. J. Fish. Res. Bd. Canada 28:935–946.

Tyler, R. M., and T. E. Targett. 2007. Juvenile weakfish *Cynoscion regalis* distribution in relation to diel-cycling dissolved oxygen in an estuarine tributary. Mar. Ecol. Prog. Ser. 333:257–269.

Tyler, R. M., D. C. Brady, and T. E. Targett. 2009. Temporal and spatial dynamics of diel-cycling hypoxia in estuarine tributaries. Estuaries and Coasts 32:123–145.

Tzeng, M. W. 2000. Patterns of ingress, age, and growth of snappers (principally *Lutjanus griseus*) near Beaufort Inlet, North Carolina. M.S. thesis, University of North Carolina at Wilmington.

Tzeng, M. W., J. A. Hare, and D. G. Lindquist. 2003. Ingress of transformation stage gray snapper, *Lutjanus griseus* (Pisces: Lutjanidae) through Beaufort Inlet, North Carolina. Bull. Mar. Sci. 72(3):891–908.

Uncles, R. J., and R. E. Smith. 2005. A note on the comparative turbidity of some estuaries of the Americas. J. Coast. Res. 21(4):845–852.

Underwood, A. J., M. G. Chapman, and S. D. Connell. 2000. Observations in ecology: you can't make progress on processes without understanding the patterns. J. Exp. Mar. Biol. Ecol. 250:97–115.

Urho, L. 2002. Characters of larvae—what are they? Folia Zool. 51(3):161–186.

Valesini, F. J., I. C. Potter, M. W. Platell, and G. A. Hyndes. 1997. Ichthyofaunas of a temperate estuary and adjacent marine embayment. Implications regarding choice of nursery area and influence of environmental changes. Mar. Biol. 128:317–328.

van Damme, C. J. G., and A. S. Couperus. 2008. Mass occurrence of snake pipefish in the Northeast Atlantic: Result of a change in climate? J. Sea Res. 60:117–125.

Van Diggelen, R., A. P. Grootjans, and J. A. Harris. 2001. Ecological restoration: state of the art or state of the science? Rest. Ecol. 9:115–118.

van Oosten, J. 1957. The skin and scales. Pp. 207–243 *in* M. E. Brown, ed. The Physiology of Fishes. Academic Press, New York.

Van Sant, S. B., M. R. Collins, and G. R. Sedberry. 1994. Preliminary evidence from a tagging study for a gag (*Mycteroperca microlepis*) spawning migration, with notes on the use of oxytetracycline for chemical tagging. Proc. Gulf. Carib. Fish. Inst. 43:417–428.

Vari, R. 1982. The seahorses (subfamily Hippocampinae). *In* J. E. Bohlke, D. M. Cohen, B. B. Collette, W. N. Eschmeyer, R. H. Gibbs, T. W. Pietsch Jr., W. J. Richards, C. L. Smith, and K. S. Thomson, eds. Fishes of the Western North Atlantic.

Part 8. Order Gasterosteiformes, Suborder Syngnathoidei. Syngnathidae (Doryrhamphinae, Hippocampinae). Mem. Sears Found. Mar. Res. 1(8):173–189.

Vasslides, J. M., and K. W. Able. 2008. Importance of shoreface sand ridges as habitat for fishes off the northeast coast of the United States. Fish. Bull. 106:93–107.

Vaughan, D. S., and T. E. Helser. 1990. Status of the red drum stock of the Atlantic coast: stock assessment report for 1989. NOAA Tech. Memo. NMFS-SEFC-263. 117 pp.

Veer, H. W. van der, and M. J. N. Bergman. 1987. Predation by crustaceans on a newly settled 0-group plaice (*Pleuronectes platessa*) population in the western Wadden Sea. Mar. Ecol. Prog. Ser. 35:203–215.

Victor, B. C. 1986. Larval settlement and juvenile mortality in a recruitment-limited coral reef fish population. Ecol. Monogr. 56:145–160.

Vincent, A., I. Ahnesjo, and A. Berglund. 1992. Operational sex ratios and behavioral sex differences. Beh. Ecol. Sociobiology 34:435–442.

Virginia Institute of Marine Sciences. 1962. Fishes and fish larvae collected from Atlantic plankton cruises of R/V *Pathfinder*, March 1961–March 1962. Spec. Sci. Rep. No. 33. College of William and Mary, Gloucester Point, Va.

Visintainer, T. A., S. M. Bollens, and C. Simenstad. 2006. Community composition and diet of fishes as a function of tidal channel geomorphology. Mar. Ecol. Prog. Ser. 321:227–243.

Vivian, D. N., J. T. Duffy-Anderson, R. G. Arndt, and K. W. Able. 2000. Feeding habits of young-of-the-year winter flounder, *Pseudopleuronectes americanus*, in the Hudson river estuary U.S.A. Bull. N.J. Acad. Sci. 45(2):1–6.

Vladykov, V. D. 1971. Homing of the American eel, *Anguilla rostrata*, as evidenced by returns of transplanted tagged eels in New Brunswick. Can. Field Nat. 85(3):241–248.

Vladykov, V. D., and J. R. Greeley. 1963. Order Acipenseroidei. Pp. 24–59 *in* Y. H. Olsen, ed. Fishes of the western North Atlantic. Sears Found. Mar. Res., Yale Univ. 1(3):1–630.

von Alt, C. J., and J. F. Grassle. 1992. LEO-15: An unmanned long term environmental observatory. OCEANS 92, Newport, R.I.

Voss, G. L. 1953. A contribution to the life history and biology of the sailfish, *Istiophorus americanus* Cuv. and Val., in Florida waters. Bull Mar. Sci. Gulf Caribb. 3:206–240.

Vouglitois, J. J. 1983. The ichthyofauna of Barnegat Bay, New Jersey: relationships between long term temperature fluctuations and the population dynamics and life history of temperate estuarine fishes during a five year period, 1976–1980. M.S. thesis, Rutgers University, New Brunswick, N.J.

Vouglitois, J. J., K. W. Able, R. J. Kurtz, and K. A. Tighe. 1987. Life history and population dynamics of the bay anchovy in New Jersey. Trans. Am. Fish. Soc. 116(2):141–153.

Wagner, M. C., and H. M. Austin. 1999. Correspondence between environmental gradients and summer littoral fish assemblages in low salinity reaches of the Chesapeake Bay, USA. Mar. Ecol. Prog. Ser. 177:197–212.

Waldman, J. R., D. J. Dunning, Q. E. Ross, and M. T. Mattson. 1990. Range dynamics of Hudson River striped bass along the Atlantic coast. Trans. Am. Fish. Soc. 119:910–919.

Waldman, J. R., K. E. Limburg, and D. L. Strayer. 2006. The Hudson River environment and its dynamic fish community. Pp. 1–7. *In* J. R. Waldman, K. E. Limburg, and D. L. Strayer, eds. Hudson River Fishes and Their Environment. Am. Fish. Soc. Symp. 51, Bethesda, Md.

Walford, L., and R. Wicklund. 1968. Monthly sea surface temperature structure from the Florida Keys to Cape Cod. Am. Geograph. Soc. Folio 15.

Wallace, R. A., and K. Selman. 1981. The reproductive activity of *Fundulus heteroclitus* females from Woods Hole, Massachusetts, as compared with more southern locations. Copeia 1:212–215.

Wallace, R. C. 1971. Age, growth, year class strength and survival rates of white perch, *Morone americana* (Gmelin), in the Delaware River in the vicinity of Artificial Island. Chesapeake Sci. 12:205–218.

Wallin, J. E., J. M. Ransier, S. Fox, and R. H. McMichael Jr. 1997. Short-term retention of coded wire and internal anchor tags in juvenile common snook, *Centropomus undecimalis*. Fish. Bull. 95:873–878.

Walline, P. D. 1985. Growth of larval walleye pollock related to domains within the southeast Bering Sea. Mar. Ecol. Prog. Ser. 21:197–203.

Walter, J. F., III, and H. M. Austin. 2003. Diet composition of large striped bass (*Morone saxatilis*) in Chesapeake Bay. U.S. National Marine Fisheries Service Fish. Bull. 101:414–423.

Walter, J. F., III, A. S. Overton, K. H. Ferry, and M. E. Mather. 2003. Atlantic coast feeding habits of striped bass: a synthesis supporting a coast-wide understanding of trophic biology. Fish. Man. Ecol. 10:349–360.

Walters, C. J., and F. Juanes. 1993. Recruitment limitation as a consequence of natural selection for use of restricted feeding habitats and predation risk taking by juvenile fishes. Can. J. Fish. Aq. Sci. 50(1):2058–2070.

Walther, B. D., and S. R. Thorrold. 2008. Continental-scale variation in otolith geochemistry of juvenile American shad (*Alosa sapidissima*). Can. J. Fish. Aq. Sci. 65 (12):2623–2635.

Walther, G., E. Post, P. Convey, A. Menzel, C Parmesan, T. J. C. Beebee, J. Fromentin, O. Hoegh-Guldberg, and F. Bairlein. 2002. Ecological responses to recent climate change. Nature 416:389–395.

Wang, J. C., and R. J. Kernehan. 1979. Fishes of the Delaware Estuaries: A Guide to the Early Life Histories. EA Communications, Towson, Md.

Wang, S. B., and E. D. Houde. 1995. Distribution, relative abundance, biomass, and production of bay anchovy *Anchoa mitchilli* in Chesapeake Bay. Mar. Ecol. Prog. Ser. 121:27–38.

Wannamaker, C. M., and J. A. Rice. 2000. Effects of hypoxia on movements and behavior of selected estuarine organisms from the southeastern United States. J. Exp. Mar. Biol. Ecol. 249:145–163.

Warfel, H. E., and D. Merriman. 1944. Studies on the marine resources of southern New England. I. An analysis of the fish population of the shore zone. Bull. Bingham Oceanogr. Collect. Yale Univ. 9(2):1–91.

Waring, G. T. 1975. Preliminary analysis of the status of butterfish in ICNAF Subarea 5 and Statistical Area 6. Int. Comm. North. Am. Fish. Res. Doc. 75/14.

Waring, G., and S. Murawski. 1982. Butterfish. MESA (Mar. Ecosyst. Anal.) N.Y. Bight Atlas Monogr. 15:105–107.

Warinner, J. E., J. P. Miller, and J. Davis. 1969. Distribution of juvenile river herring in the Potomac River. Proc. Annu. Conf. Southeast Assoc. Game Fish Comm. 23:384–388.

Warlen, S. M. 1964. Some aspects of the life history of *Cyprinodon variegatus* Lacedede 1803, in southern Delaware. M.S. thesis. University of Delaware, Newark.

Warlen, S. M., and J. S. Burke. 1990. Immigration of larvae of fall/ winter spawning marine fishes into a North Carolina estuary. Estuaries 13:453–461.

Warlen, S. M., K. W. Able, and E. Laban. 2002. Recruitment of larval Atlantic menhaden to North Carolina and New Jersey estuaries: evidence for larval transport northward along the east coast of the United States. Fish. Bull. 100(3):609–623.

Warren, R. S., P. E. Fell, R. Rozsa, A. H. Brawley, A. C. Orsted, E. T. Olson, V. Swamy, and W. A. Niering. 2002. Salt marsh restoration in Connecticut: 20 years of science and management. Rest. Ecol. 10:497–513.

Waterman, R. E. 1970. Fine structure of scale development in teleost, *Brachydanio rerio*. The Anatomical Record 168:361–380.

Watkins, E. S. 2001. Reproductive biology of Atlantic croaker, *Micropogonias undulatus*, in the Chesapeake Bay. M.S. thesis. College of William and Mary, Williamsburg, Va. 74 pp.

Watson, J. F. 1968. The early life history of the American shad, *Alosa sapidissima* (Wilson), in the Connecticut River above Holyoke, Massachusetts. M.S. thesis. University of Massachusetts, Amherst.

Webb, J. F. 1999. Larvae in fish development and evolution. Pp. 109–157 *in* B. K. Hall and M. H. Wake, eds. The Origin and Evolution of Larval Forms. Academic Press, San Diego, Calif.

Webster, M. M., and P. J. B. Hart. 2004. Substrate discrimination and preference in foraging fish. Animal Behavior 38:1071–1077.

Weeder, J. A., and S. D. Hammond. 2009. Age, growth, mortality, and sex ratio of American eels in Maryland's Chesapeake Bay. *In* J. Casselman and D. Cairns, eds. Eels at the Edge. American Fisheries Society Symposium 58, Bethesda, Md.

Weinstein, M. P. 1979. Shallow marsh habitats as primary nurseries for fishes and shellfish, Cape Fear River, North Carolina. U.S. Fish. Bull. 77:339–357.

———. 1983. Population dynamics of an estuarine dependent fish, the spot (*Leiostomus xanthurus*), along a tidal creek-seagrass coenocline. Can. J. Fish. Aquat. Sci. 40:1633–1638.

Weinstein, M. P., and H. A. Brooks. 1983. Comparative ecology of nekton residing in a tidal creek and adjacent seagrass meadow: community composition and structure. Mar. Ecol. Prog. Ser. 12:15–27.

Weinstein, M. P., and M. P. Walters. 1981. Growth, survival and production in young-of-the-year populations of *Leiostomus xanthurus* Lacepede residing in tidal creeks. Estuaries 4(3):185–197.

Weinstein, M. P., C. M. Courtney, and J. C. Kinch. 1977. The Marco Island estuary: a summary of physiochemical and biological parameters. Fla. Sci. 40:97–124.

Weinstein, M. P., S. L. Weiss, R. G. Hodson, and L. R. Gerry. 1980. Retention of three taxa of postlarval fishes in an intensively flushed tidal estuary, Cape Fear River, North Carolina. Fish. Bull. 78:419–436.

Weinstein, M. P., L. Scott, S. P. O'Neil, R. C. Siegfriend, and S. T. Szedlmayer. 1984. Population dynamics of spot, *Leiostomus xanthurus*, in polyhaline tidal creeks of the York River estuary, Virginia. Estuaries 7(4a):444–450.

Weis, J. S. 2005. Diet and food web support of the white perch, *Morone Americana,* in the Hackensack Meadowlands of New Jersey. Environ. Bio. Fish. 74:109–113.

Weisberg, S. B. 1986. Competition and coexistence among four estuarine species of *Fundulus*. Am. Zool. 26:249–257.

Weisberg, S. B., and W. H. Burton. 1993. Spring distribution and abundance of ichthyoplankton in the tidal Delaware River. U.S. Fish. Bull. 91(4):788–797.

Weisberg, S. B., P. Himchak, T. Baum, H. T. Wilson, and R. Allen.

1996. Temporal trends in abundance of fish in the tidal Delaware River. Estuaries 19(3):723–729.

Wells, B. K., J. C. Field, J. A. Thayer, C. B. Grimes, S. J. Bograd, W. J. Sydeman, F. B. Schwing, and R. Hewitt. 2008. Untangling the relationships among climate, prey and top predators in an ocean ecosystem. Mar. Ecol. Prog. Ser. 364:15–29.

Wells, H. W. 1961. The fauna of oyster beds, with special reference to the salinity factor. Ecol. Monogr. 31:266–329.

Welsh, W. W. 1915. Notes on the habits of the young of the squirrel hake and sea snail. Copeia 18:2–3.

Welsh, W. W., and C. M. Breder. 1922. A contribution to the life history of the puffer, *Spheroides maculatus* (Schneider). Zoologica (N.Y.) 2:261–276.

———. 1924. Contributions to the life histories of Sciaenidae of the eastern United States coast. Bull. U.S. Bur. Fish. 39:141–201.

Wenner, C. A. 1973. Occurrence of American eels, *Anguilla rostra*, in water overlying the eastern North American continental shelf. J. Fish. Res. Board Can. 30(11):1752–1755.

Wenner, C. A., and J. A. Musick. 1974. Fecundity and gonad observations of the American eel, *Anguilla rostrata*, in water overlying the eastern North American continental shelf. J. Fish. Res. Board. Can. 30(11):1752–1755.

Wenner, C. A., and G. R. Sedberry. 1989. Species composition, distribution and relative abundance of fishes in the coastal habitat off the southeastern United States. U.S. Dep. Commer. Natl. Ocean Atmos. Adm. Tech. Rep. Natl. Mar. Fish. Serv. 79.

Wenner, C. A., W. A. Roumillat, and C. W. Waltz. 1986. Contributions to the life history of black sea bass, *Centropristis striata*, off the southeastern United States. Fish. Bull. 84(3):723–741.

Wenner, E., H. R. Beatty, and L. D. Coen. 1996. Method for quantitatively sampling nekton on intertidal oyster reefs. J. Shellfish. Res. 115:769–775.

Werme, C. E. 1981. Resource partitioning in a salt marsh fish community. Ph.D. diss. Boston University, Boston.

Werner, E. E., and J. F. Gilliam. 1984. The ontogenetic niche and species interactions in size-structured populations. An. Rev. Eco. Sys. 15:393–425.

West, R. J., and R. J. King. 1996. Marine, brackish and freshwater fish communities in the vegetated and bare shallows of an Australian coastal river. Estuaries 19(1):31–41.

Wheatland, S. B. 1956. Pelagic fish eggs and larvae. Chap. VII *in* Oceanography of Long Island Sound, 1952–1954. Bull. Bingham Oceanogr. Collect. Yale Univ. 15:234–314.

White, D. S. 1977. Early development and pattern of scale formation in the spotted sucker, *Minytrema melanops* (Catostomidae). Copeia 2:400–403.

Whitfield, A. K. 1994. An estuary-association classification for the fishes of southern Africa. Suid-Afrikaanse Tydskrif vir Wetenskap 90:411–417.

———. 1998. Biology and ecology of fishes in southern African estuaries. Ichthyological monographs of the J. L. B. Smith Institute of Ichthyology No. 2. J. L. B. Smith Institute of Ichthyology, Grahamstown, South Africa.

Whitfield, A. K., and S. J. M. Blaber. 1978. Food and feeding ecology of piscivorous fishes at Lake St. Lucia, Zuzuland. J. Fish. Biol. 13:675–691.

Wicklund, R. 1970. A puffer kill related to nocturnal behavior and adverse environmental changes. Underw. Nat. 6(3):28–29.

Wicklund, R. I., S. J. Wilk, and L. Ogren. 1968. Observations on wintering locations of the northern pipefish and spotted seahorse. Underwater Nat. 5(2):26–28.

Wilber, D. H., and D. G. Clarke. 2001. Biological effects of suspended sediments: a review of suspended sediment impacts on fish and shellfish with relation to dredging activities in estuaries. N. Am. J. Fish. Man. 21:855–875.

Wilk, S. J. 1976. The weakfish—a wide ranging species. Atl. States Mar. Fish. Comm. Mar. Atl. Coast Fish. Leaf. No. 18.

———. 1977. Biological and fisheries data on bluefish, *Pomatomus saltatrix* (Linnaeus). NMFS Sandy Hook Lab., Tech. Ser. Rep. No. 11. 56 pp.

———. 1982. Bluefish, *Pomatomus saltatrix*. MESA (Mar. Ecosyst. Anal.) N.Y. Bight Atlas, Monogr. 15:86–89.

Wilk, S. J., and M. J. Silverman. 1976. Summer benthic fish fauna of Sandy Hook Bay, New Jersey 1976. Natl. Ocean. Atmos. Adm. Tech. Rep. Natl. Mar. Fish. Ser. Spec. Sci. Rep. Fish-698.

Wilk, S. J., W. W. Morse, D. E. Ralph, and E. J. Steady. 1975. Life history aspects of New York Bight finfishes (June 1974–June 1975). Annu. Rep. NOAA NMFS Middle Atlantic Coastal Fish. Cent. Sandy Hook Lab. Highlands, N.J.

Wilk, S. J., W. J. Clifford, and D. J. Christensen. 1979. The recreational fishery for pollock (*Pollachius virens*) in southern New England and the Middle Atlantic. Natl. Mar. Fish. Serv. Northeast Fish. Cent. Sandy Hook Lab. Ref. Doc. No. 79-31.

Wilk, S. J., W. W. Morse, and L. L. Stehlik. 1990. Annual cycles of gonad-somatic indices as indicators of spawning activity for selected species of finfish collected from the New York Bight. U.S. Fish. Bull. 88:775–786.

Wilk, S. J., R. A. Pikanowski, A. J. Pacheco, D. G. McMillan, B. A. Phelan, and L. L. Stehlik. 1992. Fish and megainvertebrates collected in the New York Bight apex during the 12-mile dumpsite recovery study, July 1986–September 1989. Natl. Ocean. Atmos. Adm. Tech. Mem. NMFS-F/NEC-90:1–78.

Williams, D. M., S. English, and M. J. Milicich. 1994. Annual recruitment surveys of coral reef fishes are good indicators of patterns of settlement. Bull. Mar. Sci. 54(1):314–331.

Williams, G. C. 1975. Viable embryogenesis of the winter flounder, *Pseudopleuronectes americanus* from 1.8° to 15°C. Mar. Biol. 33: 71–74.

Williams G. G., D. G. Williams, and R. J. Miller. 1973. Mortality rates of planktonic eggs of the cunner, *Tautogolabrus adspersus* (Walbaum), in Long Island Sound. Pp. 181–195 *in* Proceeding of a Workshop on Egg, Larval and Juvenile Stages of Fish in Atlantic Coast Estuaries, Sandy Hook Laboratory Technical Series Reports, Vol. 1, edited by A. L. Pacheco. Northeast Fisheries Science Center, Woods Hole, Mass.

Williams, J. T. 2002. Blenniidae: Combtooth blennies. Pp. 1768–1772 *in* K. E. Carpenter, ed. The Living Marine Resources of the Western Central Atlantic. Vol. 3, Bony Fishes, Part 2 (Opishthognathidae to Molidae). Food and Agriculture Organization of the United Nations, Rome.

Wilson, C. A., J. M. Dean, and R. Radtke. 1982. Age, growth rate and feeding habits of the oyster toadfish, *Opsanus tau* (Linnaeus) in South Carolina. J. Exp. Mar. Biol. Ecol. 62:251–259.

Wilson, H. V. 1891. The embryology of the sea bass (*Serranus atrarius*). Bull. U.S. Fish Comm. 9:209–277.

Windham, L., and R. G. Lathrop Jr. 1999. Effects of *Phragmites australis* (common reed) invasion on aboveground biomass and soil properties in brackish tidal marsh of the Mullica River, New Jersey. Estuaries 22(4):927–935.

Wingate, R. L., and D. H. Secor. 2008. Effects of winter temperature and flow on a summer-fall nursery fish assemblage in the Chesapeake Bay, Maryland. Trans. Am. Fish. Soc. 137:1147–1156.

Winslade, P. 1974a. Behavioural studies on the lesser sandeel *Ammodytes marinus* (Raitt). I. The effect of food availability on activity and the role of olfaction in food detection. J. Fish. Biol. 6: 565–576.

———. 1974b. Behavioural studies on the lesser sandeel *Ammodytes marinus* (Raitt). II. The effect of light intensity on activity. J. Fish. Biol. 6:577–586.

———. 1974c. Behavioural studies on the lesser sandeel *Ammodytes marinus* (Raitt). III. The effect of temperature on activity and the environmental control of the annual cycle of activity. J. Fish. Biol. 6:587–599.

Winters, G. H., J. A. Moores, and R. Chaulk. 1973. Northern range extension and probable spawning of gaspereau in the Newfoundland area. J. Fish. Res. Board Can. 30:860–861.

Witting, D. A. 1995. Settlement of winter flounder, *Pleuronectes americanus*, in a southern New Jersey estuary: temporal and spatial dynamics and the effect of decapod predation. Ph.D. diss., Rutgers University, New Brunswick, N.J.

Witting, D. A., and K. W. Able. 1993. Effects of body size on probability of predation for juvenile summer and winter flounder based on laboratory experiments. U.S. Fish. Bull. 91:577–581.

———. 1995. Predation by sevenspine bay shrimp, *Crangon septemspinosa*, on winter flounder, *Pleuronectes americanus*, during settlement: laboratory observations. Mar. Ecol. Prog. Series 123:23–31.

Witting, D. A., K. W. Able, and M. P. Fahay. 1999. Larval fishes of a Middle Atlantic Bight estuary: assemblage, structure, and temporal stability. Can. J. Fish. Aquat. Sci. 56:222–230.

Witting, D. A., R. C. Chambers, K. L. Bosley, and S. C. Wainwright. 2004. Experimental evaluation of ontogenetic diet transitions in summer flounder (*Paralichthys dentatus*), using stable isotopes as diet tracers. Can. J. Fish. Aquat. Sci. 61:2069–2084.

Wockley, R. C. 1968. A partial life history of *Trinectus maculatus* (Bloch and Schneider) the hogchoker. M.A. thesis. University of Delaware. 39 pp.

Wood, A. D., B. M. Wetherbee, F. Juanes, N. E. Kohler, and C. Wilga. 2009b. Recalculated diet and daily ration of the shortfin mako (*Isurus oxyrinchus*), with a focus on quantifying predation on bluefish (*Pomatomus saltatrix*) in the northwest Atlantic Ocean. Fish. Bull. 107(1):76–88.

Wood, A. J. M., J. S. Collie, and J. A. Hare. 2009a. A comparison between warm-water fish assemblages of Narragansett Bay and those of Long Island Sound waters. Fish. Bull. 107(1):89–100.

Wood, R. J. 2000. Synoptic scale climatic forcing of multispecies fish recruitment patterns in Chesapeake Bay. Ph.D. diss., College of William and Mary, Williamsburg, Va.

Wooten, M. C., K. T. Scribner, and M. H. Smith. 1988. Genetic variability and systematics of *Gambusia* in the southeastern United States. Copeia:293–289.

Wooton, R. J. 1976. The Biology of the Sticklebacks. Academic Press, New York. 387 pp.

Wright, P. J., H. Jensen, and I. Tuck. 2000. The influence of sediment type of the distribution of the lesser sandeel, *Ammodytes marinus*. J. Sea Res. 44:243–256.

Wuenschel, M. J., and K. W. Able. 2008. Swimming ability of eels (*Anguilla rostrata, Conger oceanicus*) at estuarine ingress: contrasting patterns of cross-shelf transport? Mar. Biol. 154:775–786.

Wuenschel, M. J., A. R. Jugovich, and J. A. Hare. 2004. Effect of temperature and salinity on the energetics of juvenile gray snapper (*Lutjanus griseus*): implications for nursery habitat value. J. Exp. Mar. Biol. Ecol. 312:333–347.

Wuenschel, M. J., K. W. Able, and D. Byrne. 2009. Seasonal patterns of winter flounder *Pseudopleuronectes americanus* abundance and reproductive condition on the New York Bight continental shelf. J. Fish Biol. 74:1508–1524.

Wuenschel, M. J., K. W. Able, D. O. Conover, S. B. Munch, D. H. Secor, J. A. Buckel, T. Lankford, and F. Juanes. In prep. Recruitment patterns of young-of-the-year bluefish in the Middle and South Atlantic Bights.

Wyanski, D. M. 1988. Depth and substrate characteristics of age 0 summer flounder (*Paralichthys dentatus*) habitat in Virginia estuaries. M.S. thesis. College of William and Mary, Williamsburg, Va. 54 pp.

Wyanski, D. M., and T. E. Targett. 2000. Development of transformation larvae and juveniles of *Ctenogobius boleosoma, Ctenogobius shufeldti,* and *Gobionellus oceanicus* (Pisces: Gobiidae) from western North Atlantic estuaries, with notes on early life history. Bull. Mar. Sci. 67(2):709–728.

Yako, L. A., M. E. Mather, and F. Juanes. 2002. Mechanisms for migration of anadromous herring; an ecological basis for effective conservation. Ecol. Appl. 12(2):521–534.

Yáñez-Arancibia, A. 1985. Fish Community Ecology in Estuaries and Coastal Lagoons: Towards an Ecosystem Integration. UNAM Press, Distrito Federal, Mexico. 654 pp.

Yáñez-Arancibia, A., F. A. Linares, and J. W. Day. 1980. Fish community structure and function in Terminos Lagoon, a tropical estuary in the southern Gulf of Mexico. Pp. 465–482 *in* V. S. Kennedy, ed. Estuarine Perspectives. Academic Press, New York.

Yáñez-Arancibia, A., P. Sanchez-Gil, M. T. Garcia, and M. De la C. Garcia-Abad. 1985. Ecology, community structure and evaluation of tropical demersal fishes in the southern Gulf of Mexico. Cah. Biol. Mar. 26:137–163.

Yáñez-Arancibia, A., A. L. Lara-Domínguez, and D. Pauly. 1994. Coastal lagoons as fish habitats. Pp. 363–376 *in* B. Kjerfve, ed. Coastal Lagoon Processes. Elsevier Science Publishers, Amsterdam.

Yokel, B. 1966. A contribution to the biology and distribution of the red drum, *Sciaenops ocellata*. M.S. thesis. University of Miami. 166 pp.

Young, B. H., K. A. McKown, V. J. Vecchio, K. Hattala, and J. D. Sicluna. 1991. A study of the striped bass in the Marine District of New York. VI. Anadromous Fish Conservation Act P.L. 89-304. Completion Report. New York. Project AFC-14. N.Y. State Department of Environmental Conservation.

Young, J. R., T. B. Hoff, W. P. Dey, and J. G. Hoff. 1988. Management recommendations for a Hudson River Atlantic sturgeon fishery based on an age-structured population model. Pp. 353–365 *in* C. L. Smith, ed. Fisheries Research in the Hudson River. Hudson River Environ. Soc., State University of New York Press, Albany.

Young, T. P. 2000. Restoration ecology and conservation biology. Biol. Conserv. 92:73–83.

Youson, J. H. 1988. First metamorphosis. Pp. 135–196 *in* W. S. Hoar and D. J. Randall, eds. Fish Physiology. Vol. 11B. Academic Press, San Diego, Calif.

Yozzo, D. J., and F. Ottman. 2003. New distribution records for the spotfin killifish, *Fundulus luciae* (Baird), in the lower Hudson River estuary and adjacent waters. NE Nat. 10(4):399–408.

Yozzo, D. J., and D. E. Smith. 1998. Composition and abundance of resident marsh-surface nekton: comparison between tidal freshwater and salt marshes in Virginia, USA. Hydrobiologia 362:9–19.

Yuschak, P. 1985. Fecundity, eggs, larvae and osteological development of the striped searobin (*Prionotus evolans*) (Pisces, Triglidae). J. Northw. Atl. Fish. Sci. 6:65–85.

Yuschak, P., and W. A. Lund Jr. 1984. Eggs, larvae and osteological development of the northern searobin, *Prionotus carolinus* (Pisces, Triglidae). J. Northw. Atl. Fish. Sci. 5:1–15.

Zastrow, C. E., E. D. Houde, and L. G. Morin. 1991. Spawning, fecundity, hatch-date frequency and young-of-the-year growth of bay anchovy *Anchoa mitchilli* in mid-Chesapeake Bay. Mar. Ecol. Prog. Ser. 73:161–171.

Zedler, J. B. 1995. Salt marsh restoration: lessons from California. Pp. 75–95 *in* J. Cairns Jr., ed. Rehabilitating Damaged Ecosystems. 2nd ed. CRC Press, Inc., Boca Raton, Fla.

———. 2000. Progress in wetland restoration ecology. Trends Ecol. Evol. 15:402–407.

———. 2001. Handbook for restoring tidal wetlands. CRC Press. Boca Raton, Fla.

Zedler, J. B., and J. C. Callaway. 1999. Tracking wetland restoration: do mitigation sites follow desired trajectories. Rest. Ecol. 7:69–73.

Zhang, W. G., J. L. Wilkin, and R. J. Chant. 2009. Modeling the pathways and mean dynamics of river plume dispersal in the New York Bight. J. Phy. Oceanog. 39(5):1167–1183.

Zich, H. E. 1977. The collection of existing information and field investigation of anadromous clupeid spawning in New Jersey. N.J. Div. Fish Game Shellfish. Misc. Rep. 41. Trenton, N.J.

Ziskowski, J. J., L. Despres-Patanjo, R. A. Murchelano, A. B. Howe, D. Ralph, and S. Atran. 1987. Disease in commercially valuable fish stocks in the Northwest Atlantic. Mar. Poll. Bull. 18(9): 496–504.

Zlokovitz, E. R., and D. H. Secor. 1999. Effect of habitat use on PCB body burden in Hudson River striped bass (*Morone saxatilis*). Can. J. Fish. Aq. Sci. 56 (suppl. 1):1–8.

Zlokovitz, E. R., D. H. Secor, and P. M. Piccoli. 2003. Patterns of migration in Hudson River striped bass as determined by otolith microchemistry. Fish Res. 63:245–259.

Zydlewski, J., S. D. McCormick, and J. G. Kunkel. 2003. Late migration and seawater entry is physiologically disadvantageous for American shad juveniles. J. Fish. Biol. 63:1521–1537.

INDEX

Many of the subjects are included in the species accounts and, therefore, are not listed in this index.
Page numbers followed by f refer to figures; those followed by t refer to tables.